Structural Alloys for Nuclear Energy Applications

Structural Alloys for Nuclear Energy Applications

Edited by

G. Robert Odette
Materials Department, University of California, Santa Barbara, CA, United States

Steven J. Zinkle
Department of Nuclear Engineering, University of Tennessee, Knoxville, TN, United States
Oak Ridge National Laboratory, Oak Ridge, TN, United States

ELSEVIER

Elsevier
Radarweg 29, PO Box 211, 1000 AE Amsterdam, Netherlands
The Boulevard, Langford Lane, Kidlington, Oxford OX5 1GB, United Kingdom
50 Hampshire Street, 5th Floor, Cambridge, MA 02139, United States

Copyright © 2019 Elsevier Inc. All rights reserved.

No part of this publication may be reproduced or transmitted in any form or by any means, electronic or mechanical, including photocopying, recording, or any information storage and retrieval system, without permission in writing from the publisher. Details on how to seek permission, further information about the Publisher's permissions policies and our arrangements with organizations such as the Copyright Clearance Center and the Copyright Licensing Agency, can be found at our website: www.elsevier.com/permissions.

This book and the individual contributions contained in it are protected under copyright by the Publisher (other than as may be noted herein).

Notices
Knowledge and best practice in this field are constantly changing. As new research and experience broaden our understanding, changes in research methods, professional practices, or medical treatment may become necessary.

Practitioners and researchers must always rely on their own experience and knowledge in evaluating and using any information, methods, compounds, or experiments described herein. In using such information or methods they should be mindful of their own safety and the safety of others, including parties for whom they have a professional responsibility.

To the fullest extent of the law, neither the Publisher nor the authors, contributors, or editors, assume any liability for any injury and/or damage to persons or property as a matter of products liability, negligence or otherwise, or from any use or operation of any methods, products, instructions, or ideas contained in the material herein.

Library of Congress Cataloging-in-Publication Data
A catalog record for this book is available from the Library of Congress

British Library Cataloguing-in-Publication Data
A catalogue record for this book is available from the British Library

ISBN: 978-0-12-397046-6

For information on all Elsevier publications
visit our website at https://www.elsevier.com/books-and-journals

Cover Credit: From top to bottom: An atom probe tomography reconstruction of a Cu-Ni-Mn-Si nanoscale precipitate in an irradiated reactor pressure vessel steel (G. Robert Odette, UCSB). A tomographic image of nanoscale helium bubbles in an advanced high performance steel that are a key to developing fusion energy (G. Robert Odette, UCSB). A nighttime view of the Diablo Canyon Nuclear Power Plant (Jim Zim, PG&E). Nighttime lights as viewed from space (NASA).

Publisher: Susan Dennis
Acquisition Editor: Kostas Marinakis
Editorial Project Manager: Michelle Fisher
Production Project Manager: Prem Kumar Kaliamoorthi
Cover Designer: Mark Rogers

Typeset by SPi Global, India

Contents

Contributors xv
Preface xvii

CHAPTER 1 Overview of Structural Materials in Water-Cooled Fission Reactors 1
Jeremy T. Busby

1.1 Introduction 1
1.2 The LWR Reactor Environment and Key Modes of Degradation.................. 9
   1.2.1 Thermal Aging and Fatigue 9
   1.2.2 Irradiation 10
   1.2.3 Water Environment 11
1.3 Overview of Key Materials of Construction for LWR Applications.............. 12
   1.3.1 Zirconium-Based Alloys 13
   1.3.2 Austenitic Stainless Steels... 13
   1.3.3 Cast-Austenitic Stainless Steels 15
   1.3.4 Ni-Base Alloys 16
   1.3.5 Low-Alloy Steels... 18
   References 21

CHAPTER 2 Overview of Reactor Systems and Operational Environments for Structural Materials in Gen-IV Fission Reactors 23
Stuart A. Maloy, Ken Natesan, David E. Holcomb, Concetta Fazio, Pascal Yvon

2.1 Introduction 23
2.2 Liquid Metal Cooled Fast Reactors 24
   2.2.1 Sodium Fast Reactors (SFR)–Description of Overall Design and Applications 24
   2.2.2 Lead Fast Reactor (LFR) 30
2.3 Helium Cooled Reactors 36
   2.3.1 Very High-Temperature Reactors (VHTR) 36
   2.3.2 Gas-Cooled Fast Reactor... 39
2.4 Other Gen-IV Fission Reactor Systems 41
   2.4.1 Molten Salt Fueled Reactor (MSR) 41
   2.4.2 Molten Salt Cooled Reactor.. 43
   2.4.3 Super-Critical Water Reactor (SCWR).. 43
   2.4.4 Summary 46
   References 46

CHAPTER 3 Overview of Reactor Systems and Operational Environments for Structural Materials in Fusion Reactors 51
Richard J. Kurtz, G. Robert Odette

- 3.1 Introduction 52
- 3.2 Overview of Basic Physics 52
  - 3.2.1 Neutron and Thermal Loads on Materials Surrounding the Fusion Reaction 54
- 3.3 Basic Aspects of Materials Degradation in the Fusion Environment 55
  - 3.3.1 Comparison with Fission Environment 58
- 3.4 Overview of MCF and ICF Conceptual Designs 60
  - 3.4.1 MCF Conceptual Power Plant Designs 61
  - 3.4.2 ICF Conceptual Power Plant Designs 64
- 3.5 First-Wall/Blanket Structural Material Options 66
  - 3.5.1 RAF/M Steels 66
  - 3.5.2 Nanostructured Ferritic Alloys 72
  - 3.5.3 Vanadium Alloys 75
  - 3.5.4 SiC_f/SiC Composites 77
  - 3.5.5 Helium Effects in Fusion Structural Materials 80
- 3.6 Materials for Divertor/Limiter Applications 85
  - 3.6.1 Tungsten and Tungsten Alloys 85
  - 3.6.2 Carbon Fiber Composites 90
  - 3.6.3 Liquid Walls 92
- 3.7 Vacuum Vessel (VV) Materials 94
- 3.8 Magnet Structural Materials 96
- References 96

CHAPTER 4 Research Tools: Microstructure, Mechanical Properties, and Computational Thermodynamics 103
Colin A. English, Jonathan M. Hyde, G. Robert Odette, Gene E. Lucas, Lizhen Tan

- 4.1 Introduction 104
  - 4.1.1 Background 104
  - 4.1.2 Characterization of Irradiated Materials 106
- 4.2 Microstructural Tools 107
  - 4.2.1 Irradiation-Induced Microstructure 107
  - 4.2.2 Microstructural Tools 109
  - 4.2.3 Electron Microscopy 109
  - 4.2.4 Atom Probe Tomography 117
  - 4.2.5 Small-Angle Neutron Scattering (SANS) 125
  - 4.2.6 Positron Annihilation Spectroscopy-Based Techniques 130
  - 4.2.7 Summary of Microstructural techniques 134
- 4.3 Subsized Specimen Tests to Measure Irradiated Mechanical Properties 134
  - 4.3.1 Introduction 134

    4.3.2 Subsized Tensile Tests 135
    4.3.3 Microhardness Tests 137
    4.3.4 Irradiation Embrittlement Tests: Transition Temperature Shifts
         and Fracture Toughness 138
    4.3.5 Irradiation Hardening-Embrittlement Relations 141
    4.3.6 Data from Mechanical Tests at the Nanoscale 142
    4.3.7 Summary 143
4.4 Computational Alloy Design and Optimization 144
    4.4.1 Introduction 144
    4.4.2 Alloy Optimization 145
    4.4.3 Alloy Selection and Design 147
    4.4.4 Kinetics and Mechanical Property Simulations 147
    4.4.5 Summary 149
    References 150

CHAPTER 5 Radiation and Thermomechanical Degradation Effects in Reactor Structural Alloys 163
Steven J. Zinkle, Hiroyasu Tanigawa, Brian D. Wirth

5.1 Introduction 164
5.2 Thermomechanical Degradation Processes 166
    5.2.1 Thermal Aging 166
    5.2.2 Thermal Creep 167
    5.2.3 Fatigue and Creep Fatigue 171
5.3 Radiation Hardening and Embrittlement 173
    5.3.1 Dose Dependence of Low-Temperature Radiation Hardening
         and Ductility Reduction 173
    5.3.2 Temperature Dependence of Radiation Hardening
         and Ductility Reduction 176
    5.3.3 Low-Temperature Radiation Embrittlement 176
5.4 Radiation-Induced Phase and Microchemical Changes 179
    5.4.1 Amorphization 179
    5.4.2 Radiation-Enhanced and Induced Segregation (and Precipitation) 181
5.5 Radiation- and Stress-Modified Corrosion and Cracking Phenomena 184
5.6 Radiation-Induced Dimensional Instability 187
    5.6.1 Cavity Swelling 188
    5.6.2 Irradiation Creep 191
    5.6.3 Irradiation Growth 194
5.7 High-Temperature Helium Embrittlement 195
5.8 Conclusions 199
    Acknowledgment 200
    References 200

CHAPTER 6 Corrosion Issues in Current and Next-Generation Nuclear Reactors 211
Gary S. Was, Todd R. Allen

6.1 Corrosion in Nuclear Systems 211
    6.1.1 Types of Corrosion 212
    6.1.2 Operating Conditions in Nuclear Systems 212
6.2 Corrosion in Water Cooled Reactors 212
    6.2.1 Subcritical Water 212
    6.2.2 Supercritical Water 219
6.3 Corrosion in Helium-Cooled Reactors 224
    6.3.1 Oxidation in a VHTR Environment 225
    6.3.2 Decarburization in a VHTR Environment 226
    6.3.3 Carburization in a VHTR Environment 227
    6.3.4 Internal Oxidation 227
    6.3.5 Additional Considerations 229
6.4 Corrosion in Molten Salt and Liquid Metal-Cooled Reactors 229
    6.4.1 Molten Salt 230
    6.4.2 Sodium 236
    6.4.3 Lead Alloys 239
    References 242
    Further Reading 246

CHAPTER 7 Zirconium Alloys for LWR Fuel Cladding and Core Internals 247
Suresh Yagnik, Anand Garde

7.1 Overview of Zr-Alloys 248
7.2 Fabrication and Microstructure 251
    7.2.1 General Comments 251
    7.2.2 Lattice Structure and Second-Phase Particles 252
    7.2.3 Basic Zr-Alloy Processing and Fabrication 253
    7.2.4 Anisotropy of Zr-Alloys 255
    7.2.5 Texture 255
7.3 Corrosion and Crud 257
    7.3.1 General Comments 257
    7.3.2 Corrosion of Zr-Alloys 258
    7.3.3 Crud Deposition on Fuel Rods 258
    7.3.4 PWR Coolant Chemistry 260
    7.3.5 BWR Coolant Chemistry 261
    7.3.6 Fuel Failures Due to Severe Corrosion and Crud Deposition 262
7.4 Hydriding and Mechanical Integrity 262
    7.4.1 General Comments 262
    7.4.2 Effect of Hydriding on Unirradiated Mechanical Properties 264

7.4.3 Effect of Hydriding on Irradiated Mechanical Properties 266
7.4.4 Effect of Hydrogen on Post-Accident Transient Mechanical Properties 272
7.5 Irradiation Effects 272
7.5.1 General Comments 272
7.5.2 Effect of Irradiation on Corrosion Resistance 274
7.5.3 Irradiation Hardening and Embrittlement 275
7.5.4 Irradiation Growth 275
7.5.5 Irradiation Creep 276
7.6 Failure Mechanisms 278
7.6.1 General Comments 278
7.6.2 Debris Fretting 279
7.6.3 Grid-To-Rod Fretting (GTRF) 280
7.6.4 Pellet-Cladding Mechanical Interaction (PCMI) 280
7.6.5 Pellet-Cladding Interaction-Stress Corrosion Cracking (PCI-SCC) 282
7.6.6 Less Common Failure Mechanisms 283
7.7 Summary/Conclusions 284
References 286

CHAPTER 8 Austenitic Stainless Steels 293
Gary S. Was, Shigeharu Ukai

8.1 Introduction 293
8.2 Application in LWRs and GenIV Reactors 294
8.2.1 Light Water Reactors 294
8.2.2 Sodium-Cooled Fast Reactors 295
8.3 Radiation-Induced Metallurgical Changes 298
8.3.1 Radiation-Induced Segregation 298
8.3.2 Dislocation Microstructure 302
8.3.3 Phase Stability 303
8.3.4 Transmutation 305
8.4 Radiation-Induced Mechanical Property Changes and Degradation Modes 308
8.4.1 Irradiation Hardening 308
8.4.2 Reduction in Fracture Toughness and Embrittlement 309
8.4.3 High-Temperature He Embrittlement 311
8.4.4 Void Swelling 311
8.4.5 Irradiation Creep and Fatigue 315
8.4.6 In-Reactor Creep Rupture Properties 318
8.5 PCI/FCCI Effects With Fission Fuels 318
8.6 Chemical Compatibility With Coolants 321
8.7 Stress-Corrosion Cracking 323
8.7.1 SCC in BWRs 323
8.7.2 IGSCC in PWRs 323

8.8 Combined Effects of Water Environment and Radiation 326
    8.8.1 Irradiation-Assisted Stress-Corrosion Cracking 326
    8.8.2 Irradiation-Accelerated Corrosion...332
    8.8.3 Corrosion Fatigue 332
    8.8.4 Hydrogen Embrittlement...334
    8.8.5 Fracture Toughness 334
8.9 Perspectives and Prospects...337
    References 340
    Further Reading 347

CHAPTER 9 Ni-Based Alloys for Reactor Internals and Steam Generator Applications..349
Malcolm Griffiths

9.1 Introduction 350
9.2 Physical Metallurgy..351
9.3 Thermomechanical Processing...354
9.4 Joining...356
9.5 Mechanical Properties 358
9.6 Fracture Modes 360
9.7 Deformation Mechanisms (Yield Stress and Creep Strength) 362
9.8 Stress Corrosion Cracking..364
9.9 Ni Alloys for Generation-IV Reactors 365
9.10 Chemical Compatibility with Coolants..368
9.11 Radiation Damage and Gas Production in Ni Alloys......................................369
9.12 Radiation Hardening/Softening and Loss of Ductility in Ni Alloys...............376
    9.12.1 CANDU Reactors 377
    9.12.2 LWR Reactors 380
    9.12.3 Fast Reactors 383
    9.12.4 Proton Irradiation Facilities...385
    9.12.5 Ion Irradiation Facilities..390
9.13 Hydrogen Embrittlement..390
9.14 Helium Embrittlement..391
9.15 Point Defects 395
9.16 Irradiation Creep and Stress Relaxation 396
9.17 Fatigue and Creep Fatigue 399
    9.17.1 Fatigue 399
    9.17.2 Creep-Fatigue Deformation...400
9.18 Conclusions..401
    Acknowledgments 402
    References 402

CHAPTER 10 Low-Alloy Steels 411
Tim Williams, Randy Nanstad

10.1 Composition, Fabrication, and Properties of LAS 412
  10.1.1 Types and composition of LAS 412
  10.1.2 Design and Fabrication of LWRs 415
  10.1.3 Microstructure and Properties 423
10.2 Principal Applications of LAS...427
  10.2.1 Reactor Pressure Vessels..427
  10.2.2 Other Pressure Vessels 428
  10.2.3 Piping...428
10.3 Performance...429
  10.3.1 Regulatory Codes and Structural Integrity Assessment (SIA).........................431
  10.3.2 In-Service Degradation..438
  10.3.3 Other Performance Issues..469
10.4 Current Developments and Future Prospects..470
  10.4.1 Improved Prediction of Through-Life Toughness...470
  10.4.2 Improved Materials 472
  10.4.3 Other Issues 473
  References 473

CHAPTER 11 Ferritic and Tempered Martensitic Steels 485
Philippe Spätig, Jia-Chao Chen, G. Robert Odette

11.1 Short Historical Development of the Ferritic/Martensitic Steel:
Composition and Constitution..485
11.2 Applications of the Ferritic/Martensitic Steels in Generation IV
Nuclear Systems and Fusion Reactors...488
11.3 Environmentally Assisted Cracking...489
11.4 Compatibility with Liquid Metal Coolants..490
11.5 Radiation Hardening and Softening, Embrittlement, Fatigue,
and Thermal Creep..493
  11.5.1 Radiation Hardening and Softening..496
  11.5.2 Irradiation Embrittlement—Fast Fracture...501
  11.5.3 Fatigue 507
  11.5.4 Thermal Creep...509
11.6 Helium Effects..510
11.7 Void Swelling and Irradiation Creep..514
  11.7.1 Void Swelling..514
  11.7.2 Irradiation Creep..516
11.8 Future Prospects for Improved Performance 518
  References 519

CHAPTER 12 Nano-Oxide Dispersion-Strengthened Steels ... 529
G. Robert Odette, Nicholas J. Cunningham, Tiberiu Stan, M. Ershadul Alam, Yann De Carlan

- 12.1 Introduction ... 530
- 12.2 A Brief History of ODS Alloys ... 531
- 12.3 Some Key Attributes of Nano-Oxide Dispersion-Strengthened (NODS) Iron-Based Alloys for Nuclear Applications—An Overview ... 532
    - 12.3.1 Unirradiated Mechanical Properties ... 532
    - 12.3.2 A Summary of Alloy Stability and Irradiation Effects ... 533
    - 12.3.3 Void Swelling and Helium Effects ... 535
    - 12.3.4 Other NODS Issues ... 535
- 12.4 Overview of the Composition and Processing Paths for NFA and NMS ... 536
    - 12.4.1 Alloy Compositions, Phase Diagrams, and Transformation Paths ... 536
    - 12.4.2 Preconsolidation Processing ... 538
    - 12.4.3 Consolidation ... 539
    - 12.4.4 Deformation Processing and Tube Fabrication ... 541
    - 12.4.5 Texturing and Damage Mechanisms During Deformation Processing ... 542
    - 12.4.6 Joining ... 545
    - 12.4.7 Alternative Compositions and Processing Paths ... 545
    - 12.4.8 Summary of Processing and Fabrication ... 547
- 12.5 Characteristics and Function of the NO ... 547
    - 12.5.1 NO Statistics ... 547
    - 12.5.2 Nature of the NO ... 548
    - 12.5.3 NO Functions and He interactions ... 550
    - 12.5.4 Summary ... 552
- 12.6 Mechanical Properties ... 552
    - 12.6.1 Static Tensile Strength and Ductility ... 552
    - 12.6.2 Creep ... 554
    - 12.6.3 Fast Fracture and Fatigue ... 556
- 12.7 Thermal Aging and Irradiation Effects ... 558
    - 12.7.1 Thermal Aging ... 559
    - 12.7.2 Overview Irradiation Effects on the Microstructure ... 560
    - 12.7.3 Irradiation Stability of the NO ... 562
    - 12.7.4 Dislocation Loops ... 562
    - 12.7.5 Solute Segregation, Clustering, and Precipitation ... 564
    - 12.7.6 Cavities and Swelling ... 566
    - 12.7.7 Effects of Irradiation on Strength and Toughness ... 570
    - 12.7.8 Effect of Irradiation on Other Properties ... 572
    - 12.7.9 Summary of Thermal Aging and Irradiation Effects ... 572
- 12.8 Modeling ... 573
- 12.9 Future Prospects ... 574
    - References ... 575

CHAPTER 13 Refractory Alloys: Vanadium, Niobium, Molybdenum, Tungsten............585
Lance L. Snead, David T. Hoelzer, Michael Rieth, Andre A.N. Nemith

13.1 Introduction 585
13.2 Practical Routes for Refractory Alloy Production...587
    13.2.1 Vanadium..587
    13.2.2 Niobium 590
    13.2.3 Fabrication of Nuclear-Grade Molybdenum...................................591
    13.2.4 Practical Routes of Tungsten and Tungsten Alloy Production 596
13.3 As-Fabricated Mechanical Properties 600
    13.3.1 Vanadium..600
    13.3.2 As-Fabricated Mechanical Properties of Niobium 606
    13.3.3 As-Fabricated Mechanical Properties of Molybdenum 607
    13.3.4 As-Fabricated Mechanical Properties of Tungsten.........................611
13.4 As-Irradiated Mechanical Properties...615
    13.4.1 As-Irradiated Mechanical Properties of Vanadium 615
    13.4.2 As-Irradiated Mechanical Properties of Niobium...........................622
    13.4.3 As-Irradiated Mechanical Properties of Molybdenum 625
    13.4.4 As-Irradiated Mechanical Properties of Tungsten..........................629
    13.4.5 Summary and Conclusions..634
    References 635

Index 641

Contributors

M. Ershadul Alam
Materials Department, University of California, Santa Barbara, CA, United States

Todd R. Allen
University of Wisconsin, Madison, Wi, United States

Jeremy T. Busby
Oak Ridge National Laboratory, Oak Ridge, TN, United States

Jia-Chao Chen
Laboratory for Nuclear Materials, Nuclear Energy and Safety, Paul Scherrer Institute, Villigen, Switzerland

Nicholas J. Cunningham
Materials Department, University of California, Santa Barbara, CA, United States

Yann De Carlan
DEN, Department of Applied Metallurgical Research (SRMA), CEA, University of Paris-Saclay, Gif-Sur-Yvette, France

Colin A. English
Materials and Reactor Chemistry, National Nuclear Laboratory, Culham Science Centre, Abingdon, Oxon, United Kingdom

Concetta Fazio
European Commission, JRC, Petten, The Netherlands

Anand Garde
Engineer Emeritus, Westinghouse, Columbia, SC, United States

Malcolm Griffiths
Department of Mechanical and Materials Engineering, Queen's University, Kingston, ON, Canada; ANT International, Mölnlycke, Sweden

David T. Hoelzer
Materials Science and Technology Division, Oak Ridge National Laboratory, Oak Ridge, TN, United States

David E. Holcomb
Oak Ridge National Laboratory, Oak Ridge, TN, United States

Jonathan M. Hyde
Materials and Reactor Chemistry, National Nuclear Laboratory, Culham Science Centre, Abingdon, Oxon, United Kingdom

Richard J. Kurtz
Pacific Northwest National Laboratory, Richland, WA, United States

Gene E. Lucas
Materials Department, University of California, Santa Barbara, CA, United States

Stuart A. Maloy
Los Alamos National Laboratory, Los Alamos, NM, United States

Randy Nanstad
R&S Consultants, Knoxville, TN, United States

Ken Natesan
Argonne National Laboratory, Lemont, IL, United States

Andre A.N. Nemith
Institute for Advanced Materials, Karlsruhe Institute of Technology, Karlsruhe, Germany

G. Robert Odette
Materials Department, University of California, Santa Barbara, CA, United States

Michael Rieth
Institute for Advanced Materials, Karlsruhe Institute of Technology, Karlsruhe, Germany

Lance L. Snead
Department of Materials Science and Chemical Engineering, State University of New York at Stony Brook, Stony Brook, NY, United States

Philippe Spätig
Laboratory for Nuclear Materials, Nuclear Energy and Safety, Paul Scherrer Institute, Villigen, Switzerland

Tiberiu Stan
Materials Department, University of California, Santa Barbara, CA, United States

Lizhen Tan
Materials Science and Technology Division, Oak Ridge National Laboratory, Oak Ridge, TN, United States

Hiroyasu Tanigawa
Japan Atomic Energy Agency, Rokkasho Fusion Institute, Aomori, Japan

Shigeharu Ukai
Hokkaido University, Sapporo, Japan

Gary S. Was
University of Michigan, Ann Arbor, MI, United States

Tim Williams
39bhr Consulting (39bhr Ltd.), Derby, United Kingdom

Brian D. Wirth
Department of Nuclear Engineering, University of Tennessee, Knoxville; Oak Ridge National Laboratory, Oak Ridge, TN, United States

Suresh Yagnik
Senior Technical Executive, EPRI, Palo Alto, CA, United States

Pascal Yvon
CEA—Saclay, Gif-sur-Yvette, France

Steven J. Zinkle
Department of Nuclear Engineering, University of Tennessee, Knoxville; Oak Ridge National Laboratory, Oak Ridge, TN, United States

Preface

The advent of commercial nuclear energy was accelerated by the development of zirconium alloy fuel cladding and multiple other innovations in the 1940s and 1950s, leading to widespread deployment of first-generation nuclear reactors in the 1960s. The materials used to design the first-generation nuclear reactors were generally selected from recently developed state-of-the art high-performance alloys, but this selection methodology was hampered by limited knowledge of potential radiation-induced, thermomechanical and chemical degradation processes. Over the subsequent decades, considerable improvements in the understanding of radiation effects and other degradation processes in materials have been acquired. Therefore, it is timely to assemble a comprehensive assessment of the operating environment and structural alloy degradation processes relevant for nuclear energy applications.

Structural materials, such as fuel cladding, core internal structural components, and the reactor pressure vessel, are vital for the safety and economics of current and proposed nuclear fission and fusion energy systems. The harsh operating environment associated with the in-core and power conversion systems requires high-performance materials that can withstand intense irradiation damage, along with corrosive high-temperature coolant environments in the presence of steady or cyclic mechanical stresses for time scales ranging from 6 to more than 80 years. This book is intended to provide a state-of-the-art summary of the reactor operating environments and structural material degradation mechanisms, along with comprehensive overviews of the leading structural alloy systems relevant for nuclear energy systems. Ceramic composites are not addressed since these important systems require consideration of additional unique processing and operational issues. The intended audience includes nuclear industry engineers, early career and experienced research professionals, and graduate students.

The first three chapters provide overviews of the reactor concepts and structural material operating environments for water-cooled fission reactors, proposed Generation IV fission reactors, and proposed fusion reactors. Chapter 1 by Jeremy T. Busby summarizes current water-cooled reactor designs, typical materials used in these reactors, and some of the key property degradation concerns. Chapter 2 by Stuart A. Maloy, Ken Natesan, David E. Holcomb, Concetta Fazio, and Pascal Yvon provides a similar overview of the six proposed Generation IV fission reactor concepts and discusses some of the key structural materials and anticipated radiation degradation in properties. Chapter 3 by Richard J. Kurtz and G. Robert Odette summarizes the basic fusion reactor design concepts and candidate materials for the first wall and blanket structure, divertor, vacuum vessel, and magnet, along with it offers a comparison of the anticipated radiation-induced degradation processes in fusion vs fission neutron spectra.

The next three chapters provide assessments of the research tools used to investigate the behavior and design new materials for nuclear reactors, along with overviews of the various property degradation mechanisms associated with displacement damage, mechanical stress, and corrosion. Chapter 4 by Colin A. English, Jonathan M. Hyde, G. Robert Odette, Gene E. Lucas, and Lizhen Tan provides an overview of the research tools currently available to perform studies on radiation effects and to examine the detailed resultant changes in microstructure and mechanical properties, along with a summary of the use of computational thermodynamics for designing new high-performance alloys. Chapter 5 by Steven J. Zinkle, Hiroyasu Tanigawa, and Brian D. Wirth provides an overview of various degradation mechanisms that are commonly observed in all alloys due to thermomechanical exposures and neutron irradiation effects, including thermal aging, creep, fatigue, radiation hardening and embrittlement, microchemical and phase changes, radiation-modified corrosion and cracking, dimensional instability due to void swelling and irradiation creep, and high-temperature helium embrittlement of grain boundaries. Chapter 6 by Gary S. Was and Todd R. Allen summarizes various corrosion and stress corrosion cracking issues for current and next-generation reactors.

The final seven chapters contain detailed assessments of the leading structural alloy systems for nuclear fission and fusion reactors, including unirradiated and irradiated mechanical properties and summaries of alloy-specific radiation-induced degradation issues. Chapter 7 by Suresh Yagnik and Anand Garde summarizes the fabrication and environmental degradation processes for Zr alloys in water-cooled reactors, including corrosion under normal and transient conditions, hydride formation, irradiation effects, and key in-service failure mechanisms. Chapter 8 by Gary S. Was and Shigeharu Ukai summarizes radiation-induced microstructural and property changes in austenitic stainless steels (predominantly 300 series stainless steels), including radiation-induced segregation, precipitation, compatibility with nuclear fuel and various coolants, and stress corrosion cracking issues. Chapter 9 by Malcolm Griffiths reviews the metallurgy and properties of commercial nickel alloys, including deformation and fracture mechanisms, stress corrosion cracking, chemical compatibility, and radiation effects, with particular emphasis on the role of the large amount of transmutant He generated during irradiation in reactors with high thermal neutron fluxes. Chapter 10 by Tim Williams and Randy Nanstad reviews the fabrication methods, properties, and in-service degradation (in particular fracture toughness) of low alloy steels used for reactor pressure vessels. Chapter 11 by Philippe Spätig, Jia-Chao Chen, and G. Robert Odette summarizes the development of tempered ferritic/martensitic steels for nuclear energy applications, including coolant corrosion and embrittlement, unirradiated and irradiated mechanical properties, and helium effects on mechanical properties and cavity swelling. Chapter 12 by G. Robert Odette, Nicholas J. Cunningham, Tiberiu Stan, M. Ershadul Alam, and Yann De Carlan reviews the processing methods and properties of oxide dispersion strengthened ferritic/martensitic steels containing high densities of nanoscale second phases that exhibit impressive improvements in strength, high-temperature stability, and radiation resistance compared to traditional ferritic/martensitic steels. Chapter 13 by Lance L. Snead, David T. Hoelzer, Michael Rieth, and Andre A.N. Nemith summarizes the fabrication procedures, typical physical and mechanical properties, and reported radiation effects for four important refractory alloy systems based on vanadium, niobium, molybdenum, and tungsten. Various chapters deal with similar issues, often from different perspectives, thus cross-references between chapters are provided in a number of cases.

Finally, the editors thank the authors for sharing their expertise and insights and for the tremendous amount of selfless time and effort devoted to the preparation of their chapters. They also appreciate the efforts of a large number of individuals at Elsevier who guided this book to completion with good humor and great patience.

OVERVIEW OF STRUCTURAL MATERIALS IN WATER-COOLED FISSION REACTORS

CHAPTER 1

Jeremy T. Busby
Oak Ridge National Laboratory, Oak Ridge, TN, United States

CHAPTER OUTLINE

- 1.1 Introduction 1
- 1.2 The Lwr Reactor Environment and Key Modes of Degradation 9
  - 1.2.1 Thermal Aging and Fatigue 9
  - 1.2.2 Irradiation 10
  - 1.2.3 Water Environment 11
- 1.3 Overview of Key Materials of Construction for LWR Applications 12
  - 1.3.1 Zirconium-Based Alloys 13
  - 1.3.2 Austenitic Stainless Steels 13
  - 1.3.3 Cast-Austenitic Stainless Steels 15
  - 1.3.4 Ni-Base Alloys 16
  - 1.3.5 Low-Alloy Steels 18
- References 21

1.1 INTRODUCTION

Nuclear power currently provides a significant fraction of non-carbon-emitting power generation in the United States and around the world. According to the Nuclear Energy Institute's latest figures, the 99 operating light water reactors (LWRs) in the United States operate with a capacity of 99,300 MW. In 2016, this fleet generated over 800,000 GWh, representing 19.7% of the US electrical consumption [1]. Cumulatively, the 449 operating reactors represent 11% of the world's electrical generation and consumption [2]. The distribution of nuclear power plants is shown in Fig. 1.1. The nuclear generation capacity for each country is listed by country in Table 1.1. In 2016, 13 countries received over a quarter of their energy via nuclear power. Clearly, nuclear energy is a key resource for today's world.

In future years, nuclear power must continue to generate a significant portion of the nation's electricity to meet growing electricity demand and clean energy goals and to ensure energy independence. The existing fleet of reactors continues to improve on operational reliability and power generation via proactive management and power

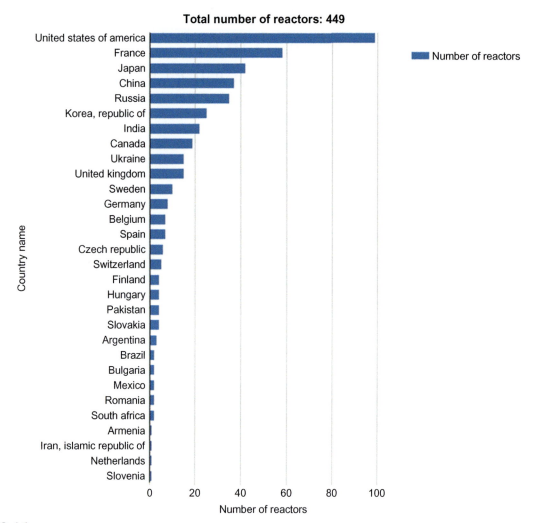

FIG. 1.1

Distribution of operating nuclear power reactors around the world [2].

From IAEA Power Reactor Information System (PRIS), https://www.iaea.org/pris/. Used with permission from IAEA.

uprates, respectively. For example, the US fleet now operates at a 92.1% capacity rate [1]. There are new plants under construction around the world and new capacity will be available in the coming years. Currently, there are 60 plants under construction around the world [2]. However, it should be noted that new plants represent a significant financial and logistics burden, and therefore the existing fleet must also continue to be managed for extended service. The balance and relative distribution of new reactors versus existing capacity is shown in the distributions in Fig. 1.2. Clearly, aging management is a key consideration for our power fleet. Today, the nuclear industry, regulators, and academia are actively exploring the potential for subsequent life extensions and how long the current generation of LWRs may operate safely and efficiently.

Nuclear reactors generate electricity in much the same way as fossil fuel plants (coal, oil, natural gas). In water-cooled reactors, water is converted to steam by a heat source at temperatures near 300 °C and the steam is used to

Table 1.1 Summary of Nuclear Generating Capacity Around the World [1]

Country	As of April 2017 Number of Nuclear Units	As of April 2017 Nuclear Capacity (MW)	2016 Nuclear Generation (GWh)	2016 Nuclear Fuel Share (%)
Argentina	3	1632	7677.4	5.6
Armenia	1	375	2194.9	31.4
Belgium	7	5913	41,430.5	51.7
Brazil	2	1884	14,970.5	2.9
Bulgaria	2	1926	15,083.5	35.0
Canada	19	13,554	95,650.2	15.6
China	36	31,384	197,829.0	3.6
Czech Republic	6	3930	22,729.9	29.4
Finland	4	2764	22,280.1	33.7
France	58	63,130	386,452.9	72.3
Germany	8	10,799	80,069.6	13.1
Hungary	4	1889	15,183.0	51.3
India	22	6240	35,006.8	3.4
Iran, Islamic Republic of	1	915	5924.0	2.1
Japan	43	40,290	17,537.1	2.2
Korea, Republic of	25	23,077	154,306.7	30.3
Mexico	2	1552	10,272.3	6.2
Netherlands	1	482	3749.8	3.4
Pakistan	4	1005	5438.9	4.4
Romania	2	1300	10,388.2	17.1
Russia	37	26,528	184,054.1	17.1
Slovakia	4	1814	13,733.4	54.1
Slovenia	1	688	5431.3	35.2
South Africa	2	1860	15,209.5	6.6
Spain	7	7121	56,102.4	21.4
Sweden	10	9740	60,647.4	40.0
Switzerland	5	3333	20,303.1	34.4
Taiwan, China	6	5052	30,461.0	13.7
United Kingdom	15	8918	65,149.0	20.4
United States[a]	99	99,319	805,327.2	19.7
Ukraine	15	13,107	76,077.8	52.3
Total	451	391,521	2,476,671.2	

[a]IAEA and EIA nuclear capacity figures vary slightly.
Source: International Atomic Energy Agency, Updated: 4/17

turn a turbine, generating electricity. The key difference between fossil and nuclear energy is, of course, how the heat is created. In a nuclear reactor, heat is generated as energetic neutrons and fission products slowdown in moderating materials and transfer heat to a coolant. The vast majority (432 out of the 449 in service [2]) of nuclear power plants in service today are water-cooled. There are three major types of water-cooled reactors being used

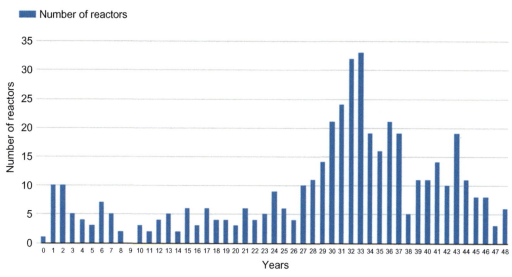

FIG. 1.2

Number of nuclear power reactors as a function of age (as of 6/2017) [2].

From IAEA Power Reactor Information System (PRIS), https://www.iaea.org/pris/. Used with permission from IAEA.

to generate electricity: pressurized water reactors (PWRs), boiling water reactors (BWRs), and pressurized heavy water reactors (PHWRs). Each of these three major water reactor types is described in more detail below. These reactors are often categorized as Generation I-III reactors. These "generations" refer to the iterative improvements on water-cooled technology (or gas-cooled reactors in the UK), predominately used for power generation. Gen I often includes the initially deployed commercial scale reactors, while Gen II captures most of the current fleet. Gen III reactors are those that are under construction or just being connected to grid. With each new generation, the nuclear industry has achieved improvements in safety, operational improvements, etc. Generation IV (Gen IV) technologies often include higher temperatures, other coolants, and more significant departures from water-cooled designs [3,4].

Advanced reactor designs, or GenIV reactors, utilize other coolants such as sodium, liquid salt, or gas and may enable other missions and opportunities for nuclear power. These reactors have been the subject of considerable research worldwide over the past 40 years. The higher operating temperature of gas or liquid salt cooled reactors may also enable the use of process heat for desalination or hydrogen production, in addition to power generation. These reactor types are described in more detail in Chapter 2. Nuclear fusion has also been studied and developed for decades, and if successful, fusion technology may offer unlimited power without any carbon emissions and very little radioactive waste. This form of nuclear power requires a high-temperature plasma environment for fusion to be sustainable; the key materials challenges are outlined in Chapter 3. Both Gen IV and fusion reactor designs will require new materials and new understanding of material degradation. While both future categories of reactors are promising, water reactor technology will likely continue to provide the bulk of nuclear power generation worldwide for many decades.

As noted above, water-coolant reactors currently make up 96% of the world's operating reactors. Today, there are 78 BWRs, 290 PWRs, and 49 PHWRs in service [2]. Of the 60 power plants currently under construction, 50 are PWR designs, while 4 BWRs and PHWRs are also under construction. Despite the differences in design and operation between BWRs, PWRs, and PHWRs, there are several common elements, most notably the use of water as the coolant and neutron moderator. Each reactor requires nuclear fuel and all water reactors utilize uranium in oxide form in a zirconium-alloy cladding. BWRs and PWRs utilize enriched uranium fuels (typically 3–5% ^{235}U), whereas PHWRs are designed to utilize natural uranium (only ~0.7% ^{235}U). Each also requires control rods, which are made with neutron-absorbing material such as cadmium, hafnium, or boron, and are inserted or withdrawn from the core to control the fission reaction, or to halt it. A pressure vessel is a robust steel vessel containing the reactor core and moderator/coolant. In addition, both PWRs and PHWRs have steam generators, where the high-pressure primary coolant bringing heat from the reactor is used to make steam for the turbine in a secondary circuit. Finally, the containment is the structure around the reactor and associated steam generators, which is designed to protect it from outside intrusion and to protect those outside from the effects of radiation in case of any serious malfunction inside. It is typically a meter-thick concrete and steel structure.

A BWR follows a straightforward concept. Cooling water is circulated through the bottom of the reactor core, as shown in Fig. 1.3. As the water flows through the core and nuclear fuel, heat is transferred and the coolant turns to steam. The steam is directly used to drive a turbine, after which it is cooled in a condenser and converted back to liquid water. This water is then returned to the reactor core, completing the loop. The BWR is the second most common design in service due to several advantages. This system requires a lower pressure and lower temperature than both PWRs and PHWRs, reducing demands on many of the key components such as the reactor pressure vessel (RPV). As the system directly produces steam, there are also fewer components and piping systems than in PWRs and PHWRs, reducing the cost and simplifying the operation. On the other hand, there are also drawbacks including the requirement to support two-phase flow in the core and the requirement for larger components to deliver similar

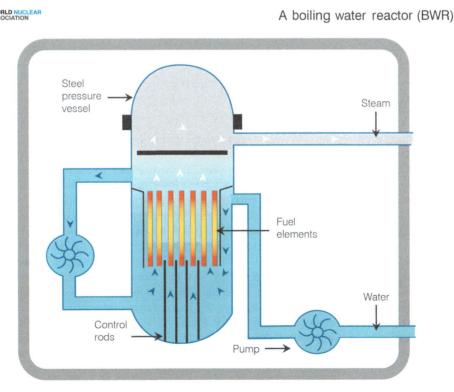

FIG. 1.3

Schematic representation of a boiling water reactor [6].

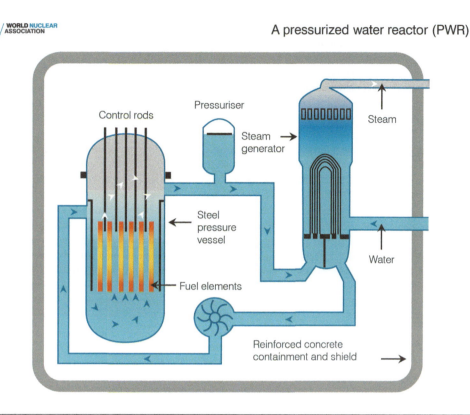

FIG. 1.4

Schematic representation of a pressurized water reactor [5].

power levels. Further, as the coolant is in contact with both the nuclear fuel and the turbine, radioactive contamination control and shielding are required in additional systems.

PWRs are the most common type of nuclear power plant in service. PWRs have several key differences from BWRs and the key features of the cycle are shown in Fig. 1.4. In a PWR, the water circulates at higher temperatures and higher pressures, staying in the liquid phase throughout the cycle. After passing through the reactor core and picking up heat, the primary coolant transfers heat in a steam generator to water in a lower pressure secondary circuit, evaporating the secondary coolant to saturated steam, which then drives a turbine as in BWRs. Pressure in the primary circuit is maintained by a pressurizer, which is a separate vessel that is connected to the primary circuit. PWRs have several distinct advantages, including stability during operation, and by incorporating a heat exchanger; the water/steam in the turbine loop is not contaminated by radioactive materials. However, owing to the high pressures required to keep the water coolant in the liquid phase, additional components (pressurizer, pumps, steam generator) are required, increasing the complexity and cost of the reactor system.

A PHWR operates in much the same manner as a PWR. The key differences are that the PHWR utilizes natural enrichment uranium oxide fuel with heavy water (deuterium oxide, D_2O) as the moderator and coolant, and the pressure vessel consists of multiple individual "calandria" rather than a large external pressure vessel surrounding the entire core. As in a PWR, the coolant is pressurized to maintain the liquid state and flows through a heat exchanger, as illustrated in Fig. 1.5. The use of heavy water as the moderator is the key to the PHWR enabling the use of natural uranium, and does so more efficiently than LWRs. Further, owing to their unique configuration, PHWRs can also be refueled without a costly shutdown period. Pressurized heavy water reactors do have some

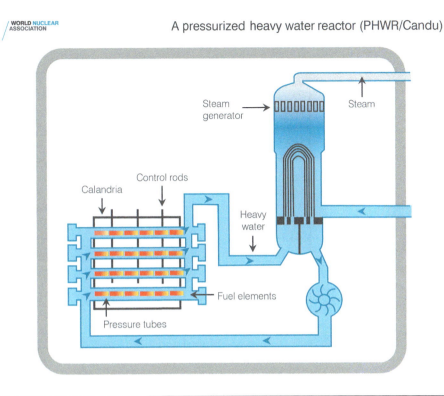

FIG. 1.5

Schematic representation of a pressurized water reactor [5].

drawbacks, primarily in the very high cost of heavy water, which generally costs hundreds of dollars per kilogram. However, this is a trade-off against reduced fuel costs as enrichment is no longer required. These designs also yield higher tritium production, which complicates operational and contamination controls.

There are clear differences between the different water reactor design types. The trade-offs in choices of coolant, temperatures, pressures, fuel enrichment, and other factors provide advantages and disadvantages. Some of the key distinguishing features between BWRs, PWRs, and PHWRs are summarized in Table 1.2. Despite these differences in materials, design, and operation, all water reactors have demonstrated high efficiency in operation, particularly in the past decade. Indeed, all water reactors now operate at over 90% capacity factors [1].

Ensuring public safety and environmental protection is a prerequisite for all nuclear power plants, whether water reactor, advanced reactor, or fusion. Materials science and engineering considerations are essential at all phases of a reactor life cycle. At the concept development stage, material properties must be balanced versus performance goals and requirements for the concept design goals. As the reactor concept matures and subsystems and components are developed, material properties are essential for component design. New materials may be required if existing alloys or materials cannot meet design requirements. Beyond subsystem design, the manufacturing of components places new demands on the materials of choice. Materials must be available, affordable, and capable of being made into actual, functioning parts and components. Then these components must be joined into subsystems and systems during actual construction. Ultimately, these materials must perform in the harsh environment of reactor service for their expected lifetime.

Unfortunately, nuclear reactors of all designs present a challenge for component service and material performance, as exemplified in water reactors. Components within a reactor must tolerate the harsh environment

Table 1.2 Comparison of Key Design Features and Operational Characteristics in Modern Water Reactors

Feature	BWR	PWR	PHWR
Number operable	78	290	49
Coolant	H_2O	H_2O	D_2O
Steam generation	Direct	Indirect	Indirect
Fuel	UO_2	UO_2	UO_2
Fuel enrichment	3–5%	3–5%	Natural
Fuel inventory	~134 tons	~104 tons	~90 tons
Refueling	Outage	Outage	continuous
Typical refueling outage	~35 days	~35 days	Continuous
Vessel type	Cylinder	Cylinder	Tubes
Active core height	~3.7 m	~4.2 m	~5.9 m
Active core dia	~4.7 m	~3.4 m	~6.0 m
Vessel thickness	~0.12 m	~0.2 m	~0.15 m
Inlet coolant temp	275 °C	275 °C	265 °C
Outlet coolant temp	288 °C	325 °C	310 °C
Operating pressure	~7.5 MPa	~15 MPa	~10 MPa
Thermal efficiency	~34%	32–33%	30–32%
Neutronic poisons		Boric acid	Light water

of high-temperature water, stress, vibration, and, for those components in the reactor core, an intense neutron field. Degradation of materials in this environment can lead to reduced performance and, in some cases, sudden failure. As described in the following chapter, the materials challenges for GenIV reactors are even more harsh than for water reactor designs. Clearly, choosing the right materials for performance under these environments and understanding materials degradation and accounting for the effects of a reactor environment in operating and regulatory limits are essential.

Materials degradation in a nuclear power plant is extremely complex due to the various materials, environmental conditions, and stress states that exist in a reactor. For example, as shown in Fig. 1.6, over 25 different metal alloys can be found within the primary and secondary systems of a modern PWR. These include austenitic stainless steels, cast-austenitic steels, low-alloy steels (LASs), and Ni-base alloys, plus all of the weldments required to join both similar and dissimilar materials. While not shown, similar diagrams and complexity exist for BWRs and PHWRs. Additional materials exist in concrete, the containment vessel, instrumentation and control equipment, cabling, buried piping, and other support facilities. Dominant forms of degradation can vary greatly between different systems, structures, and components (SSCs) in the reactor and have an important role in the safe and efficient operation of a nuclear power plant. When this diverse set of materials is placed in a complex and harsh environment, coupled with load and degradation over an extended life, an accurate estimate of the changing material behaviors and lifetime is complicated.

The following section provides insights into the materials selected for service in today's water-cooled reactor fleet, including an overview of the environment and key degradation modes, with special focus on irradiation damage and corrosion. A brief overview of the key materials of construction will be presented in Section 1.3, including zirconium-based alloys for cladding, stainless steels, and cast-austenitic stainless steels, Ni-base alloys, and, finally, LASs. Many of the subsequent chapters will explore these concepts in much greater detail.

1.2 THE LWR REACTOR ENVIRONMENT AND KEY MODES OF DEGRADATION

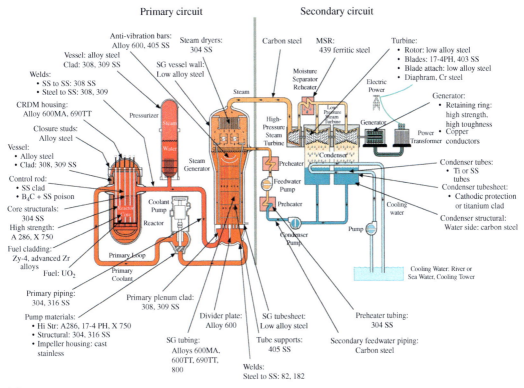

FIG. 1.6

Materials of service in a modern pressurized water reactor [6].

From NUREG 7153, Expanded Materials Degradation Assessment (EMDA).

1.2 THE LWR REACTOR ENVIRONMENT AND KEY MODES OF DEGRADATION

The reactor core is a very adverse environment, combining the effects of stress, corrosion, and irradiation. Components in this environment are also often the most critical for safe and reliable operation as the failure of a core internal component may have very severe consequences. The issues described below represent those environmental factors that may be life limiting for water reactor component life and are grouped into three key areas: thermal aging embrittlement and fatigue, irradiation-induced aging effects, and corrosion. General overviews of these degradation mechanisms are given in accompanying Chapters 5 and 6. While the material susceptibilities to these key aging effects are highly dependent upon specific material and environment combinations, these aging effects have been observed in service for many key components.

1.2.1 THERMAL AGING AND FATIGUE

The effects of elevated temperature service in metal alloys have been examined for many years. Possible effects include phase transformations that can adversely affect mechanical properties. Extended time at elevated temperature over a component lifetime may permit even very slow phase transformations to occur. This is of particular concern for cast stainless steel components where the formation of a brittle alpha phase can result in a loss of fracture toughness and lead to brittle failure. The effort required for identifying possible problems can be reduced, though, by using modern materials science modeling techniques and experience from other industries.

Fatigue refers to an aging degradation mechanism where components undergo cyclic stress. Typically, these are either low-load, high-frequency stresses or high-load, low-frequency stresses generated by thermal cycling, vibration, seismic events, or loading transients. Environmental factors may accelerate fatigue and eventually result in component failure. In an LWR, components such as the pressure vessel, pressurizer, steam generator shells, steam separators, pumps, and piping are among the components that may be affected. A general summary of thermal aging, fatigue, and creep-fatigue processes in structural alloys is given in Chapter 5.

1.2.2 IRRADIATION

Over the 40-year lifetime of an LWR, internal structural components may experience neutron flux to $\sim 10^{22}$ n/cm^2/s in a BWR and $\sim 10^{23}$ n/cm^2/s in a PWR (E > 1 MeV), corresponding to an accumulated neutron dose of ~ 7 and 70 dpa (displacements per atom), respectively. Fortunately, radiation effects in stainless steels (the most common core constituent) are also the most examined as these materials are also of interest in fast-spectrum fission and fusion reactors where higher fluences are encountered.

The neutron irradiation field can produce large property and dimensional changes in materials. This occurs primarily via one of five radiation damage processes: Radiation-induced hardening and embrittlement, phase instabilities from radiation-induced or -enhanced segregation and precipitation, irradiation creep due to unbalanced absorption of interstitials vs. vacancies at dislocations, volumetric swelling from cavity formation, and high-temperature helium embrittlement due to formation of helium-filled cavities on grain boundaries. Further, some material systems (e.g., zirconium or uranium metal, which have anisotropic structures) may also undergo irradiation growth due to unbalanced nucleation of defect clusters into different prism planes. For water reactor systems, high-temperature embrittlement and creep are not common problems due to the relatively (for creep) lower reactor operating temperature. However, radiation embrittlement, phase transformation, segregation, and swelling have all been observed in water reactor components. The following gives a brief overview of the key radiation damage mechanisms in materials; additional descriptions of these mechanisms are given in Chapter 5 for structural materials in general, and in material-specific chapters 7–13.

1.2.2.1 Radiation-induced segregation and phase transformations

Under irradiation, the large concentrations of radiation-induced defects will diffuse to defect sinks such as grain boundaries and free surfaces. These concentrations are far in excess of thermal equilibrium values and can lead to coupled diffusion with particular atoms. In engineering metals such as stainless steel, this results in radiation-induced segregation of elements within the steel. For example, in Type 316 stainless steel (SS), chromium (important for corrosion resistance) can be depleted at areas while elements like nickel and silicon are enriched to levels well above the starting homogenous composition. While radiation-induced segregation does not directly cause component failure, it can influence corrosion behavior in a water environment. Further, this form of degradation can accelerate the thermally driven phase transformations mentioned above and also result in phase transformations that are not favorable under thermal aging (such as gamma or gamma-prime phases observed in stainless steels). Additional fluence may exacerbate radiation-induced phase transformations and should be considered. The wealth of data generated for fast-breeder reactor studies and more recently in LWR-related analysis will be beneficial in this effort.

1.2.2.2 Radiation-induced swelling and creep

The diffusion of radiation-induced defects can also result in the clustering of vacancies, creating voids. If gas atoms such as He enter the void, the void becomes a bubble. While swelling is typically a greater concern for fast reactor applications where it can be life limiting, voids have recently been observed in LWR components such as baffle bolts, although the associated swelling is not considered to be a major operational issue. The motion of vacancies

1.2 THE LWR REACTOR ENVIRONMENT AND KEY MODES OF DEGRADATION

can also greatly accelerate creep rates, resulting in stress relaxation and deformation. Irradiation-induced swelling and creep effects can be synergistic and their combined influence must be considered, particularly in highly textured materials due to fabrication or even design in components such as fuel cladding. Longer reactor component lifetimes may increase the need for a more thorough evaluation of swelling as a limiting factor in LWR operation. As above, data, theory, and simulations generated for fast reactor and fusion applications can be used to help identify potentially problematic components.

1.2.2.3 Radiation-induced embrittlement

Radiation embrittlement results in an increase in the yield and ultimate tensile strength of the material. This increase in strength comes with a corresponding decrease in ductility. This hardening can be caused by the changes in the alloy's microstructure including radiation-induced segregation, phase transformations, and swelling. Ultimately, hardening and loss of ductility will result in reduced fracture toughness and resistance to crack growth.

Extended reactor lifetimes may lead to increased embrittlement issues. As noted above, over the 40-year lifetime of an LWR, internal structural components may experience ~7 dpa in a BWR and 70 dpa in a PWR. Today, the majority of the operating nuclear reactor plants in the United States have had their life extended to 60 years, and there are active discussions on extending the lifetime of the existing fleet to beyond 60 years of service. Clearly, extending the operating period of a reactor will increase the total neutron fluence to each component. Extending reactor operation to beyond 60 years will increase the demands on materials and components. While operation beyond 60 years will add additional time and neutron fluence, the primary impact will be increased susceptibility to known degradation modes, although new mechanisms are possible.

For the reactor core and primary systems, several key issues have been identified. Thermomechanical considerations such as aging and fatigue must be examined. Irradiation-induced processes must also be considered for higher fluences, particularly the influence of RIS, swelling, and/or precipitation on embrittlement. For RPVs, a number of significant issues have been identified for future research. Relatively sparse or nonexistent data at high fluences, for long radiation exposure (duration), and resulting high embrittlement create large uncertainties for embrittlement predictions. The use of test reactors at high fluxes to obtain high fluence data is not the most direct representation of the low flux conditions in RPVs. Late blooming phases, especially for high nickel welds, have been observed and additional experimental data are needed in the high fluence regime where they are expected. These and other issues related to extended service are described in more detail in Refs. [7–9].

1.2.3 WATER ENVIRONMENT

In order to transfer the large amounts of heat produced in the nuclear fission process to useful electricity, today's nuclear power fleet around the world utilizes water as the heat transfer medium and primary reactor coolant. As a result, understanding and predicting corrosion and corrosion-related processes in a water reactor environment is essential to safe and efficient operation. All of the operating commercial power plants around the world are LWRs that use water as a coolant, with temperatures as high as 320 °C. Temperatures typically range from 288 °C (550 °F) in a BWR and up to 360 °C (680 °F) in a PWR, although other water chemistry variables differ more significantly between the BWRs and PWRs. In both systems, careful control is given to minimize impurities and especially oxygen content to reduce corrosion and stress corrosion cracking (SCC) [10,11]. Chapter 6 describes the differences in water chemistry in more detail.

As discussed in further detail in Chapter 6, corrosion is a complex form of degradation that depends on temperature, material condition, material composition, water purity, water pH, chemical species present, and gas concentrations. The operating corrosion mechanism will vary from location to location within the reactor core and coolant loop and a number of different mechanisms may be operating at the same time. These may include general corrosion mechanisms such as uniform corrosion, boric acid corrosion (BAC), flow-accelerated corrosion (FAC), and/or erosion corrosion that will occur over a reasonably large area of material in a fairly homogenous manner. Localized corrosion modes occur over much smaller areas, but at much higher rates than general corrosion and include crevice corrosion, pitting, galvanic corrosion, and microbially induced corrosion (MIC). Finally, environmentally assisted cracking (EAC) includes a combination of other forms of degradation that are closely related to localized or general corrosion with the added contribution of stress, temperature, and/or irradiation. In an LWR, a number of different EAC mechanisms are observed: intergranular stress-corrosion cracking (IGSCC), transgranular stress corrosion cracking (TGSCC), PWSCC, IASCC, and low-temperature crack propagation (LTCP).

Components in the secondary (steam generator) side of a PWR are also subject to degradation. While the secondary side of the reactor does not have the added complications of an intense neutron irradiation field, the combined action of corrosion and stress can create many different forms of failure. The majority of steam generator systems in US power plants today originally used Alloy 600 (an Ni-Cr-Fe alloy) for tubes and some other components, although service experience showed many failures in tubes through the 1970s. In the last 20 years, most steam generators have been replaced with units that have Alloy 690 tubes, which show more resistance to SCC. In addition to the base material, there are weldments, joints, and varying water chemistry conditions leading to a very complex component. Indeed, the array of modes of degradation varies with location. In a single steam generator examined by Staehle and Gorman [12], 25 different modes of corrosion degradation were identified. SCC is found in several different forms and may be the limiting factor for extended service. The integrity of these components is critical for reliable power generation in extended operation, and as a result, understanding and mitigating these forms of degradation is important.

Chapter 6 provides a detailed discussion of corrosion in a water reactor environment as well as corrosion phenomena in other advanced reactor environments. The authors provide a thorough examination of the different types of environmental attack that occur in service in this aggressive environment. This includes a detailed description of the general corrosion environment and controlling factors. Uniform corrosion occurs in a roughly similar manner across the entire surface of a material and is prevalent in most reactor components to some extent, with zirconium-based alloys being a key example. Chapter 6 also touches on one of the key complications for understanding and mitigating corrosion-related degradation. Since equipment in complex engineering systems often consists of multiple materials joined by welds or other solid-state joining processes, site-specific corrosion mechanisms must also be considered. Even within a single material, if second phase strengthening is used or if surface defects are present, local corrosion effects such as galvanic, crevice, or pitting corrosion are possible.

As described in the previous section, there are ongoing efforts to consider further life extension of the US fleet and other reactors around the world. Corrosion takes many forms within the reactor core and piping systems, although IASCC and PWSCC are of high interest in extended life scenarios. Research in these areas can build upon other ongoing programs in the LWR industry as well as other reactor materials programs (such as fusion and fast reactors) to help resolve these issues for extended LWR life. In the secondary systems, corrosion is extremely complex. Understanding the various modes of corrosion and identifying mitigation strategies is an important step for long-term service.

1.3 OVERVIEW OF KEY MATERIALS OF CONSTRUCTION FOR LWR APPLICATIONS

The original water-cooled reactors were developed and constructed decades ago, using materials and alloys available at the time. Materials such as 304 and 316 stainless steel were well known with established mechanical properties and fabrication routes. Over the decades, performance knowledge relevant to water reactor operation was gained. In addition, new materials were developed and deployed for specific reactor environments. For example, nuclear fuel clad transitioned from stainless steels (which experienced SCC in BWR water chemistry) to the zirconium-based alloys still in use today, as a response to SCC issues with stainless steel and the superior neutronics of zirconium. As zirconium alloys were developed and shown to be viable, the transition away from stainless steel was natural. Power plant piping has evolved from stainless steel to Ni-base alloys such as Alloy 600 and Alloy 690. Today, a modern nuclear power plant incorporates a wide range of alloys, spanning stainless steels to LASs, as shown earlier in Fig. 1.6.

1.3 OVERVIEW OF KEY MATERIALS OF CONSTRUCTION FOR LWR APPLICATIONS

The objective of this section is to provide a brief overview for the key alloys in use, their function, and limitations. These materials will be organized by starting with Zircaloy-based cladding and moving outward from the reactor core and including austenitic stainless steels, cast-austenitic stainless steels (CASSs), LASs, and Ni-base alloy piping. While a brief overview is presented in this chapter, much greater detail on each class of material is found in later chapters.

1.3.1 ZIRCONIUM-BASED ALLOYS

Zirconium alloys are used in water reactors in several key components. Zirconium is utilized as a fuel cladding material in all water reactor systems due to the superior neutronics and its general resistance to SCC. The dominant use of Zr in water reactors is as fuel cladding and today is the result of 40 years of continuous development and improvement [13].

The ideal nuclear fuel clad must possess a large number of properties. It must be available and readily manufactured into thin-walled tubing to maximize heat transfer. It must have high strength and suitable ductility. It must be compatible with transmutation products on the inner surface of the clad and maintain good corrosion performance in high-temperature water on the outside surface of the clad. Finally, it must have a low neutron absorption cross section. Most common metal systems are quickly eliminated. Iron and nickel alloys are well known, but can be eliminated due to their high thermal neutron macroscopic absorption cross sections. Aluminum and its alloys have low strength at operating reactor temperatures. Zirconium, while not as strong as other alloy systems, has very favorable neutron absorption characteristics and undergoes manageable uniform corrosion.

As a result, zirconium was selected for fuel cladding in early commercial LWRs. In PWRs, the most widely used cladding has been Zircaloy-4 (containing Zr-Sn-Fe-Cr). Although still in use in many PWRs, Zircaly-4 is gradually being replaced by newer Nb-containing alloys: Zirlo and M5. In Russia, Zr-1%Nb and Zr-2.5%Nb alloys are traditionally used. In BWRs, the cladding has essentially remained as Zircaloy-2 (containing Zr-Sn-Fe-Cr-Ni). Chapter 7 provides considerably more detail about these materials and their development.

Following the Fukushima Daiichi accident, a worldwide examination of fuel cladding has been initiated. While zirconium alloy clad fuel operates successfully to high burnup, under severe accident conditions, the high-temperature zirconium-steam interaction is exothermic and generates hydrogen and may exacerbate an accident rather than mitigate. There is now renewed international interest in alternative fuel designs that would be more resistant to fuel failure and hydrogen production. Recent results and directions are beyond the scope of this document, but the most recent research directions are highlighted in Refs. [14–16].

1.3.2 AUSTENITIC STAINLESS STEELS

Austenitic steels are a class of nonmagnetic stainless steels that contain high levels of chromium and nickel and low levels of carbon. Known for their formability and resistance to corrosion, austenitics are the most widely used grade of stainless steel. There are over 150 different grades of stainless steel in current commercial use, of which 15 are commonly used in LWRs. These alloys are listed in Table 1.3.

The most common alloys utilized in water reactors are 304 SS and 316 SS, with specific applications of 308 SS, 309 SS, 321 SS, and 347 SS. Both 304 SS and 316 SS are widely used in nuclear reactor applications as well as 321 SS and 347 SS grades. As noted in Chapter 8, developmental alloys such as D9 and HT-UPS have been developed and utilized in advanced reactor designs. A much wider range of austenitic alloys has been explored in fast reactors, but these reactors have different neutron spectra, lower rates of helium and hydrogen production, and, most importantly, an absence of water and therefore thermal neutrons.

Decades of experience in both commercial LWRs and fast-breeder reactors have resulted in further refinement of stainless steel compositions for reactor applications.

The background on the choice of these materials is described in the *Corrosion and Wear Handbook for Water-Cooled Reactors* [17], which explains the concept of material selection for the first commercial PWR

Table 1.3 Chemical Compositions of Stainless Steels used in LWRs (wt.%) (balance is Fe)

	C	Nb	Cr	Cu	Mn	Mo	Ni	P	S	Si
Type 316	0.08 max.	—	16.0–18.0	—	2.0 max.	2.0–3.0	10.0–14.0	0.045 max.	0.03 max.	1.0 max.
Type 316L	0.03 max.	—	16.0–18.0	—	2.0 max.	2.0–3.0	10.0–14.0	0.045 max.	0.03 max.	1.0 max.
Type 304	0.08 max.	—	18.0–20.0	—	2.0 max.	—	8.0–10.5	0.045 max.	0.03 max.	1.0 max.
Type 304L	0.03 max.	—	18.0–20.0	—	2.0 max.	—	8.0–12.0	0.045 max.	0.03 max.	1.0 max.
Type 347	0.08 max.	10 × C min.	17.0–19.0	—	2.0 max.	—	9.0–13.0	0.045 max.	0.03 max.	1.0 max.
Type 308L	0.04 max.	—	18.0–21.0	0.75 max.	0.5–2.5	0.75 max.	9.0–11.0	0.04 max.	0.03 max.	0.90 max.
Type 309L	0.04 max.	—	22.0–25.0	0.75 max.	0.5–2.5	0.75 max.	12.0–14.0	0.04 max.	0.03 max.	0.90 max.
Type 403	0.15 max.	—	11.5–13.0	—	1.0 max.	—	—	0.04 max.	0.03 max.	0.5 max.
Type 410	0.15 max.	—	11.5–13.5	—	1.0 max.	—	—	0.04 max.	0.03 max.	1.0 max.
Type 630	0.07 max.	0.15–0.45	15.0–17.5	3.0–5.0	1.0 max.	—	3.0–5.0	0.04 max.	0.03 max.	1.0 max.

1.3 OVERVIEW OF KEY MATERIALS OF CONSTRUCTION FOR LWR APPLICATIONS

(Shippingport, Pennsylvania). The main criteria for the material selection were (a) high corrosion resistance in a readily available alloy; (b) good fabrication characteristics; and (c) extensive experience with the alloys in other industries, such as petrochemical and fossil power. The combined experience in other industries and their history were key factors as this reduced the risk of unexpected failures and allowed the previous research and literature review results to be leveraged. Past experience in both LWRs and fast-breeder reactors has shown that 316 SS offers higher performance and greater reliability than 304 SS, particularly in aqueous corrosion and radiation damage resistance, which will be discussed in more detail in Chapters 6 and 8 and in Refs [17–19].

The stainless steels listed in Table 1.3 are used in the primary systems in a variety of forms, including seamless piping, forgings, castings, and plates. The specific stainless steel–component combinations vary among reactor designs and manufacturers but in general share the following traits:

- Main coolant piping and elbows for PWR primary circuits are cast 316 SS (CF8M). Cast grades are discussed in detail below.
- Seamless Type 316 SS is also used for main coolant piping in some PWRs.
- The inner surfaces of the RPV, pressurizer, and steam generator channel head are clad with Type 308L SS; they are then stress relieved during a post-weld heat treatment.
- Pump and valve casings are generally made of cast stainless steel (CF8). Type 410 and Type 630 SS are used for the valve stem. Type 347 SS is used for the reactor coolant pump stem.
- Type 410 and Type 403 SS are used for the stem and parts of the control rod drive mechanism, such as the plunger.
- Main coolant piping, elbows, joints and elbows for BWR circuits are composed for 304 and 316 SS grades.
- Like PWRs, the inner surfaces of the RPV are clad with stainless steel, typically 309 or 308L SS grades.
- Core internal structures in BWRs are made of wrought 304 SS and 316 SS.

Various types of corrosion, such as general corrosion, crevice corrosion, SCC, and corrosion product transport in the primary system, were studied in the stainless steels described above in hydrogenated and oxygenated high-temperature water to determine their suitability for use in the Nautilus and in the first commercial PWRs. These forms of degradation are covered in more detail in Chapters 5, 6, and 8.

1.3.3 CAST-AUSTENITIC STAINLESS STEELS

CASS materials are prevalent in many safety-related nuclear power plant components, including the primary coolant piping system in PWRs. CASS components are also currently being designed for new power plant construction. These steels have similar chemistry and properties as above, but are cast into the near final component form rather than forging and machining. These grades are often chosen for economic reasons as the casting process is cheaper. Historically, cast stainless steel grades have performed well in nuclear reactor applications and there are relatively few key degradation modes of concern.

Today, CASSs are used in a variety of applications in both BWRs and PWRs. Common alloys in service include the CF3 and CF8 series of alloys with the CF3, CF3A, CF3M, CF8, CF8A, and CF8M being the most prominent choices. The compositions of these alloys are listed in Table 1.4, and compared to wrought alloys. The alloys are frequently used for reactor coolant and auxiliary system piping as well as pump casings, valve bodies, and other assorted fittings.

Like the other stainless steels described above, CASSs are exposed to elevated temperatures and corrosive environments of the primary cooling circuit. Overall, degradation modes for CASS in reactor applications are relatively minor when compared to other alloy systems under normal operating conditions through 40 or 60 years of life. Thermal aging and irradiation effects are not considered to be areas of concern given the relatively low temperatures and fluences for most components. There have been limited cases of SCC in CASS components in both BWRs and PWRs; however, these are attributed to irregularities in composition or microstructure rather than general vulnerabilities. In BWRs, there is an increased susceptibility to SCC in areas of cold work or

Table 1.4 Comparison of ASTM Chemistry Specifications for Cast Stainless Steel and Wrought Equivalents (compositions in wt.%) [20,21]

Grade	Type	C Max	Mn Max	Si Max	P Max	S Max	Cr	Ni	Mo
CF3	ASTM 743	0.03	1.50	1.50	0.04	0.04	17.0–21.0	8.0–12.0	
CF3A	ASTM 743	0.03	1.50	1.50	0.04	0.04	17.0–21.0	9.0–13.0	2.0–3.0
CF3M	ASTM 743	0.03	1.50	1.50	0.040	0.040	17.0–22.0	9.0–13.0	2.0–3.0
CF8	ASTM 743	0.08	1.50	2.00	0.040	0.040	18.0–21.0	8.0–11.0	
CF8A	ASTM 743	0.08	1.50	2.00	0.040	0.040	18.0–21.0	8.0–11.0	
CF8M	ASTM 743	0.08	1.50	2.00	0.040	0.040	18.0–21.0	9.0–12.0	2.0–3.0
304 SS	Wrought	0.08	2.00	1.00	0.045	0.03	18.0–20.0	8.0–11.0	–
304L SS	Wrought	0.03	2.00	1.00	0.045	0.03	18.0–20.0	8.0–12.0	–
316 SS	Wrought	0.08	2.00	1.00	0.045	0.03	16.0–18.0	10.0–14.0	2.0–3.0
316L SS	Wrought	0.03	2.00	1.00	0.045	0.03	16.0–18.0	10.0–14.0	2.0–3.0
321 SS[a]	Wrought	0.08	2.00	1.00	0.045	0.03	17.0–19.0	9.0–13.0	–
347 SS	Wrought	0.08	2.00	1.00	0.045	0.03	17.0–19.0	9.0–13.0	–

[a]0.70 wt.% Ti.

weldments. To date, there has been no record of IASCC in these components. Similarly, to date, there are no concerns for CASS components related to general or localized corrosion, fatigue, flow-accelerated corrosion (FAC), or wear for current lifetimes.

Under extended service scenarios, there may be additional degradation modes to consider. Thermal aging could lead to decomposition of key phases, resulting in increased susceptibility to embrittlement, irradiation-induced degradation, SCC, and general corrosion [22]. This section will explore those degradation modes in more detail.

1.3.4 Ni-BASE ALLOYS

Nickel and nickel alloys are used for a wide variety of applications, spanning many industries including aircraft engines to springs to piping. The vast majority of these applications involves corrosion resistance and/or heat resistance. These alloys, often called superalloys, have high strength, good ductility, and excellent corrosion resistance.

These alloys are used extensively in nuclear reactor applications for water-based designs due to their corrosion resistance, including both low uniform corrosion rate and a resistance to SCC. The coefficient of thermal expansion is also similar to that of LASs, enabling the ability to join these key material systems reliably. These alloys are also utilized in a variety of advanced reactor designs due to their high-temperature performance. Yonezawa [23] has provided an excellent review of these alloys and their properties for nuclear energy applications.

While these alloys are widely utilized, there are some drawbacks that limit an even broader use. Nickel has a very high cross section and absorbs thermal neutrons, reducing the efficiency of power generation and yielding highly radioactive components. Further, nickel is a very expensive alloying element and Ni-based alloys are also expensive as a result.

Several key alloys are utilized in water reactor designs. These alloys, their composition, and water reactor utilization are summarized in Table 1.5.

These materials are used in a wide range of components. For example, Fig. 1.7 provides an overview of where key Ni-base alloys (specifically Alloy 600) are used in BWRs. Uses include attachment welds, bolts, butters, covers, and nozzles. Other BWR components fabricated from Ni-base alloys are highlighted in Table 1.6. This class of alloys is used even more extensively in PWR systems as the pressurizer and steam generator contain a significant

1.3 OVERVIEW OF KEY MATERIALS OF CONSTRUCTION FOR LWR APPLICATIONS

Table 1.5 Summary of Key Ni-Base Alloys Utilized in Nuclear Reactor Applications

Alloy	Usage	Ni	Cr	Fe	Ti	Al	Nb	Mo
600	BWR, PWR, PHWR	75	16	8	0.3	0.2		
690	PWR, PHWR	61	29	9	0.5	0.5		
625	BWR	61	22	<5	0.3	0.3	3.5	9
718	BWR, PWR, PHWR	53	18	19	0.9	0.6	5	2.5
X-750	BWR, PWR, PHWR	72	15.5	7	2.5	0.7	1	
800	BWR, PWR, PHWR	33	21	>39.5	0.4	0.4		

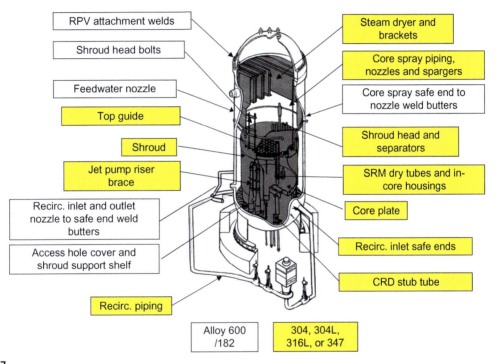

FIG. 1.7

BWR components containing Alloy 600 and alloy 182 and 82 weld metals (white boxes). Austenitic stainless steels are shown in yellow boxes [24].

From NUREG 7153, Expanded Materials Degradation Assessment (EMDA).

number of Alloy 600 or alloy 690 components. The utilization, metallurgy, and in-service degradation of Ni-base alloys are discussed in more detail in Chapter 9.

Ni-base alloys have performed very well in nuclear reactor applications, particularly in recent years. The most significant area of degradation for Ni-base alloys in water reactors is readily SCC and environmentally assisted fatigue or fracture. Thermal aging and defect-related failures may also impact weldments. These degradation modes are discussed in Chapter 9 in some detail as well as in Refs. [8,25,26].

Significant SCC was observed in these alloys in the early 1970s, particularly in BWR components under normal water chemistry. Cracking was also observed in PWR environments in laboratory tests as early as 1959, but not

Table 1.6 BWR Components Fabricated from Ni Alloys

BWR Component	Nickel Alloy Designation
BWR shroud head bolts	Alloy 600
Pressure vessel attachment pads	Alloy 182
Control rod penetrations	Alloy 600
Control rod penetration welds	Alloy 182
Core shroud support welds	Alloy 182
Pressure vessel nozzles	Alloys 182 and 82
Safe ends	Alloy 600
Weld metal deposits	Alloys 82 and 182
Jet pump beams	Alloy X-750
Fuel rod spacers	Alloy X-750

observed in service until the early 1970s. Over the decades, however, there have been trends in degradation in Ni-base alloys in water environments. The nuclear industry has responded to each trend with new materials and mitigation.

For example, consider the history and development of Ni-base alloys for steam generators. As noted above, the first significant observations of susceptibility to SCC in service occurred in the early 1970s. At least 12 power plants found that highly cold-worked regions of Alloy 600, such as the tight U-bends in steam generator tubes and cold-worked expansion of the tubes within the tube sheet, were susceptible to IGSCC. Alternative material research to mitigate this degradation was undertaken by multiple vendors and academic institutions.

By the 1980s, IGSCC became prevalent in the 1980s, leading to steam generator retirement and replacement. In addition to primary-side and secondary-side steam generator tubing degradation, cracking of other Alloy 600 PWR components has been documented, including pressurizer heater sleeves and welds, pressurizer instrument nozzles, reactor vessel closure head nozzles and welds, reactor vessel outlet nozzle welds, and reactor vessel head instrumentation nozzle and welds. In response, thermally treated Alloy 600 (600TT) was put into service after demonstrating dramatic improvement over mill-annealed 600. In parallel, Alloy 690 was also considered and qualified for service. The most critical difference between Alloy 690 and Alloy 600 chemical requirements is for Cr; Alloy 600 requires a much lower concentration, 14–17 wt.%. This change in Cr level in Alloy 690 is compensated for by a decrease in the Ni concentration.

Wrought Alloy 690 and its associated weld metals (Alloys 152, 52, 52M, and other variants) have become the common replacement and repair materials for Alloy 600 and Alloy 182/82 weld metals with lower chromium content in PWRs. While this is an active area of research for the nuclear industry, the SCC susceptibility is significantly reduced in over Alloy 600. In general, potential degradation modes of concern for Alloy 690 are similar to those for Alloy 600, including SCC, corrosion fatigue, and environment-induced fracture at high and low temperatures. Chapter 9 provides much more detail about the development, use, and degradation of these materials in LWR service.

1.3.5 LOW-ALLOY STEELS

Plain carbon steels provide adequate strength for many structural engineering purposes. Through the addition of particular other elements at low concentrations, LASs possess precise chemical compositions and provide better mechanical properties than many conventional mild or plain carbon steels. LASs offer higher strength, good fracture toughness properties, are weldable, and are relatively low cost, making them attractive for nuclear applications, particularly thick section components.

1.3 OVERVIEW OF KEY MATERIALS OF CONSTRUCTION FOR LWR APPLICATIONS

Carbon and low alloys steels serve in a variety of locations within the nuclear power generating fleet, in a variety of forms, including seamless piping, forgings, castings, plate, and bolting. Table 1.7 gives the compositions of the major LASs currently in use in water reactor systems. Many of these steels were derived originally from those developed for thick-walled vessels used for boilers and chemical plants. Chapter 10 provides a detailed assessment of the history and alloying approach for these complex and important alloy systems.

Given their ductility and cost, they are used extensively as pressure boundary materials such as pressure vessels and piping in the primary, secondary water, and service water systems of water reactors. In reactor coolant system components, such as the pressure vessel, pressurizer, and some piping, the carbon steels and LASs are clad on the inside wetted surface with corrosion-resistant materials such as austenitic stainless steels or nickel-base alloys. Further details on RPV and piping steels are provided in Chapter 10 of this book and in Refs [27,28], among many others.

The specific carbon steel or LAS/components that are used in a reactor vary between reactor designs and manufacturers. In general, however, the reactor components include the following:

- The RPV is the most significant LAS component and provides a critical role in safety of the reactor. This component must be able to contain the reactor core at elevated temperatures and pressures, even in accident conditions. Owing to harsh environmental conditions and the impracticality of replacing this component, the RPV cylinder shell material is often the lifetime-limiting component for a nuclear reactor. RPV sections are typically manufactured from rolled LAS A533 Gr. B plates, which are then welded to form a cylinder. The cylinder is clad on the internal surface with a layer of austenitic SS to reduce corrosion.
- The top and bottom heads of the pressure vessel are generally clad LAS A508 Gr. 2 forgings using the same cladding/heat treatment conditions as for the vertical sections.
- Steam generator shells for PWRs and PHWRs are LAS A533 Gr. A Class 1 or Class 2 plates, which, like the pressure vessel, are heat-treated after subassembly. The secondary side of the steam generator is not usually clad.
- Steam generator tube sheets in PWRs are generally A508 Gr. 2 Class 1 or A508 Gr. 2 Class 2, with cladding on the primary side of the bottom head. Steam generator channel heads in PWRs are generally A216 Gr.
- The pressurizer shells in PWRs are generally LAS A516 Gr.70 or A533 Gr. B plate with internal stainless steel cladding.
- Reactor coolant piping for PWR primary circuits may be seamless carbon steel with, in some designs, austenitic stainless steel cladding. Alternatively, a higher-cost option of using cast stainless steel (CF8M) has been used for the main coolant piping, as described above. The recirculation piping in BWRs is usually stainless steel (Types 304 SS, 316 SS, 304L SS, 316L SS), although unclad A333 Gr. 6 carbon steel piping may be used in the main steam and the feedwater lines.

Given their critical importance for safety, regulations require that RPV steels maintain a conservative margin for fracture toughness to ensure the integrity of an RPV during either normal operation and maintenance cycles or under accident transients such as pressurized thermal shock (PTS). There are a number of key degradation modes that must be considered for LASs and major components such as the RPV.

In general, corrosion is not a significant issue for these alloys, despite their general vulnerability to uniform corrosion, as the environment is separated from the material via cladding in the most critical components. SCC of ductile carbon steel and LAS components and their associated weldments can also occur, but such observations are quite rare. Corrosion fatigue cracking and FAC are the primary modes of degradation observed in piping applications, although both are manageable.

Irradiation effects are easily the most significant potential impact on LAS. Neutron irradiation degrades fracture toughness, in some cases severely. The past few decades have seen remarkable progress in developing a better understanding of irradiation embrittlement mechanisms as is described in both Chapters 4 and 10 as well as [29–31]. A major step forward was the development of physically based and statistically calibrated models of

Table 1.7 ASTM Compositional Specifications for Ferritic and Bainitic Carbon Steel and LAS Concentrations given as Weight Percentages (balance is Fe)

ASME/ASTM	C Max[a]	Mn	P Max	S Max	Si	Cu Max	Ni	Cr	Mo	V Max
Ferritic steels										
A105	0.035	0.6–1.05	0.035	0.04	0.1–0.35	0.4 (1)				
A106 GrB	0.3	0.29–1.06	0.035	0.035	0.1 min	0.4 (2)	0.4 max[b]	0.3 max[b]	0.12 max[b]	0.05
A216 Gr WCB	0.3	1.0 max	0.04	0.045	0.6 max	0.3 (3)	0.4 max[c]	0.4 max[c]	0.15 max[c]	0.08[c]
A302 GrB	0.25	1.15–1.50	0.035	0.035	0.15–0.4		0.5 max[d]	0.5 max[d]	0.2 max[d]	0.03[d]
A333 Gr6	0.3	0.29–1.06	0.035	0.035	0.1 max				0.45–0.6	
A508 Gr3	0.25	1.2–1.5	0.025	0.025	0.15–0.4		0.4–1.0	0.25 max	0.45–0.6	0.05
A516 Gr70	d	0.85–1.2	0.035	0.035	0.15–0.4					
A533 Type A	0.25	1.15–1.5	0.035[e]	0.035	0.15–0.4				0.45–0.6	
A533 Type B	0.25	1.15–1.5	0.035	0.035	0.15–0.4		0.4–0.7		0.45–0.6	
Bainitic steels										
1Cr1Mo0.25V	0.33	0.85	0.012	0.009	0.25			1.0	1.25	0.25
2Cr1Mo Gr22	0.026	0.49			0.28	0.05		2.42	0.98	
NiCrMoV (A469 Cl8)	0.28	0.6	0.015	0.018	0.15–0.3		3.25–4.0	1.25–2.0	0.3–0.6	0.15
NiCrMoV (A470 Cl8)	0.35	1.0	0.015	0.018	0.15–0.35		0.75	0.9–1.5	1.0–1.5	0.3
NiCrMoV (A471 Cl8)	0.28	0.7	0.015	0.015	0.15–0.35		2.0–4.0	0.7–2.0	0.2–0.7	0.05

[a]Carbon: max. varies with thickness of plate; 0.5–2 in., 0.28% max; 2–4 in., 0.30% max; 4–8 in., 0.31% max.
[b]Sum of Cu, Ni, Cr, and Mo shall be <1.00%; sum of Cr and Mo shall not exceed 0.32%.
[c]Limit for V and Nb may be increased to 0.1% and 0.05%, respectively.
[d]Sum of Cr and Ni shall not exceed 0.32%.

Charpy V-notch (CVN)-indexed transition-temperature shifts. Those semi-empirical models account for key embrittlement variables and their interactions, including the effects of copper (Cu), nickel (Ni), phosphorus (P), fluence (ϕt), flux (ϕ), and irradiation temperature (T_i). Models of the evolution of nanoscale precipitates rich in Cu, manganese (Mn), and Ni are quantitatively consistent with experimental observations of the complex interplay between those elements and other embrittlement variables. The models have not only provided early warnings of potential technical challenges, such as the contribution of Mn and Ni in high-Ni steels to embrittlement by so-called "late blooming" phases, but have also enabled the assessment of outliers in the Transition Temperature Shift Database as well as other contradictory observations. However, these models and the present understanding of radiation damage are not fully quantitative and do not take into consideration the potential contribution of all potentially significant variables and aging technical issues.

Chapter 10 provides an in-depth assessment of the key alloys, fabrication techniques, as well as an overview of base mechanical properties. This chapter also provides an overview of the main applications for water- and gas-cooled reactors as well as the structural integrity tools that have been developed to ensure their reliability over the lifetime of the reactor. Key degradation modes and implications for service are also provided.

REFERENCES

[1] Nuclear Energy Institute website, https://www.nei.org/Knowledge-Center/Nuclear-Statistics/World-Statistics/World-Nuclear-Generation-and-Capacity, updated April 2017.
[2] IAEA Database on Nuclear Power Reactors, https://www.iaea.org/pris/.
[3] A Technology Roadmap for Generation IV Nuclear Energy Systems, issued by US DOE Nuclear Energy Research Advisory Committee and Generation IV International Forum 2002.
[4] Technology Roadmap Update for Generation IV Nuclear Energy Systems, issued by Generation IV International Forum 2014.
[5] Courtesy of World Nuclear Association, http://world-nuclear.org/.
[6] R. Staehle, private communication.
[7] Life Beyond 60 Workshop Summary Report, NRC/DOE Workshop U.S. Nuclear Power Plant Life Extension Research and Development, Feb. 19–21, 2008, U.S. Nuclear Regulatory Commission and U.S. Department of Energy, Washington, 2008.
[8] Expanded Materials Degradation Assessment (EMDA), US Nuclear Regulatory Commission, NUREG/CR-7153, vols. 1–5 2013.
[9] T.R. Allen, J.T. Busby, Radiation damage concerns for extended light water reactor service, JOM J. Miner. Met. Mater. Soc. 61 (7) (2009) 29–34.
[10] Y. Solomon, An Overview of Water Chemistry for Pressurized Water Nuclear Reactors, in: Water Chemistry of Nuclear Reactor Systems, Proceedings of an international conference organized by the British Nuclear Energy Society, Bournemouth, 1978.
[11] D.D. Macdonald, Viability of hydrogen water chemistry for protecting in-vessel components of boiling water reactors, Corrosion 48 (3) (1992) 194–205.
[12] R.W. Staehle, J.A. Gorman, Proceedings of the 10th International Conference on Environmental Degradation of Materials in Nuclear Power Systems: Water Reactors, NACE International, Houston, TX, 2002.
[13] J.S. Armijo, L.F. Coffin, H.S. Rosenbaum, Development of Zirconium-Barrier Fuel Cladding, in: A.M. Garde, E.R. Bradley (Eds.), Zirconium in the Nuclear Industry: Tenth International Symposium, STP1245, ASTM International, 1994.
[14] C.R.F. Azevedo, Selection of fuel cladding material for nuclear fission reactors, Eng. Fail. Anal. 18 (8) (2011) 1943–1962.
[15] Accident tolerant fuel concepts for light water reactors, Proceedings of a Technical Meeting Held at Oak Ridge National Laboratories, United States of America, 13–16 October 2014, IAEA TECDOC No. 1797, 2016.
[16] M. Kurata, Research and development methodology for practical use of accident tolerant fuel in light water reactors, Nucl. Eng. Technol. (2016) 26–32.
[17] D.J. DePaul (Ed.), Corrosion and Wear Handbook for Water-Cooled Reactors, United States Atomic Energy Commission/McGraw-Hill, New York, 1957.

[18] P.J. Maziasz, J.T. Busby, Properties of austenitic stainless steels for nuclear reactor applications, in: Comprehensive Nuclear Materials, Elsevier, 2012, pp. 267–283.
[19] S.J. Zinkle, J.T. Busby, Structural materials for fission & fusion energy, Mater. Today 12 (11) (2009) 12–19.
[20] Specification for castings, iron-chromium, iron-chromium-nickel, corrosion-resistant, for general application, Standard ASME/ASTM SA/A 743/A 743M.
[21] Specification for castings, iron-chromium-nickel, corrosion resistant, for severe service, Standard ASME/ASTM SA/A 744.
[22] H.M. Chung, Aging and life prediction of cast duplex stainless steel components, Int. J. Press. Vessel. Pip. 50 (1–3) (1992) 179–213.
[23] T. Yonezawa, Nickel alloys: properties and characteristics, in: Comprehensive Nuclear Materials, vol. 1, Elsevier, Amsterdam, 2012, pp. 233–266.
[24] P. Andresen, private communication.
[25] S.M. Bruemmer, G.S. Was, Microstructural and microchemical mechanisms controlling inter-granular stress corrosion cracking in light-water-reactor systems, J. Nucl. Mater. 457 (2015) 165–172.
[26] P.M. Scott, An Overview of Materials Degradation by Stress Corrosion Cracking in PWR's, in: D. Féron, J.M. Olive (Eds.), Corrosion Issues in Light Water Reactors – Stress Corrosion Cracking, European Federation of Corrosion Publications, vol. 51, Woodhead Publishing, Cambridge, UK, 2007.
[27] L.M. Davies, A comparison of western and eastern nuclear reactor pressure vessel steels, Int. J. Press. Vessel. Pip. 76 (3) (March 1999) 163–208.
[28] Y. Tanaka, Reactor pressure vessel (RPV) components processing and properties, in: N. Soneda (Ed.), Irradiation Embrittlement or Reactor Pressure Vessels (RPVs) in Nuclear Power Plants, Woodhead Publishing, Cambridge, UK, 2015. ch. 2.
[29] R.E. Lott, S.S. Brenner, M.K. Miller, A. Wolfenden, Development of radiation damage in pressure vessel steels, Proc. Am. Nucl. Soc., Trans. 38 (1981) 303.
[30] G.E. Lucas, An evolution of understanding of reactor pressure vessel steel embrittlement, J. Nucl. Mater. 407 (2010) 59–69.
[31] Integrity of reactor pressure vessels in nuclear power plants: assessment of irradiation embrittlement effects in reactor pressure vessel steels, In: IAEA Nuclear Energy Series No. NP-T-3.11, IAEA, Vienna, 2009.

CHAPTER 2

OVERVIEW OF REACTOR SYSTEMS AND OPERATIONAL ENVIRONMENTS FOR STRUCTURAL MATERIALS IN GEN-IV FISSION REACTORS

Stuart A. Maloy*, Ken Natesan[†], David E. Holcomb[‡], Concetta Fazio[§], Pascal Yvon[¶]

Los Alamos National Laboratory, Los Alamos, NM, United States* Argonne National Laboratory, Lemont, IL, United States[†] Oak Ridge National Laboratory, Oak Ridge, TN, United States[‡] European Commission, JRC, Petten, The Netherlands[§] CEA—Saclay, Gif-sur-Yvette, France[¶]

CHAPTER OUTLINE

- 2.1 Introduction 23
- 2.2 Liquid Metal Cooled Fast Reactors 24
    - 2.2.1 Sodium Fast Reactors (SFR)—Description of Overall Design and Applications 24
    - 2.2.2 Lead Fast Reactor (LFR) 30
- 2.3 Helium Cooled Reactors 36
    - 2.3.1 Very High-Temperature Reactors (VHTR) 36
    - 2.3.2 Gas-Cooled Fast Reactor 39
- 2.4 Other Gen-IV Fission Reactor Systems 41
    - 2.4.1 Molten Salt Fueled Reactor (MSR) 41
    - 2.4.2 Molten Salt Cooled Reactor 43
    - 2.4.3 Super-Critical Water Reactor (SCWR) 43
    - 2.4.4 Summary 46
- References 46

2.1 INTRODUCTION

The first-generation nuclear electricity production involved a mix of graphite moderated, gas cooled and light-water cooled reactors [1]. A second generation emerged, based mostly on light-water cooled reactors with higher power levels and improved core designs, which constitutes the vast majority of the current fleet of nuclear power plants. Even safer and more efficient evolutions of these light-water cooled reactors are being brought on line but they may not be sustainable in the long run, as the uranium reserves are limited. Therefore, in 2000, an international panel of experts selected six Generation-IV (Gen-IV) reactors, which offer significant advantages over typical light-water reactors including breeding and waste management capacity, increased power conversion efficiency, passive safety features and in some cases, process heat for other applications (e.g., hydrogen production) [2]. These families of reactors include three fast reactors [sodium fast reactor (SFR), lead fast reactor (LFR) and gas-cooled fast reactor (GFR)], one thermal reactor [very high-temperature reactor (VHTR)] and two fast or thermal reactors [supercritical water reactor (SCWR) and molten salt reactor (MSR)]. The extreme environments in these families of reactors create significant challenges to various selected materials, ranging from high fluences of fast neutrons (SFR, LFR, GFR, SCWR, and MSR), more corrosive environments such as MSR or LFR and high temperatures in the helium-cooled reactor concepts (e.g., GFR and VHTR). Among all the Gen-IV systems, the leading concept is the SFR. In part, because of a substantial base of experience from previous technology programs [3]. This chapter discusses the materials challenges posed by Gen-IV reactor concepts, including summarizing those posed by prototypic irradiation effects in candidate structural alloys for these concepts, while pointing the reader to other chapters for more detailed discussions of radiation effects in specific classes of materials.

2.2 LIQUID METAL COOLED FAST REACTORS

2.2.1 SODIUM FAST REACTORS (SFR)–DESCRIPTION OF OVERALL DESIGN AND APPLICATIONS

Typical SFRs can be either a pool- or loop-type design, with the same objective of transferring heat from the core through primary and intermediate sodium circuits to a steam generator (SG) driving a power turbine. Fig. 2.1 shows a schematic representation of the two design concepts [4]. Over the past 40 years, or more, several sodium reactors of various power levels and types have been built and operated around the world resulting in substantial databases on fuel and materials performance, as well as component design and operation. Table 2.1 lists both pool and loop types that were built in various countries, some of which have operated for 30 years or more. The results from these experimental and demonstration reactors have contributed to an understanding of both generic issues and system specific issues in SFRs.

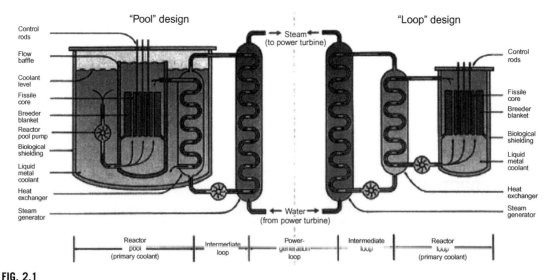

FIG. 2.1

Schematic representation of pool and loop designs of sodium fast reactors [4].

Table 2.1 A List of Experimental and Demonstration SFR (>50 MWth)

Country	Reactor	MWth	Reactor Type	Operating Life (years)	Inlet to IHX Temp. (°C)	Inlet to SG Temp. (°C)
USA	EBR-II	62.5	Pool	30	473	467[a]
	FFTF	400	Loop	12	565	
UK	DFR (NaK)	60	Loop	14	350	330
	PFR	650	Loop	20	550	540
Russia	BOR-60	55	Loop	39	545	480
	BN-350	750	Loop	26	430	415
	BN-600	1470	Pool	29	550	520
	BN-800	2100	Pool	3	547	505
France	Phenix	563	Pool	36	560	550
	SuperPhenix	2990	Pool	12	542	525
Japan	JOYO	140	Loop	32	500	470
	Monju	714	Loop	15	529	*
China	CEFR	65	Pool	4	516	495

EBR-II, experimental breeder reactor-II; CEFR, Chinese experimental fast reactor; DFR, Dounreay fast reactor; PFR, prototype fast reactor at Dounreay.
[a]No Steam generator and only sodium to air heat exchanger.

The major components in a SFR include the in-core components such as fuel cladding, duct, wrapper wire, etc., and out-of-core components that can be grouped under the reactor system and primary containment boundary, primary heat transport system, intermediate heat transport system, and power generation system.

The most important in-core component is the fuel and cladding. The fast reactor fuel pin cladding tubes, used to encapsulate cylindrical fuel pellets, are typically ∼7 mm in diameter with a wall thickness of ∼0.5 mm [5]. The fuel pin length varies with reactor design and is typically several meters long. Groups of fuel pins (typically 50–200) are placed into a duct as shown in Fig. 2.2. The duct segments the core, provides support to the fuel pins, and provides a means to distribute coolant across the reactor core. A fast reactor duct is typically hexagonal in shape with a flat-to-flat distance of ∼150 mm, and is made from ∼3-mm-thick steel.

2.2.1.1 Sodium flow conditions for out of core components

The sodium flow conditions for the out of core components vary depending on the reactor type, as shown in Table 2.1. In the past, the reactor inlet temperature to the intermediate heat exchanger (IHX) varied from 350°C to 565°C and the inlet to the SG varied from 330°C to 550°C [7]. Alloys used for the reactor vessel, primary/secondary piping, IHX, and SGs must comply with relevant national construction codes such as the American Society of Mechanical Engineers (ASME) Boiler and Pressure Vessel Code Division 5 for high-temperature reactors or the RCC-MRX (design and construction rules for mechanical components in high-temperature structures, experimental reactors, and fusion reactors) in France [8]; elevated-temperature components must comply with corresponding high-temperature construction codes such as ASME Section III Subsection NH [9]. Designs must consider time-dependent effects on mechanical properties such as creep, creep-fatigue, and creep ratcheting in addition to the environmental effects such as irradiation effects, erosion-corrosion, localized cracking, and interstitial element transfer.

2.2.1.2 Sodium flow conditions for in-core components

Typical in-core clad temperatures in a fast reactor, such as the Fast Flux Test Facility (FFTF), range from ≈375°C to 650°C with a peak temperature between 550°C and 650°C [5, 10]. A limit of 650°C is imposed on the inner surface of cladding materials to reduce effects of thermal creep. Because the fuel releases fission gas during irradiation, the

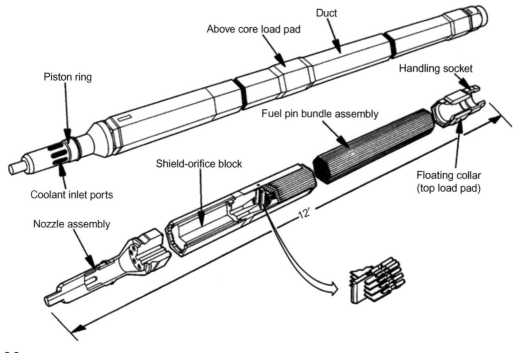

FIG. 2.2

Diagram showing typical layout of fuel pin bundle assembly and duct in a fast reactor [6].

From Maloy, S.A., Toloczko, M., Cole, J., Byun, T.S., Core materials development for the fuel cycle R&D program, Journal of Nuclear Materials, volume 415, issue 3, pp. 302–305 (2011).

pressure inside a fuel pin builds up with increasing dose. As the fuel swells, the possibility of mechanical and chemical interaction between the fuel and the cladding increases. In particular, lanthanide fission products, which are produced under irradiation, diffuse to the clad/fuel interface leading to enhanced Fuel Clad Chemical Interactions (FCCI), which increase with higher burnup. Such chemical interactions decrease the effective load-bearing cladding thickness and further degrade the cladding properties [11]. In addition, fission product gases, Xe and Kr, cause the fuel to swell, which can lead to large stresses in the cladding. The buildup of these stresses with burnup can be delayed through novel fuel designs such as using oxide fuels with low smear densities and a central hole and venting of fission gases [12]. The outer surface of the cladding must be resistant to corrosion by the sodium coolant [13].

Typical duct operating conditions in a fast reactor such as the FFTF are an average irradiation temperature of 370–575°C, while the highest flux region is at a temperature from 410°C to 475°C [6, 14]. Unless there is a clad breach, the duct does not have any interaction with the fuel and is typically under low pressures during operation. However, ducts can be exposed to bundle-duct interaction (BDI), bowing, or duct-to-duct interaction under normal reactor conditions. The cladding and duct can experience nonuniform dimensional instabilities from swelling and irradiation creep, in part due to temperature and flux/fluence variations, that lead to distortions and interactions. These clad and duct effects described here do not account for the significant changes that can occur from possible scenarios caused by a loss of coolant accident as well as flux variations under irradiation.

Of particular importance to transmutation reactors is the buildup of pressure inside the fuel pin because transmutation fuel can produce significantly more fission gas [15]. Predictions for cladding stresses due to fission gas depend on the fuel type and burnup, length of the plenum, and the diameter and wall thickness of the cladding. As an example, a standard FFTF mixed oxide (MOX) fuel pin can develop an effective stress of approximately 40 MPa after about 100 dpa of exposure [16]. Stresses from gas pressure and fuel clad mechanical interaction (FCMI) have been accommodated to up to 20 at.% burnup, while limiting gas pressure to keep hoop stress in the clad below 150 MPa. However, stresses created by burnups higher than 20%, will need to be accommodated through specific fuel element design alterations (e.g., using appropriate fuel smear densities, <75%, or increasing the plenum size or cladding wall thickness or designing a vented fuel). Alternatively, advanced cladding alloys could be developed that are able to accommodate higher stresses. In addition to normal operation, the effect of transient over temperature events on plenum pressure and material strength limits must be considered.

2.2.1.3 Predicted dose rates and total fluences

In-core components experience the highest fluences. Fuel cladding in typical sodium cooled fast reactors experience 20–40 dpa/year [17]. Thus, because of these high dose rates, most designs require replacement of fuel bundles every 2–7 years, which is sometimes before reaching the desired fuel burnup, depending on the reactor and fuel type. The high-performance structural materials that simultaneously exhibit good thermal creep strength and superior radiation resistance are needed for the demanding fuel cladding environment in SFRs. In contrast, dose rates for the SFR out of core components are <1 dpa/year. Nevertheless, the use of structural materials such as austenitic or tempered martensitic steels with at least moderate radiation tolerance is required to achieve the goal of a 60-year design lifetime.

2.2.1.4 Leading candidate alloys and materials issues for out of core components

The out-of-core nuclear structural components must comply with national construction codes such as ASME Boiler and Pressure Vessel Code Division 5 and elevated temperature components must comply with corresponding high-temperature codes such as ASME Section III subsection NH. Only 5 alloys are currently qualified in subsection NH: Type 304 and 316 austenitic stainless steels, 2.25Cr-1Mo steel, modified 9Cr-1Mo, and Alloy 800H. The candidate alloys for out of core components are listed in Table 2.2. The corresponding degradation mechanisms range from thermal aging, creep, creep-fatigue uniform and crevice corrosion. Transport/deposition of nonmetallic elements such as carbon and nitrogen when in contact with sodium can also be an issue [18]. Austenitic and high Cr ferritic/martensitic steels decarburize in the high-temperature region and carburize in the low-temperature region. The carbon activity and the extent of carburization/decarburization depend on the temperature distribution around the sodium circuit, alloy composition and thermal-mechanical treatment, as well as external carbon sources and sinks.

Table 2.2 Materials for SFR out of Core Components

Structure and Component	Material	Environment	Degradation Process or Mechanism	Factors Controlling Occurrence
Vessel	Type 316 SS (see Chapter 8)	Primary Na and Ar gas	Thermal aging Thermal creep Weld integrity	Neutron fluence Service temperature Service life
Vessel enclosure	Type 316 SS (see Chapter 8)	Primary Na and Ar gas	Thermal aging Thermal creep Weld integrity	Neutron fluence Service temperature service life
Guard vessel	Type 316 SS Fe-9Cr-Mo (see Chapters 8 and 11)	Argon gas and leaking Na	Corrosion	Temperature service life Na purity
Core support structure	Types 304 and 316 SS (see Chapters 8 and 11)	Primary Na	Irradiation thermal aging Crevice corrosion	Temperature Neutron fluence Service life

Continued

Table 2.2 Materials for SFR out of Core Components—cont'd

Structure and Component	Material	Environment	Degradation Process or Mechanism	Factors Controlling Occurrence
Primary and secondary piping	Type 316, High-Cr Steels (see Chapters 8 and 11)	Primary Na Secondary Na	Interstitial element transfer Thermal aging Fatigue resistance Radioactive mass transport (primary piping only)	Temperature and temperature differences Na purity Mechanical loads Na flow velocity
Mechanical pump (impeller, diffuser)	Type 316 (see Chapter 8)	Primary Na Secondary Na	Corrosion Fatigue resistance	Flow velocity Vibrations Applied load Na purity Temperature
Intermediate heat exchanger shell	Type 304 or 316 (see Chapter 8)	Primary Na inside	Thermal aging Fatigue and creep-fatigue Interstitial element transfer	Na purity Service life Temperature Mechanical load
SG shell	Fe-2 1/4Cr-1Mo (see Chapter 11)	Secondary Na	Interstitial element transfer Thermal aging Sodium-water reaction Thermal creep Fatigue and creep-fatigue	Na purity T. delta T Steam leak Applied load Aging time
SG tubing	Fe-2 1/4Cr-1Mo (see Chapter 11)	Steam	Na corrosion C transfer Thermal aging Caustic effect Thermal and creep-fatigue	Na purity T & Delta T Steam leak Applied load Transients Aging time
Hot leg steam piping	Fe-2 1/4Cr-1Mo (see Chapter 11)	Steam	Flow-assisted corrosion Fatigue and creep-fatigue Thermal aging	Flow velocity Steam T & P Service time Aging time Water chemistry
Cold leg steam piping	Carbon steel (see Chapter 11)	Treated water	Flow-assisted corrosion General corrosion Fatigue	Flow velocity Steam pressure Temperature Service time Aging time Water chemistry
Steam turbine	Ferritic steel (intermediate chromium) (see Chapter 11)	Steam	Steam oxidation Scale exfoliation Creep and fatigue Flow induced corrosion	Temperature Service time Applied load Steam velocity

2.2.1.5 Leading candidate alloys and materials issues for in-core components

The austenitic stainless steels were initially the alloy of choice. However, research beginning in the early 1970s showed that austenitic stainless steels undergo excessive radiation-induced void swelling prior to reaching the desired burnup [19]. Subsequently, different tracks were followed to develop a cladding alloy, which is able to withstand high irradiation doses. The US fast reactor cladding and duct development program switched its focus from austenitic steels to ferritic steels for high exposure applications. Selected ferritic steels exhibit strong resistance to swelling, adequate microstructural stability under irradiation, and retain adequate ductility at typical reactor operating temperatures [20]. Earlier in the US fast reactor development program, HT9 was selected for cladding and ducts for the Integral Fast Reactor (IFR) concept and as the next-generation driver fuel cladding for EBR-II and FFTF [21]. The nominal composition of HT9 is Fe-11.95%Cr-1%Mo-0.6Mn-0.6Ni-0.3%V-0.5%W-0.38Si-0.2%C [22]. Following processing and fabrication, the typical heat treatment of HT9 is a short normalizing anneal for 30 min at 1038°C, with a rapid air cool, followed by a tempering treatment at 760°C for 30 min before a rapid air cool [23]. France pursued austenitic steels and switched from 316 to 316 Ti, and subsequently to AIM 1 (15Cr-15Ni-Ti) which was successfully used in Phenix up to about 120 dpa [24]. Further efforts are still being made to improve austenitic grades. As for Japan, they developed oxide dispersed strengthened steels (ODS) [25], using the benefits of a ferritic/martensitic or a ferritic matrix, and improving the creep resistance with the oxide precipitates. The HT9 cladding was tested with both oxide and metal fuel in the FFTF. In the most extreme test, ACO-3, with mixed-oxide fuel was irradiated to a burnup of 238 MWd/kgM (a fast fluence of 39×10^{22} n/cm^2, or approximately 200 dpa), at a peak cladding temperature of 661°C. At this burnup, neither elongation of the duct nor any pin breach was observed [26]. The HT9 was irradiated as cladding for a metal fuel U-Zr alloy to a burnup of 143 MWd/kgM (a fast fluence of 20×10^{22} n/cm^2) at a peak cladding temperature of 640°C [27].

In addition to HT9, the 9Cr ferritic/martensitic steel T91 was considered for IFR, EBR-II, and FFTF. The two primary weaknesses of these alloys were irradiation embrittlement, or fracture toughness degradation at temperatures below $\sim$430°C and loss of creep strength above $\sim$550°C. Much of the ferritic/martensitic steel research in the US National Clad and Duct (NCD) program was to build an irradiation database on the mechanical response and maximize the radiation and thermal tolerance of HT9 and modified 9Cr-1Mo steel identified as T91. Numerous experiments were performed to explore the effects of normalizing and tempering conditions on irradiation resistance with the goal of maximizing toughness, while maintaining adequate creep strength. Testing showed that HT9 and T91 had approximately the same fracture toughness but T91 had better impact fracture resistance [28]. The two alloys were found to have similar in-reactor creep compliance.

At about the mid-point of the NCD program, attempts to develop improved ferritic/martensitic steels through compositional changes began. These efforts focused on materials with 10 wt% chromium. Examples of such materials are modified HT9 heat XA3607 (Fe-10Cr-1.26Mo-0.15C-0.06Si-0.48Mn-0.19V-0.44Ni-0.018N-0.17Ta) and 10Cr-1Mo (Fe-10.4Cr-1.5Mo-0.20C-0.45Si-0.95Mn-0.15V-0.12Ni-0.01N). However, only minimal research on these materials was performed and published [29] due to the demise of fast reactor development in the US in the early 1990s. Notably, the oxide dispersion strengthened steel MA957 underwent a significant development effort prior to the demise of the US fast reactor program [30].

Although the fast reactor program was severely curtailed in the US by the early 1990s with the shutdown of FFTF and EBR-II, research on ferritic/martensitic steels also was conducted in the Fusion Reactor Materials program. This program focused primarily on (1) improving fracture toughness resistance after low-temperature irradiation and (2) creating a new class of ferritic/martensitic steels that are less susceptible to neutron induced activation and were aptly called reduced activation ferritic/martensitic (RAFM) steels. A large number of experimental heats of material were produced by the Japanese organizations and irradiated in FFTF and high flux isotope reactor (HFIR) and in Europe and Russia. The primary effort of this research was to study the effect of composition on radiation tolerance and included variations in Cr, Ta, W, B, Al, and Nb. One of the most important outcomes of this research was the conclusion that impact toughness is optimized at 9 wt% chromium [31]. Another useful outcome was the determination that Ta improves the resistance to embrittlement under irradiation (shift of DBTT) of ferritic/martensitic steels [32]. More recent research has focused on third and fourth-generation ferritic/martensitic steels that have been developed for fossil fuel boiler applications. For example, NF616 has received interest in the irradiation damage community because of its favorable composition. Through the use of nonstandard heat treatments, NF616 has been produced with reduced grain size and a fine distribution of precipitates leading to improved thermal creep properties. Other ferritic/martensitic steels are discussed in detail in Chapter 11.

2.2.2 LEAD FAST REACTOR (LFR)

Under the category of Lead Cooled Fast Neutron Reactors (LFR) two classes of designs can be identified, namely classical fast reactors and Accelerator-Driven Systems (ADS) [33]. The ADS cores are designed to be subcritical; therefore, an additional continuous neutron source (e.g., a liquid metal neutron spallation target connected to a proton accelerator) is needed to sustain the nuclear chain reaction. The objectives of LFRs are electricity production and process heat generation, while in the case of ADS, nuclear waste transmutation is the primary goal. Moreover, some LFR variants which comply with the definitions of Small Modular Reactors (SMR) and long-life, battery-type reactors are also under consideration [34].

The proposed coolants for LFR are both pure liquid lead (Pb) and an eutectic Pb-Bi melt (LBE, lead bismuth eutectic). The difference between Pb and LBE cooled systems is in the operational temperature range due to a lower melting temperature of eutectic (125°C) compared to Pb (321°C). All of the LFR concepts that are being studied in various countries follow the pool-design principle. Fig. 2.3 [35] illustrates a LFR pool concept. Unlike SFR pool reactors, in principle, LFRs do not require an intermediate heat removal system.

However, practical experience on liquid lead cooled nuclear systems is very limited. In fact, only in Russia has an LFR been operated for the development of submarine technology [36]. More recently, within an international consortium, a liquid LBE neutron spallation target (MEGAPIE, megawatt pilot irradiation experiment) has been irradiated at the proton accelerator facility at the Paul Scherrer Institute (PSI) [37, 38]. There are a number of LFR

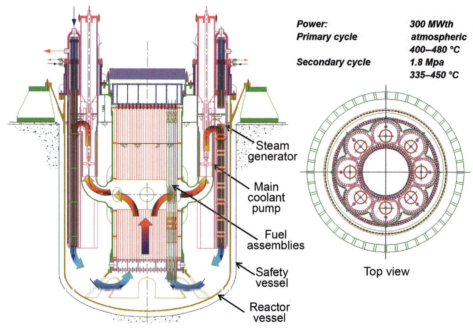

FIG. 2.3

Cross section and top view of the ALFRED concept LFR [35].

2.2 LIQUID METAL COOLED FAST REACTORS

Table 2.3 LFR Design Parameters

Region/Country	Reactor/Coolant	Power MWth	Type/Application	SG Inlet Temperature (°C)	SG Outlet Temperature (°C)	Reference
Europe	ALFRED/Pb	300	FR/SMR	400	480	[35]
Europe	ELFR/Pb	1500	FR/large reactor	400	480	[35]
Russia	SVBR/LBE	270	FR/SMR	280	440	[36]
Russia	BREST/Pb	700/2800	FR/large reactor	420	540	[36]
USA	SSTAR/Pb	45	FR/long-life battery reactor	420	570	[39]
USA	HYPERION/LBE	75	FR/SMR	?	500	[40]
Belgium	MYRRHA/LBE XT-ADS/LBE	50–100 80	ADS/research reactor ADS/technology demonstrator	300 300	400 400	[41] [42]
Europe	EFIT/Pb	400	ADS/waste transmutation	400	480	[42]

EFIT, European Facility on Industrial Scale Transmuter.

designs under development worldwide and some selected concepts are listed in Table 2.3. Almost all of these LFRs are in a conceptual design phase, which means that the parameters given in Table 2.3, and those discussed in this chapter are not frozen and can evolve as knowledge is acquired from the supporting R&D activities.

Due to some common features of the molten Pb or LBE and Na, materials selection and qualification are similar for LFR and SFR, both for in-core and the out-of-core components. The most relevant in-core components are the fuel assembly (FA) and the duct, and in case of ADS the neutron spallation target. The fuel pins in a LFR assembly are separated by either wires or spacers (both concepts are considered). The number of pins and their linear power vary depending on the reactor concept. For instance in case of the European Pb cooled ALFRED (Advanced Lead Fast Reactor European Demonstration) concept there are about 130 fuel pins in an assembly, and the linear power of a fuel pin is in the order of 25–30 kW/m [43]. Fig. 2.4 shows a schematic representation of the fuel assembly for the ALFRED concept. The fuel pin cladding diameter and thickness varies between the concepts and ranges are between 6 and 10 mm (in case of SVBR [Russian acronym for lead-bismuth fast reactor] up to 12 mm) in diameter and 0.5–0.6 mm (in case of SVBR 0.4 mm) in thickness.

2.2.2.1 LFR flow conditions

The thermal-hydraulic conditions in LFR systems depend on their design. In general, the minimum and maximum temperatures are lower in the case of LBE compared to Pb cooled systems. This is illustrated in Tables 2.4 and 2.5, where the flow conditions for the Pb-cooled ALFRED/European Lead Fast Reactor (ELFR) system and the LBE cooled system XT-ADS are listed, respectively. The operational temperatures of LFR designs are lower compared to the SFR. This design choice is mainly due to the fact that corrosion and corresponding mechanical property degradation due to the environment are more severe in the LFR compared to the SFR.

All reactor components in contact with Pb/LBE operate at about atmospheric pressure. For the out-of-core LFR components it might be necessary to adapt existing design codes, such as the ASME (USA) or the RCC-MRx (EU) [8]. In addition to the typical mechanical properties that need to be evaluated for high-temperature components, such as tensile, toughness and time-dependent creep, fatigue, and creep-fatigue, a careful assessment is needed as to

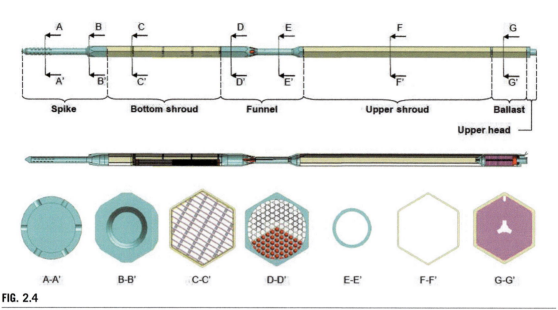

FIG. 2.4
The fuel assembly geometry of the EUROPEAN LFR reactor design ALFRED [43].

Table 2.4 Flow Conditions for ALFRED and ELFR

Component	Min/Max Temperature °C (Normal Operation)	Maximum Lead Flow Velocity m/s	Maximum Dose Rate (Radiation Damage) (dpa/year)
Reactor vessel	380/430	0.1	$<10^{-5}$
Inner vessel	380/480	0.2	0.1
Steam generator	380/480	0.6	0.006
Primary pump	380/480	10	$<10^{-5}$
FA clad	380/550	1	20
FA structure	380/530	2	20
Dummy assemblies	380/480	0.01	20
Refueling equipment	380/480	0.2	0.02
DHR heat exchanger	380/430	0.2	$<10^{-5}$

the effects of the short-, medium- and long-term exposure to liquid metals including liquid metal embrittlement [44]. Moreover, these properties need to be assessed for both the base metal and the welds. In addition to the mechanical properties, corrosion, erosion, and fretting resistance of candidate materials need to be assessed. Corrosion mechanisms and rates in Pb and LBE depend on the oxygen content in the liquid metal, temperature, flow velocity (erosion phenomena impact on the operational lifetime at pump impeller where a flow velocity of 10 m/s is foreseen), and the occurrence of flow structures such as vortices. The corrosion phenomena can be accelerated by fretting, which can occur at tube bundles of heat exchangers or between fuel bundles with spacers. The operational window (including safety margins where appropriate) has to take into account all these phenomena.

Table 2.5 Flow Condition for XT-ADS

Component	Min/Max Temperature °C (Normal Operation)	Max Lead flow Velocity m/s	Maximum Dose Rate (Radiation Damage) (dpa/year)
Reactor vessel	300/360	0.01	0.6×10^{-4}
Inner vessel	300/360	0.1	1.54
Steam generator	300/360	1.1	0.03
Primary pump	300/360	10	0.06
FA clad	300/430	1.6	29
FA structure	300/400	2.3	29
Dummy assemblies	na	0.2	na
Refueling equipment	300/360	–	na
Neutron spallation target	400/470	1.5	30

na, not available.

A challenge to be addressed in LFR design is the high corrosiveness of LBE and Pb toward structural materials [45]. Indeed, both liquid metals have high solubility values of the main steel elements (Fe, Cr, Ni), with consequent very severe chemical attacks if no corrosion protection is included [45]. Corrosion protection by oxide layer barrier formation on the steel surface or by artificial coatings has been considered in several designs, such as the formation of a native oxide layer which has been developed in Russia. Moreover, protective coatings and/or native oxide layers must be addressed in the framework of design codes and their performance understood for both normal and accidental conditions.

The critical issues for in-core components of LFRs include irradiation effects, compatibility with the liquid metal, and temperature distributions. As with the out-of-core components, the temperatures of the LFR in-core components are lower compared to SFRs, again mainly due to compatibility issues with the coolant. Typical calculated clad temperatures for Pb systems are between 380°C and 550°C, with clad peak temperature in the order of 600°C in case of ALFRED/ELFR, and 650°C in case of the Russian Pb-cooled Fast Reactor (BREST). The temperature ranges are lower in LBE systems, with the clad peak temperature below 600°C (e.g., XT-ADS). The flow velocity in the fuel subassemblies are given in the Tables 2.4 and 2.5.

Various concepts of fuel pin designs have been investigated to accommodate fission gas release and other sources of excessive stresses. The fuel pin designs of an LFR are generally comparable to those described for the SFR. However, for ADS cladding, hoop stress related issues in normal and off-normal conditions might be more severe, since the amounts of minor actinides (Np, Am, Cm) in the fuel are much higher compared to conventional fuels. For instance, a MOX pin with 2% of Americium irradiated up to about 6 at.% burn up has a 4 times higher helium content compared to a standard MOX SFR pin [15]. The typical phenomena that must be accounted for in the fuel pin design and the selection of the clad material are of mechanical and chemical in nature. Clad properties relevant to the pin design are hoop stress, tensile, and creep and fatigue limits. To address the grace time in transient and off-normal conditions, stress-to-rupture, and burst resistance data are required. In addition, chemical compatibility of clad materials with both fuels and coolants needs to be taken into account, as well as their mechanical properties degradation due to the chemical interactions and the irradiation field.

In general, oxide and nitride fuels are considered for LFR designs. In the case of ADS, that has as its main mission the reduction of the Pu and minor actinides stockpiles, the fuel under development is U-free in order to achieve this mission. Since the ADS design uses an external neutron source to sustain the chain reaction, inert matrix fuels have been specifically developed for transmutation purposes. For the clad material, several solutions are envisaged, which take advantage of existing materials data obtained in the past cladding materials development and qualification programs. For instance, in Europe [e.g., ALFRED and the ADS developed at SCK-CEN named Multi-purpose hYbrid Research Reactor for High-tech Applications (MYRRHA)] [41, 46] the austenitic 15Cr-15Ni Ti stabilized steel (for more details on austenitic steels see Chapter 8 and Ref. [24]) is considered for the first core because of its high level of technological readiness. However, due to the high solubility of Ni in the liquid metal, for the medium to long-term designs (e.g., ELFR), either ferritic or ferritic/martensitic steels (for more details on ferritic/martensitic steels see Chapter 11) are also under consideration. Relevant R&D programs on these types of steels have been carried out in Russia and Europe. The Russian approach foresees the addition of Si to the Fe-Cr steel to enhance corrosion resistance [36]. However, it remains to be demonstrated if this approach is sound in all conceivable conditions for LFR cores, especially if Si enhances radiation effects [47] and liquid metal embrittlement [48]. In Europe, the limits of ferritic or ferritic/martensitic steels, such as thermal creep and stress-to-rupture limits at temperatures above 550°C, are being addressed through the development of oxide dispersion strengthened (ODS) variants (for more details on ODS steels see Chapter 12 and Ref. [25]). Finally, alternative clad material such as SiC/SiC composites [49] might be used to increase the overall system temperatures of LFR, while controlling corrosion related phenomena.

2.2.2.2 Predicted LFR dose rates and total fluence
The dose rates on the out-of-core and in-core components for two selected reactor designs, ALFRED/ELFR and XT-ADS (as examples), are reported in the Tables 2.4 and 2.5 respectively. Similar to SFRs, the most severe radiation damage is in the fuel assemblies. Thus, in addition to reactor physics and safety considerations, LFR fuel burn up limits must take into account materials' in service properties and performance. In particular, compatibility issues might restrict fuel burn up. Moreover, in case of ADS transmutation efficiency, the trade-off between high fuel burn up, safety, and technological feasibility are always interconnected issues.

Fluences experienced by out-of-core LFR components are relatively low. However, the SG design and lifetime assessment must account for the temperature and the irradiation field as well as the presence of the liquid metal.

2.2.2.3 Leading LFR candidate metal alloys and materials issues
The candidate materials for the out of core structural components are mainly austenitic and ferritic/martensitic steels. Special materials are proposed for specific components where, for instance, high erosion rates are expected (e.g., MAX-phases for the pump impeller). Examples of materials indicated for the LFR out-of-core components are shown in Table 2.6. Key degradation mechanisms of the structural materials are those related to their compatibility in the liquid metals (for more details on corrosion and compatibility see Chapter 6). As already indicated previously, main steel elements such as Ni, Cr, and Fe are highly soluble in liquid Pb and LBE [45], when compared to their solubility in liquid Na. Therefore, steel element dissolution in the hot part of the reactor and redeposition, as corrosion products in cold parts of the reactor is possible. The consequences of these phenomena are reduction of load-bearing capabilities of corroded components in the hot part and plugging of channels in the cold part. Moreover, in given temperature ranges, thermal-hydraulic conditions, mechanical and thermal stresses, and oxygen potential in the liquid metal, both Pb and LBE can degrade the mechanical properties of the materials [45]. All of these phenomena must be accounted for when selecting an alloy and assessing a component's lifetime. Finally, several compatibility phenomena in liquid Pb and LBE are driven mainly by nonmetallic impurities in the liquid metal. The main nonmetallic impurity dissolved in Pb and LBE is oxygen and its presence might induce the buildup of an oxide layer on a component's surface [45]. These layers can be protective, that is, they mitigate dissolution phenomena of the steel elements. However, for components such as the SG and oxide layers can reduce the thermal conductivity, a phenomenon that needs to be accounted for in heat exchanger designs and dimensioning.

As shown in Table 2.6 the prime candidate alloys for Pb and LBE reactor out of core components are the austenitic stainless steel AISI 316L and the ferritic/martensitic (F/M), T91. Data of AISI 316L are included in nuclear components design codes but not for the T91 steel. Degradation phenomena related to the Pb/LBE coolant should

2.2 LIQUID METAL COOLED FAST REACTORS

Table 2.6 Materials Selected for ALFRED/ELFR

Structure and Component	Material	Environment	Degradation Process or Mechanism	Occurrence
Reactor vessel	AISI 316L (see Chapter 8)	Primary Pb and Ar/O$_2$ containing gas	Thermal aging and creep; ratchetting, weld integrity; corrosion/oxidation	Neutron fluence Temperature Service life
Inner vessel	AISI 316L (see Chapter 8)	Primary Pb	Creep-fatigue, ratchetting; weld integrity; corrosion/oxidation	Neutron fluence Temperature/ temperature gradient
Steam generator	T91 (see Chapter 11)	Primary Pb and steam	Fouling, creep-fatigue, buckling, deformation, ratchetting; weld integrity; corrosion/oxidation	Neutron fluence Temperature/ temperature gradient, flow velocity, cyclic loading Service life
Primary pump Duct and shaft Impeller	AISI 316LN or T91 (see Chapters 8 and 11) MAXTHAL or Aluminized steel	Primary Pb	Creep-fatigue, ratchetting; weld integrity; corrosion/oxidation and erosion	Neutron fluence Temperature gradient, flow velocity, vibration, cyclic loading Service life
DHR	AISI 316LN (see Chapter 8)	Primary Pb and water		Neutron fluence Temperature Service life

be addressed in the design code, in terms of the determination of materials design limits, for both types of steels. Finally, in the case of T91 for SG applications the consequences of anticipated cyclic softening under low cycle fatigue loading must be accounted.

The leading candidate alloys for LFR in-core-components in general are 9-12Cr F/M steels. The specific choices have been:

- In Europe 9Cr-T91 steel for ADS systems and future ELFR
- In Russia 12 Cr EP823 steel for both SVBR and BREST
- In USA 12 Cr HT9 steel for HYPERION and SSTAR (small sealed transportable autonomous reactor)

The nominal compositions of these steels are given in Table 2.7. The main difference between T91 and HT9 is in the Cr content, which is lower in the former, to avoid α' precipitation under irradiation [50]. The main difference between HT9 and EP823 is in the Si content, which is considerably higher in the latter, which is to improve the oxidation resistance of the steel. Design limits must be considered for whichever F/M steel is selected. First, the

Table 2.7 Composition of 9-12Cr F/M Steels (wt%) Considered as Options for the Claddings of LFR

Steel	C	Cr	Ni	Mo	Si	Mn	V	W	Nb
T91	0.1	9	0.03	1	0.2	0.3	0.2	–	0.1
EP823	0.18	11	0.7	0.7	1.2	0.6	0.3	0.6	0.3
HT9	0.2	12	0.6	1	0.4	0.6	0.3	0.5	–

temperature window in an irradiation environment is 350–550°C. The lower limit is due to severe irradiation hardening and embrittlement that occurs below 350°C [14] and the upper limit is due to relatively low thermal creep and stress-to-rupture strength of these steels above 550°C [51]. Further limitations are imposed by corrosion/oxidation and liquid metal assisted degradation of mechanical properties. In principle, passivation by oxidation of the steel surface is a favorable phenomenon, which can be achieved by adjusting the oxidation potential of the liquid metal. However, above 550°C the oxide layer growth on conventional 9-12Cr F/M steels (without addition of Si or Al) is very fast, and the thick oxide layer can spall off exposing fresh surface to liquid metal for dissolution attack and possible degradation of mechanical properties [42]. Unfortunately, the already poor F/M steel stress-to-rupture strength is further reduced by liquid metal exposure, even if the surface is protected by an oxide layer [42, 52]. The enhanced degradation of the mechanical properties of the steels in contact with the liquid metal becomes even more pronounced in the presence of liquid lead alloy combined with an irradiation field.

As a result of these numerous challenges, the technological readiness level of 9-12Cr F/M steel as a cladding material is very low. To overcome all these limitations several alternative research directions are being considered:

- The use of an austenitic stainless steel (Ti stabilized 15Cr-15Ni) as cladding material is combined with oxygen control in the liquid metal and also considered for low burn up fuels. While, Fe-Cr-Al alloys, which provide a corrosion protection layer, might be of interest for higher fuel burn up.
- The development of ODS F/M steel with the aim to improve thermal creep and stress-to-rupture strength. The ODS might also contain Al to combine the better mechanical properties with a higher corrosion/oxidation resistance.
- The development of an engineered corrosion protection layer to be used on 9Cr steel in combination with the modification of the design of the core such as to accommodate high stresses.

A comprehensive development and qualification program is needed to meet the challenge of a cladding for fuels designed for high fuel burn up. This program should include corrosion and irradiation tests (ideally combined) to screen the materials properties. The down selection of the most promising alloys identified in the screening phase should be followed by single pin, fuel bundle, and fuel assembly test programs to assess the multiple effects behavior of alloys under realistic operating conditions, as well as their performance under LFR core transient and accidental conditions.

2.3 HELIUM COOLED REACTORS
2.3.1 VERY HIGH-TEMPERATURE REACTORS (VHTR)

The VHTR is helium cooled and graphite moderated with the objective of reaching outlet gas temperatures in the range of 950–1000°C. VHTRs can operate either with a direct cycle [cf., the pebble bed modular reactor (PBMR) design], or with an IHX. The former requires the development of helium turbines, which represents a significant technological challenge. The latter design presents the benefits of offering a high-temperature fluid, which can be used not only for high efficiency electricity generation but also for a wide range of industrial applications requiring process heat.

The very high-temperature gas-cooled reactor concept benefits from the previous development of high-temperature reactors from 1970 to 1980 in the United States, France (in cooperation with General Atomics), Germany, and the United Kingdom. Five gas-cooled reactors have been built as shown in Table 2.8, accumulating a total of 60 years of operating experience. Fourth-generation gas-cooled reactors will draw on the technological and industrial progress made since then, namely the development of high-temperature gas turbines and high-temperature materials.

Both direct and indirect helium-cooled reactors can be split into two families, depending on the structure of their core: pebble bed reactors and prismatic reactors. The main characteristic of the pebble bed reactors (of German design)

2.3 HELIUM COOLED REACTORS

Table 2.8 A Listing of Experimental and Demonstration VHTR

Country	Reactor	MWth	Reactor Type	Operating Life (years)	Outlet Temp. (°C)
USA	Peach Bottom	115	Prismatic	7	715
	Fort St Vrain	842	Prismatic	15	780
UK	Dragon	20	Prismatic	14	750
Germany	AVR	46	Pebble bed	21	950
	THTR-300	750	Pebble bed	6	750
China	HTR-10	10	Pebble bed	13[a, b]	700
	HTR-PM	500	Pebble bed		750
Japan	HTTR	30	Prismatic	15[a]	950

[a] Still operating.
[b] Being built.

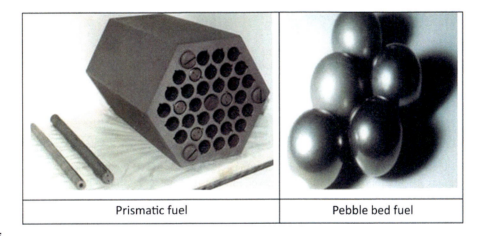

FIG. 2.5

Fuel arrangements for a prismatic fuel on left and pebble bed fuel on the right.

is the use of compacted micro-particle fuel in a graphite matrix in the form of 60-mm diameter spheres as illustrated in Fig. 2.5. The spheres are continuously inserted in at the top and extracted at the bottom of the reactor at a rate of one pebble approximately every 20 s. The pebbles are reintroduced until they reach their maximum burnup, on average, staying in the reactor for a few years. This allows continuous operation with no refueling outages.

The prismatic reactors (of American design) illustrated in Fig. 2.6, differ from the pebble bed reactors mainly in terms of core and fuel arrangements. The prismatic core is composed of graphite blocks containing the fuel compacts. These reactors are easier to operate as the fuel does not move, but they require periodic outages for refueling.

The prismatic reactors are assembled with power level modules ranging from 200 to 600 MWt. The maximum module power is limited to provide inherent passive safety features such that in case of depressurization, the reactor naturally transfers the decay heat without an active cooling system. Reactor outlet helium temperatures typically range from 750°C to 1000°C and the downstream heat transport-exchange systems provide steam and/or other high-temperature fluids. The range of power ratings, temperatures, and heat transport system configurations of prismatic provides flexibility in adapting the modules to the specific applications.

Reactors that operate at these high temperatures increase the electricity production efficiency up to 50% for an outlet temperature of 1000°C compared to the 33% achieved with current light-water reactors. The energy

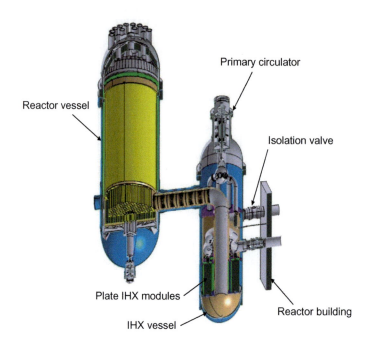

FIG. 2.6

The VHTR concept [53].

conversion system will be based on a gas turbine (Brayton cycle) and/or a SG (Rankine cycle). Another advantage of high-temperature gas reactors (HTGR's) are improved safety aspects under loss of coolant accidents because of the low power densities compared to LWRs or other Gen-IV reactor concepts [54].

The increase in coolant temperature would also enable nonelectric use of the heat, most notably, efficient hydrogen production. Thermochemical processes can be used to decompose water molecules and produce hydrogen (and oxygen) at temperatures close to 900°C. Hydrogen-generation processes are the subject of active cooperative research in Japan, the United States, and Korea. A small-scale test loop has been built in Japan, producing approximately 30 L of hydrogen per hour, and a larger test loop has been built on the General Atomics site in San Diego (in a cooperation with Sandia National Laboratory and CEA). Researchers have demonstrated the feasibility of a hydrogen production at efficiency of 35%, with a longer-term objective of increasing this 50%–55%. Another promising hydrogen production method is high-temperature steam electrolysis, which also requires very high temperatures to obtain sufficient efficiency. Other nonelectric applications include process heat for various industries including chemical, petrochemical, cement, iron and steel, as well as desalination of seawater.

Japan Atomic Energy Agency (JAEA) built a small HTTR (high-temperature test reactor [Japanese design]) that achieved a temperature of 950°C in April 2004. The United States performed fuel R&D and developed initial design plans for the construction of a 200–600 MWth first of a kind VHTR (NGNP) to produce both electricity and hydrogen. Following the start of HTR-10 VHTR, China is now planning the construction of several pebble bed reactors.

2.3.1.1 VHTR flow conditions

The coolant is helium circulating under a high pressure of 50–90 bar in the primary circuit. Depending on the concept and the application, VHTR inlet temperatures range between 400°C and 500°C and outlet temperatures between 750°C and 1000°C. The major impact of these high temperatures is on the IHX alloy selection. Indeed, new structural materials will need to be developed for temperatures above ~750°C. The helium purity is also very important, and must provide a favorable oxidizing atmosphere to create durable protective scales.

Environmental attack on the secondary circuit components by various working fluids (nitrogen, nitrogen-helium, steam in case of electricity production, and also sulfuric acid in case of the sulfur/iodine thermochemical hydrogen production process) is also important, especially in terms of corrosive effects.

2.3.1.2 VHTR predicted dose rates and total fluences

The in-core components will accumulate a dose of a few dpa per year due to the mostly thermal flux. The vessel will be slightly irradiated (see Table 2.9). The VHTR metallic components are listed in Table 2.9. For the vessel, several options are possible: the vessel can be cooled in which case, the best option among current nuclear qualified structural materials is SA508, which is the steel used for PWR vessels. In the case where the vessel is not cooled, alloys with a better creep resistance at higher temperatures might be used such as $2^{1/4}$ Cr (for instance in the HTTR) or 9Cr steel.

The other metallic components are the IHX which can be made of Alloy 800 if the outlet temperature stays below 800°C and nickel-based alloys (Inconel 617, Haynes 230, Hastelloy XR) for higher temperatures. For temperatures in excess of 950°C, nickel-based ODS would need to be developed or ceramic components have to be used.

2.3.2 GAS-COOLED FAST REACTOR

The GFR can be seen as a sustainable evolution of the high-temperature reactor, in the sense that it combines the advantages of high temperatures with a breeding capacity and a closed fuel cycle (see diagram in Fig. 2.7). The GFR has many similarities to the other helium cooled reactor designs with an IHX. The main differences will be in the core, but the vessel and the energy conversion systems will be the same. While the GFR will benefit from several components of the VHTR, one has never been built, and this concept presents significant challenges, namely the fuel assembly and the safety issues (a high power density combined with a low thermal inertia). In case of

Table 2.9 A List of Metallic Materials Considered for some Components (the Maximum Temperature is Indicated and for Higher Temperatures Ceramics have to be Considered)

Component	Min/Max Temperature °C (Normal operation)	Maximum dose (Radiation Damage) (dpa)	Materials Considered
Reactor vessel	450/500 (VHTR) 480/500 (GFR)	$<10^{-2}$ up to 40 (depending on shielding)	SA 508 (cooled vessel) 2 ¼ Cr 9 Cr (see Chapter 11)
Intermediate heat exchanger, piping	750°C 850°C 950°C 1000°C	None	Alloy 800H Haynes 230 Inconel 617 Hastelloy XR Ni based ODS (see Chapter 9), SiC
Turbine blades (cooled)	650°C 750°C	None	IN 718 Udimet 720 (see Chapter 9)
Control rod sleeves	900°C	25 dpa	Alloy 800H (see Chapter 9)

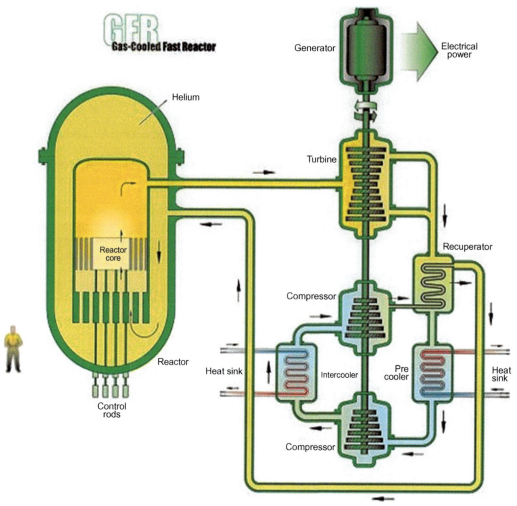

FIG. 2.7

The GFR concept [55].

depressurization, in core temperatures could reach 2000°C, which imposes severe constraints on materials. The proposed power rating of this type of reactors ranges from 600 to 3000 MWth. An eastern European consortium, the V4G4 Center of Excellence (Hungary, Czech Republic, Slovakia, and Poland), is proposing to build the first experimental reactor of this kind (ALLEGRO, experimental fast reactor cooled with helium developed by the European V4G4 Consortium) [56]. Note that the first core of this GFR will operate at low temperature (530°C) and will feature steel cladding and wrapper tubes with an oxide fuel.

2.3.2.1 GFR flow conditions

The GFR inlet temperature is between 400°C and 500°C and the outlet temperature ranges from 700°C to 850°C. The primary circuit pressure will range from 50 to 90 bar, similar to the VHTR. A major difference will be the high power density of the core, which is between and 50 and 100 MWth/m^3, in contrast to <10 MWth/m^3 for the VHTR.

2.3.2.2 Predicted GFR dose rates and total fluences

The in-core components will accumulate a lifetime dose of a 50–100 dpa, due to the fast flux. The vessel will only be slightly irradiated but higher than the vessel in a VHTR [57].

2.3.2.3 Leading candidate GFR core materials

The GFR and VHTR core materials will be very different. The components of the GFR core [fission product containment barriers (e.g., fuel cladding) and internal structures] must tolerate a combination of particularly severe conditions. These components will be simultaneously exposed to very high temperatures and intense neutron irradiation. Thus, the attributes of core materials must include significant resistance to irradiation damage-induced degradation of key material properties and high-temperature mechanical strength and toughness within the safety margins under normal operating and accident conditions. In addition, core materials must have favorable neutronic properties, post-irradiation radiotoxicity that is as low as possible (i.e., low activation), high thermal conductivity, and low permeability to fission products. The fissile material considered for the GFR is a carbide or nitride with higher density than oxides and therefore higher thermal output. This allows for a "cold" fuel better suited to retaining its fission products and therefore exerting smaller loads on the containment barriers.

The cladding has to withstand high temperatures under accident conditions and possess the qualities of adequate thermal conductivity and transparency to neutrons. Most refractory alloys are ruled out because they absorb too many neutrons and only vanadium alloys, exploiting the experience gained on these alloys for fusion applications have been investigated as a potential solution [58]. These investigations indicate, however, that the use of vanadium alloys would probably limit the helium outlet temperature to $\sim$700°C due to high-temperature performance limits like creep and impurity/atmospheric corrosion. Therefore, ceramics have been considered in terms of carbides based on their transparency to neutrons, sufficient thermal conductivity, and high melting point. However, monolithic ceramics are too brittle and thus they need to be incorporated in the form of composites. Among the ceramic composites, SiC/SiC presents the best properties and has been thoroughly investigated. There are concerns about the ability of ceramic composites to retain fission products as the matrix is porous and cracks under stress. The CEA proposed a "sandwich" concept, where attractive mechanical properties and resistance to corrosion are provided by inner and outer layers of SiC/SiC, while the hermeticity is added by a metallic liner (Ta, Nb). Some initial results on behavior and fabricability of SiC/SiC cladding show that this is a viable option [59]. For structural materials, namely the internal components, which will constitute the link between the vessel and the fuel, only ceramics are being considered (SiC, ZrC, and TiC are the most promising candidates), as metallic materials cannot achieve the necessary performance requirements.

2.4 OTHER GEN-IV FISSION REACTOR SYSTEMS
2.4.1 MOLTEN SALT FUELED REACTOR (MSR)

The defining characteristic of a Molten Salt fueled Reactor (MSR) is that its fissile material is a liquid halide salt. All of the reactor-significant transuranic elements can form chemically and radiolytically stable salts with halide elements. The safety aspects of MSRs are distinctive. The fuel can be drained from the reactor core into subcritical tanks via a melt plug in the event of a severe accident. In some designs the MSRs will be designed to have a negative salt void coefficient and negative thermal reactivity feedback that avoids a set of major design constraints imposed in solid-fuel fast reactors. On the other hand, MSRs will inherently have radionuclides in more locations than solid fuel systems. MSRs will operate at near atmospheric pressure. In general, the reactor relevant halide salts are high melting point and high boiling point materials with large heat capacities and low-to-moderate thermal conductivity.

Much of the technology base for MSRs derives from the US Aircraft Reactor Experiment molten salt breeder reactor program of the 1950–70s that produced the Molten Salt Reactor Experiment, which operated from 1965 to 1969 with over 13,000 full-power hours of operation. The book "Fluid Fuel Reactors" [60] published by

Addison-Wesley in 1958 under contract to the Atomic Energy Commission remains a useful overview of the concepts underlying MSRs and the major MSR technologies.

Liquid fuel results in distinctive design options and fuel cycles. MSRs encompass both fast and thermal spectrum systems. Both fluoride and chloride-based salts have been considered as the fuel carrier salt and either single or dual fluid (separate fissile and fertile circuits) designs are possible. MSRs can potentially support breeder, burner, and converter fuel cycles based upon ^{232}Th/^{233}U, U/Pu, or TRU fuel cycles. MSRs can incorporate fission product and/or fissile material separation into their operation. MSRs that are fueled with unenriched or low-enrichment uranium that do not include on-site fissile material separation are referred to as "denatured". In MSRs, each batch of fuel that is fed into the reactor is blended into the existing fuel inventory; consequently, the addition has a limited impact on the overall isotopic composition. Thus, MSRs can accommodate substantial variation in the isotopic content of their fuel feed. Both integral and loop type reactor configurations are possible as well as hybrid configurations in which a loop configuration is located within another containment vessel. Alternatively, the fuel salt can be contained in tubes in the core with the coolant circulating on the "shell-side."

MSRs remain technologically immature. However, several organizations are pursuing their development. Notably, the Chinese Academy of Sciences is developing a 2-MW thermal spectrum test reactor with anticipated first criticality at the end of 2020. The US Department of Energy has initiated an MSR reactor campaign seeking to support deployment in the US and the US Nuclear Regulatory Commission has begun the process to adapt its regulation to the characteristics of MSRs. The European Union has been evaluating the viability of fast spectrum MSR technology. Russia, the Czech Republic, and India also maintain small MSR R&D efforts. The Japanese government has also recently added MSRs to its energy-policy document. Multiple private companies and institutions seeking to promote and/or commercialize MSRs have been founded over the past 10 years.

2.4.1.1 MSR flow conditions

The MSR flow conditions depend on the core power and power density. Power densities under consideration range from a few hundred megawatts per cubic meter in fast reactors, to thermal spectrum reactors with power densities $\sim$10 MW/m^3 that seek to extend the lifetime of the moderator graphite. The core outlet temperature in most of the MSR design variants range from 650°C to 750°C. The core outlet temperature is limited primarily by the strength of available structural materials.

Different prospective MSR vendors are employing different strategies for selecting their structural alloy depending on their reactor characteristics and deployment strategies. Vendors that are focused on entering the market as rapidly as possible are focusing on well-known alloys with adequate performance characteristics such as stainless steel 316H. Others are considering separating the chemical compatibility and mechanical strength requirements by employing a clad structural alloy [61]. Still others are considering deploying optimized nickel-based alloys with improved elevated temperature strength and resistance to neutron embrittlement. The silicon carbide composites are also receiving increasing interest for application as structural components in MSRs.

The vendor design decisions will be substantially impacted by the regulatory requirements in place at the time that they are seeking a license. The US NRC is currently considering allowing MSRs to employ functional containment (SECY-18-0096 Functional Containment Performance Criteria for Non-Light-Water-Reactors was approved in December 2018). Functional containment is defined by NRC as a barrier, or set of barriers taken together that effectively limit the physical transport and release of radionuclides to the environment across a full range of normal operating conditions, anticipated operational occurrences (AOOs), and accident conditions. Functional containment would substantially broaden the range of possible materials, fabrication methods, and operating procedures available for MSRs. If functional containment is approved for MSRs, it would obviate the need to ASME BPVC (boiler and pressure vessel code) Section III Division 5 qualification of its salt wetted boundary material. Further, functional containment could enable treating any particular layer of containment as a consumable component intended for periodic replacement rather than vital structures required for plant safety.

For liquid-fuel MSRs, the reactor coolant boundary, and the first layer of radionuclide containment are nearly synonymous with the only distinction being in the extent to which the cover gas containment also serves as a reactor coolant boundary. The MSRs operate at near atmospheric pressure and consequently, failure of the salt-wetted containment layer does not result in significant stress on the next layer of containment. The lack of cascading failure accident sequences enables MSRs to employ independent containment layers to provide defense-in-depth. Even massive failure of an MSR's fuel salt boundary only results in fuel salt flowing out into the next layer of containment. Further, as liquid fuel is not vulnerable to structural damage, it retains its radionuclide retention properties following leaking out of the first layer of containment.

Corrosion in halide salts is dominated by oxidation of the elements of the structural alloy. Consequently, the salts are maintained in a reducing condition through careful impurity and composition control. While halide salts can be highly corrosive, studies on molten fluoride mixtures for reactor applications began in the early 1950s with primary consideration to the compatibility of the salt mixtures with structural materials. In the intervening 15 years, an extensive corrosion program has been conducted at ORNL with several families of fluoride mixtures using both commercial and developmental high-temperature alloys. As a consequence, the corrosion technology for the molten-salt reactor concept is now in an advanced stage of development. Furthermore, container materials are available that have shown extremely low corrosion rates in fluoride mixtures at temperatures considerably above the 540–650°C range proposed [62]. The corrosion characteristics of molten salts and reactor materials have recently been reviewed [70].

2.4.2 MOLTEN SALT COOLED REACTOR

Fluoride salt-cooled high-temperature reactors (FHRs) are an emerging concept that combines attractive attributes from previously developed reactor classes. FHRs are a broad reactor class that maintain strong passive safety at almost any scale and feature significant evolutionary potential for higher thermal efficiency (through higher temperatures), process heat applications, on-line refueling, thorium utilization, and alternative power cycles. FHRs use low-pressure liquid fluoride salt cooling, coated particle fuel, a high-temperature power cycle, and fully passive decay heat rejection. As is the case for other high-temperature concepts, FHRs can support either electricity generation or process heat production. However, no FHR has been built, and no FHR design has reached the stage of maturity where realistic, detailed economic analysis can be performed.

Kairos Power is seeking to commercialize FHR technology. Kairos' FHR design has its origin in the pebble bed FHR design developed at UC Berkeley under a series of DOE-NE projects [63]. The FHR Technology Development Roadmap [64] provides guidance on the state of the art in FHR technologies.

The primary coolant of FHRs will be a liquid fluoride salt. While several fluoride salts could be used in FHRs, the combination of desirable neutronic properties, good hydraulic performance, and very low activation of $(^7LiF)_2 BeF_2$ (FLiBe Li_2BeF_4) makes it the preferred primary coolant. The boiling point of FLiBe is over 1400°C and its volumetric heat capacity at 700°C is 4.67 J/cm^3-K (as compared to 4.04 J/cm^3-K for water at 100°C).

The TRISO fuel particles selected for FHR use consist of a microsphere (i.e., kernel) of uranium oxycarbide (UCO) material encapsulated by multiple layers of pyrocarbon and a SiC (silicon carbide) layer. This multiple-coating-layer system has been engineered to retain the fission products generated by fission of the nuclear material in the kernel during normal operation and all licensing basis events (LBEs) over the design life of the fuel.

2.4.3 SUPER-CRITICAL WATER REACTOR (SCWR)

The SCWR (see Fig. 2.8) is the only Gen-IV system that uses water as a coolant. This is a natural evolution of Gen 2 and 3 light-water reactors, but in order to reach higher efficiencies, the temperature and pressure are increased above the thermodynamic critical point of water (374°C, 22 MPa) to give a thermal efficiency about 46% compared

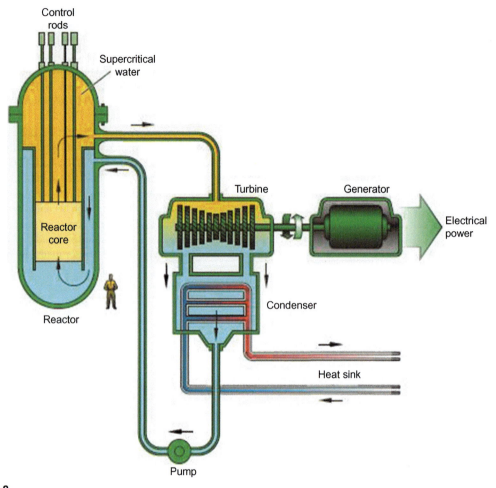

FIG. 2.8

Diagram of the SCWR [55].

to about 33% in today's light-water reactors. Even though the idea of using supercritical water as the coolant in a water-cooled reactor emerged in the 1960s [65], no reactor operating at both supercritical temperature and pressure was ever built. The supercritical water is the working fluid and directly drives the turbine without any secondary steam system. Thus the SCWR can be considered as an LWR operating at higher pressure and temperatures with a direct once-through cycle with only one phase present. The SCWR can operate at much higher temperatures and pressure than both current and expected LWRs. Besides a significantly higher efficiency, another cost advantage is expected from plant simplifications such as eliminating steam separators or primary pumps in the case of a once-through steam cycle at supercritical pressure. Also, by using supercritical water in a nuclear reactor, a boiling crisis is physically excluded, which adds an interesting safety feature.

The thermal spectrum of the core can be either fast or thermal, but in the latter case, additional water feeds have to be introduced, as the density of supercritical water is relatively low. The SCWR concepts include a pressure vessel option, as proposed for instance by Schulenberg and Starflinger et al. with the high-performance light-water reactor (HPLWR), or a pressure tube concept as proposed by Canada [66].

2.4.3.1 SCWR flow conditions
The Feedwater is heated up to 280–350°C and pumped to the reactor at a pressure of approximately 25 MPa. Neither reactors with pressure vessels nor those with multiple pressure tubes require recirculation pumps. Only extraction pumps for the condensers and feedwater pumps are needed to drive the steam cycle. In the pressure tube concept, the core produces superheated steam at ≈ 25 MPa between 500°C and 625°C. The superheated steam is supplied directly to the high-pressure turbine.

2.4.3.2 Predicted SCWR dose rates and total fluences
In the case of the pressure tube concept for the cladding materials, fluences similar to those obtained in LWRs (~ 10 dpa) are expected, while for the pressure tube, doses up to 100 dpa are predicted [67]. For the HPLWR, the dose on the cladding will be of the same order.

2.4.3.3 Leading SCWR candidate metal alloys
Since supercritical water has different chemical properties compared to liquid water, and because it requires a higher operating temperature, materials developed LWRs may not be suitable for an SCWR. The out of pile components operate under conditions for which significant operating experience and a well-developed knowledge base exist. The SCWR feed train material selection will be based on BWR and SCW Fossil-Power Plants (FPPs) best practices and advances in material development for ultrasupercritical FPPs [68]. Here, we will focus on the irradiated components. For the SCWR a wide variety of options are currently considered including both thermal- and fast-neutron spectra and pressure vessel and pressure tube configurations. The fuel is uranium or mixed uranium-plutonium oxide. The core may use thermal-neutron spectrum with light or heavy water moderation or be a fast reactor with full actinide recycle based on conventional reprocessing.

Key materials and design issues for the SCWR are those associated with high-temperature creep and strength, irradiation (irradiation-assisted creep, void swelling, and embrittlement) and stress corrosion cracking and corrosion. The challenges include requirements for increased temperature and pressure, increased heat transfer and corrosion, and water chemistry issues and stability. The materials challenges and candidate materials for SCWR development have been summarized (Guzonas and Novotny) [69]. The materials considered for the fuel cladding and the modified pressure tubes specifically for the SCWR are shown in Table 2.10. While fuel cladding will be replaced periodically as reactors are refueled, other structural components will require much longer expected lifetimes.

In structures where the temperatures will be significantly above 300°C, or irradiation doses above several dpa, candidate structural materials will be primarily ferritic or martensitic steels and low swelling austenitic stainless steels. The Fe-Cr-Ni alloys may have acceptable mechanical behavior and dimensional stability; however, there is currently an insufficient knowledge base for predicting stress corrosion cracking (SCC) or IASCC behavior under supercritical water conditions. Some alloys have demonstrated low swelling for doses of up to 50–100 dpa in both mixed-spectrum and fast reactors in the temperature regime of 450–550°C. Ferritic-martensitic steels in the 9%–12% Cr range are also more swelling-resistant than austenitic steels and acceptable behavior has been demonstrated in these alloys at neutron irradiation doses of 50–100 dpa. There has also been a strong interest in the development and application of ODS steels which provide enhanced high-temperature strength.

Table 2.10 Compositions of Candidate Fuel Cladding Alloys for the Canadian SCWR, EU HPLWR, and Japanese JSCWR Concepts [69]

Alloy	Fe	Cr	Ni	Mo	Mn	Si	Al	Zr	C	Ti	Others
310	Bal	24.5	20.2	0.29	1.17	0.33	–		0.048		
310S + Zr (H2)	Bal	25.04	20.82	0.51		0.51		0.59	0.034		P:0.016
310S FG (T3F)	Bal	24.74	21.92	–	–	0.25	–	–	0.099	0.81	N:0.0006 P:<0.005
310S FG (T6F)	Bal	25.03	22.81	2.38	–	–	–	–	–	0.41	Nb:0.26 N:0.002 P:<0.045
347	Bal	18.0	11.0	–	2.0	1.0	–	–	0.08	–	Nb: 10xC
800H	Bal	22.5	34.8	–	1.59	0.95	0.45	–	0.08	–	
625	4.9	22.6	Bal	9.8	0.43	0.47	0.47	–	0.09	0.45	Nb:3.7 Cu:0.68
214	3	16	75	–	0.5	0.2	4.5	0.1	0.05	–	Y:0.01 B:0.01
1.4970	Bal	15	15.3	1.18	1.68	0.53	–	–	0.095	0.45	
ODS PM 2000	Bal	19	–	–	–	–	5.5	–	–	0.5	Y$_2$O$_3$:0.5
316L	Bal	16.6	10	2.0–3.0	1.9	0.65	–	–	0.022		
316L (N)	Bal	18	12	2.0–3.0	2.0	1.0	–	–	0.030	–	N:0.16
316 Ti	Bal	16.6	12.1	2.03	1.15	0.45	–	–	0.032	0.38	
316L + Zr (H1)	Bal	16.54	10.71	2.22	–	0.46	–	0.56	0.006	–	P:0.016

2.4.4 SUMMARY

A summary of the operating conditions leading to materials challenges for candidate reactor materials has been provided for the six Gen-IV Reactor concepts. This includes two liquid metal cooled fast reactors (SFR and LFR), two helium cooled reactors (VHTR and GFR), the molten salt cooled reactor (MSR), and the supercritical water cooled reactor (SCWR). The reactor concept with the largest experience base is the SFR although research is underway to further the development of the others. The following chapters provide more detail on development and understanding of improved materials leading to further advancement of these advanced reactor concepts.

REFERENCES

[1] G.H. Marcus, Considering the next generation of nuclear power plants, Prog. Nucl. Energy 37 (1–4) (2000) 5–10.
[2] A Technology Roadmap for Generation IV Nuclear Energy Systems, 2002. http://www.gen-4.org/PDFs/GenIVRoadmap.pdf.
[3] Y. Sakamoto, J.-C. Garnier, J. Rouault, C. Grandy, T. Fanning, R. Hill, Y. Chikazawa, S. Kotake, Selection of sodium coolant for fast reactors in the US, France and Japan, Nucl. Eng. Des. 254 (2013) 194–217.
[4] https://en.wikipedia.org/w/index.php?title=Sodium-cooled_fast_reactor&oldid=846453107,EN.WIKIPEDIA.ORG/WIKI/BREEDER_REACTOR.
[5] A.L. Pitner, R.B. Baker, Metal fuel test program in the FFTF, J. Nucl. Mater. 204 (1993) 124–130.
[6] B.H. Sencer, J.R. Kennedy, J.I. Cole, S.A. Malloy, F.A. Garner, Microstructural analysis of an HT9 fuel assembly duct irradiated in FFTF to 155 dpa at 443°C, J. Nucl. Mater. 393 (2) (2009) 235–241.

REFERENCES

[7] K. Natesan, M. Li, Materials performance in sodium cooled fast reactors: past, present and future, in: S. Monti (Ed.), Fast Reactors and Related Fuel Cycles: Safe Technologies and Sustainable Scenarios, 2015, pp. 461–486.

[8] C. Petesch, T. Lebarbe, C. Pascal, M. Blat-Yrieix, RCC-MRX code: context, overview, on-going developments, in: S. Xu, K. Hojo, R. Cipolla (Eds.), Proceedings of the ASME Pressure Vessels and Piping Conference, vol. 1b, 2016.

[9] W.J. O'Donnell, A.B. Hull, S. Malik, Historical context of elevated temperature structural integrity for next generation plants: regulatory safety issues in structural design criteria of ASME section III subsection NH, in: K. Hasegawa, D. Scarth (Eds.), Proceedings of the ASME Pressure Vessels and Piping Conference, vol. 1, 2009, pp. 729–738.

[10] R.D. Leggett, E. Heck, P. Levine, R.F. Hilbert, Steady state irradiation behavior of mixed oxide fuel pins irradiated in EBR-II, in: International Conference on Fast Breeder Reactor Fuel Performance, Hanford Engineering Development Laboratory, Monterey, CA, 1979, p. 22.

[11] P. Martin, M. Pelletier, D. Every, D. Buckthorpe, French and United Kingdom experience of high-burnup mixed-oxide fuel in sodium-cooled fast breeder reactors, Nucl. Technol. 161 (1) (2008) 35–44.

[12] S. Qvist, Optimizing the design of small fast spectrum battery-type nuclear reactors, Energies 7 (8) (2014) 4910–4937.

[13] T. Furukawa, E. Yoshida, Material performance in sodium, in: R.J.M. Konings (Ed.), Comprehensive Nuclear Materials vol. 5: Materials Performance and Corrosion/Waste Materials, Elsevier, Amsterdam, 2012, pp. 327–341.

[14] T.S. Byun, M.B. Toloczko, T.A. Saleh, S.A. Maloy, Irradiation dose and temperature dependence of fracture toughness in high dose HT9 steel from the fuel duct of FFTF, J. Nucl. Mater. 432 (1–3) (2013) 1–8.

[15] C. Prunier, F. Boussard, L. Koch, M. Coquerelle, Some specific aspects of homogeneous americium- and neptunium-based fuels transmutation through the outcomes of the SUPERFACT experiment in Phenix fast reactor, Nucl. Technol. 119 (2) (1997) 141–148.

[16] S. Ukai, S. Ohtsuka, Irradiation creep-swelling interaction in modified 316 stainless steel up to 200 dpa, J. Nucl. Sci. Technol. 44 (5) (2007) 743–757.

[17] B. Raj, C. Materials, in: A.E. Waltar, D.R. Todd, P.V. Tsvetkov (Eds.), Fast Spectrum Reactors, Springer US, Boston, MA, 2012, pp. 299–363.

[18] Hanford Engineering Development Laboratory, In-Sodium Corrosion Behavior of Candidate Commercial Fuel Cladding and Duct Alloys, Richland, Wash: Hanford Engineering Development Laboratory, Richland, Wash, HEDL-TME-77-71.

[19] F.A. Garner, Irradiation performance of cladding and structural steels in liquid metal reactors, in: B.R.T. Frost (Ed.), Nuclear Materials, VCH Veerlagsgesellschaft mbH, Weinheim, Germany, 1996, pp. 420–543.

[20] D.S. Gelles, Development of martensitic steels for high neutron damage applications, J. Nucl. Mater. 239 (1–3) (1996) 99–106.

[21] B.A. Chin, R.J. Neuhold, J.L. Straalsund, Materials development for fast breeder reactor cores, Nucl. Technol. 57 (3) (1982) 426–435.

[22] R.L. Klueh, D.R. Harries, High-Chromium Ferritic and Martensitic Steels for Nuclear Applications, ASTM Monograph Series 3, 2001.

[23] T. Lauritzen, W. Bell, S. Vaidyanathan, Effects of irradiation on the mechanical properties of ferritic alloys HT-9 and 2.25Cr-1Mo, in: Proceedings of Topical Conference on Ferritic Alloys for Use in Nuclear Energy Technologies, Metallurgical Society of AIME, Warrendale, PA, 1984, pp. 623–630.

[24] J.L. Séran, M. Le Flem, 8 – Irradiation-resistant austenitic steels as core materials for Generation IV nuclear reactors, in: P. Yvon (Ed.), Structural Materials for Generation IV Nuclear Reactors, Woodhead Publishing, 2017, pp. 285–328.

[25] S. Ukai, S. Ohtsuka, T. Kaito, Y. de Carlan, J. Ribis, J. Malaplate, 10 – Oxide dispersion-strengthened/ferrite-martensite steels as core materials for Generation IV nuclear reactors, in: P. Yvon (Ed.), Structural Materials for Generation IV Nuclear Reactors, Woodhead Publishing, 2017, pp. 357–414.

[26] T. Uwaba, M. Ito, T. Mizuno, Irradiation performance of fast reactor MOX fuel assemblies irradiated to high burnups, J. Nucl. Sci. Technol. 45 (11) (2008) 1183–1192.

[27] R.D. Leggett, L.C. Walters, Status of LMR fuel development in the United States of America, J. Nucl. Mater. 204 (1993) 23–32.

[28] R.L. Klueh, J.M. Vitek, W.R. Corwin, D.J. Alexander, Impact behavior of 9-CR and 12-CR ferritic steels after low-temperature irradiation, J. Nucl. Mater. 155 (1988) 973–977.

[29] F.H. Huang, Influence of cold work, heat-treatment, and chromium content on the fracture properties of irradiated martensitic steels, in: A.S. Kumar, D.S. Gelles, R.K. Nanstad, E.A. Little (Eds.), Effects of Radiation on Materials: 16th International Symposium, 1994, pp. 575–590.

[30] M. Hamilton, D. Gelles, R. Lobsinger, M. Paxton, W. Brown, Fabrication Technology for ODS Alloy MA957, Pacific Northwest National Lab., Richland, WA, 2000. Report number: PNL-13165.

[31] R.L. Klueh, D.J. Alexander, P.J. Maziasz, Impact behavior of reduced-activation ferritic steels irradiated in the fast flux test facility, J. Nucl. Mater. 186 (2) (1992) 185–195.

[32] R.L. Klueh, D.J. Alexander, M. Rieth, The effect of tantalum on the mechanical properties of a 9Cr-2W-0.25V-0.07Ta-0.1C steel, J. Nucl. Mater. 273 (2) (1999) 146–154.

[33] C.c. Rubbia, The European Technical Working Group on ADS, A European Roadmap for Developing Accelerator Driven Systems (ADS) for Nuclear Waste Incineration, 2001. https://www.oecd-nea.org/pt/docs/ADSROADMAP.pdf.

[34] IAEA Report, Advances in Small Modular Reactor Technology Developments, IAEA, 2016. http://aris.iaea.org.

[35] M. Frogheri, A. Alemberti, L. Mansani, The lead fast reactor: demonstrator (ALFRED) and ELFR design, in: S. Monti (Ed.), Fast Reactors and Related Fuel Cycles: Safe Technologies and Sustainable Scenarios, 2015, pp. 233–247.

[36] IAEA, Liquid Metal Coolants for Fast Reactors Cooled by Sodium, Lead, and Lead-Bismuth Eutectic, IAEA Nuclear Energy Series No. NP-T-1.6, 2012.

[37] F. Groeschel, C. Fazio, J. Knebel, C. Perret, A. Janett, G. Laffont, L. Cachon, T. Kircher, A. Cadiou, A. Guertin, P. Agostini, The MEGAPIE 1 MW target in support to ADS development: status of R&D and design, J. Nucl. Mater. 335 (2) (2004) 156–162.

[38] Y. Dai, C. Fazio, D. Gorse, F. Groschel, J. Henry, A. Terlain, J.B. Vogt, T. Auger, A. Gessi, Summary on the preliminary assessment of the T91 window performance in the MEGAPIE conditions, Nucl. Instr. Meth. Phys. Res. Sect. A–Acceler. Spect. Detect. Assoc. Equip. 562 (2) (2006) 698–701.

[39] L. Cinotti, C.F. Smith, H. Sekimoto, L. Mansani, M. Reale, J.J. Sienicki, Lead-cooled system design and challenges in the frame of Generation IV International Forum, J. Nucl. Mater. 415 (3) (2011) 245–253.

[40] J. Zhang, R.J. Kapernick, P.R. McClure, T.J. Trapp, Lead-bismuth eutectic technology for Hyperion reactor, J. Nucl. Mater. 441 (1–3) (2013) 644–649.

[41] H.A. Abderrahim, Future advanced nuclear systems and the role of MYRRHA as a waste transmutation R&D facility, in: S. Monti (Ed.), Fast Reactors and Related Fuel Cycles: Safe Technologies and Sustainable Scenarios, 2015, pp. 193–207.

[42] C. Fazio, Proceedings of the international Demetra workshop on development and assessment of structural materials and heavy liquid metal technologies for transmutation systems preface, J. Nucl. Mater. 415 (3) (2011) 227–228.

[43] C. Petrovich, ALFRED core, Summary, synoptic tables, in: Conclusions and Recommendations, ENEA Technical Report UTFISSM-P9SZ-006, 2013.

[44] C. Fazio, F. Balbaud, 2 – Corrosion phenomena induced by liquid metals in Generation IV reactors, in: P. Yvon (Ed.), Structural Materials for Generation IV Nuclear Reactors, Woodhead Publishing, 2017, pp. 23–74.

[45] OECD-NEA, Handbook on Lead-Bismuth Eutectic Alloy and Lead Properties, Materials, Compatibility, Thermal-Hydraulics and Technologies, OECD-NEA 6195, 2007.

[46] H.A. Abderrahim, Progress on the MYRRHA project status at the end of 2017 new implementation plan towards start of construction, in: Nuclear 2018, International Conference on Sustainable Development through Nuclear Research and Education, Pitesti, Romania, 2018.

[47] G.S. Was, Fundamentals of Radiation Materials Science—Metals and Alloys, Springer, 2007.

[48] J. Van den Bosch, G. Coen, P. Hosemann, S.A. Maloy, On the LME susceptibility of Si enriched steels, J. Nucl. Mater. 429 (1–3) (2012) 105–112.

[49] J.Y. Park, 12 – SiCf/SiC composites as core materials for Generation IV nuclear reactors, in: P. Yvon (Ed.), Structural Materials for Generation IV Nuclear Reactors, Woodhead Publishing, 2017, pp. 441–470.

[50] A. Alamo, M. Horsten, X. Averty, E.I. Materna-Morris, M. Rieth, J.C. Brachet, Mechanical behavior of reduced-activation and conventional martensitic steels after neutron irradiation in the range 250–450°C, J. Nucl. Mater. 283 (2000) 353–357.

[51] C. Fazio, P. Dubuisson, Achievement and New Challenges for High Performance Materials in Europe, 2015.

[52] A.D. Kashtanov, V.S. Lavrukhin, V.G. Markov, V.A. Yakovlev, S.N. Bozin, V.N. Leonov, B.S. Rodchenkov, A.I. Filin, Corrosion-mechanical strength of structural materials in contact with liquid lead, Atomic Energy 97 (2) (2004) 538–542.

[53] F. Bertrand, I. Germain, F. Bentivoglio, F. Bonnet, Q. Moyart, P. Aujollet, Safety study of the coupling of a VHTR with a hydrogen production plant, Nucl. Eng. Des. 241 (7) (2011) 2580–2596.

[54] J.A. Lake, R.G. Bennett, J.F. Kotek, Next-generation nuclear power, Sci. Am. 286 (1) (2002) 72–81.

[55] GIF Website, https://www.gen-4.org.

REFERENCES

[56] P. Ponya, S. Czifrus, Core optimisation issues of MOX fueled ALLEGRO reactor, Ann. Nucl. Energy 108 (2017) 188–197.

[57] W. Corwin, L. Snead, S. Zinkle, R. Nanstad, A. Rowcliffe, L. Mansur, R. Swindeman, W. Ren, D. Wilson, T. McGreevy, P. Rittenhouse, J. Klett, T. Allen, J. Gan, K. Weaver, The Gas Fast Reactor (GFR) Survey of materials Experience and R&D Needs to Assess Viability, ORNL/TM-2004/99, April 30, 2004.

[58] M. Le Flem, J.-M. Gentzbittel, P. Wident, Assessment of a European V-4Cr-4Ti alloy – CEA-J57, J. Nucl. Mater. 442 (1–3) (2013) S325–S329.

[59] L. Chaffron, C. Sauder, C. Lorrette, L. Briottet, A. Michaux, L. Gelebart, A. Coupe, M. Zabiego, M. Le Flem, J.-L. Seran, Innovative SiC/SiC composite for nuclear applications, in: C. Galle (Ed.), Minos - Materials Innovation for Nuclear Optimized Systems, 2013. https://doi.org/10.1051/epjconf/20135101003.

[60] J.A. Lane, U.S.A.E. Commission, United Nations International Conference on the Peaceful Uses of Atomic, Fluid Fuel Reactors, Addison-Wesley Pub. Co., Reading, MA, 1958.

[61] R.N. Wright, T.-L. Sham, Status of Metallic Structural Alloys for Molten Salt Reactors, INL, INL/Ext-18-45171, Rev. 0, 2018.

[62] H.E. McCoy, R.L. Beatty, W.H. Cook, R.E. Gehlbach, C.R. Kennedy, J.W. Koger, A.P. Litman, C.E. Sessions, J.R. Weir, New developments in materials for molten-salt reactors, Nucl. Appl. Technol. 8 (2) (1970) 156.

[63] C. Andreades, A.T. Cisneros, J.K. Choi, A.Y.K. Chong, M. Fratoni, S. Hong, L.R. Huddar, K.D. Huff, J. Kendrick, D.L. Krumwiede, M.R. Laufer, M. Munk, R.O. Scarlat, N. Zweibau, Design summary of the mark-I pebble-bed, fluoride salt–cooled, high-temperature reactor commercial power plant, Nucl. Technol. 195 (3) (2016) 223–238.

[64] D.E. Holcomb, G.F. Flanagan, G.T. Mays, W.D. Pointer, K.R. Robb, G.L. Yoder, Jr., Fluoride salt-cooled high-temperature reactor technology development and demonstration roadmap, in: Proceedings of the 20th International Conference on Nuclear Engineering and the ASME 2012 Power Conference – 2012, vol. 2, Oak Ridge National Lab. (ORNL), Oak Ridge, TN, 2013. https://www.osti.gov/servlets/purl/1107839.

[65] J.H. Wright, J.F. Patterson, Status and application of supercritical-water reactor coolant, in: American Power Conference, Chicago, IL, 1966, pp. 139–149.

[66] M. Yetisir, M. Gaudet, D. Martin, ASME, Mechanical Aspects of The Canadian Generation IV Supercritical Water-Cooled Pressure Tube Reactor, 2012.

[67] L. Walters, M. Wright, D. Guzonas, Irradiation issues and material selection for Canadian SCWR components, J. Nucl. Eng. Radiat. Sci. 4 (3) (2018) 031005–031010.

[68] R. Viswanathan, J.F. Henry, J. Tanzosh, G. Stanko, J. Shingledecker, B. Vitalis, R. Purgert, US program on materials technology for ultra-supercritical coal power plants, J. Mater. Eng. Perform. 22 (10) (2013) 2904–2915.

[69] D. Guzonas, R. Novotny, Supercritical water-cooled reactor materials—summary of research and open issues, Prog. Nucl. Energy 77 (2014) 361–372.

[70] S. Guo, J. Zhang, W. Wu, W. Zhou, Corrosion in the molten fluoride and chloride salts and materials development for nuclear applications, Prog. Mater. Sci. 97 (2018) 448–487.

CHAPTER 3

OVERVIEW OF REACTOR SYSTEMS AND OPERATIONAL ENVIRONMENTS FOR STRUCTURAL MATERIALS IN FUSION REACTORS

Richard J. Kurtz*, G. Robert Odette[†]

Pacific Northwest National Laboratory, Richland, WA, United States *Materials Department, University of California, Santa Barbara, CA, United States*[†]

CHAPTER OUTLINE

3.1 Introduction 52
3.2 Overview of Basic Physics 52
    3.2.1 Neutron and Thermal Loads on Materials Surrounding the Fusion Reaction 54
3.3 Basic Aspects of Materials Degradation in the Fusion Environment 55
    3.3.1 Comparison with Fission Environment 58
3.4 Overview of MCF and ICF Conceptual Designs 60
    3.4.1 MCF Conceptual Power Plant Designs 61
    3.4.2 ICF Conceptual Power Plant Designs 64
3.5 First-Wall/Blanket Structural Material Options 66
    3.5.1 RAF/M Steels 66
    3.5.2 Nanostructured Ferritic Alloys 72
    3.5.3 Vanadium Alloys 75
    3.5.4 SiC$_f$/SiC Composites 77
    3.5.5 Helium Effects in Fusion Structural Materials 80
3.6 Materials for Divertor/Limiter Applications 85
    3.6.1 Tungsten and Tungsten Alloys 85
    3.6.2 Carbon Fiber Composites 90
    3.6.3 Liquid Walls 92
3.7 Vacuum Vessel (VV) Materials 94
3.8 Magnet Structural Materials 96
References 96

CHAPTER 3 OVERVIEW OF FUSION MATERIALS AND OPERATING ENVIRONMENTS

3.1 INTRODUCTION

This chapter presents an overview of the conditions that structural materials being developed for fusion power reactor applications would encounter. We begin by briefly describing the relevant plasma physics that dictate the fusion environment, which drive microstructure evolutions in structural alloys. The fundamentals of radiation damage in the fusion environment are discussed, but the reader is referred to a much more detailed treatment of this topic given in Chapter 5. Our intent is to highlight the most important degradation mechanisms operating in the fusion environment. We compare and contrast the fusion environment with that found in existing and proposed advanced fission energy systems, highlighting in particular the role that gaseous and solid transmutation products have on property evolution. The majority of the chapter is devoted to a presentation of the principal structural material choices for major subsystems of a fusion power reactor, such as the first wall/blanket, divertor, and vacuum vessel (VV). Our aim is to highlight the essential features of the material choices for each subsystem, and not to delve into the details, since other chapters address each class of structural alloys in greater depth. The emphasis is on the implication of the environment to the basic physical and mechanical properties requirements of these materials and challenges they face specific to the fusion environment.

3.2 OVERVIEW OF BASIC PHYSICS

Nuclear fusion is the process by which atomic nuclei of low atomic number join to form a heavier nucleus with the release of considerable energy. In the core of the Sun, hydrogen nuclei fuse to produce helium nuclei, but this is an extremely slow reaction and cannot be used for applications on Earth. The most favorable reaction for a practical terrestrial fusion energy source involves two heavy isotopes of hydrogen, deuterium (D), and tritium (T). The basic reaction is

$$D + T \rightarrow {}^{4}_{2}He + {}^{1}_{0}n + 17.6\,MeV$$

where approximately 20% (3.5 MeV) of the energy released is imparted to the He nucleus and remainder (14.1 MeV) resides in the neutrons. The reacting nuclei are charged particles, which repel each another. In order for nuclear fusion to occur, the two nuclei must be brought close enough so that the strong nuclear force overcomes the electrostatic repulsion between the ion cores. This is accomplished by heating the DT fuel to very high temperatures, on the order of $\sim 10^8$ K [1]. At such temperatures matter exists in a plasma state, which can be viewed as a mixture of two fluids: one of positively charged ions and another of negatively charged electrons [1].

The hot DT plasma would tend to disperse and rapidly quench any fusion reactions if it were not confined. To enable fusion to occur, the DT plasma must be confined at a high enough density for a long enough time. In stars, confinement is achieved by gravity, but gravity is a weak force compared to nuclear forces. At the temperatures needed for fusion, conventional materials cannot confine the ultra-hot plasma. Any contact between the plasma and material leads to melting and evaporation on one hand, and cooling (or quenching) of the plasma on the other hand.

There are two basic schemes used to confine fusion plasmas. One employs powerful magnetic fields to form a barrier between the hot DT fuel and the structural materials of the reaction chamber. Intense magnetic fields constrain the motion of the charged particles comprising the plasma so that they are prevented from directly interacting with the solid walls. This is known as magnetic confinement fusion (MCF) and the concept is illustrated schematically in Fig. 3.1. An alternate approach is to compress the DT fuel and heat it so rapidly that fusion occurs before the fuel expands. This method is known as inertial confinement fusion (ICF). Figure 3.2 illustrates ICF. Laser beams or x-rays produced by lasers rapidly heats the surface of a spherical DT capsule, forming a surrounding plasma envelope (Fig. 3.2A). The DT fuel is compressed by the jet-like ablation of the hot surface layer (Fig. 3.2B). As the capsule implodes the DT fuel at its core is compressed and eventually ignites (Fig 3.2C). A fusion reaction front then propagates quickly through the compressed fuel (Fig 3.2D).

3.2 OVERVIEW OF BASIC PHYSICS

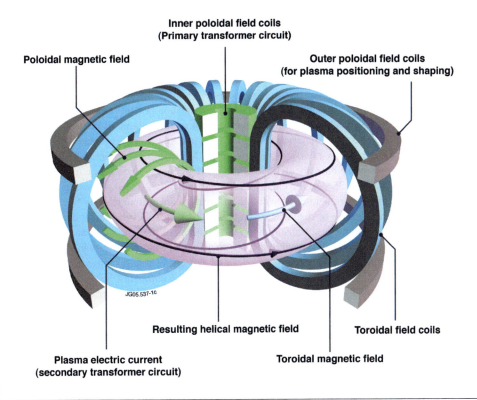

FIG. 3.1

Schematic representation of plasma confinement for a magnetic fusion energy system.

Reproduced with permission of EUROfusion/Culham Centre for Fusion Energy.

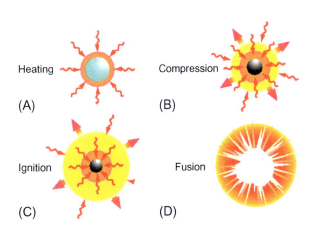

FIG. 3.2

Schematic representation of plasma confinement for an inertial fusion energy system. (A) heating of the DT capsule by laser beams or X-rays, (B) capsule compression due to ablation of the hot surface layer, (C) ignition of the fusion reaction, and (D) propagation of the fusion reaction throughout the capsule.

Obtained from http://iter.rma.ac.be/en/physics/how2doit/inertial/index.php.

3.2.1 NEUTRON AND THERMAL LOADS ON MATERIALS SURROUNDING THE FUSION REACTION

In MCF fusion, during normal operation, the plasma does not physically touch the materials directly surrounding it, but those materials are subjected to intense thermal radiation, and to bombardment by 14.1 MeV neutrons, 3.5 MeV α-particles, as well as ions, neutral particles and impurities with relatively low energies ($E < 10\,\text{keV}$) [2]. The high-energy neutrons penetrate deeply into the components enclosing the plasma and contribute additional thermal energy and irradiation damage throughout the volume of the first-wall/blanket structures by interacting with the associated materials. For an ICF system, the energy released from the ignited target is primarily in the form of energetic x-rays, ions, and neutrons. Neutrons deposit about 71% of the target yield into the chamber wall; x-rays contribute about 1.5% of the incident energy over a time interval of about 0.5 ns; and ions contribute the remaining 28% of the energy over approximately 2 μs [3]. As will be discussed in more detail below, the high-energy neutrons interact with the first-wall/blanket materials by transferring kinetic energy to lattice atoms causing them to be displaced from their lattice sites. Displacement damage is also accompanied by nuclear reactions, which cause transmutation of the atoms of the structural materials into other solid elements and gases, such as He and H. The level of displacement damage is typically quantified as displacements per atom/full-power year (dpa/FPY) [4]. A damage level of 1 dpa corresponds, on average, to displacement of every atom in a material from its most stable lattice position. Table 3.1 compares the peak displacement damage and gaseous transmutation production for typical first-wall and plasma facing materials in representative MCF and ICF fusion power systems [5].

The plasma facing and first-wall structure is only a few millimeters thick, so the volumetric heating in these materials is small in comparison to the radiative surface-heating component. Since the energies of impinging particles are relatively low their depth of penetration is $<\sim 1\,\mu\text{m}$, so their corresponding effects are primarily confined to the near surface region. The magnitude of the surface heat flux depends on the specific design features of a fusion power system, but a comparison of heat fluxes in a variety of technologies and nature is presented in Fig. 3.3 [6]. The steady-state heat flux on important components of a fusion power system ranges from $\sim$1 to 20 MW/m². Note, the steady-state heat flux is comparable to that endured by rocket nozzles and requires a highly capable heat transfer system in order to remove these high thermal loads. In addition, to steady-state heat loads, first-wall materials are also subjected to off-normal plasma events that deposit extremely large quantities of heat in small regions over a very short period of time. The events are referred to as plasma disruptions and consist of intense local discharges of the plasma energy in the form of electrons and protons [7]. As illustrated in Fig. 3.3, the duration of these events is very short and the local heat flux ranges from 1 to 10 GW/m². Plasma disruptions also cause significant electromagnetically induced forces in the structure of an MCF device and its components [8].

Table 3.1 Peak neutron wall loads and gas production for typical MCF and ICF systems [5]

Material	dpa/FPY		He appm/FPY		H appm/FPY	
	MCF	ICF	MCF	ICF	MCF	ICF
F82H	86.6	63.4	843.5	416.1	3738	1852
W	27.2	19.5	13.4	6.5	49.8	24.1
V4Cr4Ti	80.7	51.0	333.6	155.3	1704	798.6
SiC/SiC	109.3	95.1	9503	4538	3511	1739
Be	44.3	44.1	18,151	10,972	287.1	131.5
CFC	74.0	70.7	14,431	6959	2.9	4.9

3.3 BASIC ASPECTS OF MATERIALS DEGRADATION IN THE FUSION ENVIRONMENT

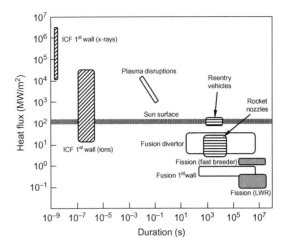

FIG. 3.3

Typical steady-state and transient heat fluxes for fission and fusion energy systems compared to the surface of the Sun and space vehicles [6].

Both MCF and ICF systems suffer radiation damage from high-energy neutrons. The instantaneous displacement damage rate in intrinsically pulsed ICF is much higher than for MCF. In addition, as noted above, the first wall in an ICF system must tolerate intense pulses of ions and x-rays that deposit enormous quantities of heat over very small time intervals, as shown in Fig. 3.3.

3.3 BASIC ASPECTS OF MATERIALS DEGRADATION IN THE FUSION ENVIRONMENT

As mentioned above, the first-wall and blanket structural materials in both MCF and ICF energy systems will be subjected to high thermal and neutron wall fluxes and will experience very complex, time-dependent mechanical loading. In addition, some structural components will also be exposed to chemically reactive tritium breeding materials, and corrosive coolants. The 14.1 MeV neutrons produced in the DT reaction carry the bulk of the fusion energy that is dissipated in the materials, components and structures surrounding the plasma or target pellet. About 10% of the incident neutron energy is deposited in the first wall, while most of the remainder is transferred to a much larger volume in the blanket [9]. The neutrons are slowed by nuclear collisions and reactions as they penetrate the reactor structure. Thus the fast neutron energy distribution, or spectrum, changes with penetration depth. However, it is important to note that the slowed neutrons are also transported (diffuse) in all directions in space. As a consequence the neutron spectrum is a combination of uncollided 14.1 MeV neutrons and neutrons with a wide range of lower but still significant kinetic energies. The ratio of 14.1 MeV to lower energy neutrons depends on the material used in the first wall and blanket and could be as low as 0.1 even at the first wall [5].

The energetic neutrons interact with the atoms of the structural materials in two basic ways. One involves transmutation, or conversion of one chemical element into others, which occurs by a wide range of reaction paths. Note, nuclear reactions, or sequences of reactions, can also produce new isotopes of the parent and preceding daughter elements. These newly created isotopes are often radioactive and represent important safety and environmental challenges that are important considerations in material selection and reactor design.

The various nuclear reaction rates depend on energy-dependent cross sections and the neutron spectrum. The most important transmutation reactions are (n, α) and (n, p), that produce He and H, which have threshold energies of several MeV. These and other transmutation reactions also produce solid impurities, such as ^{55}Fe(n, p)^{55}Mn. The second interaction mechanism involves elastic and inelastic neutron scattering collisions that do not produce transmutations, but do generate a distribution of energetic primary recoil atoms. The energies of the primary recoils resulting from both nuclear and scattering reactions range up to 1 MeV in Fe over a corresponding spectrum where about half the recoils are above ~3.1 keV [10].

The high-energy recoils, in turn, generate a cascade of atoms displaced from their lattice positions by a branching series of initially binary and later multibody atomic collisions. The collisions continue to geometrically multiply until the last generation of recoils does not have sufficient energy to displace lattice atoms from their equilibrium positions. Since lattice atom threshold displacement energies for most structural materials of interest to fusion are on the order of tens of electron volts, high-energy cascades produce a large number of vacancy and self-interstitial atom (SIA) defects [11].

The displacement dose is generally expressed on per atom units, or dpa. Energy-dependent cross sections for dpa, transmutation, and activation generation are used in conjunction with neutron flux spectra to compute the corresponding species production rates as a function of position in the entire fusion system.

Displacement cascades also cause high local temperature spikes, ballistic mixing of elements, and extended lattice defect clusters, such as nanovoids and small SIA dislocation loops as well as defect-solute complexes. All of these damage processes can lead to degradation of fusion materials. For irradiation at temperatures $\geq\sim 0.2 T_M$ diffusional transport of the defects coupled with redistribution of solute and transmuted atoms control radiation damage accumulation and microstructure evolution [11]. A way to mitigate dpa damage accumulation is to promote recombination of vacancies and SIA: vacancy + SIA → 0 defects. Recombination takes place by long-range defect diffusion, especially at defect trapping sites, as well at sinks that annihilate equal numbers of vacancies and SIA. Damage accumulation also decreases at higher temperatures because recovery processes are much faster and thermal vacancy concentrations much higher, leading to a recombination dominated defect fate.

Material properties depend on composition and microstructure at length scales ranging from angstroms to centimeters. Within limits, the microstructures of materials can be manipulated to optimize some, but usually not all, relevant properties. Thus, tailoring microstructures provide a viable approach to developing radiation-resistant materials for fusion. Typical alloys usually contain five or more constituents, two or more phases, and complex dislocation and grain boundary structures. The detailed morphology and distribution of these features control properties. Useful microstructures almost always represent a nonequilibrium thermodynamic state that evolves with time, thermally at high temperatures, and more rapidly under irradiation at lower temperature. Microstructural evolution includes not only redistribution of constituents that existed at the start of service, but also the formation of new constituents (transmutation) and features, and micro to macro damage, which leads to property degradation, increased failure probability, and decreased component lifetime.

Radiation damage strongly depends on irradiation temperature in three low, medium, and high regimes. At low temperatures, $<\sim 40\%$ of the absolute melting temperature, $0.4T_M$, the irradiated microstructure is dominated by the formation of a high density of defect clusters and precipitates that are barriers to the motion of dislocations. This results in increased strength, hardness, and reduced ductility. The strength increases with dpa to a saturation level that decreases with increasing irradiation temperature. The corresponding dpa to saturation also increases with increasing irradiation temperature [4,11–13]. Strength increases are accompanied by decreased uniform tensile ductility and strain hardening rates.

Notably, the resistance to crack propagation, or fracture toughness, decreases as strength increases, and in body-centered cubic (bcc) metals the transition from ductile-to-brittle fracture shifts to higher temperatures [14]. Helium also affects the mechanical property evolution at low irradiation temperatures. While the precise effects are not known, due to the lack of a fusion relevant neutron source for performing prototypical irradiation effects studies, considerable insight has been obtained from surrogate experiments using charged particles, spallation neutron sources, and a novel technique called in situ He implantation [15]. Such studies suggest that high levels of He lead to very large positive ductile-to-brittle transition temperature (DBTT) shifts due to both hardening and accumulation at grain boundaries, which decreases their cohesive strength, thus becoming favored pathways for crack propagation. High He also increases irradiation hardening at dpa values that would normally saturate under displacement damage conditions alone and extends it up to much higher temperatures [15].

There are three other very important effects of irradiation on alloys in the low-temperature regime. Two are related to interactions between solutes, excess vacancies, and SIA created by displacement damage. First, excess defects enhance diffusion of substitutional solutes. Radiation-enhanced diffusion (RED) can greatly accelerate thermodynamically driven radiation-enhanced precipitation (REP), which would be very sluggish under normal thermal diffusion kinetics. Hardening by the nanometer-scale precipitates that form at these low temperatures can be very significant. Second, differential coupling of solutes with persistent defect fluxes to sinks results in radiation-induced segregation (RIS). That is, various elements are enriched or depleted at sites like grain boundaries, dislocations, and cavity surfaces. The enrichment may reach levels that cross-phase boundaries resulting in radiation-induced precipitation (RIP) of nonequilibrium phases. Even lower levels of segregation promote phenomena such as grain boundary environmentally assisted cracking. For example, Cr depletion at grain boundaries is believed to make a significant contribution to what is called irradiation-assisted stress corrosion cracking, which is a critical issue in austenitic stainless steels in contact with water coolants. Finally, irradiation creep is a ubiquitous phenomenon over a wide range of temperatures. Irradiation creep mechanisms associated with biased interactions of irradiation defects, mainly involving SIA and dislocations, lead to creep at rates that scale linearly with stress and that are generally only weakly dependent on temperature.

In ceramics the predominant degradation mechanisms at low irradiation temperatures include point defect swelling, amorphization, and a substantial decrease in thermal conductivity [16]. Swelling, which is volumetric growth under irradiation, is significant in SiC at irradiation temperatures below about $0.4T_M$ with the volume expansion inversely proportional to temperature [16]. Similar to point defect swelling, the thermal conductivity of SiC is degraded with increasing fluence and the degradation is greatest at low irradiation temperatures [16]. These degradation mechanisms in SiC tend to saturate at doses of a few dpa [16].

At intermediate temperatures, between 0.3 and $0.6T_M$, irradiation-induced dpa damage decreases and, in the absence of high He, softening eventually begins to occur due to instabilities in start-of-life microstructures. Dimensional instabilities caused by volumetric swelling and irradiation creep may also occur. Irradiation creep remains important up to the temperature where thermal creep eclipses it. There is a significant incubation dose for onset of swelling that depends on many variables, such as initial microstructure, chemical composition, irradiation temperature, dose, dpa dose rate, and especially the He/dpa ratio. Significant degradation by these mechanisms typically requires more than 10 dpa [4,11–13]. The effects of He in this temperature regime are critical, as growing voids form from sufficiently large He bubbles, and the incubation time for onset of high rates of swelling is controlled by the He required for bubbles to attain a critical size [15]. However, bubbles are also sink-recombination sites for mobile point defects. Partitioning of He to a large number of stable bubbles may prevent a significant number of them from reaching the critical size needed to form voids. Higher He or He/dpa ratios may increase the number of bubbles. Thus increasing He may either increase or decrease swelling depending on the specific combination of all the material and irradiation variables. Ferritic-martensitic (bcc) steels are much more swelling resistant than face-centered cubic (fcc) austenitic stainless steels in fission reactor irradiations. In part, this is due to lower He generation rates and higher densities of He trapping sites in ferritic-martensitic steels. Other factors, such as high self-diffusion rates and low dislocation bias may contribute to the intrinsic swelling resistance of ferritic-martensitic steels relative to austenitic stainless steels. Void swelling and irradiation creep are also observed in ceramics, such as SiC, at irradiation temperatures $> \sim 0.50T_M$ [16].

At high irradiation temperatures ($> \sim 0.5T_M$) recombination is much greater since the concentration of thermal vacancies becomes dominant, thus they become the primary sink for SIA. Thought of another way, any displacement damage-induced defect aggregation recovers (anneals) at a much faster rate than it can accumulate. In this temperature regime, the consequences of displacement damage are far less important than the deleterious effects of transmutation products, such as He. A fraction of the He diffuses to, and is trapped at, grain boundaries, where it forms bubbles. The bubbles act as nucleation sites for creep cavities that are above a critical size and that grow unstably due to stresses normal to the boundary [11,12,14]. Rapid growth and coalescence of a large population of bubble nucleated grain boundary cavities leads to reduced creep rupture times and ductility, in some cases by well over an order of magnitude. High-temperature He intergranular creep embrittlement begins to emerge at He concentrations ranging from as low as 1 up to ~500 appm [14]. The dose corresponding to these He concentrations depends sensitively on the material composition, microstructure, and neutron spectrum [12]. Ferritic-martensitic steels are much more He embrittlement resistant than austenitic stainless steels in fission reactor irradiations for reasons that are similar to those that make them swelling resistant. Generally, the lifetime-temperature limits for fusion first-wall/blanket structural materials in the high-temperature regime are primarily controlled by chemical compatibility with breeding materials and coolants and thermal creep as well as creep-rupture and creep-fatigue interaction, which also are likely to be affected by high He [14]. These degradation mechanisms are discussed in more detail in the next section.

3.3.1 COMPARISON WITH FISSION ENVIRONMENT

Structural materials for proposed advanced fission power systems will face many of the same technical challenges that fusion materials will face with respect to the need for resistance to high levels of neutron-induced displacement damage and, for cladding, the capability to operate under severe thermomechanical conditions of high temperatures and stresses. However, while the experience gained from materials technology for advanced fission reactors is highly pertinent to fusion materials development, there are significant differences. Fig. 3.4 illustrates the substantially more demanding service environment that proposed advanced fission and fusion power systems will encounter compared to currently operating nuclear power plants. It is notable that the upper operating temperature limits of

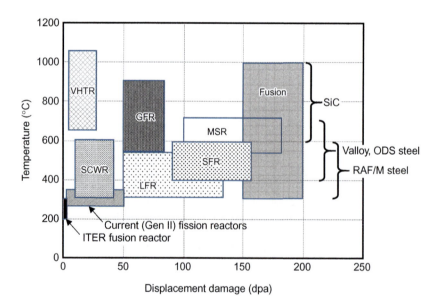

FIG. 3.4

Operating temperature and displacement damage regimes for current fission reactors, and proposed advanced fission and fusion energy systems. The advanced fission systems are very high-temperature reactor (VHTR), supercritical water reactor (SCWR), lead fast reactor (LFR), gas fast reactor (GFR), sodium fast reactor (SFR), and molten salt reactor (MSR) [12].

advanced fission and fusion systems are hundreds of degrees higher than conventional fission, and the projected damage limits are 2–4 times, or more, greater in some cases. Most significantly, the thermomechanical structural challenges of fusion systems are singular and unique.

While there are similarities in the fission and fusion operating environments, the 14.1 MeV neutrons produced in the DT fusion reaction are unique and must be considered. For typical fusion first-wall conditions the primary knock-on atom (PKA) spectrum has energies up to about ~1 MeV and an average value of 48 keV [17]. The corresponding maximum and average PKA energies in a typical light water reactor spectrum are ~300 and ~10 keV, respectively [17,18]. The PKA spectra for these two cases are remarkably similar for energies ≤ 100 keV with the fusion spectrum having a significantly higher fraction of PKAs above ~200 keV. However, differences in the PKA spectra do not have a major influence on production of surviving vacancies and SIAs, which escape cascade recombination, as well as the production of clusters of these defects. The PKA spectral effects are mitigated by the fact that high-energy recoils produce lower energy, more fission-like subcascades [18]. There are likely residual effects of time and spatial proximity of the subcascades, but these are expected to be second order. A much more important factor than fission versus fusion displacement damage is that 14.1 MeV fusion neutrons are above the threshold energy for (n, α) and (n, p) transmutation reactions, which cause significant production of He and H.

Assuming an optimistic end-of-life dose of about 200 dpa for first-wall structural materials, the peak neutron wall load estimates in Table 3.1 indicate that the lifetime of an MCF first-wall constructed of a reduced activation ferritic/martensitic (RAF/M) steel, such as F82H, would be about 2.3 years. Of course, first-wall lifetime is not dictated solely by displacement damage alone. The peak MCF transmutation production rate of He and H in F82H is 844 and 3738 appm/FPY, respectively. Note, the transmutation production of He is more than 11 times higher in SiC than it is in F82H. The effects of such high levels of gas production on the microstructural stability of first-wall materials are not fully understood, since no fusion-relevant neutron source exists for carrying out experiments to end-of-life doses. Novel experimental techniques [19] have been developed to explore He effects at fusion relevant production rates and considerable progress has been made in understanding basic mechanisms, but they do not permit the types of measurements of bulk properties needed for material qualification and licensing. Furthermore, the effects of solid transmutation remain largely unexplored, which could significantly affect property evolution in some cases, such as in SiC and W, where the rate of solid transmutation is substantial.

While the fusion nuclear environment is extremely challenging, the fusion thermomechanical environment may be even more problematic. As illustrated in Fig. 3.3, the steady-state heat flux on the MCF first wall and divertor is very high, and time varying. Transient heat loads associated with plasma disruptions deposit gigawatts of thermal energy into very small regions of the first wall over a very short time. In addition, plasma-facing heat transfer structures are thicker (e.g., a few millimeters) and geometrically much more complex than typical fast reactor fuel cladding. Although the steady-state heat load on an MCF first wall is comparable to that on thin-wall fast breeder reactor cladding (see Fig. 3.3), the thermal stresses in plasma facing and first-wall components are considerably larger and more complex. Stresses will also arise from differential thermal expansion and radial temperature gradients that occur over distances of meters and in three dimensions. These stresses and strains must be accommodated in massive interconnected structures that must maintain precise dimensional tolerances. The thermomechanics is further complicated by spatially varying dimensional instabilities (swelling and creep) and property evolution. This leads to extremely complex time varying stresses and strains. From Fig. 3.4 it is apparent that thermodynamically attractive fusion power reactor structures must operate at temperatures near the thermal creep regime. The combination of severe time varying temperatures and stresses suggests that material degradation mechanisms, such as thermal creep, creep-fatigue, and creep crack growth, will all be much more significant than in a fission system. In summary, it is no exaggeration to state that the ultra-severe thermomechanical environment of fusion structures is unprecedented and represents a grand challenge feasibility issue regarding the potential for large-scale MCF energy production.

Another significant challenge that is unique to fusion is the need to control the flow of large quantities of tritium. This will be a huge technical challenge because of the very large inventory of tritium involved, and the potential for significant retention of hydrogen isotopes in irradiated materials. Hydrogen isotopes are highly mobile and easily trapped at preexisting and radiation-induced defects. The processes by which tritium is absorbed at surfaces and adsorbed at internal interfaces, how tritium diffuses in irradiated metals and ceramics, and how it interacts with all microstructural features including radiation generated voids, helium bubbles, and point defect clusters are poorly understood [20]. To illustrate the potential impact of neutron irradiation, estimates suggest that a dose of as little as 20 dpa in steels could increase the tritium sink density sufficiently to quadruple the retained tritium in the blanket [21]. The potential for retention of large amounts of tritium in first-wall/blanket materials represents a potential safety issue that will require further research on the basic mechanisms of tritium transport, trapping, and release in irradiated materials.

Finally, the experience of the commercial fission power industry to address the challenge of high-level radioactive waste has motivated development of high-performance materials that minimally impact the environment. Radioactive isotope inventory and release paths are important considerations in designing for safety. Consequently, a worldwide materials development strategy has emerged that is focused on low or reduced activation materials. The basic concept is that materials removed from service should be recyclable and/or clearable and will not require long-term geological disposal, thereby minimizing the impact on the environment. On the basis of safety, waste disposal, and performance considerations the main materials systems under investigation include (1) Fe-Cr-based alloys that are RAF/M steels and nanostructured ferritic alloys (NFAs), (2) VCrTi alloys, (3) SiC composites, and (4) W alloys. The leading candidate materials are RAF/M steels, which achieve reduced activation compositions by replacement of high activation elements, such as Mo and Nb with V and W. Attainment of truly low activation steels and vanadium-based alloys will require reduction of impurities that restrict the free-release of in-vessel components [22].

3.4 OVERVIEW OF MCF AND ICF CONCEPTUAL DESIGNS

This section presents an overview of conceptual design approaches for both MCF and ICF power systems. The objective here is to give a broad overview of the major design features and components of MCF and ICF systems. From a materials science perspective, there are many similarities between the two approaches. Both MCF and ICF have first-wall and blanket materials that suffer radiation damage from high-energy neutrons generating displacement damage and transmutation products. Consequently, both approaches seek to develop materials that are more radiation damage resistant to provide for longer service life. Coupled with radiation tolerance is the need to develop and use low-activation structural materials to increase safety and environmental attractiveness. Both technologies desire materials capable of high-temperature operation to permit high power conversion efficiency, and materials that can withstand high heat fluxes and transfer thermal energy to the coolant. Further, structural materials in both environments must be resistant to aggressive chemical species. Finally, MCF and ICF system development needs extensive experiments and modeling efforts to determine acceptable service condition limits, including but not limited to radiation damage.

While there are many similarities between MCF and ICF, unique challenges and opportunities confront ICF development. First, the instantaneous neutron damage rate is orders of magnitude higher in ICF, which drives a need to understand how this affects material response. Second, the ICF first wall must tolerate intense pulses of ions and x-rays from the target in addition to damage from neutrons. Third, for laser-based ICF neutron damage to final optics is a concern, and chamber conditions for beam propagation and target injection must be reestablished between implosions. Fourth, most ICF designs are not constrained by liquid metal breeder/coolant magnetohydrodynamic (MHD) effects since powerful magnetic fields do not exist in ICF. Finally, a possibility exists with ICF that use of a thick liquid first wall would stop ions and x-rays, and may significantly attenuate the neutrons, thereby reducing the structural materials challenges to levels more comparable to present day fission reactor systems.

3.4 OVERVIEW OF MCF AND ICF CONCEPTUAL DESIGNS

3.4.1 MCF CONCEPTUAL POWER PLANT DESIGNS

MCF is presently considered to be the most promising approach to demonstrating commercial fusion power production and the tokamak confinement concept is the leading approach. Other confinement concepts that have received attention include stellarator, spherical torus, field-reversed configuration, reversed-field pinch, and spheromak, but currently the worldwide tokamak research effort accounts for over 90% of the total MCF program [23].

A tokamak is a device that uses powerful magnets to confine the plasma in the shape of a torus. The main components of a tokamak power core are shown in the cutaway view given in Fig. 3.5. A large set of equally spaced toroidal field (TF) and poloidal field (PF) coils are the principal magnets that confine the plasma. In addition, there are sets of divertor, equilibrium field, central solenoid, and vertical position coils used to create, shape, position, and stabilize the plasma within the D-shaped elliptical cross-section toroidal vessel. The red region in Fig. 3.5 is where the plasma resides, which is surrounded by the first-wall and blanket structure. The generic elements of an outboard radial build of a tokamak are depicted schematically in Fig. 3.6. Due to high particle and heat fluxes, the first wall and blanket must be actively cooled. The first wall is the material surface nearest the plasma and is integral to the blanket that is behind it.

The first wall may have a thin layer of armor material to limit surface erosion and sputtering. The three main functions of the first-wall/blanket structure are to (1) provide a physical boundary for the plasma and reduce the number of energetic neutrons reaching outboard components; (2) breed tritium fuel, which is not available in nature; and (3) convert the kinetic energy of neutrons and secondary gamma rays into heat at high temperature for energy conversion.

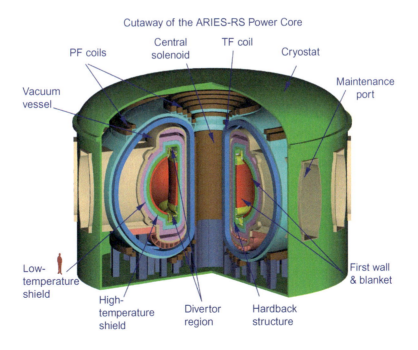

FIG. 3.5

Cutaway view of the ARIES-RS (Reversed Shear) power core, which is typical of many tokamak design concepts.

Reproduced from L.A. El-Guebaly, Fifty years of magnetic fusion research (1958–2008): brief historical overview and discussion of future trends, Energies 3 (2010) 1067.

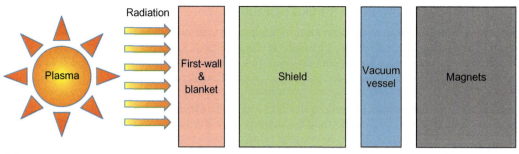

FIG. 3.6

Schematic representation of the generic elements of a radial build of a tokamak.

Interaction of neutrons with blanket materials produces significant volumetric heating. The incident neutron and radiative energy is recovered by coolants flowing in the first wall and blanket. The heat is transported out to where it can be thermally converted to electricity. Energetic neutrons also interact with lithium-bearing compounds in the blanket to produce tritium through nuclear reactions. The tritium is then recovered to refuel the power plant.

The two basic blanket concepts for breeding tritium are solid breeders that use lithium ceramics or liquid breeders that use liquid metals or molten salts. Typical solid breeding ceramics include Li_2O, Li_4SiO_4, Li_2TiO_3, and Li_2ZrO_3. Liquid metal breeders are either Li or PbLi and molten salts under consideration include LiF-BeF_2 also known as Flibe or LiF-BeF_2-NaF, which is referred to as Flinabe. Solid breeder blankets are always separately cooled, typically with He or water. Liquid breeder blankets may be self-cooled, separately cooled, or dual cooled depending on the details of the blanket design and choice of structural materials. In a self-cooled system, the liquid breeder is circulated at high enough speed to also serve as the coolant. For a separately cooled breeding blanket, the breeder is circulated only at a low speed for tritium extraction and He is often used as the coolant. In a dual-coolant blanket design, the first-wall and blanket structures are separately cooled with He and the breeding zone is self-cooled by the liquid metal.

An advantage of a solid breeding blanket is the selection of a coolant (e.g., He) that avoids problems related to safety, such as potential corrosion and MHD effects. A concern with solid breeder blankets is intrinsically low thermal conductivity, which might be further reduced under neutron irradiation. The low thermal conductivity of solid breeders has implications for the operating temperature window of these materials. The low thermal conductivity imposes limitations on power density and the achievable tritium-breeding ratio. Molten salt liquid breeders also suffer from low thermal conductivity compared to liquid metal breeders. Liquid metals are most attractive for high-performance blankets because they possess high tritium breeding capability with no need for neutron multipliers, unlike solid breeder concepts. They also have high thermal conductivity and immunity to radiation damage. Since ^6Li is the isotope that makes tritium there is also the potential for unlimited lifetime if the ^6Li content can be continuously or frequently replenished. Challenges associated with liquid metal breeders include potential reactivity with water and air, which is primarily a concern for liquid Li; embrittlement of the V-alloy structural material through interstitial impurity ingress from the liquid Li; and corrosion of steel and deposition of activated corrosion products along with generation of ^{210}Po for PbLi. Generic to flowing liquid metal blankets is MHD-induced pressure drop. The consequence of MHD pressure drop is increased stress on the flow channel. MHD effects can be minimized by electrically isolating the liquid metal flow from conductive structural materials or by reducing the flow velocity of the liquid metal.

Outside the first-wall/blanket region, there is a structural material zone that mechanically stiffens and supports the blanket. Typically, this region and those beyond it are permanent components of a tokamak. This region is depicted as a shield in Figs. 3.5 and 3.6, and it is designed so that (1) the blanket can be periodically replaced, (2) to resist electromagnetic forces originating in the plasma, and (3) to protect the VV and magnets from neutrons.

The next major component is the VV, which resides between the shield and magnet structures, see Fig. 3.6. The VV separates the high vacuum region, which is required to create and sustain the fusion reaction, and the atmospheric pressure conditions existing external to the vessel. Frequently the VV is double walled and it may be

3.4 OVERVIEW OF MCF AND ICF CONCEPTUAL DESIGNS

designed to provide additional shielding by including neutron-absorbing materials. The VV is considered the primary safety boundary for tritium containment and other radioactive nuclides.

The magnets are furthest from the plasma because their materials of construction are very sensitive to radiation damage. The magnets are superconducting to minimize electrical power needs. Both low-temperature and high-temperature superconductors may be used in future fusion power systems. Materials for low-temperature superconducting magnets include Nb_3Sn for high-field applications and NbTi for low-field situations. High-temperature superconductors are less technologically mature than low-temperature superconductors, but $YBa_2Cu_3O_5$ is currently the leading choice for fusion magnet applications.

The final major component of interest here is the divertor, which may represent the greatest technological challenge to overcome. Alpha particles from the fusion reaction, ions, and electrons escaping magnetic confinement are collected in the divertor. The primary functions of the divertor are plasma exhaust and heat removal. Typically, tokamaks use a single divertor at the bottom, or double divertors located above and below the plasma region as shown in Fig. 3.5. Divertor materials face some of the most challenging conditions in a fusion power system because they are exposed to very high heat and particle loads. The small area available to absorb the incident energy exacerbates this situation.

Over the last 50 years there have been many fusion power plant conceptual design studies. As noted by El-Guebaly, such studies provide important information to understand the implications of future technological developments, to give guidance to physics and engineering research for advanced systems, and to assess the impact of current technical challenges on the feasibility of fusion power [23]. Given the large number and evolving nature of power plant design studies it is not practical to give details of the work here. The interested reader should consult El-Guebaly [24] for a recent survey of these studies. It is sufficient to note that a wide range of structural material/coolant combinations for the major components of an MCF power system has been considered. Table 3.2

Table 3.2 Materials for major components in ARIES tokamak power plant studies

Component	Power plant study		
	ARIES-RS [25]	ARIES-ST [26]	ARIES-AT [27]
First-wall			
Structure	V-4Cr-4Ti	RAFM steel/ODS alloy	SiC composite
Coolant	Li	He	PbLi
Blanket			
Structure	V-4Cr-4Ti	RAFM/ODS steel	SiC composite
Coolant	Li	He + PbLi	PbLi
Breeder	Li	PbLi	PbLi
Shield			
Structure	V-4Cr-4Ti	Ferritic steel	SiC composite
Coolant	Li + He	He	PbLi
Vacuum vessel			
Structure	High Mn steel	Al	Ferritic steel
Coolant	He	H_2O	H_2O
Divertor			
Structure	V-4Cr-4Ti	ODS ferritic alloy	SiC composite
Coolant	Li	He	PbLi
Armor	W	W	W

summarizes the material choices for major components of MCF power systems from three recent design studies [25–27]. Note, the material selections depend heavily on design goals. For example, the ARIES-ST study focused on more conservative design goals so the structural material choices reflect a lower level of technological risk and attractiveness from a system performance perspective. On the other hand, the ARIES-AT study established more ambitious design objectives and focused on advanced technology alternatives. It is apparent from the information in Table 3.2 that less technologically mature structural materials selections were made in this study, representing greater risk, but providing for a potentially more attractive fusion power system. It is important to note that all these conceptual designs represent materials and structural feasibility issues that are far beyond current technological capabilities. Low-activation structural materials under development for fusion have been utilized in these design studies, but as will be seen below an enormous amount of work remains to be done to fully develop, qualify materials, and compile a database that is tightly coupled to advanced structural analysis methods so that a viable fusion power system can be designed, constructed, and safely operated.

3.4.2 ICF CONCEPTUAL POWER PLANT DESIGNS

As summarized in Section 3.2, ICF, involves the implosion of a small (~3 mm diameter) spherical DT capsule to produce the fusion reaction [28]. In order for this process to work, the fuel capsule must be compressed to more than 100 times solid density before ignition is achieved [29]. Al-Ayat et al. [29] note that an ICF power plant would function in a manner analogous to an internal combustion engine. The DT fuel capsule is injected into a reaction chamber; the fuel is compressed and heated to ignition by a laser or other driver, which is similar to the role of a piston; the spent fuel is exhausted and the cycle is repeated many times [29]. Repetition rates on the order of about 10 times per second are needed for ICF to be a practical fusion power source [29]. At this rate, an ICF power system theoretically would be able to produce a gigawatt of electrical energy [29].

The essential components of an ICF power plant are illustrated schematically in Fig. 3.7. They include (1) a factory to manufacture the cryogenic DT capsules, which are sent to the reactor and injected into the reaction

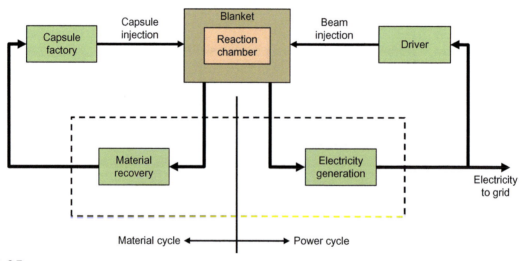

FIG. 3.7

Essential high-level components of an ICF power plant.

Adapted from R.A. Al-Ayat, E.I. Moses, R.A. Hansen, Inertial fusion technology, in: S.B. Krivit, J.H. Lehr, T.B. Kingery (eds.), Nuclear Energy Encyclopedia: Science, Technology, and Applications, John Wiley, Hoboken, NJ, 2011, p. 421.

chamber, (2) a reaction chamber that is equipped with instrumentation to precisely track the position, flight direction and velocity of the injected capsule, (3) a driver, which is a laser or particle accelerator, that creates short bursts of light or particles that are delivered to the capsule surface in the proper spatial and temporal arrangement to cause the fusion reaction, and (4) the first wall and blanket system, which extracts tritium from the blanket and the complex mix of reaction chamber exhaust gases, and that also transfers thermal energy deposited in the blanket to a useful electrical energy conversion system. The balance of plant completes two essential cycles in the ICF power reactor. In the material cycle, shown in Fig. 3.7, tritium bred in the blanket surrounding the reaction chamber is returned to the capsule factory to maintain the fuel supply, and in the power cycle a fraction of the electricity generated is recirculated to power the driver.

An ICF power plant must attain several high-level requirements for it to compete successfully with MCF. First, both the driver efficiency and fusion energy gain must be high. An economically attractive ICF power plant is obtained when the product of capsule energy gain and driver efficiency is about 10 or greater [29]. Since driver efficiencies are generally fairly low, ranging from 7% to 20%, this requires fusion energy output to be approximately 50–150 times greater than the driver energy input to the DT capsule [29]. Second, as noted above, fabrication, insertion, and implosion of the target must be repetitively accomplished at a rate of around 10 per second [29]. This implies that the disturbances created from the previous implosion event must sufficiently dissipate so as to not interfere with the next implosion event. Third, the fusion energy deposited in the blanket must efficiently produce electricity and breed new tritium from lithium compounds, and allow for the extraction of that tritium from the breeding medium to enable refueling of the reactor [29]. Finally, all of the components of an ICF power system must function with sufficient availability and reliability to make it economically competitive and environmentally attractive.

From a structural materials perspective the ICF blanket, heat transfer system, VV, neutron shielding, and electricity generating components share much in common with similar MCF components [30]. The major differences with MCF reside in the capsule factory, the laser or heavy-ion driver, and the reaction chamber [30]. The first wall of the reaction chamber faces the most severe materials challenges from exposure to the ICF thermomechanical and particle impingement environment. Critical issues that must be addressed in ICF power plant design include a rapidly evolving surface due to the very high transient temperatures attained from the absorption of x-rays and plasma debris, and bulk radiation damage from the high-energy neutrons. Particle impingement causes evaporation and recondensation of the surface and thermomechanical fatigue of the structural material due to the cyclic nature of the loading.

As summarized in Table 3.3, several reaction chamber concepts have been developed to address some of the materials challenges created by ICF conditions. A unique aspect of the ICF environment is the possibility that the plasma facing surface may be a liquid since, in contrast to MCF, high vacuum is not required [28]. From a materials perspective, perhaps the most challenging reaction chamber configuration is the dry wall design, with an innermost solid surface that must accommodate the energy released from the capsule explosion that is not transported by the 14.1 MeV neutrons. Some protection of the first-wall surface is provided by a low-pressure chamber inert gas that reduces the energy flux carried by impinging ions [28]. A possible alternative is to impose a "cusp" magnetic field on the chamber that guides the ions through poloidal holes toward an external collector [31]. The ions do not contact the chamber wall, but deposit their energy in external "dumps," which consist of tubes filled with Ga mist [31]. Perhaps the most materials friendly first-wall protection scheme is use of thin liquid layers that coat the solid surface, thus mitigating the large thermal transients associated with x-ray and ion absorption. However, evaporation and recondensation of the thin liquid layer and possible interference with restoration of the chamber environment between implosions are major technical challenges with this concept. Finally, lithium-bearing liquid walls, order 50 cm thickness, have been proposed [29]. A thick liquid flowing between the target capsule and the solid first wall not only provides protection from x-rays and ions, but would also mitigate neutron induced radiation damage. From a structural materials development standpoint, the thick liquid wall concept offers the greatest potential to ameliorate particle-induced damage, but shifts the technical challenges toward thermal hydraulics of liquid metal flows and restoration of the environment between implosion events.

Table 3.3 Summary of ICF reaction chamber concepts

Reaction chamber type	Key concepts	Critical issues	Conceptual design studies
Dry wall	Solid first-wall externally cooled by gas or liquid	Protection of surface from X-ray and ions by using low density gas in the chamber or a distributed magnetic on the surface	SOMBRERO HAPL LIFE KOKI
Wetted wall	Thin liquid layer that seeps from a porous wall of fabric to protect the surface of the first-wall	Evaporation and re-condensation of wetted surface Seeping out phenomenon to keep thin wetted layer	Osiris KOYO-F
Thick-liquid wall	Magnetically guided flow or oscillating thick flow designed to form a cavity	Cavity formation and stability of thick flow Restoration of the cavity for capsule and laser injection for next repetition	HYLIFE-II Z-IFE SENRI

R.A. Al-Ayat, E.I. Moses, R.A. Hansen, Inertial fusion technology, in: S.B. Krivit, J.H. Lehr, T.B. Kingery (eds.), Nuclear Energy Encyclopedia: Science, Technology, and Applications, John Wiley, Hoboken, NJ, 2011, p. 421.

3.5 FIRST-WALL/BLANKET STRUCTURAL MATERIAL OPTIONS

This section provides an overview of the principal low or reduced activation structural materials for fusion applications. Based on safety, waste disposal, and decay heat considerations candidate structural materials for first-wall and blanket components have chemical compositions based on the following elements: Fe, Cr, W, V, Ta, Ti, Si, and C. Given this set of elements, structural materials of interest include RAF/M steels, NFAs, vanadium alloys, fiber-reinforced silicon carbide composites, and tungsten alloys. While the technological maturity of these candidate materials varies considerably, they are all in very early stages of development as far as being sufficiently durable to provide a functional basis for fusion structures. Generally a more technologically mature material involves fewer feasibility issues, but might be less economically attractive due to possibly lower system performance. However, it is important to emphasize that upper use temperature based on a simple consideration like strength is not anywhere near an appropriate metric of realistic performance capabilities in a practical electricity producing system.

The intent here is to highlight the essential features of each of these structural material systems, but not delve into details, which are covered in other chapters. The emphasis is on the implication of the environment on the requirements for basic physical and mechanical properties of these materials and challenges they face specific to the fusion environment. The same approach is followed in Sections 3.6 and 3.7 dealing with divertors and other components.

3.5.1 RAF/M STEELS

RAF/M steels are the most technologically mature first-wall and breeding blanket structural material, and are the leading candidates for nearer term applications, such as test blanket modules for ITER and advanced plasma devices, such as the proposed Fusion Nuclear Science Facility and DEMO. These steels, more precisely described as normalized and tempered martensitic steels (TMS), evolved from a high-strength, creep resistant Grade T91 modified 9Cr-1Mo alloy developed in the 1970s for fossil-fired steam plants. Substituting W for Mo and Ta for Nb in

3.5 FIRST-WALL/BLANKET STRUCTURAL MATERIAL OPTIONS

Table 3.4 Chemical composition specifications for RAF/M steels under development for fusion in various countries

Element	China	Europe	India	Japan	Russia	USA
Name	CLAM	EUROFER	INRAFM	F82H	RUSFER	9Cr-2WVTa
Fe	Bal.	Bal.	Bal.	Bal.	Bal.	Bal.
Cr	8.8–9.2	8.5–9.5	8.9–9.1	7.5–8.5	12.0	8.5–9.0
W	1.3–1.7	1.0–1.2	0.90–1.10	1.8–2.2	1.3	2.0
V	0.15–0.25	0.15–0.25	0.20–0.24	0.15–0.25	0.40	0.25
Ta	0.10–0.20	0.10–0.14	0.06–0.08	0.01–0.06	0.15	0.07
Mn	0.35–0.55	0.20–0.60	0.40–0.60	0.05–0.20	0.60	0.45
C	0.080–0.12	0.09–0.12	0.10–0.12	0.08–0.12	0.16	0.1

Values are in weight percent.

T91 lowered the residual activity induced by fusion neutrons. Some of the advantages enjoyed by these steels over other low-activation alternatives, such as V alloys and SiC composites, include (1) lower cost, (2) the existence of more mature, or many might say the only basically feasible, structural design, and fabrication technologies, (3) compatibility with a wide range of coolants, and (4) resistance to radiation-induced swelling [32,33]. Table 3.4 lists RAF/M steel chemical composition specifications for six of the seven countries involved in the ITER project [34–36]. With the exception of Russia, all countries are developing a RAF/M steel with about 8%–9% Cr and around 0.10% C. The concentrations of the other alloying elements are similar, with somewhat larger variation in the W level.

The operating temperature window for RAF/M steels is estimated by some to be about 200°C wide [37]. The lower operating temperature limit is about 350°C and the upper operating temperature limit is approximately 550°C. The lower limit is governed by irradiation-induced hardening and embrittlement. We first discuss hardening and embrittlement for fission reactor displacement dominated irradiations that do not produce significant amounts of He. A separate section is dedicated to He effects. Figure 3.8 shows the temperature dependence of the tensile yield strength for several TMS for both unirradiated controls and fission reactor irradiated conditions over a range of dpa levels [38]. The data clearly show that TMS irradiation harden significantly below 350°C, with yield stress (σ_y) increases ($\Delta\sigma_y$) of up to 600 MPa. Experimental evidence indicates that hardening of RAF/M steels saturates by a dose of ~10 dpa for fission irradiations from 300°C to 335°C [39]. The irradiation effects database in this temperature regime extends to nearly 80 dpa. As the irradiation temperature increases to ~400°C, or more, hardening falls to low levels and softening may occur at even higher temperatures [40]. Large amounts of irradiation hardening are accompanied by severe loss of uniform strain ductility that decreases to <1% in tensile tests due to plastic instabilities. Plastic instabilities are caused by a combination of both higher true flow stresses and lower strain hardening rates [41].

Irradiation hardening is also directly associated with irradiation embrittlement manifested as a loss of fast fracture resistance, or degraded fracture toughness. A characteristic of bcc metals, even in the unirradiated condition, is a strong temperature dependence of both σ_y and fracture toughness. As the test temperature decreases bcc metals undergo a transition in fracture mode from high-energy driven ductile tearing to low-energy brittle cleavage fracture. This transition is typically indexed by the so-called DBTT. There are various measures and indices for the DBTT, which are test specific. For standard Charpy V-notch (CVN) impact tests, the DBTT = T_c, is often indexed at 41 J of absorbed energy. For true fracture toughness, characterizing the loading conditions leading to the extension of a sharp fatigue crack, the DBTT = T_o indexed at a median $K_{Jc}(T_o) = 100$ MPa m$^{1/2}$. The entire toughness-temperature curve, $K_{Jc}(T)$, often has a universal master curve shape that is placed on an absolute temperature scale at T_o. Typical T_c and T_o values for high-quality RAF/M steels are between −80 and −120°C [42].

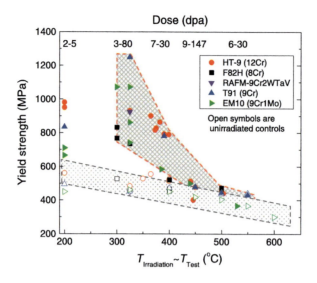

FIG. 3.8

Evolution of yield strength with irradiation temperature in F/M steels. Two RAF/M steels are included on the plot (F82H and 9Cr2WTaV). Note the irradiation temperature and tensile testing temperature were approximately the same [38].

Reproduced from O. Anderoglu, T.S. Byun, M. Toloczko, S.A. Maloy, Mechanical performance of ferritic martensitic steels for high dose applications in advanced nuclear reactors, Metall. Mater. Trans. A Phys. Metall. Mater. Sci. 44 (2013) S70, with permission from Springer.

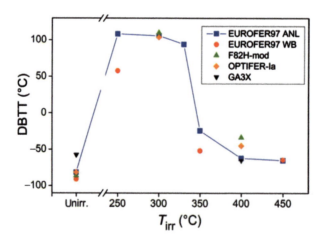

FIG. 3.9

Effect of irradiation temperature on the DBTT for several RAF/M steels. The average neutron dose was 16.3 dpa except for the data point at $T_{irr} = 330°C$ where the dose was 15 dpa [39].

Low-temperature neutron irradiation embrittlement results in increases in the DBTT, or transition temperature shifts (TTS). The TTS depend on irradiation temperature as illustrated in Fig. 3.9 for several RAF/M steels based on sub-sized CVN tests [39]. The magnitude of the TTS is largest up to ~330°C, decreasing sharply at higher temperatures. There is a general and well understood relationship between TTS and irradiation hardening, typically measured as increases in either $\Delta\sigma_y$ or the plastic flow ($\Delta\sigma_{fl}$) stress at finite plastic strains as: $TTS = C\Delta\sigma_y$.

3.5 FIRST-WALL/BLANKET STRUCTURAL MATERIAL OPTIONS

In the hardening dominated embrittlement regime, the value of C depends on both the test method and measure of irradiation hardening. In the case of the master curve toughness T_o, TTS $C_k \approx 0.7°C/MPa$ for the average $\Delta\sigma_{fl}$ between 0% and 10% strain; for sub-sized CVN tests, typical $C_c \approx 0.4°C/MPa$ for $\Delta\sigma_y$ [42]. The dose dependence of the TTS and yield or flow stress changes is also generally similar. These relations hold only for hardening dominated embrittlement.

However, at temperatures above 375–400°C, nonhardening embrittlement mechanisms may emerge. Indeed, as discussed below at temperatures >500°C purely thermal aging can lead to a combination of softening ($\Delta\sigma_y < 0$) and embrittlement (TTS > 0). The underlying mechanisms include precipitation and coarsening and gross microstructural instabilities of dislocation and lath substructures. However, there is considerable evidence that irradiation acts synergistically with thermal processes shifting regimes of nonhardening embrittlement to lower temperatures and shorter times [42]. Such effects remain largely unexplored. However, a far more serious source of nonhardening embrittlement in fusion service is the accumulation of He on grain boundaries. This leads to a very low-energy intergranular fracture path. As noted previously, high He also appears to lead to higher hardening that extends to higher temperatures. This is discussed in Section 3.5.5 on He effects.

It is important to emphasize that the DBTT is not a material property or a quantitative measure of the loading conditions leading to fracture in specimens, let alone in complex structures. Even fracture toughness, which in principle can be used to assess fracture stress and strain margins, depends on the absolute size and geometry of the cracked body, and a variety of factors that characterize the loading conditions, including the rate, multiaxiality, and structural compliance. The classical issue of how to properly apply test coupon derived data to the evaluation of complex structures represents a grand challenge in the case of fusion structures.

At intermediate operating temperatures, irradiation-induced creep and volumetric swelling control the dimensional stability of RAF/M steels. Irradiation creep is a constant volume deformation process in a material subjected to a stress below the static yield stress that is enabled by the displacement damage-induced supersaturation of vacancies and SIA. Void swelling is an isotropic volume increase of a material caused by nucleation and growth of internal voids, resulting from biased dislocation sinks for SIA, that leave an excess flux of vacancies to accumulate at neutral void sinks. Void swelling typically involves a prolonged incubation dpa prior to the formation of a significant population of voids. The incubation dpa is generally associated with pressurized bubbles stably growing by the addition of He up to a critical size where they convert to unstably growing voids. TMS are much more swelling resistant than fcc austenitic stainless steels. The higher resistance is believed to be largely due to a combination of lower He generation rates, higher self-diffusion coefficients, and a smaller dislocation bias for SIA in TMS. The intrinsically lower He generation rate in TMS is due to the absence of Ni.

Swelling occurs in the absence of stress; however, the swelling rate is enhanced by stress adding a component of positive strain in the loading direction. A general empirical equation relating the irradiation creep strain rate to the applied stress, σ, is

$$\dot{\varepsilon}_c = (B_o + D\dot{S})\sigma^n$$

where n is a stress exponent with a value of around unity, B_o is the creep compliance, $\dot{S}$ is the swelling rate as a function of dpa, and D is the creep-swelling coupling coefficient. Values of the creep compliance have been experimentally determined for a number of TMS and generally range from $B \approx 0.43$ to 1.7×10^{-6} MPa^{-1}dpa^{-1} with an average of 0.67×10^{-6} MPa^{-1}dpa^{-1} for alloys with 7.8%–12% Cr [34]. The creep compliance is not a strong function of material composition or microstructure. On the basis of the overall irradiation creep database for a number of materials, B_o is expected to have a weak to moderate dependence on temperature and damage rate. The corresponding creep-swelling coupling coefficients range from $D \approx 0.4$ to 1.1×10^{-2} MPa^{-1} [38]. However, in the absence of high He, swelling is not expected to be a major issue in RAF/M steels.

The upper operating temperature limit of about 550°C for RAF/M steels is governed by the high-temperature stability of the tempered martensitic substructure, corrosion, and thermal creep strength [40,43], which is well above the displacement damage regime. Thermal annealing of unirradiated F82H for up to 10^5 h shows that the

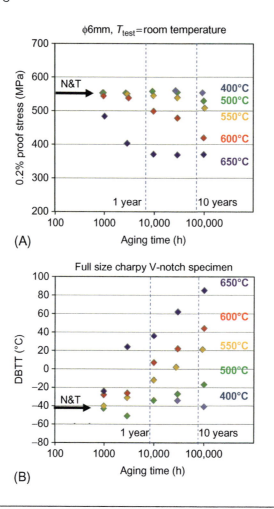

FIG. 3.10

Effect of long-term aging of F82H on (A) 0.2% yield stress and (B) DBTT at temperatures from 400°C to 650°C [40].

yield strength decreases only slightly at 550°C, but softening is much more significant at 600°C, see Fig. 3.10A [40]. However, even the slight softening observed at 550°C is accompanied by thermal embrittlement as manifested by a ~60°C shift in DBTT to around room temperature, see Fig. 3.10B [40]. Both softening and embrittlement increase with aging temperature. The thermal aging resistance of Eurofer97 is reported be superior to F82H [40], probably due to the lower tungsten concentration of the former, which gives that steel better resistance to formation of embrittling Laves phases.

A significant database has been developed on creep rates and creep rupture times of unirradiated RAF/M steels Eurofer97 and F82H [43,44]. The ~90 MPa at 600°C creep rupture is accompanied by very large total elongation strains, so ductility is not an issue. A more practical measure of creep strength relates to dimensional stability, which is particularly important for fusion structures. The primary plus secondary creep strain limit is typically taken as 1% although this may be higher or lower for a particular component. The creep strength for 1% strain in Eurofer97 and F82H at 20,000 h is ~120 and ~80 MPa at 550°C and 600°C, respectively. However, the corresponding stress-temperature combinations for longer times of, order 100,000 h, cannot be determined fully reliably by extrapolation.

Characterization of the tertiary creep behavior of RAF/M steels has been performed along with related Grade 91 ferritic-martensitic steel commonly used in the nonnuclear power generation industry. Interestingly, the creep deformation behavior of RAF/M steels displays a very long tertiary stage compared to time spent in the primary and secondary stages of creep [45]. These studies show that the primary thermal creep damage mechanism for RAF/M steels is microstructural degradation, which is characterized by gradual coarsening of the subgrain size and strengthening particles, such as MX and $M_{23}C_6$ precipitates [45]. Ashby and Dyson developed a simple parameter to characterize the predominant failure mechanism for the tertiary stage of creep that is the ratio of the failure strain divided by the Monkman-Grant constant [46]. Ashby and Dyson refer to this parameter as the creep damage tolerance factor, λ, and for engineering alloys it has values between 1 and 20 [47]. Creep damage due to void growth exhibits λ values $<\sim 2.5$, whereas microstructural degradation gives $\lambda > 5$ [47]. Typical λ values for RAF/M and Grade 91 steels are around 4–5 [45,48,49]. In contrast, the limited data available for NFAs indicate much lower values for the creep damage tolerance parameter. Wilshire and Burt report λ values for MA957 tested between 600°C and 700°C between 1 and 2, indicating a predominantly void growth mechanism contributes to tertiary creep damage in this material [50].

The synergistic effects of high temperatures and irradiation may exacerbate purely thermal effects, especially those caused by diffusion driven instabilities. The potential for nonhardening embrittlement caused by precipitation and coarsening of brittle phases, segregation of trace elements at grain boundaries such as phosphorous or depletion of beneficial elements such as carbon, and instabilities in the dislocation and lath-packet substructures leading to larger effective subgrain sizes may be significant at irradiation temperatures where hardening is not observed (e.g., above $\sim 425°C$). A need exists for high-dose irradiations at temperatures beyond the hardening regime [51]. This is particularly true when the potential for He-induced grain boundary decohesion is considered (see discussion below).

First-wall structures and components will also be exposed to complex, time varying, thermomechanical loadings at high operating temperatures that will limit their lifetimes, even in the absence of radiation damage. In this environment, severe mechanical damage can occur by the interaction of creep deformation with cyclic loading, known as fatigue. For example, high-temperature strain controlled fatigue typically causes significant strength decreases in 9Cr TMS and can lead to macroscopic component-scale dimensional instabilities, such as ratcheting. The effects of cyclic plastic loading are generally far more severe than under pure monotonic creep, especially in regimes with very damaging long tensile hold times and out of phase thermal and mechanical fatigue. The development of creep cracks, especially on grain boundaries, could lead to very short component lifetimes. Creep-fatigue interaction is a complex process that depends on a large number of mechanical and metallurgical variables, such as test temperature, strain-rate, hold-time, environment, and the microstructure of the material [52]. The micromechanical mechanisms that control creep-fatigue are poorly understood [53]. Current high-temperature design and regulatory codes are very approximate at best, and typically based more on experience than mechanistic insight. For many applications, these codes are adequate because they are very conservative. However, it is not at all clear that such conservatism extends to very complex and intricate fusion structures that experience time-dependent dimensional changes, gradients, plastic strains, and short and longtime stress redistributions. Thus, it is imperative to replace simple structural design rules with sophisticated multi-physics computational design tools if fusion is to become a viable power production technology.

These considerations are further amplified by the potential for synergistic interactions between the effects of irradiation, thermal, chemical, and mechanical environments. As noted above, long-term aging at elevated temperatures causes softening and embrittlement of RAF/M steels, but the creep life is also significantly reduced due to static recovery of the tempered martensitic lath structures and heterogeneous subgrain growth [54]. Recent research has also shown that prior low-cycle fatigue damage dramatically reduces creep rupture life, decreases tertiary creep ductility, and increases the minimum creep rate [55]. Very little low-cycle fatigue exposure ($\sim 1\%$ of fatigue life at 550°C) was needed to promote subgrain coarsening, which resulted in a 60% reduction in creep-rupture life [55]. Finally, the degrading effects of high concentrations of He coincident with the other environmental challenges may represent the greatest grand challenge to fusion structures as described in more detail in Section 3.5.5.

Of course corrosion and compatibility are also key issues. But since they are highly system specific it is difficult to deal with them in any simple and generic fashion. Hence, here we focus on only one of the leading high-performance blanket designs under consideration for advanced plasma devices that uses liquid PbLi eutectic for breeding tritium and partially cooling the RAF/M steel structure. Since liquid metal is in direct contact with RAF/M steel components, it is essential to know the corrosion characteristics of those components in PbLi because deposition of activated corrosion products could have serious implications for plant operations and safety. The corrosion performance of RAF/M steel in slowing flowing lithium-lead at $<\sim480°C$ is reasonably well understood, but the effects of magnetic fields and neutron irradiation on corrosion are not well characterized. In addition, corrosion of other important product forms, such as weld metal and heat-affected-zone microstructures, have not been determined. A typical corrosion study generally involves isothermal exposure of a structural material in static coolant, or in a flowing coolant in a temperature gradient, which is a more realistic experiment. Corrosion rates are often empirically correlated to coolant temperature and flow velocity. These types of correlations do not fully capture the fundamental physical mechanisms involved in the corrosion process and, therefore, are not helpful for predicting material performance outside the range of experimental conditions.

Development of quantitative corrosion models, however, is challenging because many interacting variables affect the corrosion process, including temperature, chemical composition of the structural material and the coolant, coolant flow velocity and velocity profile, coolant impurity concentration, and radiation. Significant uncertainties in basic quantities, such as the saturation concentration of Fe in PbLi, increase the complexity of the problem. Various correlations for the temperature dependence of the Fe saturation concentration in PbLi give values ranging over more than three orders of magnitude at 500°C [56]. The experimental data on mass loss for RAF/M steels in flowing PbLi at temperatures above 450°C exhibit a similar wide range of values from 20 up to 900 μm/year, with the higher values corresponding to studies conducted in a magnetic field [57]. Recent modeling studies suggest that magnetic field effects on RAF/M steel corrosion are related to MHD effects on the PbLi flow velocity. Flow pathways oriented parallel with the magnetic field can develop high-velocity jets in the flow, which substantially enhances corrosion [56]. While quantitative agreement with experiment remains elusive, models provide a physical description of blanket conditions that qualitatively rationalizes experimental results.

3.5.2 NANOSTRUCTURED FERRITIC ALLOYS

While RAF/M steels possess many positive attributes for fusion first-wall/blanket structural applications, their limited operating temperature window and lifetime neutron dose restrictions may preclude their use in high-performance fusion power reactors. Radiation tolerant microstructures must have a high density of stable sinks to trap transmutation helium and provide sites for point defect recombination [58]. Experimental observations and associated modeling suggest that structural material microstructures may need sink strengths of $\sim 10^{16}/m^2$ in order to effectively survive long-term exposure in the fusion nuclear environment [59]. One strategy to enhance radiation resistance is to introduce a uniform dispersion of stable nanometer-scale precipitates, or nanofeatures (NFs). The sink strength of a NF distribution is proportional to the product of the average precipitate diameter and number density [60]. In the case of nanometer-scale precipitates, this implies particle number densities of order $5 \times 10^{23}/m^3$ are needed.

Thus a new class of oxide dispersion strengthened Fe-based materials is being developed for fusion applications called NFAs. Typical NFA compositions, in weight percent, are iron plus 12%–20% Cr, 1%–3% W, 0.1%–1% Ti, and 0.15%–0.5% Y_2O_3 [61]. These materials differ from transformable oxide dispersion strengthened steels, which are alloyed with C and generally contain about 9% Cr [62]. Adding Cr to Fe increases the high-temperature oxidation and corrosion resistance, while W provides solid solution strengthening. When Y_2O_3, Ti, and O combine they form a NF dispersion that significantly increases creep strength by impeding dislocation motion. The NFs also improve radiation resistance by increasing the sink strength into the desired $10^{16}/m^2$ range, while trapping He in fine-scale bubbles [59].

3.5 FIRST-WALL/BLANKET STRUCTURAL MATERIAL OPTIONS

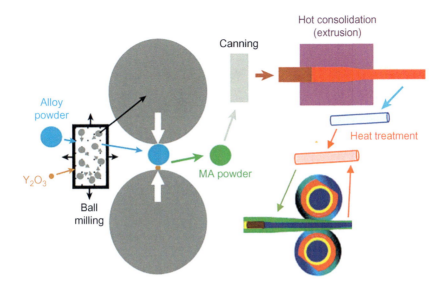

FIG. 3.11

Schematic representation of the sequence of processing steps needed to produce NFAs from metallic and yttria powders.

Mechanical alloying is typically used to process NFAs as illustrated schematically in Fig. 3.11. Powdered Y_2O_3 is mixed with metallic alloy powders and ball milled under an inert atmosphere, generally Ar. The mechanically alloyed powder is then canned and vacuum degassed before consolidation by hot isostatic pressing or hot extrusion at temperatures ranging from ~800°C to 1200°C [61] to $\geq 98\%$ of the theoretical density. NF diameters range from 1 to 4 nm at number densities on the order of 10^{23}–$10^{24}/m^3$. In contrast to HIPing, extrusion typically produces elongated and textured grains [61]. NFA generally require further thermal-mechanical processing to achieve more isotropic properties and to fabricate product forms.

Alloying with titanium suppresses coarser scale Y_2O_3, resulting in the formation of much smaller Y-Ti-O precipitates with 2–3 nm average diameters. The precipitates are mostly of $Y_2Ti_2O_7$ complex oxides, possibly with a core-shell structure and nonstoichiometric compositions with excess Ti. Other coarser scale precipitate features in NFAs include larger Ti-oxides and carbonitride phases. Recent studies indicate the NFs and NFA microstructures in MA957 are essentially completely thermally stable at anticipated service temperatures and times below about 950°C [63–65]. However, the thermal stability of NFs and NFA processed at lower temperatures remains to be explored.

The mechanical properties of unirradiated NFAs are far superior to RAF/M steels. Typical NFA yield and ultimate tensile stresses are 1 GPa, or more, and are accompanied by significant ductility [61]. The typical creep strength of NFAs is also much higher than RAF/M steels as shown in Fig. 3.12 [66–71]. Here creep-rupture time Larson-Miller data for typical NFAs, such as MA957, are compared to corresponding data for RAF/M steels Eurofer97 and F82H. However, the creep strength of as-extruded NFAs is highly anisotropic and is often much weaker in the transverse direction. Hence, deformation processing to produce more equiaxed microstructures is critical. Notably, NFA have much better low cycle and creep-fatigue properties than RAF/M steels, and they rarely cyclically soften [56]. On the other hand, the fracture behavior of NFAs may be inferior to that in RAF/M steels and is also often strongly anisotropic, due to the elongated and textured grain structures in extruded product forms [72]. However, fine-grained thermal-mechanically treated NFA have much improved and more isotropic toughness, especially in terms of low DBTTs. The NFA upper-shelf tearing toughness is generally lower than in RAF/M. Nevertheless, properly processed NFA manifest extensive stable crack growth (are highly ductile) at loads controlled uncracked net section tensile stresses.

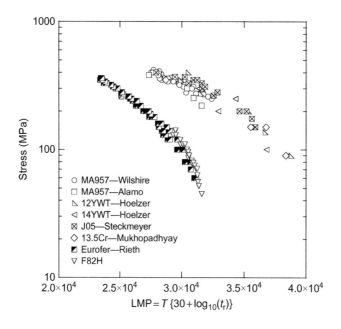

FIG. 3.12

Larson-Miller Parameter plot (C=30) comparing the creep-rupture strength of typical NFAs (MA957, J05, and 13.5Cr) with conventional RAF/M steels Eurofer and F82H [66–71].

There are many unmet challenges to developing NFA, including joining, fabricating complex and defect-free product forms, high cost, limited data on key properties, achieving uniform and reproducible microstructures and properties, and so on. All of these issues are being addressed at least in limited ways. For example, it has been shown that, in principle, NFA can be joined by friction stir welding without serious degradation of the NFs [73]. Engaging these difficult challenges is motivated by the remarkable irradiation tolerance of NFA.

Issues of irradiation effects on NFA at fusion relevant conditions with high He are discussed in Section 3.5.5. NF and NFA stability under irradiation with low He fission reactor conditions are extremely promising, although this depends on a number of variables, such as temperature, dpa and dpa rate, feature size, structure and composition, as well as the type of irradiation [74,75]. However, most studies suggest that fine NFs are unaffected, or only moderately affected, by irradiation up to order 120 dpa, or more in charged particle irradiations. This observation applies to practical conditions of temperature and dpa rate that allow NF self-repair at rates that exceed the corresponding rates of dissolution and damage [74–77]. The irradiation stability of NF can be understood in terms of the enormous thermodynamic driving forces for precipitation; thus reattachment of solutes dissolved by short-range ballistic ejection and deposited near a NF has a high probability. NF amorphization under irradiation might also be expected, but again this has not been observed for the small $Y_2Ti_2O_7$ precipitates.

Due to the high sink density, displacement damage in NFA is expected to accumulate at a lower rate at low to intermediate temperatures compared to RAF/M steels. This may be a partial explanation of the smaller irradiation-induced yield stress increases observed in NFA compared to RAF/M alloys. However this difference is also partly due to pre and post irradiation dispersed barrier strengthening superposition effects. Further precipitation of Cr-rich α' phases extends the temperature range of irradiation hardening compared to RAF/M steels. Notably, hardening in NFA is not accompanied by large reductions in uniform tensile and total strain ductility. Irradiation embrittlement effects, primarily manifested as DBTT shifts, are not fully characterized; however, they are expected to generally follow similar hardening trends. NFA are highly swelling resistant even in the presence of high helium. A primary irradiation damage concern is irradiation creep that occurs in NFA at rates comparable to those in a wide range of other steels.

3.5.3 VANADIUM ALLOYS

Vanadium (V) alloys have been considered as the structural material for an advanced blanket concept that utilizes liquid Li as the coolant and tritium breeding material. A self-cooled vanadium/lithium blanket has the potential for outstanding thermal performance owing to the favorable physical properties of V alloys and heat transfer capabilities of liquid Li. In addition, a neutronically attractive vanadium/lithium blanket will be able to breed tritium without the need for neutron multiplying elements such beryllium or lead. Compared to Fe-base alloys and SiC composites, V-base alloys have the respective advantages not being ferromagnetic, in MCF applications, while possessing significant ductility and fracture toughness. Vanadium alloys are also appealing because of their low induced activation along with reasonably good high-temperature strength, and ability to tolerate high heat flux conditions [78–80].

Most of the worldwide research effort on vanadium alloys has focused on developing an alloy containing 4%–5% Cr and 4%–5% Ti with the balance V as the reference composition [78–80]. Chromium is added to provide elevated temperature solid solution strengthening, and Ti to improve the ductility and workability by removal of interstitial impurities from the matrix by formation of Ti-rich precipitates [78–80]. A significant technical challenge for vanadium alloy development for the demanding fusion environment is their strong affinity for gaseous impurities, such as C, O, N, and H. Vanadium is highly reactive with these elements, and when in sufficiently high concentrations in the matrix, they can strengthen the material considerably, while drastically reducing ductility and elevating the DBTT [78–80]. The use of Li as a breeder/coolant is necessary because it has stronger affinity for O than V, thereby reducing the pick-up and corresponding embrittlement associated with this element. On the other hand, V has a greater affinity for C and N than Li, so transfer of these impurities from the coolant to V is a concern [79].

More generally, it is well known that the mechanical properties of bcc metals and alloys are strongly affected by the presence of interstitial species, such as C, O, and N [81]. Vanadium alloys exhibit engineering stress-strain curves with the usual features of a yield drop, Lüders strains, as well as static and, especially, dynamic strain aging (DSA), depending on the test temperature and strain rate. The tensile curves for V-4Cr-4Ti plotted in Fig. 3.13 show that DSA occurs over a fairly broad temperature range from ~400°C to 700°C at a strain rate of 10^{-3}/s [81]. DSA occurs under these test conditions because the interstitials are mobile and continuously form and reform atmospheres, sequentially pinning dislocations. At low temperatures, interstitial mobility is too low for sufficient solute concentrations to develop in the vicinity of moving dislocations; and at higher temperatures a lower concentration of more mobile interstitial solute atmospheres exert a low drag force [81].

The effects of interactions between V alloys and Li have received considerable attention, particularly with respect to long-term thermal creep and corrosion [68]. Studies of thermal creep in both vacuum and liquid lithium [67,68] indicate that the high-temperature strength of V-4Cr-4Ti may not be adequate. This has motivated the search for improved V-alloy strength through thermomechanical processing and introduction of ultrafine particles of Y_2O_3 or YN [82,83].

Figure 3.14 shows the effect of irradiation temperature and neutron dpa dose on the yield strength of V-4Cr-4Ti. Vanadium alloys harden significantly at irradiation temperatures below ~400°C. Hardening increases with dose, saturating at several dpa. Above 400°C, hardening decreases and is negligible at ~600°C, at least at lower dose, but persists at higher dpa. Hardening below ~400°C is associated with the formation of point defect clusters, interstitial cluster complexes, enhanced by interstitials released by the dissolution of coarser scale Ti-C-O precipitates that had formed during heat treatments [84]. These observations have motivated exploring the effects of various alloy additions and heat treatments [79,80,84] to more effectively control the matrix interstitial content.

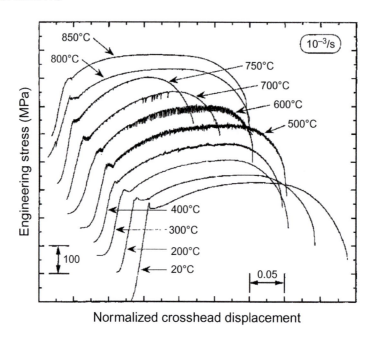

FIG. 3.13

Tensile curves for annealed V-4Cr-4Ti at a strain rate of 10^{-3}/s showing the temperature regime for DSA. The curves are curves offset on stress and strain axes for clarity [81].

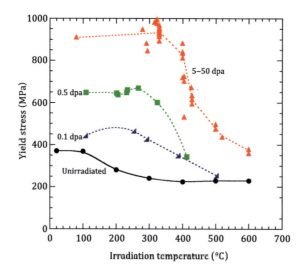

FIG. 3.14

Effect of irradiation temperature and dose on yield strength of V-4Cr-4Ti.

Adapted from S.J. Zinkle, J.T. Busby, Structural materials for fission and fusion energy, Mater. Today 12 (2009) 12.

More recent vanadium alloy research has included a focus on development of fabrication technologies with emphasis on joining by gas tungsten arc and laser welding methods. Making welds with adequate strength, ductility, and fracture toughness has been challenging, but successful joints have been prepared using high-purity filler metal in combination with a high-purity cover gas to minimize pick-up of interstitial impurities [79,80]. The unirradiated mechanical properties of such joints are similar to base metal properties. Some differences in low-temperature irradiation hardening behavior has been found, which could be reduced by postweld heat treatment (PWHT), but more irradiation affects data are needed [80].

Another extremely critical practical issue facing use of V alloy-Li coolant systems is the need for an insulator coating to electrically isolate the V structure from the flowing Li. Electrical isolation is required in an MCF system to mitigate the MHD pressure drop that results when a conductive liquid metal, such as Li, is flowing in a strong magnetic field, and is in direct contact with a conductive structure. The insulating layer must have a very low defect (crack and flaked) fraction ($<10^{-6}$) in order to be effective [85]. The number of ceramic materials that closely match the thermal expansion coefficient of V, and that are stable in liquid Li for long-times at high-temperature, while acting as good insulators are limited [85]. Due to the low defect tolerance, research has focused on self-healing ceramic layers that form in situ, and a two-layer approach that uses a thin protective outer layer of pure V to protect the insulating layer from the liquid Li. Experiments performed in convective flow loop experiments show that Er_2O_3 and Y_2O_3 are potential coating candidates, but much more work remains to be done to resolve the critical coating issue [85].

In summary, recent review papers [78–80] have discussed the maturity level of vanadium alloys for fusion applications. Several technical issues were identified in these broad surveys that need to be addressed for vanadium alloys to become a compelling choice for first-wall/blanket structural materials. The crucial issues include (1) the need for more data on thermal and irradiation creep, (2) a better understanding of the effects of irradiation on fracture properties, (3) much more data on the effects of helium on microstructure and property evolution at all temperatures, particularly in concert with neutron irradiation, (4) the composition of the reference alloy and thermomechanical heat treatments to improve strength and to turn interstitial impurities, such as carbon, oxygen, and nitrogen into an asset rather than a detriment, (5) an improved understanding of the effects of impurity transfer between the lithium breeder/coolant and vanadium structural components, and (6) the development of an effective MHD insulator coating. The last critical issue is of particular importance because without an effective MHD coating the vanadium/lithium blanket concept is not feasible for MCF systems.

3.5.4 SiC$_F$/SiC COMPOSITES

Silicon carbide composites are, in principle, very attractive materials for fusion power system structural and insulating applications because of desirable attributes, such as (1) the ability to operate at higher temperatures than metals, which offers high plant efficiency, (2) stability under neutron irradiation and exposure to some aggressive chemical environments, (3) a very low-level of long-lived radioisotopes that enhances environmental attractiveness, and (4) the ability to tailor properties through design of composite architectures for specific applications [16,86]. Silicon carbide composites are fabricated from largely brittle constituents, most often including SiC fibers. These composites achieve useful strength and fracture resistance through incorporation of low shear strength or debonding coatings between the fibers and the matrix that are around 100 nm thick. The brittle fibers fracture at various distances below the crack surfaces due to the build-up of sliding stresses. Intact segments of the fibers bridge the crack with closing tractions, shielding the remote loading stress intensity factor at the tip. Pyrolytic carbon and boron nitride are the most commonly used fiber coatings [87]. Multilayer interfacial coatings have also been used consisting of alternating layers of C and SiC where each layer is between 25 and 50 nm thick.

Significant progress has been made toward understanding the response of SiC to neutron irradiation, and the development of composites resistant to that environment [16,86–89]. A recent evaluation concluded that high-temperature swelling and irradiation creep would likely control the upper operating temperature limit of SiC/SiC composites for fusion applications [16]. The lower operating temperature limit is dictated by irradiation-induced

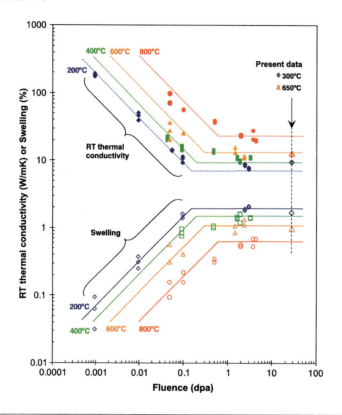

FIG. 3.15

Temperature and fluence dependence of swelling and thermal conductivity of CVD SiC/SiC over the temperature range 200°C to 800°C [88].

degradation of thermal conductivity, which must remain sufficient to accommodate very high heat fluxes [16]. Figure 3.15 shows the significant reduction of thermal conductivity caused by neutron irradiation in the temperature range from 200°C to 800°C. Note, these effects appear to saturate at doses of ∼0.5 dpa. Further, thermal conductivity degradation is most pronounced at low irradiation temperatures [88]. The data in Fig. 3.15 also show the connection between thermal conductivity degradation and point defect-induced swelling [88]. The defect clusters that accumulate in the lower temperature regime are responsible for both degradation effects [88]. Note, irradiations below about ∼150°C amorphize SiC, altering all its properties [16]. In contrast, significant void swelling occurs at high irradiation temperatures, that when coupled with irradiation creep, restricts the use of SiC at temperatures <∼1200°C [16].

However, even absent the effect of irradiation, it is problematic that large complex fusion structures can be designed and operated that are almost entirely composed of SiC/SiC composites. While the composites are stronger and tougher than their constituents, SiC/SiC composites are still inherently brittle. Indeed, the highest toughness of these composites is similar to the lower-shelf toughness of RAF/M steels and NFA. Further, composite toughness is provided by damage in the form of cracks that may in themselves compromise the functionality of the structures, like hermeticity.

Perhaps the most appropriate role for SiC/SiC is in components that bear minimum loads and have special functional requirements. For example, SiC is the material of choice for advanced blanket designs where it electrically and thermally isolates the flow of hot liquid PbLi breeder/coolant from the adjacent structural steel. In this

3.5 FIRST-WALL/BLANKET STRUCTURAL MATERIAL OPTIONS

application, the thermal and/or electrical insulating properties determine the maximum acceptable upper use temperature, while the lower operating temperature limit is established by swelling-induced secondary stresses [16]. The electrical conductivity of monolithic SiC depends on the impurity concentration since it is a wide-band gap semiconductor [16]. The pyrolytic carbon fiber coating and its response to irradiation largely determine the electrical conductivity of irradiated SiC/SiC composites [16,87]. Under neutron irradiation the electrical conductivity degrades markedly. Recent research shows that acceptably low values of electrical conductivity can be achieved for insulating applications by careful control of chemical purity, composite architecture, and irradiation conditions [16].

Radiation damage can also affect the mechanical properties of SiC and SiC/SiC composites. The most advanced generation of composites exhibit no significant strength degradation up to several tens of dpa, and they are capable of operating at temperatures up to 1000°C, or more, as illustrated in Fig. 3.16. More recent development efforts have focused on improving the engineering properties of composites. A focus of recent research is exploration of advanced processing technologies, development of an industrial basis for composite production, and extensive characterization of properties over a wide range of conditions needed to develop an engineering database to support design activities.

While SiC/SiC composites are very attractive materials for fusion power systems, a fairly large number of technical challenges must be resolved before fusion applications are feasible. Recent reviews have highlighted the crucial issues that must be addressed to make these materials realistic candidates for fusion structural applications [16,86–90]. Data and models to predict time-dependent deformation processes, such as slow crack growth have been developed, but need to be extended to account for loading conditions when the composite fibers are not aligned with the loading axis. Another critical issue related to the development of improved time-dependent deformation models includes determination of the composite strength limit. It has been suggested that the strength limit needs to be correlated with failure behavior, including construction of a strength anisotropy map and lifetime

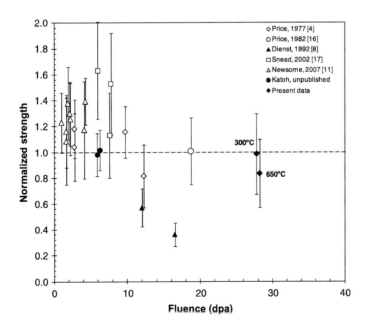

FIG. 3.16

Ratio of irradiated strength to unirradiated strength for CVD SiC versus neutron fluence. Irradiation temperatures ranged from 300°C to 1300°C. Error bars signify one standard deviation [88].

evaluations for fatigue and creep loading conditions [89]. Structural applications will require effective joining technologies that can tolerate severe heat loads [87]. A variety of joining technologies are being investigated, but there is only limited information about joint response to neutron irradiation. Hermeticity is also an important issue for first-wall/blanket applications requiring containment of high-pressure coolants. Consequently, development of sealing layers resistant to the effects of irradiation and cyclic thermal and mechanical loads is needed. Other crucial areas need thorough investigation include (1) assessment of the effects of transmutation products on all relevant properties, (2) definition of irradiation-induced creep to design relevant neutron fluence, and (3) establishment of structural design criteria for materials with limited ductility and fracture toughness. The potential effects transmutation products are particularly significant. Evaluation of various blanket concepts shows that He production near the first wall will vary from 30 to 170 appm/dpa, depending on design details, which are many times greater than for metallic materials. In addition, solid transmutations may also significantly affect composite properties. In the fusion neutron spectrum burnout of SiC occurs nonstoichiometrically, and burn-in of substantial concentrations of impurities, such as aluminum, magnesium, lithium, beryllium, and phosphorus occurs. Estimates for an MCF reactor at a neutron fluence of 100 dpa show that about ~3900 appm of solid impurities will be generated with Mg comprising about 60% of this total [87]. The consequences of introducing such large levels of impurities on the properties of SiC composites are unknown.

3.5.5 HELIUM EFFECTS IN FUSION STRUCTURAL MATERIALS

As stated in Section 3.3.1 the high flux of 14 MeV neutrons in fusion devices produces substantially more He and H than in similar irradiations performed in fission reactors. A greater proportion of high-energy neutrons than a fission neutron energy spectrum gives rise to a higher gas generation rate in the fusion environment. The neutron capture cross-sections for He and H generation by (n, α) and (n, p) reactions are substantial at neutron energies that are above those where the fission neutron spectra fall to low values [91]. A summary of the He and H generation rates for typical plasma facing/first-wall/blanket structural and functional materials was given in Table 3.1. Because He can profoundly affect material properties this section is devoted to highlighting these effects.

Helium is essentially insoluble in solids and this is ultimately is the reason for its detrimental effects on properties. Low solubility is equivalent to a very strong tendency for He to form clusters or bubbles in regions of high excess volume. Accumulation of enough He at grain boundaries is known to compromise the integrity of fusion materials. Helium can significantly degrade the tensile, creep, and fatigue properties at elevated temperatures. Helium bubbles at grain boundaries may lead to crack initiation and premature failure under stress. The extent of degradation depends on temperature, He concentration, He production rate, stress level, and composition and microstructure of the material. At low temperatures He may also influence irradiation hardening and cause fatigue life reductions by interfering with dislocation movement. At intermediate temperatures, He can shorten the incubation time for onset of swelling through stabilization of small vacancy clusters, which can transition from small bubbles into unstably growing voids.

Because He can substantially affect properties, it is essential to carefully quantify its effects in order to develop a safe and reliable fusion power system. Currently, it is difficult to properly explore the effects of He under prototypic conditions due to a lack of appropriate neutron irradiation sources. To design, develop, and validate He resistant microstructures, appropriate experimental methods are needed to introduce He into a material at fusion relevant He-to-dpa ratios. A novel technique [92] has been developed to inject He into a substrate while under neutron irradiation. A nickel bearing coating on a substrate causes production of energetic He atoms due to a two-step thermal neutron reaction sequence under mixed spectrum neutron irradiation conditions, and the He is injected into the substrate. With this approach it is possible to explore the effects of He on microstructural development by implanting He at almost any desired He-to-dpa ratio to a uniform depth of a few microns. It is not possible to obtain bulk mechanical property information by this technique, so it does not alleviate the need for a fusion-relevant neutron source. Dual ion beam irradiation experiments have also been utilized to simultaneously create displacement damage and implant He [93]. Such irradiations cannot exactly duplicate the damage caused by

3.5 FIRST-WALL/BLANKET STRUCTURAL MATERIAL OPTIONS

neutrons because of higher dpa rates ($\geq 10^3$–10^5) and spatial nonuniformity of dpa and He, but these experiments are relatively easy to perform and they do not activate the specimen, which greatly simplifies postirradiation examination. Ion irradiation studies can provide valuable insights into the interaction of He with important microstructural features.

Under irradiation conditions where He production and displacement damage occur concurrently the nucleation and growth of He bubbles depends on diffusion and clustering of He atoms, vacancies, and SIAs. At low irradiation temperatures, $\leq 0.3 T_M$, where the migration of vacancies is sluggish relative to He atoms and SIAs, the He bubble density will be high and bubble diameters will be observed on the nanometer scale. Theoretical estimates of bubble densities range from 10^{25} m^{-3} at $0.2 T_M$ down to 10^{21} m^{-3} at $0.5 T_M$ [94]. He bubbles constitute an additional barrier to the motion of dislocations, which increases the tensile yield and flow stresses, and reduces the ductility (to some extent) and fracture toughness is severely degraded. In addition, sufficient accumulation of He at grain boundaries enhances low-energy intergranular fracture. Recently, numerous molecular dynamics (MD) simulations of dislocation interactions with He clusters and small bubbles in pure Fe have been performed [46,95–99]. The simulation results for equilibrium He bubbles ranging from ~0.4 to 4 nm diameter show that the obstacle strengthening parameter, α, is between 0.05 and 0.45 for the smallest to largest bubbles investigated as shown in Fig. 3.17. The strengthening parameter is defined as

$$\alpha = \tau_c (L-d)/\mu b$$

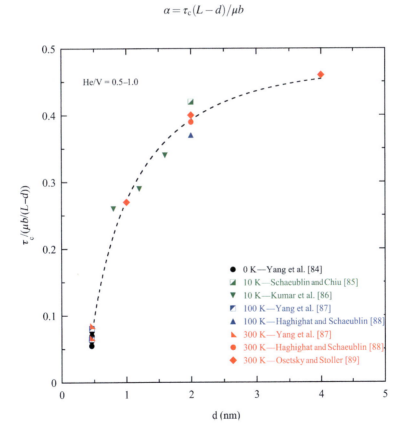

FIG. 3.17

Dependence of obstacle strengthening parameter, α, on bubble diameter and temperature for He clusters/bubbles with He-to-vacancy ratios between 0.5 and 1.0.

where τ_c is the critical resolved shear stress from the MD simulation for He bubbles with a diameter d that are spaced L apart, μ is the shear modulus, and b is the Burgers vector. The results plotted in Fig. 3.17 are for He bubbles with He-to-vacancy ratios between 0.5 and 1.0. Recent computer calculations of the conditions for mechanical equilibrium between small He bubbles and the Fe matrix show that the He-to-vacancy ratio for equilibrium bubbles falls within the range 0.3–1.0 [100].

An estimate of the strengthening contributed at ~300°C by a typical population of He bubbles about ~1.5 nm diameter and number density ~10^{23} m^{-3} can be obtained from the dispersed barrier hardening model, which is

$$\Delta\sigma = M\alpha\mu b\sqrt{Nd}$$

where M is the Taylor factor with a value of 3.06 and N is the bubble number density. From Fig. 3.17 the value of α for a 1.5 nm bubble is about 0.35. Using $\mu = 80$ GPa and $b = 0.248$ nm for Fe gives a $\Delta\sigma$ estimate for bubbles of about 260 MPa. This can be compared to the amount of hardening caused by neutron displacement damage alone for irradiated RAF/M steels with $\Delta\sigma$ ~450 MPa after a dose of only ~10 dpa [42]. This estimate is consistent with data obtained from spallation neutron/proton irradiations that involve very high He at dpa levels greater than typical saturation values, which show much larger $\Delta\sigma$ values as well as associated shifts in the DBTT [101].

Very little is known about He-induced hardening and embrittlement in other low-activation fusion materials. The effects of He on the tensile properties of several V alloys have been studied from 400°C to 900°C [78]. Figure 3.18 presents the temperature dependence of the normalized tensile ductility for He concentrations ranging from 14 to 480 appm for several V alloys. Here normalized ductility is the ratio of total elongation for a He-implanted specimen over the total elongation for a He-free specimen tested at the same temperature.

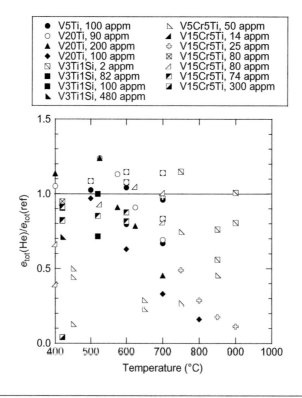

FIG. 3.18

Temperature dependence of the normalized total elongation for several V alloys preimplanted with He [78].

Normalized ductility near unity indicates no effect of He. It is apparent that He degrades the tensile ductility at all concentrations with the most pronounced effects observed at temperatures below about 450°C and above 650°C.

Helium generation in SiC will be more than an order of magnitude greater than that generated in metallic alloys (see Table 3.1). The implications of such high quantities of He on the properties of SiC and SiC composites are largely unknown. While SiC composites are stronger and tougher than their constituents, they are still inherently brittle and introduction of large levels of He could seriously further degrade composite mechanical properties. Some experimental work has been carried out to explore the effect of He on the mechanical properties of monolithic SiC [102]. High-purity, chemically vapor-deposited SiC bend bars were implanted at 590°C to He concentrations of 100 and 1000 appm. The samples were then neutron irradiated at 800°C to a dose of about 8 dpa to assess displacement damage interactions with preexisting He. Clearly, this does not simulate the fusion environment, but it does provide some indication of the effect of He on irradiated SiC mechanical properties. This study showed that bend strength, hardness, swelling, and indent fracture toughness significantly increased following the lower-temperature He implantation [102]. Since vacancies are not mobile in SiC at 590°C, no bubbles were observed [102]. Annealing the He implanted specimens at 1000°C showed that the mechanical property changes measured following implantation were caused by the $\sim$0.1 dpa displacement damage associated with the implantation [15]. While neutron irradiation of preimplanted condition produced larger mechanical property changes compared with neutron irradiated-only control case was not statistically significant [102]. While these results suggest little effect of He on mechanical properties at 800°C, considerably more work under prototypical conditions and concurrent irradiations to much larger doses and He concentrations are needed to properly assess the impact of He on the full range of SiC properties.

The strong tendency for He to stabilize vacancy clusters in fusion materials has significant implications for degradation mechanisms, such as swelling and high-temperature creep cavitation. Here we use the term cavity to refer to both bubbles and voids. A He-stabilized cavity is a bubble, with an internal gas pressure $\sim 2\gamma/r$, where γ is the surface free energy. Bubbles grow by absorbing He atoms and enough vacancies to reach the $2\gamma/r$ condition. Bubbles emit vacancies such that the excess vacancy arrival rate (see below) is exactly balanced by vacancy emission plus SIA absorption. A void is a cavity that is largely a collection of vacancies, where the He pressure may be $\ll 2\gamma/r$. The main difference between bubbles and voids is that smaller bubbles grow only by the addition of more He, while larger voids grow continuously due to an excess flux of vacancies impinging on them. Small bubbles initially grow by absorbing He atoms up to a critical He content (or size), where they convert to unstably growing voids [55]. Unstable void growth drives swelling.

Vacancies and interstitials are produced in equal numbers, but their annihilation at individual sinks is not equal. At intermediate irradiation temperatures, $0.3 \leq T_M \leq 0.6$, both SIAs and vacancies are mobile, and SIAs slightly preferentially annihilate at dislocations, or other biased sinks, giving rise to an excess flux of vacancies that are absorbed at the growing voids. However, for a given excess vacancy flux, voids only grow when they are larger than a critical size due to vacancy emission that is mediated by the Gibbs-Thomson effect, at a rate that exponentially increases with decreasing size [55]. Large critically sized vacancy clusters can form by via statistical fluctuations because there is a nonzero probability of having vacancy clusters of size $n+1$ given a population of size n. However, such homogeneous nucleation rates are much too low to explain the experimentally observed void densities under neutron irradiation conditions [103]. On the low-temperature end of the swelling regime void growth is kinetically constrained by slow vacancy-SIA recombination, and at the high-temperature end by vacancy emission and large critical bubble sizes.

As a consequence of the role of bubbles in void formation, RAF/M steels are resistant to swelling in fission reactors where little He is generated, but at high He-to-dpa ratios voids form on critically-sized bubbles, if present [15,55,104]. Thus, an effective strategy is to keep all bubbles below the critical size by promoting the formation of a high density of bubbles. An additional benefit derived from this approach is that bubbles are excellent sinks for both point defects thus effectively enhancing recombination, thereby reducing excess vacancy fluxes. Bubbles also strongly trap He generated at higher doses. In this case, the critical He bubble size increases and the rate of swelling decreases, if voids do manage to form, with increasing bubble number density [15,55,104].

Thus, swelling can be almost completely suppressed by distributing the He in a large number of small bubbles. This is the main material design principle guiding the development of NFAs. Void formation is prevented in NFAs because they contain $\sim 10^{23-24}$ m^{-3} of $\sim$2.5 nm diameter Y-Ti-O nano-oxides. The high nano-oxide number density benignly traps He in subcritical bubbles and substantially reduces the amount of He reaching grain boundaries, thus reducing toughness loss at low-to-intermediate temperatures and potential degradation of creep rupture properties at high temperatures [15,55,104]. These nano-oxides, along with a high dislocation density, and small grain sizes also impart resistance to displacement damage in these alloys. A significant additional advantage of NFAs is that they can operate at temperatures well above the displacement damage-swelling regime [55], where the primary damage mechanism of concern is creep embrittlement due to accumulation of He on grain boundaries.

At temperatures above $\sim 0.5T_M$ metals and alloys deform by time-dependent creep. At high stresses and lower temperatures, many alloys fail in a transgranular manner by nucleation and growth of microvoids in the matrix as discussed in [105]. In the intermediate temperature range the primary creep failure mode is intergranular, associated with nucleation, growth, and coalescence of cavities on grain boundaries normal to the tensile axis. Cavity nucleation occurs by two mechanisms: breaking weak particle interfaces, or condensation of vacancies, or both. Creep cavity growth occurs by vacancy creation and diffusion along the grain boundary, which is equivalent to plating the flux of atoms flowing in the opposite direction. Very high stresses on the order of $E/100$ are required for homogeneous creep cavity nucleation, where E is Young's modulus. Creep stresses are generally about an order of magnitude lower. Thus in the absence of He, large stress concentrations are required that are typically provided by particles on sliding grain boundaries and dislocation pile-ups at slip bands intersecting a grain boundary. Creep cavities grow and ultimately coalesce by a vacancy creation—diffusion atom plating mechanism and final rupture of the remaining ligaments. For unhindered creation and diffusional growth rupture times would be very short. However, particles on grain boundaries can greatly reduce cavity growth rates, as can back stresses created by the need for creep to accommodate local grain boundary swelling [94,105–109].

Helium bubbles preferentially form on grain boundaries, dislocations, and particle/matrix interfaces primarily due to greater available excess volume at such these features compared to a defect-free matrix [110]. Growing stressed grain boundary creep cavities form on critically sized bubbles in a way that is completely analogous to matrix voids. In the case of voids, the excess vacancy flux produces a chemical stress, while a mechanical stress drives the process in the case of creep cavities. Again, as in the case of void swelling at higher He concentrations, the population of critical bubbles for a particular applied stress largely governs postirradiation creep rupture times. The corresponding in situ creep rupture dpa are largely governed by the dose needed to grow a significant population of critically sized grain boundary bubbles [97]. Consequently, He can reduce the creep rupture time by orders of magnitude, even for relatively low He concentrations, as illustrated for cold-worked austenitic stainless steels shown in Fig. 3.19. The number of bubble nucleated creep cavities typically far exceeds those in the absence of He. As a result less cavity growth is required to reach the coalescence condition, hence, creep ductility is also greatly reduced at high He levels.

There are no data on creep rupture of RAF/M or NFA steels at high He levels from reactor neutron irradiations. However, numerous high-energy He ion implantation studies, including in-beam experiments, have been performed to assess He effects in RAF/M steels [111] and in 14YWT [112]. Fairly high He concentrations (up to 1000 appm) on creep lifetime, minimum creep rate, and rupture elongation show little deterioration of these properties [111,112]. This has been attributed to trapping of most of the He in the complex matrix microstructure so that less is apportioned to grain boundaries as well as the low creep stresses that are characteristic of RAF/M compared to austenitic alloys. While these experimental observations are encouraging, the He-implantation experiments are not prototypical of the fusion neutron environment. Additional data on properties, such as creep and creep fatigue crack growth at high temperatures with high He levels, are not available. Hence, possible effects of high He on the upper operating temperature limits of fusion alloys remains an open question.

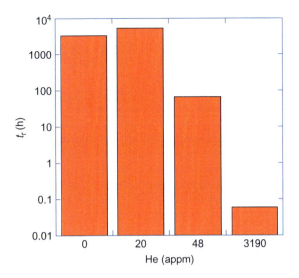

FIG. 3.19

The creep-rupture time for 20% cold-worked Type 316 austenitic stainless steel tested at 550°C and 310 MPa for different He levels after irradiation in the high flux isotope reactor at temperatures ranging from 545°C to 605°C [15].

3.6 MATERIALS FOR DIVERTOR/LIMITER APPLICATIONS

The MCF divertor and limiter are the main plasma-facing components (PFCs) that will be subjected to intense particle and thermal loads during operation. PFC materials serve two main functions: transporting enormous heat loads away from the surface and protecting the first wall from the high incident particle flux. Linke [113] and Cottrell [2] enumerate the general requirements for PFC materials including high thermal conductivity, adequate mechanical properties, low-activation by neutrons, and ready availability at acceptable cost. In addition, PFC materials must resist physical and chemical sputtering, in large part to avoid contaminating the plasma, possess a high melting temperature, and display good thermal shock resistance [113]. In this section, we briefly discuss the principal materials being considered for plasma-facing applications.

3.6.1 TUNGSTEN AND TUNGSTEN ALLOYS

Tungsten (W), W-alloys, and W-composites are currently the leading candidate materials for PFCs in fusion devices due to their excellent thermophysical properties. Rieth and coworkers have recently summarized the large European W research program [114,115]. The main advantages of W-based materials for PFC applications are a high melting point, good thermal conductivity, low sputtering yields and erosion rates, high intrinsic mechanical strength, low thermal expansion, and high resistance to swelling. The corresponding disadvantages are a high DBTT and relatively low recrystallization temperature. The resulting temperature window for divertor applications is about 800–1200°C, where the lower end is dictated by the DBTT [116] and the upper limit is controlled by the recrystallization temperature [117]. Note, even at temperatures above the DBTT the fracture toughness of tungsten is low.

Powder metallurgy (PM) is the main processing path for W [118]. The microstructure and properties of PM W are sensitive to the details of the processing method. The grain size, shape, and texture are governed by sintering conditions, the type and extent of postsintering deformation, and the intermediate and final annealing temperatures [118]. Hot isostatically pressed and sintered products generally must be forged and rolled to achieve reasonable toughness, which at ambient temperature is still only about $\approx 8\,\mathrm{MPa\,m^{1/2}}$ in favorable orientations in simple rolled plates. Control of interstitial impurities is also vital to obtaining acceptable mechanical properties [118]. Low solubility impurities, such as O, segregate to grain boundaries, which can result in embrittlement. In addition to the inherent effect of larger grains in reducing toughness, grain boundary area is reduced after recrystallization leading to an increased concentration of impurities. Various schemes of producing so-called nano-grained W, including with refractory fine second-phase carbide and oxide particles intended to stabilize them as processed microstructures have had only limited success in increasing the corresponding toughness. The toughness of W can be increased in two ways as follows: (a) additions of Re to promote dislocation mobility, and, (b) severe plastic deformation to ultrahigh strains that result in nanometer-scale grains and high dislocation densities [119,120]. However, neither of these toughening strategies appears to be viable for fabrication of W components. More recent research is focusing on fabrication of W composites that are toughened by engineered reinforcement architectures [121]. Greatly improved fracture resistance is achieved by incorporating ductile phases or strong reinforcing fibers that bridge a growing crack and lower the effective stress intensity factor at the crack tip. Application of the large body of work on toughening of brittle ceramics and intermetallic compounds to reinforce W for PFC applications is just beginning, but holds great promise [122–124].

The PM-deformation processing approach described above may be sufficient for fabricating simple W product forms like armor plates. However, conventional machining of W components with complex shapes is not feasible for divertor applications. Thus, new manufacturing paths must be explored. One promising approach is metal injection molding (MIM). MIM involves blending W powders with an organic binder that is then molded into a near net shape, in a way that is analogous to processing of many plastic parts. The binder is burned off and the green powder shape is then sintered to near theoretical density. Both pure W and alloys containing fine particles to reduce grain size can be processed by MIM. The toughness of MIM-W is low with a DBTT of about 800°C. It is not known if the toughness of MIM-W above the DBTT is sufficient for structural applications. However, since MIM-W is not deformed it is more resistant to recrystallization. Small and simple MIM-W coupons have been exposed to multiple ultrahigh heat flux pulses that are associated with off-normal transient plasma conditions as described below. These exposures lead to extensive surface cracking and roughening. The corresponding effects of normal high heat fluxes on MIM-W components in this condition are not known. Additive manufacturing that involves spray forming of powder that is melted in place, or deposited as liquid droplets is another approach to fabricating tungsten components. However it is likely that the deposits will face similar or more severe toughness challenges than conventional powder consolidation deformation approaches.

Another fabrication challenge is creating robust joints between tungsten dissimilar metal substrates like NFA typically by diffusion bonding. The joints are a potential life limiting failure path that is subject to both large coefficient of thermal expansion mismatch stresses and extremely brittle intermetallic phases. Joint debonding not only represents a mechanical failure path but also affects the effective heat transfer capacity of tungsten divertor components.

Perhaps partly due to its inherent brittleness, the motivation to explore neutron irradiation effects on W has been limited. Table 3.1 shows that the peak neutron wall load and gas production for W in typical MCF and ICF power systems are considerably smaller than for first-wall materials, such as RAF/M steels, and alternative PFC materials, such as carbon fiber composites. However, production of solid transmutation elements is much more significant in irradiated W than in these other materials. The principal W solid transmutation products are Re and Os. As shown in Fig. 3.20 by 80 dpa initially pure W is alloyed with ~13% Re and ~8.5% Os, with potentially significant consequences to the mechanical and thermal properties. Notably, due to large thermal-epithermal neutron cross sections, the corresponding transmutation rates in fast and especially mixed thermal and fast spectrum fission reactor irradiations are much greater than for fusion by factors of about 5 and 50, respectively [125,126]. Thus, fission reactor data cannot be used to directly evaluate irradiation effects in fusion environments solely based on temperature and dpa. Osmium accumulates in significant quantities due to transmutation of Re.

3.6 MATERIALS FOR DIVERTOR/LIMITER APPLICATIONS

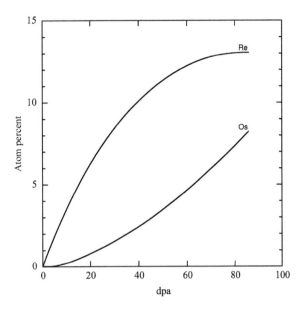

FIG. 3.20

Transmutation of tungsten to rhenium and osmium in a fusion neutron spectrum (Starfire).

Adapted from L.R. Greenwood, F.A. Garner, Transmutation of Mo, Re, W, Hf, and V in various irradiation test facilities and STARFIRE, J. Nucl. Mater. 212–215 (1994) 635.

Hasegawa and coworkers have compiled fission neutron irradiation data from low dpa irradiations of pure W and W alloys. Unfortunately, these data are for a limited temperature range [127–129]. Pertinent irradiation data for Mo are also available at somewhat higher dpa. These irradiations are all in the hardening regime below about $0.3T_m$. For example, Fig. 3.21 shows representative hardness data as a function of dpa for both pure W and a W-26Re alloy irradiated in various fission reactors at 500°C and 600°C. At higher temperatures, the irradiation microstructures are composed of a mixture of dislocation loops, voids, and intermetallic precipitates. At doses below about 0.4 dpa, the W-Re alloy exhibits slightly lower hardening than pure W. This is likely due to enhanced void and loop contributions in this case. However, at higher dpa, the W-26Re alloy irradiation hardens enormously more than pure W due to enhanced precipitation of nonequilibrium χ (WRe_3) phase [128]. More generally, neutron-irradiated W contains equiaxed σ (WRe)-phase, as well as χ-phase in the form of plates or acicular needles. The precipitates and defects result in enormous hardening in fast reactor irradiations up to 11 dpa at 500–600°C. While the W-26Re alloy composition is an extreme case, hardening for pure W is significant and increases in a roughly linear manner with Re content once the dose reaches about 1 dpa [127]. Formation of nonequilibrium χ-phase is likely due to RIS to dislocation loops, accelerated by RED. Segregation results in precipitation of Re from highly subsaturated solutions. To make matters worse, Os has a lower solubility than Re in equilibrium with the ternary W-Re-Os σ-phase, and also lowers the solubility of Re. For example, based on the 1600°C ternary W-Re-Os phase diagram [130] and Fig. 3.20, which shows a plot of Re and Os generation for a fusion spectrum, there would be about 0.25 volume fraction of σ-phase at 80 dpa. Hardening decreases with increasing temperature, primarily due to coarser precipitate microstructures. However, large amounts of even more brittle intermetallic phases will exacerbate the inherently low toughness of W. While much research remains to be carried out, two conclusions are obvious for fusion applications of tungsten: (a) alloying W with Re is counterproductive; and, (b) monolithic W certainly will be embrittled by neutron irradiation limiting its temperature and dpa window.

CHAPTER 3 OVERVIEW OF FUSION MATERIALS AND OPERATING ENVIRONMENTS

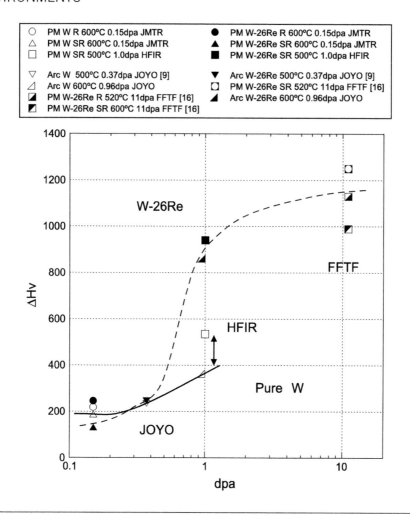

FIG. 3.21

Dose dependence of irradiation hardening of tungsten and tungsten-26% rhenium between 500°C and 600°C in various fission reactor environments [127].

Tungsten PFCs will also be subjected to severe and variable surface heat loads, damage from energetic ions, sputtering and erosion by ions (or neutral particles). These conditions may induce failure modes that differ from that elsewhere in the system. For example, impurities from erosion can directly compromise plasma stability. Relative to other PFC materials the energy threshold for sputtering of W by deuterium or tritium ions is high ($\sim$100 eV) [2]. For He ions the peak sputtering yield is about $\sim$ 0.03 at 1000 eV [2]. The maximum sputtering rates are lower for the hydrogen isotopes. The net erosion rate of W under ITER like conditions has been estimated to be approximately 0.1 mm/FPY, which is far less than the 3 mm/FPY estimated erosion rate for a low-Z material like beryllium. Tungsten atoms eroded from the surface are ionized in the plasma resulting in radiative cooling that is proportional to Z^2, where Z is the atomic number of the ion. Fortunately, the high sputtering threshold for W coupled with a propensity for ejected W ions to quickly redeposit on the surface tends to reduce its influence on plasma performance [2].

3.6 MATERIALS FOR DIVERTOR/LIMITER APPLICATIONS

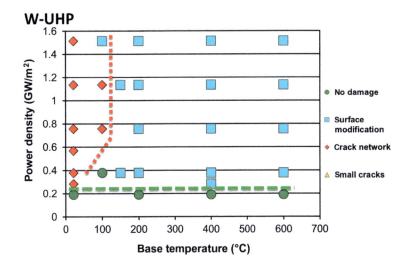

FIG. 3.22

Thermal shock response of ultrahigh purity W to 100 pulses, with an absorption coefficient of 0.55 and a pulse duration of 1 ms. The dashed line marks the boundary between undamaged and damaged W and the dotted line shows the conditions that produced cracking [131].

From M. Wirtz, J. Linke, G. Pintsuk, et al., Comparison of the thermal shock performance of different tungsten grades and the influence of microstructure on the damage behaviour, Phys. Scr. T145 (2011) 014058. © The Royal Swedish Academy of Sciences. Reproduced by permission of IOP Publishing. All rights reserved.

A substantial threat to the integrity of PFCs is the effects of transient heat loads. As shown in Fig. 3.3 steady-state heat loads in the divertor region are on the order of 10 MW/m^2, while transient heat loads associated with plasma events, such as edge localized modes, disruptions, and vertical displacement events, result in short-time surface heat fluxes of 1–10 GW/m^2. Thermal shock experiments have been performed to characterize the effects of severe transient heat loads that approximate the conditions of plasma disruptions. Fig. 3.22 shows the effects of exposing a polished surface of an ultrahigh purity W coupon to a focused high power electron beam for 100 pulses of 1 ms duration [131]. A 3-s dwell time between pulses allows the specimen to cool to a base temperature before the next pulse. Below about 0.2 GW/m^2 no surface damage is observed at all the base temperatures. However, above 0.2 GW/m^2 either cracking occurred at base temperatures below 100°C or surface modifications above 100°C [131]. The damage depends on both the test conditions, as well as the microstructure and properties of the W alloy. Alloys with higher strength exhibited a higher damage threshold and greater resistance to cracking [115]. On the other hand, W alloys with better thermal properties experienced increased depth penetration of the crack network [115].

Tritium (T) retention in W is orders of magnitude lower than for carbon fiber composites and Be [132]. However, one safety issue is trapping of the large quantities of T impinging on the surface. Another safety issue is the potential for release of oxide particles with activated tungsten and entrained tritium under accident conditions associated with failure of the VV boundary. Loss of coolant could result in peak surface temperatures up to ~1200°C due to nuclear decay heat. Air ingress would result in formation of volatile WO$_3$ at an estimated evaporation rate on the order of 10–100 kg/h for a reactor at 1000°C with 1000 m^2 of surface area. Thus, research has been carried out on alloys with elements added to make W self-passivating, thereby mitigating potential accidental releases [133].

Another surface modification that has been observed in tungsten alloys exposed to pure He plasma, such as in the PISCES-B linear-divertor-plasma simulator, is shown in Fig. 3.23 [134]. Under conditions where the average helium ion energy is below the threshold for physical sputtering of W, a tendril-like nanostructure grows on the

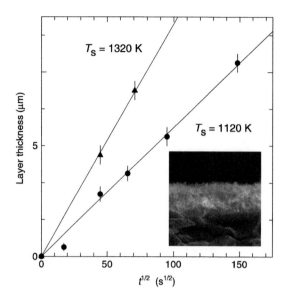

FIG. 3.23

Time dependence of the nanostructured layer thickness on tungsten exposed to a helium plasma with a flux of about $5 \times 10^{22}/m^2$ s at the indicated temperatures. Inset shows a scanning electron microscope image of the nanostructured layer formed after a 9000 s exposure.

Adapted from M.J. Baldwin, R.P. Doerner, Helium induced nanoscopic morphology on tungsten under fusion relevant plasma conditions, Nucl. Fusion 48 (2008) 035001.

surface. As shown in Fig. 3.23, the so-called "W-fuzz" growth rate obeys diffusion-controlled kinetics, but there is no detailed understanding of how these features nucleate and grow. Tendril formation occurs over a limited range of exposure temperatures and He ion energies, and can be suppressed by the presence of other low-Z species in the plasma. The implications of tendril formation may be both positive and negative for divertor operation. There is experimental evidence that surface fuzz may increase the resistance of W to thermal transients and reduce sputtering rates presumably due to an increased redeposition probability. On the other hand, fuzz growth may contribute to an increase of W dust and could possibly lead to enhanced tritium retention associated with nano-scale helium bubbles embedded in the tendril structure [134,135].

3.6.2 CARBON FIBER COMPOSITES

Carbon-based composites are the traditional material of choice in PFCs of experimental tokamaks. Carbon matrix-carbon fiber (CFC) composites have many desirable properties including low Z, low activation, and good thermal conductivity relative to W [2,136]. Graphite sublimates rather than melts making it an attractive choice for the separatrix region on the divertor [137] with increased tolerance for off-normal events, such as plasma disruptions. Carbon fiber composites generally offer superior mechanical and physical properties compared to monolithic carbon. Composite architectures can be designed to mitigate some of the deleterious effects of neutron irradiation while achieving potentially adequate fracture toughness under cyclic thermal loading [2].

A significant challenge for use of CFCs in fusion applications is erosion, which manifests itself in three forms: physical sputtering; chemical erosion associated with chemical reactions between the hydrogen ions and carbon that form volatile hydrocarbon species; and radiation-enhanced sublimation [2,136]. The threshold ion energy for physical C sputtering is from 10 to 30 eV [2] with the maximum rate between 100 and 1000 eV [2]. Chemical erosion is a greater concern than physical sputtering for C-based PFCs since it occurs at much lower ion energies than for physical sputtering [2]. The main engineering issue is associated with control and accountancy of the reactor tritium inventory. Tritiated volatile hydrocarbons redeposit at other sites in the reaction chamber, hampering tritium recovery and increase the risk of uncontrolled radioactivity releases in the event of an accident.

Large time-varying thermal loads also challenge the structural integrity of carbon-based composites during normal operation, and extreme heat fluxes during off-normal events associated with plasma instabilities. Thermal fatigue is the principal concern during normal operation and thermal shock during off-normal events. High heat flux test facilities have been developed to simulate fusion-relevant thermal loading conditions, with power densities ranging from MW/m^2 levels to several GW/m^2 [137]. These facilities utilize electron or hydrogen ion beams that are operated in a pulsed or rastered mode, with durations ranging from microseconds to nearly continuous, to simulate the relevant range of thermal loading conditions [137]. Small-scale mockups of various CFC configurations (flat tiles or monoblocks) have survived up to 1000 cycles at heat fluxes between 19 and 25 MW/m^2 [137]. Thermal shock tests on CFC armor to 60 MW/m^2, which simulates a vertical displacement off-normal event, show that thermal erosion occurs by carbon sublimation and brittle fragmentation [137] due to surface microcracking to depths of several hundred microns [3]. Cracking is associated with large temperature gradients and thermal stresses that result from short-pulse plasma disruptions. Fragmentation leads to particle ejection and accumulation of C dust particles in the plasma chamber.

The effects of neutron irradiation on CFCs can be very significant. Perhaps the most important effect is a dramatic decrease in thermal conductivity that is caused by enhanced phonon scattering by irradiation-induced defects. The thermal conductivity degradation depends on the irradiation temperature and dpa as illustrated in Fig. 3.24. It is evident that at low irradiation temperatures the thermal conductivity of CFCs decreases markedly, at doses as low as ~ 0.01 dpa. At higher irradiation temperatures thermal conductivity degradation rate is less dramatic. There is also evidence that the decrease in thermal conductivity saturates at ~ 1 dpa. High-temperature annealing partially recovers CFC thermal conductivity [138].

Neutron irradiation can also affect the dimensional stability and mechanical properties of CFCs [138,139]. Dimensional instabilities are dominated by the fiber reinforcement component of the composite [138]. Irradiation

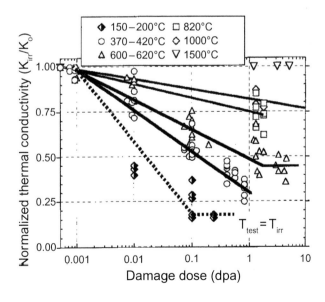

FIG. 3.24

Neutron dose and irradiation temperature depend on the normalized thermal conductivity of several CFCs [138].

tends to cause the fibers to contract in the longitudinal direction and swell in the radial direction due to the highly anisotropic structure of graphite crystals [138,139]. The dimensional changes lead to lock-up of the fibers, reducing or eliminating the effectiveness of the fiber crack bridging on increasing composite toughness. CFCs can also undergo substantial nonuniform macroscopic dimensional changes at high neutron doses. Neutron irradiation can also lead to changes in the elastic modulus of CFC composites and increased strength [126]. One study reported that initially the elastic modulus increases up to a dose of ~ 9.5 dpa then decreases significantly by 32 dpa [139]. The dimensional changes tend to increase with increasing irradiation temperature and to be more isotropic in 2D and 3D CFCs versus 1D composite architectures [138].

Retention of tritium is also a significant issue for carbon-based PFC materials due to trapping by both intrinsic and irradiation-induced defects. High levels of retained tritium have been reported following low-dose irradiation of fine grain graphites, saturating at ~ 7000 appm at doses above ~ 0.1 dpa [138]. Substantially less tritium appears to be retained in CFC compared to graphite, but the quantities may still be significant [138]. The amount of retained tritium depends on the neutron dose, but saturation levels are reached at less than about 1 dpa [138].

The response of irradiated CFCs under high plasma disruption loading conditions has been investigated. A 0.3 dpa irradiation at 290°C was found to double the CFC erosion during high-heat flux testing to simulate plasma disruptions [138]. This behavior was attributed to the corresponding reduction in thermal conductivity.

As in the case of all other materials of interest for the fusion divertor and first-wall/blanket applications, and as noted in the section on SiC, He generation may play a significant to dominant role in the effects of irradiation on the mechanical, dimensional, and physical properties of CFCs. Notably, fusion spectrum neutron irradiation of carbon generates He at a rate of ~ 380 appm/dpa [138]. However, very little is known about the consequences of up to more than 42,000 appm helium generated over a 3 FPY divertor lifetime.

3.6.3 LIQUID WALLS

The enormous challenges associated with solid PFCs have motivated research on the possible use of flowing liquid walls. This is particularly true for ICF power system designs where the pulsed heat loading delivered by the exploding target is on par with MCF plasma disruptions [140,141]. Some of the concepts developed for ICF were briefly summarized in Section 3.4.2 above. Liquid walls have several characteristics that make them particularly appealing for high heat and neutron wall loadings. Compared to solid structures, liquid walls would be (1) effectively immune to irradiation effects [142] and (2) much more tolerant of high surface heat fluxes, including during transients. A sufficiently thick (~ 0.4 m) liquid layer would also reduce neutron damage to the underlying solid structures. Liquid walls offer several other possible advantages [143] such as: (1) improvements in plasma stability and confinement, (2) higher power density through reduction of thermal stresses and wall erosion, (3) decreased volume of radioactive waste, and (4) the potential for increased plant availability through extended lifetimes, reduced failure rates, and more rapid maintenance operations. A schematic representation of the liquid wall concept is shown in Fig. 3.25 [143].

Liquid wall options have also been considered for MCF applications. Similar to ICF, concepts have been explored where the thickness of the liquid ranges from ~ 2 cm to more than ~ 40 cm [143]. The liquids receiving the most attention are lithium based so that tritium breeding is possible. Liquid metals, such as pure Li, a SnLi alloy, or a molten salt, Li_2BeF_4 (also known as Flibe) are the leading candidates. A number of concepts have been developed to power the liquid flow and maintain its adherence and uniformity against a solid backing wall. One concept uses gravity and fluid momentum to create the desired flow pattern. Fluid is injected at the top of the chamber at an angle tangential to the curved backing wall [143]. As long as the centrifugal force pushing the liquid against the backing wall exceeds the gravitational force the liquid remains continuously attached. Another electromagnetically restrained concept, which only works for liquid metals, involves controlling the fluid flow by inducing a current in the poloidal direction. The current in the flow interacts with the toroidal magnetic field to create internal forces that cause the liquid to adhere to the solid wall [143].

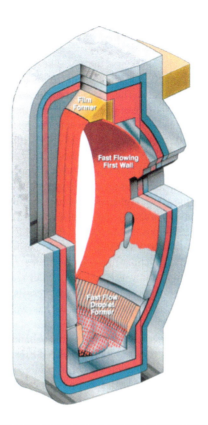

FIG. 3.25

Schematic representation of a liquid wall concept. Liquid (red) enters at the top of the chamber and flows downward over the structural material (gray) behind the liquid layer [143].

However, liquid walls also present several scientific and engineering challenges that need to be addressed if they are to become realistic replacements for solid wall concepts. Perhaps the major challenge is maintaining hydrodynamic control of a free-surface flow inside a geometrically complex chamber that incudes penetrations, submerged walls, and inverted surfaces [143]. These considerations are particularly acute for thick liquid walls. In the case of liquid metal walls, magnetohydrodynamic interaction, which can create eddy current forces tending to pull the liquid metal off the backing wall or impose a substantial drag on a free-surface flow [143]. Abdou et al. also note that the effects of liquid walls on the plasma core need to be characterized, along with the interaction between the plasma edge and the free surface of the liquid [143].

The influence of various atomic-level processes, such as sputtering, evaporation, segregation, and particle retention on limiting the operating envelope for liquid plasma facing materials, has been considered [142]. The high evaporation rate of pure Li appears to limit its use to temperatures below about 450°C [142]. A tin-lithium alloy is attractive because its vapor pressure is lower than pure Li, which should permit higher operating temperatures. However, the higher sputtering yield of tin-lithium could introduce unacceptable concentrations of impurities into the plasma [142]. Flibe also suffers from concerns about sputtering yield as well as degradation due to particle bombardment [142] and corrosion compatibility issues. The overall conclusion of an assessment of candidate liquid walls is that a completely suitable concept for their use in plasma-facing service does not exist [142]. However, further research is needed to confirm this finding.

3.7 VACUUM VESSEL (VV) MATERIALS

An important structural component of an MCF system is the VV. The VV is outboard of the first-wall/blanket and shield elements, and inboard from the magnets. Its primary function is to provide the high-vacuum environment needed to maintain a high-quality burning plasma. The VV also provides a critical safety barrier to confine tritium and other radioactive species so as to limit exposure of plant workers and the public in the event of an accident. As a primary safety barrier, the VV requires a high level of mechanical integrity. In recent MCF design studies [144,145] the thickness of the VV ranges from 5–10 cm (thin design) to 25–40 cm (thick design). A thick VV is intended to shield the magnets from radiation. It would be water cooled and operate between 150°C and 200°C. A thin He-cooled VV would operate between 400°C and 500°C. The VV also provides support for the first-wall/blanket structures and must tolerate electromagnetic loads during various types of plasma disruptions and potential seismic events [21]. An advantage of operating the VV at higher temperature is improved dust control and reduced tritium accumulation, along with reduced radiation hardening and loss of fracture resistance in bcc steels [146]. The VV could weigh in the neighborhood of ~1000 tons, not including the additional ~2000 tons of shielding material. El-Guebaly and coworkers [146] have proposed VV material selection criteria that are primarily based on activation and fabrication concerns. Neutron activation impacts in-service inspections and maintenance, remote handling, decay after heat, decommissioning and waste disposal. Other requirements include in-service compatibility with operating temperatures in the ranges of thin and thick vessels, especially considering the effects of neutron irradiation on ductility, toughness, and potential swelling, as well as welding and re-welding without the need for complex PWHTs.

The same austenitic stainless steel that was selected for the blanket/shield (316LN-IG) will be used for the ITER VV. The sister alloy used in the French breeder program (316LN-SPH) was modified for VV application by reducing Co, Mn, and Ta related to neutron activation concerns, while still remaining within the specification requirements of the French safety RCC-MR Code [21]. On the basis of considerable fabrication experience and the existence of a comprehensive database, 316LN-IG may also be an attractive choice for long-term fusion systems. However this alloy exhibits two potential drawbacks. First, the composition includes a minimum of ~1.6% Mo, which precludes shallow land burial. Presumably replacing Mo with W could resolve this issue. Second, VV sections behind assembly gaps and close to penetrations and ports will be subjected to displacement damage approaching 20 dpa at ~200°C in ARIES designs, which might cause unacceptable reductions in uniform elongation and fracture toughness, as well as possible dimensional changes and internal stresses associated with differential swelling [146]. Note, there is no direct experience in fabrication of $\geq$1000-ton VV-scale heavy-section stainless steel pressure vessels. Moreover, it will likely be necessary to stress relieve any welds in such heavy-section components, which might be problematic.

Other candidate alloys for VV applications include 3Cr-3WV bainitic steel, 8%–9%Cr RAF/M steels (ORNL-FS), 430 ferritic steel (430-FS), 316 SS austenitic stainless steel, modified DIN-4970 austenitic stainless steel (RAAS), and AMCR-0033, which is a Mn-stabilized austenitic stainless steel (Mn-AS). Corresponding activation calculations are shown in Figs. 3.26 and 3.27 for the thick VV and thin VV design options, respectively. All of these steels with nominal impurity levels, except 316-SS, meet the requirements for Class C low-level waste disposal (shallow land burial) for the thick VV option, while none of them meet this requirement for a thin VV. Meeting Class C requirements for the thin VV option will require reductions in impurity levels. Fortunately, this appears to be possible for four of these steels (MF82H, 3Cr-3WV, ORNL FS, 430-SS) [133].

Most recent MCF VV designs have focused on ferritic, and especially RAF/M steels mainly as a way of reducing alloy qualification and fabrication procedure development costs [21]. However, a thick VV will operate in a temperature range of maximum radiation hardening and embrittlement for these alloys. In addition, RAF/M welds will require PWHTs. Given the wall thickness and overall size of the VV, proper PWHT presents an enormous challenge. A major conclusion of the El-Guebaly study is that 3Cr-3WV bainitic steel, with adequate mechanical strength, ductility, and fracture toughness, appears to be an attractive option for both VV designs that can meet low-level waste disposal criteria while avoiding complex PWHT tempering steps [146]. Low-chromium

3.7 VACUUM VESSEL (VV) MATERIALS

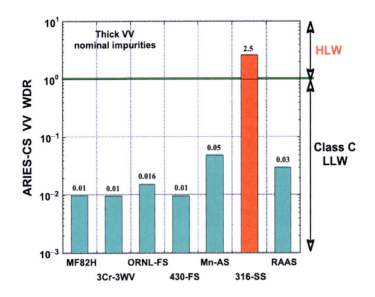

FIG. 3.26

Class C waste disposal requirements of the seven candidate steels for ARIES-CS thick VV with "nominal" impurities [146].

From L.A. El-Guebaly, T. Huhm, A.F. Rowcliffe, et al., Design challenges and activation concerns for ARIES vacuum vessel materials Fusion Sci. Technol. 64 (2013) 449. Copyright 2013 by the American Nuclear Society, La Grange Park, IL.

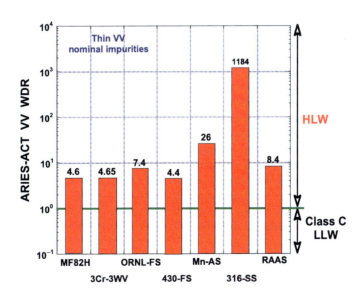

FIG. 3.27

Waste disposal requirements of the seven candidate steels for ARIES-ACT-1 thin VV with "nominal" impurities [146].

From L.A. El-Guebaly, T. Huhm, A.F. Rowcliffe, et al., Design challenges and activation concerns for ARIES vacuum vessel materials Fusion Sci. Technol. 64 (2013) 449. Copyright 2013 by the American Nuclear Society, La Grange Park, IL.

steels generally do not require PWHT because the fusion and heat-affected zones do not transform to a brittle martensitic phase on cooling [147]. However, stress relief treatments would probably still be needed. Further data on low-temperature irradiation hardening and embrittlement as well as the potential for irradiation-assisted stress corrosion cracking need to be assessed for this relatively new, steel [146].

In summary, there are significant drawbacks using a RAF/M steel for the VV because of (1) the difficulty of carrying out PWHTs and (2) problems with managing irradiation-induced shifts in DBTT. The ITER solution of using 316LN-IG is not acceptable because of the long-term activation products stemming from Mo and the potential for mechanical property degradation. Consequently, the newer 3Cr-3WV bainitic steel appears to offer a more attractive choice for VV applications, but alloy performance under fusion relevant operating conditions needs to be determined.

3.8 MAGNET STRUCTURAL MATERIALS

The design for the ITER Central Solenoid (CS) magnet system involves six identical coils, each consisting of a superconducting Nb_3Sn cable-in-conduit conductor operating at up to 13 T [148]. The conductor jacket is the main structural element of the CS coils. During operation these coils experience large electromagnetic loads, which are resisted by the conductor jacket. Due to the pulsed operation of ITER the jacket must be designed to accommodate an estimated 60,000 electromagnetic load cycles. The jacket will contain low-temperature ($\sim$4.2 K), pressurized He and must be designed to tolerate off-normal forces (e.g., earthquakes). Since the magnets are furthest from the plasma, due to the high sensitivity of Nb_3Sn to radiation damage and heating, the properties of greatest importance for the conductor jacket are adequately high tensile strength and fracture toughness, as well as low fatigue crack growth rates. In addition, the jacket material must meet stringent dimensional specifications and be compatible with the coil winding. In the case of the Nb_3Sn superconductor the jacket material must accommodate a necessary aging heat treatment of 650°C for about $\sim$200 h. Further, the jacket material requires a coefficient of thermal expansion similar to Nb_3Sn to avoid degrading superconducting performance associated with thermal contraction strains [149]. For ITER the total length of jacket that must be manufactured while maintaining all these requirements is approximately 43 km [150].

The ITER CS jacket sections are fabricated by hot-extrusion followed by cold drawing into seamless square outer cross-section pipes with a circular inner diameter [137]. The leading material developed for this application is a high-Mn-bearing austenitic stainless steel designated JK2LB [149]. The nominal composition is Fe-20.5Mn-12Cr-8Ni-0.03C-0.5Mo-0.1Co-0.09N-0.001B [150]. Specifications have also been developed defining an acceptable microstructure. The grain size must be $\geq$ASTM 4, the material must be fully austenitic (no ferrite present), and specifications dictate acceptable levels of carbides, carbonitrides, and inclusions of various sizes. The minimum 0.2% yield strength and ultimate tensile strength required at 4.2 K is 850 MPa and 1150 MPa, respectively. The fracture toughness must be $>$130 MPa-m$^{1/2}$ [150]. These mechanical properties apply to a JK2LB jacket following the superconductor aging heat treatment. Other alloys have been considered for the conductor jacket in the past such as Incoloy 908 and 316LN. Both of these alloys possess acceptable characteristics, but also some undesirable characteristics such as the propensity for stress-accelerated grain boundary oxidation in the case of Incoloy 908 and a significant thermal-expansion mismatch with Nb_3Sn in the case of 316LN.

REFERENCES

[1] F. Romanelli, Plasma physics and engineering, in: S.B. Krivit, J.H. Lehr, T.B. Kingery (Eds.), Nuclear Energy Encyclopedia: Science, Technology, and Applications, John Wiley, Hoboken, NJ, 2011, p. 371.
[2] G.A. Cottrell, A survey of plasma facing materials for fusion power plants, Mater. Sci. Technol. 22 (8) (2006) 869.
[3] J.P. Blanchard, R. Raffray, Laser fusion chamber design, Fusion Sci. Technol. 52 (2007) 440.

REFERENCES

[4] N. Baluc, Materials for fusion power reactors, Plasma Phys. Controlled Fusion 48 (2006) B165.

[5] M.E. Sawan, Damage parameters of structural materials in fusion environment compared to fission reactor irradiation, Fusion Eng. Des. 87 (2012) 551.

[6] S. Sharafat, private communication.

[7] L.M. Waganer, Fusion technology, in: S.B. Krivit, J.H. Lehr, T.B. Kingery (Eds.), Nuclear Energy Encyclopedia: Science, Technology, and Applications, John Wiley, Hoboken, NJ, 2011, p. 389.

[8] P. Testonia, F. Cau, A. Portone, et al., F4E studies for the electromagnetic analysis of ITER components, Fusion Eng. Des. 89 (2014) 1854.

[9] D.R. Harries, in: Ferritic martensitic steels for use in near term and commercial fusion reactors, Proceedings of the Topical Conference on Ferritic Alloys for Use in Nuclear Energy Technologies, Snowbird, Utah, June 19–23, 1983.

[10] L.R. Greenwood, Neutron interactions and atomic recoil spectra, J. Nucl. Mater. 216 (1994) 29.

[11] A. Molvik, A. Ivanov, G.L. Kulcinski, et al., A gas dynamic trap neutron source for fusion material and subcomponent testing, Fusion Sci. Technol. 57 (2010) 369.

[12] S.J. Zinkle, J.T. Busby, Structural materials for fission and fusion energy, Mater. Today 12 (2009) 12.

[13] S.J. Zinkle, N.M. Ghoniem, Prospects for accelerated development of high performance structural materials, J. Nucl. Mater. 417 (2011) 2.

[14] G.R. Odette, T. Yamamoto, H.J. Rathbun, et al., Cleavage fracture and irradiation embrittlement of fusion reactor alloys: Mechanisms, multiscale models, toughness measurements and implications to structural integrity assessment, J. Nucl. Mater. 323 (2003) 313.

[15] Y. Dai, G.R. Odette, T. Yamamoto, The effects of helium in irradiated structural alloys, Comprehensive Nuclear Materials, vol. 1, 2012, p. 141.

[16] Y. Katoh, L.L. Snead, Operating temperature window for SiC ceramics and composites for fusion energy applications, Fusion Sci. Technol. 56 (2009) 1045.

[17] R.E. Stoller, L.R. Greenwood, Subcascade formation in displacement cascade simulations: implications for fusion reactor materials, J. Nucl. Mater. 271–272 (1999) 57.

[18] S.J. Zinkle, A. Moeslang, Evaluation of irradiation facility options for fusion materials research and development, Fusion Eng. Des. 88 (2013) 472.

[19] G.R. Odette, P. Miao, D.J. Edwards, T. Yamamoto, H. Tanagawa, R.J. Kurtz, A comparison of cavity formation in neutron irradiated nanostructured ferritic alloys and tempered martensitic steels at high He/dpa ratio, Trans. Am. Nucl. Soc. (2008) 98.

[20] J.P. Sharpe, R.D. Kolasinski, M. Shimada, et al., Retention behavior in tungsten and molybdenum exposed to high fluences of deuterium ions in TPE, J. Nucl. Mater. 390–391 (2009) 709.

[21] C.E. Kessel, M.S. Tillack, V.S. Chan, et al., Fusion Nuclear Science Pathways Assessment, Princeton Plasma Physics Report PPPL-4736, (February, 2012).

[22] R.L. Klueh, E.T. Cheng, M.L. Grossbeck, E.E. Bloom, Impurity effects on reduced-activation ferritic steels developed for fusion applications, J. Nucl. Mater. 280 (2000) 353.

[23] L.A. El-Guebaly, Power plant projects, in: S.B. Krivit, J.H. Lehr, T.B. Kingery (Eds.), Nuclear Energy Encyclopedia: Science, Technology, and Applications, John Wiley, Hoboken, NJ, 2011, p. 405.

[24] L.A. El-Guebaly, Fifty years of magnetic fusion research (1958-2008): brief historical overview and discussion of future trends, Energies 3 (2010) 1067.

[25] M.S. Tillack, S. Malang, L. Waganer, et al., Configuration and engineering design of the ARIES-RS tokamak power plant, Fusion Eng. Des. 38 (1997) 87.

[26] M.S. Tillack, X.R. Wang, J. Pulsifer, et al., Fusion power Core engineering for the ARIES-ST power plant, Fusion Eng. Des. 65 (2003) 215.

[27] A.R. Raffray, L. El-Guebaly, S. Malang, et al., Advanced power core system for the ARIES-AT power plant, Fusion Eng. Des. 80 (2006) 79.

[28] S. Nakai, K. Mima, Laser driven Inertai fusion energy: Present and prospective, Rep. Prog. Phys. 67 (2004) 321.

[29] R.A. Al-Ayat, E.I. Moses, R.A. Hansen, Inertial fusion technology, in: S.B. Krivit, J.H. Lehr, T.B. Kingery (Eds.), Nuclear Energy Encyclopedia: Science, Technology, and Applications, John Wiley, Hoboken, NJ, 2011, p. 421.

[30] G.M. McCracken, P.E. Stott, Fusion: The Energy of the Universe, Academic, Waltham, MA, 2013.

[31] J.D. Sethian, D.G. Colombant, J.L. Giuliani, et al., The science and technologies for fusion energy with lasers and direct-drive targets, IEEE Trans. Plasma Sci. 38 (4) (2010) 690.

[32] N. Baluc, R. Schaeublin, P. Spaetig, M. Victoria, On the potentiality of using ferritic/martensitic steels as structural materials for fusion reactors, Nucl. Fusion 44 (2004) 56.
[33] R.J. Kurtz, A. Alamo, E. Lucon, et al., Recent progress toward development of reduced activation ferritic/martensitic steels for fusion structural applications, J. Nucl. Mater. 386–388 (2009) 411.
[34] Q. Huang, N. Baluc, Y. Dai, et al., Recent progress of R&D activities on reduced activation ferritic/martensitic steels, J. Nucl. Mater. 442 (2013) S2.
[35] B. Raj, K.B.S. Rao, A.K. Bhaduri, Progress in the development of reduced activation ferritic-martensitic steels and fabrication technologies in India, Fusion Eng. Des. 85 (2010) 1460.
[36] V.M. Chernov, M.V. Leonteva-Smirnova, M.M. Potapenko, Structural materials for fusion power reactors—the RF R&D activities, Nucl. Fusion 47 (2007) 839.
[37] S.J. Zinkle, N.M. Ghoniem, Operating temperature windows for fusion reactor structural materials, Fusion Eng. Des. 51–52 (2000) 55.
[38] O. Anderoglu, T.S. Byun, M. Toloczko, S.A. Maloy, Mechanical performance of ferritic martensitic steels for high dose applications in advanced nuclear reactors, Metall. Mater. Trans. A Phys. Metall. Mater. Sci. 44 (2013) S70.
[39] E. Gaganidze, J. Aktaa, Assessment of neutron irradiation effects on RAFM steels, Fusion Eng. Des. 88 (2013) 118.
[40] H. Tanigawa, K. Shiba, A. Moeslang, et al., Status and key issues of reduced activation ferritic/martensitic steels as the structural material for a DEMO blanket, J. Nucl. Mater. 417 (2011) 9.
[41] G.R. Odette, M.Y. He, E.G. Donahue, et al., Modeling the multiscale mechanics of flow localization-ductility loss in irradiation damaged bcc alloys, J. Nucl. Mater. 171 (2002) 307–311.
[42] T. Yamamoto, G.R. Odette, H. Kishimoto, et al., On the effects of irradiation and helium on the yield stress changes and hardening and non-hardening embrittlement of 8Cr tempered martensitic steels: Compilation and analysis of existing data, J. Nucl. Mater. 356 (2006) 27.
[43] M. Rieth, M. Schirra, A. Falkenstein, et al., EUROFER 97 Tensile, Charpy, Creep and Structural Tests, Forschungszentrum Karlsruhe Report FZKA 6911, 2003.
[44] P. Fernandez, A.M. Lanchaa, J. Lapena, et al., Creep strength of reduced activation ferritic/martensitic steel Eurofer'97, Fusion Eng. Des. 75–79 (2005) 1003.
[45] J. Vanaja, K. Laha, R. Mythili, et al., Creep deformation and rupture behaviour of 9Cr–1W–0.2V–0.06Ta reduced activation ferritic–martensitic steel, Mater. Sci. Eng. A 533 (2012) 17.
[46] Y.N. Osetsky, R.E. Stoller, Molecular Dynamics Modeling of Dislocation-Obstacle Interactions and Mechanisms of Hardening and Strengthening in Irradiated Metals, Fusion Materials Semiannual Progress Report for Period Ending December 31, 2011, DOE/ER-0313/51, U.S. Department of Energy, 2011, p. 119.
[47] M.F. Ashby, B.F. Dyson, Creep damage mechanics and micromechanisms, Adv. Fract. Res. 1 (1984) 3.
[48] T. Shrestha, M. Basirat, I. Charit, et al., Creep rupture behavior of grade 91 steel, Mater. Sci. Eng. A 565 (2013) 382.
[49] B.K. Choudhary, E.I. Samuel, Creep behaviour of modified 9Cr–1Mo ferritic steel, J. Nucl. Mater. 412 (2011) 82.
[50] B. Wilshire, H. Burt, Tertiary creep of metals and alloys, Z. Metallkd. 96 (2005) 6.
[51] R.L. Klueh, K. Shiba, M.A. Sokolov, Embrittlement of irradiated F82H in the absence of irradiation hardening, J. Nucl. Mater. 386–388 (2009) 191.
[52] M. Li, S. Majumdar, K. Natesan, Modeling creep-fatigue behavior of mod. 9Cr-Mo steel, J. ASTM Int. 8, 2011, JAI103824.
[53] P.J. Karditsas, Design issues and implications for structural integrity of fusion power plant components, Fusion Eng. Des. 84 (2009) 2104.
[54] H.G. Armaki, R. Chen, K. Maruyama, M. Igarashi, Premature creep failure in strength enhanced high Cr ferritic steels caused by static recovery of tempered martensite lath structures, Mater. Sci. Eng. A 527 (2010) 6581.
[55] A. Sarkar, V.D. Vijayanand, P. Parameswaran, et al., Influence of prior fatigue cycling on creep behavior of reduced activation ferritic-martensitic steel, Metall. Mater. Trans. A Phys. Metall. Mater. Sci. 45A (2014) 3023.
[56] S. Smolentsev, S. Saedi, S. Malang, M. Abdou, Numerical study of corrosion of ferritic/martensitic steels in the flowing PbLi with and without a magnetic field, J. Nucl. Mater. 432 (2013) 294.
[57] I. Bucenieks, R. Krishbergs, E. Platacis, et al., Corrosion phenomena of Eurofer steel in Pb-17Li stationary flow at magnetic field, Magnetohydrodynamics 42 (2006) 237.
[58] S.J. Zinkle, L.L. Snead, Designing radiation resistance in materials for fusion energy, Annu. Rev. Mater. Res. 44 (2014) 241.
[59] S.J. Zinkle, Challenges in developing materials for fusion technology-past, present and future, Fusion Sci. Technol. 64 (2013) 65.

REFERENCES

[60] G.S. Was, Fundamentals of Radiation Materials Science: Metals and Alloys, Springer, New York. ISBN 978-3-540-49471-3, 2007.

[61] G.R. Odette, M.J. Alinger, B.D. Wirth, Recent developments in irradiation-resistant steels, Ann. Rev. Mater. Res. 38 (2008) 471.

[62] G.R. Odette, Recent progress in developing and qualifying nanostructured ferritic alloys for advanced fission and fusion applications, JOM 66 (2014) 2427.

[63] A. Hirata, T. Fujita, Y.R. Wen, et al., Atomic structure of nanoclusters in oxide-dispersion-strengthened steels, Nat. Mater. 10 (2011) 922.

[64] M.J. Alinger, G.R. Odette, D.T. Hoelzer, The development and stability of Y–Ti–O nanoclusters in mechanically alloyed Fe–Cr based ferritic alloys, J. Nucl. Mater. 329–333 (2004) 382.

[65] N. Cunningham, Y. Wu, D. Klingensmith, G.R. Odette, On the remarkable thermal stability of nanostructured ferritic alloys, Mater. Sci. Eng. A 613 (2014) 296.

[66] B. Wilshire, T.D. Lieu, Deformation and damage processes during creep of Incoloy MA957, Mater. Sci. Eng. A 386 (2004) 81.

[67] A. Alamo, V. Lambard, X. Averty, M.H. Mathon, Assessment of ODS-14%Cr ferritic alloy for high temperature applications, J. Nucl. Mater. 329–333 (2004) 333.

[68] A. Steckmeyer, V.H. Rodrigo, J.M. Gentzbittel, et al., Tensile anisotropy and creep properties of a Fe–14CrWTi ODS ferritic steel, J. Nucl. Mater. 426 (2012) 182.

[69] D. Mukhopadhyay, F. Froes, D.S. Gelles, Development of oxide dispersion strengthened ferritic steels for fusion, J. Nucl. Mater. 258 (1998) 1209.

[70] M. Rieth, M. Schirra, A. Falkenstein, et al., Eurofer97: Tensile, Charpy, Creep, and Structural Tests, FZKA-6911, Forschungszentrum Karlsruhe, 2003.

[71] D.T. Hoelzer, J.P. Shingledecker, R.L. Klueh, Creep Behavior of MA957 and 14YWT (SM10 heat), Fusion Materials Semiannual Progress Report for the Period Ending June 30, 2010, DOE-ER-0313/48, 2010, p. 50.

[72] A.L. Rouffié, P. Wident, L. Ziolek, et al., Influences of process parameters and microstructure on the fracture mechanisms of ODS steels, J. Nucl. Mater. 433 (2013) 108.

[73] Y. Wu, E.M. Haney, N.J. Cunningham, G.R. Odette, Transmission electron microscopy characterization of the nanofeatures in nanostructured ferritic alloy MA957, Acta Mater. 60 (2012) 3456.

[74] T.R. Allen, J. Gan, J.I. Cole, et al., Radiation response of a 9 chromium oxide dispersion strengthened steel to heavy ion irradiation, J. Nucl. Mater. 375 (2008) 26.

[75] J. Ribis, S. Lozano-Perez, Nano-cluster stability following neutron irradiation in MA957 oxide dispersion strengthened material, J. Nucl. Mater. 444 (2014) 314.

[76] N.A. Bailey, E. Stergar, M. Toloczko, P. Hosemann, Atom probe tomography analysis of high dose MA957 at selected irradiation temperatures, J. Nucl. Mater. 459 (2015) 225.

[77] E. Aydogan, N. Almirall, G.R. Odette, et al., Stability of nanosized oxides in ferrite under extremely high dose self ion irradiations, J. Nucl. Mater. 486 (2017) 86–95.

[78] R.J. Kurtz, K. Abe, V.M. Chernov, et al., Recent progress on development of vanadium alloys for fusion, J. Nucl. Mater. 329–333 (2004) 47.

[79] T. Muroga, J.M. Chen, V.M. Chernov, et al., Review of advances in development of vanadium alloys and MHD insulator coatings, J. Nucl. Mater. 367–370 (2007) 780.

[80] J.M. Chen, V.M. Chernov, R.J. Kurtz, T. Muroga, Overview of the vanadium alloy researches for fusion reactors, J. Nucl. Mater. 417 (2011) 289.

[81] A.F. Rowcliffe, S.J. Zinkle, D.T. Hoelzer, Effect of strain rate on the tensile properties of unirradiated and irradiated V-4Cr-4Ti, J. Nucl. Mater. 283–287 (2000) 508.

[82] P.F. Zheng, T. Nagasaka, T. Muroga, J.M. Chen, Y.F. Li, Creep properties of V-4Cr-4Ti strengthened by cold working and aging, Fusion Eng. Des. 86 (2011) 2561.

[83] P.F. Zheng, T. Nagasaka, T. Muroga, J.M. Chen, Investigation on mechanical alloying process for vanadium alloys, J. Nucl. Mater. 442 (2013) S330.

[84] T. Muroga, T. Nagasaka, H. Watanabe, M. Yamazaki, The effect of final heat treatment temperature on radiation response of V-4Cr-4Ti, J. Nucl. Mater. 417 (2011) 310.

[85] T. Muroga, B.A. Pint, Progress in the development of insulator coating for liquid Lithium blankets, Fusion Eng. Des. 85 (2010) 1301.

[86] Y. Katoh, L.L. Snead, I. Szlufarska, et al., Radiation effects in SiC for nuclear structural applications, Curr. Opin. Solid State Mater. Sci. 16 (2012) 143.
[87] L.L. Snead, T. Nozawa, M. Ferraris, et al., Silicon carbide composites as fusion power reactor structural materials, J. Nucl. Mater. 417 (2011) 330.
[88] Y. Katoh, T. Nozawa, L.L. Snead, et al., Stability of SiC and its composites at high neutron fluence, J. Nucl. Mater. 417 (2011) 400.
[89] T. Nozawa, T. Hinoki, A. Hasegawa, et al., Recent advances and issues in development of silicon carbide composites for fusion applications, J. Nucl. Mater. 386–388 (2009) 622.
[90] Y. Katoh, L.L. Snead, C.H. Henager Jr, et al., Current status and critical issues for development of SiC composites for fusion applications, J. Nucl. Mater. 367–370 (2007) 659.
[91] L.K. Mansur, M.L. Grossbeck, Mechanical property changes induced in structural alloys by neutron irradiations with different to displacement ratios, J. Nucl. Mater. 130 (1988) 155–157.
[92] R.J. Kurtz, G.R. Odette, P. Miao, et al., The transport and fate of helium in martensitic steels at fusion relevant He/dpa ratios and dpa rates, J. Nucl. Mater. 367–370 (2007) 417.
[93] T. Yamamoto, Y. Wu, G.R. Odette, et al., A dual ion irradiation study of helium-dpa interactions on cavity evolution in tempered martensitic steels and nanostructured ferritic alloys, J. Nucl. Mater. 449 (2014) 190.
[94] H. Trinkaus, B.N. Singh, Helium accumulation in metals during irradiation—where do we stand? J. Nucl. Mater. 323 (2003) 229.
[95] L. Yang, X.T. Zu, F. Gao, et al., Dynamic interactions of helium-vacancy clusters with edge dislocations in α-Fe, Physica B 405 (2010) 1754.
[96] R. Schaeublin, Y.L. Chiu, Effect of helium on irradiation-induced hardening of iron: a simulation point of view, J. Nucl. Mater. 362 (2007) 152.
[97] N.N. Kumar, P.V. Durgaprasad, B.K. Dutta, G.K. Dey, Modeling of radiation hardening in ferritic/martensitic steel using multi-scale approach, Comput. Mater. Sci. 53 (2012) 258.
[98] L. Yang, Z.Q. Zhu, S.M. Peng, et al., Effects of temperature on the interactions of helium–vacancy clusters with gliding edge dislocations in α-Fe, J. Nucl. Mater. 441 (2013) 6.
[99] S.M.H. Haghighat, R. Schaeublin, Influence of the stress field due to pressurized nanometric he bubbles on the mobility of an edge dislocation in iron, Philos. Mag. 90 (2010) 1075.
[100] R.E. Stoller, Y.N. Osetsky, An atomistic assessment of helium behavior in iron, J. Nucl. Mater. 455 (2014) 258.
[101] Y. Dai, J. Henry, Z. Tong, et al., Neutron/proton irradiation and he effects on the microstructure and mechanical properties of ferritic/martensitic steels T91 and EM10, J. Nucl. Mater. 415 (2011) 306.
[102] L.L. Snead, R. Scholz, A. Hasegawa, et al., Experimental simulation of the effect of transmuted helium on the mechanical properties of silicon carbide, J. Nucl. Mater. 307–311 (2002) 1141.
[103] R.E. Stoller, G.R. Odette, in: F.A. Garner, N.H. Packan, A.S. Kumar (Eds.), A Comparison of the Relative Importance of Helium and Vacancy Accumulation in Void Nucleation, American Society of Testing and Materials, Philadelphia, PA, 1987, p. 358. ASTM STP 955.
[104] G.R. Odette, D.T. Hoelzer, Irradiation-tolerant nanostructured ferritic alloys: transforming helium from a liability to an asset, JOM 62 (2010) 84.
[105] W.D. Nix, Mechanisms and controlling factors in creep fracture, Mater. Sci. Eng. A 103 (1988) 103.
[106] B.F. Dyson, Constraints on diffusional cavity growth rates, Met. Sci. 10 (1976) 349.
[107] W. Beere, Models for constrained cavity growth in polycrystals, Acta Metall. 28 (1980) 143.
[108] H. Trinkaus, On the modeling of the high-temperature embrittlement of metals containing helium, J. Nucl. Mater. 118 (1983) 39.
[109] H. Trinkaus, H. Ullmaier, High temperature embrittlement of metals due to helium: is the lifetime dominated by cavity growth or crack growth, J. Nucl. Mater. 303 (1994) 212–215.
[110] R.J. Kurtz, H.L. Heinisch, The effects of grain boundary structure on binding of He in Fe, J. Nucl. Mater. 329–333 (2004) 1199.
[111] N. Yamamoto, Y. Murase, J. Nagakawa, An evaluation of helium embrittlement resistance of reduced activation martensitic steels, Fusion Eng. Des. 81 (2006) 1085.
[112] J. Chen, P. Jung, T. Rebac, et al., Helium effects on creep properties of Fe–14CrWTi ODS steel at 650 °C, J. Nucl. Mater. 453 (2014) 253.
[113] J. Linke, Plasma facing materials and components for future fusion devices—development, characterization and performance under fusion specific loading conditions, Phys. Scr. T123 (2006) 45.

[114] M. Rieth, S.L. Dudarev, S.M. Gonzalez de Vicente, et al., Recent progress in research on tungsten materials for nuclear fusion applications in Europe, J. Nucl. Mater. 432 (2013) 482.
[115] M. Rieth, S.L. Dudarev, S.M. Gonzalez de Vicente, et al., A brief summary of the progress on the EFDA tungsten materials program, J. Nucl. Mater. 442 (2013) S173.
[116] P. Norajitra, L.V. Boccaccini, E. Diegele, et al., Development of a helium-cooled divertor concept: Design-related requirements on materials and fabrication technology, J. Nucl. Mater. 329–333 (2004) 1594.
[117] M. Rieth, B. Dafferner, Limitations of W and W–1%La_2O_3 for use as structural materials, J. Nucl. Mater. 342 (2005) 20.
[118] E. Lassner, W.D. Schubert, Tungsten—Properties, Chemistry, Technology of the Element, Alloys, and Chemical Compounds, Springer, New York, 1999.
[119] S. Wurster, B. Gludovatz, R. Pippan, High temperature fracture experiments on tungsten-rhenium alloys, Int. J. Refract. Met. Hard Mater. 28 (2010) 692.
[120] M. Faleschini, H. Kreuzer, D. Kiener, R. Pippan, Fracture toughness investigations of tungsten alloys and SPD tungsten alloys, J. Nucl. Mater. 367–370 (2007) 800.
[121] J. Riesch, T. Höschen, C. Linsmeier, et al., Enhanced toughness and stable crack propagation in a novel tungsten fibre-reinforced tungsten composite produced by chemical vapour infiltration, Phys. Scr. T159 (2014) 014031.
[122] G.R. Odette, R.L. Chao, J.W. Sheckherd, G.E. Lucas, Ductile phase toughening mechanisms in a TiAl-TiNb laminate composite, Acta Metall. Mater. 40 (1992) 2381.
[123] B.N. Cox, Extrinsic factors in the mechanics of bridged cracks, Acta Metall. Mater. 39 (1991) 1189.
[124] L.S. Sigl, P.A. Mataga, B.J. Dalgleish, et al., On the toughness of brittle materials reinforced with a ductile phase, Acta Metall. Mater. 36 (1988) 945.
[125] L.R. Greenwood, F.A. Garner, Transmutation of Mo, Re, W, Hf, and V in various irradiation test facilities and STARFIRE, J. Nucl. Mater. 212–215 (1994) 635.
[126] M.R. Gilbert, S.L. Dudarev, D. Nguyen-Manh, et al., Neutron-induced dpa, transmutations, gas production, and helium embrittlement of fusion materials, J. Nucl. Mater. 442 (2013) S755.
[127] A. Hasegawa, T. Tanno, S. Nogami, M. Satou, Property change mechanism in tungsten under neutron irradiation in various reactors, J. Nucl. Mater. 417 (2011) 491.
[128] A. Hasegawa, M. Fukuda, T. Tanno, S. Nogami, Neutron irradiation behavior of tungsten, Mater. Trans. 54 (2013) 466.
[129] M. Fukuda, A. Hasegawa, T. Tanno, et al., Property change of advanced tungsten alloys due to neutron irradiation, J. Nucl. Mater. 442 (2013) S273.
[130] J.-C. He, A. Hasegawa, M. Fujiwara, et al., Fabrication and characterization of W-Re-Os alloys for studying transmutation effects of W in fusion reactor, Mater. Trans. 45 (2004) 2657.
[131] M. Wirtz, J. Linke, G. Pintsuk, et al., Comparison of the thermal shock performance of different tungsten grades and the influence of microstructure on the damage behaviour, Phys. Scr. T145 (2011) 014058.
[132] J. Roth, E. Tsitrone, A. Loarte, et al., Recent analysis of key plasma wall interactions issues for ITER, J. Nucl. Mater. 1 (2009) 390–391.
[133] C. García-Rosalesa, P. López-Ruiz, S. Alvarez-Martín, et al., Oxidation behaviour of bulk W-Cr-Ti alloys prepared by mechanical alloying and HIPing, Fusion Eng. Des. 89 (2014) 1611.
[134] M.J. Baldwin, R.P. Doerner, Helium induced nanoscopic morphology on tungsten under fusion relevant plasma conditions, Nucl. Fusion 48 (2008) 035001.
[135] G.D. Tolstolutskaya, V.V. Ruzhytskiy, I.E. Kopanets, et al., Displacement and helium-induced enhancement of hydrogen and deuterium retention in ion-irradiated 18Cr10NiTi stainless steel, J. Nucl. Mater. 356 (2006) 136.
[136] D.M. Duffy, Modeling plasma facing materials for fusion power, Mater. Today 12 (2009) 38.
[137] J. Linke, High heat flux performance of plasma facing materials and components under service conditions in future fusion reactors, Trans. Fusion Sci. Technol. 61 (2012) 246.
[138] V. Barabash, G. Federici, M. Roedig, et al., Neutron irradiation effects on plasma facing materials, J. Nucl. Mater. 138 (2000) 283–287.
[139] L.L. Snead, Y. Katoh, K. Ozawa, Stability of 3-D carbon Fiber composite to high neutron fluence, J. Nucl. Mater. 417 (2011) 629.
[140] W.R. Meier, A.R. Raffray, S.I. Abdel-Khalik, IFE chamber technology: status and future challenges, Fusion Sci. Technol. 44 (2003) 27.
[141] R.W. Moir, Liquid wall inertial fusion energy power plants, Fusion Eng. Des. 32–33 (1996) 93.
[142] R. Bastasz, W. Eckstein, Plasma-surface interactions on liquids, J. Nucl. Mater. 290–293 (2001) 19.

[143] M. Abdou, The APEX Team, A. Ying, et al., On the exploration of innovative concepts for fusion chamber technology, Fusion Eng. Des. 54 (2001) 181.
[144] The ARIES Project: http://aries.ucsd.edu/ARIES/.
[145] M.S. Tillack, X.R. Wang, S. Malang, F. Najmabadi, ARIES-ACT-1 power Core engineering, Fusion Sci. Technol. 64 (2013) 427.
[146] L.A. El-Guebaly, T. Huhm, A.F. Rowcliffe, et al., Design challenges and activation concerns for ARIES vacuum vessel materials, Fusion Sci. Technol. 64 (2013) 449.
[147] R.L. Klueh, Reduced-activation bainitic and martensitic steels for nuclear fusion applications, Curr. Opin. Solid State Mater. Sci. 8 (2004) 239.
[148] P. Libeyrea, D. Bessettea, A. Devred, et al., Conductor jacket development to meet the mechanical requirements of the ITER central solenoid coils, Fusion Eng. Des. 86 (2011) 1553.
[149] K. Hamada, H. Nakajima, K. Kawano, et al., Optimization of JK2LB chemical composition for ITER central solenoid conduit material, Cryogenics 47 (2007) 174.
[150] H. Ozeki, K. Hamada, Y. Takahashi, et al., Establishment of production process of JK2LB jacket section for ITER CS, IEEE Trans. Appl. Supercond. 24 (2014) 4800604.

CHAPTER 4

RESEARCH TOOLS: MICROSTRUCTURE, MECHANICAL PROPERTIES, AND COMPUTATIONAL THERMODYNAMICS

Colin A. English*, Jonathan M. Hyde*, G. Robert Odette[†], Gene E. Lucas[†], Lizhen Tan[‡]

Materials and Reactor Chemistry, National Nuclear Laboratory, Culham Science Centre, Abingdon, Oxon, United Kingdom
Materials Department, University of California, Santa Barbara, CA, United States[†] Materials Science and Technology Division, Oak Ridge National Laboratory, Oak Ridge, TN, United States[‡]

CHAPTER OUTLINE

4.1 Introduction 104
    4.1.1 Background 104
    4.1.2 Characterization of Irradiated Materials 106
4.2 Microstructural Tools 107
    4.2.1 Irradiation-Induced Microstructure 107
    4.2.2 Microstructural Tools 109
    4.2.3 Electron Microscopy 109
    4.2.4 Atom Probe Tomography 117
    4.2.5 Small-Angle Neutron Scattering (SANS) 125
    4.2.6 Positron Annihilation Spectroscopy-Based Techniques 130
    4.2.7 Summary of Microstructural Techniques 134
4.3 Subsized Specimen Tests to Measure Irradiated Mechanical Properties 134
    4.3.1 Introduction 134
    4.3.2 Subsized Tensile Tests 135
    4.3.3 Microhardness Tests 137
    4.3.4 Irradiation Embrittlement Tests: Transition Temperature Shifts and Fracture Toughness 138
    4.3.5 Irradiation Hardening-Embrittlement Relations 141
    4.3.6 Data From Mechanical Tests at the Nanoscale 142
    4.3.7 Summary 143
4.4 Computational Alloy Design and Optimization 144
    4.4.1 Introduction 144
    4.4.2 Alloy Optimization 145
    4.4.3 Alloy Selection and Design 147

4.4.4 Kinetics and Mechanical Property Simulations 147
4.4.5 Summary 149
References 150

4.1 INTRODUCTION
4.1.1 BACKGROUND

The purpose of this chapter is to provide an overview of the research tools that are available to evaluate the response of materials that are subject to atomic displacements caused by fast particle irradiation and transmutation products, especially He. The major focus is on metals employed in structural components in operating or planned nuclear fission and fusion reactors. Prior to the discussion of research tools that are of interest, we briefly outline:

- The irradiation parameter space that we are interested in for fission and fusion systems.
- The material systems of interest and the major degradation mechanisms that must be understood and predictively modeled.

The relevant irradiation parameter space for operating fission reactors is shown in Fig. 4.1 in terms of the irradiation temperature and dose (characterized by displacements per atom). The importance of the major degradation mechanisms is also illustrated in the figure. The parameter space for the planned Generation IV [1] and fusion reactors (see, e.g., Ref. [2]) is given in Table 4.1 [3]. It can be seen that this further extends the irradiation temperature and dose range of interest. Transmutation products are also an issue and this is especially the case for He and H in fusion environments where gas production occurs and a rate of order 10 and 100 appm/dpa, respectively. These degradation mechanisms and their consequences are addressed in much more detail in the other chapters of this book.

Exposure of structural materials to displacement damage doses and transmutation products gives rise to complex changes in their macroscopic dimensional stability, mechanical properties, and corrosion resistance. These changes result from modification the materials microstructures from atomic to mesoscopic length scales that depend on the irradiation exposure and that are governed by the interactions of a large number of mechanisms and variables acting in combination over long periods of time. A schematic illustration of the key processes is presented in Fig. 4.2.

It is to be emphasized that the choice of research tools, and indeed the properties that can be studied, is strongly influenced by the source of the irradiated materials, that is, whether it be fast neutron or ion beam irradiation. Samples irradiated in operating reactors or material test reactors (MTRs) may have the advantage of larger volumes that generally enable a large range of properties to be studied, although a very expensive infrastructure is required to irradiate and examine radioactive samples. Indeed, the literature on the development of materials for reactor applications has been dominated by the long standing history of using fast neutrons to study the effects of radiation displacement damage on material properties (see, e.g., Refs. [4, 5] and papers published in early conferences on void swelling, e.g., Ref. [6]). Over time, a wide variety of reactor environments became available for materials irradiations including high power MTRs, in operating Generation II and III power reactors and in prototype Fast reactors. Of particular note is the marked decline, since the 1990s, in MTR irradiation facilities, especially those capable of reaching 100 dpa or more in practical time scales and, on a similar timescale, a severe decline in nationally funded nuclear R&D programs (see, e.g., Ref. [7]).

There is a long-standing history of using ion beams and electrons to study radiation damage. Research workers at the Atomic Energy Research Department of North American Aviation at Downey California were probably the first to recognize the advantages of charged particle irradiation (CPI) for the studies of radiation damage in materials (quoted in Ref. [8]). Initially these studies focused on fundamental studies of point defect properties (see Ref. [9] for details of the use of ion beams for fundamental work carried out during the 1950s and 1960s).

4.1 INTRODUCTION

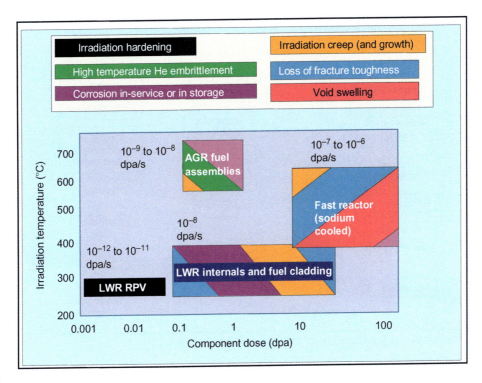

FIG. 4.1

Schematic representation of the damage level vs irradiation temperature that components experience in different operating reactor systems. In addition, the major degradation mechanisms are also indicated.

Table 4.1 Indicative Levels of the Maximum Temperature, Dose Range and Level of Transmutation for Fission, and Fusion Reactors [3]

	Fission (Gen I, i.e., Early Magnox or LWR)	Fission (Gen III LWR)	Fission (Gen IV, i.e., Fast reactor)	Fusion (DEMO)
Structural material T_{max}	<300–350°C	<300–400°C	300–1000°C	550–1000°C
Max dose for core internal structures	~0.01–1 dpa	80–120 dpa	~30–200 dpa	~200 dpa
Max helium concentration	0.1–1 appm	~10–30 appm	~3–40 appm	~1500 appm (~10,000 appm for SiC)
Max hydrogen concentration				~6750 appm
Neutron energy E^{max}	<1–2 MeV High fluxes of thermal neutrons	<1–3 MeV High fluxes of thermal neutrons	<1–3 MeV In core very low fluxes of thermal neutrons	< 14 MeV

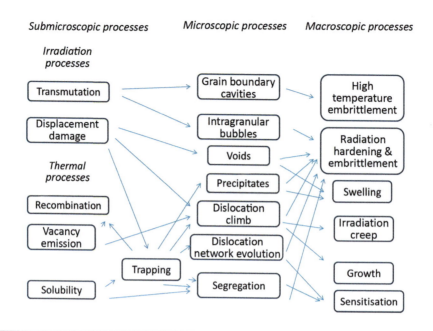

FIG. 4.2

Schematic representation of the interaction of radiation damage processes.

CPI was also used for early irradiation damage studies on the behavior of inert gases in metals [10, 11]. In 1966, the discovery [12] of the phenomenon of void swelling in fast reactor stainless steel fuel cladding and core materials prompted further development of the use of CPI to study such effects [6, 13, 14]. From the 1970s there was the need for the innovative use of ion beams to develop alloys that can withstand high levels of displacement damage and transmutation gas (He and H) anticipated in a fusion reactor [15, 16]. This need was underlined by the fact that there are no fusion power or test reactors in which materials properties can be evaluated in samples irradiated under fusion conditions (very high dose of 14 MeV neutrons).

No single irradiation source covers the range of irradiation temperatures (<300°C–1000°C), levels of transmutation gas production (maximum of 2000 appm He or 6750 appm H) or dose range (from <1 to 200 dpa) required. Even in combination, all these irradiation sources have limitations. For example, a source for high-energy 14 MeV neutrons for fusion relevant irradiation studies has long been sought but is not yet available [17]. Further limitations arise from the permitted volume of irradiated material and uncertainties in critical variables like irradiation temperature even in MTR irradiations.

4.1.2 CHARACTERIZATION OF IRRADIATED MATERIALS

Understanding, predicting, and characterizing the changes in material properties under irradiation requires the ability to characterize the irradiation-induced microstructure at and above the atomic scale, and the ability to relate such changes to the observed degradation in performance sustaining properties. Thus, any discussion of research tools necessarily involves:

- microstructural characterization techniques
- mechanical property characterization techniques

In addition, computational thermodynamics (and related multiscale, multiphysics modeling) can also be considered characterization tools since they enable the study of the thermodynamically and kinetically mediated mechanisms (influenced by the irradiation environment) that affect microstructural evolution. Hence, this chapter also contains a brief review of computational thermodynamics.

Developing relevant databases and modeling frameworks for the accurate prediction of end of life properties of reactor components at the design stage, or even during plant operation is extremely challenging. The approach often adopted includes the following elements:

- Use of available data to construct physically based models to extrapolate property predictions to yet-to-be-experienced service conditions.
- Collection of in-service irradiation effects data in well-designed surveillance programs.
- Filling data gaps and providing mechanism, microstructural and related materials' property data from accelerated MTR and CPI to help inform, improve and calibrate predictive physical models.

Note, it should be emphasized that highly accelerated MTR and CPI do not directly simulate longer time service conditions. However, they have the advantage of being well controlled, at least in principle, and highly flexible in their coverage of variable ranges and combinations beyond those available in surveillance databases.

There is a clear need to comprehensively characterize materials following irradiation. The available toolkit of atomic, and even subatomic, tools to characterize the nano-microstructure of materials, has expanded tremendously, and continues to grow at a tremendous pace. Further, irradiations intrinsically involve severe limits on specimen numbers and sizes. Thus the recent progress in small, to now micro-nano scale testing are extremely important.

In the following sections these research tools are reviewed, highlighting their individual strengths and weaknesses, where appropriate selected examples of application are presented. While it is not possible to provide an in-depth analysis of each technique, the summaries presented include references to critical reviews and examples of major applications.

4.2 MICROSTRUCTURAL TOOLS
4.2.1 IRRADIATION-INDUCED MICROSTRUCTURE

Microstructural development in a huge range of neutron, electron, and ion-irradiated metals and alloys has been the subject of study over a wide range of dose, dose rate, and irradiation temperature. It is essential to have microstructural tools that are capable of comprehensively characterizing the large variety of irradiation-induced features that are created during irradiation. At the most basic level, ion and neutron irradiation produce (mobile) point defects and small defect clusters. These defects are created following the initial interaction of a neutron (or ion) with a lattice atom. If sufficiently energetic, the initial event may result in a displacement cascade, where a high density of displacements is created in a region several nm in diameter. Both freely migrating point defects and small clusters of vacancies and interstitials are typically created in the cascade event. At the temperature range of most operating reactors, both vacancy and interstitial point defects, and very small clusters, are mobile. Migrating defects escaping from the initial cascade, or creation site, interact with point defect sinks, such as preexisting dislocations, recombine with each other, either directly or at solute traps, or form vacancy or interstitial clusters. In all cases, the defects lose their individual identity. Complex interactions with solutes occur during the migration of the defects and irradiation resulting in enhanced diffusion or chemical segregation. Interstitial solutes are mobile and rapidly reach sink or cluster sites where they lose their individual identity. Vacancy-interstitial recombination eliminates both defects (self-healing). The sequence of events governing these processes are illustrated the flow chart in Fig. 4.3.

The irradiation environment of a component depends not only on its location in the reactor (i.e., outside or inside the core) but also on the reactor operating regime. For example, the pressure vessel of a light water reactor (LWR) experiences a relatively low dose rate (and thus lifetime dose) at a relatively low temperature $\sim 0.3 T_m$, and

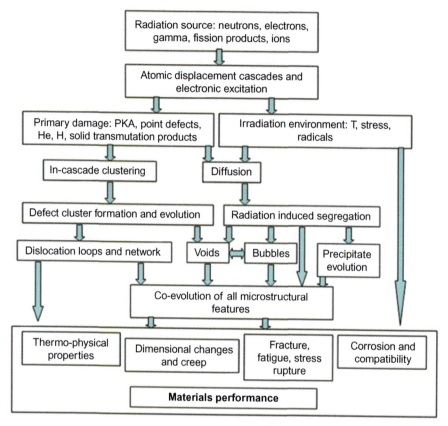

FIG. 4.3

Schematic representation of the evolution of the irradiation produced microstructure [18].

the microstructure is dominated by a high density of small cluster-solute complexes and precipitates [19]. In contrast, reactor internals in fusion and fission reactors experience much larger doses and high to very high temperatures. The microstructure developed under each of these conditions is very different consisting of dislocation loops, lines, bubbles, and voids, as well as precipitates of various types. It is also noted in Fig. 4.3 that segregation of solutes to sinks, such as grain boundaries (GBs), may also occur. The processes described in Fig. 4.3 illustrate the need for microstructural characterization at, or near, the atomic scale (and upwards), and the importance of being able to identify not only the fate of point defects but also the redistribution of solute elements in the material.

The development of understanding into specific degradation mechanisms frequently requires acquiring quantitative data on specific microstructural features, such as voids, dislocation loops, or small precipitates, or solute concentration profiles adjacent to microstructural features such as GBs. There needs to be sufficient understanding of the response of a given microstructural techniques to the feature of interest that the uncertainties associated with a given measurement can be assessed.

Furthermore, it is important to ensure that microstructural data are obtained from relevant regions of the samples being examined. For instance, failure paths are also associated with crack propagation in irradiation-assisted stress corrosion cracking (IASCC) of austenitic steels or delayed hydride cracking in Zr alloys. These require the characterization of the microstructure, including oxide or hydride phases around crack tips. Producing samples from specific microstructural features or specific regions of a sample, for example, following ion irradiation, have historically been highly challenging. However, the advent of focused ion beam (FIB) techniques [20] now enables selection of regions of materials for microstructural analysis with a resolution of ∼10 nm. FIB systems can now be used for the preparation of samples extracted from specific regions of the microstructure, for example, adjacent to the crack tip. Such samples are best suited to techniques such as transmission electron microscopy (TEM) or atom probe tomography (APT) where small volumes of material can be examined at a high spatial resolution. For example, it is possible to use FIB milling to prepare TEM samples in metals and alloys containing crack tips [21–23].

It should also be emphasized that the study of irradiation-induced defects also necessitates the study of the start of life microstructure. There are two main reasons. Firstly, without a baseline, it is not possible to relate changes in the microstructure to changes in properties under irradiation and, secondly, the very fine-scale irradiation-induced features are very sensitive to the start of life condition. For instance local variations in the concentrations of minor solute elements can dramatically impact the development of irradiation-induced solute clusters and precipitates. It is therefore imperative to characterize the microstructure and understand variability.

4.2.2 MICROSTRUCTURAL TOOLS

The available microstructural techniques can be categorized as direct or indirect. With direct techniques an image of the sample is obtained with resolvable microstructural features (e.g., TEM or APT). The interpretation of such images is relatively simple although quantitative interpretation may require an understanding of the contrast theory or the effect of aberrations. In indirect techniques a volume-averaged response is obtained and detailed interpretation may require detailed modeling.

A summary of microstructural tools used to characterize the effects of irradiation is presented in Table 4.2 with additional details provided. For each technique, a brief overview of the principles is presented, supported by references to major state-of-the-art reviews (where appropriate), selected examples demonstrating application for the study of radiation damage and a brief summary. It is important to note that in certain areas, for example, reactor pressure vessel (RPV) embrittlement, it was found necessary to develop and/or apply new techniques to enable the required microstructural characterization to take place.

4.2.3 ELECTRON MICROSCOPY

4.2.3.1 Principles

TEM and scanning transmission electron microscopy (STEM) are exceptionally powerful techniques for providing direct characterization of a number of critical features of the irradiation-induced microstructure. There are numerous books dedicated to the field of TEM and STEM, see for example, Refs. [42, 43].

In the transmission electron microscope a beam of high-energy electrons, typically 100–1000 keV, are focused on a thin sample of the material under investigation. Electrons transmitted through the sample are detected. These electrons can be used to form an image of the sample showing internal microstructural features such as GBs, second phases, precipitates, and on a finer scale dislocations and dislocation loops. The elastic and inelastic scattering theory required to interpret the images and their associated diffraction patterns was established in the 1960s [44, 45]. Some important developments have occurred since the early seminal work, for example, the development of the weak-beam diffraction technique [46], which led to improvements in the characterization of small point defect clusters (>∼2 nm in diameter [47]). A most complete reference to the imaging techniques is given in the book by Jenkins and Kirk which focuses on the characterization of radiation damage by TEM [47].

A STEM consists of a modified electron microscope, which permits a fine electron probe to be scanned across the specimen. Dedicated STEMs are equipped with high brightness field emission sources (field emission gun–FEG), which allow probes down to 1.5 nm or less to be formed. The fine probe can be stopped in selected positions and the X-ray signal generated by the interaction of the electron beam and the specimen is collected and analyzed. Dedicated FEG-STEM instruments thus offer the benefit of high spatial resolution chemical analysis by energy-dispersive X-ray (EDX) spectroscopy or electron energy loss spectroscopy (EELS) from thin foils. This enables not only the detection of the chemical composition on a very fine scale, but also the mapping of the compositions of features of interest. EDX is sensitive to a wider range of elements than EELS and reliable quantification is possible. However, EDX

Table 4.2 Summary of Microstructural Techniques Used to Characterize Irradiation Damage [24, 25]

Application	Technique	Feature
Basic characterization of the starting microstructure	Optical microscopy	Gross features—grain size shape
	SEM (Scanning Electron Microscopy)	Hydrides, large second phase particles or inclusions gross features of deformation structure
	EPMA (Electron Probe Micro-Analysis)	Gross features of the chemical compositions
	(TEM/STEM)	See below
Direct Observation of radiation-induced microstructure	TEM (Transmission Electron Microscopy)	Small and large dislocation loops, dislocation network, bubbles and voids—microstructure associated with features such as crack tips[a]
	STEM (Scanning TEM)	Fine scale precipitation, minor and major elements to internal sinks such as grain boundaries, characterization of second phase particles[a]
	FIM (Field Ion Microscopy)	Point defects and point defect clusters
	APT (Atom Probe Tomography)	Composition and sizes of nanometer sized precipitates, segregation to grain boundaries[a]
	AES (Auger Electron Spectroscopy)	Ideally suited to characterizing segregation of elements such as P or C to grain boundaries [25–30]
Distribution of microstructural features within bulk sample (indirect)	SANS (Small-Angle Neutron Scattering)	Characterization of irradiation-induced precipitates, bubbles, and voids[a]
	PAS (Positron Annihilation Spectroscopy)	Local variations in electron density, vacancy point defects, and clusters[a]
Minor techniques	Nuclear probes (Mossbauer, PAC/PAD, Muon Spin Resonance)	Determining the local environment around specific target atoms [31, 32]
	Internal Friction	Interaction of radiation damage and interstitial solutes with dislocations [33, 34]
	Synchrotron extended X-ray absorption fine structure (EXAFS)	Detection of defect interactions with each individual alloying element in irradiated steels at the atomic level [35–37]
	SAXS (Small-Angle X-ray Scattering) and XRD (X-ray diffraction) at synchrotrons	Detection of nanoscale voids, bubbles particles, and vacancy defect clusters [38, 39]
	TEP (Thermo-electric power). Resistivity	Monitoring the effect of irradiation on solute and vacancy content [40]
	SIMS (Secondary Ion Mass Spectroscopy) and Nano-SIMS	Quantification of trace elements, often better than ppm. NanoSIMS provides high spatial resolution (down to 50 nm) [41]

[a]For more details, see Section 4.2.3.

is relatively slow and sensitivity to low Z elements is poor. EELS is faster to perform than EDX mapping, and the higher energy resolution can provide information on coordination and bonding [48]. However, with EELS multicomponent quantification is difficult; the technique is not suitable for all elements and is sensitive to sample thickness (samples should be <100 nm in thickness [48]). Lozano-Perez [25] has also pointed out that advances in computing power and the design of detectors has enabled the acquisition of images where an EDX or EELS spectrum can be obtained serially for each pixel of a scanned image (spectrum imaging). Multivariate analyses can then be used to extract chemically relevant components which can provide information on both composition and bonding [49].

The advent of aberration corrected electron microscopes such as the JEOL JEM-2200FS and the FEI Titan series improves resolution. In STEM mode the reduced probe size allows EELS and EDX spectra to be taken from a very small area of the specimen [50] near, and in some cases at, atomic resolution. Further, Nellist and Pennycook [51], in a review carried out in 2000 of the principles of annular dark-field imaging, point out that the combination of atomic-resolution Z-contrast STEM and EELS is a powerful means of characterizing the atomic and electronic structure of materials, nanoscale systems and interfaces at near atomic resolution. Further, Sigle stresses [52] that with the advent of new instrumentation (spherical aberration correctors, electron monochromators, energy filters, CCD detectors) it has now become possible to obtain chemical information with a spatial resolution well below 1 nm and an energy resolution better than 0.2 eV in analytical TEM (see, e.g., Ref. [53]).

Several authors have argued that recent advances in the capability of modern electron microscopes enable such techniques to be applied to the characterization of irradiation-induced microstructures with greater insight that ever before. Parish et al. [54] have argued that traditional methods of electron microscopy are well-suited to the materials. However, the development of radiation-resistant nanostructured materials requires a characterization by modern nanocharacterization methods. Parish et al. also stress that the coordinated use of STEM X-ray mapping and transmission Kikuchi diffraction will drive future advances in grain boundary engineering, radiation-induced segregation (RIS), and related phenomena.

Multivariate statistical methods (MVSA) are increasingly being used to extract more accurate compositional information [55]. In recent years there has been a development of data evaluation and interpretation (see Refs. [52, 54]). More specifically, Parish et al. highlight that STEM in conjunction with MVSA-based data mining will allow far more detailed and comprehensive interrogation of the redistribution of solutes in irradiated materials which gives rise to phenomena (such as radiation-induced precipitation, RIS, and radiation-induced phase instability). Development of TEM techniques continues apace, particularly in areas such as correlative microscopy. For instance, the combined use of electron tomography and APT enables information from one technique to overcome limitations of the other (and vice versa) [56].

4.2.3.2 Application
Introduction
TEM can detect the strain field of small precipitates, defect clusters, and dislocation loops and lines. TEM images can be used to estimate the number density, size, and structure of such small clusters with diameters above the visibility limit of ~1–2 nm. In addition TEM is well suited to characterizing coarser features such as dislocations (and hence dislocation density), dislocation loops, voids, larger precipitates, and grain size. In recent years, there has been an explosion of information from TEM as FIB techniques have been employed to prepare thin foils that could not easily be produced by standard electropolishing methods [25, 57–59]. Lim and Burke [60] have pointed out that there are FIB-induced damages such as amorphous and irradiation-induced dislocation loops that are very similar to those induced by neutron irradiation in low alloy steels. However, the use of low-energy FIB for the final stage of foil fabrication or cleaning with conventional broad argon ion milling after FIB fabrication is considered a means of removing such damage (see, e.g., Ref. [61]).

TEM has been used in the study of radiation damage from the early investigations of the irradiation-induced microstructure (see, e.g., Refs. [62–64]). These studies followed the successful observation of vacancy clusters in quenched metals [65]. TEM and STEM have made major contributions to understanding of radiation damage mechanisms and effects in Generation II–IV reactors (and fusion) [66]. A major advantage of using the TEM is that the sample volume is such that even highly radioactive samples may be examined, provided appropriate sample preparation facilities are available. The combination of TEM and STEM is routinely employed to examine the internal microstructure [cavities, dislocation lines, dislocation loops, solute redistribution (segregation and precipitation)] in irradiated materials. Most importantly it has frequently been possible to characterize the development of a particular feature, e.g., dislocation loops or cavities, as a function of irradiation dose and temperature. In addition, the advent of FIB to prepare samples from specific regions in a sample has allowed the examination of interface regions such as metal/oxide interfaces.

Fast reactors and ODS steels

In the 1970s and 1980s the major use of TEM in radiation damage studies was to characterize the microstructure at high doses as part of the alloy development programs for the liquid-metal fast reactors. A particularly powerful example of the use of TEM is the initial observation of void swelling and understanding of the phenomenon of void swelling [67], see Fig. 4.4.

TEM images utilizing phase contrast were subsequently used extensively to measure the level of void swelling in samples irradiated to high doses in either ion accelerators or prototype fast reactors (see, e.g., Refs. [66, 68–74]). Such quantitative measurements require an accurate assessment of the TEM foil thickness where the observations were made. There are two established techniques: energy loss spectroscopy [75, 76] and convergent beam electron diffraction [77].

The power of the combination of TEM and STEM can be seen from an early study by Williams et al. [78], who studied void swelling, solute segregation, and microstructural instability in neutron irradiated 12Cr-15Ni-Si austenitic alloys. The Si was in the range of 0.14%–1.42% and the alloys were irradiated to approximately 20 dpa at temperatures 400–645°C. The authors observed through TEM that Si additions reduced void swelling, but that more significantly after irradiation only the low Si alloy remained fully austenitic. Transformation to ferrite occurred in the other alloys because of extensive irradiation-induced redistribution of the constituents of the austenitic alloys, by segregation and precipitation. Further, even in intragranular regions away from GBs, regions were observed with enhanced chromium and iron levels, but where nickel was depleted, especially at voids. The overall composition change resulted in local transformation of austenite to ferrite.

FIG. 4.4

Electron micrograph showing the first observations of voids in stainless steel following irradiation in the Dounreay Fast Reactor to $\sim 4.7 \times 10^{22}\,\mathrm{n\,cm^{-2}}$ (fission) at 510°C [67].

The emphasis in characterizing steels irradiated to high doses (Generation IV) has, more recently, focused on irradiated and unirradiated oxide dispersion strengthened (ODS) steels that are prime candidates for void swelling-resistant applications in components subject to high temperatures and high fluences in proposed fast and fusion reactors [79, 80]. In addition to low swelling, ODS steels have good thermomechanical properties even after irradiation [81]. A wide variety of TEM methods have been applied to characterizing the nanoclusters that are actually primarily $Y_2Ti_2O_7$ pyrochlore complex oxides, as shown by Wu [82]. De Castro [83] et al. have employed scanning-transmission electron microscope three-dimensional (3D) tomography to characterize the Y- and Cr-rich secondary phases present in a reduced activation Fe–12Cr ODS. More conventional TEM imaging techniques have been used to characterize the dislocation and cavity distributions in both neutron-irradiated (see, e.g., Refs. [84, 85]) and ion-irradiated ODS alloys (see, e.g., Ref. [86, 87]). Wells used correlative TEM and APT tomography to demonstrate the one to one association of nanoclusters and bubbles in He implanted Fe-14Cr ODS MA957 [88].

Generation II–III

There has also been a significant interest in the microstructures developed in austenitic steels irradiated as constituent parts of core internals in LWRs, examined as part of R&D programs. Most studies have investigated irradiation conditions close to those experienced by core internal components in LWRs. The microstructural studies have also been driven by the need to investigate IASCC and void swelling at high displacement doses (and somewhat higher component temperatures than 300°C). Solute levels at GBs, particularly Cr and Si, are considered to be important factors in determining propensity for IASCC. This has led to a significant number of studies of grain boundary composition in steels irradiated at $\sim$300°C and doses $<\sim$80 dpa. Observations have been made on material extracted from core internal components (e.g., bolts, thimble tubes) as well as from accelerated irradiations in high flux MTRs or (Russian) fast reactors [89–94]. Significant irradiation-induced hardening occurs in annealed stainless steels and this is believed to lead to enhanced IASCC, perhaps in part due to the accompanying propensity to flow localization and channel formation. This, in turn, has led to characterization of the underlying dislocation loop (and precipitate) structures (see, e.g., Refs. [95, 96]). An appreciation of the effects of gamma heating in raising component temperatures to close to 400°C raised further concerns that appreciable void swelling may occur at high doses; an additional factor was that in the components of interest there was significant He and H produced under irradiation. These concerns led to the characterization of bubble and void distributions. For example, voids were observed in Tihange baffle-former bolt made with cold-worked 316 SS after irradiation at 345°C to 12 dpa [97–99] (see Fig. 4.5).

Fukuya and coworkers [96] observed fine cavities in thimble tubes irradiated up to 73 dpa. These bubbles had a diameter of 1 nm, at doses higher than 1 dpa. The He/dpa ratio was 8–22 (appm/dpa). These cavities are helium bubbles given their morphology, size, and density [100–102].

In contrast TEM has been less successful in characterizing the irradiation damage in RPV steels. Here, the dominant damage consists of small point defect and solute clusters or precipitates ($\sim$2–3 nm in diameter). FIB techniques have largely resolved issues such as ferromagnetic specimens and surface contamination when properly applied. In addition, in steels the complexity and inhomogeneity of the start of life microstructures renders observation of small features difficult. TEM observations of dislocation loops in RPV steels (rather than high-purity Fe or model alloys) are very limited [103], and the lack of evidence for micro-void formation in irradiated steels (as opposed to simple model alloys) is also significant. However, when imaged under strong two beam conditions ($Sg=0$ where Sg is the deviation from the Bragg condition), Cu-containing precipitates give rise to characteristic black-white (B-W) contrast [104]. Further, FEGSTEM was used to measure directly the copper concentration in the matrix. Matrix Cu measurements may be used as a measure of the amount of Cu available for subsequent precipitation and hence embrittlement (through measurements at the start of life), or as a basis for the prediction of small solute cluster development at high fluence [105].

The advent of aberration corrected microscopes has enabled the FEGSTEM to be used to characterize the small solute clusters formed in irradiated RPV steels. For example Lim and Burke [60] examined three neutron irradiated model alloys to $\sim$1 dpa using a FEI aberration-corrected Titan G2 80-200 S/TEM X-FEG with Super X EDX and a

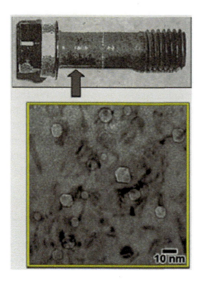

FIG. 4.5

Polyhedral voids observed in Tihange baffle-former bolt made with cold-worked 316 SS after irradiation at 345°C to 12 dpa [97].

From F.A. Garner. Volume 4 Chapter 4.02, "Radiation Damage in Austenitic Steels" in Comprehensive Nuclear Materials, Editor-in-Chief: Rudy J.M. Konings. ISBN: 978-0-08-056033-5 Volume 4 Chapter 4.02 Pages 33–95, 200 Originally from Edwards, D. J.; Simonen, E. P.; Garner, F. A.; Greenwood, L. R.; Oliver, B. A.; Bruemmer, S. M. J. Nucl. Mater. 2003, 317, 32–45.

Gatan Quantum 965 EELS system. STEM-EELS analyses on these model alloys show that the nanometer-scaled solute features that nucleated and formed during the neutron irradiation are strongly dependent on the alloying elements; in this case, Ni, Mn, and Cu. It is important to note that without the presence of Ni in the alloy, Mn, and Si were segregated to well-resolved dislocation loops, and no other nanometer-scaled solute features were observed. For the alloys with Ni, Ni-Mn-Si-enriched solute clusters were detected.

Zheng and Kaoumi [106] studied the cavities formed in a ferritic/martensitic HT9 steel irradiated with 5 MeV Fe^{++} ions to 16.6 displacements per atom (dpa) at 432°C. The sample was co-implanted with 3.22 appm He (at the peak damage level). Characterization of the irradiated samples in the FEI Titan 80–300 probe aberration corrected microscope revealed the segregation of Ni at cavity surfaces (cavity radius ~2.5 nm). RIS was observed to occur throughout the irradiated region and small Ni/Si/Mn-rich clusters precipitated on dislocation lines.

The irradiation-induced dissolution of SPPs, Fe-Cr, and Fe-Ni type, has been studied by Harte et al [107] in both proton and neutron-irradiated Zircaloy-2. The authors considered that the development of TEM in its high spatial resolution through aberration-corrected probes and the development of EDX detectors of large total solid angles have been essential in obtaining such measurements.

There is the need to develop corrosion-resistant components in LWRs which are irradiated with fast neutrons and high-energy gammas and are exposed to high-temperature water. Specifically, core components need to be resistant to IASCC, and zirconium alloys used for fuel cladding need to be oxidation resistant. These requirements, coupled to the ability to fabricate location and feature specific TEM samples using the FIB, has led to widespread interest in advanced TEM characterization of the thin oxide films formed during high-temperature exposure in an autoclave (or in samples irradiated in a power or materials test reactor).

Several authors report detailed TEM characterizations of zirconium oxide films formed during aqueous corrosion of a nuclear-grade zirconium alloys [108–113]. The grain size and shape, oxide phases, texture, cracks, and incorporation of precipitates on the oxide and the phases apparent in the interfaces have also been studied. Most importantly researchers frequently combine different TEM techniques. For example, Garner et al. used automated crystal orientation mapping in the TEM and transmission-electron backscatter diffraction (t-EBSD) to characterize the phase and micro-texture of the films [114]. They reported that the former enabled tetragonal ZrO_2 grains as small as 5 nm to be identified, and unambiguous indexing with t-EBSD enabled verification of the TEM observations. Hua and co-workers [115] report a methodology combining TEM, STEM, t-EBSD and EELS to analyze the structural and chemical properties of the metal-oxide interface of corroded Zr alloys. Different combinations of techniques allowed different features of the microstructure to be characterized. For example, TEM, STEM, and diffraction results were used to characterize the suboxide grains at the metal interface. t-EBSD was used to analyze the structure of the suboxide grains, and for mapping of a relatively large region of the metal-oxide interface, revealing the location and size distribution of the suboxide grains. EELS provided a means of characterizing the oxygen concentration across the interface. Reliable quantification of the oxygen content at the metal-oxide interface has been obtained by EELS analyses [116, 117] after careful calibration of inelastic partial scattering cross-sections against a ZrO_2 powder standard and estimation of the effective inelastic mean free path of high-energy electrons in oxidized Zr alloys. The authors estimate that an accuracy of ± 5 at.% can be achieved for the oxygen concentration with a spatial resolution better than 2 nm. This enables local oxygen stoichiometry variations at the metal-oxide interface to be analyzed.

Stress corrosion cracking (SCC) is an important but complex phenomena which potentially affects the integrity of internal components in LWRs. Susceptibility to SCC is influenced by irradiation by high-energy neutrons and X-rays, and specific mechanisms controlling SCC can vary with changes in alloy characteristics, applied/residual stress or environmental conditions [118]. Analytical TEM has been used extensively to characterize the structure and chemical nature of the corrosion features formed near crack tips and along attacked GBs in metals undergoing SCC. Indeed, Bruemmer and Thomas [119] stress that such characterization is essential if mechanistic understanding is to be obtained. The local crack electrochemistry, crack-tip mechanics and material metallurgy are the main factors controlling crack growth. Bruemmer and coworkers [120–123] employed both conventional BF and precipitate DF imaging, and, fine-probe (0.7-nm diameter) compositional analysis by EDS and PEELS, and fine-probe parallel-beam diffraction with electron probes as small as 5 nm in diameter. In addition, stereoscopic TEM photographs were used to observe finely porous structures along with attacked GBs (see Fig. 4.6).

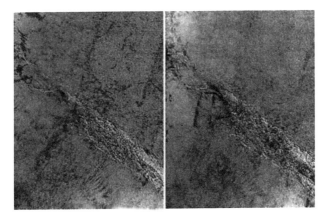

FIG. 4.6

Stereoscopic image of IG attack tip area showing fine porosity in thin corroded layer at grain boundary and along boundary plane ahead of oxidized material [119].

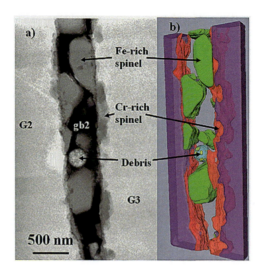

FIG. 4.7

(A) HAADF image showing a portion of the open crack along a grain boundary approximately 3 μm from the tip and (B) 3D reconstruction using surface segmentation obtained from an HAADF tomographic series of 69 images. Sample: SUS304 5%CW [124].

More recently, Lozano-Perez and coworkers have demonstrated the benefits of using 3D FIB slicing and electron tomography [124]. They illustrate how all the relevant microstructural features can now be studied in detail and their relative orientation with respect to the strain direction and grain boundary plane accurately measured. Electron tomography offers the best spatial resolution; the 3D shapes were obtained using high-angle annular dark field (HAADF) from the selected crack tips (Fig. 4.7).

Transmission EBSD (t-EBSD) has been utilized to characterize [125] the cracks formed during SCC of type 316 stainless steels. The main advantage of t-EBSD is a significantly higher spatial resolution compared to the conventional EBSD. This is because of the smaller region of the specimen that interacts with the incident electron beam. Lozano-Perez has developed a systematic method to examine cracked samples across length scales ranging from millimeters to subnanometer [126]. More specifically, cracks have been analyzed using electron microscopy, 3D Atom-Probe, and NanoSIMS in order to explore the interaction between SCC, chemistry, and microstructure. (S)TEM microanalysis is the dominant electron microscopic technique, and here data are acquired using HAADF, EELS, and EDX.

4.2.3.3 Summary

The discussion in the previous two sections has demonstrated that the both TEM- and STEM-based techniques have, and will continue to be, very powerful tools for characterizing the irradiation-induced microstructure. Traditional contrast mechanisms for characterization of microstructures containing dislocation loops, lines, and phase contrast objects such as voids or bubbles are understood well. Quantitative information can be obtained on the size and number density of point defect clusters (etc.) with diameters above the visibility limit of $\sim$2 nm. However, the analysis is limited to relatively small areas ($\sim\mu m^2$) less than 0.1–0.5 μm thick. The resolution may be adversely affected by the magnetic nature of the material or by oxide formation on the surface.

The majority of metallic material used in reactor components is complex multi-alloyed containing a range of second phases. Redistribution of the solutes to internal sinks or to influence the second phases is a vital part of understanding the response to irradiation. The discussion earlier has shown that the STEM is an excellent technique for providing microchemical information on the spatial distribution of solute elements, for example, the composition of small precipitates, matrix chemistry, or segregation to GBs. Elemental mapping allows insight into compositional fluctuations in high dose irradiations and crack tip chemistry in SCC or interface chemistry in the oxidation of zirconium processes, etc. Further, the advent of aberration corrected microscopes has and will enable far more detailed characterization of solute redistribution and clustering in complex technological materials with almost atomic resolution.

Advanced specimen techniques, such as FIB milling, can be used to minimize the sample volume or to locate a specific microstructural feature. However, care must be taken to ensure that image artifacts are not introduced (e.g., small dislocation loops). Specific care is required in the analysis of EDX spectra from irradiated steels. X-rays produced by the decay of ^{55}Fe have an energy very similar to the characteristic X-rays from Mn.

Finally, both techniques are most powerful when applied in combination with techniques such as SANS and APT.

4.2.4 ATOM PROBE TOMOGRAPHY
4.2.4.1 Principles

APT enables the characterization of the 3D spatial distribution and chemical identity of atoms with near-atomic resolution within a very small volume of material [127–130]. There have been many APT instrument developments in the last three decades, mainly based on the pioneering studies of Cerezo and coworkers who introduced a position-sensitive detector (enabling reconstruction of the atomic chemistry in 3D) [131] and the reflectron [132, 133] for greatly improved mass to charge resolution, and Kelly, who invented the local electrode atom probe (LEAP) [134, 135]. APT limitations include the fact that the volume of material analyzed is extremely small [typically (50–100) x (50–100) x (100–500) nm^3] and that approximately 30%–50% of the atoms are not detected. Vacancy- and interstitial-type defects cannot be characterized although solute segregation can signal the presence of dislocations, GBs, and dislocation loops.

APT is performed on needle-shaped samples, prepared by electropolishing [136] or FIB techniques [137], with tip radii of ~50–70 nm. The tip is cooled to cryogenic temperatures (typically 50K) in a high vacuum (10^{-10} torr) and held at a positive bias potential, typically 2–6 kV, with respect to a field emission electrode. The electric field is highly concentrated at the specimen apex enabling individual atoms on the surface of the tip to be field evaporated during application of a very short pulsed negative potential on the local electrode, or if the tip temperature is increased. The former is known as voltage mode, the latter can be achieved by focusing a laser pulse on the specimen apex. Typically the APT settings are adjusted so that about one ion is detected per 100 pulses. The ions created by the field evaporation are the radially accelerated toward a multichannel plate detector. A schematic representation of the APT instrument (without reflectron) is shown in Fig. 4.8.

The precisely measured time of flight between the voltage or laser pulse and the x-y position-sensitive detector hit depends on the ion mass-to-charge ratio. Thus the time-of-flight spectrum of detector hits is converted to a mass-to-charge spectrum, with sharp peaks at the relevant isotopic masses and charge states of the field emitted ions. In principle, this allows the atomic isotopes to be determined. However, there are many complications to this simple picture, such as peak overlap, the presence of multi atom ion species, and various ion charge states.

The x-y detector position signal is back projected to a common focal point below the assumed hemispherical tip surface. The normal intersections of the field lines with the assumed tip surface are used to identify the corresponding x-y position of the evaporated ion. The significant assumptions in this simple geometric model lead to distortions of the atomic positions even for pure metals. If the precise shape of the tip is known, it is possible to partially correct these distortions. However, distortions are amplified in the presence of small second-phase features and in layered structures [139, 140]. Repeated voltage, or laser, pulsing, leads to the detection of up to tens of millions of ions, generating a 3D x-y-z map of the relative positions and chemical identities of the atoms in the tip. The z-position typically assumes uniformly distributed evaporation of the ions from the surface, but the actual evaporation will depend on the local microstructure. Furthermore not all atoms are detected as the inherent channel plate detection efficiency is ~50%–80% and the detection efficiency is further reduced for atom probes that employ a wide angle reflectron to improve the mass resolution [133].

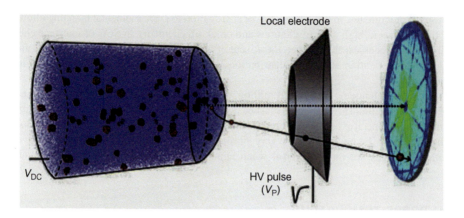

FIG. 4.8

Schematic representation of Local Electrode Atom Probe [138].

The complexities associated with the field emission processes coupled with inherent physical limitations of the technique limit the accuracy of the technique and near atomic scale resolution is often not achieved in practice. For example, small precipitates with lower solute evaporation potentials than the surrounding matrix cause a local flattening of the tip that focuses atoms into the precipitate region (local magnification effect). Trajectory aberrations (TAs) can result in matrix atoms being incorrectly positioned within clusters [141–143]. Uncertainties in atomic positions and the composition of nm-scale features due to TAs are amplified by other un-modeled phenomena, such as surface diffusion (field enhanced), multiple hits, and pre-emission under the standing potential. Many of these effects depend strongly on the tip temperature. The uncertainties are further exacerbated by the inconsistencies in the approaches taken to analyze APT data [144].

Simple analytical techniques to correct raw APT data (containing atom positions and mass-to-charge ratios) are a very desirable objective. For example, De Geuser proposed an iterative atom position relocation to remove the higher than physical densities to approximately correct for local magnification effects [145] and Hyde proposed use of a simple analytical function to reverse the local magnification effects associated with small precipitates and demonstrated that the resultant atom densities in precipitates were consistent with the matrix [144]. Morley proposed a fitted matrix-precipitate mixing zone width to correct precipitate compositions [141]. In parallel, more accurate physical models of the evaporation processes and evolution of tip shape are being developed and will lead to improved reconstruction algorithms [146–151].

It is important to note that within the reconstructed data there may be clear evidence of artifacts (for instance a very high local atom density), but additional information from other sources is required to enable correctly positioned atoms and those influenced by TA and other APT artifacts to be distinguished. Hence, other independent methods such as TEM-EDS, SANS, and PA must be used to characterize the nanoscale features, along with APT. For instance, Edmonson used complimentary scanning transmission electron microscopy with energy dispersive spectroscopy (STEM-EDX) combined with STEM-EDS modeling to conclude that the maximum Fe content in nanoscale Ni-Mn-Si-rich precipitates is ~6% [152]. Further, a recent paper by Shu et al. used a multi technique method to provide convincing evidence that there is little or no Fe in Cu precipitates in simple Fe-Cu and Fe-Cu-Mn alloys [153] supporting new approaches to correcting for APT artifacts such as that proposed by Hatzoglou et al. [154].

4.2.4.2 Application
Introduction

APT has been widely used to characterize irradiation damage in steels. The original instruments did not have position sensitive detectors, so compositional variations had to be interpreted from the sequence of ions evaporated from a very small borehole region, typically ~2nm in diameter, below the tip apex. These instruments could, however, also be used as "field ion microscopes" (FIMs) to examine atoms at the tip apex. In 1973, Seidman used FIMs to directly observe point defects in irradiated tungsten [155] and in 1978 Brenner et al. used field ion microscopy to identify the presence of "Cu-stabilized microvoids" in an Fe-0.34w.%Cu alloy irradiated to 3×10^{19} n cm^{-2} at 288°C [156]. However, they could not determine chemical composition of regions containing microvoids because of "visual difficulty of spotting small imperfections on the microscope screen".

The continued development of APT [157] in the 1980s enabled the identification of ultrafine copper-enriched clusters and precipitates formed during in-service irradiation of RPV steels. These clusters act as barriers to dislocation movement and hence are responsible for most of the observed irradiation-induced hardening [158, 159]. The technique was also used to study grain boundary segregation in RPV steels [160] which can lead to temper embrittlement, manifested as a failure-mode transition from transgranular to intergranular brittle fracture [161]. However, obtaining statistically significant data was hugely challenging since only a very small fraction of samples prepared by electropolishing contained a grain boundary sufficiently near the tip apex for analysis.

The development of atom probes with position-sensitive detectors, energy compensating reflectrons, and in particular local electrodes (LEAPs), coupled with FIB specimen preparation has significantly increased both the quality and the quantity of information that can be obtained by APT as well as the range of materials than can be studied [162]. Coupling FIB with EBSD to locate particular features like GBs is an especially powerful combination. APT is now routinely used to study a wide range of irradiation-induced phenomena including RPV embrittlement, IASCC, and the irradiation response of Gen IV and fusion materials.

RPV embrittlement

Miller and Russell reviewed the contribution of APT to the study of RPV embrittlement in 2007 [163] and Marquis and Hyde used the same issue in 2010 as a case study to demonstrate the capabilities of APT [164]. APT has provided direct observations of (i) irradiation-induced intragranular clusters containing Cu, Mn, Ni, and Si; (ii) - irradiation-induced segregation to, and precipitation on, dislocations (including loops); and (iii) irradiation-induced/enhanced segregation of solutes to GBs. These data are illustrated in Fig. 4.9 which shows the results of a LEAP analysis of an irradiated high Cu (~0.3 wt%) RPV steel. Within the analyzed volume more than 900 2–3-nm diameter Cu-enriched clusters were identified both in the matrix and on dislocation lines. The atom maps of P and C clearly show the presence of dislocation lines and a grain boundary. In addition carbides can also

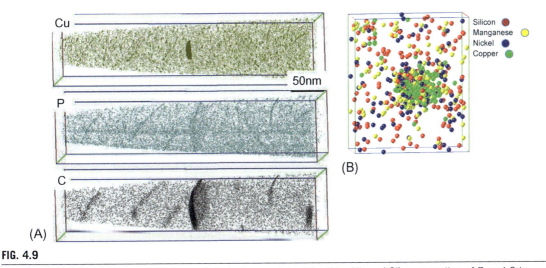

FIG. 4.9

(A) LEAP data showing the nanometer scaled clusters-enriched Cu (Mn, Ni, and Si), segregation of P and C to dislocations, and a grain boundary, the presence of a large Cu precipitate and a carbide on the grain boundary and (B) example of single irradiation-induced cluster ~3 nm in diameter [164].

be identified as well as a large associated Cu precipitate (which presumably formed during the post weld heat treatment). The size distribution of the irradiation-induced clusters, and their volume fraction, has been shown to directly correlate to changes in mechanical properties.

During the last decade, there has been an increasing focus on very low Cu RPV steels which are more typical of modern RPV steels. In 1997, Odette used thermodynamic arguments to show that precipitation of Mn and Ni could occur in RPV steels even if the Cu levels were very low. He suggested that the associated embrittlement would be observed at high fluence compared to that needed for rapid Cu clustering, but that the associated precipitate volume fractions and corresponding embrittlement could be significant and could thus impact the long-term operation of nuclear reactors [165]. Odette dubbed these late blooming phases, and more recently they have been described as Mn-Ni-Si precipitates (MNSPs). Notably, APT has been instrumental in proving that Cu-free or very low Cu solute atom clusters consisting mainly of Mn, Ni, and Si atoms do indeed form under neutron irradiation [166–171]. Here, we briefly outline the sequence of events that have emerged from APT (and SANS) studies, on the evolution of MNSPs above a very high fluence that show the dominant role of Ni.

In low Cu steels the total precipitate volume fraction grows slowly at first (with the formation of Mn-Ni-Si solute-defect cluster complexes formed in displacement cascades) and then more rapidly, above a high threshold dose, once MNSPs have formed. The MNSPs can grow to significant volume fractions as illustrated in Fig. 4.10A. MNSPs can also form in Cu-bearing steels ($>\sim$0.07 wt%) but are typically not seen until the matrix Cu is sufficiently depleted. In this case precipitates first evolve with a Cu-enriched core and Mn-Ni-Si shell, Fig. 4.10B, but at higher fluence can grow as discrete MNSP appendages (Fig. 4.10C). At the early stages of development, the observed compositions appear to reflect the initial solute concentrations in the matrix rather than predictions from thermodynamic models [intermetallic $G(T_3)$- or $G2(T_6)$-phase MNSPs as seen in Fig. 4.10D].

APT data in Fig. 4.10E show that in Cu-bearing steels, the total precipitate volume fraction (f) approximately scales as $f \approx 0.37(3Cu + Ni)$ at a fluence of 1.4×10^{20} n/cm^2 [172] at 290°C and intermediate flux; here the MNSPs have just started to develop. In comparison, at the very high fluence of 1.1×10^{21} n/cm^2 ($\sim$300°C), the volume fraction f is dominated by the Ni (due to the presence of MNSPs) content $f \approx 0.77(2Ni + Cu)$. Finally, Fig. 4.10F shows that the APT data can be used to predict yield stress changes in the very high fluence alloy conditions.

Zircaloys

Zirconium alloys are used for nuclear fuel cladding in LWRs because of their low neutron cross section and good mechanical behavior. However, corrosion in a pressured water environment leads to oxidation and associated hydrogen pick-up. This has implications for fuel efficiency and safety, particularly at high burn-ups or extended service life. Understanding the corrosion mechanisms is challenging as the corrosion is observed to occur in a cyclic manner eventually leading to a more rapid and continuous linear regime [135, 173]. APT has been used, in combination with a wide variety of other techniques, to help underpin a mechanistic understanding of the degradation processes. Interpretation of APT data from Zircaloys is particularly challenging as Zr has four isotopes (with an abundance of >10%) and can be field evaporated in various charge states and as complexes (issues exacerbated by the necessity of using laser pulsing). Similarly, there are multiple isotopes of the key trace elements and multiple overlaps in the mass to charge spectrum. Nonetheless, several researchers have performed high-quality informative analyses on Zircaloys, often using complementary techniques to confirm, or help interpret, the observations.

Gabory [174] used EELS and APT to study the evolution of the metal/oxide interface during Zircaloy-4 and ZIRLO autoclave oxidation. They found that the oxide/metal interface exhibits an intermediate layer, with an oxygen content between 45 and 55 O (at.%) and a layer corresponding to an oxygen-saturated solid solution in the metal matrix side (Fig. 4.11).

These data are consistent with previous work of Ni et al. who also used a combination of TEM and APT to study ZIRLO, Zircaloy-4, and a Zr-Nb-Ti model alloy [175]. They found thin intermediate oxide layers with compositions close to ZrO (Fig. 4.12) in almost all the pre-transition samples studied but after transition the suboxide was generally absent. They suggested that the suboxide cannot by itself act as a protective layer and conclude that it is

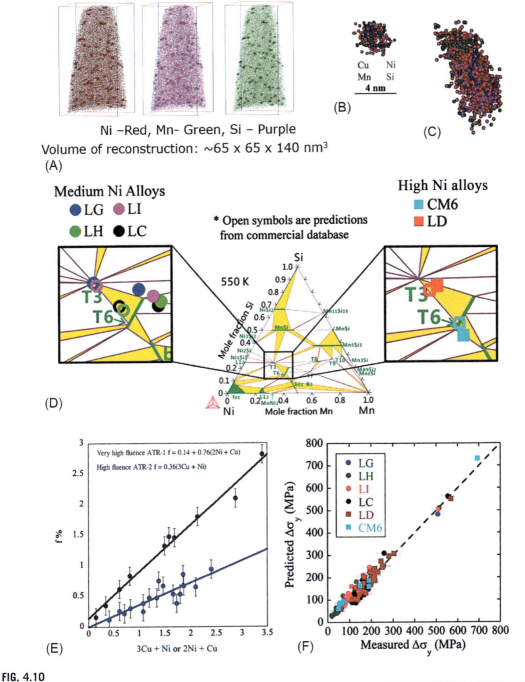

FIG. 4.10

(A) MNSPs observed in Ringhals Unit 3 weld irradiated to 5.6×10^{19} n cm^{-2} [171], (B) CRP core Mn-Ni-Si shell structure in high Ni-Cu content steel (LD) at a 1.3×10^{20} n/cm^2 [134], (C) discrete MNSP appendage on a CRP in steel LD at a very high fluence 1.1×10^{21} n/cm^2 [134], (D) predictions from thermodynamic models, (E) effect of Ni content on observed precipitate volume fraction at fluences of 1.4×10^{20} n/cm^2 (blue) and 1.1×10^{21} n/cm^2 (green) at 290°C [172], (F) correlation between predicted yield stress changes at the very high fluence based on APT data [134].

(D) Wells PB, Yamamoto T, Miller B, Milot T, et al. "Evolution of manganese–nickel–silicon-dominated phases in highly irradiated reactor pressure vessel steels", Acta Mater. 2014;80:205–219. Originally adapted from Xiong W, Ke H, Wells P, Barnard L, Krishnamurthy R, Odette GR, Morgan D. Thermodynamic models of low temperature Mn–Ni–Si precipitation in reactor pressure vessel steels, MRS Commun 2014. https://doi.org/10.1557/mrc.2014.21.

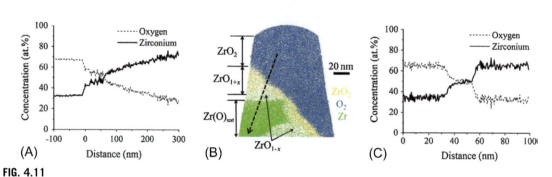

FIG. 4.11

Zircaloy-4 pre-transition oxide region (A) composition as a function of distance to the oxide/metal interface as measured by EELS; (B) 10 nm slice from an APT reconstruction showing the presence of different oxide phases; and (C) concentration profile from APT taken along the arrow indicated in (B) [174].

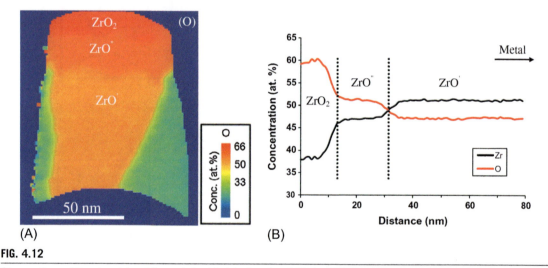

FIG. 4.12

Observations of the chemistry at the metal-oxide interface in a 100-day ZIRLO_ RXA sample (A) 2D oxygen composition map from the APT analysis and (B) a 1D line scan across the same data set illustrating the identification of separate ZrO" and ZrO' layers [175].

the development of interlinked porosity down to the metal-oxide interface that is the reason for the transition in oxidation kinetics.

Dong et al. [176] performed APT on various Zr alloys including Zircaloy-4 subjected to 360°C water in an autoclave. They found that the distribution of alloying elements is modified in the oxygen-rich region of the metal next to the oxide front. Segregation and clustering of Fe and Sn were observed along GBs in ZrO_2, at ZrO_2/ZrO and ZrO/Zr(O)$_{sat}$ interfaces.

APT studies have also been performed on unirradiated Zircaloy-2 (LK3™) [177–179] and subsequently following BWR irradiations [180]. Wide grain to grain variations in solute levels were observed, but repeat experiments enabled these variations to be quantified. In the irradiated materials, re-precipitation of Fe and Cr (from dissolution of the second-phase particles) was observed. The size distribution of the Fe- and Cr-rich clusters and precipitates was found to be consistent with <a> loops (1–5 nm in diameter) suggesting

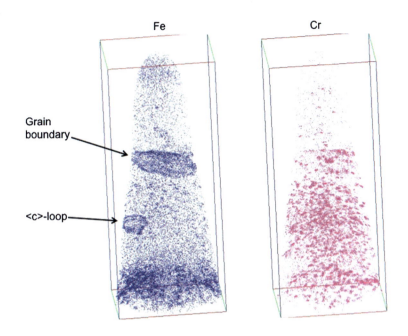

FIG. 4.13

Distribution of Fe and Cr in an APT analysis. Fe (left) is seen to segregate to a grain boundary and an assumed $<c>$-component loop, but not to clusters. Cr (right) is not residing in the grain boundary or the $<c>$-loop but has formed a large number of nano-sized clusters. The size of the box is $\sim 400 \times 170 \times 170$ nm^3 [180].

that solute precipitation may occur in the strain field of such defects. At higher doses ring-shaped Fe (and Sn) clustering was observed in the metal, probably at $<c>$ component vacancy loops (Fig. 4.13). This is an important observation as development of $<c>$ loops has been linked with breakaway growth that can occur after prolonged irradiation exposure and may therefore be linked to the behaviors of Fe and Sn. In contrast to the autoclave materials, the irradiated material did not appear to have extended oxygen gradients in the metal underneath the oxide front.

Austenitic stainless steels, SCC and IASCC

SCC is a significant concern for both PWRs and BWRs. SCC was predicted from autoclave tests [181] and has now been seen in operating nuclear power plants [182]. The cracking phenomena are dependent on a complex interaction between the microstructure, stress, and environment [183, 184]. Techniques such as FIB lift outs and electron tomography can be used to measure crack openings and provide insight on 3D crack morphology. APT has also been used to show the complex surface oxidation processes that occur [185] during SCC. This is illustrated in Fig. 4.14, which shows APT data from a cold-worked 316 stainless steel following autoclave testing (1500 h at 340°C) [186]. The total oxide thickness was about 400 nm. The reconstructions show the inner Cr-rich surface oxide (top of atom map) and Cr-rich sub-interface oxides following regions of high dislocation density. At the metal-oxide interface (Fig. 4.14B) there is a Ni-rich region. APT has demonstrated that the network of precursors which forms ahead of the oxide-metal interface provides fast diffusion paths for oxygen.

IASCC in austenitic stainless steels is frequently associated with RIS at GBs [183]. This phenomenon has been extensively studied using neutron irradiation but developing a mechanistic understanding of the

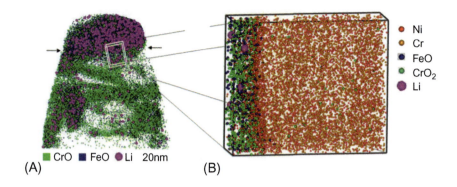

FIG. 4.14

(A) APT reconstruction showing the presence of Li atoms within the cap and sub-interface oxides. The arrows indicate the location of the cap-oxide-to-metal interface. The Li atom distribution is superimposed on the oxide atom maps. (B) Sub-volume (5 × 15 × 18 nm; 40,000 detected atoms) taken from the cap-oxide-metal interface showing selected species [186].

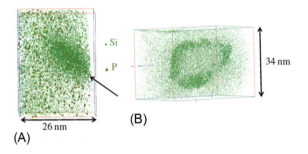

FIG. 4.15

(A) Ni–Si lenticular rounded cluster, only Si and P atoms are represented. P atoms are located at the center of the cluster. (B) Ni–Si cluster in the form of loop (torus-shaped cluster), only Si atoms are represented. Si atoms, in regions where the Si concentration is higher than 3% at., are emphasized [187].

degradation has proved elusive. As noted in Section 4.2.3, it can be advantageous to use ion (rather than neutron) irradiations for mechanistic studies since the variables can be easily controlled, experiments are less expensive and sample activation can be avoided (while recognizing that damage rates are different and that there may be differences between bulk and near surface behavior). Etienne et al. [187] took this approach to study radiation-induced precipitation and segregation in a cold-worked 316 austenitic stainless steel irradiated with 10 MeV Fe^{5+} ions. Several different intragranular features were observed including cylindrical features (dislocations), rounded clusters (small dislocation loops), and torus-shaped clusters (large dislocation loops) formed by RIS (Fig. 4.15). Similar Ni-Si features had previously been seen in neutron irradiated 316 SS [188]. The authors also measured the grain boundary Cr concentration and found, for a given dose, that RIS was higher than that reported in the literature for 316 SS irradiated at lower dose rates with neutron and proton irradiations. This may be due to the use of different techniques since the majority of data in the literature came from STEM/EDX measurements, a technique which may not measure peak RIS. Alternatively, it may be insufficient boundaries have been analyzed by APT to provide statistically significant data. However, an enhancement of segregation at higher dose rate is to be expected.

Generation IV and fusion materials. The next generation of nuclear reactors needs to be capable of operating at much higher temperatures than those used in Gen-II reactors (see Fig. 4.1) and be resistant to radiation damage. ODS steels are candidate materials but their production is complex, involving ball milling and extrusion or hot isostatic pressing, and this leads to a complex microstructure consisting of submicron grains and a very fine distribution of nanoscale oxide particles. There are two broad classes of the alloys including 9Cr transformable ODS steels, with a wide range of microstructures associated with various heat treatment paths and fully ferritic 14Cr steels often called nanostructured ferritic alloys (NFA). ODS and NFA are processed by ball milling metallic powders to add dispersed Y and O, which along with Ti form complex oxides nanoscale precipitates, like pyrochlore $Y_2Ti_2O_7$ during hot consolidation (see Section 4.2.3.2). The nanoclusters are expected to promote recombination of irradiation-produced point defects and to trap transmutation-produced He in small, high-pressure harmless bubbles [189]. These nano-oxides have been extensively studied by APT [190–193]. The APT results are generally consistent with TEM observations. APT has revealed oxide core shell structures with a Y-Ti-O enriched center region surrounded be Cr and Ti zones of segregation. Other elements like C, N, V, Mn and Si have also been observed in either the core or shell, depending on the alloy composition. APT has also shown that the nano-oxides are extremely stable under both high-temperature long-time thermal aging and intense irradiation conditions [194].

However, there has been a long-standing significant discrepancy between the compositions of the nano-oxides measured by APT and other techniques. In general, APT showed much lower Y and O oxide contents than TEM found. The reasons for this are complex, but it is now accepted that these results are due to significant APT artifacts, due to a combination of TAs, solute emission sequences that are not consistent with reconstruction assumptions and surface diffusion [195]. In part these issues are more severe in the case of the nano-oxides since their dielectric nature perturbs the tip fields in their vicinity. However, further discussion is beyond the scope of this chapter.

Summary

APT provides near atomic resolution, combined with the chemical analysis of individual atoms. APT is one of few techniques that provide direct information on the composition, size, and number density of solute-enriched clusters as well as segregation of solutes to/from microstructural features such as GBs and dislocations. Experiments performed under optimized conditions have provided a huge contribution to our understanding of the effects of irradiation on steels and corrosion processes.

However, interpretation of APT data is nontrivial and the effects of the imperfect detection efficiency and a number of potential artifacts must be carefully considered [130, 196, 164]. Many approaches to improvements of simulations of the field evaporation process are underway, but much more work is needed to improve APT reconstructions. Most importantly, new protocols for measuring, reconstructing, and interpreting APT data are needed, including objective and transparent indications of the limits and uncertainties associated with this powerful technique. Progress may be helped by technological improvements, particularly in relation to improving the detection efficiency (e.g., by developing superconducting detectors [197, 198]).

4.2.5 SMALL-ANGLE NEUTRON SCATTERING (SANS)

4.2.5.1 Principles

While there were limited early defect studies using neutron and X-ray scattering ([199–201]), small-angle neutron scattering (SANS) has been used extensively to study nm-scale features in irradiated materials beginning in the mid-1980s. Much of the SANS was initially directed at characterizing 1–5-nm Cu-rich precipitates in RPV steels and model alloys, primarily in Germany, the UK and the US [202, 203].

SANS measurements are carried out by directing a nearly monoenergetic collimated beam of cold neutrons into a sample and recording the scattering using a position-sensitive detector at low angles. The neutrons typically interact with a few cubic millimeters of the sample and hence the scattering provides information on the bulk microstructure in contrast to techniques such as TEM and APT, which can only examine microscopic quantities of material.

SANS provides information about the sizes of scattering features and, if their "contrast" with the matrix is known, their number densities and volume fractions. For example, the technique has been widely applied to characterize the irradiation-induced solute clusters in RPV steels [204–207]. Additional information on the nature of the scattering features can be obtained if measurements are taken in a magnetic field. Often irradiation results in changes to the formation of nanometer scale clusters, or precipitates, and so it is important to analyze the difference in scattering observed between an irradiated and an unirradiated sample.

Full interpretation of SANS data generally requires supporting microstructural data from techniques such as APT or STEM/EDX. For example, in the case of RPV steels, the in-magnetic field measurements can be used to provide a constraint on the potential range of features that are causing the observed scattering. However, the constraint is often not strong enough to uniquely identify the composition of the scattering features and so alternative microstructural techniques must be employed.

In the application to irradiated pressure vessel steels the neutron wavelength, λ, is selected to give scattering from only inhomogeneous structural features that differ from the matrix they are embedded in, typically ~ 0.5 nm. The scale of the inhomogeneity detected depends on the scattering angle that can be measured by the position sensitive detector. In many facilities inhomogeneities down to 0.7–1.0 nm can be detected. SANS samples can conveniently be cut from Charpy specimens, that is, the slices can typically be 10×10 mm^2 and 1–2 mm thick.

Neutrons are scattered by individual nuclei. The scattering lengths depend on the nuclear and magnetic structure of the nuclei, which vary greatly among different elements and isotopes [208, 209]. A scattering vector, q, is defined as the difference between the wave vectors of the incident and scattered beams, as $q = 4\pi \sin(\theta)/\lambda$, where 2θ is the scattering angle which determines the constitutive and destructive interactions between the scattered and transmitted coherent neutron waves. In SANS, only very small θ are examined.

A typical SANS setup is shown in Fig. 4.16. Notably small nanometer scale features scatter up to high (relatively) values of q. Thus SANS is a powerful tool in characterizing nano-precipitates that form under irradiation, since such small features are generally not present in the unirradiated condition. Characterizing the hardening precipitates in pressure vessel steels features as small as ~ 1 nm in diameter requires measurements out to q values of at least 3 nm^{-1}.

The scattering intensity also depends on the "form factor" of the scattering features. For simple shapes the form factor can be evaluated analytically. For instance, if the scattering features are spheres with a Guinier radius R_g and the magnitude of q is small (i.e., small scattering angles) then the Guinier approximation holds and $I(q) = I(0)e^{\frac{-R_g^2 q^2}{3}}$ where I(q) is the observed scattering intensity as a function of q. Hence a plot of log(I(q)) versus q^2 will be a straight line with slope $\frac{-R_g^2}{3}$ and the intercept at $q = 0$ gives the relative volume fraction of scatterers. Real systems will contain a range of sizes of scattering features and so the simple Guinier approximation is inadequate.

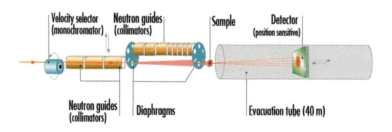

FIG. 4.16

Schematic showing layout of D11 at the institute Laue-Langevin (small-angle neutron scattering instrument) [210].

From http://www.ill.eu/d11. Used with permission from Institut Laue-Langevin.

The shape of the size distribution can be assumed, for instance a log-normal, and then a least-squares refinement applied to obtain the best fit [211]. Indirect Fourier transforms can also be used but these can result in un-physical negative size distributions [212, 213]. These limitations can be avoided by adopting a Monte Carlo approach as developed by Martelli et al. [214] and applied by Wagner et al. [215]. Alternatively, the Maximum Entropy technique can be used to determine the most probable distribution of cluster sizes that is consistent with the observed scattering [216–218]. This method does not require any prior assumption as to the form of the size distribution. If information is also available on the composition of the features then absolute volume fractions can be obtained.

Neutrons undergo both nuclear and magnetic scattering, where the latter depends on the difference in the magnetic scattering length density squared in the feature and matrix. For a zero magnetic field, the measured scattering intensity is proportional to the nuclear contrast (N) plus two-thirds of the magnetic contrast (M). However, in the case of ferromagnetic steels, SANS measurements should be made in a strong saturating magnetic field for two reasons since the field: (a) suppresses scattering from domain walls; and, (b) allows the separation of magnetic and nuclear scattering. The latter is critical since magnetic scattering provides another contrast factor, independent of nuclear scattering. Specifically, the total scattering cross section, $\left(\frac{d\Sigma(q)}{d\Omega}\right)_T$, depends on both q and the angle ϕ of the detector position with respect to the magnetic field as

$$\left(\frac{d\Sigma(q)}{d\Omega}\right)_T = \left(\frac{d\Sigma(q)}{d\Omega}\right)_N + \sin(\phi)^2 \left(\frac{d\Sigma(q)}{d\Omega}\right)_M$$

Here, N and M refer to nuclear and magnetic scattering, respectively. The cross-section magnitudes depend on the corresponding nuclear and magnetic scattering length density differences. The M and N scattering measurements provide useful information regarding the nature of the scattering features as expressed by the A ratio, defined as

$$A_{\text{Ratio}} = \frac{N+M}{N}$$

The value of the *A ratio* depends on the composition of the features, their atomic density, and their magnetic state, both relative to the matrix. In many cases the scattering features are assumed to be nonmagnetic, in contrast to the saturated ferromagnetic state of the ferritic steel matrix. In this case the number density, size distribution, and volume fraction of the features can be accurately determined from the $\left(\frac{d\Sigma(q)}{d\Omega}\right)_M$ data. In contrast, $\left(\frac{d\Sigma(q)}{d\Omega}\right)_N$ explicitly depends on the elemental and isotopic composition of the feature. Thus for nonmagnetic features A can be used to constrain the composition and atomic density of the feature. If the feature is composed of only one element, then A is a fixed value for a given atomic density (for instance, A ≈ 8.7 for a bcc Cu precipitates in an Fe matrix). If the scatterer contains two elements, A can be determined as a function of its composition, again assuming the atomic density in the nonmagnetic features is known; thus in this case A measures the feature composition. However, if the feature contains more than two elements, or the density of the feature is not known, then the measured value of A constrains possible compositions of the feature but the actual composition cannot be uniquely determined.

In practice, there are a number of complications to this simple basis for SANS analysis: (a) the features may not be completely nonmagnetic; (b) the atomic density in the features may not be known; (c) typical assumptions about the feature geometries and chemical structures (dilute homogeneous defect free spheres) may not be valid.

Most of these issues can be dealt with, at least to some extent, while others are under active investigation. However, as is almost always the case, proper interpretation of SANS measurements requires, or benefits from, complementary techniques like APT- and TEM-based microanalysis. Further the interpretations should be consistent with basic physical considerations like thermodynamics when near equilibrium conditions can be demonstrated. The technique has been used to study a range of radiation damage phenomena including RPV embrittlement, void, and bubble populations in ferritic steels irradiated to high displacement doses as part of fusion materials development programs as well as for the study of different thermomechanical treatments. These are explored in the following section.

4.2.5.2 Application

SANS has been used extensively to study precipitate and solute cluster formation in irradiated RPV steels (and relevant model alloys) (see, e.g., Refs.[216, 219–221]). The most complete analysis has been carried out by Odette and coworkers who have undertaken a very extensive and systematic study of the effect of material and irradiation variables (IVAR) on the microstructure and hardening of RPV steels and model alloys. These experiments were mainly carried out at the IVAR facility in the University of Michigan reactor [222] but also involve a number of other irradiation experiments including in the BR2 reactor at the SCK Laboratory in Belgium and in the Advanced Test Reactor at Idaho National Laboratory. A large matrix of alloys was irradiated in these experiments, including a series of split-melt A533B bainitic steels with systematic variations of Cu, Ni, and Mn. The fluxes ranged from 8×10^{11} n cm^{-2} s^{-1} to 2.3×10^{14} n cm^{-2} s^{-1}, the fluences 2×10^{17} to 10^{21} n/cm^2 and the irradiation temperatures from 270°C to 310°C. Thus this body of research has provided a detailed map of material and irradiation variable effects, singly and in combination, on both RPV microstructure and properties [223]. For example, Fig. 4.17 illustrates the application of SANS to a set split melt of RPV steels with systematic variations of Ni at a fixed bulk Cu ($\sim$0.4 wt%) and Mn ($\sim$1.5%) for an irradiation (IVAR T6) at the same flux (9.7×10^{11} ncm^{-2} s^{-1}), fluence (3.3×10^{19} ncm^{-2}), and temperature (290°C) [223]. The strong and systematic effect of Ni on increasing the precipitate number density (N_p), average radius ($<r>$), and hence volume fraction (f_v) are clear. The M/N ($= A - 1$) decreases with Ni, which at first seems surprising, since increasing Ni in the precipitates increases the M/N. However, the results are consistent with the fact that increased Ni leads to increased Mn, with a net result of decreasing M/N. Since there are actually four elements in the precipitates (Cu, Ni, Mn, and Si), their precise composition cannot be specified by SANS alone.

Fig. 4.18 [223] shows the precipitate volume fraction (f_v) as a function of the neutron fluence of two 0.4 wt% Cu, 1.4 wt% Mn alloys with 0.8 and 1.25%Ni (LC and LD, respectively), irradiated at 290°C at fluxes ranging from 8×10^{10} (IVAR-L) to 1×10^{14} (BR2-HH) ncm^{-2} s^{-1}. The SANS results demonstrate the strong effects of increasing flux to increase the fluence for a specified f_v. Similar SANS data are available for other compositional variables including Cu, Mn, Si, and P as well as for systematic variations in the irradiation temperature, for various flux and fluence levels.

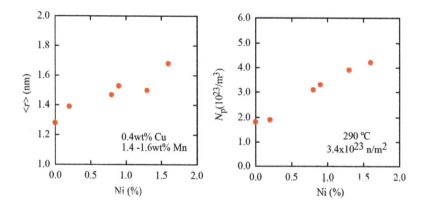

FIG. 4.17

SANS analyses of a set split-melt of RPV steels with systematic variations of Ni at a fixed bulk Cu ($\sim$0.4 wt%) and Mn ($\sim$1.5%) for an irradiation (IVAR T6) at the same flux (9.7×10^{11} ncm^{-2} s^{-1}), fluence (3.3×10^{19} ncm^{-2}), and temperature (290°C) [223].

From E. D. Eason, G. R. Odette, R. K. Nanstad, T. Yamamoto, A Physically Based Correlation of Irradiation-Induced Transition Temperature Shifts for RPV Steels. ORNL/TM-2006/530. Courtesy of Oak Ridge National Laboratory, U.S. Dept. of Energy.

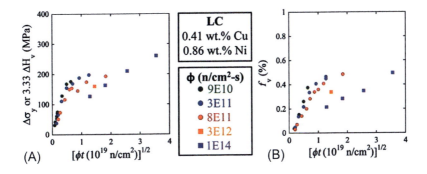

FIG. 4.18

The effects of flux (φt) on a high Cu, medium Ni RPV steel: (A) the yield stress increase ($\Delta\sigma_y$); and, (B) the precipitate volume fraction (f_v) [224].

From E. D. Eason, G. R. Odette, R. K. Nanstad, T. Yamamoto, A Physically Based Correlation of Irradiation-Induced Transition Temperature Shifts for RPV Steels. ORNL/TM-2006/530. Courtesy of Oak Ridge National Laboratory, U.S. Dept. of Energy.

These microstructural studies coupled to experimental determinations of the irradiation hardening were particularly influential in providing strong mechanistic support [224] to the formulation of reduced order physically motivated fitted models to predict the embrittlement of US RPV steels [223].

More recently a series of SANS studies been documented in the thesis of Wells [225] for a very high fluence RPV steel irradiations in the advanced test reactor that produced nanometer-scale precipitates dominated by Ni, Mn, and Si even in Cu-bearing steels.

SANS has also been employed to characterize precipitates distributions in unirradiated materials, frequently to determine the effect of different thermomechanical treatments (see, e.g., Refs. [226–228]). Although the majority of studies have been on neutron irradiated samples, Albertini and coworkers [229] presented a most interesting study on samples of the low activation MANET steel implanted with 1200 appm of He. There was a good agreement between TEM and SANS data on the resultant He bubble radius and number density. The authors also presented a methodology for evaluating the He content of the bubbles. SANS has also been used to characterize bubbles in helium implanted alloys [230] and as a powerful tool in characterizing nano-oxides in ferritic and martensitic alloys that provide dispersions strengthening up to high temperatures and remarkable radiation tolerance [231–234]. These studies include characterizing the evolution of the nano-oxides during processing steps (powder atomization, ball milling, high-temperature consolidation and deformation) and after long-term thermal aging and high-dose irradiations (see Chapter 6).

Finally it should be noted that modern synchrotron light sources are rapidly emerging as a powerful tool for characterizing irradiated microstructures.

4.2.5.3 Summary

SANS provides direct information on the size of scattering features in bulk specimens and indirect information on their volume fraction and number density. It has better resolution for small precipitates than TEM (~1 nm) and the analyses are performed over greater volumes. In performing the analysis of SANS data it is necessary to make assumptions about the nature of the scattering centers, for example, spherical precipitates of a uniform composition are commonly assumed despite evidence from other microstructural data for more complex structures (e.g., core-shell structures). Analysis requires either an assumption on the form of the size distributions or use of techniques such as maximum entropy.

The technique provides limited information on precipitate composition, although the magnetic A-ratio measurements permit some insight into the chemical composition of the scattering centers.

It has most commonly been used for the characterization of the small solute clusters formed in irradiated RPV steels by several groups worldwide that have employed a similar methodology, albeit with differences in the methods employed to determine size distributions and assumptions on the magnetic nature of the small solute clusters. The work of Odette and coworkers has demonstrated that SANS is a very powerful technique for systematically studying trends in the development of a precipitate population with material and irradiation parameters.

4.2.6 POSITRON ANNIHILATION SPECTROSCOPY-BASED TECHNIQUES
4.2.6.1 Principles
Introduction

An energetic positron injected into a metal is rapidly thermalized, and undergoes random walk diffusion until it annihilates in the bulk crystal, or at a trapping site. Positrons are trapped in open volume regions where atomic nuclei are further apart, and where the electron densities are lower, such as vacancy defects. Clusters of solute atoms in a metal also have different affinities for positrons and they too can be trap sites. The positron annihilates with an electron in a few hundred ps and, in doing so, generates two 511-keV γ-rays traveling in nearly opposite directions. Positron annihilation (PA) techniques can be used to characterize the timing of the annihilation event (lifetime) and/or the energies (and momenta) of the resultant γ-rays. The former is possible since the lifetime of a positron before annihilation is so short that under normal conditions only one positron is in the lattice at any one time. The latter reflects the fact that precise energies of the individual γ-rays depend on the momentum of the electron that is annihilated by the positron.

Thus several basic PA characterization methods have been developed:

(a) Analysis of positron annihilation lifetime spectra (PALS) associated with various trapping sites [235].
(b) Analysis of the distribution of the PA γ-ray energies around the nominal energy of 511 keV, by positron annihilation lineshape analysis (PALA) or coincidence Doppler broadening (CDB). CBD is based on using two high-resolution high-purity Ge detectors to record the energy of both γ-rays from the same PA event. A related technique is angular correlation of annihilation radiation (ACAR), which measures the 3D distribution of angles between the two γ-rays around the nominal value of 180°.

All these techniques (and others) have been refined to improve their sensitivity and to better understand and model (e.g., by DFT methods) the nature of the annihilation sites. Note there are a wide range of potential traps, especially in irradiated alloys. Thus PA techniques generally cannot quantify multiple features in a complex irradiated microstructure without a validated trapping model. Consequently PA methods are best used as differencing techniques, for instance to characterize the observable changes between irradiated and unirradiated alloy conditions, or the effect of fluence from a series of irradiations on a single material, and so on. Annealing has also been frequently used to characterize the recovery of the various trapping sites. More than the other common microstructural techniques PA very sensitive to point defects and small defect and solute clusters (or precipitates) that form under irradiation. PA is generally most useful as a source of complimentary qualitative information for features that cannot be characterized by TEM, SANS, and APT.

PALA, CDB, orbital electron momentum spectra and ACAR

Contributions from Compton scattering and general background must first be subtracted to isolate the γ-ray spectrum from other PA events. Conduction electrons have, on average, lower momenta than core, or tightly bound valence, electrons. Consequently the spread of energies associated with PA with conduction electrons will be lower than that for annihilations with the other more tightly bound electrons. As vacancy, or other open volume defect concentrations increase, the number of annihilations with conduction electrons increases, thence the corresponding width of the γ-ray energy spectral peak decreases as schematically shown in Fig. 4.19A. An analysis of the γ-ray spectrum line shape can

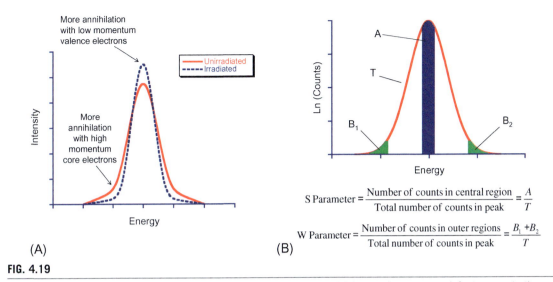

FIG. 4.19

(A) Schematic of width of the γ-ray energy spectral peak decreasing with increasing vacancy defect concentration and (B) methodology to calculate S and W parameter.

be used to detect vacancy-type defects. The width of the annihilation γ-ray energy lineshape is measured by a defined lineshape parameter, S (Fig. 4.19B). The S parameter is usually defined as the ratio of the total counts in a narrow central region of the energy peak to the total counts in the whole peak. Thus as the vacancy concentration increases the 'S' parameter increases. A second parameter, W, is defined by the number of counts in the two wings of the γ-ray spectrum, again divided by the total number of counts in the entire peak (Fig. 4.19B). Thus compared to vacancies, W is sensitive to higher energy core and valance electrons. Annihilations at solute cluster traps tend to be with higher energy core and valance electrons, thus they have a much higher W than vacancies. The corresponding solute annihilations result in much broader Doppler energy spectra. Vacancies have a very high S and low W, associated with free electrons, and can be plotted on a S-W coordinate system. Notably, individual elements also have specific position signatures on a S-W plot. Thus solute clusters with affinity for positrons can be detected by their S-W positions. Odette et al. [236, 237] also show that if the annihilation solute has a magnetic moment, S-W points measured in a magnetic field shift when it is reversed in polarity. This implies that PA can detect ferromagnetic elements like Fe and Ni in solute clusters. Notably, in an Fe matrix the positron affinities and trapping strengths are high for Cu, Ni, and Si compared to Fe. Thus the emergence of precipitates enriched in these elements shift the position of points on a S-W plot compared to when they are in solution. The S-W position stabilizes if all of the positrons are completely trapped at a particular feature. In typical steels, PA is sensitive to trap concentrations in the range 10^{-7}–10^{-4} (traps per atom).

CBD spectrum data can also be represented by a momentum (P_L) vs frequency spectrum, $n(P_L)$. The $n(P_L)$ is usually normalized by the corresponding $n_m(P_L)$ spectrum for the matrix, Fe in most cases of interest. Each element has a relatively distinct $n(P_L)$ spectrum [238]. Thus properly weighted combinations of n(PL) spectra can be used to estimate the composition of clusters. Proper weighting requires accounting for positron affinity and the degree to which it is confined in the solute cluster.

The momentum of the annihilation electron also affects the angular distribution between the coincidentally emitted γ-rays. An ACAR experiment yields a 2D projection of the electron positron momentum distribution that can help establish the nature of the trapping sites. However, the ACAR technique is complex, time consuming and needs high-quality models to help interpretation of the experimental data.

Positron annihilation lifetime spectroscopy

The most common positron source is the ^{22}Na isotope which undergoes β^+ decay accompanied by a 1.275-MeV γ-ray. PALS measurements are typically made by sandwiching a ^{22}Na source, usually in the form of ^{22}NaCl, between two flat faced samples and placing detectors on opposite sides of the sample. The technique is based on measuring the time difference between detection of the 1.275-MeV γ-ray and detection of one of the two 511 keV γ-rays from the subsequent annihilation event. The resultant data will consist of a spectrum of lifetimes characterized by a steep increase to a peak, followed by an exponential decay to longer lifetimes. The lifetime spectra for various types of annihilation site typically overlap. The background must be subtracted from the PA signal. The analysis of the measured spectra is based on the decomposition into exponentially decaying components for individual traps, and matrix sites as

$$\tau_{av} = I_1\tau_1 + I_2\tau_2 + I_3\tau_3 + \ldots$$

Here I_i is the relative intensity and τ_i the lifetime of the component i. The spectra are usually decomposed into one, two, or three I and τ, using fitting codes such as PALSfit [239]. Maximum entropy techniques have also been used to determine the most probable underlying distribution of lifetimes [240].

As the number of different traps and τ_i increase, PALS data interpretation becomes increasingly difficult. PA lifetime measurements are related to the size of the defects. In most metals free positron lifetimes range from 100 to 150 ps, with vacancy or dislocation-trapped positron lifetime somewhat higher (50%). The lifetimes for positrons trapped in voids are about 450 ps, in smaller vacancy clusters lifetimes lie between this value and the value for single vacancies. In iron lifetimes of $\sim$110 ps for bulk (τ_b) and $\sim$ 170 ps for a monovacancy (τ_v) have been experimentally determined. The lifetimes for vacancy clusters and small precipitates in iron have been calculated [241–243]. Thus, for small, submicroscopic voids ("nano-voids") the lifetime value is a measure of their size.

Variable energy positron annihilation spectroscopy (VEPAS)

In the VEPAS (variable energy positron annihilation spectroscopy) technique, positrons are used as a monoenergetic beam (in the range of a few eV to tens of keV) in order to obtain a defined small penetration depth for research on surfaces or defects under surfaces. These slow positrons allow depth profiling and may be important in the study of ion-irradiated materials [244, 245]. Iwai and Tsuchida connected a variable energy positron beam to an ion irradiation chamber [246]. The authors describe how the experimental setup enables positron Doppler broadening spectroscopy data to be obtained during and immediately after irradiation in the chamber under controlled temperature conditions.

4.2.6.2 Application

Positrons have been employed for a number of decades to investigate vacancies in metals. For example, MacKenzie et al. [247] observed that in a number of metals the mean life of positrons increases reversibly with increasing temperatures, and this was associated with preferential trapping of positrons at thermal vacancies. Seeger [248] reviewed the application of PA to the study of equilibrium point defects in metals, and provided a detailed discussion in terms of ion core size, valency, and vacancy relaxation of the trapping criteria for different metals. Since these early studies PA has been used to study the early development of vacancies and small vacancy clusters in irradiated metals. Studies using positrons have made important contributions to the understanding of microstructural development in pure metals, and notable in ferritic steels used in RPVs. Examples are given in the following.

PAS is obviously well suited to characterizing small vacancy clusters in irradiated metals, and, importantly, to detect such clusters at sizes too small to be observed by TEM. For example, Eldrup and Singh [249, 250] have studied neutron-irradiated Cu and Fe. They found that for all irradiation temperatures in Fe and for high temperatures in Cu a lifetime of 500 ps (τ_3) signaled the presence of voids. In Fe at low temperatures they detected that the formation of a high density of nano-voids [from an intense ($I_2 = 80\%$) component with a lifetime of 325 ps (τ_2)]. In contrast, in Cu small vacancy clusters or non-perfect stacking fault tetrahedra are formed.

A particularly important result was obtained comparing the microstructures developed in irradiated model alloys and RPV steels. Valo [251] examined various steels neutron irradiated to a fluence of 2.5×10^{23} nm^{-2} (E>1MeV). In contrast to the model alloys, no evidence was found for vacancy clusters in the steels. It was postulated that higher alloying and impurity concentrations in the RPV steels may be important in preventing the formation of detectable vacancy clusters. Nagai et al studied the effect of irradiation on the CDB signal obtained from various FeCu model alloys [252]. They irradiated the model alloys at ~300°C to a dose of 8.3×10^{18} n/cm^2 (E>1MeV). They presented information on P_L, the e^+/e^- momentum along the direction of γ-ray emission. Despite very low Cu contents, a significant fraction of the annihilations occurs in the vicinity of Cu atoms. This provides evidence that the vacancy traps are associated with Cu. The distribution was consistent with development of Cu-mono-vacancy-Cu complexes.

A S-W plot for Fe-0.9 Cu (VH) and a Fe-0.9Cu-1Mn (VD) (wt%) alloys irradiated at different fluxes (H/L) to increasing fluence (1–5) is shown in Fig. 4.20 from the work of Shu et al [153]. The figure shows that the H5 and H5A (lightly annealed) conditions (blue triangles) fall close to a line connecting Cu and Mn. The S-W positions of the low fluence irradiations, L1-4 and H1, of the Fe-Cu-Mn alloys remain close to Cu, with a small decrease in the low momentum fraction and a slight downward shift in W in the direction of the Fe controls. The CRPs in Fe-Cu-Mn are very small at low fluence (<1 nm), thus the positrons are unlikely to be fully confined. This hypothesis is consistent with the small degree of magnetic splitting that is also observed in L1-4 and H1 conditions that occur upon reversal of a magnetic field if the feature is itself magnetic [253].

The previous examples focused on characterizing samples irradiated to relatively low displacement doses typical of RPV steels. PA has also been applied to ODS steels, in both unirradiated [254] and following irradiation to high displacement doses. More specifically, the microstructures of four different commercial oxide-dispersion-strengthened steels (MA 956, ODM 751, MA 957, and ODS Eurofer) were studied after irradiation with 0.5 MeV He$^+$ to doses of ~45 dpa at a temperature ~62°C [255]. The implantation depth was calculated to be around 1.2 μm. Positron Doppler broadening spectroscopy with a slow positron beam was applied for detection of defects to a depth ~1.6 μm. The data showed that the ODS Eurofer is the most radiation resistant from the group of all investigated steels. Most interestingly, the authors considered on the basis of only slight changes in PA lifetimes in all investigated steels that probably only mono-vacancies were accumulated even though the dose was ~45 dpa.

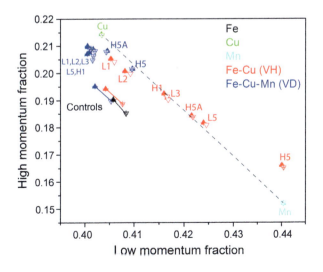

FIG. 4.20

The spin polarized PAS CDB S-W plot for the VH and VD alloys irradiated to various fluences along with the corresponding points both for the unirradiated controls and Cu, Mn, and Fe [153].

4.2.6.3 Summary

In recent years there have been significant developments in both the positron techniques available and the interpretation of data. Bulk samples are employed ($\sim 10 \times 10 \times 1$ mm^3) and no special surface preparation is required. However, irradiated materials normally contain many different types of trap and the positron trapping from individual defects cannot easily be deconvoluted. Currently PA lifetime measurements provide only semiquantitative information in complex irradiated steels with a range of trapping sites. Absolute values are sensitive to the experimental arrangements (e.g., background noise, definitions of parameters such as "S," etc.) which makes comparisons between research groups complex. Orbital electron momentum spectra (OEMS) measurements may be more quantitative, especially if there is a dominant solute cluster trapping site. Further, in irradiated steels and iron alloys the high-energy γ-radiation, especially from ^{60}Co, may limit the sensitivity of the technique [256], by giving rise to high peak count rates and a significant background.

Further, interpretation often needs to be underpinned by modeling. More quantitative interpretations of PA data will require well-calibrated, multi-feature trapping models as well as DFT-based calculations of the affinity and trapping characteristics of complex traps like multi-constituent solute clusters and defect solute complexes.

However, the methodology has the potential to provide information on matrix features that has not proved possible using other microstructural methods; for example, the presence of small vacancy defects in irradiated RPV steels. Finally, PAS is often used with PIA, and in association with H_v, TEM, APT, and SANS.

4.2.7 SUMMARY OF MICROSTRUCTURAL TECHNIQUES

In this section we have described some of the microstructural features that have to be analyzed, emphasizing the need to characterize irradiation-induced features on an atomic scale upward. In the complex multi-alloy materials employed in reactor components, it is essential to be able to analyze irradiation-induced solute re-distribution.

An overview of the many techniques that have been employed is given and four "mature" techniques described in detail. The latter are electron microscopy, SANS, APT, and PA. The strengths and limitations of these techniques are described and their application to radiation damage studies illustrated. An important recent development is the increasing use of focused ion beam (FIB) milling in specimen preparation. The use of the FIB has enabled in-depth analysis of the microstructure in specific regions, for example, at a crack tip or grain boundary.

Lastly, the focus has been on technique description, rather than the understanding of radiation damage mechanisms. It should be noted that in several phenomena, greatest insight has come from combining techniques. Further, in spite of the very great advances in capability for characterizing the irradiation-induced microstructure there are still features that are difficult to characterize, for example point defect clusters in RPV embrittlement.

4.3 SUBSIZED SPECIMEN TESTS TO MEASURE IRRADIATED MECHANICAL PROPERTIES

4.3.1 INTRODUCTION

The lifetimes of structural materials in nuclear power systems are generally limited by a variety of aging phenomena, many directly related to irradiation service [257]. Aging results in changes in the fundamental flow and fracture properties, including: tensile yield and ultimate stress and ductility; fracture toughness; fatigue life and crack growth rates; thermal and irradiation creep rates, rupture times and strains; and void swelling, which represents a dimensional instability that interacts with irradiation creep. Oxidation and corrosion may result in loss of strength and toughness, erode load-bearing sections and, most seriously, cause, or enhance, cracking [258, 259]. These degradation processes often involve strong synergistic interactions, especially in a corrosive environment that accelerates fracture under static and cyclic loading (e.g., SCC and corrosion fatigue). Thus it is necessary to monitor changes in the mechanical properties of key reactor components to ensure that the plant can continue to operate with large safely margins [260]. Monitoring may involve surveillance, component sampling studies, or separate test reactor irradiations. In most cases there are limits to the sizes and numbers of specimens that can be included in the test capsules.

As one example, the mechanical property tests required for the conservative estimate of the safety margin of a RPV of a LWR are tensile, Charpy-V notch, drop-weight, and fracture mechanics tests [261]. The need to ensure a good testing practice has led to the development of standards for measuring the mechanical properties of interest. There are a large number of national and international standards and code body organizations, but most are similar to the American Society of Testing and Materials (ASTM) and the American Society of Mechanical Engineers (ASME). A summary of standards important to nuclear technology reported by English et al. can be found in Ref. [260] with further discussion in Ref. [262].

The ways that irradiation can produce dramatic changes in mechanical properties strongly depend on the synergistic combination of the material and irradiation conditions [263]. A prime example is embrittlement in RPV steels that depend on the synergistic combination Cu, Ni, Mn, Si, P, flux, fluence, and irradiation temperature [264,265]. As a consequence a very large matrix of alloy and irradiation conditions is required to develop validated predictive models of property changes.

As noted above, mechanical testing for nuclear applications is most often challenged by the limited space available in surveillance and test reactor irradiations. Most often this necessitates use of nonstandard, subsized test specimens. Indeed, an extreme case of these limits is the design of test pieces for the proposed International Fusion Materials Irradiation Facility (IFMIF) [266]. Space limits are only one motivation, and others include the amount of available materials and radioactivity issues. The importance of test miniaturization is reflected in six ASTM Symposia on this topic, generally referred to subsized test techniques (SSTTs), beginning 1983 [267] and most recently held in January 2014 [268].

However, a complete discussion of this topic is beyond the scope of this chapter, and we briefly focus on only four tests: subsized tensile, microhardness, subsized Charpy, and precracked static fracture toughness. These have been chosen primarily because they have been extensively used for the determination of the effect of irradiation on the mechanical properties of irradiated materials. Likewise the blossoming field of nanoscale testing, that is applicable to ion-irradiation studies, is only briefly noted (Section 4.3.6). Further, there are well-defined standards for the testing of full-scale mechanical property test pieces (see, e.g., https://www.astm.org/Standards/physical-and-mechanical-testing-standards.html) which are also beyond the scope of this chapter.

4.3.2 SUBSIZED TENSILE TESTS

Standard ASTM tensile test standards (ASTM E8/E8M) typically involve round bar specimens with gauge section lengths and diameters of 50 mm and 12.5 mm, respectively, and total lengths of 70 mm or more. Smaller round bar specimens scale in proportion to these dimensions. Standard ASTM sheet specimens have similar gauge and width dimensions, but with a thickness of 0.5 mm. For a variety of reasons such sheet specimens are used in most irradiation experiments, but with greatly reduced gauge lengths and widths but similar or only slightly smaller thicknesses. For example, the dimensions of the so-called SSJ2 and SSJ3 specimens are total length/gauge section length/width and thickness of 16/5/1.2/0.25 (SSJ2) or 0.5 (SSJ3) mm. Subsized tensile specimens are adequate for testing alloys with reasonably homogenous microstructures and grain sizes that do not exceed ≈ 0.125–0.25 of the specimen thickness, which is true in most cases. The yield stress is the property least sensitive to the specimen dimensions, while the total elongation is the most sensitive (it is not really a property). However, for the common thinning dominated tensile fracture mode, the ultimate tensile and uniform elongation stress may also be affected somewhat by the tensile specimen size and geometry. Note reductions in area are a useful measure of ductility, but are not usually reported.

Tensile test results are usually specified only in terms of the yield and ultimate stresses and uniform and total elongations. However, engineering stress strain curves $s(e) = P/A_o(dl/l_o)$ contain a great deal more information about deformation properties. Here P is the instantaneous load at an engineering strain of dl/l_o where l_o is the initial gauge section length, and dl is the corresponding increase in the uniform gauge length. The $s(e)$ curves can be directly converted to a true stress strain curves, $\sigma(\varepsilon)$, that are needed in finite element analysis, only up to the point of necking. In many cases, this is an adequate basis specifying alloys constitutive properties in the unirradiated condition. However, high-dose neutron irradiations typically reduce the uniform elongation strain at necking to 1% or less. In this case $\sigma(\varepsilon)$ can be determined by an iterative inverse method based on FEM simulations of the engineering stress-strain curve for trial true stress strain inputs [269, 270]. The trial $\sigma(\varepsilon)$ are systematically adjusted until the predicted and measured $s(e)$ converge. The FEM simulations are based on J_2 flow theory, thus are generally applicable to a range of monotonic loading stress states. It is also useful to verify the FEM simulations inverse analysis $\sigma(\varepsilon)$ by comparing predicted versus measured values for other observables, like strain distributions and large geometry changes, ideally in three dimensions. In addition to standard in situ optical and post-test optical and SEM observations, digital image correlation strain mapping and confocal microscopy are very useful in validating the FEM-based inverse method $\sigma(\varepsilon)$ results [271]. Some typical results are shown in Fig. 4.21 for a variety of steels irradiated in the ATR-2 reactor to 6.5 dpa at 295°C [270, 271].

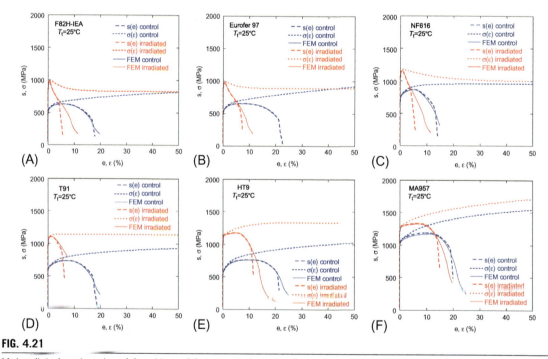

FIG. 4.21

Unirradiated engineering $s(e)$ and true $\sigma(\varepsilon)$ stress-strain curves for various alloys in 6.5 dpa at 295°C condition. The $\sigma(\varepsilon)$ is derived from the $s(e)$ curves using the FEM-based inverse method [270]. Note that three alloys strain soften, one is perfectly plastic and two still strain harden in the irradiated condition.

Pre and post irradiation tensile tests are readily carried over a wide range of temperatures and strain rates, including in the viscoplastic creep regime. Strain rate jump tests, involving changes in imposed displacement rates during a single test, are useful both for basic dislocation dynamics studies as well as providing approximate creep rate vs stress data based on saturation of the σ vs ε curve prior to necking and fracture. Finally, in reactor creep and creep rupture studies are conducted with pressurized thin-walled tubes. Laser profilometry is used to measure the diameter of the tubes for a sequential series of irradiation-discharge fluence increments. In some cases in situ release of tag gases was used to signal in reactor rupture times [272].

4.3.3 MICROHARDNESS TESTS

Hardness and microhardness tests (H) are a simple and powerful way to probe constitutive properties of alloys in both irradiated and unirradiated conditions. Microhardness is the indentation load (P) divided by a metric of the indentation area (A). For a variety of reasons it is difficult to precisely determine the actual contact area, due to effects such as elastic-reversed plasticity and pile-ups. Thus standardized test procedures use easily measured indentation length scales to determine H. The most common method to measure H is the Vickers microhardness (H_v) test using a diamond pyramid indenter, with programmed indentation and dwell times, at loads from about 100 to 10^4 g. Vickers microhardness is based on optically measuring the diagonals of the post load isosceles trapezoidal indent. Note very low loads my result in size-dependent H_v. Further, the indent area must be sufficient to homogenize the plastic response of the material over relevant microstructural (e.g., grain) length scales, if H_v is to emulate a tensile test results.

The most common application of H_v data in irradiation studies is to estimate the $\Delta\sigma_y$ that would be measured in a tensile test. Typically this has involved developing empirical correlations in the form: $\Delta\sigma_y = C_H \Delta H_v$. When, as is usual, the units of H_v are kg_f/mm^2 typical values of C_H range from $\approx$ –3 to 3.6. However, it is important to recognize that indentation hardness involves significant plastic strains; hence, the relation between C_H and H_v intrinsically depends on the strain hardening (σ_{sh}) characteristics of the material. He et al. carried out FEM studies for a wide range of constitutive laws to derive a universal relation between H_v and strain hardened flow stress (σ_{fl}) [273]. This and subsequent research showed that the relation of H_v and σ_{fl} averaged between $\sim$0% and 10% plastic strain is consistent for a wide range of $\sigma(\varepsilon)$ for unirradiated and irradiated alloys ranging from perfectly plastic to large increments of strain hardening. The slightly nonlinear H_v-σ_{fl} relation also depends on the elastic modulus (E). The corresponding average strain hardening $\sigma_{fl} = \sigma_{fl} - \sigma_y$ is generally close to the difference between the ultimate tensile and yield stress for ferritic steels used in nuclear applications.

The corresponding $\Delta\sigma_y = \Delta\sigma_{fl}$ if there is no change in the strain hardening ($\Delta\sigma_{sh} = 0$). However, irradiation typically reduces σ_{sh}, hence $\Delta\sigma_y > \Delta\sigma_{fl}$. Thus C_H depends on both the $\Delta\sigma_{sh}$ and $\Delta\sigma_{sy}$. Theoretical values of C_H are much higher if the σ_{sh} decrease due to irradiation is large. For example a simple model that $\sigma_y \approx 3*H - \sigma_{sh}$, yields a $C_H = 3.26$ MPa/$kg_f mm^{-2}$ if σ_{sh} does not change from an unirradiated value of 130 MPa, while C_H increases to $\approx$ 4.1 if σ_{sh} is reduced to 20 MPa after irradiation. Such large reductions in σ_{sh} are associated with high dose irradiations of 9Cr steels at $\approx$ 300°C, while reductions in the strain hardening in RPV steels are much lower even for large $\Delta\sigma_y$. Perhaps the main point is that there is not a simple and unique relation between $\Delta\sigma_y$ and ΔH. Fig. 4.22 shows H_v versus σ_y data for a large set of irradiated (unfilled circles) and unirradiated (half-filled squares) RPV steels [274]. The fitted slope is close to the theoretical value of 3 and the intercept is −136 MPa, close to the average value of the measured $\sigma_{sh} = 140$ MPa.

Finally we note that indentation measurements can provide a variety of other types of information. For example, ball indentation methods can estimate flow stresses over a range of plastic strains [275, 276]. Indentation pile-up measurements also correlate with strain hardening behavior: sharp high pile ups indicate low strain hardening while diffuse, low pile ups are associated with high strain hardening [277, 278], especially when associated with sink-in regions. Flow localization is signaled by crowned pile-ups as well as associated slip-step traces intersecting the unindented surface. Further indentation can be used over a wide range of temperatures to derive constitutive properties, including in the viscoplastic creep regime.

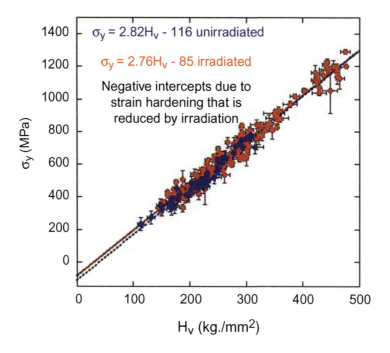

FIG. 4.22

Vickers hardness (H_v) versus the tensile yield stress (σ_y) for a large number of RPV steels in unirradiated and irradiated conditions [274]. The fitted σ_y line intercepts of 116 (unirradiated) and 85 (irradiated) MPa are the strain hardening increments associated with a Vickers indentations. The relation between $\Delta\sigma_y$ and ΔH_v is affected by the loss of strain hardening following irradiation.

4.3.4 IRRADIATION EMBRITTLEMENT TESTS: TRANSITION TEMPERATURE SHIFTS AND FRACTURE TOUGHNESS

An overarching requirement for safety critical nuclear system components is that the probability of fracture is essentially nil under normal operating, start-up, accident transient, maintenance, and fuel cycle handling conditions. While there are many circumstances that are pertinent, perhaps the two key examples are light water RPVs (see Chapter 10) and fusion reactor first wall and blanket structures (see Chapter 3). Structural alloys used for these applications have very high fracture resistance at the start of life. However various aging processes, most notably expose to neutron irradiation, results in microstructural changes that lead to embrittlement of structural alloys as manifested as a combination of ductile to brittle transition temperature increase shifts (ΔT) and corresponding reductions in fracture toughness at a specified temperature.

Historically, ΔT were monitored by Charpy V-Notch (CVN) impact tests. CVN tests involve measuring the energy (E) absorbed in breaking a shallow notched bar with a pendulum hammer as a function of test temperature (T), based on the guidance provided by standards such as ASTM E23. As illustrated in Fig. 4.23, for RPV applications, the ΔT is temperature indexed at 41 J (T_{41}) for a standard CVN specimen with a 10-mm square cross section and 55 mm length, impacted in three-point bending. As shown in Fig. 4.23 Charpy $E(T)$ curves have a sigmoidal shape composed of: (a) a low-temperature fully elastic brittle cleavage fracture region; (b) a higher temperature fully ductile, crack tearing upper shelf energy (USE) region; and, (c) an intermediate temperature transition region of increasing energy with mixed mode fracture. Irradiation shifts the TTS up in temperature (ΔT_{41}) and reduces the upper shelf energy (ΔUSE).

4.3 SUBSIZED SPECIMEN TESTS

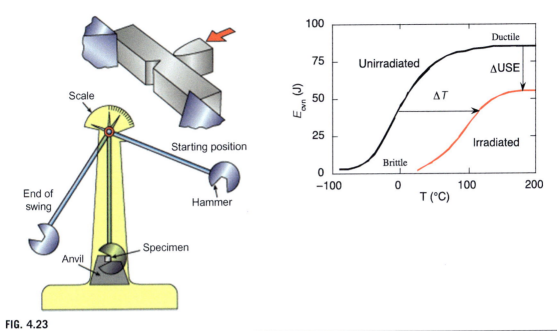

FIG. 4.23

(A) A schematic illustration of the Charpy impact test; (B) an illustration of irradiation embrittlement effects on Charpy $E(T)$ curves.

A wide variety of subsized CVN specimens have been tested in various irradiation programs. Common examples include specimens with all dimensions self-similarly scaled to 1/3 the dimensions of standard Charpy bars, and so-called KLST specimens with width (W), thickness (B), and length (L) dimensions of $4 \times 3 \times 27$ mm^3 Even smaller scaled specimens with 1/6 standard CVN dimensions have been tested. In all cases, the basic $E(T)$ attributes of standard CVN tests are preserved. The main differences between full and subsized CVN tests are: (a) the absorbed energy is lower, often roughly scaling with a Bb^2, where b is the ligament dimension. Thus for a 1/3 sized CVN, the scaling factor is nominally $\approx 1/27$ (note other scaling correlations have been proposed); (b) the E(T) curves are shifted down in temperature; (c) the transition region $E(T)$ slopes are steeper; and, (d) The ΔT due to irradiation is also somewhat smaller.

However, standard CVN tests do not measure the fracture toughness-temperature $K_{Ic/Jc}(T)$ curve needed for quantitative safety assessments. Here the subscripts refer to either elastic (K_{Ic}), or elastic-plastic (K_{Jc}), fracture toughness. Fracture toughness can be used to determine the load, or stress (elastic), or load and load point displacement (elastic-plastic stress and strain) that results in the extension of an initially sharp fatigue precrack. The elastic fracture stress intensity factor (K_I) for an applied stress σ and crack length a is $K_I = f(a/W)_t \sigma \sqrt{a}$. Here $f(a/W)$ is a known function of the nondimensional test specimen geometry (e.g., a and W in the simplest case) that is available in handbooks [279]. Thus, the elastic fracture toughness (K_{Ic}) is determined by measuring the fracture stress (σ_{ft}) in a precracked test specimen. The basic concept of elastic fracture mechanics is that the corresponding fracture stress for a structure is $\sigma_{fs} = K_{Ic}/[f(a/W)_s \sqrt{a}]$. Elastic-plastic fracture toughness (K_{Jc}) further accounts for the extra energy expended when permanent deformation occurs beyond the overall yield stress of a cracked body (specimen or component), marked by a nonlinear plastic deviation from the elastic loading line.

Current ASME regulations use Charpy TTS to position, or index, an ASME reference elastic toughness curve $K_r(T)$ on an absolute temperature scale [280]. The reference curve represents the lower bound of a very dated set of K_I data. It should be noted that the ability to transfer K_{Jc} data, measured in coupon scale tests, to other cracked body and loading geometries and size scales is limited, and subject to a number of restrictions. However, this complex topic is beyond the scope of this brief section.

Within a set of prescribed conditions, cleavage fracture $K_{Jc}(T)$ curves can be measured and used in design and safety assessments. The limiting conditions are related to two basic issues:

(1) That the specimen size and geometry is such that the triaxial stress and strain fields near the tip of a crack that blunts under applied loading, remain self-similar (e.g., have the same shape) when distances are scaled (normalized) by the crack tip opening displacement (δ), where $\delta = K_I^2/[2\sigma_y E']$, where $E' = E/[1-\nu^2]$, E is the elastic modulus, and ν is Poisson's ratio [281]. This so-called small-scale yielding (SSY) condition, imposes constraints on the specimen thickness (B) and ligament (b) dimensions as well as a/W. The formal requirements imposed by the ASTM E1921 master curve (MC) method standard practice are that $\delta/b > 30$, $B/W > 0.5$, and $a/W \approx 0.5$ (e.g., a deep crack in bending). The net effect is that the measured nonvalid toughness for small specimens are higher (nonconservative) than would be found from valid tests on larger test pieces [282–284]. Deviations from SST conditions result in lower crack tip stress fields due to the loss of triaxial constraint. A significant body of research has shown that the δ requirement in E1921 is nonconservative, and more appropriate value for ensuring SSY is $\delta/b > 100$–200 [283]. Thus specimens with $W \geq 1$ cm are needed to measure K_{Jc} up to 100 MPa $\sqrt{m}$. However, both higher and lower toughness must be measured in tests for a median $K_{Jc} = 100$ MPa $\sqrt{m}$; this requires larger specimens, scaling with K_{Jc}^2 (thus $W = 4$ cm for $K_{Jc} = 200$ MPa), to maintain SSY fields.
(2) Accounting for the inherently statistical nature of cleavage fracture in the transition due to weakest link statistics [283, 284]. The critical cleavage event is when the high (3–5 × σ_y) normal stress fields at the tip of a blunting crack cause microcracks to form at brittle trigger particles (e.g., large grain boundary carbides) that then can propagate to extend the precrack. This occurs when critical stress (σ^*), which scales with the inverse square root of the potential trigger particle size, encompasses a critical volume of the material (V^*). The size distribution of potential trigger particles in a highly stressed volume can be described by weakest link Weibull statistics for trigger particles at the largest end of the overall size distribution. The net effect is that at a given temperature $K_{Jc}(T)$ may statistically vary by factors of 3 or more.

ASTM E1921 provides a method to calculate mean and median K_{Jc} and the corresponding upper and lower bounds at a specified confidence interval [285]. The standard also prescribes a way to adjust the measured K_{Jc} for various crack front lengths (B) to a reference size (B_r) as $K_{Jcr} = (K_{Jc}(B) - K_{Jmin})[B/B_r]^{1/4} + K_{Jmin}$, where $K_{Jmin} = 20$ MPa $\sqrt{m}$ is a minimum toughness.

As schematically illustrated in Fig. 4.24, the ASTM Standard Practice E1921, for measuring cleavage toughness in the transition, permits indexing an invariant MC cleavage toughness $K_{Jc}(T-T_o)$ shape on an absolute temperature scale by determining the temperature (T_o) corresponding to a median toughness of 100 MPa $\sqrt{m}$. The shape of the MC is given by

$$K_{Jc}(T) = 30 + 70\exp[0.019^*(T-T_o)] \, (MPa\sqrt{m})$$

The E1921 standard requires a minimum of six valid K_{Jc} tests to determine T_o, although a greater number result in improved statistical upper and lower bound confidence intervals. Thus as noted by Odette, indexing master $K_{Jc}(T)$ curves based on measuring T_o requires a relatively small number of relatively small specimens [285]. Clearly the MC method is an immensely powerful tool for assessing irradiation embrittlement. However, irradiation experiments generally require use of precracked specimens that are even smaller than $W = 1$ cm.

Measured $K_{Jm}(T)$ LSY curves are shifted to lower temperatures and have higher slopes than SSY conditions. The corresponding ΔT_o shifts are also smaller. Fortunately, deviations from SSY can be corrected, up to some extent, for large scale yielding (LSY) to estimate the corresponding toughness for fully constrained conditions as illustrated in Fig. 4.25 [282]. The adjustments involve use of a micromechanical model that postulates that cleavage occurs when σ^* encompasses V^* as noted earlier. Values of σ^* and V^* are determined by calibration tests using

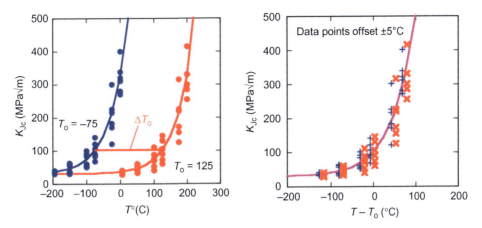

FIG. 4.24

(A) A schematic illustration of unirradiated and irradiated $K_{Jc}(T)$ master curves plotted on an absolute temperature scale showing a $\Delta T_o = 200°C$; (B) The same data plotted on a T_o indexed temperature scale as $K_{Jc}(T - T_o)$.

From E1921-97. Standard test method for determination of reference temperature, TO, for ferritic steels in the transition range. Annual Book of ASTM Standards, vol. 3.01. West Conshohocken, PA, 1998.

larger specimens, generally in the unirradiated condition, using finite element method simulations of the crack tip fields. These local fracture properties are then used to determine the ratio of $K_{Jc}(LSY)/K_{Jc}(SSY) > 1$, again based on FEM simulations for LSY conditions. These model-based adjustments allow the use very small, precracked specimens to estimate the MC T_o. Further the statistical effects in difference in the crack front length B are accounted for by the $[B/B_r]^{1/4}$ scaling cited earlier. This approach to determining ΔT_o assumes that σ^* and V^* are unaffected by irradiation, which is often, but not always the case. If these local fracture properties are affected by irradiation more detailed model and analysis procedures are needed. Very small specimen-based estimates of ΔT_o also require that the fracture mode remains cleavage and δ not be too large, such that ductile tearing or plastic hinging and collapse occurs.

It has been shown that after constraint loss adjustments, LSY toughness data, for tests on three point 18.3-mm long bend bar tests, with $b = 1.65$, $B = 1.65$, $W = 3.3$ mm, can be used to estimate the MC T_o [282, 283, 286], see Fig. 4.25. Tiny pre-cracked specimens with half of these dimensions also have been used to measure invalid LSY toughness $K_{Jm}(T)$ curves at very low temperatures and high loading rates [287]. These tests with very sharp transitions do provide a measure of ΔT. However, it is not possible to accurately evaluate MC ΔT_o based on such tests. Finally, small, precracked bend bar specimens also can be used to measure fatigue crack growth da/dN curves.

The preceding discussion applies to cases where TTS are primarily due to irradiation hardening. However in some cases TTS are associated with nonhardening embrittlement due to weakening of GBs, as the result of phenomena such as radiation-enhanced segregation of P and He leading to intergranular fracture paths. These effects result in a change in σ^* and V^*.

4.3.5 IRRADIATION HARDENING-EMBRITTLEMENT RELATIONS

As in the case of the correlation between H_v and $\Delta\sigma_y$ and $\Delta\sigma_{fl}$ described previously, considerable progress has been made in establishing other property-property relations. These relations are based on extensive finite element simulations of the tests combined with large verification databases. A prime example is the relation between irradiation hardening and the ΔT_o MC shift (also for Charpy shifts developed more than 30 years ago [288]).

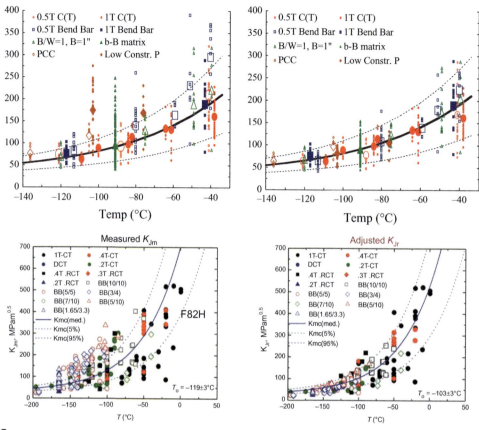

FIG. 4.25

(A) As measured K_{Jm} values of a unirradiated 9Cr tempered martensitic steel, for tests on a large range of fracture specimen sizes and geometries, along with a MC with a $T_o = -103$ and 5 and 95% confidence interval bounds; (B) the same data following size adjustments for constraint loss and statistical effects of the crack front length, B, along with a MC with a $T_o = -104°$.

From G.R. Odette, T. Yamamoto, H.J. Rathbun, M.Y. He, M.L. Hribernik, J.W. Rensman, Journal of Nuclear Materials vol. 323 (2003) 313.

In the case of steels irradiated at around 300°C, the ΔT_o is dominated by irradiation hardening. Finite element simulations have shown that the process zone fracture region in front of a blunting precrack experiences an average of $\approx 10\%$ plastic strain under both SSY and LSY conditions. This is the same strain range $\Delta\sigma_{sh}$ as well as the $\Delta\sigma_y$. Approximately average relations are $\Delta T_o \approx 0.8\Delta\sigma_{fl} \approx 0.7\Delta\sigma_y \approx 2.3\Delta H_v$. More detailed models show that the relation between irradiation hardening and embrittlement is not a simple linear function.

4.3.6 DATA FROM MECHANICAL TESTS AT THE NANOSCALE

Increasingly, mechanical tests at a very small scale are being used to characterize the mechanical properties of irradiated materials, primarily material subject to ion irradiation. There are two main techniques, nanoindentation and testing of samples (μms in size) machined using the FIB.

Nanoindentation has been used since the mid-1980s to monitor the hardness of ion-irradiated material [289], recently there has been an increase in the number of papers reporting nanoindentation data of ion-irradiated metals. Many authors employ a diamond Berkovich tip used with the continuous stiffness measurement (CSM) method [290], which enables the measurement of both hardness and elastic modulus as a function of indenter depth. Nanoindentation is a highly sensitive qualitative tool for assessing hardening effects of ion irradiation, but it requires rather sophisticated analysis before valid extrapolations to macroscopic tensile properties that can be made. In fact, one has to be particularly careful with the data evaluation, as a range of effects can contribute to the actual data obtained [291]. Pharr and Oliver [292] have pointed out that these effects include indentation size effect, crystal anisotropy, pile-up and sink-in, and the possibility of residual stresses between the ion-implanted layer and substrate [293]. The limited penetration depth of the nanoindentation gives rise to concerns about the effect of the surface preparation technique employed. For example, mechanical polishing may produce a work-hardened surface region that can cause misleading hardness results [294]. This is a particular concern at low indenter loads.

Attention has also been focused on comparing results from nanoindentation testing with microhardness data and also to properties like yield strength. This requires taking the indentation size effect into account [295–297], whereby it is known that smaller indents give higher hardness compared with a macroscopic indent. Nix and Gao [296] discussed a model describing this effect based on the presence of dislocations required to accommodate the plastic deformation prescribed by the rigid indenter geometry. Empirical correlations may then be used to derive yield strength data. However, Kiener et al. [290] consider that there is an influence of radiation on indentation size effects and that a size effect relationship obtained on an unirradiated material cannot be applied directly to an irradiated material. Further, Kasada et al. have suggested a model based on Nix and Gao [296] which accounts for the effect of the softer unirradiated region beyond the irradiation range [298]. A number of authors have employed these models to extrapolate the experimentally obtained nanoindentation hardness to the bulk-equivalent hardness of ion-irradiated materials [297, 299]. Overall, it can be seen that making a comparison between nanoindentation hardness, microhardness, or bulk mechanical uniaxial data is complex. The problems and promise of relating nanoindentation results to tensile yield stress changes are also highlighted in a recent publication by Krumwiede [300].

With the advent of the FIB, it is possible to develop samples suitable for miniaturized uniaxial tests; either by pillar compression or microtensile testing [290]. Keiner et al. consider that testing such samples offer several benefits over nanoindentation experiments which offset the more complex sample preparation [290]. However, Grieveson et al. [301] have demonstrated that there are marked differences between the deformation behavior of micropillars in annealed and heavy-ion-irradiated material. They consider that these differences are due to the ease or difficulty of dislocation nucleation in these small test volumes, rather than differences in flow stresses. They recommend caution should be exercised when interpreting micropillar test data on irradiated materials in terms of bulk mechanical properties. Relating micro beam bending data to macroscopic properties has also proven to be an unmet challenge [302].

4.3.7 SUMMARY

Subsized tensile specimen tests are a well-established method for measuring stress-strain constitutive behavior of alloys in both the irradiated and irradiated conditions. Inverse methods can be used to derive post necking true stress-strain curves, avoiding the typical challenge of nearly complete loss of uniform strain in highly irradiated metallic materials. Relevant ductility parameters can also be derived from subsized tensile tests. Irradiation-induced changes in yield and ultimate necking stress, strain hardening, and various ductility measures are best measured in tensile tests. In part, this is due to the fact that the uncertainties associated with the irradiated and control yield stress estimates are generally smaller. Use of alternative methods to estimate the yield stress changes, like shear punch microhardness tests, involve conversion uncertainties that have inherent experimental scatter. Note, such added uncertainties are especially significant, since the errors in establishing estimated changes in yield stress are greatly amplified for relatively modest differences between fairly large numbers.

Microhardness tests directly measure flow stresses between 0% and 10% plastic strain, based on empirically validated theoretical finite element-based analysis. Thus the relation between microhardness and yield stress depends on the strain hardening behavior of the material. Therefore there is no single unique change in yield stress-hardening relation. However, general trends in strain hardening established in tensile tests, or even ball indentation or indentation pile measurements, can be exploited to relate microhardness measurements to yield stress estimates.

Within well-understood limits, subsized fracture specimens can be used to evaluate shifts in MC cleavage reference temperatures (ΔT_o). This generally requires micromechanical adjustments for constraint loss effects at smaller sizes, as well as accounting for statistical stressed volume effects. Uncorrected toughness temperature curves are shifted to lower temperatures and estimates of transition temperature shifts are smaller for small versus full size specimens, in both CVN and pre-cracked fracture toughness tests.

In principle, nanoindentation tests can be related to corresponding properties measured in tensile and microhardness tests. However, such relations depend on a number of test and material factors, and a robust general method to convert Berkovich nano hardness to tensile yield and flow stress has not been fully demonstrated. Further, applying nano indentation to evaluating macroscopic strength changes, caused by heavy ions with a depth of only a few microns, remains a very difficult challenge. Micro-pillar compression tests and microbeam bending tests have attributes of conventional larger scale tests but they have not been demonstrated to be successful measuring radiation effects on macroscopic mechanical properties.

Finally, due to space limitation we have not covered a number of small specimen methods, for example disc bulge and subsized fatigue tests, to estimate irradiation effects on mechanical properties. Nor have we even broached the critical topic of environmental effects on properties and cracking. And the issue of differences between in reactor versus post irradiation tests on properties like fatigue remains almost completely unexplored.

4.4 COMPUTATIONAL ALLOY DESIGN AND OPTIMIZATION
4.4.1 INTRODUCTION

Application of computational tools to alloy design and optimization is an important topic and has received broad attention in research/education institutes and industries. A variety of computational modeling and simulation methods from atomistic to mean-field have been developed to help understand materials behavior at different length and time scales. In many cases, computational models at different scales have to be coupled together to solve specific problems. Some comprehensive overviews can be found in Refs. [303–306].

Most of the alloys used in nuclear reactors are complex materials with about five to more than ten alloying elements. The microstructures of such materials depend on alloy compositions, heat treatment parameters, environmental conditions, etc. Computational thermodynamics using the Calphad (CALculation of PHAse Diagram) method [307] has been widely used in modeling the phase properties of multicomponent engineering alloys under different thermal and environmental conditions.

The essence of the Calphad method is to simultaneously evaluate all available thermochemical and constitutional data for the considered system, and then obtain a set of model equations with adjustable parameters for describing the Gibbs energies of all phases as functions of temperature, composition, and pressure. The optimal values of the unknown parameters, which provide the best match between the calculated quantities and their experimental counterparts, are usually obtained by thermodynamic optimization using the weighted nonlinear least squares minimization procedure. Ab initio calculation data also provide valuable inputs in the case of unavailable experimental data. The model selection of a phase is based on the physical and chemical properties of the phase such as crystallography, type of bonding, ordering, and defect structure. The model selection and parameter optimization require all sources of data to be self-consistent and also consistent with thermodynamic principles. By using a stepwise approach, the Gibbs energies of phases in the unary, binary, and higher order systems can be modeled and compiled into large self-consistent multicomponent alloy thermodynamic database. Once the Gibbs energies of phases in a multicomponent system are obtained, phase properties such as phase fraction, composition, transition temperature, and thermodynamic properties can be calculated by numerically finding the minimum energy of the system. As a result of this approach, several commercial software packages were developed for calculating phase properties at a given temperature and pressure and their chemical compositions, such as Thermo-Calc, JMatPro, Pandat, MatCalc, etc. [308]. This method has played a significant role in the development of nuclear fuel and structural materials [309]. A few examples are presented further to demonstrate how the computational tools can efficiently help designing and optimizing alloys for nuclear reactor applications.

4.4.2 ALLOY OPTIMIZATION

Alloy optimization refers to microstructural optimization of existing alloys. For example, Fig. 4.26A shows the calculated phase amount (mole fraction) in alloy A1 [Fe-0.09C-15Cr-1Mn-2.5Mo-1Si-1.4Al-7Ni (wt%)] as a function of temperature. If this alloy is fabricated using one of the standard heat treatments with mill annealing at 1066°

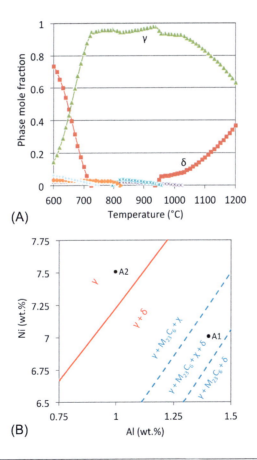

FIG. 4.26

Calculated (A) phase mole fraction vs temperature in alloy A1 [Fe-0.09C-15Cr-1Mn-2.5Mo-1Si-1.4Al-7Ni (wt.%)] and (B) phase diagram showing the effects of Al and Ni contents on phase stability in the Fe-0.09C-15Cr-1Mn-2.5Mo-1Si-xAl-yNi (UNS S15700) system at 1066°C (solid red line) and 954°C (dashed blue lines).

C followed by austenite conditioning at 954°C [310], a significant amount of δ-ferrite that is detrimental to properties would be developed in the matrix. The formation of δ-ferrite can be prevented by controlling alloy compositions. As suggested in Fig. 4.26B, within the specifications of alloy UNS S15700 with Al (0.75–1.5 wt%) and Ni (6.5–7.75 wt%), only the alloys within the up-left corner of the figure with low-Al and high-Ni contents would eliminate the δ-ferrite formation. For example, a modified alloy A2 with 1Al and 7.5Ni is in the single γ-austenite region during mill annealing and would develop $M_{23}C_6$ and X in the γ matrix during austenite conditioning. In contrast, the original alloy A1 has γ and δ during mill annealing, which are retained during austenite conditioning with additional $M_{23}C_6$ and X precipitates.

Ferritic-martensitic (FM) steels, another important category of structural materials used for nuclear reactors, are known for their greater resistance to radiation-induced void swelling, high thermal conductivity, and low thermal expansion coefficient compared to austenitic stainless steels. Grade 92 is the third generation FM steels, which was developed to have strength superior to Grade 91 by substituting the majority of Mo with W. However, the creep strength/resistance of Grade 92 is believed to be impaired by the coarsening of the large amount of $M_{23}C_6$ [311, 312] and Laves phase [313] and the formation of Z-phase, for example Cr(V,Nb)N, which consumes the ultrafine (V,Nb)N-type MX nanoprecipitates that benefit alloy strength [314]. The solid lines in Fig. 4.27 exhibit the calculated phase fraction at the tempering and general application temperatures using the MJT heat of Grade 92 as an example [315]. In addition to $M_{23}C_6$, Laves phase, and σ-phase at lower temperatures, MX transforms into Z-phase at temperatures below ~700°C. To enhance the performance of Grade 92, chemical composition has been adjusted to mitigate the formation of Z-phase and reduce the amount of $M_{23}C_6$ and Laves phase, as illustrated by the dashed lines in Fig. 4.27. Preliminary creep testing at 600°C on the modified-Gr92 exhibited significant increases in creep life by about 700 times and creep strength by about 40% compared to conventional Grade 92.

In addition to composition modification, computational thermodynamics has also been used to guide thermomechanical treatment of alloys for performance improvements. One of the successful examples is its application to alloy 800H that is usually solution annealed at 1177°C followed by water quench. It has been found that a low level of cold work, for example, <10%, followed by annealing at elevated temperatures can significantly increase the population of twin boundaries in face-centered cubic alloys with a low level of stacking fault energies [316]. In the

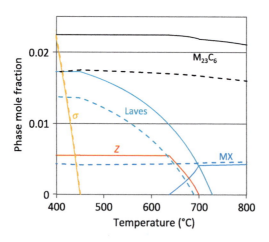

FIG. 4.27

Calculated phase fraction vs. temperature in T92 (solid lines) for the MJT heat [315] and modified Grade 92 with an adjusted composition (dashed lines).

Data from K. Kimura, K. Sawada, H. Kushima, Creep rupture ductility of creep strength enhanced ferritic steels, Proceedings of the ASME 2010 Pressure Vessels & Piping Division/K-PVP Conference, July 18–22, 2010, Bellevue, Washington, USA, paper ID PVP2010-25297.

cases of precipitate-strengthened alloys, such processing heat treatment would also alter the formation of precipitates. The annealing temperature of 1050°C after ~6% cold work on alloy 800H was advised by computational thermodynamics calculations to favor the formation of TiC and prevent the formation of a large amount of $M_{23}C_6$ [317]. The special treatment increased, by ~36% the length fraction of special low-Σ, low energy, boundaries (improving the corrosion/oxidation resistance [318]) and introduced TiC nanoprecipitates in matrix (improving the alloy strength [319] and irradiation resistance by reducing the radiated-induced formation of γ'-Ni_3(Ti,Al) [320, 321]).

4.4.3 ALLOY SELECTION AND DESIGN

Oxidation resistance is an important property for alloys to be used in power plants. It can be qualitatively evaluated by computational thermodynamics. The microstructures of oxide scales that form on FM steels, austenitic stainless steels, and Ni-base superalloys exposed to supercritical water were evaluated by Tan and co-workers [322–324] by simulating the oxide phases that form at different oxygen chemical potentials in the scale. The calculated results were consistent with the layout of the experimentally observed scales. Fig. 4.28 shows an example of using computational thermodynamics to help screen the oxidation resistance of candidate steels exposed to steam or pressurized water. As shown in the calculated oxide stability diagram of Fe-Ni-20Cr (at.%) at 1000°C in Fig. 4.28A, to prevent the formation of FeO, the Fe content needs to be less than ~53 at.%. The shaded box (pink) on the right specifies a series of alloys (~47–53 at.% Fe) favoring the formation of Cr_2O_3, spinel-1 (Fe-rich M_3O_4 with some Cr and limited Ni), spinel-2 (M_3O_4 with Fe, Ni, and Cr), and M_2O_3 (primarily Fe_2O_3). Alloy 800H is one of the alloys in this region, and its oxidation behavior is exemplified in Ref. [326]. In contrast, the oxide scale formed on the alloys in the blue shaded box on the left (~40–47 at.% Fe) is expected to be dominantly spinel-2 at the surface. Alloy DS is one of the alloys in this region. Because the spinel-2 with greater Cr and Ni contents is more oxidation resistant and has greater resistance to exfoliation than Fe_2O_3 and Fe_3O_4 [325], alloy DS is expected to have oxidation resistance superior to alloy 800H. This expectation is confirmed by the steam oxidation testing results as shown in Fig. 4.28B [327].

The previous examples demonstrated that computational thermodynamics can provide invaluable insight and guidance for complex alloy systems, and can enable more efficient alloy modification and selection. The success of computational thermodynamics is highly dependent on the reliability of thermodynamic databases. Multicomponent thermodynamic databases need to be validated simultaneously by low-order phase diagrams (such as binary and ternary systems) and phase properties of multicomponent alloys. However, experimental validation is an indispensable step for practical alloy development. New alloy development process would be significantly accelerated by using computational thermodynamics-assisted alloy design and the following kinetic and mechanical property simulations, which would greatly reduce experimental efforts.

4.4.4 KINETICS AND MECHANICAL PROPERTY SIMULATIONS

In addition to constructing phase diagrams at equilibrium states, the Calphad method is now becoming more and more integrated with kinetics simulations for studying phase transformation in alloys under thermal ageing and external forces. Diffusion and precipitation processes in multicomponent alloy systems are important kinetics simulation topics. A few integrated simulation tools, such as Dictra and TC-Prisma (products of Thermo-Calc), JMatPro, MatCalc, Panprecipitation (product of CompuTerm), etc., have been developed to conduct these types of simulations. A variety of models have evolved to be able to simultaneously simulate precipitation stages in multicomponent, multiphase alloys for any given thermal treatment, time, temperature, and alloy chemistry [328].

Although thermodynamic property data have been modeled to describe equilibrium conditions, these data also provide information on thermodynamic quantities in the metastable state, from which the chemical driving forces can be derived and then used in models describing kinetic processes. It has been found that the accuracy and reliability of thermodynamic databases are essential for underpinning reliable and predictive kinetic simulation

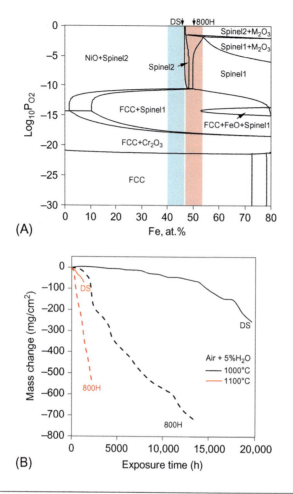

FIG. 4.28

Oxide stability diagram of Fe-Ni-20Cr (at.%) at 1000°C [326] and mass changes of alloys 800H and DS exposed to steam at 1000°C and 1100°C (adapted from [327]). P_{O_2} is the partial pressure of O_2.

Adapted from L. Tan, X. Ren, K. Sridharan, T.R. Allen, Effect of shot-peening on the oxidation of alloy 800H exposed to supercritical water and cyclic oxidation, Corros. Sci. 50 (2008) 2040–2046; G.Y. Lai, High Temperature Corrosion and Materials Applications, ASM International, Materials Park, OH, USA, 2007.

results [329]. Experimental data points from precipitation in a 316L stainless steel aged at 815°C [330] are presented in Fig. 4.29. Fig. 4.29A also shows the calculated evolution of the precipitates using one of the latest commercial databases whereas Fig. 4.29B shows the calculated results using the database being developed at Oak Ridge National Laboratory (OCTANT: ORNL Computational Thermodynamics for Applied Nuclear Technology) [328]. Both plots were calculated using the MatCalc software and identical kinetics simulation parameters. Fig. 4.29A shows no Sigma phase at this temperature, while Fig. 4.29B shows the precipitation of Sigma phase before Laves phase. The effect of the thermodynamic databases on the precipitation kinetics simulation is evident.

Precipitate simulation results are the major inputs for mechanical property simulations that have been incorporated into tools such as JMatPro and MatCalc. For example, creep strength at given service temperatures and times could be simulated or predicted based on computational thermodynamics and precipitation simulations. The difference between the predicted creep strength to the experimental data could be within about 30 MPa

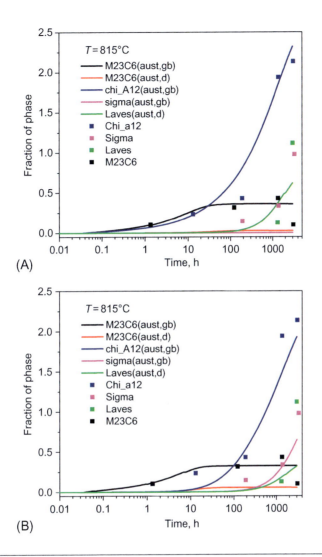

FIG. 4.29

Calculated time-dependent fraction of phases in a 316L stainless steel at 815°C using (A) a latest commercial database and (B) the being developed OCTANT database [329]. The experimental data are illustrated in different symbols [330].

Adapted from Y. Yang, J. T. Busby, Thermodynamic modeling and kinetics simulation of precipitate phases in AISI 316 stainless steels, J. Nucl. Mater. 448 (2014) 282–293. and B. Weiss and R. Stickler, Phase instabilities during high temperature exposure of 316 austenitic stainless steel, Metallurgical Transactions, vol. 3, 1972, p. 851–66.

[331]. It is expected that the prediction fidelity would be improved by optimizing databases as well as the development of models and software.

4.4.5 SUMMARY

Multiscale modeling methods have been developed for specific applications to understand a broad spectrum of microstructural evolution and resulting property changes. This section focused on the applications of computational thermodynamics, primarily based on the Calphad method, and kinetics simulations to the processing, selection, optimization, and design of multicomponent multiphase engineering alloys. The accuracy and reliability of thermodynamic database are essential to the calculation fidelity of not only the phase stabilities in equilibria states, but also the kinetics and followed mechanical property simulations. The advance of the computational tools as well as the databases including both thermodynamics and kinetics would facilitate problem solving and significantly accelerate the cost-effective development of advanced alloys that benefit the safety margins, design flexibility, and economics of advanced nuclear reactors.

REFERENCES

[1] Technology Roadmap Update for Generation IV Nuclear Energy Systems January 2014. Issued by the OECD Nuclear Energy Agency for the Generation IV International Forum January 2014.
[2] S. Paidassi, EU fusion technology: towards a commercial power reactor, in: Fusion Engineering and Design, 49–50, November 2000, pp. 53–63.
[3] Workshop on "Science Applications of a Triple Beam Capability for Advanced Nuclear Energy Materials", Chairs Michael Fluss & Wayne E. King, LLNL-MI-413125, 2009.
[4] J.R. Hawthorne, Treatise on materials science & technology, in: Embrittlement of Engineering Alloys, 25, 1983, pp. 461–524.
[5] Radiation Damage of Structural Materials, Changes in the properties of Zr alloys as a result of irradiation, in: Materials Science Monographs, 79, 1994, pp. 264–288.
[6] Proceedings of a conference on "Radiation-induced voids in metals": International Conference, Albany, June 1971, James W. Corbett, Louis C. Ianniello Eds. CONF-710601. AEC Symposium Series No. 26, 1972.
[7] HOUSE OF LORDS, Select Committee on Science and Technology 3rd Report of Session 2010–12 Nuclear Research and Development Capabilities, November 2011 HL 221. ISBN 978 0 10 847395 1.
[8] A.K. Seeger, Proc. R. Soc. London A371 (1980) 165.
[9] M.W. Thompson, Defects and Irradiation Damage in Metals, Cambridge University Press, Cambridge, 1969.
[10] R.S. Barnes, G.B. Relding, J. Nucl. Energy (A) Reactor Sci. 10 (1959) 32.
[11] R.S. Barnes, D.J. Mazey, Phil Mag. 5 (1960) 1247.
[12] C. Cawthorne, E.J. Fulton, UKAEA Unclassified Report AERE-R5269, Nature 216 (1967) (1966) 1.
[13] R.S. Nelson, D.J. Mazey, J.A. Hudson, J. Nucl. Mater. 37 (1970) 1.
[14] J.H. Worth, International conference on the use of cyclotrons in chemistry, in: C.B. Amphlett (Ed.), Metallurgy and Biology, Butterworths, London, 1970, p. 282.
[15] J.L. Brimhall, E.P. Simonen, J. Nucl. Mater. 68 (1977) 235–243.
[16] S.J. Zinkle, A. Möslang, Evaluation of irradiation facility options for fusion materials research and development, Orig. Res. Art. Fus. Eng. Des. 88 (2013) 472–482.
[17] P. Garin, M. Sugimoto, Status of IFMIF design and R&D, Fusion Eng. Des. 83 (7–9) (2008) 971–975.
[18] I. Chant, K.L. Murty, Structural materials issues for the next generation fission reactors, JOM 62 (9) (September 2010) 67–74.
[19] C.A. English, J.M. Hyde, Radiation damage of reactor pressure vessel steels, in: R. Konings (Ed.), Comprehensive Nuclear Material, Elsevier, 2012. ISBN: 9780080560274.
[20] L.A. Giannuzzi, F.A. Stevens, Introduction to Focused Ion Beams: Instrumentation, Theory, Techniques and Practice, Springer Press, 2004. ISBN 978-0-387-23116-7.
[21] J. Li, T. Malis, S. Dionne, Recent advances in FIB-TEM specimen preparation techniques, Mater Charact (2006).
[22] H. Saka, G. Nagaya, Plan-view transmission electron microscopy observation of a crack tip in silicon, Philos Mag Lett. (1995).
[23] S. Lozano-Perez, Y. Huang, R. Langford, J.M. Titchmarsh, Preparation of TEM techniques containing stress corrosion cracks in austenitic alloys using FIB, in: Electron Microscopy and Analysis 2001, Inst Phys Conference series, No 168, Section 5, 2001, p. 191.
[24] J.M. Hyde and C.A. English, "Microstructural characterisation techniques for the study of reactor pressure vessel (RPV) embrittlement", in "Part III Techniques for the Study of Reactor Pressure Vessel (RPV) Embrittlement", Woodhead Publishing Series in Energy, in press.

REFERENCES

[25] S. Lozano-Perez. "Characterisation techniques for assessing irradiated and ageing materials in nuclear power plant systems, structures and components." In Understanding and Mitigating Ageing in Nuclear Power Plants: edited by Ph G Tipping published by Woodhead Publishing Ltd Abington Hall Cambridge, 2010, p. 38.

[26] W.E.S, Unger SURFACE ANALYSIS | Auger Electron Spectroscopy Reference Module in Chemistry, in: Molecular Sciences and Chemical Engineering, 2013.

[27] I.A. Vatter, J.M. Titchmarsh, Comparison of FEGSTEM and AES measurements of equilibrium segregation of phosphorus In 9% Cr ferritic steels, Surf. Interface Anal. 25 (1997) 760–776.

[28] F. Christien, Y. Borjon-Piron, R. Le Gall, S. Saillet, Quantifying the effect of fracture surface topography on the scattering of grain boundary segregation measurement by Auger electron spectroscopy, Mater Charact 61 (2010) 73–84.

[29] R.D. Carter, D.L. Damcott, M. Atzmon, G.S. Was, S.M. Bruemmer, E.A. Kenik, Quantitative analysis of radiation-induced grain-boundary segregation measurements, J. Nucl. Mater. 211 (July 1994) 70–84.

[30] S. Fisher, G. Knowles, B. Lee, Auger electron spectroscopy of magnox pressure vessel steels, in: R.K. Nanstad, M.L. Hamilton, F.A. Garner, S. Kumar (Eds.), Effects of Radiation on Materials, 18th International Symposium ASTM STP 1325, American Society for Testing and Materials, West Conshohocken, PA, 1999.

[31] G.J. Long, F. Grandjean (Eds.), Mössbauer Spectroscopy Applied to Magnetism and Materials Science, 1, Plenum, New York, 1993.

[32] T. Wichert, Nuclear Probes for investigating radiation damage, J. Nucl. Mater. 216 (1994) 199–219.

[33] K. Van Ouytsel, R. De Batist, A. Fabry, F. Poortmans, R. Schaller, A mechanical spectroscopy study of neutron irradiation and thermal ageing of reactor pressure vessel steels, J. Phys. IV, Colloque C8, Journal de Physique III 6 (1996). Decembre.

[34] J.A. Caro, P.R. Bloch, J. de Miguel, W. Benoit, The flux dependence of internal friction under irradiation, J. Phys. F 14 (1984) 55.

[35] M. Li, D. Olive, Y. Trenikhina, H. Ganegoda, J. Terry, S.A. Maloy, Study of irradiated mod.9Cr–1Mo steel by synchrotron extended X-ray absorption fine structure, J. Nucl. Mater. 441 (October 2013) 674–680.

[36] F. Maury, N. Lorenzelli, C.H. de Novion, P. Lagarde, EXAFS study of the copper precipitation in an FeCuMn dilute alloy, Scr. Metall. Mater. 25 (1991) 1839–1844.

[37] S. Cammelli, C. Degueldre, G. Kuri, J. Bertsch, D. Lützenkirchen-Hecht, R. Frahm, Study of atomic clusters in neutron irradiated reactor pressure vessel surveillance samples by extended X-ray absorption fine structure spectroscopy, J. Nucl. Mater. 385 (2009) 319–324.

[38] P. Ehrhart, Investigation of radiation damage by X-Ray diffraction, J. Nucl. Mater. 216 (1994) 170–198.

[39] G.S. Was, M. Hash, R.G. Odette, Hardening and microstructure evolution in proton-irradiated model and commercial pressure-vessel steels, Philos. Mag. 85 (2005) 703–722.

[40] M. Perez, V. Massardier, X. Kleber, Thermoelectric power applied to metallurgy: principle and recent applications, Int. J. Mat. Res. (formerly Z. Metallkd.) 100 (2009) 10.

[41] S. Lozano-Perez, M. Schröder, T. Yamada, T. Terachi, C.A. English, C.R.M. Grovenor, Using NanoSIMS to map trace elements in stainless steels from nuclear reactors, Appl. Surf. Sci. 255 (2008) 1541–1543.

[42] P.J. Goodhew, F.J. Humphreys, R. Beanland, Electron Microscopy and Analysis, third ed., Taylor and Frances, 2001.

[43] D.B. Williams, C.B. Carter, Transmission Electron Microscopy: A Textbook for Materials Science, Plenum Press, NY, 1996. 0-306-45324-X.

[44] D.B. Williams, C.B. Carter, Parts I–IV, Plenum Press, New York, 1996.

[45] P. Hirsch, A. Howie, R. Nicholson, D.W. Pashley, M.J. Whelan, Electron Microscopy of Thin Crystals, Butterworths/Krieger, London Malabar/FL, 1965/1977. ISBN 0-88275-376-2.

[46] D.J.H. Cockayne, I.L.F. Ray and M.J. Whelan, "Investigations of dislocation strain fields using weak beams", Phil. Mag. 20 1265-1270.

[47] M.L. Jenkins, M.A. Kirk, Characterisation of radiation damage by transmission electron microscopy, IOP Publishing Ltd, 2001. ISBM 0 7503 0748 X (hbk).

[48] R. Egerton, "Electron Energy-Loss Spectroscopy in the Electron Microscope", Springer, New York, 1986/1996.

[49] M. Boseman, M. Watanabe, D.T.L. Alexander, V.J. Keast, Mapping chemical and bonding information using multivariate analysis of electron energy loss spectrum images, Ultramicroscopy 106 (2006) 1024–1032.

[50] C.J.D. Hetherington, Strategies for one angstrom resolution, Electron Microsc Anal. (2003) 219–224.

[51] P.D. Nellist, S.J. Pennycook, The Principles and Interpretation of Annular Dark-Field Z-Contrast Imaging, Adv. Imag. Electron Phys. 113 (2000).

[52] W. Sigle, Analytical transmission electron microscopy, Annu. Rev. Mat. Res. 35 (2005) 239–314.
[53] P.G. Kotula, D.O. Klenov, H.S. von Harrach, Microsc. Microanal. 18 (4) (2012) 691–698.
[54] C.M. Parish, K. Wang, P.D. Edmondson, Viewpoint: Nanoscale chemistry and crystallography are both the obstacle and pathway to advanced radiation-tolerant materials, Scr. Mater. 143 (2018) 169–175.
[55] J.M. Titchmarsh, EDX spectrum modelling and multivariate analysis of sub-nanometer segregation, Micron 30 (1999) 159–171.
[56] I. Arslan, E.A. Marquis, M. Homer, M.A. Hekmaty, N.C. Bartelt, Towards better 3-D reconstructions by combining electron tomography and atom-probe tomography, Ultramicroscopy 108 (2008) 1579–1585.
[57] J Mayer, L.A. Giannuzzi, T Kamino, J Michael, TEM sample preparation and FIB-induced damage, MRS Bull. VOLUME 32, MAY 2007 p400.
[58] L.A. Giannuzzi, F.A. Stevie, Introduction to Focused Ion Beams: Instrumentation, Theory, Techniques, and Practice, Springer, New York, 2005.
[59] S. Lozano-Perez, A guide on FIB preparation of samples containing stress corrosion crack tips for TEM and atom-probe analysis, Micron 39 (2008) 320–328.
[60] J.J.H. Lim, M.G. Burke, Revealing nanometre-scaled solute clusters in neutron irradiated low alloy steels, Microsc. Microanal. 23 (Suppl 1) (2017).
[61] K. Naoko, Reducing focused ion beam damage to transmission electron microscopy samples. J. Electron Microsc. (Tokyo) 53 (5) (2004) 451–458, https://doi.org/10.1093/jmicro/dfh080.
[62] R.E. Smallman, K.H. Westmacott, J. Appl. Phys. 30 (1959) 603.
[63] B.L. Eyre, Phil. Mag. 7 (1962) 2107.
[64] H.G.F. Wilsdorf, D. Kuhlmann-Wilsdorf, Dislocation behavior in quenched and in neutron irradiated stainless steel, J. Nucl. Mater. 5 (1962) 178–192.
[65] J. Silcox, P.B. Hirsch, Phil Mag. 4 (1959) 72.
[66] B.L. Eyre and J.R. Matthews, "Technological impact of microstructural evolution during irradiation", Volume 205, October 1993, Pages 1–15, https://doi.org/10.1016/0022-3115(93)90066-8
[67] C. Cawthorne, E.J. Fulton, Nature 216 (1976) 25.
[68] P.J. Maziasz, J. Nucl. Mater. 191–194 (1992) 701.
[69] H.R. Brager, J.L. Straalsund, J. Nucl. Mater. 46 (1973) 134.
[70] S. Hamada, M. Suzuki, P.J. Maziasz, A. Hashinuma, M.P. Tanaka, N.H. Packan, R.E. Stoller, A.S. Kumar (Eds.), Effects of Radiation on Materials, vol. I, 1989, p. 172. ASTM-STP 1046.
[71] H.R. Brager, F.A. Garner, E.R. Gilbert, J.E. Flinn, W.G. Wolfer, M.L. Bleiberg, J.W. Bennett (Eds.), Radiation Effects in Breeder Reactor Structural Materials, TMS-AIME, New York, 1977, p. 727.
[72] S.J. Zinkle, R.L. Sindelar, J. Nucl. Mater. 155–157 (1988) 1196.
[73] K.Q. Bagley, J.I. Brammman, C. Cawthorne, in: S.F. Pugh, M.H. Loretto, D.I.R. Norris (Eds.), Voids Formed by Irradiation of Reactor Materials, BNES, 1971. p. 1.
[74] K. Farrell, N.H. Packan, in: H.R. Brager, J.S. Perrin (Eds.), Effects of Radiation on Materials, 1982, p. 953. ASTM-STP 782.
[75] A. Bardal, K. Lie, Measuring the thickness of aluminium alloy thin foils using electron energy loss spectroscopy, Mater Charact 44 (2000) 329–343.
[76] R.F. Egerton, Electron Energy Loss Spectroscopy in the Electron Microscope, second ed., Plenum Press, New York & London, 1996.
[77] S.M. Allena, E.L. Hallb, Foil thickness measurements from convergent-beam diffraction patterns: An experimental assessment of errors, Philos.l Mag. A 46 (2) (1982) 243–253.
[78] T.M.Williams, R.M.Boothby and J.M.Titchmarsh, in Proc Int Conf on Materials for Nuclear Reactor Core applications BNES London (1987), p. 293.
[79] R. Lindau, A. Moeslang, M. Schirra, P. Schlossmacher, M. Klimenkov, J. Nucl. Mater 307-311 (2002) 769–772.
[80] L. Hsiung, M. Fluss, S. Tumey, J. Nucl. Mater 409 (2011) 72–79.
[81] R.L. Klueh, D.R. Harries, High-Chromium Ferritic and Martensitic Steels for Nuclear Applications, ASTM, West Conshohocken, PA, 2001.
[82] Y. Wu, J. Ciston, S. Krämer, N. Bailey, G.R. Odette, P. Hosemann, The crystal structure, orientation relationships and interfaces of the nanoscale oxides in nanostructured ferritic alloys, Acta Mater. 111 (2016) 108–115.
[83] V. de Castro, P. Rodrigo, E.A. Marquis, S. Lozano-Perez, Oxide dispersion strengthened Fe–12Cr steel in three dimensions: An electron tomography study, J. Nucl. Mater. 444 (2014) 416–420.

REFERENCES

[84] O. Weiss, E. Gaganidze, J. Aktaa, J. Nucl. Mater. 426 (2012) 52.

[85] M. Klimenkov, A. Moslang, E. Materna-Morris, J. Nucl. Mater. 453 (2014) 54.

[86] D. Brimbal, L. Beck, O.r. Troeber, E. Gaganidze, P. Trocellier, J. Aktaa, R. Lindau, Microstructural characterization of Eurofer-97 and Eurofer-ODS steels before and after multi-beam ion irradiations at JANNUS Saclay facility, J. Nucl. Mater. 465 (2015) 236–244.

[87] M.J. Swenson, J.P. Wharry, The comparison of microstructure and nanocluster evolution in proton and neutron irradiated Fe 9%Cr ODS steel to 3 dpa at 500°C, J. Nucl. Mater. 467 (2015) 97–112.

[88] Y. Wu, J. Ciston, S. Kraemer, N. Bailey, G.R. Odette, P. Hosemann, Acta Mater. 111 (2016) 108–115.

[89] A.J. Jacobs, G.P. Wozadlo, K. Nakata, T. Yoshida, I. Masaoka, The Correlation of grain boundary composition in irradiated stainless steel with IASCC resistance, in: G. Theus, J.R. Weeks (Eds.), Proceedings of the 3rd International Conference on the Environmental Degradation of Materials in Nuclear Power Systems—Water Reactors, TMS, Warrendale, PA, 1988. p. 673.

[90] R. Pathania, K. Gott, P. Scott, An Overview of the Cooperative IASCC Research (CIR) Program, in: 13th International Conference on environmental Degradation of Materials in Nuclear Power Systems, Canadian Nuclear Society, Whistler, B.C., 2007.

[91] D.J. Edwards, E.P. Simonen, S.M. Bruemmer, Comparison of Microstructural Evolution in LWR and Fast-reactor Irradiations of AISI 304 and 316 Stainless Steels, in: Proc. of International Symposium Fontevraud VI: Contributions of Materials Investigations to Improve the Safety and Performance of Light-Water Reactors, French Nuclear Energy Society, 2006, p. 603.

[92] E.P. Simonen, D.J. Edwards, S.M. Bruemmer, Microstructural Evolution in Neutron-Irradiated Stainless Steels: Comparison of LWR and Fast-Reactor Irradiations, in: 12th International Conference on Environmental Degradation of Materials in Nuclear Power Systems—Water Reactors, Salt Lake City, 2005, p. 429.

[93] E. Simonen, S. Bruemmer, Radiation-induced segregation behavior in austenitic stainless steels: fast reactor versus light water reactor irradiations, in: D.J. Edwards (Ed.), 13th International Conference on environmental Degradation of Materials in Nuclear Power Systems, Canadian Nuclear Society, Whistler, B.C., 2007.

[94] S. Bruemmer, New issues concerning radiation-induced material changes and irradiation-assisted stress corrosion cracking in light-water reactors, in: 10th International Conference on environmental Degradation of Materials in Nuclear Power Systems, Lake Tahoe, Nevada, August 5 to 9, 2001.

[95] C. Pokor, Y. Brechet, P. Dubuisson, J.-P. Massoud, A. Barbu, Irradiation damage in 304 and 316 stainless steels experimental investigation and modeling. Part I: Evolution of the microstructure, J. Nucl. Mater. 326 (2004) 19–29.

[96] K. Fukuya, K. Fujii, H. Nishioka, Y. Kitsunai, J. Nucl. Sci. Tech. 43 (2006) 159–173.

[97] F.A. Garner, Chapter 4.02: Radiation damage in austenitic steels, in: R.J.M. Konings (Editor-in-Chief) Comprehensive Nuclear Materials, vol. 4, 2012, pp.33–95 (Chapter 4.02).

[98] D.J. Edwards, E.P. Simonen, S.M. Bruemmer, Evolution of fine-scale defects in stainless steels neutron-irradiated at 275°C, J. Nucl. Mater. 317 (2003) 13–31.

[99] D.J. Edwards, E.P. Simonen, F.A. Garner, L.R. Greenwood, B.M. Oliver, S.M. Bruemmer, Influence of irradiation temperature and dose gradients on the microstructural evolution in neutron-irradiated 316SS, J. Nucl. Mater. 317 (2003) 32–45.

[100] J.P. Foster, D.L. Porter, D.L. Harrod, T.R. Mager, M.G. Burke, J. Nucl. Mater. 224 (1995) 207.

[101] S. Hamada, P.J. Maziasz, M.P. Tanaka, M. Suzuki, A. Hishinuma, J. Nucl. Mater. 155–157 (1988) 838.

[102] L.E. Thomas, J.M. Beeston, J. Nucl. Mater. 107 (1982) 159.

[103] K. Fujii, K. Fukuya, Characterization of defect clusters in ion-irradiated A533B steel, J. Nucl. Mater. 336 (2005) 323–330.

[104] W.J. Phythian, A.J.E. Foreman, C.A. English, J.T. Buswell, M. Hetherington, K. Roberts, S. Pizzini, The structure and hardening mechanism of copper precipitation in thermally aged or irradiated Fe-Cu and Fe-Cu-Ni model alloys, in: R.E. Stoller, A.S. Kumar, D.S. Gelles (Eds.), Effects of Radiation on Materials: 15th International Symposium, ASTM STP 1125, ASTM Philadelphia, 1992, pp. 131–150.

[105] C.A. English, J.M. Hyde, S. Dumbill, S.R. Ortner, Determining Matrix Cu in RPV Steels, in: IAEA/LMNPP Specialists Meeting On "Irradiation Embrittlement And Mitigation" held Gloucester, England, U.K. 14th-17th, May 2001.

[106] C. Zheng, D. Kaoumi, Radiation-induced swelling and radiation-induced segregation & precipitation in dual beam irradiated Ferritic/Martensitic HT9 steel, Mater Charact 134 (2017) 152–162.

[107] A. Harte, M. Topping, P. Frankel, D. Jadernas, J. Romero, L. Hallstadius, E.C. Darby, M. Preuss, Nano-scale chemical evolution in a proton-and neutron-irradiated Zr Alloy, J. Nucl. Mater. 487 (2017) 30–42.

[108] A. Yilmazbayhan, E. Breval, A.T. Motta, R.J. Comstock, J. Nucl. Mater. 349 (2006) 265–289.
[109] D. Pêcheur, J. Godlewski, P. Billot, J. Thomazet, Microstructure of oxide films formed during the waterside corrosion of the Zircaloy-4 cladding in lithiated environment, in: E.R. Bradley, G.P. Sabol (Eds.), Zirconium in the nuclear industry: eleventh international symposium, ASTM STP 1295, ASTM International, West Conshohocken, PA, 1996.
[110] M. Oskarsson, E. Ahlberg, U. Sodervall, U. Andersson, K. Pettersson, J. Nucl. Mater. 289 (2001) 315.
[111] M. Oskarsson, E. Ahlberg, U. Andersson, K. Pettersson, J. Nucl. Mater. 297 (2001) 77.
[112] W.F. Hsieh, J.J. Kai, C.H. Tsai, X.J. Guo, Mater. Chem. Phys. 50 (1997) 37.
[113] J.-Y. Park, S.J. Yoo, B.-K. Choi, Y.H. Jeong, Oxide microstructures of advanced Zr alloys corroded in 360°C water loop, J. Alloys Compd. 437 (1–2) (2007) 274–279.
[114] A. Garner, A. Gholinia, P. Franke, M. Gass, I. MacLaren, M. Preuss, The microstructure and microtexture of zirconium oxide films studied by transmission electron backscatter diffraction and automated crystal orientation mapping with transmission electron microscopy, Acta Mater. 80 (2014) 159–171.
[115] J. Hua, A. Garner, N. Ni, A. Gholinia, R.J. Nicholls, S. Lozano-Perez, P. Frankel, M. Preuss, C.R.M. Grovenor, Identifying suboxide grains at the metal–oxide interface of a corroded Zr–1.0%Nb alloy using (S)TEM, transmission-EBSD and EELS, Micron 69 (2015) 35–42.
[116] N. Ni, S. Lozano-Perez, J. Sykes, C. Grovenor, Quantitative EELS analysis of zirconium alloy metal/oxide interfaces, Ultramicroscopy 111 (2011) 123–130.
[117] N. Ni, D. Hudson, J. Wei, P. Wang, S. Lozano-Perez, G.D.W. Smith, J.M. Sykes, S.S. Yardley, K.L. Moore, S. Lyon, R. Cottis, M. Preuss, C.R.M. Grovenor, How the crystallography and nanoscale chemistry of the metal/oxide interface develops during the aqueous oxidation of zirconium cladding alloys, Acta Mater. 60 (2012) 7132–7149.
[118] P. Scott, Review article: A review of irradiation assisted stress corrosion cracking, J. Nucl. Mater. 211 (1994) 101–122.
[119] S.M. Bruemmer, L.E. Thomas, High-resolution analytical electron microscopy characterization of corrosion and cracking at buried interfaces, Surf. Interface Anal. 31 (7) (2001) 571–581.
[120] V.Y. Gertsman, S.M. Bruemmer, Study of Grain Boundary Character Along Intergranular Stress Corrosion Crack Paths In Austenitic Alloys, Acta Mater. 49 (2001) 1589–1598.
[121] L.E. Thomas, S.M. Bruemmer, Corrosion J. 56 (2000) 572.
[122] Thomas, L. E. and Bruemmer, S. M., Proc. 9th Int. Conf. on Environmental Degradation of Materials in Nuclear Power Systems–Water Reactors. TMS, 2000, p. 72.
[123] S.M. Bruemmer, L.E. Thomas, Proc. Int. Conf. on Chemistry and Electrochemistry of Corrosion and Stress Corrosion. TMS; 2001.
[124] S. Lozano-Perez, P. Rodrigo, L.C. Gontard, Three-dimensional characterization of stress corrosion cracks, J. Nucl. Mater. 408 (2011) 289–295.
[125] M. Meisnar, A. Vilalta-Clementea, A. Gholinia, M. Moody, A.J. Wilkinsoa, N. Huinc, S. Lozano-Perez, Using transmission Kikuchi diffraction to study intergranular stress corrosion cracking in type 316 stainless steels, Micron 75 (2015) 1–10.
[126] S. Lozano-Perez, Novel characterization of stress corrosion cracks, J. Phys.: Conf. Ser. 126 (2008).
[127] T.T. Tsong, Atom-Probe Field Ion Microscopy, Cambridge University Press, 1990.
[128] M.K. Miller, A. Cerezo, M.G. Hetherington, G.D.W. Smith, Atom Probe Field Ion Microscopy, Oxford Science Publications—Clarendon Press, 1996.
[129] M.K. Miller, Atom Probe Tomography: Analysis at The Atomic Level, Kluwer Academic, New York, 2000.
[130] B. Gault, M.P. Moody, J.M. Cairney, S.P. Ringer, Atom probe microscopy, in: Springer Series in Materials Science, Springer, 2012.
[131] A. Cerezo, T.J. Godfrey, G.D.W. Smith, Rev. Sci. Instrum. 59 (1988) 862.
[132] P. Panayi, Great Britain Patent Application GB2426120A, November 15, 2006.
[133] P. Panayi, P.H. Clifton, G. Lloyd, G. Shellswell, A. Cerezo, A wide angle achromatic reflectron for the atom probe, in: IVNC 2006/IFES 2006, 2006. p. 63.
[134] P.B. Wells, T. Yamamoto, B. Miller, T. Milot, et al., Evolution of manganese–nickel–silicon-dominated phases in highly irradiated reactor pressure vessel steels, Acta Mater. 80 (2014) 205–219.
[135] B. Cox, J. Nucl. Mater. 336 (2005) 331.
[136] M.K. Miller, A. Cerezo, M.G. Hetherington, G.D.W. Smith, Atom Probe Field Ion Microscopy, Oxford Science Publications—Clarendon Press, Oxford, 1996.
[137] D.W. Saxey, J.M. Cairney, D. McGrouther, T. Honma, S.P. Ringer, Atom probe specimen fabrication methods using a dual FIB/SEM, Ultramicroscopy 107 (9) (September 2007) 756–760.

REFERENCES

[138] M. P. Moody, A. V. Ceguerra, A. J. Breen, X. Y. Cui, B. Gault, L. T. Stephenson, R. K. W. Marceau, R. C. Powles and S. P. Ringer, "Atomically resolved tomography to directly inform simulations for structure–property relationships, Nat. Commun. 5, Article number: 5501, https://doi.org/10.1038/ncomms6501.

[139] B. Gault, M.P. Moody, J.M. Cairney, S.P. Ringer, Atom Probe Microscopy, first ed., Springer, New York, 2012.

[140] F. Vurpillot, A. Cerezo, D. Blavette, D.J. Larson, Microsc. Microanal. 10 (2004) 384–390.

[141] A. Morley, G. Sha, S. Hirosawa, A. Cerezo, G.D.W. Smith, Ultramicroscopy 109 (2009) 535–540.

[142] F. Vurpillot, A. Bostel, D. Blavette, Appl. Phys. Lett. 76 (2000) 3127, https://doi.org/10.1063/1.126545.

[143] E.A. Marquis, F. Vurpillot, Microsc. Microanal. 14 (2008) 561.

[144] J.M. Hyde, P. Styman, H. Weekes, et al., Analysis of radiation damage in light water reactors: development of standard protocols for the analysis of atom probe data, Microsc. Microanal. 23 (2) (2017 Apr) 366–375, https://doi.org/10.1017/S1431927616012678.

[145] F. DeGeuser, W. Lefebvre, F. Danoix, F. Vurpillot, B. Forbord, D. Blavette, An improved reconstruction procedure for the correction of local magnification effects in three-dimensional atom-probe, Surf. Interface Anal. 39 (2007) 268–272.

[146] F. Vurpillot, A. Bostel, A. Menand, D. Blavette, Trajectories of field emitted ions in 3D atom-probe, Eur. Phys. J. 6 (1999) 217–221, https://doi.org/10.1051/epjap:1999173.

[147] C. Oberdorfer, S.M. Eich, G. Schmitz, A full-scale simulation approach for atom probe tomography, Ultramicroscopy 128 (2013) 55–67.

[148] F. Vurpillot, C. Oberdorfer, Modelling Atom Probe Tomography: A Review, Ultramicroscopy 132 (2013) 19–30.

[149] F. Vurpillot, C. Oberdorfer, Modeling atom probe tomography: a review, Ultramicroscopy, Volume 159, Part 2, 2015, 202–216.

[150] http://www.uni-stuttgart.de/imw/mp/forschung/atom_probe_RD_center/software.en.html. Accessed 13 June 2018.

[151] E.A. Marquis, B.P. Geiser, T.J. Prosa, D. Larson, Evolution of tip shape during field evaporation of complex multilayer structures, J. Microsc. 241 (3) (2011) 225–233.

[152] P.D. Edmondson, C.M. Parish, R.K. Nanstad, Using complimentary microscopy methods to examine Ni-Mn-Si-precipitates in highly-irradiated reactor pressure vessel steels, Acta Mater. 134 (2017) 31–39.

[153] S. Shu, B.D. Wirth, P.B. Wells, D.D. Morgan, G.R. Odette, Multi-technique characterization of the precipitates in thermally aged and neutron irradiated Fe-Cu and Fe-Cu-Mn model alloys: Atom probe tomography reconstruction implications, Acta Mater. 146 (2018) 237–252.

[154] C. Hatzoglou, B. Radiguet, P. Pareige, Experimental artefacts occurring during atom probe tomography analysis of oxide nanoparticles in metallic matrix: Quantification and correction, J. Nucl. Mater. 492 (2017) 279–291.

[155] D.N. Seidman, The direct observation of point defects in irradiated or quenched metals by quantitative field ion microscopy, J. Phys. F 3 (Feb 1973).

[156] S.S. Brenner, R. Wagner, J.A. Spitznagel, Metall. Trans. A. 9A (1978) 1761.

[157] M.K. Miller, J. Physiol. Paris 47 (1986) 493.

[158] J.T. Buswell, C.A. English, M.G. Hetherington, W.J. Phythian, G.D.W. Smith, G.M. Worrall, Effects of Radiation on Materials: 14th International Symposium, in: N.H. Packan, E.E. Stoller, A.S. Kumar (Eds.), ASTM STP 1046, Vol. II, 1990. p. 127.

[159] M.K. Miller and M.G. Burke, in "Effects of Radiation on Materials: 14th International Symposium, Eds N.H. Packan, E.E. Stoller and A.S. Kumar (1990) ASTM STP 1046, Vol. II, p. 107.

[160] M.K. Miller, K.F. Russell, Appl. Surf. Sci. 94–95 (1996) 378–383.

[161] C.A. English, S.R. Ortner, G. Gage, W.L. Server, S.T. Rosinski, Review of phosphorus segregation and intergranular embrittlement in reactor pressure vessel steels, in: Effects of Radiation Materials, ASTM STP 1405, 2001, pp. 151–173.

[162] E.A. Marquis, J.M. Hyde, D.W. Saxey, S. Lozano-Perez, V. de Castro, D. Hudson, C. Williams, R. Hu, S. Humphry-Baker, G.D.W. Smith, Atomic-scale characterisation of nuclear reactor materials, Mater. Today 12 (11) (November 2009) 30–37. http://www.sciencedirect.com/science/article/pii/S1369702109702962#bib23.

[163] M.K. Miller, K.F. Russell, J. Nucl. Mater. 371 (2007) 145–160.

[164] E.A. Marquis and J.M. Hyde, "Atomic scale analysis of solute behaviours by atom-probe tomography", Mater. Sci. Eng. R: Rep., volume 69, issues 4–5, July 2010, pp. 37-62.

[165] G.R. Odette, in: Proc. Mater. Res. Soc. Symp., vol. 373, 1995, pp. 137–148.

[166] M.G. Burke, R.J. Stofanak, J.M. Hyde, C.A. English, W.L. Server, Microstructural aspects of irradiation damage in A508 Gr 4N forging steel: composition and flux effects, in: Effects of Radiation on Materials: 21st International Symposium ASTM-SP 1447, American Society for Testing and Materials, Tuscon, AZ, 2004, pp. 194–207.

[167] M.K. Miller, K.A. Powers, R.K. Nanstad, P. Efsing, Atom probe tomography characterizations of high nickel, low copper surveillance RPR welds irradiated to high fluences, J. Nucl. Mater. 437 (2013) 107–115.

[168] M.K. Miller, K.F. Russell, J. Kocik, E. Keilova, Embrittlement of low copper VVER 440 surveillance samples neutron-irradiated to high fluences, J. Nucl. Mater. 282 (2000) 83–88.

[169] M.K. Miller, M.A. Sokolov, R.K. Nanstad, K.F. Russell, APT characterization of high nickel RPV steels, J. Nucl. Mater. 351 (2006) 187–196.

[170] H. Ke, P. Wells, P.D. Edmondson, N. Almirall, L. Barnard, G.R. Odette, D. Morgan, Thermodynamic and kinetic modeling of Mn-Ni-Si precipitates in low-Cu reactor pressure vessel steels, Acta Mater. 138 (2017) 10–26.

[171] P.D. Styman, J.M. Hyde, D. Parfitt, K. Wilford, M.G. Burke, C.A. English, P. Efsing, Post-Irradiation Annealing of Ni-Mn-Si-Enriched Clusters in a Neutron-Irradiated RPV Steel Using Atom Probe Tomography, J. Nucl. Mater. 459 (2015) 127–134.

[172] G.R. Odette, T. Yamamoto, P.B. Wells, N. Almirall, et al., Update on the ATR-2 reactor pressure vessel steel high fluence irradiation project, 2017. UCSB ATR-2, 2017-2.

[173] A. Yilmazbayhan, A.T. Motta, R.J. Comstock, G.P. Sabol, B. Lai, Z. Cai, J. Nucl. Mater. 324 (2004) 6–22.

[174] B. de Gabory, Y. Dong, A. Motta, E. Marquis, Atom probe tomography study of the evolution of the metal/oxide interface during zirconium alloy oxidation, J. Nucl. Mater. 462 (2015) 304–309.

[175] N. Ni, D. Hudson, J. Wei, P. Wang, S. Lozano-Perez, G.D.W. Smith, J.M. Sykes, S.S. Yardley, K.L. Moore, S. Lyon, R. Cottis, M. Preuss, C.R.M. Grovenor, How the crystallography and nanoscale chemistry of the metal/oxide interface develops during the aqueous oxidation of zirconium cladding alloys, Acta Mater. 60 (20) (2012) 7132–7149.

[176] Y. Dong, A.T. Motta, E.A. Marquis, Atom probe tomography study of alloying element distributions in Zr alloys and their oxides, J. Nucl. Mater. 442 (1–3) (2013) 270–281.

[177] G. Sundell, M. Thuvander, H.-O. Andrén, Corros. Sci. 65 (2012) 10–12.

[178] P. Tejland, M. Thuvander, H.-O. Andrén, S. Ciurea, T. Andersson, M. Dahlbäck, L. Hallstadius, ASTM STP, 1529 (2011), pp. 595-619

[179] P. Tejland, H.-O. Andrén, J. Nucl. Mater. 430 (2012) 64–71.

[180] G. Sundell, M. Thuvander, P. Tejland, M. Dahlbäck, L. Hallstadius, H.-O. Andrén, Redistribution of alloying elements in Zircaloy-2 after in-reactor exposure, J. Nucl. Mater. 454 (1–3) (2014) 178–185.

[181] K. Arioka, et al., Corrosion 62 (2006) 568.

[182] Couvant, T., et al., In International Symposium on Contribution of Materials Investigations to Improve the Safety and Performance of LWRs, SFEN (French Nuclear Energy Society), Fontevraud (2006) 1-2.

[183] G.S. Was, P. Andresen, JOM 44 (1992). p. 8.

[184] S.M. Bruemmer, E.P. Simonen, P.M. Scott, P.L. Andresen, G.S. Was, J.L. Nelson, J. Nucl. Mater. 274 (1999) 299–314.

[185] Lozano-Perez, S., et al., In 14th International conference on environmental degradation of materials in nuclear power systems-water reactors, Virginia Beach, USA, (2009).

[186] S. Lozano-Perez, D.W. Saxey, T. Yamada, T. Terachi, Atom-probe tomography characterization of the oxidation of stainless steel, Scr. Mater. 62 (2010) 855–858.

[187] A. Etienne, B. Radiguet, N.J. Cunningham, G.R. Odette, P. Pareige, Atomic scale investigation of radiation-induced segregation in austenitic stainless steels, J. Nucl. Mater. 406 (2) (2010) 244–250.

[188] A. Etienne, P. Pareige, B. Radiguet, J.-P. Massoud, C. Pokor, J. Nucl. Mater. 382 (2008) 64–69.

[189] T. Yamamoto, G.R. Odette, P. Miao, D.T. Hoelzer, J. Bentley, N. Hashimoto, H. Tanigawa, R.J. Kurtz, J. Nucl. Mater. 367–370 (2007) 399–410.

[190] C.A. Williams, E.A. Marquis, A. Cerezo, G.D.W. Smith, Nanoscale characterisation of ODS–Eurofer 97 steel: An atom-probe tomography study, J. Nucl. Mater. 400 (1) (2010) 37–45.

[191] N.J. Cunningham, Y. Wu, A. Etienne, E.M. Haney, G.R. Odette, E. Stergar, D.T. Hoelzer, Y.D. Kim, B.D. Wirth, S.A. Maloy, Effect of bulk oxygen on 14YWT nanostructured ferritic alloys, J. Nucl. Mater. 444 (1–3) (2014) 35–38.

[192] M.J. Swenson, J.P. Wharry, Nanocluster irradiation evolution in Fe-9%Cr ODS and ferritic-martensitic alloys, J. Nucl. Mater. 496 (2017) 24–40.

[193] J.P. Wharry, M.J. Swenson, K.H. Yano, A review of the irradiation evolution of dispersed oxide nanoparticles in the b.c.c. Fe-Cr system: current understanding and future directions. J. Nucl. Mater. 486 (2017) 11–20, https://doi.org/10.1016/j.jnucmat.2017.01.009.

[194] K.D. Zilnyk, K.G. Pradeep, P. Choi, H.R.Z. Sandim, D. Raabe, Long-term thermal stability of nanoclusters in ODS-Eurofer steel: An atom probe tomography study, J. Nucl. Mater. 492 (2017) 142–147.

[195] C. Hatzoglou, B. Radiguet, F. Vurpillot, P. Pareige, A chemical composition correction model for nanoclusters observed by APT - Application to ODS steel nanoparticles, J. Nucl. Mater. 505 (2018) 240–248.
[196] J.M. Hyde, M.G. Burke, B. Gault, D.W. Saxey, P. Styman, K.B. Wilford, T.J. Williams, Atom probe tomography of reactor pressure vessel steels: An analysis of data integrity, Ultramicroscopy 111 (2011) 676–682.
[197] R.H. Hadfield, M.J. Stevens, S.S. Gruber, A.J. Miller, R.E. Schwall, R.P. Mirin, S.W. Nam, Single photon source characterization with a superconducting single photon detector, Opt. Express 13 (2005) 10846–10853.
[198] M.K. Miller, T.F. Kelly, The Atom TOMography (ATOM) Concept, Microsc. Microanal. 16 (2010) 1856–1857.
[199] H.A. Mook, J. Appl. Phys. 45 (1974) 43.
[200] R.W. Hendricks, J. Schelten, W. Schmatz, Philos. Mag. 36 (1974) 819.
[201] R.N. Sinclair, C.G. Windsor, in: C. Janot et al., (Ed.), Proc. Workshop on Atomic Transport and Defects in Metals by Neutron Scattering, Springer, Berlin, 1986, p. 213.
[202] M. Große, F. Eichhorn, J. Böhmert, G. Brauer, H.-G. Haubold, G. Goerigk, ASAXS and SANS investigations of the chemical composition of irradiation-induced precipitates in nuclear pressure vessel steels, Nucl. Instrum. Methods Phys. Res., Sect. B 97 (1–4) (1995) 487–490.
[203] C.A. English, A.J. Fudge, R.J. McElroy, W.J. Phythian, J.T. Buswell, C.J. Bolton, P.J.H. Heffer, R.B. Jones, T.J. Williams, Approach and methodology for condition assessment of thermal reactor pressure vessels, Int. J. Press. Vess. Pip. 54 (1–2) (1993) 49–87.
[204] R.G. Carter, N. Soneda, K. Dohi, J.M. Hyde, C.A. English, W.L. Server, Microstructural characterization of irradiation-induced Cu-enriched clusters in reactor pressure vessel steels, J. Nucl. Mater. 298 (2001) 211–224.
[205] F. Bergner, M. Lambrecht, A. Ulbricht, A. Almazouzi, Comparative small-angle neutron scattering study of neutron-irradiated Fe, Fe-based alloys and a pressure vessel steel, J. Nucl. Mater. 399 (2010) 129–136.
[206] E. Meslin, M. Lambrecht, M. Hernández-Mayoral, F. Bergner, L. Malerba, P. Pareige, B. Radiguet, A. Barbu, D. Gómez-Briceño, A. Ulbricht, A. Almazouzi, Characterization of neutron-irradiated ferritic model alloys and a RPV steel from combined APT, SANS, TEM and PAS analyses, J. Nucl. Mater. 406 (2010) 73–83.
[207] E.D. Eason, G.R. Odette, R.K. Nanstad, T. Yamamoto, M.T. EricksonKirk, A physically based correlation of irradiation-induced transition temperature shifts for RPV steels, ORNL/TM-2006/530, Oak Ridge National Laboratory, Oak Ridge, TN, 2007.
[208] Neutron News, Vol. 3, No. 3, 1992, pp. 29-37.
[209] http://www.ncnr.nist.gov/resources/n-lengths/.
[210] https://www.ill.eu/users/instruments/instruments-list/d11/description/instrument-layout/ (14/6/2018).
[211] B.D. Wirth, P. Asoka-Kumar, R.H. Howell, G.R. Odette, P.A. Sterne, Positron annihilation spectroscopy and small angle neutron scattering characterization of nanostructural features in irradiated Fe–Cu–Mn alloys, MRS Proc. 650 (2001). R6.5.
[212] O. Glatter, Determination of particle-size distribution functions from small angle scattering data by means of the indirect transformation method, J. Appl. Cryst. 13 (1980) 7–11.
[213] F. Bergner, M. Lambrecht, A. Ulbricht, A. Almazouzi, Comparative small angle neutron scattering study of neutron-irradiated Fe, Fe-based alloys and a pressure vessel steel, J. Nucl. Mater. 399 (2010) 129–136.
[214] S. Martelli, P.E. di Nunzio, Particle size distribution of nanospheres by monte carlo fitting of small angle X-ray scattering curves, Part. Part. Syst. Charact. 19 (2002) 247–255.
[215] A. Wagner, F. Bergner, A. Ulbricht, C.D. Dewhurst, Small-angle neutron scattering of low-Cu RPV steels neutron-irradiated at 255°C and post-irradiation annealed at 290°C, J. Nucl. Mater. 441 (2013) 487.
[216] R.G. Carter, N. Soneda, K. Dohi, J.M. Hyde, C.A. English, W.L. Server, Microstructural characterization of irradiation-induced Cu-enriched clusters in reactor pressure vessel steels, J. Nucl. Mater. 298 (2001) 211–224.
[217] J.A. Potton, G.J. Daniell, B.D. Rainford, Particle size distributions from SANS data using the maximum entropy method, J. Appl. Cryst. 21 (1988) 663–668.
[218] J.M. Hyde, C.A. English, An analysis of the structure of irradiation induced Cu-enriched clusters in low and high nickel welds, MRS Proc. 650 (2001). R6.6.
[219] R.B. Jones and C.J.Bolton, "Neutron radiation embrittlement studies in support of continued operation and validation of sampling of magnox reactor steel pressure vessels and components", 24th Water Reactor Safety Meeting 21-23 October 1996.
[220] M. Hasegawa, Y. Nagai, T. Toyama, Y. Nishiyama, M. Suzuki, A. Alamazouzi, E van Walle, R. Gerard, Evolution of irradiation-induced cu precipitation and defects in surveillance test specimens of pressure vessel steels of nuclear power reactors: positron annihilation and 3 dimensional atom probe study, in: B.L. Eyre, I. Kimura (Eds.), Proc Int. Symp on Research For Ageing Management of Light Water Reactors, held Fukui City, Japan October 2007, pub. INSS, Japan, 2008, pp. 327–344.

[221] A. Wagner, F. Bergner, A. Ulbricht, C.D. Dewhurst, Small-angle neutron scattering of low-Cu RPV steels neutron-irradiated at 255°C and post-irradiation annealed at 290°C, J. Nucl. Mater. 441 (2013) 487–492.

[222] G.R. Odette, T. Yamamoto, R.D. Klingensmith, On the effect of dose rate on irradiation hardening of RPV steels, Phil. Mag. 85 (2005) 779.

[223] E. D. Eason, G. R. Odette, R. K. Nanstad, T. Yamamoto, A Physically Based Correlation of Irradiation-Induced Transition Temperature Shifts for RPV Steels. ORNL/TM-2006/530.

[224] G.R. Odette, T. Yamamoto, E.D. Eason, R.K. Nanstad, On the metallurgical and irradiation variable dependence of the embrittlement of RPV steels: Converging physically based predictions and critical unresolved issues, in: B.L. Eyre, I. Kimura (Eds.), Proc Int. Symp on Research For Ageing Management Of Light Water Reactors, held Fukui City, Japan October 2007, pub. INSS, Japan, 2008, pp. 279–306.

[225] P. Wells, The Character, Stability and Consequences of Mn-Ni-Si Precipitates in Reactor Pressure Vessel Steels, Ph.D. Thesis, UCSB 2016.

[226] G. Albertini, F. Carsughi, R. Coppola, F. Fiori, F. Rustichelli, M. Stefanon, Small-angle neutron scattering microstructural investigation of MANET steel, J. Nucl. Mater. 233–237 (1996) 253–257.

[227] R. Coppola, R. Lindau, M. Magnani, R.P. May, A. Moslang, J.W. Rensmane, B. van der Schaaf, M. Valli, Microstructural investigation, using small-angle neutron scattering, of neutron irradiated Eurofer97 steel, Fusion Eng. Des. 75–79 (2005) 985–988.

[228] M.H. Mathon, M. Perrut, S.Y. Zhong, Y. de Carlan, Small angle neutron scattering study of martensitic/ferritic ODS alloys, J. Nucl. Mater. 428 (2012) 147–153.

[229] G. Albertini, F. Carsughi, R. Coppola, W. Kesternich, G. Mercurio, F. Rustichelli, D. Schwahn, H. Ullmaier, Study of He-bubble growth in MANET steel by small-angle neutron scattering, J. Nucl. Mater. 191 - 194 ((1992) 1327–1330.

[230] R. Coppola, M. Klimiankou, M. Magnani, A. Möslang, M. Valli, Helium bubble evolution in F82H-mod – correlation between SANS and TEM, J. Nucl. Mater. 329–333 (2004) 1057–1061. Part B.

[231] Z. Száraz, G. Török, V. Kršjak, P. Hähner, SANS investigation of microstructure evolution in high chromium ODS steels after thermal ageing, J. Nucl. Mater. 435 (1–3) (2013) 56–62.

[232] A. Chauhan, F. Bergner, A. Etienne, J. Aktaa, Y. de Carlan, C. Heintze, D. Litvinov, M. Hernandez-Mayoral, E. Oñorbe, B. Radiguet, A. Ulbricht, Microstructure characterization and strengthening mechanisms of oxide dispersion strengthened (ODS) Fe-9%Cr and Fe-14%Cr extruded bars, J. Nucl. Mater. 495 (2017) 6–19.

[233] A. Muñoz, M.A. Monge, B. Savoini, R. Pareja, A. Radulescu, SANS characterization of particle dispersions in W-Ti and W-V alloys, Int. J. Refract. Met. Hard Mater. 61 (2016) 173–178.

[234] M.H. Mathon, M. Perrut, S.Y. Zhong, Y. de Carlan, Small angle neutron scattering study of martensitic/ferritic ODS alloys, J. Nucl. Mater. 428 (1–3) (2012) 147–153.

[235] M. Eldrup, J. Phys. (Paris) IV Colloq. 5 (1995) C1.

[236] P. Asoka-Kumar, B.D. Wirth, P.A. Sterne, G.D. Odette, Philos. Mag. Lett. 82 (2002) 609.

[237] S.C. Glade, B.D. Wirth, G.R. Odette, P. Asoka-Kumar, Positron annihilation spectroscopy and small angle neutron scattering characterization of nanostructural features in high-nickel model reactor pressure vessel steels, J. Nucl. Mater. 351 (1–3) (2006) 197–208.

[238] Y. Nagai, Z. Tang, M. Hasegawa, Chemical analysis of precipitates in metallic alloys using coincidence Doppler broadening of positron annihilation radiation, Rad. Phys. Chem. 58 (5–6) (2000) 737–742.

[239] P. Kirkegaard, M. Eldrup, O.E. Mogensen, N.J. Pedersen, Program system for analysing positron lifetime spectra and angular correlation curves, Comput. Phys. Commun. 23 (1981) 307–335.

[240] A. Shukla, M. Peter, L. Hoffmann, Nucl. Instr. and Meth. A 335 (1993) 310.

[241] P. Hautojärvi, L. Pöllönen, A. Vehanen, J. Yli-Kauppila, J. Nucl. Mater. 114 (1983) 250.

[242] A. Vehanen, P. Hautojärvi, J. Johansson, J. Yli-Kauppila, P. Moser, Phys. Rev. B25 (1982) 762.

[243] G. Brauer, M. Sob, J. Kocik, Report ZfK-647, 1990

[244] P.J. Schultz, K.G. Lynn, Rev. Mod. Phys. 60 (1988) 701.

[245] N. Djourelova, B. Marchand, H. Marinova, N. Moncoffre, Y. Pipon, P. Nédélec, N. Toulhoat, D. Sillouf, Variable energy positron beam study of Xe-implanted uranium oxide, J. Nucl. Mater. 432 (2013) 287–293.

[246] T. Iwai, H. Tsuchida, In situ positron beam Doppler broadening measurement of ion-irradiated metals—current status and potential, Nucl. Instrum. Methods Phys. Res. B 285 (2012) 18–23.

[247] I.K. MacKenzie, T.L. Khoo, A.B. McDonald, B.T.A. McKee, Phys. Rev. 19 (1967) 94G–8.

[248] A. Seeger, J. Phys. F 3 (1973) 248, https://doi.org/10.1088/0305-4608/3/2/003.

[249] M. Eldrup, B.N. Singh, Risø-R-1241 (EN), Risø National Laboratory, 2001. p. 21.

REFERENCES

[250] M. Eldrup, B.N. Singh, Mater. Sci. Forum 363–365 (2001). 79.
[251] M. Valo, R. Krause, K. Saarinen, P. Hautojarvi, J.R. Hawthorne, Irradiation response and annealing behaviour of pressure vessel model steels and iron ternary alloys measured with positron techniques, in: R.E. Stoller, A.S. Kumar, D.S. Gelles (Eds.), Effects of Radiation on Materials: 15th International Symposium, ASTM 1125, ASTM, Philadelphia, 1992, p. 172.
[252] Y. Nagai, Z. Tang, M. Hasegawa, T. Kana, M. Saneyasu, Phys. Rev. B 63 (2001) 134110.
[253] S.C. Glade, B.D. Wirth, G.R. Odette, P. Asoka-Kumar, Positron annihilation spectroscopy and small angle neutron scattering characterization of nanostructural features in high-nickel model reactor pressure vessel steels, J. Nucl. Mater. 351 (2006) 197–208.
[254] M.J. Alinger, S.C. Glade, B.D. Wirth, G.R. Odette, T. Toyama, Y. Nagai, M. Hasegawa, Positron annihilation characterization of nanostructured ferritic alloys, Mater. Sci. Eng. A 518 (1–2) (2009) 150–157.
[255] J. Simeg Veternikova, V. Slugen, S. Sojak, M. Skarba, E. Korhonen, S. Stancek, J. Degmova, V. Sabelova, I. Bartosova, Application of slow positron beam for study of commercial oxide-dispersion-strengthened steels J. Simeg Veternikova et al. J. Nucl. Mater. (2013)https://doi.org/10.1016/j.jnucmat.2013.12.003.
[256] G.H. Dai, P. Moser and J.C. Van Duysen, Proc. 9th Int. Conf. Positron Annihilation, Szenibathely, Hungary, Aug 1991.
[257] G.E. Lucas, Effects of radiation on the mechanical properties of structural materials, J. Nucl. Mater. 216 (1994) 322–325.
[258] K. Takakura, K. Nakata, IASCC Crack Growth behaviour of neutron irradiated stainless steels, in: B.L. Eyre, I. Kimura (Eds.), Proc Int. Symp on Research for Ageing Management of Light Water Reactors, held Fukui City, Japan October 2007, pub. INSS, Japan, 2008, pp. 181–200.
[259] Delayed Hydride Cracking in Zirconium Alloys in Pressure Tube Nuclear Reactors IAEA, VIENNA, 2004 IAEA-TECDOC-1410 ISBN 92-0-110504-5 ISSN 1011–4289.
[260] C.A.English, G.Gage, R.M.Boothby, S.R.Ortner, P. Hahner, H. Stamm. AMES Report No 9: Methodology for Characterising Materials, 1997 EUR 17328 EN.
[261] J. Fohl, AMES Report No 5 Part II "Survey of existing, planned and required standards" EUR 16313 EN 1995.
[262] R.K. Nanstad, W.L. Server, M.A. Sokolov, M. Brumovsky, Evaluating the fracture toughness of reactor pressure vessel materials subject to embrittlement, in: N. Soneda (Ed.), Irradiation Embrittlment of Reactor Pressure Vessel in Nuclear Power Plants, Elsevier, 2015. https://doi.org/10.1533/9780857096470.3.295 Chapter 10, p. 295.
[263] G.R. Odette, R.K. Nanstad, Predictive reactor pressure vessel steel irradiation embrittlement models: Issues and opportunities, JOM 61 (7) (2009) 17–23.
[264] G. Robert Odette, Takuya Yamamoto, Ernest D. Eason, Randy K. Nanstad, "On the Metallurgical and Irradiation Variable Dependence of the Embrittlement of RPV Steels: Converging Mechanistic Based Predictions and Critical Unresolved Issues," Institute for Nuclear Safety System.
[265] E.D. Eason, G.R. Odette, R.K. Nanstad, T. Yamamoto, A physically-based correlation of irradiationinduced transition temperature shifts for RPV steels, J. Nucl. Mater. 433 (2013) 240–254.
[266] A. Moslang, V. Heinzel, H. Matsui, M. Sugmoto, The IFMIF test facilities design, Fusion Eng. Des. 8 (2006) 863–871.
[267] The Use of Small-Scale Specimens for Testing Irradiated Material. First International Symposium on Small Specimen Test Techniques, Sponsored by Committee E10 on Nuclear Technology and Applications Co-chairs W Corwin, G.E.Lucas. ASTM STP888. DOI: 10.1520/STP888-EB, ISBN-EB: 978-0-8031-4960-1.
[268] Sixth International Symposium on Small Specimen Test Techniques, Sponsored by Committee E10 on Nuclear Technology and Applications January 29-31, 2014 Co-Chairs: M. Sokolov, E. Lucon January 28–31, 2014.
[269] G.R. Odette, M.Y. He, E.G. Donahue, G.E. Lucas, ASTM STP 1418 (2002) 221.
[270] S.A. Maloy, T.A. Saleh, O. Anderoglu, T.J. Romero, G.R. Odette, T. Yamamoto, S. Li, J.I. Cole, R. Fielding, J. Nucl. Mater. (2015). https://doi.org/10.1016/j.jnucmat.2015.07.039.
[271] T. Yamamoto, G.R. Odette, S. Li, S. Maloy, T. Saleh, Derivation of true stress-true strain constitutive laws for irradiated ferritic steels, DOE/ER-0313/58(2015) pp. 239–245.
[272] M.B. Toloczko, B.R. Grambau, F.A. Garner, K. Abe, Comparison of thermal creep and irradiation creep of HT9 pressurized tubes from ≈ 490 to 605°C, effects of irradiation on materials, in: 20th International Symposium, American Society of Testing and Materials, ASTM_STP, 1405 2001, pp. 557–567.
[273] Y. He, G.R. Odette, T. Yamamoto, D. Klingensmith, A universal relationship between indentation hardness and flow stress. J. Nucl. Mater. 367–370 (2007) 556 560, https://doi.org/10.1016/j.jnucmat.2007.03.044.
[274] G.R. Odette, T. Yamamoto, P.B. Wells, N. Almirall, et al., Update on the ATR 2 reactor pressure vessel steel high fluence irradiation project, UCSB ATR-2 (2016). 2016-1.
[275] P. Au, G.E. Lucas, J.W. Scheckherd, G.R. Odette, Non-destructive Evaluation in the Nuclear Industry, American Society for Metals, Metals Park, OH, 1980, p. 597.

[276] F.M. Haggag, G.E. Lucas, Metall. Trans. A. 14A (1983) 1607.
[277] D. Tabor, The Hardness of Metals, Clarendon Press, Oxford, UK, 1951.
[278] C. Santos, G.R. Odette, G.E. Lucas, T. Yamamoto, J. Nucl. Mater. 258–263 (1998) 452.
[279] H. Tada, et al., The Stress Analysis of Cracks Handbook, second ed., Paris Productions, St. Louis, MO, 1985.
[280] ASTM, E23-16b, Standard Test Methods for Notched Bar Impact Testing of Metallic Materials, ASTM International, West Conshohocken, PA, 2016. https://www.astm.org.
[281] T.L. Anderson, Fracture mechanics fundamentals and applications, second ed., CRC Press, Boca Raton, Florida, 1995.
[282] G.R. Odette, T. Yamamoto, H.J. Rathbun, M.Y. He, M.L. Hribernik, J.W. Rensman, J. Nucl. Mater. 323 (2003) 313.
[283] H.J. Rathbun, G.R. Odette, M.Y. He, T. Yamamoto, "Influence of statistical and constraint loss size effects on cleavage fracture toughness in the transition—a model based analyses" Eng. Fract. Mech. vol.73 pp. 2723–2747.
[284] H.J. Rathbun, G.R. Odette, T. Yamamoto, G.E. Lucas, "Statistical stressed volume and constraint loss size effects on cleavage fracture toughness in the transition—a single variable experiment and database" Eng. Fract. Mech. vol.73 pp.134–158.
[285] E1921-97, Standard test method for determination of reference temperature, T0, for ferritic steels in the transition range, Annual Book of ASTM Standards, vol. 3.01. West Conshohocken, PA, 1998.
[286] T. Yamamoto, G.R. Odette, M.A. Sokolov, On the fracture toughness of irradiated F82H: Effects of loss of constraint and strain hardening capacity, J. Nucl. Mater. 417 (2011) 115–119.
[287] G.R. Odette, M. He, D. Gragg, et al., Some recent innovations in small specimen testing, J. Nucl. Mater. 307–311 (2002) 1643–1648.
[288] G.R. Odette, P.M. Lombrozo, R.A. Wullaert, The relation between irradiation hardening and embrittlement, in: 12th International Symposium on the Effects of Irradiation on Materials-12, American Society for Testing and Materials, 1985, pp. 841–860. ASTM-STP-870.
[289] S.J. Zinkle, W.C. Oliver, Mechanical property measurements on ion-irradiated copper and Cu-Zr, J. Nucl. Mater. 141–143 (1986) 548–552.
[290] W.C. Oliver, G.M. Pharr, J. Mater. Res. 7 (1992) 1564–1583.
[291] D. Kiener, A.M. Minor, O. Anderoglu, Y. Wang, S.A. Maloy, P. Hosemann, Application of small-scale testing for investigation of ion-beam-irradiated materials, J. Mater. Res. 27 (21) (Nov 14, 2012).
[292] G.M. Pharr, W.C. Oliver, MRS Bull. 17 (1992) 28–33.
[293] C.D. Hardie, S.G. Roberts, Nanoindentation of model Fe–Cr, J. Nucl. Mater. 433 (2013) 174–179.
[294] H. Bluckle, Met. Rev. 4 (1959) 49.
[295] J.G. Swadener, E.P. George, G.M. Pharr, J. Mech. Phys. Solids 50 (2002) 681.
[296] F. Schulz, H. Hanemann, Z. Metallkd. 33 (Heft 3) (1941).
[297] W.D. Nix, H. Gao, J. Mech. Solids 46 (3) (1998) 411.
[298] R. Kasada, Y. Takayama, K. Yabuuchi, A. Kimura, A new approach to evaluate irradiation hardening of ion-irradiated ferritic alloys by nano-indentation techniques, Fusion Eng. Des. 86 (2011) 2658–2661.
[299] P. Hosemann, J.G. Swadener, D. Kiener, G.S. Was, S.A. Maloy, N. Li, An exploratory study to determine applicability of nano-hardness and micro-compression measurements for yield stress estimation, J. Nucl. Mater. 375 (2008) 135–143.
[300] D.L. Krumwiede, T. Yamamoto, T.A. Saleh, et al., Direct comparison of nanoindentation and tensile test results on reactor-irradiated materials, J. Nucl. Mater. 504 (2018) 135–143.
[301] E.M. Grieveson, D.E.J. Armstrong, S. Xu, S.G. Roberts, Compression of self-ion implanted iron micropillars, J. Nucl. Mater. 430 (2012) 119–124.
[302] C.D. Hardie, G.R. Odette, Y. Wu, et al., Mechanical properties and plasticity size effect of Fe-6%Cr irradiated by Fe ions and by neutrons, J. Nucl. Mater. 482 (2016) 236–247.
[303] R. Lesar, Introduction to Computational Materials Science – Fundamentals to Applications, Cambridge University Press, New York, 2013.
[304] E.B. Tadmor, R.E. Miller, Modeling Materials: Continuum, Atomistic and Multiscale Techniques, Cambridge University Press, New York, 2011.
[305] K.G.F. Janssens, D. Raabe, E. Kozeschnik, M.A. Miodownik, B. Nestler, Computational Materials Engineering—An Introduction to Microstructure Evolution, Elsevier Academic Press, 2007.
[306] S.J. Zinkle, N.M. Ghoniem, Prospects for accelerated development of high performance structural materials, J. Nucl. Mater. 417 (2011) 2–8.
[307] H. Lukas, S.G. Fries, B. Sundman, Computational Thermodynamics: The Calphad Method, Cambridge University Press, 2007.

REFERENCES

[308] Commercial software of Thermo-Calc, http://www.thermocalc.com/ JMatPro http://www.sentesoftware.co.uk/ Pandat http://www.computherm.com/ MatCalc, http://matcalc.tuwien.ac.at/.

[309] T.M. Besmann, Computational thermodynamics: application to nuclear materials, in: R.J.M. Konings, T.R. Allen, R. Stoller, S. Yamanaka (Eds.), Comprehensive Nuclear Materials, Elsevier, Spain, 2012.

[310] AK Steel PH 15-7 Mo Stainless Steel (UNS S15700), AK Steel Product Data Bulletin, 2007. (http://www.aksteel.com/pdf/markets_products/stainless/precipitation/15-7_Mo_Data_Bulletin.pdf).

[311] F. Abe, Precipitate design for creep strengthening of 9% Cr tempered martensitic steel for ultra-supercritical power plants, Sci. Technol. Adv. Mater. 9 (2008). 013002.

[312] L. Tan, J.T. Busby, P.J. Maziasz, Y. Yamamoto, Effect of thermomechanical treatment on 9Cr ferritic-martensitic steels, J. Nucl. Mater. (2013), https://doi.org/10.1016/j.jnucmat.2013.01.323.

[313] C. Panait, A.-F. Gourgues-Lorenzon, J. Besson, A. Fuchsmann, W. Bendick, J. Gabrel, M. Pieete, Long-term aging effect on the creep strength of the T92 steel, in: 9th Liege Conference: Materials for Advanced Power Engineering 2010, Liege, Belgium, 2010.

[314] H.K. Danielsen, J. Hald, Influence of Z-phase on long-term creep stability of martensitic 9 to 12% Cr steels, VGB PowerTech 5 (2009) 68–73.

[315] K. Kimura, K. Sawada, H. Kushima, Creep rupture ductility of creep strength enhanced ferritic steels, Proceedings of the ASME 2010 Pressure Vessels & Piping Division/K-PVP Conference, July 18–22, 2010, Bellevue, Washington, USA, paper ID PVP2010-25297.

[316] V. Randle, Mechanism of twinning-induced grain boundary engineering in low stacking-fault energy materials, Acta Mater. 47 (1999) 4187–4196.

[317] L. Tan, L. Rakotojaona, T.R. Allen, R.K. Nanstad, J.T. Busby, Microstructure optimization of austenitic alloy 800H (Fe-21Cr-32Ni), Mater. Sci. Eng. A 528 (2001) 2755–2761.

[318] L. Tan, K. Sridharan, T.R. Allen, The effect of grain boundary engineering on the oxidation behavior of Incoloy alloy 800H, J. Nucl. Mater. 348 (2006) 263–271.

[319] L. Tan, K. Sridharan, T.R. Allen, R.K. Nanstad, D.A. McClintock, Microstructure tailoring for property improvements by grain boundary engineering, J. Nucl. Mater. 374 (2008) 270–280.

[320] R.K. Nanstad, D.A. McClintock, D.T. Hoelzer, L. Tan, T.R. Allen, High temperature irradiation effects in selected generation IV structure alloys, J. Nucl. Mater. 392 (2009) 331–340.

[321] L. Tan, J.T. Busby, H.J.M. Chichester, K. Sridharan, T.R. Allen, Thermomechanical treatment for improved neutron irradiation resistance of austenitic alloy (Fe-21Cr-32Ni), J. Nucl. Mater. 437 (2013) 70–74.

[322] L. Tan, T.R. Allen, Y. Yang, Corrosion of austenitic stainless steels and nickel-base alloys in supercritical water and novel control methods, in: S.K. Sharma (Ed.), Green Corrosion Chemistry and Engineering—Opportunities and Challenges, Wiley-VCH Verlag & Co. KGaA, Weinheim, Germany, 2012, p. 211.

[323] L. Tan, X. Ren, T.R. Allen, Corrosion behavior of 9-12%Cr ferritic-martensitic steels in supercritical water, Corros. Sci. 52 (2010) 1520–1528.

[324] L. Tan, T.R. Allen, Y. Yang, Corrosion behavior of alloy 800H (Fe-21Cr-32Ni) in supercritical water, Corros. Sci. 53 (2011) 703–711.

[325] L. Tan, X. Ren, K. Sridharan, T.R. Allen, Effect of shot-peening on the oxidation of alloy 800H exposed to supercritical water and cyclic oxidation, Corros. Sci. 50 (2008) 2040–2046.

[326] L. Tan, Y. Yang, T.R. Allen, J.T. Busby, Computational thermodynamics for interpreting oxidation of structural materials in supercritical water, in: J.T. Busby, B. Ilevbare, P.L. Andresen (Eds.), Proc. of the 15th Int. Conf. on Environmental Degradation of Materials in Nuclear Power Systems-Water Reactors, TMS, 2011, pp. 1909–1917.

[327] G.Y. Lai, High Temperature Corrosion and Materials Applications, ASM International, Materials Park, OH, USA, 2007.

[328] E. Kozeschnik, Modeling Solid-State Precipitation, Momentum Press, 2012.

[329] Y. Yang, J.T. Busby, Thermodynamic modeling and kinetics simulation of precipitate phases in AISI 316 stainless steels, J. Nucl. Mater. 448 (2014) 282–293.

[330] B. Weiss, R. Stickler, Phase instabilities during high temperature exposure of 316 austenitic stainless steel, Metall. Trans. A 3 (1972) 851–866.

[331] D.-I.I. Holzer, Modeling and Simulation of Strengthening in Complex Martensitic 9-12%Cr Steel and a Binary Fe-Cu Alloy, Ph.D. Thesis, Graz University of Technology, 2010.

CHAPTER 5

RADIATION AND THERMOMECHANICAL DEGRADATION EFFECTS IN REACTOR STRUCTURAL ALLOYS

Steven J. Zinkle[*,†], Hiroyasu Tanigawa[‡], Brian D. Wirth[*,†]

Department of Nuclear Engineering, University of Tennessee, Knoxville, TN, United States Oak Ridge National Laboratory, Oak Ridge, TN, United States† Japan Atomic Energy Agency, Rokkasho Fusion Institute, Aomori, Japan‡*

CHAPTER OUTLINE

- 5.1 Introduction 164
- 5.2 Thermomechanical Degradation Processes 166
  - 5.2.1 Thermal Aging 166
  - 5.2.2 Thermal Creep 167
  - 5.2.3 Fatigue and Creep Fatigue 171
- 5.3 Radiation Hardening and Embrittlement 173
  - 5.3.1 Dose Dependence of Low-Temperature Radiation Hardening and Ductility Reduction 173
  - 5.3.2 Temperature Dependence of Radiation Hardening and Ductility Reduction 176
  - 5.3.3 Low-Temperature Radiation Embrittlement 176
- 5.4 Radiation-Induced Phase and Microchemical Changes 179
  - 5.4.1 Amorphization 179
  - 5.4.2 Radiation-Enhanced and Induced Segregation (and Precipitation) 181
- 5.5 Radiation- and Stress-Modified Corrosion and Cracking Phenomena 184
- 5.6 Radiation-Induced Dimensional Instability 187
  - 5.6.1 Cavity Swelling 188
  - 5.6.2 Irradiation Creep 191
  - 5.6.3 Irradiation Growth 194
- 5.7 High-Temperature Helium Embrittlement 195
- 5.8 Conclusions 199
- Acknowledgment 200
- References 200

5.1 INTRODUCTION

The severe operating environment in current and proposed future nuclear reactors can lead to a variety of property degradations in structural materials during extended service. As discussed in Chapters 1–3, the chief operational concerns in current and proposed reactors are associated with high radiation damage levels that often exceed 10 displacements per atom (dpa), along with high stresses (often multiaxial) that may exceed 50–100 MPa at elevated temperatures (300–800°C) in a variety of coolants that are often corrosive. Components in nuclear energy systems typically need to reliably function for time periods of 5 to >60 years. Of greatest importance for structural alloys are phenomena that produce progressive degradation in the allowable applied stress (e.g., creep-fatigue, radiation embrittlement) or produce unacceptable changes in the dimensions of engineering components. Fig. 5.1 shows an example of lateral bowing of fuel rods in an early light-water reactor due to excessive axial expansion associated with irradiation growth of the zirconium alloy cladding [1]. This type of bowing is undesirable as it can negatively affect the thermal hydraulic and neutronic behaviors of the fuel assembly, or if severe enough can cause deformation of the entire fuel assembly that can impede the insertion of the control rods, which are important safety considerations.

Fig. 5.2 shows another example of undesirable property degradation in an operating nuclear reactor component [2]. In this figure, embrittlement and fracture are evident in an austenitic stainless steel reflector assembly duct in the BOR-60 sodium cooled fast-fission reactor exposed to a maximum dose of 34 dpa. The failure occurred during scheduled maintenance involving withdrawal of the duct, and was due to high removal stresses associated with void swelling and warping (the latter effect was due to damage gradients in the reflector).

Operating temperature is of primary importance for the thermomechanical and radiation degradation of materials. Thermal creep is generally very pronounced for operating temperatures that exceed $0.5T_M$ (where T_M is the melting temperature of the structural alloy), and may become significant at operating temperatures as low as ~ 0.35–$0.4T_M$ [3]. Property degradation due to thermal aging may also occur, particularly at elevated temperatures. The temperature dependence of radiation degradation processes is relatively more complex due to the potential presence of multiple radiation effects. As shown in Fig. 5.3 [4], at a given operating temperature there may be as many as four simultaneously occurring radiation degradation phenomena. Three important temperature stages are listed near the bottom of Fig. 5.3: Stage I (onset temperature for self-interstitial migration), Stage III (onset temperature for vacancy migration), and Stage V (onset temperature for thermal dissolution of small vacancy clusters that are directly produced by energetic

FIG. 5.1

Example of bowing of a fuel rod in an irradiated fuel assembly due to irradiation growth of the zirconium alloy cladding in an early light-water reactor that did not provide sufficient accommodation for axial expansion [1].

5.1 INTRODUCTION   165

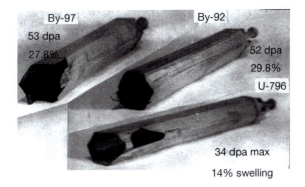

FIG. 5.2

Embrittlement and fracture in a BOR-60 reflector assembly duct made from austenitic stainless steel (Russian alloy X18H10T, analogous to Type 321 stainless steel) following irradiation to a maximum dose of 34 dpa near 400°C [2].

From F.A. Garner, D.J. Edwards, S.M. Bruemmer, S.I. Porollo, Yu. V. Konobeev, V.S. Neustroev V.K. Shamardin and A.V. Kozlov, "Recent Developments Concerning Potential Void Swelling of PWR Internals Constructed from Austenitic Stainless Steels", PROCEEDINGS, Fontevraud 5, Contribution of Materials Investigation to the Resolution of Problems Encountered in Pressurized Water Reactors, 23–27 September, 2002, paper #22. Used with permission from F.A. Garner.

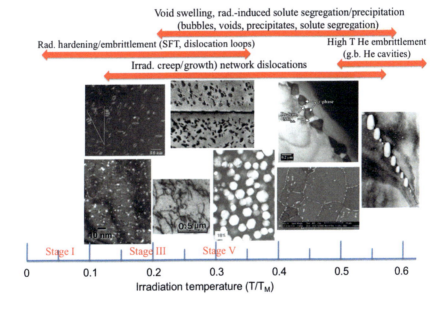

FIG. 5.3

Overview of the temperature regimes where major radiation-induced degradation mechanisms are most pronounced, along with representative microstructures.

Modified from S.J. Zinkle, L.L. Snead, Designing radiation resistance in materials for fusion energy, Annu. Rev. Mat. Res. 44 (2014) 241–267.

displacement cascades). Although the specific values for these three recovery stages and for the various radiation damage degradation phenomena depend on material-specific parameters such as vacancy migration and binding energies, in many cases these temperature regimes can be approximately correlated with the T_M fraction [5]. Considering that practical nuclear reactor operating temperatures often span temperature ranges of 100–200°C (cf. Chapters 1–3), a structural material for reactor core applications must typically be resistant to multiple radiation damage phenomena.

In the following sections, an overview is presented of the major degradation processes in structural alloys associated with elevated temperatures, applied mechanical stress, and neutron irradiation. In the absence of irradiation, key thermomechanical processes include phase instabilities due to prolonged thermal aging (with or without applied mechanical stress), thermal creep, and creep-fatigue phenomena. Radiation-induced degradation processes include low-temperature radiation hardening and embrittlement, radiation-modified solute segregation and precipitation, radiation- and stress-modified corrosion and cracking, dimensional instabilities associated with void swelling, irradiation creep and radiation growth, and high-temperature helium embrittlement. Additional material-specific details regarding most of these degradation phenomena are provided in the relevant accompanying Chapters 7–13. Degradation mechanisms associated with corrosion and stress corrosion cracking (SCC) are reviewed in Chapter 6.

5.2 THERMOMECHANICAL DEGRADATION PROCESSES
5.2.1 THERMAL AGING

In general, steels and nickel-base alloys in use or under consideration for nuclear reactor structural applications exhibit good stability during long-term thermal aging, particularly for light-water reactor operating conditions. Similarly, the Zr alloy compositions of interest for water-cooled reactor applications have good thermal stability during operation near 300°C [6]. Prolonged operation at higher temperatures that are representative of the conditions for some Generation IV fission reactors or proposed fusion reactor concepts may lead to microstructural changes (and accompanying property degradation) due to thermal aging. Experimental studies of thermal aging in austenitic steels, nickel-base alloys, and ferritic/martensitic steels are briefly summarized in the following; additional details may be found in Chapters 7–13 for specific alloy systems.

Some modest reduction in tensile elongation has been reported for Type 316 austenitic steel due to $M_{23}C_6$ precipitation on grain boundaries after aging for 3–7 years at 625°C [7]. At low temperatures relevant for light-water reactor power plants (~300°C), enhanced precipitation of $M_{23}C_6$ and Laves phases are predicted to occur only for exposure times above 5–10 years [8]. More significant thermal aging effects have been observed in austenitic steel weldments following aging for times >1 year at ~500°C, where a reduction in fracture toughness [9] and degradation in low-cycle fatigue properties [10] has been observed. These mechanical property degradations have been linked to phase transformations in the delta ferrite regions of the weld.

Nickel-base alloys have been developed for multiple commercial applications including fossil and nuclear energy systems and generally exhibit very good thermal stability up to high temperatures. A recent overview of general characteristics and properties of nickel-base alloys of interest for nuclear energy applications has been carried out by Yonezawa [11], and a review is provided in Chapter 9 for relevant reactor applications. Some phase instabilities are observed in high chromium alloys; thermal aging at 400–650°C can produce significant reductions in fracture toughness and poor aqueous corrosion resistance, with the poorest performance observed for the alloys containing higher Cr levels [12]. Nickel alloy 718 exhibits acceptable phase stability up to temperatures of ~650°C, but at higher temperatures exhibits coarsening of the gamma double prime and gamma prime precipitates responsible for its strength [13]. Nickel alloys designed for higher temperature applications such as Alloy 617 and 230 have generally displayed good phase stability during long-term thermal aging up to 800–850°C [14–16]. Some carbide precipitate coarsening ($M_{23}C_6$, etc.) has been observed during thermal aging at higher annealing temperatures of 900–1000°C, with accompanying loss of strength [14, 15].

Ferritic/martensitic steels containing 9%–12%Cr generally have good phase stability during long-term aging. At low temperatures (e.g., ~300°C relevant for water-cooled commercial nuclear power reactor operation), precipitation of the Cr-rich alpha prime phase is generally not observed in the absence of irradiation due to low thermal diffusion [17–20]. As discussed in Section 5.3, alpha prime embrittlement is nevertheless of concern during low temperature (~e.g., 300°C) neutron irradiation due to radiation-enhanced diffusion that can stimulate the precipitation of alpha prime [21]. At intermediate temperatures near 475°C, strengthening and embrittlement due to alpha prime precipitation is of concern for ferritic/martensitic alloys containing more than ~10% Cr. At

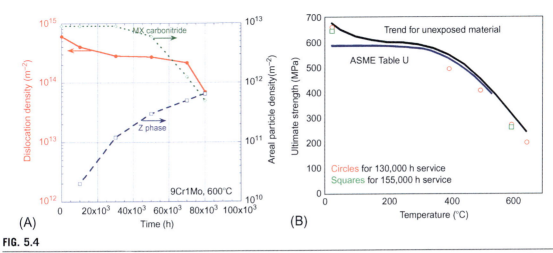

FIG. 5.4

(A) Effect of thermal creep (70 MPa, 600°C) exposure time on several microstructural features [27] and (B) effect of long-term (>10^5 h) thermal exposure at 400–650°C on the tensile strength of 9%Cr-1Mo ferritic/martensitic steel [29].

(A) Data from K. Sawada, H. Kushima, M. Tabuchi, K. Kimura, Microstructural degradation of Gr.91 steel during creep under low stress, Mater. Sci. Eng. A, 528 (2011) 5511–5518. (B) Based on the sources R.W. Swindeman, M.L. Santella, P.J. Maziasz, B.W. Roberts, K. Coleman, Issues in replacing Cr-Mo steels and stainless steels with 9Cr-1Mo-V steel, Int. J. Pressure Vessels and Piping, 81 (2004) 507–512.; R.W. Swindeman, V.K. Sikka, P.J. Maziasz, D.A. Canonico, Evaluation of T91 after 130,000 hours in service, in: H.S. Mehta, et al. (Eds.) Fatigue, Environmental Factors, and New Materials, Pressure Vessels & Piping-Vol. 374, ASME, 1998, pp. 305–312.

high temperatures (>500°C), alpha prime precipitation is generally not of concern due to increased Cr solubility. However, other phase instabilities can emerge in 9%–12%Cr ferritic/martensitic steels during long-term operation at elevated temperatures, demonstrating the complexities in long-term phase stability in ferritic/martensitic steels. These phase instabilities typically only emerge after operation at temperatures above 500°C for >10 years, and include substructure coarsening and precipitation of Laves and M_6C phases [22, 23] and dissolution of M_2X fine-scale precipitates [24] in 9%Cr steels at 600–650°C that can produce loss of the tensile thermal creep strength. $M_{23}C_6$ precipitates gradually coarsen during long-term aging at temperatures above ~550°C [25]. MX carbonitrides are stable for aging times up to ~3 years at 550–700°C [24–28] and then gradually diminish and Z-phase precipitates are observed at long annealing times. Fig. 5.4 shows the change in several microstructural features in 9%Cr-1Mo steel during long-term thermal creep testing at 600°C [27], and reductions in tensile strength following long-term thermal aging [29]. The ultimate tensile strength at 400–650°C of 9%Cr-1Mo has been reported to be unchanged following thermal aging for 30,000 h, whereas the strength decreased to ~83% of the as-tempered value following thermal aging for >130,000 h at 400–650°C [29].

5.2.2 THERMAL CREEP

Prolonged exposure of structural materials to mechanical stress at elevated temperatures leads to permanent plastic deformation at stresses that are a fraction of the yield stresses obtained during short-term mechanical testing [3, 30, 31]. Fig. 5.5 shows examples of typical thermal creep deformation curves [32], where three deformation regimes are evident: an initial short, transient, primary creep regime where the deformation rate is initially high and progressively decreases, an extended secondary ("steady-state") creep regime where the creep deformation rate is approximately constant, and a tertiary creep regime where the creep rate begins to rapidly increase prior to sample rupture. The relative magnitudes of the different creep regimes can vary significantly for different materials. For example, oxide dispersion strengthened alloys typically exhibit a very long secondary creep regime and a very short tertiary creep regime [33]. The onset temperature for pronounced thermal creep deformation typically varies between ~0.3 and

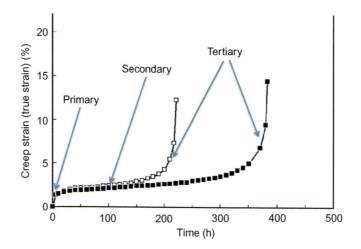

FIG. 5.5

Examples of thermal creep deformation vs exposure time behavior for two austenitic alloys tested at 750°C and a constant engineering stress of 170 MPa [32]. The transient primary, secondary ("steady state"), and tertiary creep regimes are highlighted.

$0.5 T_M$ (T_M = melting temperature) for different materials, with materials specifically engineered for thermal creep resistance (e.g., nickel-base superalloys) generally exhibiting the highest onset temperature for creep.

From an engineering design perspective, time-dependent degradation phenomena such as thermal creep and cyclic fatigue require the application of more restrictive material-dependent design rules for the allowable stress compared to tensile strength design rules [34–39]. Depending on the particular application, the relevant design specification for allowable thermal creep stress (S_t) may be based on avoidance of component failure with an appropriate safety factor (e.g., S_t equal to 2/3 of the minimum stress to produced creep rupture for the design lifetime) or it may be based on limiting the plastic deformation to a certain level (e.g., 1% deformation) or avoidance of the onset of tertiary creep over the design lifetime. It is generally not allowable to extrapolate thermal creep data to times beyond the experimental creep database, although for preliminary screening "point design" evaluations such invalid extrapolations are commonly used.

Several different physical mechanisms are responsible for the predominant thermal creep behavior, depending on material, temperature, applied stress, and deformation rate [3, 30, 31, 40]. At relatively low temperatures and intermediate stresses, thermal creep is typically controlled by the climb of dislocations past obstacles. Dislocation (power law) creep exhibits a deformation rate that is independent of grain size and proportional to σ^n, where σ is the applied stress and n is ~3–6. At higher temperatures and lower applied stresses, two different types of diffusional creep mechanisms are observed. At very high temperatures, Nabarro-Herring creep associated with lattice vacancy diffusion occurs with a linear stress dependence and an inverse square dependence on grain size. Grain boundary sliding (Coble creep) associated with grain boundary atomic diffusion is often observed at low stresses and at temperatures intermediate between the dislocation creep and Nabarro-Herring creep regimes, with a characteristic linear stress dependence and an inverse cubic dependence on grain size. Several additional thermal creep mechanisms may also occur in specific material systems; some material specific data on thermal creep are summarized in Chapters 7–13. In general, the thermal creep behavior can be described by

$$\frac{d\varepsilon}{dt} = \frac{DGb}{kT} A \frac{(\sigma - \sigma_{th})^m}{d^n} \quad (5.1)$$

where $d\varepsilon/dt$ is the plastic deformation rate, D is the diffusion coefficient (self-diffusion for dislocation and Nabarro-Herring creep, grain boundary diffusion for Coble creep), G is the shear modulus, b is the Burgers vector of the predominant glide dislocation, k is Boltzmann's constant, T is the absolute temperature, A is a material

5.2 THERMOMECHANICAL DEGRADATION PROCESSES

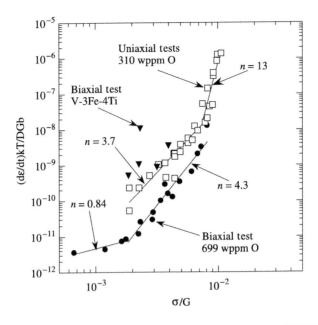

FIG. 5.6

Thermal creep behavior at 600–800°C for V-4%Cr-4%Ti and V-3%Fe-4%Ti alloys [41].

constant, σ is the applied stress, σ_{th} is a threshold stress to initiate flow (dependent on microstructural features such as precipitates), d is the grain diameter, and m and n are material constants that depend on which creep mechanism is predominant. Fig. 5.6 shows an example of thermal creep rate data for refractory vanadium alloys, where several distinct thermal creep regimes exhibiting different dependencies on applied stress are visible [41].

The Larson-Miller parameter [42] is a widely employed convenient approach to summarize thermal creep data obtained at different times and temperatures, and to provide a general overview of the thermal creep behavior of a given material. The typical form of the Larson-Miller parameter (LMP) is LMP=$T(C+\log(t))$, where T is the test temperature, t is the test time, and C is a constant with typical values between 20 and 30. The analysis is based on the assumption that creep strain rate follows the same Arrhenius relationship for all of the test data $A \cdot \exp(-Q/RT)$. Fig. 5.7 compares thermal creep rupture data on low-alloy steels plotted as rupture stress vs time for different test temperatures and a master creep rupture plot utilizing a Larson-Miller parameter [43]. Whereas data plotted in traditional rupture stress vs. time plots provide important detailed information, the Larson-Miller plots are convenient for evaluating overall behavior, particularly when comparing different classes of materials over a range of test times and temperatures; from Fig. 5.7B, it is readily apparent that there is not a major difference in thermal creep behavior for the seven sets of plotted material conditions that encompass over 1800 individual creep tests. It should be noted that a Larson-Miller analysis is invalid if the constitutive creep equation is not a simple exponential with a constant activation energy, or if the evaluated data encompass creep conditions involving different predominant creep mechanisms. Therefore, LMP trends should not be extrapolated well beyond the range of times and temperatures used to construct the plot, as different creep mechanisms (with significantly different dependencies on stress and exposure time) may emerge.

In general, the best approach for analyzing thermal creep data is to calculate the contributions from known thermal creep mechanisms and assemble the calculated behavior into a deformation mechanism "Ashby" map (applied stress vs temperature for a given deformation strain rate) that displays the regimes where different creep mechanisms may be dominant [31, 44]. Although there is typically significant uncertainty in the value of the thermal creep parameters, this method helps to identify likely dominant creep mechanisms for the test conditions of interest, and then further specific deformation tests can be performed to verify the dominant creep mechanism. Fig. 5.8 shows an example of such a deformation map that was calculated for a vanadium alloy and was experimentally validated using

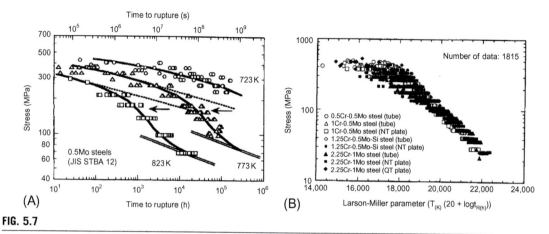

FIG. 5.7

Comparison of thermal creep rupture data for low-alloy steels: (A) rupture stress vs time for 0.5%Cr steels at three test temperatures, and (B) Larson-Miller plot comparing the creep rupture behavior for several low-alloy steels at multiple test times and temperatures [43].

From K. Kimura, H. Kushima, F. Abe and K. Yagi: in 'Microstructural stability of creep resistant alloys for high temperature plant applications', (ed. A. Strang, J. Cawley and G. W. Greenwood), 185–196; 1998, London, Institute of Materials, Copyright © Institute of Materials, Minerals and Mining, reprinted by permission of Taylor & Francis Ltd, http://www.tandfonline.com on behalf of Institute of Materials, Minerals and Mining.

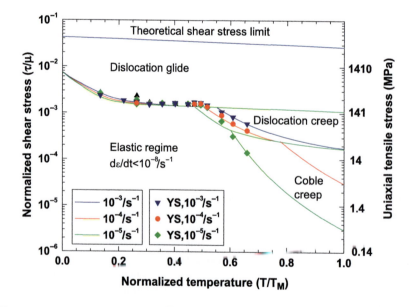

FIG. 5.8

Deformation mechanism map for V-4%Cr-4%Ti at strain rates of 10^{-3}–10^{-5}/s [45]. The filled data points denote yield strengths measured at different strain rates that were used to experimentally validate the predicted predominant deformation modes.

mechanical test data performed at different strain rates [45]. As noted in Section 5.2.1, an additional potential issue during long-term thermal creep testing is the possibility of thermal aging (along with stress-enhanced aging) that could produce microstructural changes and accompanying creep strength degradation. For example, the thermal creep strength of 9–12Cr FM steel at test temperatures > 600°C has been observed to decrease (compared to the extrapolation of lower duration creep tests) for exposure times above ~30,000 h [27].

5.2.3 FATIGUE AND CREEP FATIGUE

Fatigue design is recognized as an essential part of structural design as the majority of product failures in structural applications are typically caused by fatigue failure. It has been recognized since the mid- to late-1800s that the stress (or strain) to produce failure under cyclic loading conditions progressively decreases with increasing number of fatigue cycles up to a high-cycle "fatigue limit". Fatigue property is typically evaluated in a plot of cyclic stress (or strain) amplitude vs number of cycles to failure ("S-N diagram"). Fig. 5.9 compares the fatigue failure behavior for three austenitic and tempered martensitic steels [46].

High-cycle fatigue, where the number of cycles to failure is above 10^5–10^6 cycles, was discovered to be responsible for the fracture of railway steam engine's wheel axles in the early days of the industrial revolution [47]. High-cycle failure occurs from applied loads that are smaller than the yield stress. A fatigue limit, that is, the stress (or strain) amplitude below which no high-cycle fatigue failure occurs, is historically observed to be 35%–60% of the yield strength in steels and most structural alloys whereas several nonferrous alloys (Al, Cu, Mg) do not exhibit distinct high-cycle fatigue limits and an equivalent "endurance limit" at a specified high number of fatigue cycles is usually specified [48]. Some recent very high-cycle fatigue studies have questioned whether a true fatigue limit exists for any material [49]. Low-cycle fatigue, where the number of cycles to failure is below 10^5 cycles, involves significant plastic deformation and has received significant attention for public safety applications as the first commercial jet airplane crashes in 1954 were discovered to be due to low-cycle fatigue fracture [50].

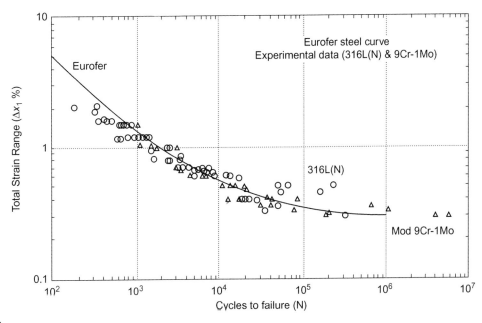

FIG. 5.9

Fatigue behavior of 316L(N) austenitic steel, Mod. 9Cr-1Mo, and reduced activation EUROFER 9Cr-1WVTa tempered martensitic steel *(solid curve)* at 823 K [46].

For high-temperature applications, thermal fatigue due to differential thermal expansion in a structure with a temperature gradient or under cyclic thermal loading has to be considered as the potential failure mode [51]. Thermal expansion stress, which is categorized as secondary stress in design criteria, is a strain-controlled stress; the plastic deformation due to thermal expansion must be limited in order to avoid inducing progressive deformation (ratcheting [52]) under cyclic loading. The thermal fatigue failure is considered within this limit by evaluating relevant operation scenarios.

It is well known that the number of cycles to failure in high-temperature fatigue tend to decrease when hold time effects are considered. Creep fatigue is the failure mode, and this could dominate the life of high-temperature components that experience the combination of steady operation and startup/shutdown cycles [53]. It has been considered that creep damage is the most significant damage in most of the high-temperature components that are operated under constant high-pressure load (considered as primary stress) with few startup/shutdown conditions. However, the creep-fatigue damage becomes significant where secondary stress (thermal expansion stress) is dominant compared to primary stress, for example, in the case of sodium-cooled fast breeder reactors. The difficulty in evaluating creep fatigue is that it is often impractical to conduct "generalized" creep-fatigue tests that can be extrapolated to the real operation cycles. Thus, mechanistic understanding of damage accumulation is essential to extrapolate the limited variation of test results to design criteria.

The fatigue damage-accumulation process is categorized into three stages, that is, cyclic hardening/softening, crack initiation, and crack propagation; each requires cyclic plastic deformation at the stress concentration location. High-cycle fatigue crack initiation is typically observed at the defects or flaws inside the structure (or specimen), and the formation of a micro-crack around the defects is typically the most important factor for determining fatigue life. Low-cycle fatigue crack initiation is mostly observed at the surface (or near surface volume) where the constraint factor is low (plane stress condition) and the largest plastic deformation is expected due to bend mode or structural discontinuity (due to notch or surface flaw). These conditions induce fatigue slip bands and form an intrusion on the surface, especially at grain boundaries or at surface defects (inclusions). Fatigue crack propagation is the major process that defines low-cycle fatigue life, which forms fatigue striation on the fracture surface.

Creep-fatigue damage is described as the combination of thermal fatigue damage and creep damage. Creep damage is generally defined as cavity formation and growth at a grain boundary, and it is described as the function of stress, temperature, and time. The typical creep-fatigue interaction mechanisms are crack initiation and/or propagation enhancement due to cavitation damage [54], which describes creep-fatigue damage for relatively high strain and short holding time. The consequence of this mechanism is often associated with an intergranular fracture surface. The other mechanism is creep damage enhancement by cyclic loading [54], which explains creep-fatigue damage accumulated by a longer hold time under a smaller inelastic cyclic strain range. The phenomena such as cyclic softening, or decrease of the deformation resistance due to the inelastic deformation in the opposite direction, are known as the key phenomena that consequently increase the inelastic deformation during holding time.

The creep-fatigue interaction diagram (Fig. 5.10), which is currently used in major design codes for creep-fatigue design criteria [55], assumes that creep-fatigue failure occurs when the sum of fatigue damage described by the linear cumulative damage rule [56] and creep damage described by the time fraction rule [57] reaches an empirically determined material-specific limit [58].

This stress-based time-fraction approach is very practical as it is based on existing data, and proved to be sufficiently conservative with a code safety factor, but it cannot be extrapolated to other conditions nor to other materials. The strain-based ductility exhaustion approach has been studied as an alternative creep damage evaluation method [59], as the stress-based time-fraction approach is overly conservative for some important structural materials such as Modified 9Cr 1Mo [60]. In this approach, the damages are described as the integral of the ratio of accumulated inelastic strain, creep strain, and plastic strain induced by fatigue vs fracture strain.

It should be noted that the creep-fatigue failure in the real application is much more complex, as there is typically an impact of multiaxial loading effects due to a discontinuous structure or metallurgical complexity at weld/joints, and environmental effects such as corrosion, oxidation, and/or irradiation. Thus, mechanistic understanding of creep-fatigue phenomena is essential to make it possible to extrapolate the controlled test results to design criteria for the real components.

5.3 RADIATION HARDENING AND EMBRITTLEMENT

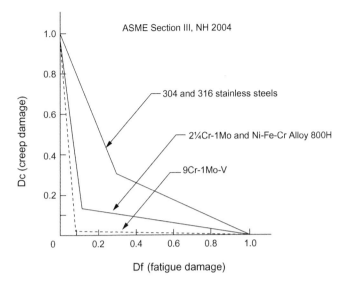

FIG. 5.10

Creep-fatigue interaction diagram of ASME Section III, NH [55].

Reprinted from ASME 2004 Edition, BPVSC, Section III-NH, by permission of The American Society of Mechanical Engineers. All rights reserved.

5.3 RADIATION HARDENING AND EMBRITTLEMENT

Neutron irradiation of structural alloys at low temperatures results in the creation of high densities of fine-scale defect clusters, including dislocation loops, stacking fault tetrahedra (for fcc materials), gas-filled bubbles, precipitates, and solute-defect clusters [5]. These defect clusters act as obstacles to the motion of gliding dislocations, and therefore typically produce a significant increase in the alloy strength [61]. Although this increase in strength ("radiation hardening") is nominally beneficial (e.g., compared to the strength decrease that may occur in alloys during long-term thermal aging), the radiation hardening is typically accompanied by reductions in tensile elongation [62, 63] and fracture toughness [64–66]. Material-specific data are summarized in Chapters 7–13.

5.3.1 DOSE DEPENDENCE OF LOW-TEMPERATURE RADIATION HARDENING AND DUCTILITY REDUCTION

The yield strength of metals and alloys increases rapidly with increasing dose due to low-temperature radiation hardening and typically approaches a near-constant saturation value after doses of 0.1–10 dpa (as discussed in Chapter 10, an important exception is the delayed formation of so-called "late-blooming phases" in irradiated reactor pressure vessel steels, where additional hardening occurs at doses above ~1 dpa). As shown in Fig. 5.11 for the case of copper [67], pure metals harden rapidly and generally approach their saturation hardness value after doses of ~0.1–1 dpa [62, 63]. For engineering alloys, the rate of increase of yield strength tends to be less rapid than observed for pure metals and the dose required to achieve saturation in hardening is typically ~1–10 dpa. Fig. 5.12 shows an example of the dose dependence of the low-temperature radiation hardening behavior of austenitic stainless steels [68].

Low-temperature irradiation hardening is accompanied by a progressive reduction in work-hardening capacity, as manifested by decreases in uniform elongation, total elongation, and engineering strain-hardening rate. Fig. 5.13

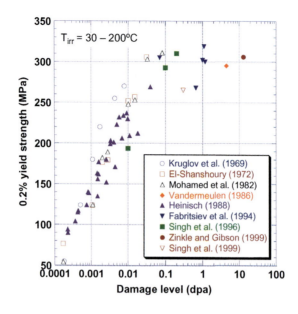

FIG. 5.11

Dose dependence of the yield strength of copper irradiated with fission neutrons near room temperature [67].

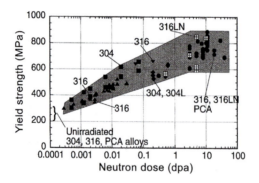

FIG. 5.12

Dose dependence of the yield strength of annealed austenitic stainless steel irradiated near room temperature [68].

From A.F. Rowcliffe, S.J. Zinkle, J.F. Stubbins, D.J. Edwards, D.J. Alexander, Austenitic stainless steels and high strength copper alloys for fusion components, J. Nucl. Mater., 258–263 (1998) 183–192. Original from J.E. Pawel et al., J. Nucl. Mater. 239 (1996) 126.

shows an example of the engineering stress-strain curves for neutron-irradiated nickel near room temperature [69]. Fig. 5.14 summarizes the dose-dependent uniform elongation for several pure metals following neutron irradiation near room temperature [69–71]. Analyses of the tensile curves have found that there are several potential factors that contribute to the loss of strain-hardening capacity. The leading conventional explanation of the low strain-hardening capacity is based on extreme flow localization due to dislocation channeling [62, 69, 72–76], wherein the leading dislocations emitted from a dislocation source efficiently annihilate defect clusters in their glide path. Subsequent dislocations emitted from the same source encounter a relatively weak defect-free slip channel when they glide along the same slip planes as the initial dislocation. This leads to highly localized flow and tends to minimize interactions between dislocations emitted from different sources, thereby minimizing normal dislocation

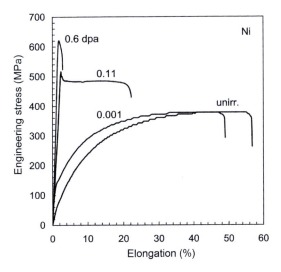

FIG. 5.13

Uniaxial tension stress-strain curves for nickel irradiated with fission neutrons at several doses near room temperature [69].

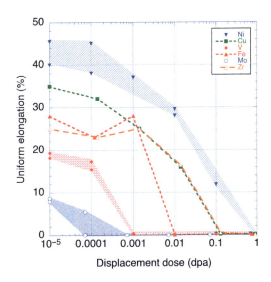

FIG. 5.14

Dose dependence of uniaxial tension uniform elongation of several pure metals after neutron irradiation near room temperature [69–71].

multiplication processes that are largely responsible [40] for work hardening in unirradiated metals. Additional proposed physical mechanisms that may contribute to the reductions in engineering strain-hardening rate and uniform elongation include mechanical (necking) instabilities associated with high matrix strength [61, 63, 70, 77–79]. For example, Byun and coworkers [61, 63, 70, 78, 79] have observed that the material specific plastic instability stress (true stress at the onset of necking) in dozens of metals and alloys is dose-independent for unirradiated and low-temperature-irradiated samples.

5.3.2 TEMPERATURE DEPENDENCE OF RADIATION HARDENING AND DUCTILITY REDUCTION

The high levels of radiation hardening and accompanying reduction in tensile ductility are typically most pronounced below a characteristic material- and dose rate-dependent temperature; above this characteristic temperature, the radiation hardening and reduction in tensile ductility usually rapidly diminish with increasing temperature [66, 74, 80–85]. As a rule of thumb, the characteristic temperature is usually ~0.3T_M for most neutron-irradiated materials. Fig. 5.15 shows an example of the dose- and irradiation temperature-dependent hardening in neutron-irradiated V-4%Cr-4%Ti solid solution alloys [84]. In this case, the high hardening levels at irradiation temperatures below ~400°C (~0.31T_M) were accompanied by uniform elongations below 2%. At irradiation temperatures above ~400°C, the hardening rapidly decreased and the tensile elongations rapidly increased with increasing temperature, with both the strength and ductility approaching unirradiated values at higher temperatures.

In most irradiated materials, the measured radiation hardening at a given dose is either nearly constant or slowly decreases with increasing temperature below the critical temperature of ~0.3T_M [66, 74, 80–85]. In some materials, a localized peak in radiation hardening may occur at an intermediate temperature due to processes such as radiation-enhanced or -induced precipitation [21, 86, 87] or helium bubble formation [68, 88] that introduce additional obstacles to gliding dislocations. Fig. 5.16 compares the dose dependence of the radiation-hardening behavior for 9%Cr ferritic/martensitic steel irradiated near 80°C and near 300°C [89]. Whereas the hardening at low doses (<1.5 dpa) is highest for the lower irradiation temperature, the hardening at higher doses (>2 dpa) was slightly higher for the higher irradiation temperature. The difference in dose dependence may be attributable to direct production in neutron displacement cascades of small hardening centers at the lower irradiation temperature compared to nucleation and growth of hardening centers at the higher irradiation center. In general, detailed microstructural characterization (cf. Chapter 4) is needed to provide quantitative insight into the features responsible for observed property changes.

5.3.3 LOW-TEMPERATURE RADIATION EMBRITTLEMENT

Several different methods are traditionally used to quantify the degree of embrittlement in structural alloys. The two simplest methods are based on evaluation of ductility in uniaxial tension tests: tensile elongation (uniform and/

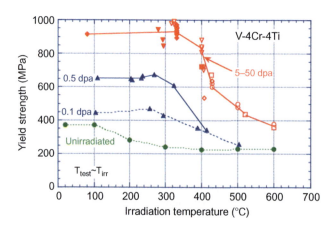

FIG. 5.15

Effect of neutron-irradiation dose and temperature on the yield strength of V-4%Cr-4%Ti alloys [84]. The tensile tests were performed at the irradiation temperature.

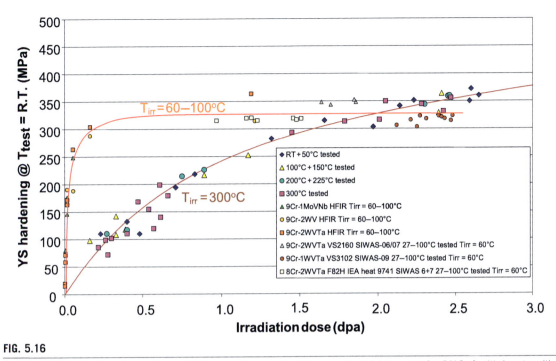

FIG. 5.16

Comparison of the dose dependence of the room temperature increase in yield strength for 9%Cr ferritic/martensitic steels irradiated at ~80°C vs ~300°C [89].

or total elongation) and reduction in area [90]. In most unirradiated and irradiated materials, there is a good correlation between these different measures of tensile ductility. However, for cases where extreme flow localization occurs such as in many pure metals and structural alloys irradiated at low temperatures, it is important to note that the total elongation and reduction in area might remain high while the uniform elongation is near zero. A typical example of high reduction in area and low uniform elongation following irradiation of a V alloy at temperatures below 400°C is given in Fig. 5.17 [91, 92]. Similar examples of high reduction in area for low-temperature irradiated alloys exhibiting low uniform elongation have been reported for austenitic [93, 94] and ferritic/martensitic [95, 96] steels after low-temperature irradiation, due to flow localization (dislocation channeling) effects.

A third general property to quantify embrittlement is based on fracture toughness [97]. In general, low-temperature irradiation causes a progressive reduction in the fracture toughness of structural alloys, regardless of crystal structure [86, 94, 98–105]. In body centered cubic materials, the temperature-dependent fracture toughness exhibits a relatively sharp ductile to brittle transition temperature (DBTT) between the low-toughness cleavage (or intergranular fracture) regime at low temperatures and the higher toughness regime (typically microvoid coalescence) at high temperatures. Initiation of cleavage (and hence the DBTT) occurs when the matrix strength exceeds the material-dependent threshold stress for cleavage (including a stress enhancement constraint factor $M > 3$ to account for stress enhancements near defects such as cracks) [97, 106, 107]. Therefore, the pronounced radiation hardening that typically occurs following low-temperature neutron irradiation can produce significant shifts in the DBTT. Fig. 5.18 shows the correlation between DBTT shift and increase in yield strength for several steels [108]. In general, for nuclear reactor applications the DBTT should remain below room temperature to avoid potential rapid cleavage fracture in case of a pressurized thermal shock scenario for light-water reactors (or equivalent accident scenarios involving rapid ingress of room temperature coolant).

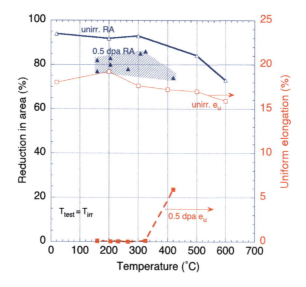

FIG. 5.17

Comparison of tensile ductility parameters, uniform elongation (e_u) and reduction in area (RA) for unirradiated and fission neutron-irradiated V-4%Cr-4%Ti alloys [91, 92].

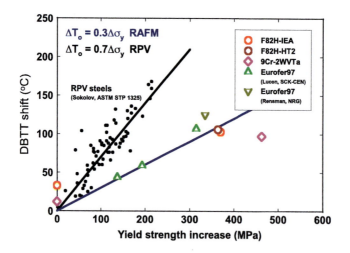

FIG. 5.18

Comparison of the neutron irradiation-induced increase in DBTT for several reduced activation ferritic/martensitic (RAFM) steels and for reactor pressure vessel (RPV) steels [108].

Adapted from M.A. Sokolov, H. Tanigawa, G.R. Odette, K. Shiba, R.L. Klueh, Fracture toughness and Charpy impact properties of several RAFMS before and after irradiation in HFIR, J. Nucl. Mater., 367–370 (2007) 68–73.

Although in most cases materials with low values of fracture toughness also exhibit low values of tensile uniform elongation, this relationship is not universal. For example, copper alloys irradiated with neutrons at low temperature generally exhibit very low values of uniform elongation after doses >0.1 dpa, whereas the corresponding fracture toughness may exhibit only a slight reduction from its high unirradiated value [103, 109, 110]. Conversely, body centered cubic alloys such as V-4%Cr-4%Ti or HT-9 (12%Cr ferritic steel) can exhibit moderate uniform elongation ($e_u \sim 2$ to 5%) but low toughness after low-temperature neutron irradiation whereas other materials such as F82H (9%Cr ferritic/martensitic steel) can exhibit low uniform elongation but high toughness [74, 111].

5.4 RADIATION-INDUCED PHASE AND MICROCHEMICAL CHANGES

While the topic of radiation effects in structural materials encompasses a vast literature dating from 1942 [112], it is now well understood that the root cause of radiation damage and the effects of radiation discussed in this chapter and book stem from the production of individual and clustered vacancies and self-interstitials in collision cascades resulting from the collisions between energetic particles and target atoms [113, 114]. The subsequent diffusion and clustering of these defects, along with the associated transport of solutes and impurities that drive microstructural changes at the nanometer-to-micrometer length scale lead to accelerated creep [115], volumetric swelling [116], segregation of alloying elements [117, 118], and embrittlement [99]. Furthermore, the forced atomic mixing taking place in the displacement cascades themselves can lead to disordering of chemically ordered phases, including complete loss of crystal structure, or amorphization, in some cases of low-temperature irradiation of ceramic compounds and intermetallic precipitates [5, 118, 119] and to dissolution of precipitates [118, 120].

A substantial literature exists on the energy loss mechanisms of high-energy particle irradiation in solids, resulting in both electronic and nuclear energy loss mechanisms, and corresponding atomic defect creation, but those defect creation processes will not be reviewed here. Instead, we focus on the subsequent diffusional evolution of the vacancy and self-interstitial type defects produced in displacement cascades, and the corresponding microstructural and microchemical changes that are responsible for phase instability or radiation-enhanced or -induced segregation. MD simulations of displacement cascades indicate that the primary recoiling atoms induced by an elastic collision with a neutron or high-energy irradiating particle rapidly lose kinetic energy through a branching chain of atomic displacement collisions, as well as nondisplacing interactions with electrons, producing a displacement cascade that evolves over very short times of 100 ps or less, and small volumes with a characteristic length scale of 50 nm or less [113, 114, 121, 122]. The key conclusions of MD simulations of high-energy (>20 keV) simulations are that (i) intracascade recombination of vacancies and self-interstitial atoms (SIA) results in $\sim$25%–30% of the defect production expected from displacement theory, (ii) many-body collision effects produce a spatial correlation (separation) of the vacancy and self-interstitial defects, (iii) substantial clustering of the self-interstitials and to a lesser extent, the vacancies occurs within the cascade volume, and (iv) high-energy displacement cascades tend to break up into lobes, or subcascades, which may either enhance recombination (thereby decreasing damage production), or increase damage through the interaction of supersonic shock fronts for cascade energy divided by displacement energy above about 300. Large concentrations of defects, along with local cascade mixing, are believed responsible for amorphization of intermetallic compounds, and are reviewed in Section 5.4.1, with more details provided in Chapter 7 for Zr alloy-specific phases.

Ultimately, the long-range diffusion of vacancy and self-interstitial defects that escape the cascade region is responsible for nano-/microstructural evolution, and this transport occurs across a range of timescales from nanoseconds to times greatly exceeding seconds. Enhanced defect transport leads to a redistribution of solute and impurity elements as a consequence of radiation-enhanced diffusion and segregation. Solute redistribution can alter the local chemistry and result in the precipitation of phases not expected within a homogeneously distributed alloy. Section 5.4.2 will review the current understanding of radiation-enhanced diffusion and radiation-induced segregation (RIS) phenomena, with further material-specific results provided in Chapters 7–13.

5.4.1 AMORPHIZATION

Amorphization by particle irradiation refers to the complete loss of long-range crystalline structure, and is often called irradiation-induced amorphization. Amorphization in ceramics and materials is well documented, with the first reference of the metamict state dating to the late 1800s [123, 124], although it was not until well into the 20th century before the cause was conclusively identified as the result of self-irradiation by alpha decay [124, 125]. As noted by Motta [119], the amorphization of crystalline solids, particularly intermetallic compounds, is a complex phenomenon, which depends on numerous irradiation parameters including temperature, as well as the details of the preexisting material microstructure.

For intermetallic compounds, Motta notes that irradiation-induced amorphization is primarily a low-temperature phenomenon, and, as such, is generally not a concern for structural alloys in nuclear environments. However, it is important to note that amorphization of ternary intermetallic precipitates [e.g., $Zr(Cr,Fe)_2$ or $Zr_2(Ni,Fe)$] in the zirconium alloy cladding of nuclear fuel rods has been observed and is believed to impact the corrosion behavior [119, 126]. Fig. 5.19 provides a schematic representation of the dose to amorphization of a crystalline solid, as a function of irradiation temperature and the type of particle irradiation [119]. The first item to note in Fig. 5.19 is that there is a critical dose to induce amorphization, which increases dramatically with increasing temperature; conversely, it can be stated that amorphization will not occur at temperatures higher than T_e for electron irradiation, T_n for neutron irradiation, or T_{l-i}/T_{h-i} for light ion vs heavy ion irradiation, respectively. Thus, these temperatures denote the critical amorphization temperature and are dependent on the type of irradiating particle, such that amorphization by cascade-producing irradiation has a higher critical temperature than that of electron irradiation, which only forms isolated Frenkel pairs. As well, there is a slight increase in the critical dose vs temperature indicated in Fig. 5.19, and noted with the temperature T_{ann}, which is usually related to an annealing process that makes the damage accumulation process more difficult (thus the increase in critical dose) without fully impeding it. The effect of dose rate is also indicated in Fig. 5.19; the critical temperature for amorphization by electron irradiation is schematically illustrated with a family of three red curves (solid line, dotted line, and dotted-dashed, respectively), in which T_e is shown to increase with increasing dose rate.

The severity of the damage, both in terms of higher dose rate and more dense displacement cascades, also strongly influences the irradiation-induced amorphization behavior. Fig. 5.19 indicates that the critical amorphization temperature of ion irradiation (denoted by green and darker green lines, respectively, for light ion vs heavy ion irradiation) is higher than that of neutrons, which are typically at much lower dose rates than ion irradiation. This shift of critical amorphization temperature from neutron to ion irradiation is caused by the increased dose rate of ion irradiation experiments, while the temperature increase from light-to-heavy ion irradiation involves a combination of dose rate and more dense displacement cascades.

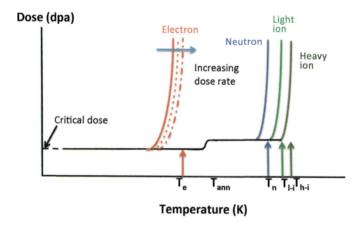

FIG. 5.19

Schematic illustration of the amorphization dose versus temperature, as well as the influence of different types of irradiating particles and dose rate [119].

5.4 RADIATION-INDUCED PHASE AND MICROCHEMICAL CHANGES

Motta concludes his review [119] by noting that the presence of extended defects in the initial material microstructure, along with deviations from ideal compound stoichiometry of intermetallic compounds, can facilitate amorphization. As well, amorphization does not appear to depend on long-range diffusion, since it is only observed to occur during low-temperature irradiation, but it does occur by local atomic rearrangements driven by particle irradiation that destroy the long-range crystalline order while maintaining some short-range order [119]. Once a material has been made amorphous, that amorphous structure persists during further irradiation.

Likewise, the irradiation-induced amorphization of complex ceramics is a low-temperature phenomena, for which amorphization is not possible at higher irradiation temperatures. Wang and Ewing [125] note that the critical dose for amorphization of ceramics depends on a combination of (i) structural complexity (e.g., number of distinct cation sites), (ii) chemical complexity (e.g., number of different types of atoms), (iii) the ionicity of the average bonding (e.g., relative percentage of ionic versus covalent bonding), and (iv) the melting point of the ceramic, which provides a measure of the bond strength. They also note that amorphization in ceramics can occur as a result of electronic excitation, for example, swift heavy ion tracks, as well as nuclear stopping (e.g., defect production in elastic collisions). But, the overall dose and temperature dependence of amorphization is generally similar to that of intermetallic alloys as indicated in Fig. 5.19.

5.4.2 RADIATION-ENHANCED AND INDUCED SEGREGATION (AND PRECIPITATION)

As noted earlier, the primary product of radiation is a greatly increased local concentration of vacancy and self-interstitial defects, and it is the long-range diffusion and transport of these defects that is responsible for driving radiation-enhanced diffusion and chemical segregation. It has long been recognized that the diffusion of atoms within a solid is driven by the motion of point defects and small clusters. Russell [118] provides an assessment of the enhanced diffusion of alloying elements in metallic systems as a result of increased point defect concentrations observed under high-energy particle irradiation, and the interested reader is referred to the comprehensive treatment by Sizmann [127]. Fig. 5.20 provides a schematic illustration of the effect of temperature and dose rate on solute diffusivity under conditions of relatively high-defect sink concentrations (precipitates, dislocations, grain boundaries, etc.), in which three different regimes of behavior are observed. At high temperatures, the defect mobilities are sufficiently high that the locally high point defect concentrations can rapidly diffuse to defect sinks and the thermal equilibrium defect populations dominate the diffusion behavior. In this high-temperature regime, the

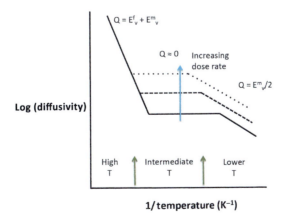

FIG. 5.20

Schematic illustration of radiation-enhanced diffusion under sink-dominant conditions as a function of dose rate and irradiation temperature, following Fig. 8 of Russell [118].

activation energy of diffusion is essentially the same as for self-diffusion, and corresponds to the sum of the vacancy formation and migration energy, as $Q \approx E_v^f + E_v^m$, and there is no significant irradiation-enhanced diffusion. Fig. 5.20 indicates that for sink-dominant conditions there is a regime of essentially temperature-independent diffusivity at intermediate temperatures, in which the radiation-enhanced diffusivity greatly exceeds the thermal diffusivity (as indicated by extrapolating the high-temperature diffusivity along its Arrhenius slope). In this intermediate regime, the point defects are expected to annihilate at preexisting microstructural features like grain boundaries or dislocations, which controls the nearly temperature-dependent diffusion behavior with greatly enhanced diffusivity. For cases where the sink density is low such as in irradiated pure metals (recombination-dominant conditions), the diffusion activation energy at intermediate temperatures is given by $Q \approx E_v^m$ [127, 128]. In the lower temperature regime, the vacancies have limited mobility and their concentration increases to such a high level that recombination with interstitials begins to dominate over diffusion to fixed sinks, and the diffusivity again begins to decrease with decreasing temperature. In this low temperature, recombination-dominant regime, the activation energy for diffusion is approximately given by one-half of the vacancy migration energy, as $Q \approx E_v^m/2$ [127]. The effect of increasing dose rate is to generally increase point defect concentrations and generally increase the diffusion coefficient in both the intermediate and low-temperature regime, although the transition between the two regimes will shift to higher temperatures (decreased inverse temperature) with increasing dose rate. One classic example of radiation-enhanced diffusion is the precipitation of copper into copper-rich precipitates, which is responsible for the irradiation embrittlement of low-alloy ferritic steels that are used as the pressure vessel of light-water reactors, and the interested reader is referred to numerous seminal articles by Odette and coauthors [64, 99].

RIS is a nonequilibrium process that occurs at point-defect sinks during high-energy particle irradiation of alloys at intermediate temperature (~ 0.3–$0.5 T_m$, where T_m is the melting temperature) [129]. As indicated in Fig. 5.20, the excess concentrations of point defects in irradiated metals drive increases in solute diffusivity and, at intermediate temperatures, the defects diffuse long distances to extended defect sinks including surfaces, grain boundaries, and dislocations. Segregation occurs when a given alloying component has a preferential association with the defect flux, which can lead to either local enrichment or depletion of elements based on the relative interaction with the defect flux. RIS in austenitic alloys, as well as face centered cubic (FCC) Fe-Ni-Cr model alloys, has received considerable attention over the past 3 decades, as austenitic steels are a widely used in light-water reactor core material. Solute segregation has numerous important consequences on radiation effects and component performance. For example, precipitates may form in local regions of enriched solute concentration, even though the precipitate phase is thermodynamically unfavored at the average matrix composition. Conversely, precipitates that would normally be thermodynamically stable (at the average bulk concentration) may dissolve in regions that experience solute depletion. Likewise, nonequilibrium segregation to cavity surfaces may alter the surface energy and influence void nucleation and growth, or cause local embrittlement [118, 130, 131].

Two mechanisms have been proposed as being significant contributors to RIS in Fe-Cr-Ni alloys. One is the preferential exchange of an alloying element with the vacancy flux, resulting in a net solute flux toward or away from the boundary (e.g., the vacancy mechanism portion of the Inverse Kirkendall effect). Preferential exchange of a solute with the vacancy flux was initially proposed by Marwick et al. [132, 133], and serves as the basis for the Perks model [134]. The second mechanism is the preferential association of undersized atoms with the interstitial flux (interstitial binding) [117, 135] and was applied to Fe-Cr-Ni alloys [136]. In the interstitial binding model, both interstitial binding and preferential association of solutes with the vacancy flux contribute to the segregation, and low-temperature RIS behavior does support the possibility of Ni atom binding with interstitial atoms [137]. Both RIS mechanisms are summarized in recent review articles [138, 139].

Analysis of several different Fe-Cr-Ni alloys irradiated with protons at temperatures from 200–600°C and to doses from 0.1 to 3.0 dpa provided substantial insight into the RIS mechanism. In these austenitic alloys, a vacancy-mediated Inverse Kirkendall mechanism can largely explain all of the observed RIS behavior. Fig. 5.21 summarizes the temperature and radiation dose dependence of grain boundary segregation of Ni, Cr, and Fe in an Ni-18Cr-9Fe model alloy following proton irradiation at a dose rate of $\sim 7 \times 10^{-6}$ dpa/s [140]. Fig. 5.21A shows the average

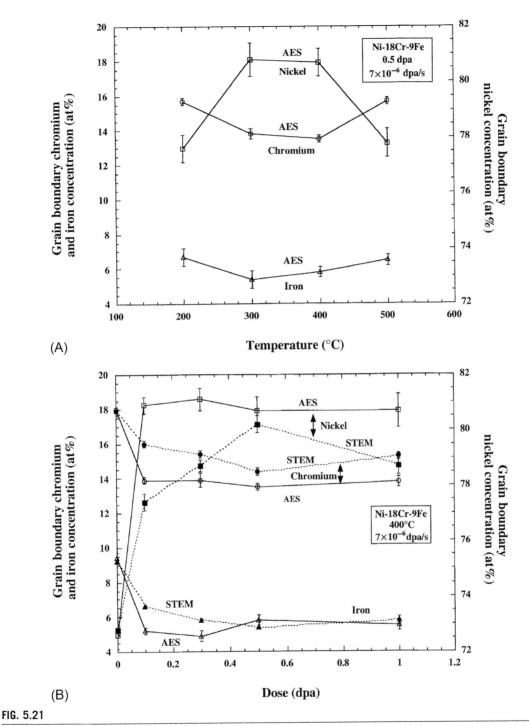

FIG. 5.21

Summary of RIS data of the grain boundary Fe, Cr and Ni concentration following 3.4 MeV proton irradiation of a model Ni-18%Cr-9% Fe alloy. (A) Irradiation temperature dependence at ~0.5 dpa, and (B) radiation dose dependence at 400°C irradiation temperature.

Reproduced from G.S. Was, T.R. Allen, J.T. Busby, J. Gan, D. Damcott, D. Carter, M. Atzmon, E.A. Kenik, Microstructure and microchemistry of proton-irradiated austenitic alloys under conditions relevant to LWR core components, J. Nucl. Mater. 270 (1999) 96–114.

grain boundary concentration of Fe, Cr, and Ni as a function of irradiation temperature at ~0.5 dpa, while Fig. 5.21B shows the dose dependence at 400°C and indicates a trend toward a saturating grain boundary concentration. As shown in Fig. 5.21, Cr consistently depletes at the grain boundaries, while Ni generally increases in concentration with increasing dose. The temperature dependence is nonmonotonic, with the amount of Ni enrichment typically peaking around 300–400°C, while the Cr concentration goes through a minimum at the same temperature. The behavior shown in Fig. 5.21 for this Ni-based model alloy is quite consistent with observations in austenitic steels, in which the RIS is characterized by grain boundary chromium depletion and nickel enrichment, with iron either enriching or depleting depending on the alloy composition. Segregation asymptotically approaches saturation with increasing dose and exhibits a characteristic bell-shaped curve with temperature in the intermediate temperature range ($0.3 < T_m < 0.5$). RIS has been successfully modeled with the Inverse Kirkendall model, in which differences in vacancy migration energies determine the direction and degree of segregation. RIS in iron-base ferritic-martensitic alloys is considerably less well understood and studied.

Finally, it is important to briefly summarize the effect of high-energy particle irradiation on precipitates, as numerous current and proposed structural materials for nuclear applications are precipitate strengthened, including the very high number density of nanoscale yttrium-titanium-oxygen precipitates in recent oxide dispersion strengthened alloys [141–143]. As noted earlier, high-energy particle radiation can enhance diffusion, which can accelerate precipitation as in the case of copper in reactor pressure vessel steels, but such irradiation can also induce the re-solution of precipitated atoms back into the metallic matrix. In contrast to fission gas bubbles in nuclear fuel, the re-solution mechanism associated with precipitates in structural alloys appears to be entirely based on elastic collisions. Certain and coworkers showed that at low temperatures, well below Stage III, high-dose rate Ni^{2+} ion irradiation to a dose of 100 dpa at −75°C could produce complete dissolution of the Y-Ti-O nanoparticles, although these particles were stable for higher temperature radiation and were not observed to undergo significant coarsening for irradiation temperatures up to 600°C and 100 dpa [144]. Russell provides a theoretical framework for assessing precipitate stability under irradiation including recoil resolution effects [118, 145], and more recently, Xu and coworkers have introduced a precipitate re-solution model into a cluster dynamics model of precipitate stability that reproduces the low-temperature dissolution and high-temperature coarsening behavior of copper precipitates in model Fe-Cu model alloys following ion irradiation [146].

Overall, the evolution of precipitates or dispersoids during neutron irradiation is influenced by the dynamic competition between ballistic recoil dissolution of precipitates by energetic displacement cascades, radiation-induced solute segregation (depletion or enrichment) to microstructural features such as grain boundaries or dislocation loops, and radiation-enhanced diffusion (accelerated kinetics). Material-specific behaviors regarding precipitate evolution for key structural alloys in nuclear energy systems are summarized in Chapters 7–12.

5.5 RADIATION- AND STRESS-MODIFIED CORROSION AND CRACKING PHENOMENA

Corrosion and SCC issues in water-cooled nuclear power plants have been extensive, with intergranular stress corrosion cracking (IGSCC) and irradiation-assisted stress corrosion (IASCC) observed in numerous nickel- and iron-based piping and heat exchanger materials [147–161]. Table 5.1 provides a partial summary of the corrosion and cracking issues observed in nuclear reactors, as tabulated by Gordon [161]. The vast majority of corrosion-related issues observed in light-water reactors are associated with IRSCC. The term IRSCC has been defined such that the intergranular cracking associated with SCC is either accelerated (either in terms of crack initiation or crack growth rate), or in some cases uniquely observed in nuclear reactor conditions where cracking is not observed in unirradiated alloys at comparable temperature, stress, and coolant conditions [159]. One well-publicized safety-related issue involved IASCC of nickel-base nozzles in the reactor pressure vessel head;

5.5 RADIATION- AND STRESS-MODIFIED CORROSION

Table 5.1 A Chronological Summary of Corrosion and Cracking Issues in Water-Cooled Nuclear Reactors

Corrosion Event	Time of Detection
Alloy 600 IGSCC in a laboratory study	Late 1950s
IGSCC BWR stainless steel fuel cladding	Late 1950s
BWR IGSCC of type 304 stainless steel during construction	Late 1960s
IGSCC of furnace sensitized type 304 during BWR operation	Late 1960s
IGSCC in U-bend region of PWR steam generator	Early 1970s
Denting of PWR alloy 600 steam generator tubing	Mid 1970s
PWSCC of PWR alloy 600 steam generator tubing	Mid 1970s
Pellet-cladding interaction failures of BWR zircaloy fuel cladding	Mid 1970s
IGSCC of BWR welded small diameter type 304 piping	Mid 1970s
IGSCC of BWR large diameter type 304 piping	Late 1970s
IGSCC of BWR alloy X-750 jet pump beams	Late 1970s
IGSCC of BWR alloy 182/600 in nozzles	Late 1970s
Accelerating occurrence of IGSCC of BWR internals	Late 1970s
PWSCC in PWR pressurizer heater sleeves	Early 1980s
General corrosion of carbon steel containments	Early 1980s
Crevice-induced IGSCC of type 304L/316L in BWRs	Mid 1980s
FAC of single-phase carbon steel systems in PWRs	Mid 1980s
PWSCC in PWR pressurizer instrument nozzles	Late 1980s
IGSCC of BWR low carbon and stabilized stainless steels	Late 1980s
IGSCC of BWR internal core spray piping	Late 1980s
Axial PWSCC of alloy 600 of PWR top head penetration	Early 1990s
Circumferential PWSCC of j-groove welds	Early 1990s
PWSCC of PWR hot leg nozzle alloy 182/82	Early 2000s
PWSCC-induced severe boric acid corrosion of a PWR head	Early 2000s
SCC of stainless steels in PWRs	Early 2000s

Reproduced from B.M. Gordon, Corrosion and corrosion control in light water reactors, JOM, 65 (2013) 1043–1056.

the through wall cracking in the control rod drive mechanism nozzles led to general corrosion (wastage) of the bainitic steel reactor pressure vessel head [162].

As noted by Scott [158], the operating environments that led to corrosion of nickel- and iron-based alloys are dramatically different in BWR and PWRs, as shown in the simplified Pourbaix diagram of Fig. 5.22. As noted in Fig. 5.22, BWRs typically operate at a pH of 5.5–5.6 with a normal water chemistry potential in the range of 50–300 mV, while the potential is significantly reduced to −250 to −400 mV under BWR hydrogen water chemistry. PWR conditions are different in the primary vs secondary side, but are nominally in a region where the iron or nickel oxide phases would be expected. Also noted in this diagram is a hatched box denoting the typical region of pH and potential where pressurized water SCC (PWSCC) has been observed in nickel-based alloys, most notably Alloy 600 that makes up the steam generator tubes, and is the most severe example of SCC in nuclear reactors in terms of observed economic impact [158]. Two other hatched regions highlighted in Fig. 5.22 correspond to other IGSCC causes observed in PWR steam generator tubes, namely caustic intergranular cracking that is typically observed at higher pH values; this phenomenon has been reduced by taking care to restrict sodium impurities [158]. At lower pH values, cracking driven by sulfate ions has also been observed to occur [158]. The IGSCC, also referred to

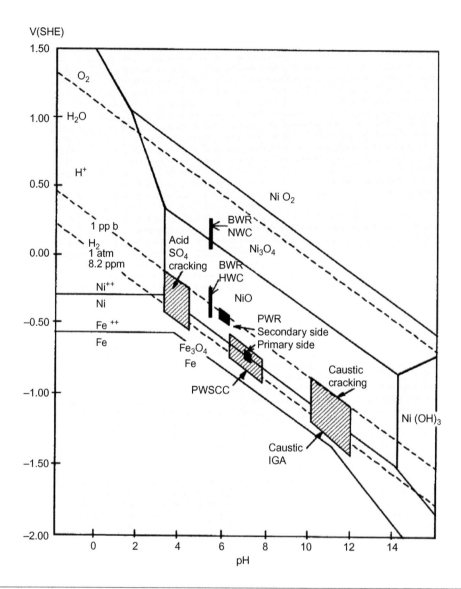

FIG. 5.22

Simplified Pourbaix diagram at ~300°C for nickel and iron, which shows the pH and electrochemical potential combinations for PWR primary and secondary side water, BWR Normal Water Chemistry (NWC) and BWR Hydrogen Water Chemistry (HWC), in addition to the SCC modes of alloy 600.

From P.M. Scott, An overview of materials degradation by stress corrosion in PWRs In D. Féron Ed., Corrosion Issues in Light Water Reactors, Elsevier, 2007.

as PWSCC, of Alloy 600 steam generator U-tubes has without a doubt been the most significant corrosion problem to affect PWRs [161]. Alloy 600 is a nickel-base alloy that contains 14%–17% Cr and 6%–10% Fe, and was initially selected for steam generator U-tubes because it had generally good resistance to chloride cracking compared to stainless steel [163], as well as being used for the control rod drive mechanism penetrations of the upper reactor pressure vessel head. Extensive cracking of alloy 600 steam generator tubes was initially mitigated through plugging the cracked tubes, followed by eventual steam generator replacement in which Ni-base alloy 690, which has a higher chromium content, replaced alloy 600.

In addition to the PWSCC of alloy 600, there have been cracking failures in the iron- and nickel-base stainless alloys in both BWR and PWR core components, which experience significant neutron fluence at temperatures between 270°C and 340°C in water environments that span from oxygenated to hydrogenated water [161]. Such failures are generally referred to as IASCC, and have been observed to occur in baffle former bolts, instrumentation tubes, and BWR core shrouds [164]. While the detailed mechanisms of IASCC are not yet well established, it is evident that changes in the microstructure and mechanical properties of alloys subject to neutron irradiation (reviewed earlier in this chapter as well as Chapters 7–13) play a role in IASCC degradation [159, 160]. In particular, as noted previously in Section 5.4, RIS in iron- and nickel-based austenitic alloys is known to decrease the local chromium concentration at grain boundaries while locally increasing the concentration of nickel, in addition to increasing the concentration of nm-sized defect clusters throughout the microstructure that produce radiation hardening. Was and Busby have also postulated that the dislocation channeling phenomena that is typically observed in irradiated materials could enhance the localized deformation and influence the crack nucleation behavior [159, 161]. In addition, radiation can affect the IG cracking behavior by modifying the water chemistry through radiolysis [159]. The effect of radiolysis on modifying the concentration of ionic and molecular species in water is well established [165], most notably increasing the concentration of oxidant species that directly increases the corrosion potential [159]. Indeed, numerous data indicate that the extent of intergranular cracking, as well as the average crack growth rate, increases with an increasing value of the corrosion potential [159]. However, the extent of intergranular cracking is also known to correlate to other radiation-induced material changes, including increasing with increasing yield strength and increasing with reductions in Cr content at the grain boundaries [159]. Thus, while it is clear that the irradiation environment definitely accelerates the IGSCC phenomenon, further elucidating the controlling mechanisms requires improved fundamental understanding of the IG cracking process associated with SCC for both irradiated and nonirradiated conditions [152].

Irradiation-modified corrosion and stress corrosion issues are considered to be of secondary importance in reactors that are cooled by liquid metals or helium. It is worth noting that, in contrast to the pronounced stress corrosion issues with LWRs, no significant adverse stress corrosion effects associated with high-purity sodium have been observed in austenitic or ferritic/martensitic steels for sodium fast reactor operating conditions [166]. Further research is needed to determine whether irradiation-modified corrosion mechanisms may occur in molten salt-cooled reactor concepts.

While the fundamental responsible mechanisms of IASCC for water-cooled reactor environments are not fully resolved, the empirical radiation dose dependence is well established. Fig. 5.23, which has been reproduced from the review article by Bruemmer and coauthors [152], indicates that there is a threshold fluence of about 5×10^{20} N/cm^2 (neutron energy > 1 MeV) for the IASCC of type 304 stainless steel. This threshold fluence corresponds to the onset of significant radiation-induced microstructure and chemical segregation changes and radiation hardening of the material, but as noted previously, definitive understanding of the specific controlling mechanism(s) is not yet established.

In summary, PWSCC and IASCC have been observed in nickel- and iron-bearing austenitic alloys, in addition to low-alloy carbon steels, for doses above 1–10 dpa at LWR irradiation conditions. Notably, ferritic-martensitic alloys appear to be much less susceptible to RIS [137] as well as intergranular SCC in nuclear environments. Further details on the corrosion and cracking of nuclear core materials can be found in Chapter 6 (Corrosion) and Chapter 8 (Stainless steels).

5.6 RADIATION-INDUCED DIMENSIONAL INSTABILITY

In addition to phase changes previously discussed in Section 5.4, there are three major radiation-induced phenomena that can lead to dimensional distortions in structural materials: cavity (void) swelling, irradiation creep, and

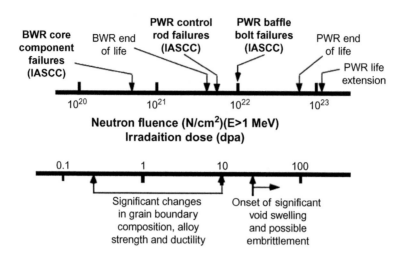

FIG. 5.23

The influence of radiation dose (neutron fluence) on type 304 stainless steel BWR and PWR component cracking susceptibility.

Reproduced from S.M. Bruemmer, E.P. Simonen, P.M. Scott, P.L. Andresen, G.S. Was, J.L. Nelson, Radiation-induced material changes and susceptibility to intergranular failure of light-water-reactor core internals, J. Nucl. Mater. 274 (1999) 299–314.

irradiation growth. Cavity swelling can occur in all irradiated materials at intermediate temperatures, irrespective of applied stress. Irradiation creep occurs in all materials irradiated with an applied stress, and is most pronounced at intermediate temperatures. Irradiation growth only occurs in materials with anisotropic crystal structures (with or without applied stress), and is most pronounced at low to intermediate temperatures.

5.6.1 CAVITY SWELLING

Swelling is generally observed in all materials at irradiation temperatures between 0.3 and $0.6T_m$ for damage levels above a few dpa, due to the aggregation of vacancies that are induced by displacement damage to concentration levels that are far beyond the thermal equilibrium concentration [5, 116, 167–169]. The phenomenon was first observed in type 316 austenitic stainless steel irradiated in the Dounreay fast reactor in 1967 [170], and it has been one of the most intensively researched radiation-induced phenomena since then, as void swelling has been recognized to be a potential major contributor to dimensional instability in structural materials in nuclear energy systems. The swelling typically produces isotropic volumetric expansion; for example, linear expansion ($\Delta l/l_0$) is typically one-third the volumetric expansion ($1/3 \Delta V/V_0$) in cubic systems, although in some cases for BCC and HCP materials ordered arrays of cavities can be produced over a limited range of irradiation temperatures that can create anisotropic swelling [171]. The term "void" is historically reserved for gas-free (or nearly gas-free) collections of vacancies. The term "bubble" is used to describe vacancy + gas clusters that may be near thermodynamic equilibrium or overpressurized due to excess amount of gas atoms relative to the number of vacancies. The term "cavity" is the generic term to describe either voids or bubbles.

The dose-dependent swelling behavior can be generally categorized into two regimes: a transient or incubation regime where no or very low swelling occurs, and a steady-state swelling regime that is observed after the transient regime is terminated. It is generally recognized that volumetric swelling over ~5% in structural components is difficult to accommodate in engineering designs [172], and swelling levels exceeding ~10% are typically accompanied by pronounced embrittlement [173]. The steady-state swelling rate appears to be relatively insensitive to initial thermomechanical condition, slight compositional variations or to irradiation conditions once it enters the posttransient rapid swelling regime [174]. Fig. 5.24 summarizes the general dose-dependent swelling behavior for neutron-irradiated austenitic stainless steel at a variety of temperatures in the void swelling regime. The duration of the low-swelling transient regime is typically longest at very low or very high irradiation temperatures. Due to the relatively high volumetric swelling rates (0.2%–1%/dpa) observed in most materials in the posttransient steady-state swelling regime, research on void swelling-resistant materials has focused on extending the duration of the low-swelling transient regime [4, 141, 175, 176].

Swelling behavior has been observed to depend on the crystal system of the irradiated material. In body centered cubic (bcc) metals, it is generally recognized that void swelling is suppressed [169], and tends to saturate by forming an ordered void lattice as a consequence of "self-organization" [171, 177]. Higher diffusivity of point defects [178] and lower biased self-interstitial atom absorption at dislocations [179, 180] are recognized as possible contributing mechanisms. The superior swelling resistance of ferritic and ferritic/martensitic steels, as shown in Fig. 5.25 [104], is due to its longer transient regime as a consequence of these mechanisms. It should be noted that suppressive swelling is not universally observed for all bcc metals, for example, large void swelling has been observed in neutron-irradiated V-5Fe bcc alloy whereas pure vanadium and most other V alloys exhibit good void swelling resistance [181].

For the case of fcc metals, swelling appears to be an unsaturable process [182, 183], and it becomes the life-limiting phenomenon of reactor core components that use austenitic stainless steels, such as type 316 or 304 stainless steel, at temperatures above ~400°C. Consequently, a vast amount of experimental and theoretical studies have been conducted on fcc metals, providing most of our current mechanistic understandings on swelling [173].

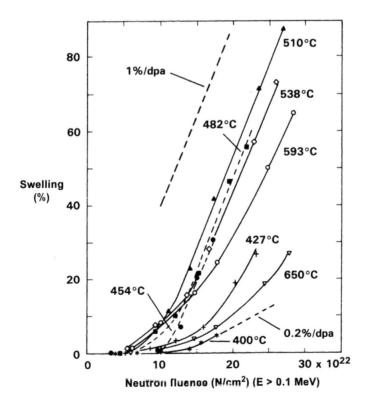

FIG. 5.24

Swelling of 20% cold-worked Type 316 stainless steel irradiated in EBR-II [174].

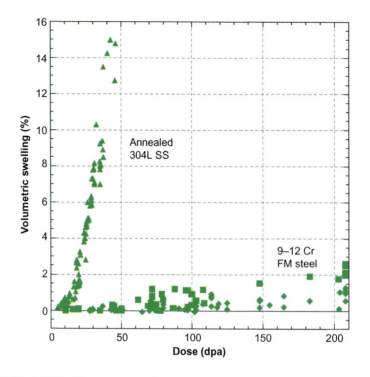

FIG. 5.25

Comparison of swelling in neutron-irradiated-type 304-L austenitic (FCC) stainless steel and 9–12 Cr ferritic/martensitic (BCC) steel at 400–550°C [104].

The termination of the transient regime is associated with completion of the void nucleation phase and onset of rapid void growth. Void nucleation is the key process, and the input parameters that govern the void nucleation process are temperature, dpa rate, and applied stress. A sufficient supersaturation of vacancies is the key requirement for the onset of void nucleation. Therefore, the irradiation temperature has to be low enough to achieve a steady-state vacancy concentration well above the thermal equilibrium value. Vacancy diffusion is another key process, and can be modified by solute additions. Solutes such as phosphorus in stainless steels are known to bind solute interstitials and enhance vacancy diffusion, which allows vacancies to recombine with interstitials or be absorbed into point defect sinks, and as a result, suppress swelling [184]. The material starting state is also an important factor. Cold working, usually chosen to be 20%–25%, is known as a useful approach to initially suppress swelling by inducing a high dislocation density, that is, a high point defect sink distributed throughout the matrix that thereby suppresses void nucleation [185, 186]. It should be noted that too much cold working tends to enhance swelling [187] by inducing a high free-energy status in the matrix that evokes enhanced irradiation diffusion and, as a consequence, vigorous recovery and recrystallization that eliminate induced void sinks. Applied stress, mainly a shear stress component but not a hydrostatic component of stress, could shorten the transient regime [188–190]. This phenomenon could be interpreted as the result of stress-induced Frank loop unfaulting and/or dislocation glide and localization, which eliminates the Frank loop as a point defect sink and accelerates the onset of void nucleation by reducing the overall point defect sink density. It should be noted that not only the external (primary) stress but also the thermal (secondary) stress that is induced by temperature gradients can enhance the same effect [191].

Experimental studies on austenitic and ferritic/martensitic steels have reported that the amount of void swelling tends to decrease with increasing displacement damage rate for both neutron-reactor-relevant

($\sim 10^{-8}$ to 10^{-6} dpa/s) [173, 192, 193] and ion-irradiation-relevant ($\sim 10^{-4}$ to 10^{-2} dpa/s) [194, 195] damage rates. The low swelling rate transient regime is extended to higher doses with increasing damage rate, whereas the post-transient steady-state swelling rate regime does not appear to be significantly affected. These observations suggest that accelerated damage rate data obtained from typical studies in test reactors or ion accelerators may underpredict the void swelling that would occur under actual (lower damage rate) service conditions in power reactors.

Under a high-energy neutron irradiation such as 14 MeV fusion neutron or spallation proton-neutron irradiation (and also under mixed spectrum neutron irradiation in materials containing certain elements such as Ni, B, or Be), transmutation reactions such as (n, α) or (n, p) take place in irradiated materials with relatively high production cross-sections. This causes substantial amounts of helium and hydrogen to be produced in the matrix, for example, 10 appm He/dpa under 14-MeV fusion neutron irradiation for iron. Since helium is insoluble in solids and can readily diffuse in the matrix even at relatively low temperature, it can easily combine with vacancies and form bubbles (helium-filled gas pressurized small cavities), and serve as nuclei for cavity swelling. This process can accelerate the onset of void swelling and increase cavity density. Conversely, if extremely high bubble densities are produced, it can also extend the transient regime by forming tiny bubbles at high densities in the various trapping sites, such as dislocation or grain boundaries or precipitate surfaces. In general, maximum cavity swelling occurs when the cavity sink strength is comparable to the biased interstitial sink strength [141, 175, 196]. Since helium is primarily coupled with vacancies, all microstructural evolutions under irradiation (precipitation, dislocation loop formation, network dislocation evolution, etc.) are affected and it induces various effects on mechanical properties as well [197].

Based on current understanding of void swelling, maintaining a high point defect sink strength is essential to exhibit a longer low-swelling transient regime. For example, more complex alloys, such as bcc steel-based oxide dispersion strengthened steel (ODS) or nanostructured ferritic alloy (NFA), exhibit very low swelling compared to a normalized and tempered ferritic/martensitic steel due to their very high point-defect sink strengths associated with nanoscale dispersoids, high dislocation densities, and fine-scale grain size [198, 199]. Thus, designing a stable high-sink-strength material with a radiation-resistant matrix phase will be the key to develop materials that exhibit negligible radiation-induced property degradation [4].

5.6.2 IRRADIATION CREEP

Irradiation creep is a stress-induced dimensional change that occurs in all crystalline materials under irradiation at intermediate temperatures; its effects are typically most pronounced in the temperature regime ~ 0.2–$0.45 T_M$. In this temperature regime, the magnitude of deformation associated with irradiation creep can be orders of magnitude larger than that for thermal creep. At very low temperatures, irradiation creep is generally suppressed due to low point defect mobility and high radiation-induced sink strengths that cause extensive point defect recombination. As operating temperatures approach or exceed $0.5 T_M$, the exponential increase in thermal creep deformation coupled with a decrease in irradiation creep due to lower point defect supersaturation leads to a transition to thermal creep-dominated behavior. Fig. 5.26 compares the temperature-dependent behavior of irradiation creep and conventional thermal creep in austenitic stainless steel. Irradiation creep is characterized by a relatively weak dependence on temperature and by a progressive increase in deformation with increasing damage level at a given temperature. Irradiation creep was first observed in 1955 in uranium [200], and in stainless steel in 1959 [201]. As irradiation creep involves plastic deformation, irradiation volume is conserved (i.e., expansion along the direction of applied tensile stress, shrinkage along the lateral direction).

Irradiation creep and void swelling are interrelated processes. Thus, irradiation creep can be categorized into four stages: the transient regime, the creep regime in the absence of swelling, swelling-enhanced creep, and creep disappearance [173]. The irradiation creep transient is mainly involved with the transient evolution of the dislocation density to achieve quasiequilibrium and is particularly pronounced in the case where the initial dislocation density is high, but this phase can be ignored in most cases. The irradiation creep disappearance stage occurs when void swelling becomes pronounced. Thus, most irradiation creep studies have concentrated on irradiation creep

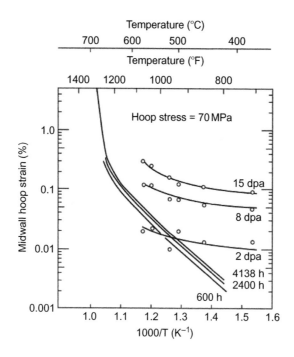

FIG. 5.26

Comparison of thermal creep *(solid lines)* and irradiation creep *(open circle data points)* of cold-worked Type 316 stainless steel [173].

Original from Gilbert, E. R.; Bates, J. F. J. Nucl. Mater.1977, 65, 204–209.

deformation with and without swelling. The irradiation creep in these regimes can be described using the following general equation [115]:

$$\dot{\varepsilon}/\sigma = B_0 + D\,dS/d\phi, \tag{5.2}$$

where $\dot{\varepsilon}$ is the plastic strain rate per unit dose, σ is the applied effective stress, B_0 is the creep compliance for irradiation creep deformation, D is the creep-swelling coupling coefficient for irradiation creep deformation, and $dS/d\phi$ is the swelling rate per unit dose.

The creep compliance B_0 has been observed to be roughly independent of composition, starting state, dpa rate, and irradiation temperature in austenitic steel, with a typical value of $\sim 1 \times 10^{-6}$/MPa-dpa [169, 173]. The creep compliance has been reported to become larger in the case of very low-temperature ($< \sim 100°C$) irradiation where transient point defect diffusion effects occur [202], or when helium effects are significant [203, 204]. In the case of BCC ferritic steel, the creep compliance is lower ($\sim 0.5 \times 10^{-6}$/MPa-dpa) [169, 205]. Apparent creep compliance values for other structural material systems range from $B_0 \sim 2.5-10 \times 10^{-6}$/MPa-dpa) for vanadium alloys [206, 207] to much larger values for Zr alloys [208], although in the latter case there is a large contribution from irradiation growth (see Section 5.6.3 and Chapter 7). The swelling creep coupling coefficient D is assumed to be constant in a given crystal system ($D \sim 0.006$/MPa for steels) [169], with some evidence the value for ferritic steel may be larger than that of austenitic steel [205]. The stress exponent of irradiation creep is generally observed to be linear for austenitic stainless steel, which is different from thermal creep where exponent values larger than 1 are often observed. The observed stress exponent for irradiation creep has generally be found to be near 1 in BCC alloys such as ferritic steel and V alloys [206, 207, 209] but has been reported to be as high as 1.5 in some ferritic steels [210]. The irradiation creep, Eq. 5.2, suggests that the creep rate is proportional to dpa both in the absence of swelling and in the steady-state swelling-enhanced regime.

The irradiation creep can be understood as an anisotropic redistribution of mass in response to shear stress. In austenitic stainless steel, stress-induced preferential nucleation (SIPN) of Frank loops (faulted loops) is considered as a potential mechanism [211, 212], but the SIPN process alone cannot explain the observed irradiation creep magnitude. Stress-induced preferential growth and unfaulting of Frank loop accompanied with network dislocation climb and glide are necessary to be taken into account.

It should be noted that irradiation creep is regarded as a nondamaging phenomenon at the microstructural level, that is, irradiation creep does not enhance void nucleation or growth at grain boundaries, which is a typical damage mechanism associated with thermal creep. In other words, irradiation creep can mitigate thermal creep damage by relaxing stress and allow deformation without failure. Furthermore, the typical magnitude of irradiation creep in structural materials for moderate operating stresses (~50–100 MPa) can be seen from Eq. 5.2 to produce <1% plastic deformation for doses below ~100 dpa if void swelling is small. Therefore, in terms of dimensional changes irradiation creep is generally only of concern for high-dose, high-stress conditions or cases where significant void swelling occurs (although in the latter case void swelling by itself tends to cause unacceptable dimensional changes).

However, irradiation creep during reactor operation can potentially cause several problems, most notably stress relaxation (creep down) of gaskets and in-core prestressed springs or fasteners such as bolts [173, 213–215]. Fig. 5.27 shows an example of the dramatic stress relaxation observed in a nickel alloy commonly used for hold-down springs in light-water reactors [215]. This can result in excessive rubbing wear (fretting) or loosening of bolted structures, and is a major concern in operating reactors. Some components may also experience a potential stress reversal between shutdown and operating conditions if thermal differential stresses are relieved by irradiation creep during operation. There is also potential concern for enhanced damage associated with irradiation creep-fatigue interactions [216].

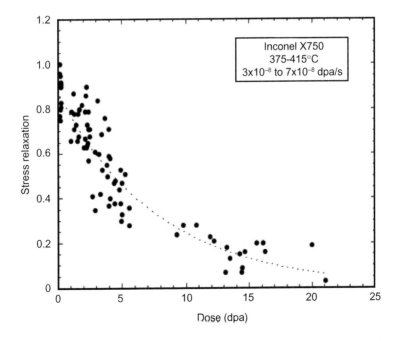

FIG. 5.27

Stress relaxation of Inconel X750 springs due to fast fission reactor irradiation [215].

5.6.3 IRRADIATION GROWTH

Irradiation growth is observed in anisotropic crystalline materials such as Zr or graphite during irradiation at low to intermediate temperatures [1, 217, 218]. As with irradiation creep, the anisotropic dimensional changes occur with no net volume change [217] (i.e., the growth in certain crystal directions is accompanied by shrinkage in other crystal directions). The magnitude of irradiation growth is most pronounced at low-to-intermediate irradiation temperatures and, unlike irradiation creep, does not require applied stress to be activated. For irradiation creep to be perceptible in polycrystalline materials, preferential crystallographic texture must be present. Significant texture typically develops in anisotropic crystalline materials during standard processing steps involving deformation and annealing. For example, Zr alloys develop <a> texture along the longitudinal direction in rolled or extruded components with <c> texture in the radial direction for tubing or in the short thickness direction for plate [6]. Fig. 5.28 shows the effect of orientation (texture) on the irradiation growth behavior of recrystallized Zircaloy-2 due to neutron irradiation [217]. Pronounced elongation occurred along the longitudinal direction (corresponding to predominantly a-axis texture) whereas contraction occurred along the transverse direction (predominantly c-axis texture). It is worth noting that the irradiation growth behavior in graphite is opposite to that observed in zirconium alloys, that is, for graphite irradiation growth produces expansion along the c-axis and contraction along the a-axis [219].

The empirical dose dependence of irradiation growth typically consists of three regimes: a low-dose transient regime with relatively high growth rate, an intermediate-dose regime where the growth rate is near zero, and a

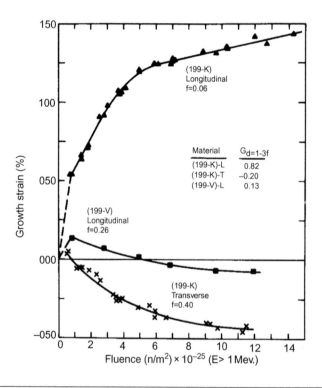

FIG. 5.28

Effect of texture on the irradiation growth of recrystallized Zircaloy-2 sheet at 330 K. Alloy 199-K was slowly cooled from 1070 K (single alpha-phase regime) whereas alloy 199-V was rapidly cooled from 1290 K (initially beta phase, cooled into alpha phase) [217]. The Kearns texture parameter f corresponds to the fraction of basal poles aligned along the direction, and the anisotropy factor $G_d = 1-3f$ (positive for a-axis texture, negative for c-axis texture).

high-dose "breakaway growth" regime where the growth rate increases [208, 217, 220–222]. The onset of the breakaway growth regime in irradiated Zr alloys has been correlated with the appearance of a significant population of <c> loops at high doses. The mechanism for irradiation growth in Zr alloys has been attributed to anisotropic diffusion of self-interstitial atoms [208, 221], although it is possible other anisotropic physical processes may also contribute. Additional information on irradiation growth phenomena for Zr alloys is given in Chapter 7.

The temperature dependence of irradiation growth in Zr alloys has been observed to be relatively weak between 330 and 700 K [208, 217]. Irradiation growth data for Zr alloys at higher temperatures has apparently not been reported, although for all materials it is expected the irradiation growth should vanish at high temperatures due to few radiation-induced microstructural changes.

In summary, irradiation growth only occurs in anisotropic materials and typically produces moderate anisotropic dimensional changes (~0.1%–1% length changes) with no significant overall volumetric change. The magnitude of these dimensional changes is generally less pronounced than irradiation creep or swelling. However, it is important to provide allowances for the dimensional changes due to irradiation growth in engineering designs, particularly due to the anisotropic growth and shrinkage aspects of the dimensional changes.

5.7 HIGH-TEMPERATURE HELIUM EMBRITTLEMENT

Neutron irradiation can induce transmutation reactions in materials, including generation of helium from (n,α) reactions. The helium is typically initially created as an interstitial impurity. Due to the high mobility of interstitial helium and high affinity for trapping by vacancy defects [197, 223–226], the helium is usually quickly trapped in lower mobility helium vacancy clusters in the matrix. At high temperatures, these helium-vacancy complexes are sufficiently mobile to migrate to grain boundaries, where the localized accumulation of He-filled cavities weakens the grain boundaries and can lead to premature failure of components that are mechanically stressed. Since it is a diffusion-activated effect, high-temperature helium embrittlement typically emerges at temperatures above $\sim 0.5 T_M$ and becomes progressively more pronounced with increasing temperature and applied stress level. The phenomenon of high-temperature helium embrittlement can produce dramatic reductions in ductility (uniform and total elongation and reduction in area), and therefore is an important engineering design concern for the reliable and safe operation of neutron-irradiated structural alloy components at high temperatures [197, 223, 224, 227–229]. The predominant fracture mode for high-temperature helium embrittlement is intergranular, as opposed to ductile dimpled transgranular fracture surfaces that typically occur in unirradiated structural alloys. High-temperature helium embrittlement is of particular concern in Ni-containing alloys such as austenitic stainless steel in mixed spectrum reactors due to the relatively large helium production from thermal neutrons via the two-step ^{58}Ni $(n,\gamma)^{59}$Ni$(n,\alpha)^{56}$Fe reaction [230].

Fig. 5.29 shows a data compilation of the decrease in thermal creep ductility in irradiated austenitic stainless steels with increasing transmutant helium content for creep testing at 900 K ($\sim 0.53 T_M$) [231]. A significant reduction in creep ductility is observed for He contents above ~0.1 appm. The decrease in creep ductility is accompanied by a transition from transgranular fracture in unirradiated and low-He samples to intergranular fracture in high-He samples [197, 223, 227, 231–234]. The decrease in elongation due to premature intergranular fracture directly leads to a reduction in creep rupture lifetime. An example of the dramatic decrease in creep lifetime with increasing He content is shown in Fig. 5.30 for neutron-irradiated austenitic steel tested at 310 MPa at a relatively low temperature of 823 K ($0.48 T_M$), where the creep lifetime decreased by a factor of 10,000 as He levels increased from 30 to 3000 appm [235]. In these tests, it is possible that radiation-induced solute segregation at grain boundaries may have partially contributed to the dramatic grain boundary weakening.

For a given concentration of helium and constant strain rate, the observed severity of high-temperature embrittlement increases rapidly with increasing test temperature as expected for a process that is controlled by diffusion of He-vacancy complexes to grain boundaries. Fig. 5.31 summarizes the normalized tensile elongation in vanadium alloys containing 2–200 appm He as a function of test temperature [236–238]. At the lowest He content (2 appm He),

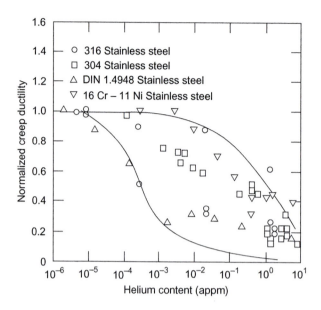

FIG. 5.29

Normalized ductilities for several austenitic stainless steels vs helium content after mixed spectrum fission reactor irradiation at 325 K and thermal creep testing at 191 MPa at 900 K [231]. The curves represent predicted upper and lower bounds for ductility due to varying grain size and other factors.

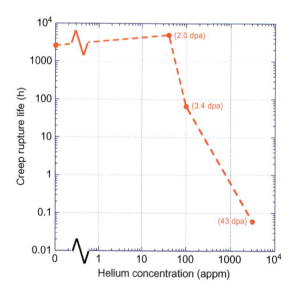

FIG. 5.30

Reduction in rupture lifetime vs helium content in mixed spectrum reactor irradiated Type 316 stainless steel after thermal creep testing at 310 MPa at 823 K [235]. The numbers in parentheses indicate the irradiation doses of the tested specimens.

Data from E.E. Bloom, F.W. Wiffen, The effects of large concentrations of helium on the mechanical properties of neutron-irradiated stainless steel, J. Nucl. Mater., 58 (1975) 171–184.

5.7 HIGH-TEMPERATURE HELIUM EMBRITTLEMENT

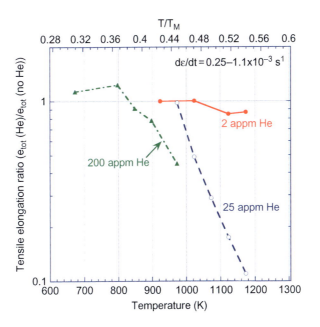

FIG. 5.31

Normalized tensile elongation ratio as a function of test temperature for V-3%Ti-1%Si with 2 appm He [236], V-15%Cr-5%Ti with 25 appm He [237], and V-20%Ti with 200 appm He [238].

no evidence of pronounced ductility degradation is observed up to a maximum test temperature of $\sim 0.54 T_M$, whereas at the highest He content (200 appm) some ductility degradation is observed even at test temperatures near $0.4 T_M$. Similar strong reductions in ductility with increasing test temperature have been observed in numerous other studies on helium-containing materials such as irradiated stainless steel [232, 235, 239, 240], nickel alloys [241], and copper [242–244]. In general, pronounced He embrittlement of grain boundaries typically requires test temperatures $>0.5 T_M$ although some embrittlement can be observed at lower temperatures for high He contents and slow strain rates (e.g., Figs. 5.31 and 5.32).

The severity of high-temperature helium embrittlement becomes more pronounced with decreasing strain rate for a given He content and test temperature. This effect is easily rationalized as slower strain rates correspond to longer times to reach a given plastic deformation, thereby providing increased diffusion times for He-vacancy clusters to migrate to grain boundaries. Fig. 5.32 summarizes the effect of strain rate on neutron-irradiated Type 304 stainless steel containing 7-appm He tested at 823 K ($0.48 T_M$) [245]. Whereas no evidence of ductility degradation relative to the unirradiated sample was observed for typical tensile test strain rates $>10^{-3}$/s (in part due to the relatively low He content and low test temperature of $0.48 T_M$), significantly lower ductility occurred in the irradiated samples at slow strain rates (10^{-9}–10^{-4}/s). Similar observations of more pronounced embrittlement with decreasing strain rate have been reported in several studies of neutron-irradiated nickel-base alloys at elevated test temperatures [241, 246]. As reactor applications typically require structural material lifetimes of 2 to >40 years (i.e., effective strain rates of $\sim 10^{-9}$/s) for satisfactory economics, it is important to recognize that accelerated strain rate tests frequently used in research tests provide lower bound estimates of the magnitude of high-temperature helium embrittlement.

Several studies on austenitic stainless steels have demonstrated that application of stress during the He introduction at high temperatures enhances the magnitude of high-temperature helium embrittlement [223, 240, 247–251]. For example, the creep rupture lifetime at 1023 K [$\sim(0.60 T_M)$] for Type 316 stainless steel was about a factor of 10 lower in samples where the He was introduced continuously during high-temperature creep testing at 50–120 MPa compared to samples that were He hot-preimplanted with zero stress prior to creep testing [249, 250].

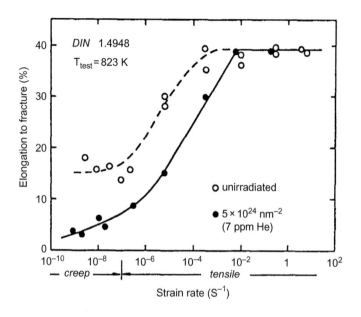

FIG. 5.32

Effect of strain rate on the measured high-temperature ductility of unirradiated and neutron-irradiated Type 304 stainless steel [245]. Although the ductility decreases at low strain rates for both cases, the decrease is more pronounced for He-containing samples.

From B. van der Schaaf, M.I. de Vries, J.D. Elen, Irradiation embrittlement of Type 304 stainless steel welds and plate at 823 K (550 C), in: M.L. Bleiberg, J.W. Bennett (Eds.) Radiation effects in breeder reactor structural materials, TMS-AIME, New York, 1977, pp. 307–316. Used with permission of The Minerals, Metals & Materials Society.

Therefore, simulation studies that do not apply stress during the He implantation will generally provide only a lower bound estimate of the amount of high-temperature He embrittlement compared to actual reactor applications. In general, the observed stress effects have been attributed to stress-enhanced diffusion of He-vacancy clusters to grain boundaries and subsequent stress-enhanced nucleation and growth of the grain boundary cavities [248]. For the case of He co-implantation during creep testing, the relatively small He-vacancy clusters continuously created during coimplantation can more easily migrate to grain boundaries under the applied stress compared to the larger He-vacancy clusters that nucleate and grow in the matrix during stress-free implantation prior to creep testing. The higher flux of He-vacancy clusters reaching grain boundaries for the coimplantation case promotes enhanced growth of grain boundary cavities. Since He arrives earlier at grain boundaries for the coimplantation case, pronounced stress-assisted growth of the grain boundary cavities can occur. Most modeling studies have assumed the effect of applied stress is mainly controlled by nucleation and growth of grain boundary cavities as opposed to stress-assisted transport of He-vacancy clusters to the grain boundaries [224, 252–255].

Material parameters such as matrix precipitate density and thermal creep strength exert a large influence on susceptibility to high-temperature He embrittlement. Experimental studies on austenitic stainless steels containing high concentrations of precipitates (which serve as matrix traps to inhibit the diffusion of He-vacancy clusters to grain boundaries) have shown suppressed embrittlement compared to precipitate-free steels [228, 229, 252, 256, 257]. Several studies have observed that ferritic/martensitic steels exhibit superior resistance to high-temperature helium embrittlement compared to austenitic stainless steels [197, 227, 229, 258, 259]. This has been attributed to the relatively low thermal creep strength of ferritic/martensitic steels (i.e., the rupture lifetime is controlled by zero-helium thermal creep rather than the transport and nucleation and growth of He bubbles at grain boundaries) [258], along with a higher size threshold ("critical radius") in ferritic/martensitic steels for the conversion of subcritical grain boundary bubbles to rapidly growing cavities [197, 223, 227, 229, 250, 252, 258]. A higher critical radius means much higher helium content is needed to induce the conversion of grain boundary bubbles to the rapidly growing cavities associated with high-temperature He embrittlement.

Since high-temperature helium embrittlement involves migration of He to grain boundaries, He production per unit time is in general a more relevant correlation parameter than the commonly used appm He/dpa parameter used for void swelling correlations [260]. This is because helium migration to grain boundaries at elevated temperatures is largely controlled by thermal diffusion, not by radiation damage defects, due to the relatively low point defect supersaturation concentrations that occur for elevated temperature irradiations [127, 212, 224, 261]. Therefore, most accelerated-rate He production simulation studies may provide underestimates of the degree of high-temperature helium embrittlement for a given He content.

High-temperature helium embrittlement is also a major contributor to cracking in conventional tungsten inert gas repair weldments in irradiated materials such as austenitic stainless steel. The phenomenon of cracking in the weldment heat-affected zone in irradiated materials has been shown to be due to rapid growth of He-containing grain boundary cavities associated with the high temperature and stress gradient present during welding [262–265]. Severe cracking has been observed in the heat-affected zones for He contents above ~1 appm. The severity of cracking is reduced for low heat inputs (~1 kJ/cm) due to reduced stress levels and shorter time at high temperature.

In summary, significant high-temperature helium embrittlement is typically observed in structural alloys for He contents above 1–100 appm He, test temperatures above $0.5T_M$, and strain rates $\sim 10^{-6}$/s or lower. The severity of high-temperature He embrittlement is most pronounced when mechanical stress and elevated temperatures are applied during He generation (as typically occurs in reactor service conditions). Matrix precipitates or dispersoids can be effective in reducing the magnitude of He embrittlement by trapping He and thereby suppressing the amount of He reaching grain boundaries. Structural materials with relatively low thermal creep strengths such as ferritic/martensitic steels are less susceptible to this embrittlement than high creep strength materials such as nickel-base alloys or austenitic stainless steels.

5.8 CONCLUSIONS

Nuclear fission and fusion reactors represent extreme operating environments for structural materials due to the elevated temperatures, steady or cyclic applied mechanical stresses, and high neutron irradiation fluxes. Both thermomechanical- and irradiation-induced degradation phenomena must be considered when evaluating the suitability of potential structural materials for reactor components. Long-term thermal aging, thermal creep, and creep-fatigue phenomena can lead to significant property degradation in many materials. Therefore, highly stable microstructures are essential for potential structural alloys. All structural materials are susceptible to five general temperature-dependent radiation-induced degradation processes. At low temperatures (below $\sim 0.4T_M$), radiation hardening and embrittlement is of potential concern for doses above ~ 0.1 to 1 dpa. At high temperatures (above $\sim 0.5T_M$), high-temperature He embrittlement typically emerges for He contents above ~ 1–10 appm (~ 1–10 dpa, depending on material and neutron spectrum). Collectively, low-temperature radiation hardening and high-temperature He embrittlement (along with chemical compatibility and thermal creep considerations) typically define the upper and lower operating temperature limits for prospective structural materials. At intermediate temperatures, radiation-enhanced or -induced solute segregation and precipitation can produce degradation in mechanical properties for doses above ~ 1–10 dpa, and void swelling and irradiation creep can produce unacceptable dimensional changes and stress relaxation for doses above ~ 10 dpa. Irradiation creep is proportional to dose and applied stress, and volumetric swelling is proportional to dose following a low-swelling transient dose that varies from <0.1 to >100 dpa, depending on material and irradiation spectrum. Two material-specific radiation effects may also need to be considered in certain classes of structural materials: Amorphization of ceramic or intermetallic precipitates can occur for doses above ~ 1 dpa at low temperatures (below $\sim 0.3T_M$ for the precipitate phase) with corresponding volumetric changes (swelling) of the precipitates and typically a decrease in alloy strength. For anisotropic crystal structure alloys such as Zr alloys, irradiation growth leads to anisotropic dimensional changes of ~1% that need to be considered in engineering designs. In all cases, microstructural stability is crucial for minimizing property degradation.

ACKNOWLEDGMENT
This work was sponsored in part by the Office of Fusion Energy Sciences, US Department of Energy.

REFERENCES
[1] D.G. Franklin, R.B. Adamson, Implications of Zircaloy creep and growth to light water-reactor performance, J. Nucl. Mater. 159 (1988) 12–21.
[2] V.S. Neustroev, Z.E. Ostrovsky, A.A. Teykovtsev, V.K. Shamardin, V.V. Yakolev, Experimental studies of the failure of irradiated ducts in the BOR-60 reactor (in Russian), in: Proceedings of the 6th Russian Conference on Reactor Materials Science, Dimitrovgrad, Russia, 2000.
[3] F.R.N. Nabarro, H.L. de Villiers, The Physics of Creep, Taylor and Francis, London, 1995.
[4] S.J. Zinkle, L.L. Snead, Designing radiation resistance in materials for fusion energy, Annu. Rev. Mat. Res. 44 (2014) 241–267.
[5] S.J. Zinkle, Radiation-induced effects on microstructure, in: R.J.M. Konings (Ed.), Comprehensive Nuclear Materials, 1, Elsevier, Amsterdam, 2012, pp. 65–98.
[6] C. Lemaignan, Zirconium alloys: properties and characteristics, in: R.J.M. Konings (Ed.), Comprehensive Nuclear Materials, Vol 2: Material Properties/Oxide Fuels for Light Water Reactors and Fast Neutron Reactors, Elsevier, Amsterdam, 2012, pp. 217–232.
[7] L.P. Stoter, Thermal aging effects in AISI type 316 stainless-steel, J. Mater. Sci. 16 (1981) 1039–1051.
[8] Y. Yang, J.T. Busby, Thermodynamic modeling and kinetics simulation of precipitate phases in AISI 316 stainless steels, J. Nucl. Mater. 448 (2014) 282–293.
[9] B.S. Dutt, G. Sasikala, G. Shanthi, S. Venugopal, M.N. Babu, P.K. Parida, A.K. Bhaduri, in: M. Guagliano, L. Vergani (Eds.), Mechanical behaviour of SS 316 (N) weld after long term exposure to service temperatures, 11th International Conference on the Mechanical Behavior of Materials, Elsevier Science Bv, Amsterdam, 2011.
[10] S. Goyal, R. Sandhya, M. Valsan, K.B.S. Rao, The effect of thermal ageing on low cycle fatigue behaviour of 316 stainless steel welds, Int. J. Fatigue 31 (2009) 447–454.
[11] T. Yonezawa, Nickel alloys: properties and characteristics, in: R.J.M. Konings (Ed.), Comprehensive Nuclear Materials, Elsevier, Amsterdam, 2012, pp. 233–266.
[12] Y. Daigo, Y. Watanabe, K. Sugahara, T. Isobe, Compatibility of nickel-based alloys with supercritical water applications: aging effects on corrosion resistance and mechanical properties, Corrosion 62 (2006) 174–181.
[13] M. Sundararaman, P. Mukhopadhyay, S. Banerjee, Some aspects of the precipitation of metastable intermetallic phases in Inconel-718, Metall. Trans. A. 23 (1992) 2015–2028.
[14] J. Veverkova, A. Strang, G.R. Marchant, G.M. McColvin, H.V. Atkinson, High Temperature Microstructural Degradation of Haynes Alloy 230, Minerals, Metals & Materials Soc, Warrendale, 2008.
[15] K. Mo, G. Lovicu, H.M. Tung, X.A. Chen, J.F. Stubbins, High temperature aging and corrosion study on Alloy 617 and Alloy 230, J. Eng. Gas Turbines Power Trans ASME 133 (2011) 9.
[16] S. Nandi, G.J. Reddy, K. Singh, Effect of ageing on creep behaviour of IN 617 Nickel base super alloy for advanced USC power plant applications, Trans. Indian Inst. Met. 69 (2016) 271–276.
[17] T.M. Angeliu, E.L. Hall, M. Larsen, A. Linsebigler, C. Mukira, The long-term aging embrittlement of Fe-12Cr steels below 773 K, Metall. Mater. Trans. A 34 (2003) 927–934.
[18] J.S. Lee, C.H. Jang, I.S. Kim, A. Kimura, Embrittlement and hardening during thermal aging of high Cr oxide dispersion strengthened alloys, J. Nucl. Mater. 367-370 (2007) 229–233.
[19] M. Terada, M.F. Hupalo, I. Costa, A.F. Padilha, Effect of alpha prime due to 475 degrees C aging on fracture behavior and corrosion resistance of DIN 1.4575 and MA 956 high performance ferritic stainless steels, J. Mater. Sci. 43 (2008) 425–433.

REFERENCES

[20] W. Xiong, M. Selleby, Q. Chen, J. Odqvist, Y. Du, Phase equilibria and thermodynamic properties in the Fe-Cr system, Crit. Rev. Solid State Mater. Sci. 35 (2010) 125–152.

[21] M.H. Mathon, Y. De Carlan, G. Geoffroy, X. Averty, A. Alamo, C.H. de Novion, A SANS investigation of the irradiation-enhanced alpha-alpha 'phases separation in 7-12 Cr martensitic steels, J. Nucl. Mater. 312 (2003) 236–248.

[22] R.W. Swindeman, V.K. Sikka, P.J. Maziasz, D.A. Canonico, Evaluation of T91 After 130,000 Hours in Service, Oak Ridge National Lab, Oak Ridge, TN, 1998.

[23] K. Shiba, H. Tanigawa, T. Hirose, H. Sakasegawa, S. Jitsukawa, Long-term properties of reduced activation ferritic/martensitic steels for fusion reactor blanket system, Fusion Eng. Des. 86 (2011) 2895–2899.

[24] M. Yoshizawa, M. Igarashi, K. Moriguchi, A. Iseda, H.G. Armaki, K. Maruyama, Effect of precipitates on long-term creep deformation properties of P92 and P122 type advanced ferritic steels for USC power plants, Mater. Sci. Eng. A 510-11 (2009) 162–168.

[25] A. Aghajani, C. Somsen, G. Eggeler, On the effect of long-term creep on the microstructure of a 12% chromium tempered martensite ferritic steel, Acta Mater. 57 (2009) 5093–5106.

[26] K. Sawada, H. Kushima, K. Kimura, Z-phase formation during creep and aging in 9-12% Cr heat resistant steels, ISIJ Int. 46 (2006) 769–775.

[27] K. Sawada, H. Kushima, M. Tabuchi, K. Kimura, Microstructural degradation of Gr.91 steel during creep under low stress, Mater. Sci. Eng. A 528 (2011) 5511–5518.

[28] C.G. Panait, A. Zielinska-Lipiec, T. Koziel, A. Czyrska-Filemonowicz, A.F. Gourgues-Lorenzon, W. Bendick, Evolution of dislocation density, size of subgrains and MX-type precipitates in a P91 steel during creep and during thermal ageing at 600 degrees C for more than 100,000 h, Mater. Sci. Eng. A 527 (2010) 4062–4069.

[29] R.W. Swindeman, M.L. Santella, P.J. Maziasz, B.W. Roberts, K. Coleman, Issues in replacing Cr-Mo steels and stainless steels with 9Cr-1Mo-V steel, Int. J. Press. Vess. Piping 81 (2004) 507–512.

[30] O.D. Sherby, A.K. Miller, Combining phenomenology and physics in describing the high temperature mechanical behavior of crystalline solids, J. Eng. Mater. Technol. 101 (1979) 387–395.

[31] H.J. Frost, M.F. Ashby, Deformation Mechanism Maps: The Plasticity and Creep of Metals and Ceramics, Pergamon Press, New York, 1982.

[32] Y. Yamamoto, M.L. Santella, M.P. Brady, Effect of alloying additions on phase equilibria and creep resistance of alumina-forming austenitic stainless steels, Metall. Mater. Trans. A 40A (2009) 1868–1880.

[33] B. Wilshire, T.D. Lieu, Deformation and damage processes during creep of Incoloy MA957, Mater. Sci. Eng. A 386 (2004) 81–90.

[34] R. Seshadri, D.L. Marriott, On relating the reference stress, limit load and the ASME stress classification concepts, Int. J. Press. Vess. Piping 56 (1993) 387–408.

[35] A.A.F. Tavassoli, Present limits and improvements of structural materials for fusion reactors—a review, J. Nucl. Mater. 302 (2002) 73–88.

[36] K. Maruyama, K. Yoshimi, Influence of data analysis method and allowable stress criterion on allowable stress of Gr.122 heat resistant steel, J. Press. Vessel Technol. Trans. ASME 129 (2007) 449–453.

[37] F. Masuyama, Creep rupture life and design factors for high-strength ferritic steels, Int. J. Press. Vess. Piping 84 (2007) 53–61.

[38] J.C. Vaillant, B. Vandenberghe, B. Hahn, H. Heuser, C. Jochum, T/P23, 24, 911 and 92: new grades for advanced coal-fired power plants-properties and experience, Int. J. Press. Vess. Piping 85 (2008) 38–46.

[39] N. Kasahara, K. Satoh, K. Tsukimori, N. Kawasaki, in: Proposals of guidelines for high temperature structural design of fast reactor vessels, Proceedings of the ASME Pressure Vessels and Piping Conference 2010, vol. 1: Codes and Standards, Amer Soc Mechanical Engineers, New York, 2010, pp. 315–322.

[40] U.F. Kocks, A.S. Argon, M.F. Ashby, Thermodynamics and kinetics of slip, Prog. Mater. Sci. 19 (1975) 1–291.

[41] R.J. Kurtz, K. Abe, V.M. Chernov, D.T. Hoelzer, H. Matsui, T. Muroga, G.R. Odette, Recent progress on development of vanadium alloys for fusion, J. Nucl. Mater. 329-333 (2004) 47–55.

[42] F.R. Larson, J. Miller, A time-temperature relationship for rupture and creep stresses, Trans. Am. Soc. Mech. Eng. 74 (1952) 765–775.

[43] R.L. Klueh, Elevated-temperature ferritic and martensitic steels and their application to future nuclear reactors, Int. Mater. Rev. 50 (2005) 287–310.

[44] M.F. Ashby, A first report on deformation-mechanism maps, Acta Metall. 20 (1972) 887–897.

[45] M. Li, S.J. Zinkle, Deformation mechanism maps of unirradiated and irradiated V-4Cr-4Ti, J. ASTM Internat. 2 (2005). JAI12462.

[46] A.A.F. Tavassoli, B. Fournier, M. Sauzay, High temperature creep-fatigue design, Trans. Indian Inst. Met. 63 (2010) 235–244.
[47] A. Wöhler, Uber Die Festigkeitversuche mit Eisen und Stahl, Engineering (London) XI (1871) 199–441.
[48] F.C. Campbell (Ed.), Fatigue, in: Elements of Metallurgy and Engineering Alloys, ASM International, Materials Park, OH, 2008, pp. 243–264 (Chapter 14).
[49] B. Pyttel, D. Schwerdt, C. Berger, Very high cycle fatigue—is there a fatigue limit? Int. J. Fatigue 33 (2011) 49–58.
[50] P.A. Withey, Fatigue failure of the de Havilland comet I, Engr. Failure Anal. 4 (1997) 147–154.
[51] L.F. Coffin Jr., A study of the effects of cyclic thermal stresses on a ductile metal, Trans. ASME 76 (1954) 931–950.
[52] D.R. Miller, Thermal-stress ratchet mechanism in pressure vessels, Trans. ASME 81 (1959) 190–196.
[53] T.L. Sham, S.J. Zinkle, Creep and fatigue issues for structural materials in demonstration fusion energy systems, Trans. Indian Inst. Met. 63 (2010) 331–337.
[54] R. Raj, Mechanisms of creep-fatigue interaction, in: R. Raj (Ed.), Flow and Fracture at Elevated Temperatures, ASM, Metals Park, OH, 1985, pp. 215–249.
[55] ASME, Boiler and Pressure Vessel Code, Section III, Subsection-NH, American Society of Mechanical Engineers, 2005.
[56] M.A. Miner, Cumulative damage in fatigue, J. Appl. Mech. 12 (1945) A159–A164.
[57] E.L. Robinson, Effect of temperature variation on the long-time rupture strength of steels, Trans. ASME 74 (1952) 777–781.
[58] S.S. Manson, J.C. Freche, C.R. Ensign, Application of a double linear damage rule to cumulative fatigue, in: Fatigue Crack Propagation, ASTM STP 415, American Society for Testing and Materials, Philadelphia, 1967, pp. 384–412.
[59] H.G. Edmunds, D.J. White, Observations of effect of creep relaxation on high-strain fatigue, J. Mech. Eng. Sci. 8 (1966) 310–321.
[60] T. Asayama, Y. Tachibana, Creep-fatigue data and evaluation procedures for Grade 91 steel and Hastelloy XR for VHTR, in: STP-NU-018, ASME Standards Technology, LLC, New York, 2008, p. 162.
[61] T.S. Byun, K. Farrell, M. Li, Deformation in metals after low-temperature irradiation: part II—irradiation hardening, strain hardening, and stress ratios, Acta Mater. 56 (2008) 1056–1064.
[62] M. Victoria, N. Baluc, C. Bailat, Y. Dai, M.I. Luppo, R. Schäublin, B.N. Singh, The microstructure and associated tensile properties of irradiated fcc and bcc metals, J. Nucl. Mater. 276 (2000) 114–122.
[63] T.S. Byun, K. Farrell, Plastic instability in polycrystalline metals after low temperature irradiation, Acta Mater. 52 (2004) 1597–1608.
[64] G.R. Odette, G.E. Lucas, Recent progress in understanding reactor pressure vessel embrittlement, Radiat. Eff. Defects Solids 144 (1998) 189–231.
[65] M. Rieth, B. Dafferner, H.-D. Röhrig, Embrittlement behaviour of different international low activation alloys after neutron irradiation, J. Nucl. Mater. 258-263 (1998) 1147–1152.
[66] S.J. Zinkle, N.M. Ghoniem, Operating temperature windows for fusion reactor structural materials, Fusion Eng. Des. 51-52 (2000) 55–71.
[67] S.J. Zinkle, M. Victoria, K. Abe, Scientific and engineering advances from fusion materials R&D, J. Nucl. Mater. 307-311 (2002) 31–42.
[68] A.F. Rowcliffe, S.J. Zinkle, J.F. Stubbins, D.J. Edwards, D.J. Alexander, Austenitic stainless steels and high strength copper alloys for fusion components, J. Nucl. Mater. 258-263 (1998) 183–192.
[69] N. Hashimoto, T.S. Byun, K. Farrell, Microstructural analysis of deformation in neutron-irradiated fcc materials, J. Nucl. Mater. 351 (2006) 295–302.
[70] T.S. Byun, K. Farrell, M. Li, Deformation in metals after low-temperature irradiation: Part I - Mapping macroscopic deformation modes on true stress-dose plane, Acta Mater. 56 (2008) 1044–1055.
[71] S.J. Zinkle, B.N. Singh, Microstructure of neutron-irradiated iron before and after tensile deformation, J. Nucl. Mater. 351 (2006) 269–284.
[72] M.S. Wechsler, Dislocation channeling in irradiated and quenched metals, in: R.E. Reed-Hill (Ed.), The Inhomogeneity of Plastic Deformation, Am. Society for Metals, Metals Park, OH, 1972, pp. 19–54.
[73] A. Luft, Microstructural processes of plastic instabilities in strengthened metals, Prog. Mater. Sci. 35 (1991) 97–204.
[74] S.J. Zinkle, G.E. Lucas, Deformation and fracture mechanisms in irradiated FCC and BCC metals, in: Fusion Materials Semiann, Oak Ridge National Lab, Oak Ridge, TN, 2003, pp. 101–125. Progress Report for Period ending June 30, 2003, DOE/ER-0313/34.

[75] K. Farrell, T.S. Byun, N. Hashimoto, Deformation mode maps for tensile deformation of neutron-irradiated structural alloys, J. Nucl. Mater. 335 (2004) 471–486.
[76] N. Hashimoto, T.S. Byun, K. Farrell, S.J. Zinkle, Deformation microstructure of neutron-irradiated pure polycrystalline metals, J. Nucl. Mater. 329-333 (2004) 947–952.
[77] G.R. Odette, M.Y. He, E.G. Donahue, P. Spätig, T. Yamamoto, Modeling the multiscale mechanics of flow localization-ductility loss in irradiation damaged BCC alloys, J. Nucl. Mater. 307-311 (2002) 171–178.
[78] T.S. Byun, K. Farrell, Irradiation hardening behavior of polycrystalline metals after low temperature irradiation, J. Nucl. Mater. 326 (2004) 86–96.
[79] T.S. Byun, K. Farrell, N. Hashimoto, Plastic instability behavior of bcc and hcp metals after low temperature neutron irradiation, J. Nucl. Mater. 329-333 (2004) 998–1002.
[80] G.E. Lucas, The evolution of mechanical property changes in irradiated austenitic stainless steel, J. Nucl. Mater. 206 (1993) 287–305.
[81] S.J. Zinkle, S.A. Fabritsiev, Copper alloys for high heat flux structure applications, Atom. Plasma Mater. Inter. Data Fusion 5 (1994) 163–192.
[82] J.E. Pawel, A.F. Rowcliffe, D.J. Alexander, M.L. Grossbeck, K. Shiba, Effects of low temperature neutron irradiation on deformation behavior of austenitic stainless steels, J. Nucl. Mater. 233-237 (1996) 202–237.
[83] K. Shiba, M. Suzuki, A. Hishinuma, Irradiation response on mechanical properties of neutron irradiated F82H, J. Nucl. Mater. 233-237 (1996) 309–312.
[84] S.J. Zinkle, H. Matsui, D.L. Smith, A.F. Rowcliffe, E. van Osch, K. Abe, V.A. Kazakov, Research and development on vanadium alloys for fusion applications, J. Nucl. Mater. 258-263 (1998) 205–214.
[85] B. Van der Schaaf, C. Petersen, Y. De Carlan, J.-W. Rensman, E. Gaganidze, X. Averty, High dose, up to 80 dpa, mechanical properties of Eurofer 97, J. Nucl. Mater. 386-388 (2009) 236–240.
[86] G.R. Odette, R.K. Nanstad, Predictive reactor pressure vessel steel irradiation embrittlement models: Issues and opportunities, JOM 61 (2009) 17–23.
[87] C.D. Hardie, C.A. Williams, S. Xu, S.G. Roberts, Effects of irradiation temperature and dose rate on the mechanical properties of self-ion implanted Fe and Fe-Cr alloys, J. Nucl. Mater. 439 (2013) 33–40.
[88] J.E. Pawel, A.F. Rowcliffe, G.E. Lucas, S.J. Zinkle, Irradiation performance of stainless steels for ITER application, J. Nucl. Mater. 239 (1996) 126–131.
[89] J. Rensman, NRG Irradiation Testing: Report on 300°C and 60°C Irradiated RAFM Steels, NRG report 20023/05.68497/P, NRG, Petten, The Netherlands, (2005).
[90] G.E. Dieter, Mechanical Metallurgy, second ed., McGraw-Hill, New York, 1976.
[91] L.L. Snead, S.J. Zinkle, D.J. Alexander, A.F. Rowcliffe, J.P. Robertson, W.S. Eatherly, Temperature dependence of the tensile and impact properties of V-4Cr-4Ti irradiated to low neutron doses, in: Fusion Materials Semiann, Oak Ridge National Lab, 1997, pp. 81–98. Prog. Report for period ending Dec. 31 1997, DOE/ER-0313/23.
[92] S.J. Zinkle, A.F. Rowcliffe, C.O. Stevens, High temperature tensile properties of V-4Cr-4Ti, in: Fusion Materials Semiann, Oak Ridge National Lab, 1998, pp. 11–14. Prog. Report for period ending June 30 1998, DOE/ER-0313/24.
[93] T.S. Byun, K. Farrell, E.H. Lee, L.K. Mansur, S.A. Maloy, M.R. James, W.R. Johnson, Temperature effects on the mechanical properties of candidate SNS target container materials after proton and neutron irradiation, J. Nucl. Mater. 303 (2002) 34–43.
[94] S.A. Maloy, M.R. James, G. Willcutt, W.F. Sommer, M.A. Sokolov, L.L. Snead, M.L. Hamilton, F.A. Garner, The mechanical properties of 316L/304L stainless steels, Alloy 718 and Mod 9Cr-1Mo after irradiation in a spallation environment, J. Nucl. Mater. 296 (2001) 119–128.
[95] S.J. Zinkle, J.P. Robertson, R.L. Klueh, Thermophysical and mechanical properties of Fe-(8-9%)Cr reduced activation steels, in: Fusion Materials Semiann, Oak Ridge National Lab, 1998, pp. 135–143. Prog. Report for period ending June 30 1998, DOE/ER-0313/24.
[96] J. Henry, X. Averty, Y. Dai, P. Lamagnere, J.P. Pizzanelli, J.J. Espinas, P. Widert, Tensile properties of 9Cr-1Mo martensitic steel irradiated with high energy protons and neutrons, J. Nucl. Mater. 318 (2003) 215–227.
[97] S.T. Rolfe, J.M. Barsom, Fracture and Fatigue Control in Structures: Applications of Fracture Mechanics, Prentice-Hall, Englewood Cliffs, NJ, 1977.
[98] F.H. Huang, Brittle-fracture potential of irradiated zircaloy-2 pressure tubes, J. Nucl. Mater. 207 (1993) 103–115.
[99] G.R. Odette, G.E. Lucas, Embrittlement of nuclear reactor pressure vessels, JOM 53 (2001) 18–22.

[100] G.R. Odette, T. Yamamoto, H.J. Rathbun, M.Y. He, M.L. Hribernik, J.W. Rensman, Cleavage fracture and irradiation embrittlement of fusion reactor alloys: mechanisms, multiscale models, toughness measurements and implications to structural integrity assessment, J. Nucl. Mater. 323 (2003) 313–340.

[101] M.A. Sokolov, H. Tanigawa, G.R. Odette, K. Shiba, R.L. Klueh, Fracture toughness and Charpy impact properties of several RAFMS before and after irradiation in HFIR, J. Nucl. Mater. 367-370 (2007) 68–73.

[102] O.K. Chopra, A.S. Rao, A review of irradiation effects on LWR core internal materials – Neutron embrittlement, J. Nucl. Mater. 412 (2011) 195–208.

[103] M. Li, S.J. Zinkle, Physical and mechanical properties of copper and copper alloys, in: R.J.M. Konings (Ed.), Comprehensive Nuclear Materials, Elsevier, Amsterdam, 2012, pp. 667–690.

[104] S.J. Zinkle, G.S. Was, Materials challenges in nuclear energy, Acta Mater. 61 (2013) 735–758.

[105] G.R. Odette, G.E. Lucas, The effects of intermediate temperature neutron irradiation on the mechanical behavior of 300-series austenitic stainless steels, J. Nucl. Mater. 179-181 (1991) 572–576.

[106] G.R. Odette, G.E. Lucas, Analysis of cleavage fracture potential of martensitic stainless steel fusion structures, J. Nucl. Mater. 117 (1983) 264–275.

[107] G.R. Odette, On the ductile to brittle transition in martensitic stainless steels — Mechanisms, models and structural implications, J. Nucl. Mater. 212-215 (1994) 45–51.

[108] M.A. Sokolov, A. Kimura, H. Tanigawa, S. Jitsukawa, Fracture toughness characterization of JLF-1 steel after irradiation in HFIR to 5 dpa, J. Nucl. Mater. 367-370 (2007) 644–647.

[109] S. Tähtinen, M. Pyykkönen, P. Karjalainen-Roikonen, B.N. Singh, P. Toft, Effect of neutron irradiation on fracture toughness behavior of copper alloys, J. Nucl. Mater. 258-263 (1998) 1010–1014.

[110] S.J. Zinkle, Applicability of copper alloys for DEMO high heat flux components, Phys. Scr. T167 (2016) 10.

[111] S.J. Zinkle, D.J. Alexander, J.P. Robertson, L.L. Snead, A.F. Rowcliffe, L.T. Gibson, W.S. Eatherly, H. Tsai, Effect of fast neutron irradiation to 4 dpa at 400°C on the properties of V-(4-5)Cr-(4-5)Ti alloys, in: Fusion Materials Semiannual, Oak Ridge National Lab, 1996, pp. 73–78. Progress Report for Period ending December 31, 1996, DOE/ER-0313/21.

[112] E.P. Wigner, Report for Month Ending December 15, 1942, Physics Division. US Atomic Energy Commission Report CP-387, University of Chicago, 1942.

[113] R.S. Averback, T. Diaz de la Rubia, Displacement damage in irradiated metals and semiconductors, Solid State Phys. 51 (1998) 281–402.

[114] D.J. Bacon, Y.N. Osetsky, Modelling atomic scale radiation damage processes and effects in metals, Int. Mater. Rev. 47 (2002) 233–241.

[115] K. Ehrlich, Irradiation creep and interrelation with swelling in austenitic stainless-steels, J. Nucl. Mater. 100 (1981) 149–166.

[116] L.K. Mansur, Void swelling in metals and alloys under irradiation: an assessment of the theory, Nucl. Technol. 40 (1978) 5–34.

[117] R.A. Johnson, N.Q. Lam, Solute segregation in metals under irradiation, Phys. Rev. B 13 (1976) 4364–4375.

[118] K.C. Russell, Phase stability under irradiation, Prog. Mater. Sci. 28 (1984) 229–434.

[119] A.T. Motta, Amorphization of intermetallic compounds under irradiation—a review, J. Nucl. Mater. 244 (1997) 227–250.

[120] G. Martin, Phase-stability under irradiation—ballistic effects, Phys. Rev. B 30 (1984) 1424–1436.

[121] J.A. Brinkman, On the nature of radiation damage in metals, J. Appl. Phys. 25 (1954) 961–970.

[122] R.E. Stoller, Primary radiation damage formation, in: R.J.M. Konings (Ed.), Comprehensive Nuclear Materials, Elsevier, Amsterdam, 2012, pp. 293–332.

[123] W.C. Brøgger, Salmonsens Store Illustrerede Konverstaion—Slexicon, vol. 1, Brødrene Salmonsens Forlag, Copenhagen, Denmark, 1893, pp. 742–743.

[124] R.C. Ewing, The metamict state—1993—the centennial, Nucl. Instrum. Meth. B 91 (1994) 22–29.

[125] L.M. Wang, R.C. Ewing, Ion-beam-induced amorphization of complex ceramic materials minerals, MRS Bull. 17 (1992) 38–44.

[126] P.R. Okamoto, L.E. Rehn, J. Pearson, R. Bhadra, M. Grimsditch, Brillouin-scattering and transmission electron-microscopy studies of radiation-induced elastic softening, disordering and amorphization of intermetallic compounds, J. Less Comm. Metals 140 (1988) 231–244.

[127] R. Sizmann, The effect of radiation upon diffusion in metals, J. Nucl. Mater. 69&70 (1978) 386–412.

[128] G.J. Dienes, A.C. Damask, Radiation enhanced diffusion in solids, J. Appl. Phys. 29 (1958) 1713.

[129] P.R. Okamoto, L.E. Rehn, Radiation-induced segregation in binary and ternary alloys, J. Nucl. Mater. 83 (1979) 2–23.

[130] T.R. Allen, J.I. Cole, J. Gan, G.S. Was, R. Dropek, E.A. Kenik, Swelling and radiation-induced segregation in austenic alloys, J. Nucl. Mater. 342 (2005) 90–100.
[131] F.A. Garner, Irradiation performance of cladding and structural steels in liquid metal reactors, in: B.R.T. Frost (Ed.), Materials Science and Technology: A Comprehensive Treatment, VCH, New York, 1994, pp. 419–543.
[132] A.D. Marwick, Segregation in irradiated alloys–Inverse Kirkendall effect and effect of constitution on void swelling, J. Phys. F 8 (1978) 1849–1861.
[133] A.D. Marwick, R.C. Piller, M.E. Horton, Radiation-induced segregation in Fe-Cr-Ni alloys, in: Dimensional Stability and Mechanical Behavior of Irradiated Metals, British Nuclear Energy Society, London, 1984, pp. 11–15.
[134] J.M. Perks, A.D. Werweck, C.A. English, A Computer Code to Calculate Radiation-Induced Segregation in Concentrated Ternary Alloys, AERE-R-12121, UKAEA Atomic Energy Research Establishment, Harwell, 1986.
[135] H. Wiedersich, P.R. Okamoto, N.Q. Lam, Theory of radiation-induced segregation in concentrated alloys, J. Nucl. Mater. 83 (1979) 98–108.
[136] N.Q. Lam, A. Kumar, H. Wiedersich, Kinetics of radiation-induced segregation in ternary alloys, in: H.R. Brager, J.S. Perrin (Eds.), 11th Int. Symp. on Effects of Radiation on Materials, ASTM STP 782, American Society for Testing and Materials, Philadelphia, 1982, p. 985.
[137] G.S. Was, J.P. Wharry, B. Frisbie, B.D. Wirth, D. Morgan, J.D. Tucker, T.R. Allen, Assessment of radiation-induced segregation mechanisms in austenitic and ferritic-martensitic alloys, J. Nucl. Mater. 411 (2011) 41–50.
[138] M. Nastar, F. Soisson, Radiation-induced segregation, in: R.J.M. Konings (Ed.), Comprehensive Nuclear Materials, Elsevier, Amsterdam, 2012, pp. 471–496.
[139] A.J. Ardell, P. Bellon, Radiation-induced solute segregation in metallic alloys, Curr. Opin. Solid State Mater. Sci. 20 (2016) 115–139.
[140] G.S. Was, T.R. Allen, J.T. Busby, J. Gan, D. Damcott, D. Carter, M. Atzmon, E.A. Kenik, Microstructure and microchemistry of proton-irradiated austenitic alloys under conditions relevant to LWR core components, J. Nucl. Mater. 270 (1999) 96–114.
[141] G.R. Odette, M.J. Alinger, B.D. Wirth, Recent developments in irradiation-resistant steels, Annu. Rev. Mat. Res. 38 (2008) 471–503.
[142] G.R. Odette, Recent progress in developing and qualifying nanostructured ferritic alloys for advanced fission and fusion applications, JOM 66 (2014) 2427–2441.
[143] S.J. Zinkle, J.L. Boutard, D.T. Hoelzer, A. Kimura, R. Lindau, G.R. Odette, M. Rieth, L. Tan, H. Tanigawa, Development of next generation tempered and ODS reduced activation ferritic/martensitic steels for fusion energy applications, Nucl. Fusion 57 (2017) 17.
[144] A. Certain, S. Kuchibhatla, V. Shutthanandan, D.T. Hoelzer, T.R. Allen, Radiation stability of nanoclusters in nanostructured oxide dispersion strengthened (ODS) steels, J. Nucl. Mater. 434 (2013) 311–321.
[145] H.J. Frost, K.C. Russell, Particle stability with recoil resolution, Acta Metall. 30 (1982) 953–960.
[146] D.H. Xu, A. Certain, H.J.L. Voigt, T. Allen, B.D. Wirth, Ballistic effects on the copper precipitation and re-dissolution kinetics in an ion irradiated and thermally annealed Fe-Cu alloy, J. Chem. Phys. 145 (2016) 9.
[147] S.M. Bruemmer, Grain boundary chemistry and intergranular failure of austenitic stainless steels, Mater. Sci. Forum 46 (1989) 309–334.
[148] G.S. Was, Grain boundary chemistry and intergranular fracture in austenitic nickel-base alloys, Mater. Sci. Forum 46 (1989) 335.
[149] P.L. Andresen, Effects of temperature on crack growth rate in sensitized type 304 stainless steel and alloy 600, Corrosion 49 (1993) 714–725.
[150] P. Scott, A review of irradiation assisted stress corrosion cracking, J. Nucl. Mater. 211 (1994) 101–122.
[151] T.K. Yeh, D.D. Macdonald, A.T. Motta, Modeling water chemistry, electrochemical corrosion potential, and crack-growth rate in the boiling water-reactor heat-transport circuits .1. The damage-predictor algorithm, Nucl. Sci. Eng. 121 (1995) 468–482.
[152] S.M. Bruemmer, E.P. Simonen, P.M. Scott, P.L. Andresen, G.S. Was, J.L. Nelson, Radiation-induced material changes and susceptibility to intergranular failure of light-water-reactor core internals, J. Nucl. Mater. 274 (1999) 299–314.
[153] P.M. Scott, F.N. Speller award lecture: stress corrosion cracking in pressurized water reactors—interpretation, modeling, and remedies, Corrosion 56 (2000) (2000) 771–782.
[154] R.W. Staehle, J.A. Gorman, Quantitative assessment of submodes of stress corrosion cracking on the secondary side of steam generator tubing in pressurized water reactors: part 1, Corrosion 59 (2003) 931–994.

[155] P. Kritzer, Corrosion in high-temperature and supercritical water and aqueous solutions: a review, J. Supercrit. Fluids 29 (2004) 1–29.
[156] R.W. Staehle, J.A. Gorman, Quantitative assessment of submodes of stress corrosion cracking on the secondary side of steam generator tubing in pressurized water reactors: part 2, Corrosion 60 (2004) 5–63.
[157] R.W. Staehle, J.A. Gorman, Quantitative assessment of submodes of stress corrosion cracking on the secondary side of steam generator tubing in pressurized water reactors: part 3, Corrosion 60 (2004) 115–180.
[158] P.M. Scott, An overview of materials degradation by stress corrosion in PWRs, in: EUROCORR 2004: Long Term Prediction and Modeling of Corrosion, Societe de Chimie Industrielle, Paris/Nice, 2004, p. 18.
[159] G.S. Was, J.T. Busby, Role of irradiated microstructure and microchemistry in irradiation-assisted stress corrosion cracking, Philos. Mag. 85 (2005) 443–465.
[160] G.S. Was, Y. Ashida, P.L. Andresen, Irradiation-assisted stress corrosion cracking, Corrosion Rev. 29 (2011) 7–49.
[161] B.M. Gordon, Corrosion and corrosion control in light water reactors, JOM 65 (2013) 1043–1056.
[162] T.A. Lang, in: Significant corrosion of the davis-besse nuclear reactor pressure vessel head, PVP2003-2167, ASME 2003 Pressure Vessels and Piping Conference, Cleveland, OH, 2003, pp. 169–176.
[163] P.M. Scott, P. Combrade, Corrosion in pressurized water reactors, in: S.D. Cramer, B.S. Covino Jr. (Eds.), ASM Handbook vol. 13C: Corrosion: Environments and Industries, ASM International, 2006, pp. 362–385.
[164] G.S. Was, J.T. Busby, P.L. Andresen, Effect of irradiation on stress-corrosion cracking and corrosion in light water reactors, in: S.D. Cramer, B.S. Covino Jr. (Eds.), ASM Handbook Volume 13C: Corrosion: Environments and Industries, ASM International, 2006, pp. 386–403.
[165] A.O. Allen, Radiation chemistry of aqueous solutions, J. Phys. Colloid Chem. 52 (1948) 479–490.
[166] T. Asayama, Y. Abe, N. Miyaji, M. Koi, T. Furukawa, E. Yoshida, Evaluation procedures for irradiation effects and sodium environmental effects for the structural design of Japanese fast breeder reactors, J. Press. Vessel Technol. Trans. ASME 123 (2001) 49–57.
[167] A.D. Brailsford, R. Bullough, The rate theory of swelling due to void growth in irradiated metals, J. Nucl. Mater. 44 (1972) 121–135.
[168] J.W. Corbett, L.C. Ianniello, Radiation-Induced Voids in Metals, National Technical Information Service, Springfield, VA, 1972, pp. 1–884.
[169] F.A. Garner, M.B. Toloczko, B.H. Sencer, Comparison of swelling and irradiation creep behavior of fcc-austenitic and bcc-ferritic/martensitic alloys at high neutron exposure, J. Nucl. Mater. 276 (2000) 123–142.
[170] C. Cawthorne, E.J. Fulton, Voids in irradiated stainless steel, Nature 216 (1967) 575–576.
[171] N.M. Ghoniem, D. Walgraef, S.J. Zinkle, Theory and experiment of nanostructure self-organization in irradiated materials, J. Comp. Aided Mater. Des. 8 (2001) 1–38.
[172] R.F. Mattas, F.A. Garner, M.L. Grossbeck, P.J. Maziasz, G.R. Odette, R.E. Stoller, The impact of swelling on fusion reactor first wall lifetime, J. Nucl. Mater. 122&123 (1984) 230–235.
[173] F.A. Garner, Radiation damage in austenitic steels, in: R.J.M. Konings (Ed.), Comprehensive Nuclear Materials, Elsevier, Amsterdam, 2012, pp. 33–95.
[174] F.A. Garner, D.S. Gelles, Neutron-induced swelling of commercial alloys at very high exposures, in: N.H. Packan, R.E. Stoller, A.S. Kumar (Eds.), Effects of Radiation on Materials, 14th Intern. Symp., ASTM STP 1046, American Society for Testing and Materials, Philadelphia, 1989, pp. 673–683.
[175] W.G. Wolfer, Advances in void swelling and helium bubble physics, J. Nucl. Mater. 122&123 (1984) 367–378.
[176] L.K. Mansur, Theory of transitions in dose dependence of radiation effects in structural alloys, J. Nucl. Mater. 206 (1993) 306–323.
[177] F.W. Wiffen, The effect of alloying and purity on the formation and ordering of voids in bcc metals, in: J.W. Corbett, L.C. Ianello (Eds.), Radiation Induced Voids in Metals, CONF-710601, National Technical Information Service, Springfield, VA, 1972, pp. 386–396.
[178] R.W. Balluffi, Vacancy defect mobilities and binding energies obtained from annealing studies, J. Nucl. Mater. 69&70 (1978) 240–263.
[179] J.J. Sniegowski, W.G. Wolfer, J.W. Davis, D.J. Michel (Eds.), On the physical basis for the swelling resistance of ferritic steels, Proc. Topical Conference on Ferritic Alloys for use in Nuclear Energy Technologies, TMS-AIME, New York, 1984, pp. 579–586.
[180] Z. Chang, P. Olsson, D. Terentyev, N. Sandberg, Multiscale calculations of dislocation bias in fcc Ni and bcc Fe model lattices, Nucl. Instrum. Meth. B 352 (2015) 81–85.

[181] H. Nakajima, S. Yoshida, Y. Kohno, H. Matsui, Effect of solute addition on swelling of vanadium after FFTF irradiation, J. Nucl. Mater. 191-194 (1992) 952–955.

[182] F.A. Garner, Recent insights on the swelling and creep of irradiated austenitic alloys, J. Nucl. Mater. 122 (1984) 459–471.

[183] F.A. Garner, M.L. Hamilton, T. Shikama, D.J. Edwards, J.W. Newkirk, Response of solute and precipitation strengthened copper alloys at high neutron exposure, J. Nucl. Mater. 191-194 (1992) 386–390.

[184] F.A. Garner, H.R. Brager, The role of phosphorus in the swelling and creep of irradiated austenitic alloys, J. Nucl. Mater. 133 (1985) 511–514.

[185] W.G. Johnston, J.H. Rosolowski, A.M. Turkalo, T. Lauritzen, Nickel-ion bombardment of annealed and cold-worked type-316 stainless-steel, J. Nucl. Mater. 48 (1973) 330–338.

[186] H.R. Brager, The effects of cold working and pre-irradiation heat treatment on void formation in neutron-irradiated type 316 stainless steel, J. Nucl. Mater. 57 (1975) 103–118.

[187] A.M. Dvoriashin, S.I. Porollo, Y.V. Konobeev, F.A. Garner, Influence of cold work to increase swelling of pure iron irradiated in the BR-10 reactor to similar to 6 and similar to 25 dpa at similar to 400 degrees C, J. Nucl. Mater. 283 (2000) 157–160.

[188] J.F. Bates, E.R. Gilbert, Experimental-evidence for stress enhanced swelling, J. Nucl. Mater. 59 (1976) 95–102.

[189] H.R. Brager, F.A. Garner, E.R. Gilbert, J.E. Flinn, W.G. Wolfer, Stress-affected microstructural development and the creep-swelling interrelationship, in: M.L. Bleiberg, J.W. Bennett (Eds.), Radiation Effects in Breeder Reactor Structural Materials, TMS-AIME, New York, 1977, pp. 727–755.

[190] M.M. Hall, J.E. Flinn, Stress state dependence of in-reactor creep and swelling. Part 2: experimental results, J. Nucl. Mater. 396 (2010) 119–129.

[191] N. Akasaka, I. Yamagata, S. Ukai, Effect of temperature gradients on void formation in modified 316 stainless steel cladding, J. Nucl. Mater. 283 (2000) 169–173.

[192] J.L. Seran, J.M. Dupouy, The Swelling of Solution Annealed 316 Cladding in RAPSODIE and PHENIX, in: H.R. Brager, J.S. Perrin (Eds.), 11th Int. Symp. on Effects of Radiation on Materials, ASTM STP 782, American Society for Testing and Materials, Philadelphia, 1982, pp. 5–16.

[193] T. Okita, T. Sato, N. Sekimura, F.A. Garner, L.R. Greenwood, The primary origin of dose rate effects on microstructural evolution of austenitic alloys during neutron irradiation, J. Nucl. Mater. 307 (2002) 322–326.

[194] T. Okita, T. Sato, N. Sekimura, T. Iwai, F.A. Garner, The synergistic influence of temperature and displacement rate on microstructural evolution of ion-irradiated Fe-15Cr-16Ni model austenitic alloy, J. Nucl. Mater. 367 (2007) 930–934.

[195] J.G. Gigax, T. Chen, H. Kim, J. Wang, L.M. Price, E. Aydogan, S.A. Maloy, D.K. Schreiber, M.B. Toloczko, F.A. Garner, L. Shao, Radiation response of alloy T91 at damage levels up to 1000 peak dpa, J. Nucl. Mater. 482 (2016) 257–265.

[196] L.K. Mansur, E.H. Lee, Theoretical basis for unified analysis of experimental data and design of swelling-resistant alloys, J. Nucl. Mater. 179-181 (1991) 105–110.

[197] Y. Dai, G.R. Odette, T. Yamamoto, The effects of helium in irradiated structural alloys, in: R.J.M. Konings (Ed.), Comprehensive Nuclear Materials, Elsevier, Amsterdam, 2012, pp. 141–193.

[198] G.R. Odette, P. Miao, D.J. Edwards, T. Yamamoto, R.J. Kurtz, H. Tanigawa, Helium transport, fate and management in nanostructured ferritic alloys: in situ helium implanter studies, J. Nucl. Mater. 417 (2011) 1001–1004.

[199] T. Yamamoto, Y. Wu, G.R. Odette, K. Yabuuchi, S. Kondo, A. Kimura, A dual ion irradiation study of helium-dpa interactions on cavity evolution in tempered martensitic steels and nanostructured ferritic alloys, J. Nucl. Mater. 449 (2014) 190–199.

[200] T.T. Konobeevsky, N.F. Pravdyuk, V.I. Kutaitsev, in: Effect of irradiation on structure and properties of fissionable materials, UN Conf. on Peaceful Uses of Atomic Energy, vol. 7, Geneva, United Nations, New York, 1955, pp. 433–440.

[201] J.W. Joseph Jr., Stress relaxation in stainless steel during irradiation, USAEC Report DP-369, E.I. DuPont de Nemours and Co., 1959.

[202] M.L. Grossbeck, L.K. Mansur, Low-temperature irradiation creep of fusion reactor structural materials, J. Nucl. Mater. 179 (1991) 130–134.

[203] M.L. Grossbeck, J.A. Horak, Irradiation creep in type 316 stainless steel and US PCA with fusion reactor He/dpa levels, J. Nucl. Mater. 155 (1988) 1001–1005.

[204] F.A. Garner, M.B. Toloczko, M.L. Grossbeck, The dependence of irradiation creep in austenitic alloys on displacement rate and helium to dpa ratio, J. Nucl. Mater. 258-263 (1998) 1718–1724.

[205] A. Uehira, S. Mizuta, S. Ukai, R.J. Puigh, Irradiation creep of 11Cr–0.5Mo–2W,V,Nb ferritic–martensitic, modified 316, and 15Cr–20Ni austenitic S.S. irradiated in FFTF to 103–206 dpa, J. Nucl. Mater. 283 (2000) 396–399.
[206] M.M. Li, D.T. Hoelzer, M.L. Grossbeck, A.F. Rowcliffe, S.J. Zinkle, R.J. Kurtz, Irradiation creep of the US Heat 832665 of V-4Cr-4Ti, J. Nucl. Mater. 386-88 (2009) 618–621.
[207] K. Fukumoto, H. Matsui, M. Narui, M. Yamazaki, Irradiation creep behavior of V-4Cr-4Ti alloys irradiated in a liquid sodium environment at the JOYO fast reactor, J. Nucl. Mater. 437 (2013) 341–349.
[208] F. Onimus, J.L. Bechade, Radiation effects in zirconium alloys, in: R.J.M. Konings (Ed.), Comprehensive Nuclear Materials, vol. 4, Elsevier, Amsterdam, 2012, pp. 1–31.
[209] F.A. Garner, M.B. Toloczko, L.R. Greenwood, C.R. Eiholzer, M.M. Paxton, R.J. Puigh, Swelling, irradiation creep and growth of pure rhenium irradiated with fast neutrons at 1030-1330 degrees C, J. Nucl. Mater. 283-287 (2000) 380–385.
[210] A. Kohyama, Y. Kohno, K. Asakura, M. Yoshino, C. Namba, C.R. Eiholzer, Irradiation creep of low-activation ferritic steels in FFTF/MOTA, J. Nucl. Mater. 212 (1994) 751–754.
[211] J.R. Matthews, M.W. Finnis, Irradiation creep models–an overview, J. Nucl. Mater. 159 (1988) 257–285.
[212] L.K. Mansur, Theory and experimental background on dimensional changes in irradiated alloys, J. Nucl. Mater. 216 (1994) 97–123.
[213] L.C. Walters, W.E. Ruther, In-reactor stress relaxation of Inconel X750 springs, J. Nucl. Mater. 68 (1977) 324–333.
[214] J.P. Massoud, P. Dubuisson, P. Scott, N. Ligneau, E. Lemaire, The effects of neutron radiation on materials for core internals of PWRs. A joint research programme, paper 62, in: Colloque International Fontevraud 5, SFEN Publications, 2002, p. 417.
[215] T.R. Allen, J.T. Busby, Radiation damage concerns for extended light water reactor service, JOM 61 (2009) 29–34.
[216] R. Scholz, R. Mueller, Irradiation creep-fatigue interaction of type 316L stainless steel, J. Nucl. Mater. 233 (1996) 169–172.
[217] V. Fidleris, The irradiation creep and growth phenomena, J. Nucl. Mater. 159 (1988) 22–42.
[218] R.A. Holt, Mechanisms of irradiation growth of alpha-zirconium alloys, J. Nucl. Mater. 159 (1988) 310–338.
[219] P.A. Tempest, M.V. Speight, A model to describe the irradiation-induced dimensional changes in polycrystalline graphites and carbons, J. Nucl. Mater. 97 (1981) 225–230.
[220] G.J.C. Carpenter, R.H. Zee, A. Rogerson, Irradiation growth of zirconium single-crystals—a review, J. Nucl. Mater. 159 (1988) 86–100.
[221] C.H. Woo, Modeling irradiation growth of zirconium and its alloys, Radiat. Eff. Defects Solids 144 (1998) 145–169.
[222] R.B. Adamson, Effects of neutron irradiation on microstructure and properties of Zircaloy, in: G.P. Sabol, G.D. Moan (Eds.), Zirconium in the Nuclear Industry, 12th International Symposium, ASTM STP 1354, American Society for Testing and Materials, West Conshohocken, PA, 2000, pp. 15–31.
[223] H. Ullmaier, The influence of helium on the bulk properties of fusion reactor structural materials, Nucl. Fusion 24 (1984) 1039–1083.
[224] H. Trinkaus, B.N. Singh, Helium accumulation in metals during irradiation—where do we stand? J. Nucl. Mater. 323 (2003) 229–242.
[225] K. Morishita, R. Sugano, Mechanism map for nucleation and growth of helium bubbles in metals, J. Nucl. Mater. 353 (2006) 52–65.
[226] D. Stewart, Y.N. Osetsky, R.E. Stoller, Atomistic studies of formation and diffusion of helium clusters and bubbles in BCC iron, J. Nucl. Mater. 417 (2010) 1110–1114.
[227] L.K. Mansur, M.L. Grossbeck, Mechanical property changes induced in structural alloys by neutron irradiations with different helium to displacement ratios, J. Nucl. Mater. 155-157 (1988) 130–147.
[228] H. Schroeder, High temperature helium embrittlement in austenitic stainless steels-correlations between microstructure and mechanical properties, J. Nucl. Mater. 155-157 (1988) 1032–1037.
[229] H. Schroeder, H. Ullmaier, Helium and hydrogen effects on the embrittlement of iron- and nickel-based alloys, J. Nucl. Mater. 179-181 (1991) 118–124.
[230] A.A. Bauer, M. Kangilaski, Helium generation in stainless steel and nickel, J. Nucl. Mater. 42 (1972) 91–95.
[231] B. Van der Schaaf, P. Marshall, The effect of boron on the development of helium induced creep embrittlement in Type 316 stainless steel, in: Dimensional Stability and Mechanical Behaviour of Irradiated Metals and Alloys, British Nuclear Energy Society, London, 1983, pp. 143–148.
[232] D. Kramer, H.R. Brager, C.G. Rhodes, A.G. Pard, Helium embrittlement in type 304 stainless steel, J. Nucl. Mater. 25 (1968) 121–131.

[233] E.E. Bloom, J.O. Stiegler, Effect of fast neutron irradiation on the creep rupture properties of type 304 stainless steel at 600 C, in: A.L. Bement Jr. (Ed.), Irradiation Effects on Structural Alloys for Nuclear Reactor Applications, American Society for Testing and Materials, Philadelphia, 1970, pp. 451–467.

[234] E.E. Bloom, Irradiation strengthening and embrittlement, in: N.L. Peterson, S.D. Harkness (Eds.), Radiation Damage in Metals, American Society for Metals, Metals Park, OH, 1976, pp. 295–329.

[235] E.E. Bloom, F.W. Wiffen, The effects of large concentrations of helium on the mechanical properties of neutron-irradiated stainless steel, J. Nucl. Mater. 58 (1975) 171–184.

[236] K. Ehrlich, H. Böhm, Irradiation effects in vanadium-base alloys, in: Radiation Damage in Reactor Materials, IAEA, Vienna, 1969, pp. 349–355.

[237] A.T. Santhanam, A. Taylor, S.D. Harkness, R.J. Arsenault (Ed.), Charged-particle simulation studies of vanadium and vanadium alloys, Proc. Int. Conf. on Defects and Defect Clusters in BCC Metals and Their Alloys, Nuclear Metallurgy, vol. 18, National Bureau of Standards, Gaithersburg, MD, 1973, pp. 302–320.

[238] M.P. Tanaka, E.E. Bloom, J.A. Horak, Tensile properties and microstructure of helium injected and reactor irradiated V-20Ti, J. Nucl. Mater. 103&104 (1981) 895–900.

[239] M. Kangilaski, J.S. Perrin, R.A. Wullaert, Irradiation-induced embrittlement in stainless steel at elevated temperature, in: A.L. Bement Jr. (Ed.), Irradiation Effects on Structural Alloys for Thermal and Fast Reactors, American Society for Testing and Materials, Philadelphia, 1969, pp. 67–91.

[240] D.R. Harries, Neutron irradiation-induced embrittlement in type 316 and other austenitic steels and alloys, J. Nucl. Mater. 82 (1979) 2–21.

[241] G.H. Broomfield, High-temperature tensile properties of unirradiated and thermal reactor irradiated Nimonic PE16, in: A.L. Bement Jr. (Ed.), Irradiation Effects on Structural Alloys for Thermal and Fast Reactors, ASTM STP 457, American Society for Testing and Materials, Philadelphia, 1969, pp. 38–66.

[242] P. Vela, B. Russell, The stress-rupture characteristics of copper and alloys containing boron and helium, J. Nucl. Mater. 22 (1967) 1–15.

[243] G.J.C. Carpenter, R.B. Nicholson, High-temperature embrittlement of metals by rare gas bubbles, in: Radiation Damage in Reactor Materials, IAEA, Vienna, 1969, pp. 383–400.

[244] S.A. Fabritsiev, A.S. Pokrovsky, S.J. Zinkle, D.J. Edwards, in: M.L. Hamilton, A.S. Kumar, S.T. Rosinski, M.L. Grossbeck (Eds.), Neutron irradiation induced high temperature embrittlement of pure copper and high strength copper alloys, 19th Int. Symp. on Effects of Radiation on Materials, ASTM STP 1366, American Society for Testing and Materials, West Conshohocken, PA, 2000, pp. 1226–1242.

[245] B. van der Schaaf, M.I. de Vries, J.D. Elen, Irradiation embrittlement of Type 304 stainless steel welds and plate at 823 K (550 C), in: M.L. Bleiberg, J.W. Bennett (Eds.), Radiation Effects in Breeder Reactor Structural Materials, TMS-AIME, New York, 1977, pp. 307–316.

[246] H. Böhm, K.D. Closs, Effects of strain rate on high temperature mechanical properties of irradiated Incoloy 800 and Hastelloy X, in: M.L. Bleiberg, J.W. Bennett (Eds.), Radiation Effects in Breeder Reactor Structural Materials, TMS-AIME, New York, 1977, pp. 347–356.

[247] A.A. Sagues, H. Schroeder, W. Kesternich, H. Ullmaier, Influence of helium on high-temperature mechanical-properties of an austenitic stainless-steel, J. Nucl. Mater. 78 (1978) 289–298.

[248] D.N. Braski, H. Schroeder, H. Ullmaier, The effect of tensile stress on the growth of helium bubbles in an austenitic stainless steel, J. Nucl. Mater. 83 (1979) 265–277.

[249] H. Schroeder, P. Batfalsky, The dependence of high-temperature mechanical properties of austenitic stainless steels on implanted helium, J. Nucl. Mater. 117 (1983) 287–294.

[250] H. Ullmaier, Helium in fusion materials: High temperature embrittlement, J. Nucl. Mater. 133-134 (1985) 100–104.

[251] S.N. Korshunov, Y.V. Martynenko, I.D. Skorlupkin, V.G. Stolyarova, Effect of mechanical deformation on development of helium porosity, Techn. Phys. 54 (2009) 527–534.

[252] H. Trinkaus, On the modeling of the high-temperature embrittlement of metals containing helium, J. Nucl. Mater. 118 (1983) 39–49.

[253] W. Beere, The growth of sub-critical bubbles on grain boundaries, J. Nucl. Mater. 120 (1984) 88–93.

[254] G.R. Odette, A model for in-reactor stress rupture of austenitic stainless steels, J. Nucl. Mater. 122&123 (1984) 435–441.

[255] H. Trinkaus, Modeling of helium effects in metals: high temperature embrittlement, J. Nucl. Mater. 133&134 (1985) 105–112.

[256] W. Kesternich, A possible solution of the problem of helium embrittlement, J. Nucl. Mater. 127 (1985) 153–160.

[257] H. Schroeder, W. Kesternich, H. Ullmaier, Helium effects on the creep and fatigue resistance of austenitic stainless steels at high temperatures, Nucl. Eng. Design/Fusion 2 (1985) 65–95.
[258] H. Schroeder, U. Stamm, High temperature helium embrittlement–Austenitic versus martensitic stainless steels, in: N.H. Packan, R.E. Stoller, A.S. Kumar (Eds.), Effects of Radiation on Materials, 14th Intern. Symp., ASTM STP 1046, American Soc. for Testing & Materials, Philadelphia, 1990, pp. 223–245.
[259] N. Yamamoto, Y. Murase, J. Nagakawa, An evaluation of helium embrittlement resistance of reduced activation martensitic steels, Fusion Eng. Des. 81 (2006) 1085–1090.
[260] B.N. Singh, A.J.E. Foreman, Some limitations of simulation studies using the ppm to dpa ratio as the helium generation rate, J. Nucl. Mater. 179-181 (1991) 990–993.
[261] H. Trinkaus, Energetics and formation kinetics of helium bubbles in metals, Radiat. Eff. Defects Solids 78 (1983) 189–211.
[262] H.T. Lin, M.L. Grossbeck, B.A. Chin, Cavity microstructure and kinetics during gas tungsten arc welding of helium-containing stainless steel, Metall. Mater. Trans. A 21 (1990) 2585–2596.
[263] W.R. Kanne, G.T. Chandler, D.Z. Nelson, E.A. Francoferreira, Welding irradiated stainless-steel, J. Nucl. Mater. 225 (1995) 69–75.
[264] C.A. Wang, M.L. Grossbeck, B.A. Chin, Threshold helium concentration required to initiate cracking during welding of irradiated stainless-steel, J. Nucl. Mater. 225 (1995) 59–68.
[265] K. Asano, S. Nishimura, Y. Saito, H. Sakamoto, Y. Yamada, T. Kato, T. Hashimoto, Weldability of neutron irradiated austenitic stainless steels, J. Nucl. Mater. 264 (1999) 1–9.

CORROSION ISSUES IN CURRENT AND NEXT-GENERATION NUCLEAR REACTORS

Gary S. Was*, Todd R. Allen[†]

University of Michigan, Ann Arbor, MI, United States *University of Wisconsin, Madison, Wi, United States*[†]

CHAPTER OUTLINE

- 6.1 Corrosion in Nuclear Systems 211
  - 6.1.1 Types of Corrosion 212
  - 6.1.2 Operating Conditions in Nuclear Systems 212
- 6.2 Corrosion in Water Cooled Reactors 212
  - 6.2.1 Subcritical Water 212
  - 6.2.2 Supercritical Water 219
- 6.3 Corrosion in Helium-Cooled Reactors 224
  - 6.3.1 Oxidation in a VHTR Environment 225
  - 6.3.2 Decarburization in a VHTR Environment 226
  - 6.3.3 Carburization in a VHTR Environment 227
  - 6.3.4 Internal Oxidation 227
  - 6.3.5 Additional Considerations 229
- 6.4 Corrosion in Molten Salt and Liquid Metal-Cooled Reactors 229
  - 6.4.1 Molten Salt 230
  - 6.4.2 Sodium 236
  - 6.4.3 Lead Alloys 239
- References 242
- Further Reading 246

6.1 CORROSION IN NUCLEAR SYSTEMS

Because the transfer of heat from a nuclear power plant requires a cooling fluid operating at high temperature and high pressure, understanding and mitigating corrosive reactions is critical to plant operation. All of the operating commercial power plants in the United States are light-water reactors (LWRs) that use water as a coolant, with temperatures as high as 320°C. The direct cost attributed to corrosion in these nuclear plants, according to a 2002 US Federal Highway Administration report [1], was $4.2 billion. Proposed advanced concepts would use coolants other than water such as helium, sodium, or lead, and at much higher temperatures than current generation of LWRs. However, each heat transfer fluid interacts with structural materials so that corrosion remains a concern for all coolants being considered. The control of corrosion in all of its forms will be critical to the success of these advanced plants.

6.1.1 TYPES OF CORROSION

There are several different types of environmental attack that occur under the general description of corrosion. Although they are classified slightly differently by various sources, a good general description is available in Ref. [2] and is graphically summarized in Fig. 6.1. Uniform corrosion occurs in a roughly similar manner across the entire surface of a material and is prevalent in engineering systems to some extent. Because equipment in complex engineering systems often consists of multiple materials joined by welds or other solid-state joining processes, site-specific corrosion mechanisms must also be considered. Even within a single material, if second phase strengthening is used or if surface defects are present, local corrosion effects such as galvanic, crevice, or pitting corrosion are possible. In this chapter, the focus will be on describing general corrosion as well as selected site-specific or environmentally enhanced corrosion mechanisms that are critical to the performance of current and proposed nuclear energy systems.

6.1.2 OPERATING CONDITIONS IN NUCLEAR SYSTEMS

To improve the performance of nuclear systems in the areas of safety, proliferation resistance, economics, and waste management, six advanced nuclear concepts were proposed by the international community under the Generation IV program. While each plant has perceived advantages, they all operate at higher temperatures and more aggressive radiation environments than current LWRs. The concepts and their operating conditions are listed in Table 6.1 [3]. This chapter outlines the unique corrosion challenges in operating and future nuclear systems, including those associated with the coolants proposed in Table 6.1, broken down into three major sections: corrosion in water, corrosion in impure helium, and corrosion in liquid metals and salts. Compositions of principal alloys referred to in this chapter are given in Table 6.2.

6.2 CORROSION IN WATER COOLED REACTORS
6.2.1 SUBCRITICAL WATER

The temperature range of the water coolant in LWRs is generally between 280°C in boiling water reactors (BWRs), in which the water boils in the core, to 320°C subcooled water in pressurized water reactors (PWRs). Although LWRs have been in operation for over 50 years, corrosion remains a significant concern that will only become more important as plants age. Corrosion occurs in all of the major systems exposed to a water environment, including the reactor core, steam generator, turbine, condenser and piping, valves and fittings, and in a wide variety of alloys such as carbon and low alloy steels used in piping and turbine components, stainless steel used in core internals, primary flow circuits and the condenser, nickel-base alloys in the steam generator and in reactor vessel penetrations and welds, and zirconium alloys that serve as fuel cladding.

Early corrosion problems in nuclear reactor systems stemmed from "epidemics" that were generally precipitated by improper water chemistry control, including ingress of chlorides that induced pitting in steam turbine discs and blades and pitting and stress corrosion cracking (SCC) in stainless steel, poor secondary side pH control that resulted in wastage and crevice corrosion in steam generator tubes, and denting of steam generator tubes due to high corrosion rates of tube support plates. Problems were also caused by poor microstructure or alloy chemistry control such as high corrosion rates of zirconium fuel cladding, and SCC of stainless steel BWR piping due to sensitization or weld knife-line attack [4]. More recently, corrosion degradation has emerged in the form of SCC of stainless steel

6.2 CORROSION IN WATER COOLED REACTORS

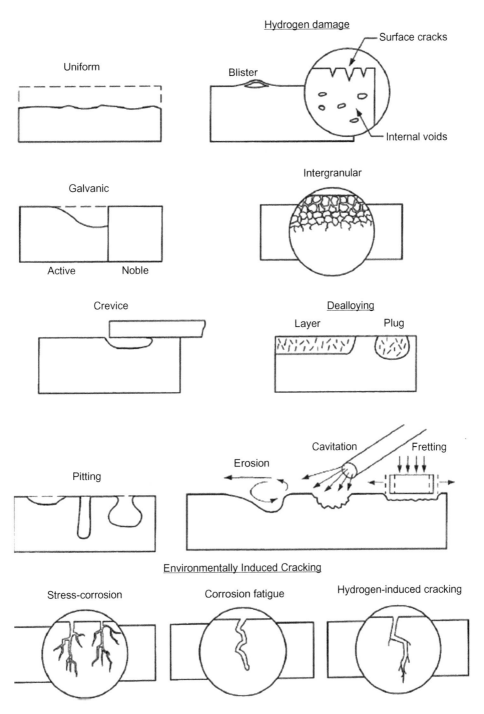

FIG. 6.1

Various forms of corrosion [2].

Redrawn from D. A. Jones, Principles and Prevention of Corrosion, Prentice-Hall, Upper Saddle River, NJ, 1996.

Table 6.1 Operating Conditions Anticipated for Generation IV Reactor Concepts [3]

Reactor Type	Coolant Inlet Temp (°C)	Coolant Outlet Temp (°C)	Maximum Dose (dpa[a])	Pressure (Mpa)	Coolant
Supercritical water-cooled reactor (SCWR)	290	500	15–67	25	Water
Very high-temperature gas-cooled reactor (VHTR)	600	1000	1–10	7	Helium
Sodium-cooled fast reactor (SFR)	370	550	200	0.1	Sodium
Lead-cooled fast reactor (LFR)	600	800	200	0.1	Lead
Gas-cooled fast reactor (GFR)	450	850	200	7	Helium/SC CO_2
Molten salt reactor (MSR)	700	1000	200	0.1	Molten salt
Pressurized water reactor (PWR)	290	320	100	16	Water

[a]dpa is displacement per atom and refers to a unit that radiation material scientists used to normalize radiation damage across different reactor types. For one dpa, on average each atom has been knocked out of its lattice site once.

steam lines, nickel-base steam generator tubes and reactor vessel penetrations, flow-assisted corrosion in low alloy steels, nodular corrosion, shadow corrosion, crud-induced localized corrosion, and fretting of zirconium alloy fuel cladding [5]. Still more recently, irradiation has played an increasingly important role in irradiation-assisted stress corrosion cracking (IASCC) and irradiation-accelerated corrosion (IAC) [6–8]. As plants age, the most important future corrosion issues will center around SCC, and the accelerating role of irradiation in both IASCC and corrosion.

6.2.1.1 Stress corrosion cracking

The SCC is a form of localized degradation that is manifest by the formation and growth of cracks due to a combination of aggressive medium, an applied tensile stress, and a susceptible material. The SCC can be either transgranular SCC (TGSCC) or intergranular SCC (IGSCC) in nature according to the path followed by the crack. The IGSCC is the most prevalent form of SCC in reactor materials today [9]. Environmental factors aggravating IGSCC are high conductivity and high corrosion potential which follows the oxygen concentration of the water. Tensile stresses are required and cracks nucleate when the applied tensile stress exceeds a critical value. Similarly, crack growth occurs when the stress intensity factor exceeds a critical value. While alloys are developed for their general corrosion resistance, they tend to exhibit a higher degree of susceptibility to localized corrosion processes, such as SCC. Further, the degree of susceptibility tends to follow the extent of alloying. However, "pure" metals are not exempt from susceptibility, they are just more resistant.

Much progress has been made in addressing the SCC issue through water chemistry control, microstructure/chemistry control, and residual stress reduction. A classic example of the confluence of environment, microstructure, and stress to cause IGSCC was 304 stainless steel BWR pipes that began to crack in the 1980s [10]. In that case, the carbon content was such that welding resulted in the precipitation of grain boundary carbides that depleted the region of chromium and resulted in tensile residual stresses that were sufficient to nucleate and grow intergranular stress corrosion cracks, resulting in failure of both small and large diameter pipes [11]. Measures taken to address this issue included use of low carbon stainless steel, 304 L, to minimize or delay sensitization. Stress relief of welds removed another factor required to avoid cracking. Water chemistry changes included stricter control of water conductivity and reduction in the corrosion potential by addition of small amounts of hydrogen (hydrogen water chemistry, HWC).

Table 6.2 Nominal or Limiting Compositions of Alloys

Alloy	C	Al	Si	Ti	V	Cr	Mn	Fe	Ni	Cu	Nb	Mo	W	Zr	Co	Other Elements
Zircaloy-2 From: Matweb						0.1		0.1	0.05					98.5		O: 0.12, Sn: 1.4
Inconel 617 From: special Metals	0.05–0.15	0.8–1.5	1.0 max	0.6 max		20–24	1.0 max	3 max	44.5 min	0.5 max		8–10			10–15	S: 0.015 max, B: 0.006 max
Haynes 230 From: haynesintl.com	0.1	0.3	0.4			22	0.5	3 max	57 as bal.			2	14		5 max	La: 0.02, B: 0.015 max
Hastelloy X From: haynesintl.com	0.1		1 max			22	1 max	18	47 as bal.			9	0.6		1.5	B: 0.008 max
Hastelloy N	0.08 max		1 max			7	0.8 max	5 max	71 as bal.	0.35 max		16	0.5 max		0.2 max	Al + Ti = 0.35 max
SUS304 www.yamco-yamashin.com	0.08 max		1.0 max			18–20	2.0 max		8–10.5							P: 0.045 max, S: 0.030 max
SUS316 www.yamco-yamashin.com	0.08 max		1.0 max			16–18	2.0 max		10–14			2–3				P: 0.045 max, S: 0.030 max
SUS321 From: Matweb	0.08 max		1.0 max	0.4		17–19	2.0 max	65.5–73.6	9–12							P: 0.045 max, S: 0.03 max
2.25Cr-1Mo ASTM A387 Grade 22	0.05–0.15 max		0.5 max			2–2.5	0.3–0.6	Bal.				0.9–1.1				P: 0.025 max, S: 0.025 max
9Cr-1Mo ASTM A387 Grade 91	0.08–0.12	0.02 max	0.2–0.5	0.01 max	0.18–0.25	8–9	0.3–0.6		0.4 max			0.85–1.05		0.01 max	0.06–0.10	P: 0.02, S: 0.01, N: 0.03–0.07
Alloy SX Norem International			4.8–6			16.5–19.5	2	Bal.		1.5–2.5		0.3–1.5				
Alloy EP 823 JNM, vol. 296, 201 p231	0.14		1.8		0.43	12	0.67	Bal.	0.89		0.4	0.7	1.2			
Alloy EP 823 (materials Science, vol 40. No. 2, 2004)	0.17		2.04		0.2	13.46	0.74	Bal.	0.28		0.2	1.6	0.19			N: 0.094
JPCA JNM, 431 (2012) p98	0.053		0.52	0.24		14.53	1.48	Bal.	15.6			2.5			<0.005	B: 0.004, P <0.005, S: 0.0017, N: 0.0012
JPCA JNM, 307–311 (2002) p353	0.06		0.50	0.24		14.2	1.77	Bal.	15.6			2.3				B: 0.003, N: 0.0039, P: 0.027, S: 0.005
1.4970 JNM, 358 (2006) p40–41	0.09		0.46	0.46		14.6	1.70	Bal.	15.00			1.25				B: 0.0045, N: 0.01
1.4970 JNM, 317 (2003), p161	0.11		0.3	0.48		15.13	1.4	Bal.	14.97			1.14				P: 0.005, S: 0.01, B: 0.0029

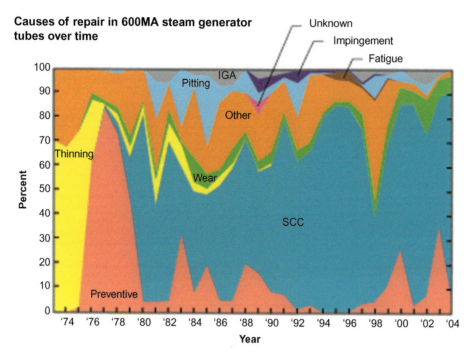

FIG. 6.2

Failure modes of mill-annealed alloy 600 over a 38-year period [9].

Redrawn from Steam Generator Progress Report, EPRI Steam Generator Database, Electric Power Research Institute, Palo Alto, CA, February 23, 2012, Fig. 2–11.

Today, the major SCC issues are with the nickel-base alloys used in steam generators and vessel penetrations and the stainless steels in the reactor core. Since steam generator tubes comprise some 75% of the surface area of the primary circuit in contact with the coolant, their performance is critical to that of the reactor. Fig. 6.2 shows that while many degradation modes of alloy 600 steam generator tubes exist, SCC has been the dominant one over the past 25 years. The SCC of nickel-base alloys is very sensitive to composition and water chemistry. Fig. 6.3 shows the propensity for cracking of austenitic alloys as a function of nickel content in both pure water and 0.1% NaCl. Note that SCC in chloride (red curve) is at a maximum at the extremes in nickel content, but SCC in pure water (blue) occurs only at high nickel concentrations.

Unfortunately, alloy 600 was originally used in steam generators and vessel penetrations and is highly susceptible to SCC in pure water. Neither microstructure modification (thermal treatment) nor water chemistry control has been successful in eliminating SCC in alloy 600. The susceptibility of alloy 600 is likely due to the low chromium concentration of the alloy, rather than its high nickel concentration. Chromium is required for passivation of the alloy, which protects it from attack by the environment. Heat treatments can cause the precipitation of chromium carbides that reduce the chromium concentration and make is susceptible to SCC in oxidizing (BWR and secondary side of PWR) environments. In a reducing environment (primary side of PWR), alloy 600 also exhibits IGSCC, and recent studies have pointed toward internal oxidation as the culprit [12,13]. Internal oxidation occurs because the chromium content is insufficient to form a protective passive layer. Results show that cracking is worst at the Ni-NiO phase boundary at which Ni metal is stable and does not substantially participate in the formation of the oxide.

Staehle has identified seven different SCC modes defined by their potential-pH combinations as shown in Fig. 6.4 [14]. In fact, while not evident from the information in Fig. 6.4, the severity of SCC in alloy 600 appears

6.2 CORROSION IN WATER COOLED REACTORS

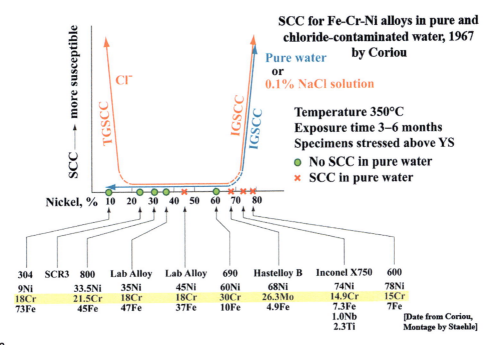

FIG. 6.3

Schematic depiction of SCC severity of austenitic alloys as a function of nickel content in pure water (*blue* curve) and 0.1% sodium chloride (*red* curve). YS is yield stress. Alloy concentrations shown are nominal and in weight %.

Courtesy: R. W. Staehle.

to track the Ni-NiO stability line in which cracking is at a maximum at this equilibrium condition and drops off both above and below it. Since the susceptibility range of potential and pH shown in Fig. 6.4 spans what is typically achievable in service, control of SCC in alloy 600 has proven very difficult. As a result, alloy 600 components have been replaced by alloy 690, which has a nominal composition of 60Ni-30Cr-10Fe and exhibits much higher resistance to SCC in pure water, chlorides, and alkaline solutions (Fig. 6.3).

The history with alloy 600 is that it was difficult to crack in pure high-temperature water in the laboratory, prompting the belief that it would be resistant in service. In fact, even after years in operation, few incidents of cracking were reported. However, the incubation period was found to be of the order of 11–12 years, Fig. 6.5, and eventually, all plants began to experience cracking that has eventually led to widespread replacement with steam generators using alloy 690. Alloy 690 contains more chromium than alloy 600, about 30% Cr in alloy 690 compared with 16% in alloy 600. The higher Cr content is believed to be responsible for greater protection of the alloy. The concern today is that the superior resistance to SCC may be due to a longer incubation period and that, in fact, it will eventually begin to crack. Current efforts are focused on trying to determine if there are microstructures, processing routes, or water chemistries for which Ni-based alloys are particularly susceptible to SCC. Recently, it was found that a single cold rolling operation for a 20%–30% reduction in thickness increases the crack growth rate in pure water by over a factor of 100 [15]. Similar efforts to identify susceptibility to crack initiation are on the increase.

6.2.1.2 Irradiation-assisted stress corrosion cracking

The IASCC of austenitic stainless steels and some nickel-base alloys have presented a significant problem in ensuring the integrity of LWR core components. The IASCC is generic as it cuts across all LWR designs and materials as illustrated in Table 6.3. The specific effects of irradiation on SCC are classified into two categories: effects on

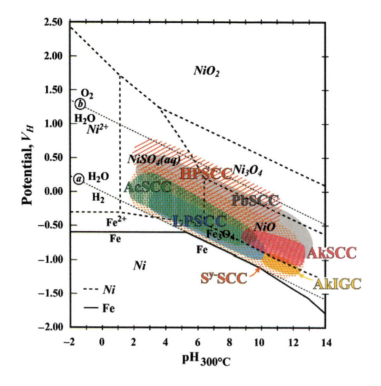

FIG. 6.4

Stress corrosion cracking modes for alloy 600 as a function of potential and pH at 300°C. *AkSCC*, alkaline SCC; *AkIGC*, alkaline IGC; *S^y-SCC*, sulfide SCC; *PbSCC*, lead SCC; *LPSCC*, low-potential SCC; *HPSCC*, high-potential SCC; *AcSCC*, acidic SCC.

Courtesy R. W. Staehle.

water chemistry and effects on microstructure [6–8]. Water chemistry effects include radiolysis and its effects on corrosion potential as it impacts IASCC. Microstructure effects include radiation-induced segregation (RIS), dislocation loop formation, swelling, and creep, and generation of H and He. Leading processes proposed to explain the roles of radiation in the SCC process are: radiolysis and crack tip strain rate, grain boundary chromium depletion, irradiation hardening, localized deformation, and RIS of minor elements.

Irradiation causes a significant change in local composition near grain boundaries and other defect sinks [6–8]. The enrichment of nickel and silicon, and the depletion of chromium can affect the susceptibility to IASCC, especially under oxidizing conditions, where chromium is needed for passivation. Silica is soluble in high-temperature water, which could explain the detrimental effect of high Si alloys. Irradiation also alters the microstructure, and under LWR conditions, faulted dislocation loops represent the primary irradiation-induced microstructure defect. The loops impede the motion of dislocations, resulting in an increase in the yield strength by factor of up to five. Radiation hardening correlates with IASCC propensity, and also induces highly localized deformation in the form of dislocation channels, which could contribute to IASCC. Irradiation also induces creep that can relax macroscopic stresses and can also enhance local dynamic deformation and local stresses. Other factors, such as swelling and formation of new phases, may accentuate IASCC at high fluence. With the many effects of irradiation that overlap spatially and temporally, more work is needed to identify the roles radiation plays in the mechanism(s) of IASCC and develop a comprehensive prediction methodology. In particular, the evolution

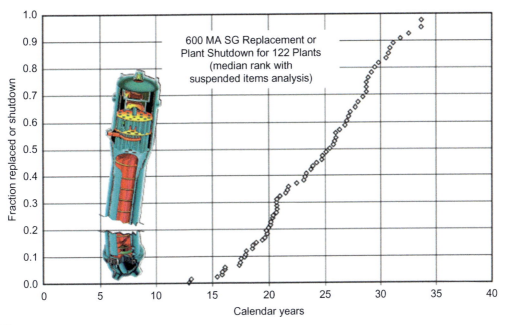

FIG. 6.5

Fraction of replaced or shutdown steam generators containing mill annealed alloy 600 tubing as a function of calendar years of operation.

Courtesy R. W. Staehle.

of localized deformation in an irradiation damage microstructure and the emergence of phases at high dose are the areas that require a better understanding to ensure the structural integrity of core components to a life of 40 or 60 years and beyond.

6.2.1.3 Irradiation-accelerated corrosion

The last major issue in LWR operated to high doses involves the interplay of radiation and corrosion in IAC. The mechanism of this process is unknown, but it has been shown to have significant impacts on corrosion rates. Zirconium irradiated in-reactor in moist carbon dioxide-air mixtures had oxygen weight gains five times more than that in the unirradiated state [16]. In-reactor corrosion rates of zirconium alloys were found to be 10 times greater than those conducted out-of-reactor, attributable, at least in part, to the greater permeability of the oxide irradiated in-reactor [17]. More recently, Lewis and Hunn [18] found that proton irradiation of a 316 stainless steel foil in room temperature water for 4 h produced an oxide that was 20× thicker than the unirradiated control. Further, old data on in-reactor exposure of Zircaloy-2 showed increases in the oxide weight gain of 40× and a strong, linear dependence on neutron flux [19]. These data suggest that displacement damage to the solid during corrosion produces a significantly larger effect than that due to radiolysis alone.

6.2.2 SUPERCRITICAL WATER

The use of water as a coolant at very high temperatures and pressures in the supercritical water (SCW) regime enables a higher efficiency of energy production and a simplified system design. However, the properties of SCW differ substantially from subcritical water, providing a very different environment for the behavior of

Table 6.3 IASCC Service Experience [6]

Component	Material	Reactor Type	Possible Sources of Stress
Fuel cladding	304 SS	BWR	Fuel swelling
Fuel cladding	304 SS	PWR	Fuel swelling
Fuel cladding[a]	20%Cr/25%Ni/Nb	AGR	Fuel swelling
Fuel cladding ferrules	20%Cr/25%Ni/Nb	SGHWR	Fabrication
Neutron source holders	304 SS	BWR	Welding & Be swelling
Instrument dry tubes	304 SS	BWR	Fabrication
Control rod absorber tubes	304/304L/316L SS	BWR	B_4C swelling
Fuel bundle cap screws	304 SS	BWR	Fabrication
Control rod follower rivets	304 SS	BWR	Fabrication
Control blade handle	304 SS	BWR	Low stress
Control blade sheath	304 SS	BWR	Low stress
Control blades	304 SS	PWR	Low stress
Plate type control blade	304 SS	BWR	Low stress
Various bolts[b]	A-286	PWR & BWR	Service
Steam separator dryer bolts[b]	A-286	BWR	Service
Shroud head bolts[b]	600	BWR	Service
Various bolts	X-750	BWR & PWR	Service
Guide tube support pins	X-750	PWR	Service
Jet pump beams	X-750	BWR	Service
Various springs	X-750	BWR & PWR	Service
Various springs	718	PWR	Service
Baffle former bolts	316 SS Cold Work	PWR	Torque, differential swelling
Core shroud	304/316/347/L SS	BWR	Weld residual stress
Top guide	304 SS	BWR	Low stress (bending)

[a] Cracking in AGR fuel occurred during storage in spent fuel pond.
[b] Cracking of core internal occurs away from high neutron and gamma fluxes.

structural alloys. Fig. 6.6 shows that, in addition to a sharp decrease in water density, the dissolution constant also decreases while the solubility of gases in H_2O increases. The SCW reactor is expected to operate over a temperature range that spans from the subcritical to as high as 620°C under normal operation. As experienced in subcritical water, corrosion and SCC increase with temperature, so it is expected that temperature alone will greatly aggravate both degradation modes. Additional concerns include the nonpolarity of SCW, its radiolysis products, the insolubility of dissolved metal ions, and the high solubility for gases, including oxygen. However, much less research and development to date has been devoted to the direct (pressure, temperature) and indirect (fluid properties) effects associated with supercritical conditions, although such efforts have recently received greater attention [20,21].

6.2.2.1 Corrosion

While many alloy systems have been studied to some degree (e.g., austenitic stainless steels, nickel-base, ferritic-martensitic, zirconium-based, titanium-based alloys), most research has focused on stainless and ferritic-martensitic steels. As in other high-temperature fluids, resistance to environmental degradation under SCW conditions depends critically on the formation and long-term stability of protective surface layers on the

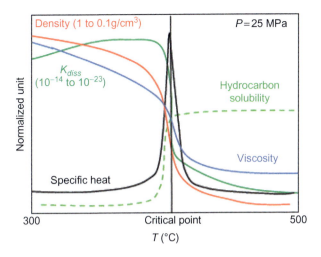

FIG. 6.6

Variation in the behavior of water upon crossing the critical point at 374.1°C.

structural/containment materials. For iron-based systems of interest for most nuclear applications, corrosion resistance is associated with the ability of iron- and chromium-based surface oxides to function as barriers to the transport of reactants (oxygen and metal ions). These oxides are the same as those typically formed in other oxidizing environments, such as steam. The main issues related to the behavior of these oxides as protective layers in SCW are similar to those for advanced-steam-cycle fossil energy plants [22]. However, wall thickness of critical core components such as fuel cladding and coolant tubes are an order of magnitude smaller than boiler tubes in fossil plants, placing a greater burden on the development of thin protective oxide layers that do not spall.

While oxide growth rates and product morphological details are specific to the oxygen content of the fluid, temperature, and steel composition and possibly other factors [20], the oxide structure on steels exposed to SCW follows pattern shown schematically in Fig. 6.7, and is similar to that observed under steam conditions for ferritic and ferritic-martensitic (F-M) steels [22]. Because it is generally accepted that chromia-containing spinels are better permeation barriers to cations (metal) and anions (oxygen, OH, etc.) relative to iron oxides [23], it is the underlying oxide layer that can proffer the best corrosion resistance in the SCW environment. This has been observed in recent work on steels under nuclear SCW conditions [20,24,25] in terms of increasing corrosion resistance with increasing chromium content of the alloy. In fact, Fig. 6.8 shows that nickel-base alloys show the greatest resistance to oxidation in SCW, followed by austenitic stainless steels, and then ferritic-martensitic alloys. Alloys other than steels for SCW nuclear service are also being considered and certain Zr alloys have exhibited corrosion resistance on a par with the most promising ferritic alloys exposed to SCW [26].

6.2.2.2 Stress corrosion cracking

The most daunting challenge for materials in the SCW environment is resistance to SCC and IASCC. While nickel-base alloys and austenitic stainless steels are very resistant to corrosion in SCW, they are most susceptible to SCC. Intergranular SCC occurs readily in high purity, deaerated SCW at 400°C and above in both austenitic stainless steels and nickel-base alloys [27]. Fig. 6.9 shows that cracking severity increases exponentially with temperature in two stainless steels and two nickel-base alloys [20]. Over this same temperature range, ferritic-martensitic alloys are resistant to SCC [20,21].

222 CHAPTER 6 CORROSION IN NUCLEAR SYSTEMS

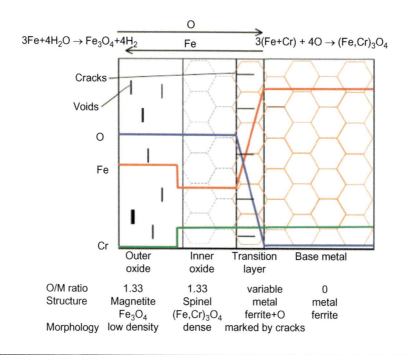

FIG. 6.7

Structure of the oxide layer on steels exposed to supercritical water [18].

From C. Cabet, J. Jang, J. Konys, P.F. Tortorelli, Environmental Degradation of Materials in Advanced Reactors, in Materials Challenges for Advanced Nuclear Energy Systems, Y. Geurin, G. S. Was, S. J. Zinkle (Eds.), MRS Bulletin, 34 (1) (2009) 35.

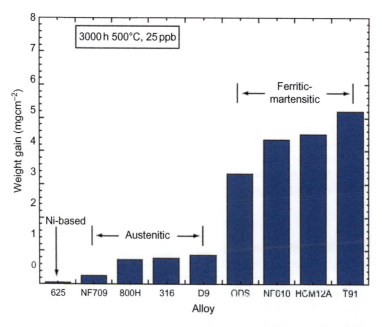

FIG. 6.8

Oxide weight gain in nickel-based, iron-based austenitic, and ferritic-martensitic alloys at 500°C.

From T. Allen, Y. Chen, X. Ren, K. Sridharan, L. Tan, G. Was, E. West, D. Guzonas, Materials Performance in Supercritical Water, Chapter 100, in Comprehensive Nuclear Materials, R. Konings, T. Allen, R. Stoller, S. Yamanaka (Eds.), Elsevier, Netherlands, in press.

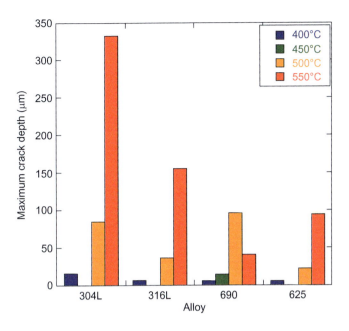

FIG. 6.9

Temperature dependence of SCC crack depth in two austenitic stainless steels and two nickel-base alloys tested in constant extension rate tests.

From S. Teysseyre, G. S. Was, Stress Corrosion Cracking of Austenitic Alloys in Supercritical Water, Corrosion 62 (2006) 1100.

6.2.2.3 Irradiation-assisted stress corrosion cracking

Radiation effects on IGSCC are only just now being investigated for SCW conditions, yet results show that indeed, irradiation significantly increases the extent of SCC in stainless steels and nickel-base alloys [28,29]. Proton irradiations of type 316L stainless steel and Ni-based alloy 690 showed a significant increase in intergranular cracking relative to the unirradiated cases, Fig. 6.10. The increased cracking could not be attributed to radiation-induced segregation or hardening alone, so combinations of factors or other defect mechanisms must be at play [29]. Both the density of cracks and crack depth increased over the unirradiated case following irradiation to 7 dpa and testing in SCW at 400°C. One set of data exists on the effect of neutron irradiation on cracking in SCW, where an austenitic stainless steel was irradiated to doses of over 40 dpa and showed extreme embrittlement [30]. Under the same irradiation and testing conditions, ferritic-martensitic alloys were found to be resistant to cracking.

6.2.2.4 Summary: SCW

Degradation of alloys in the SCW environment occurs by corrosion, SCC, and IASCC. Austenitic alloys are most resistant to oxidation but susceptible to both SCC and IASCC. Ferritic-martensitic alloys exhibit significantly higher oxidation rates but are resistant to SCC and IASCC in SCW. As such, much development will be required to either overcome the SCC susceptibility of austenitic alloys or the high corrosion rates of ferritic-martensitic alloys. Zirconium alloys must also be explored further for their potential suitability as fuel cladding.

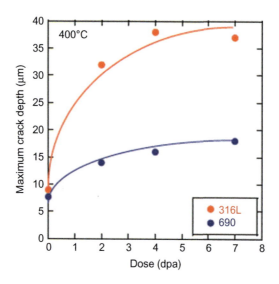

FIG. 6.10

Effect of irradiation at 400°C with 2 MeV protons on the IG cracking behavior of alloys 316L and 690 tested in 400°C SCW.

From R. Zhou, E. A. West, Z. Jiao, G. S. Was, Irradiation-Assisted Stress Corrosion Cracking of Austenitic Alloys in Supercritical Water, J. Nucl. Mater. 395 (2009) 11–22.

6.3 CORROSION IN HELIUM-COOLED REACTORS

The very high-temperature reactor (VHTR) has emerged as a leading advanced nuclear reactor concept. High efficiency of electricity production (>50%), high service lifetime (>60 years), combined with a broad range of process heat applications, such as hydrogen production, distinguish it from other "Generation IV" nuclear reactor systems [31–33]. In this concept, helium gas with outlet temperatures up to 1000°C will pass through an intermediate heat exchanger where it will transfer heat to a secondary coolant. Such temperatures require the use of nickel-based alloys rich in chromium (about 22 wt%) and strengthened by the addition of Mo, Co, and W (e.g., Inconel 617 and Haynes 230) [34].

The helium coolant of a VHTR inevitably contains parts per million (ppm) level of CO, CO_2, H_2, H_2O, and CH_4 as impurities, which arise mainly from reactions between the hot graphite core and in-leakage of O_2, N_2, and water vapor from seals, welds, and degassing of reactor materials such as fuel, thermal insulation, and in-core structural materials [35,36]. Typical concentrations of helium impurities in various gas-cooled reactors are given in Table 6.3. Depending on the impurity concentration, temperature, and alloy composition, the impurities react with the metallic surfaces of the heat exchanger resulting in oxide formation or reduction and/or carburization or decarburization. Chromium is oxidized at oxygen partial pressures above , and is reduced at partial pressures below this value. Similarly, chromium carbide is stable above a critical carbon activity, , and decarburization is expected to occur below the critical value. The equilibrium Cr-O-C stability diagram is shown in Fig. 6.11 with the dashed line representing as set by the reaction

$$C(s) + \tfrac{1}{2} O_2 (g) - CO(g) \tag{6.1}$$

Oxidation, decarburization, and carburization are processes that can degrade the mechanical properties of the alloy, for example, oxidation may reduce the load-bearing cross section of the component and/or internal oxide

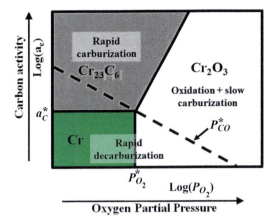

FIG. 6.11

Cr-C-O stability diagram.

precipitates may act as the preferential crack initiation sites [37], which can, potentially, decrease the creep and fatigue life of the alloy. Significant reductions in the creep-rupture ductilities of alloy 800H [38], alloy 617 [39], and Hastelloy X [40] have been reported in a carburizing environment in comparison to pure helium and air environments. A coarse and semicontinuous film of carbides forms along the grain boundaries during carburization and may act as a preferential crack initiation and propagation path, which could decrease the operating life of the alloy. Grain boundary migration and sliding has been identified as the dominant creep deformation mechanism in the candidate alloys, such as alloy 617 at 1000°C [41,42], and the dissolution of carbides due to decarburization may lead to significant loss of the creep strength. Therefore, a detailed knowledge of the oxidation mechanisms and rates of microstructure degradation is important to estimate the lifetime of the component and define mitigation strategies for improved high-temperature performance of alloys in the VHTR environment.

6.3.1 OXIDATION IN A VHTR ENVIRONMENT

Oxidation of Ni-Cr alloys in impure helium is governed by the competition between two gas-metal reactions. Assuming Cr as the main reactive element that forms oxides and carbides, the reactions are as follows:

$$2Cr_{solution} + 3H_2O \rightarrow Cr_2O_3 + 3H_2 \qquad (6.2)$$

$$2Cr_{solution} + 3CO_2 \rightarrow Cr_2O_3 + 3CO \qquad (6.3)$$

Fig. 6.12 shows a chromium oxide scale on a sample of alloy 617 exposed to He-containing CO and CO_2 in the ratio $CO/CO_2 = 9$ for 500h at 1000°C. The rate of growth of the oxide is parabolic in time, most likely indicating solid-state diffusion control. Accordingly, oxide growth is sensitive to temperature and, at high temperatures envisioned for the VHTR, the oxidation rate is fairly rapid. This, combined with the significant volatility of Cr_2O_3 at temperatures of 900°C and above [43], render the chromia layer relatively unprotective under VHTR conditions.

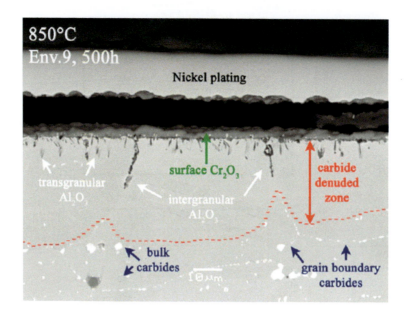

FIG. 6.12

Backscattered electron image of the surface microstructure of alloy 617 oxidized at 850°C for 500 h in He-containing CO and CO_2 in the ratio $CO/CO_2 = 9$.

6.3.2 DECARBURIZATION IN A VHTR ENVIRONMENT

In addition to reactions (6.2) and (6.3) a reaction, often referred as a "micro-climate" reaction [44,45], occurs above a "critical temperature" and decarburizes the alloy. In the impurity concentration range relevant to the VHTR (see Table 6.4), the value of the critical temperature increases with the CO concentration in helium, and is independent of the concentrations of other impurities [46,47]. Globally, the microclimate reaction takes the form

$$2Cr_2O_3 + Cr_{23}C_6 \rightarrow 27Cr + 6CO \qquad (6.4)$$

Hydrogen triggers the reaction by reducing the chromia film to provide water vapor as per the reaction

$$2Cr_2O_3\,(solid) + 6H_2\,(gas) \rightarrow 6H_2O + 4Cr \qquad (6.5)$$

Table 6.4 Typical Concentration of Impurities in Impure Helium [29]

Reactor	H_2 (ppm)	H_2O (ppm)	CO_2 (ppm)	CO (ppm)	CH_4 (ppm)	N_2 (ppm)
Dragon	20	1	<0.4	12	3	3
AVR	300	30	100	100		
PNP	500	1.5		15	20	5
HHT	50	5	5	50	5	5
HTGR–SC	200	10	<1	20	20	15
AGCNR	400	2	0.2	40	20	<10

From C. Cabet, A. Terlain, P. Lett, L. Guetaz, J.M. Gentzbittel, High temperature corrosion of structural materials under gas-cooled reactor helium. Mater. Corr. 57 (2) (2006) 147–153.

which in turn reacts with the carbide to produce CO, Cr and regenerates the hydrogen as per the reaction:

$$6 H_2O \text{ (Gas)} + Cr_{23}C_6 \text{ (Solid)} \rightarrow 23Cr + 6H_2 + 6CO \tag{6.6}$$

Thus, the net effect is that the alloy decarburizes via production of CO, while hydrogen and water vapor act as the transfer media and are neither produced nor depleted. In effect, decarburization is proposed to be occurring through the medium of gas catalysis at the oxide-metal interface. In the case where a microclimate cannot be established, decarburization occurs via the reaction [48,49]

$$Cr_2O_3 \text{ (solid)} + 3C_{\text{Solution}} \rightarrow 2Cr + 3CO \tag{6.7}$$

Both models agree that decarburization occurs via consumption of the surface oxide (Cr_2O_3) film but they differ on whether the decarburization is a gas mediated reaction or solid-solid reaction.

The consumption of Cr by oxidation leads to Cr depletion below the film, resulting in carbide dissolution. Fig. 6.13 shows the chromium-depleted region below the chromium oxide layer, and the resulting decarburization, noted by the disappearance of carbides near the surface.

Decarburization occurs for conditions that fall within the area of chromia instability and in the absence of methane. The main reaction is the reduction of chromia by carbon due to the low p_{CO} in the helium. However, water vapor can still oxidize Cr, thereby regenerating chromia. A kinetic competition therefore occurs between chromia reduction and growth that depends on environmental factors. Even when Cr_2O_3 is maintained on the surface, it isn't necessarily protective: the scale is highly porous [50] and allows ingress by reactive gases. The alloy bulk carbon content decreases with the formation of CO. A deep carbide-free zone also develops because the decrease in carbon concentrations causes carbides to dissolve. The dissolution of carbides can have a dramatic and deleterious effect on the creep-rupture life [50].

6.3.3 CARBURIZATION IN A VHTR ENVIRONMENT

Exposure to helium with a high CO partial pressure or containing methane produces carburization and prevents an oxide film from forming (or reduces the existing one). Carburization can occur due to the reaction between methane and chromium according to the following reaction:

$$7Cr_{\text{solution}} + 3CH_4 \rightarrow Cr_7C_3 + 6H_2 \tag{6.8}$$

If the CO/CO_2 ratio is high enough, surface and bulk carburization of the samples occurs. The surface carbide, Cr_7C_3 is metastable and nucleated due to preferential adsorption of carbon on the chromia surface. The Cr_7C_3 precipitates grow at the gas-scale interface via outward diffusion of Cr cations through the chromia scale until the activity of Cr at the reaction site falls below a critical value. The decrease in activity of chromium triggers a reaction between chromia and carbide: $Cr_2O_3 + Cr_7C_3 \rightarrow 9Cr + 3CO$, which resulted in a porous surface scale.

Recent observations [43] suggest that Cr migrates through the oxide film and reacts with C from dissociated CO molecules on the surface of the oxide to form chromium carbide. Some of the carbon deposited on the surface also moves into the bulk where it precipitates as coarse carbides. Transport of carbon occurs despite the low solubility of C in chromium oxide due to the very defective nature of the chromia. These carbides are associated with significantly reduced impact energy, tensile strength, and rupture elongation [51].

6.3.4 INTERNAL OXIDATION

Internal oxidation of *Al* has been observed just below the surface at both intra- and intergranular locations. Deeply penetrating "finger-like" internal oxides also occur, and this mode of degradation becomes more significant relative to surface oxidation at temperatures below 850°C. While alumina is a very stable oxide, the *Al* concentration is

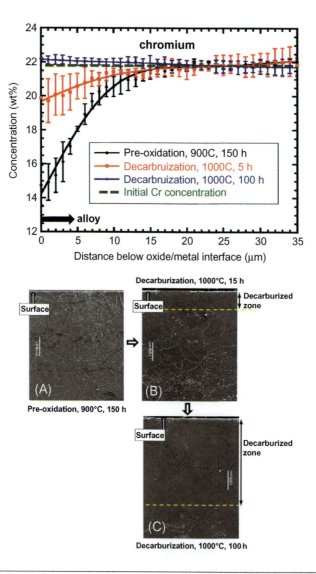

FIG. 6.13

(A) Concentration profile of Cr near the surface of alloy 617 pre-oxidized at 900°C for 150h and decarburized for 5 and 100h at 1000°C in He-containing CO and CO_2 in the ratio $CO/CO_2 = 9$. The initial concentration of Cr was 21.8wt% and is superimposed on the plot. (B) Micrographs of carbide structure for the treatments shown in part (A).

From D. Kumar, R. R. Adharapurapu, T. M. Pollock, G. S. Was, High-Temperature Oxidation of Alloy 617 in Helium Containing Part-Per-Mission Levels of CO and CO_2 as Impurities. Metall. Trans. A. 42A (2011) 1245–1265.

below the level that can support a continuous oxide layer, However, Al does oxidize as an incomplete layer and also at internal sites, increasingly dominated by grain boundaries as temperature is reduced. Fig. 6.14 shows the formation of aluminum oxide along the grain boundaries in alloy 617 exposed to 850°C He-containing CO and CO_2 in the ratio $CO/CO_2 = 9$ for 500h. Clearly, the combination of mechanical loading and grain boundary oxidation poses a great risk for fatigue strength and creep life of the components.

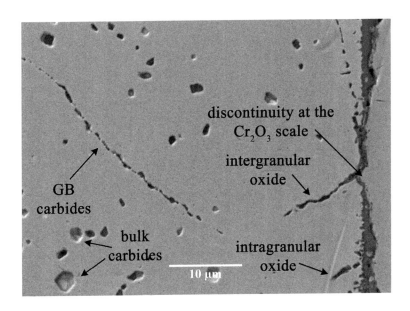

FIG. 6.14

Backscattered electron image of the cross-sectional microstructure of alloy 617 following exposure to He-containing CO and CO_2 in the ratio $CO/CO_2 = 9$ at 850°C for 500 h. The intergranular oxides are predominantly Al_2O_3.

6.3.5 ADDITIONAL CONSIDERATIONS

Carburization/decarburization are driven by thermodynamics and depend on the local conditions such as temperature and alloy composition. Temperature gradients along a heat exchanger tube can result in carburization at one location and decarburization in another. Likewise, oxidation potential gradients due to the presence of dissimilar metals in contact with a common environment can preferentially transport ions from one surface to another. Control of helium chemistry, especially in terms of the CO partial pressure, is a major issue in mitigating high-temperature corrosion of VHTR alloys. However, the atmosphere must also be compatible with carbon-based in-core structures, which obviously degrade at higher oxidizing potentials. With respect to alloy composition, small amounts of Ti and Al increase the overall reaction rate and internal oxidation [23]. Optimization of the alloy chemistry within the overall alloy specification could reduce this problem. Of perhaps greatest concern is the long-term oxidation behavior and the role of environmental factors, especially temperature.

6.4 CORROSION IN MOLTEN SALT AND LIQUID METAL-COOLED REACTORS

Molten salts and liquid metals have been proposed as primary and secondary coolants for a number of nuclear designs. Molten salts as a primary coolant were first demonstrated in the United States as the high-temperature fluid fuel for the molten salt reactor (MSR) developed at Oak Ridge National Laboratory [52]. The fluid fuel in these reactors, consisting of UF_4 and ThF_4 dissolved in fluorides of beryllium and lithium, was circulated through a reactor core moderated by graphite. Recently, molten salts as a heat transfer fluid for transferring heat from a high-temperature gas-cooled reactor to a process heat application [53] as well as a primary coolant in a graphite moderated reactor using a TRISO fuel form [54], known as a fluoride-cooled high temperature reactor (FHR), have also been suggested. Molten salts have been considered as a breeding material for helium-cooled fusion systems in helium-cooled molten salt breeder systems [55] and as the fluid for electrochemically based nuclear fuel recycling systems [56]. Chlorides are used for separating hafnium impurities from zirconium to be used as cladding in LWRs and for the conversion of uranium dioxide to uranium fluoride as part of the enrichment process in developing nuclear fuel [57]. Each application will use a salt optimized for its use including the ability to conduct heat, breed tritium, or separate fuels but, in each case, management of the corrosion of system materials by the salt will be critical to the optimized operation.

Both liquid sodium [58] and well as liquid lead alloys [59] have been proposed as coolants for fast spectrum nuclear systems. Liquid sodium was the typical choice of coolant in many national fast reactor development programs because of its excellent heat transfer properties and neutron transparency but was operationally challenging because of the high-energy release reaction between sodium and oxygen, making prevention of leaks of sodium an important challenge. Russian programs developed liquid lead alloys, both pure lead and lead-bismuth eutectic (LBE) alloys, as alternative higher temperature coolant options. Corrosion control is important in using both types of coolants, although corrosion control strategies for sodium and lead differ.

The fundamentals of liquid metal corrosion for materials to be used in nuclear systems were identified by the 1950s and captured in a report produced by Oak Ridge National Laboratory [60]. Many of the factors are similar for molten salts. The primary corrosion concerns are the dissolution of alloying elements from the components into the liquid salt or liquid metal and/or uptake of elements (such as oxygen and carbon) from the liquid metal (which can affect the form of dissolution). Attack can be a general dissolution or preferential attack along grain boundaries or inclusions depending on the free energy change associated with the dissolution, the wettability of specific portions of a surface, and how light elements partition between the liquid metal and solid. In a simple static system, dissolution attack would stop once the chemical potential of dissolved elements in the solution matches that in the alloy. In a more complex system with dissimilar materials in contact with the salt or liquid metal or with a temperature gradient, material transfer can occur. Both dissimilar-metal corrosion and differential-temperature corrosion will be greater than that experienced by simple static tests and understanding and controlling the complexity of the interactions within the system are critical to acceptable performance level.

As noted, by Manly [60], several variables affect liquid metal corrosion such as

- temperature
- temperature gradient
- cyclic temperature fluctuation with associated dissolution and subsequent deposition
- metallic surface area to salt volume ratio
- purity of liquid metal
- flow velocity, or Reynolds number
- surface condition of container material
- number of materials in contact with the same liquid metal
- condition of the container material, such as the presence of a grain boundary precipitate, the presence of a second phase, the state of stress of the metal, and the grain size.

Zhang et al. have recently published a detailed description of models for liquid metal corrosion [61].

In the following sections, selected key attributes of the above variables on corrosion response will be presented to highlight the challenges associated with liquid salt and liquid metal coolants.

6.4.1 MOLTEN SALT
6.4.1.1 Salts in nuclear systems
Salts are chosen for a specific application based on the optimization of a number of specific properties such as melting point, vapor pressure, density, heat capacity, viscosity, thermal conductivity, and cost [53]. Delpech and coworkers have outlined the uses of a number of salts in nuclear systems [57]:

- LiF-BeF$_2$ (66-33 mol%) (FLiBe) was studied at Oak Ridge National Laboratory for the development of the MSR Experiment and the Molten Salt Breeder reactor and has also been considered as a breeder blanket material for fusion power plants. In recent times, FLiBe has also been chosen as the primary coolant for FHR.

- LiF-NaF-KF (46.5-11.5-42 mol%) (FLiNaK) is a candidate for secondary cooling loops for high-temperature gas-cooled reactors.
- LiF-ThF$_4$-UF$_4$ is being studied as a salt for a fast spectrum MSR by the Centre National de la Recherche Scientifique (National Center for Scientific Research, CNRS) in France.

In addition, two other salts currently under consideration for nuclear system use are as follows:

- KCl-MgCl$_2$ and 58%KF-42%ZrF$_4$ as candidates for secondary cooling loops for high-temperature gas-cooled reactors.
- Molten LiCl-KCl in electrochemical processes for recycling nuclear fuel [56].

6.4.1.2 General corrosion

The basic corrosion process in fluoride salts is dissolution of alloying components, corresponding to the free energy of formation of specific compounds. Passivation that occurs on some metallic alloys in oxidizing environments, is not possible as the fluoride salts will dissolve oxides. As illustrated in Table 6.5, which describes the stability of fluoride compounds for the elements that typically exist in steel components, Ni is more stable relative to its fluoride than Fe which is more stable than Cr.

When exposed to the fluoride salt, Cr will preferentially dissolve before Fe or Ni. The corrosion reactions follow the driving force described by the free energy as follows [62]:

$$2HF + M = MF_2 + H_2 \text{ where } M = Ni, Cr, Fe \tag{6.9}$$

and

$$XF_2 + Cr = CrF_2 + X \text{ where } X = Ni, Fe \tag{6.10}$$

Because chromium dissolution is the major cause of material loss when steels are exposed to fluoride salts, alloys with higher chromium concentration tend to show greater weight loss, Fig. 6.15.

The corrosion can follow distinctive microstructural features that provide selective pathways for dissolution. For instance, in steels where grain boundaries provide fast diffusion pathways for Cr, possibly exacerbated by the presence of grain boundary chromium carbides, the dissolution of Cr may be faster along grain boundaries, as shown

Table 6.5 Relative Thermodynamic Stability of Fluoride Compounds Formed by Elements Employed as Alloying Additions [56]

Element	Most Stable Fluoride Compound	Standard Free Energy of Formation per Gram-Atom of Fluorine (kcal/g-atom of F) at 800°C	Standard Free Energy of Formation per Gram-Atom of Fluorine (kcal/g-atom of F) at 600°C
Al	AlF$_3$	−87	−92
Ti	TiF$_3$	−85	−90
V	VF$_2$	−80	−84
Cr	CrF$_2$	−72	−77
Fe	FeF$_2$	−66	−69
Ni	NiF$_2$	−59	−63
Nb	NbF$_5$	−58	−60
Mo	MoF$_5$	−57	−58
W	WF$_5$	−46	−48

From J. H. DeVan, R. B. Evans III, Corrosion Behavior of Reactor Materials in Fluoride Salt Mixtures, ORNL-TM-328, September 1962. Courtesy of Oak Ridge National Laboratory, U.S. Dept. of Energy.

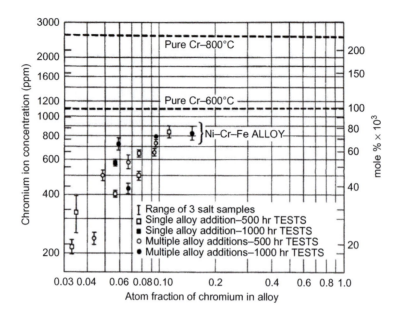

FIG. 6.15

Chromium concentration of fluoride salt circulated in thermal-convection loops as a function of chromium content of the loop. Salt mixture: NaF-LiF-KF-UF$_4$ (11.2-45.3-41.0-205 mol%). Loop temperature: hot leg, 815°C, cold leg, 650°C.

From J. H. DeVan, R. B. Evans III, Corrosion Behavior of Reactor Materials in Fluoride Salt Mixtures, ORNL-TM-328, September 1962. Courtesy of Oak Ridge National Laboratory, U.S. Dept. of Energy.

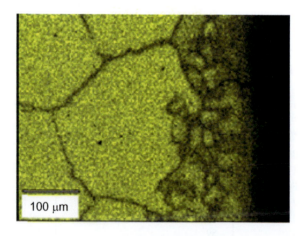

FIG. 6.16

EDS map of chromium concentration in a cross section of alloy 800H exposed to FLiNaK for 500 h at 850°C. The chromium dissolves from the surface and along connected grain boundary pathways [60].

From L. Olson, K. Sridharan, M. Anderson, T. Allen, Nickel-plating for active metal dissolution resistance in molten fluoride salts, J. Nucl. Mater. 411 (2011) 51–59.

in Fig. 6.16. Oak Ridge National Laboratory developed the low-Cr, Ni-base alloy Hastelloy N specifically for corrosion resistance to molten fluoride salts [63].

One method for controlling the dissolution of metallic components is through redox control, attempting to control the fluorine potential through the gas phase, metallic additions, or salt chemistry [64]. In gas-phase control, the fluorine potential is controlled by contacting the salt with an equilibrium mixture of H_2/HF according to the following reaction:

$$1/2 H_2(g) + 1/2 F_2 = HF(g) \tag{6.11}$$

Metallic additions can control the fluorine potential by reacting with fluorine preferentially. For example, in an FLiNaK salt, Zirconium would react as follows:

$$Zr + F_2 = ZrF_2 \tag{6.12}$$

preferentially over a reaction between the chromium and the salt. Additions of beryllium to control the fluoride potential in FLiBe have also been investigated [65,66]. Finally, if a salt containing a cation with two valence states is added to the salt, the reaction between the two valence states can be used to control the fluorine potential. As an example, in the MSR experiment, the reaction between UF_3/UF_4 was utilized:

$$UF_3 + 1/2 F_2 = UF_4 \tag{6.13}$$

Of course, the use of redox control may then lead to other undesirable reactions such as plating out of other elements onto surface, so a detailed understanding of the system design is ultimately necessary to prove the efficacy of a particular redox-salt system.

An alternate to redox control as a means of corrosion management is to coat the metallic components with a material resistant to dissolution such as Ni. Electroplating of nickel onto the surface of alloy 800H has been shown to reduce significantly the dissolution of chromium into the fluoride salt melt [67]. However, the coating does not provide a perfect barrier because the chromium does slowly diffuse through the nickel and eventually reach the salt [67]. Oak Ridge National Laboratory has examined the possibility of cladding metallic surfaces through a number of techniques including electroplating, electroless plating, physical vapor deposition, chemical vapor deposition (CVD), thermal spray techniques, weld overlay, laser cladding, and coextrusion and determined that, in the near term, the most promising approaches are (1) a laser-based surface cladding technique that is primarily applicable to surfaces with line-of-sight access and (2) CVD using the nickel-carbonyl process that is applicable to the cladding of surfaces with no line-of-sight access, such as long, narrow tubes used in heat exchangers [68]. For protection of stainless steels used in recycling processes using LiCl-KCl salts, plasma-sprayed yttria-stabilized zirconia coatings have been investigated [69].

6.4.1.3 Dissimilar-metal and temperature gradient-driven corrosion

Because the dissolution rate and thermodynamic driving forces of cations depends on local conditions, metallic ions can be transported throughout closed-loop systems. This can occur due to temperature gradients where dissolution normally occurs in a hot leg and deposition in a cold leg of a system [61,70,71]. Fig. 6.17 provides an illustration of a temperature-driven mass transport process and Fig. 6.18 is mass loss data from a loop test showing the mass transport in a corrosion loop.

Similarily, corrosion potential gradients due to the presence of dissimilar metals in contact with a common salt can preferentially transport ions from one surface to another as illustrated in Fig. 6.19.

An example is the transport of chromium from a steel component through a salt with deposition onto a graphite crucible. This mechanism has been shown to increase significantly the chromium dissolution rate in capsule experiments [60] and might have implications for FHR concepts that use metallic containers and graphite moderating materials.

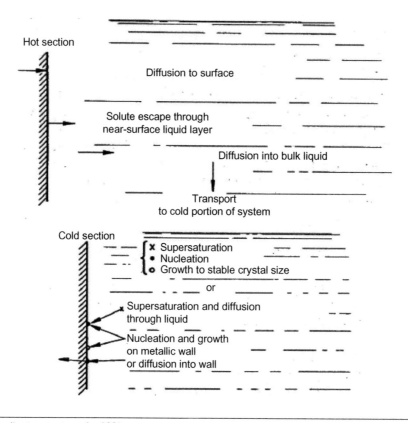

FIG. 6.17

Temperature gradient mass transfer [63].

From J. W. Koger, A. P. Litman, Mass Transfer between Hastelloy N and Haynes Alloy No. 25 in a Molten Fluoroborate Mixture, ORNL-TM-3488, 1971. Courtesy of Oak Ridge National Laboratory, U.S. Dept. of Energy.

6.4.1.4 Impurity-driven corrosion

Impurities in the salt can also affect the dissolution rates of alloying elements, with chromium being of primary concern. For example, if water remains in the FLiNaK salt as an impurity, additional HF can be formed to further drive the corrosion reactions listed in Eq. (6.9). As an example [72]

$$2KF + H_2O = K_2O + 2HF \tag{6.14}$$

To mitigate general corrosion, the moisture content of the salts is purposely kept low. As another example, if excess Fe is left in the salt, it can react to place more chromium into the salt [71], for example,

$$FeF_2 + Cr = CrF_2 + Fe \tag{6.15}$$

Fig. 6.20 shows how the rate of chromium dissolution increased due to the presence of excess iron in the salt.

6.4.1.5 Fission product-driven SCC

In molten-salt-cooled reactors that dissolve the fuel into the flowing salt, fission products will accumulate in the salt as the reactor operates and reactor with the containment material. As an example, at high enough concentrations, the tellurium concentration can react with structural steel as follows [57]:

$$xTe + yNi = Ni_yTe_x \text{ or } Cr_x + Te_y = Cr_xTe_y \tag{6.16}$$

These compounds form at grain boundaries and are brittle, leading to an intergranular attack. Salt control and alloy development can help mitigate this type of attack, at least in some cases [73].

6.4 CORROSION IN MOLTEN SALT AND LIQUID METAL-COOLED REACTORS

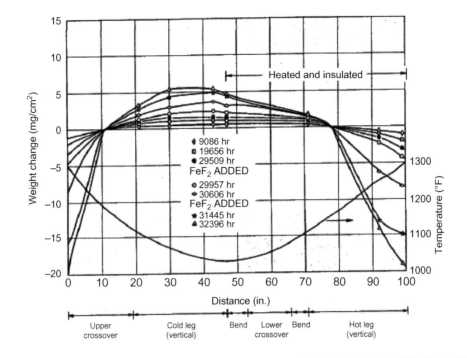

FIG. 6.18

Weight changes of Hastelloy N specimens exposed to LiF-Be-F$_2$-UF$_4$ (65.5-34.0-0.5 mol%) with FeF$_2$ added as a function of position and time [64].

From J. W. Koger, Effect of FeF2 Addition on Mass Transfer in a Hastelloy N-LiF-BeF2-UF4 Thermal Convection Loop System, ORNL-TM-4188, 1972. Courtesy of Oak Ridge National Laboratory, U.S. Dept. of Energy.

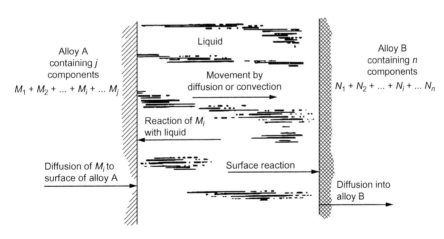

FIG. 6.19

Dissimilar metal mass transfer [63].

From J. W. Koger, A. P. Litman, Mass Transfer between Hastelloy N and Haynes Alloy No. 25 in a Molten Fluoroborate Mixture, ORNL-TM-3488, 1971. Courtesy of Oak Ridge National Laboratory, U.S. Dept. of Energy.

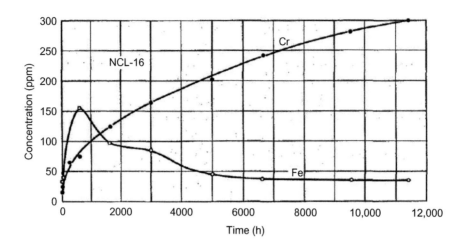

FIG. 6.20

Concentration of iron and chromium in a fuel salt loop [64].

From J. W. Koger, Effect of FeF2 Addition on Mass Transfer in a Hastelloy N-LiF-BeF2-UF4 Thermal Convection Loop System, ORNL-TM-4188, 1972. Courtesy of Oak Ridge National Laboratory, U.S. Dept. of Energy.

6.4.2 SODIUM

6.4.2.1 Sodium-cooled reactors

Sodium-cooled test reactors have been built and operated to build a foundation for fast spectrum nuclear systems that could either burn excess actinides or create needed actinides for a closed fuel cycle that recycles the fuel elements [74]. To date, 10 test fast reactors with thermal power ranging from 8 to 400 MW(th) and six commercial-size prototypes with electrical output ranging from 250 to 1200 MW(e) have been constructed and operated [58]. Sodium was selected as a preferred coolant due to its low vapor pressure, large heat capacity and thermal conductivity, high boiling point, minimal interactions with neutrons, and lack of damage from radiation [65].

6.4.2.2 General corrosion

The basic corrosion mechanism in liquid sodium is similar to that in molten salt, primarily driven by dissolution of alloying element into the sodium as well as corrosion reactions with impurities in the sodium, notably oxygen [61,65,75,76]. The primary elements leaving a metal in the sodium, and thus being responsible for corrosion weight loss, are Cr, Mn, Ni, and Si. Significant amounts of carbon can also be transferred from locations within a non-isothermal loop, but while this carbon transport can affect material properties (sometimes dramatically), it is only a small component of the corrosion weight loss.

The dissolution of metallic elements from steels in oxygen-gettered sodium is dependent on the solubility. The solubility of key metallic elements is shown in Fig. 6.21.

The solubility of Mn, Ni, and Mo is higher than Cr and Fe. Both Mn and Ni are austenite-stabilizing elements and their preferential dissolution will leave a ferritic phase on the surface of an austenitic steel [61]. Further dissolution then depends on diffusional processes that move ions through the ferrite layer. While the formation of the ferrite layer occurs early in the exposure time, interactions of cations with dissolved oxygen also play an important role in long-term corrosion.

For sodium-cooled reactors constructed to date, the concentration of oxygen in sodium is normally maintained at levels 2–3 ppm by the operation of a cold trap; however, for some maintenance activities, a level up to 10 ppm is

6.4 CORROSION IN MOLTEN SALT AND LIQUID METAL-COOLED REACTORS

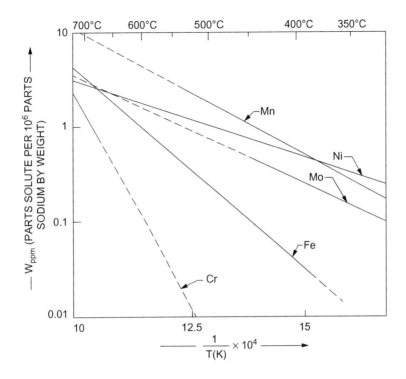

FIG. 6.21

Solubility of the alloying elements of stainless steel in liquid sodium [69].

From C. K. Mathews, Liquid Sodium-The Heat Transport Medium in Fast Breeder Reactors. Bull. Mat. Sci. 16 (6) (1993) 477–489.

authorized [58]. This control is critical, as noted in Ref. [58], an increase in the oxygen concentration in the primary circuit of the BN-350 reactor corresponds to an increase in 80 kg/year of additional corrosion products.

As oxygen concentration is increased, thermodynamic considerations indicate that dissolution is dependent on the relative free energy of formation of oxides in the sodium. For instance, as oxygen level increases, chromium will form a stable chromium oxide, $NaCrO_2$, rather than be reduced to metal. As shown in Fig. 6.22, in the temperature range of the sodium used in a typical fast reactor, only the formation of a chromium oxide is anticipated from thermodynamic considerations. Separately from the data in Fig. 6.22, Mathews reported the threshold oxygen concentration for the formation of $NaCrO_2$ at approximately 17 wppm at 500°C while the threshold oxygen concentration for the formation of Na_4MoO_5 is approximately 843 wppm at 427°C [76].

Thus, even with the range of uncertainty of the oxygen concentration at which Cr will form an oxide, Cr has little solubility as a metal and the transport of chromium from the metal surface to form an oxide in the sodium is a stronger driving force. Molybdenum, on the other hand, is unlikely to form an oxide at typical sodium reactor operating temperatures, and with limited solubility, is likely to stay in the steel. Nickel oxides are not thermodynamically stable so Ni loss from the steel surface is driven by the difference in the solubility of nickel in sodium and the amount of nickel in the alloy [65]. Iron does not form a thermodynamically favorable oxide yet corrosion data for pure iron indicates that the iron dissolution increases with the square of the oxygen concentration, indicating a kinetic effect in which oxygen promotes the local dissolution of iron into sodium [65].

6.4.2.3 Temperature gradient-driven corrosion

Because the temperature of the sodium changes throughout the system, there is transport of elements from hot to cold portions of the system, based on the difference in solubility as a function of temperature [77,78]. The selective leaching and deposition of metallic ions in contact with flowing sodium can lead to metal loss and associated

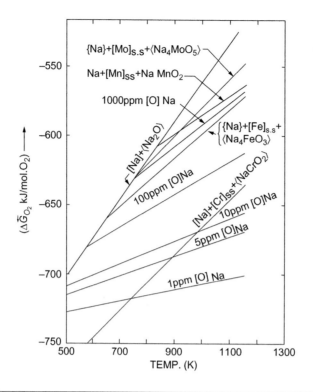

FIG. 6.22

Oxygen potentials for Na-(M)$_{SS}$-NaM$_x$O$_y$ and Na-Na-Na$_2$O systems [70].

From C. W. Mathews Pure & Appl. Chem., Vol. 67, No. 6, pp. 1011–1018, 1995. Printed in Great Britain. 1995 IUPAC.

reduction in load-carrying capability, transport of radioactivity, flow blockages, and carbon transport with the associated changes in properties [58]. As an example, transport of carbon and nitrogen from the reactor vessel (proposed to be constructed from 316 stainless steel), and associated decrements in strength was noted as an open question for General Electric as part of the Nuclear Regulatory Commission review of the proposed PRISM sodium-cooled reactor design [79].

Carbon can exist in sodium in both dissolved and undissolved forms [76]. Migration of carbon from one portion of the loop to another occurs due to differences in the activity of carbon, just as with the other species transported across a circuit. The carbon activity changes primarily due to temperature differences. In systems consisting solely of austenitic steels, carbon is transported from the hot leg to the cold leg. This decarburization and subsequent carburization can change the properties of the steel in the altered layers. In a system consisting of both austenitic and ferritic steels, the transfer of carbon is from the ferrite to the austenite [65].

6.4.2.4 Factors controlling corrosion and mass transfer in sodium systems

Operational variables can affect corrosion rates. The corrosion rate of steels increases with sodium coolant velocity between 0 and 3 m/s, but reaches a constant value for velocities at or above about 3 m/s [65]. The overall corrosion rate, as measured by metal loss per year, depends strongly on oxygen content and temperature, Fig. 6.23.

Although there is evidence that alloy composition can affect the net corrosion rates, from a practical standpoint, the Monju fast reactor adopted the following corrosion rate formula that only uses oxygen and temperature as critical variables and covers both austenitic and ferritic steels [75]:

$$\log_{10} R = 0:85 + 1.5\log_{10} C_o - 3:9 \times 10^3/(T+273) \tag{6.17}$$

6.4 CORROSION IN MOLTEN SALT AND LIQUID METAL-COOLED REACTORS

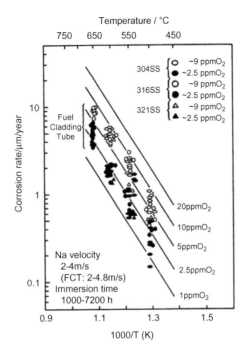

FIG. 6.23

Corrosion rate of austenitic steels in sodium [68].

From T. Furukawa, S. Kato, E. Yoshida, Compatibility of FBR materials with sodium, J. Nucl. Mater. 392 (2009) 249–254.

where R is the corrosion rate (mm/year), C_o the oxygen concentration (mass ppm) ($5 < C_o < 25$), T the temperature (°C) ($400 < T < 650$), and materials: SUS304, SUS316, SUS321, and 2.25Cr-1Mo. This equation takes the same form as that proposed in the pioneering work by Thorley and Tyzack for metal loss in austenitic flowing sodium [80]:

$$\log_{10} S \text{ (mils/year)} = 2:44 + 1.5 \log_{10}|O| - 75{,}312/(2.3RT) \tag{6.18}$$

where $[O]$ is the concentration of oxygen in sodium (ppm) (valid between 5 and 100 ppm); R the gas constant (JK^{-1} mol^{-1}); T the temperature (K); and 1 mil = 25 μm. Most flowing loop data is fit to an equation of this form.

Recent studies by Moisseytsev and colleagues [81] indicate that new higher strength steels, if deployed in sodium-cooled fast reactor systems, could reduce materials costs by approximately 45%. Confirmation of corrosion performance of these advanced alloys would be needed before taking advantage of them in newer reactor designs.

6.4.3 LEAD ALLOYS

Lead-cooled fast reactor systems are either Pb or Pb-Bi alloy-cooled reactors with a fast-neutron spectrum. Technology options include a long refueling interval transportable system ranging from 50 to 150 MWe, a modular system sized at 300–400 MWe, and a large monolithic plant at 1200 MWe [82]. In addition, European programs are actively studying the use of lead-alloy-based accelerator driven transmutation systems for the processing of radio-nuclide products.

Lead provides two advantages over sodium as a coolant. First, the boiling point of lead or LBE is much higher, providing greater safety margins and the ability to operate at higher temperature. Second, unlike sodium, lead alloys do not react exothermically with water and air. Lead has certain drawbacks relative to sodium, the most notable of which is greater corrosivity toward structural steels than that of sodium. In addition, because of the greater mass of lead alloys as compared to sodium, additional study of the seismic stability of the reactor plant is needed.

Lead and LBE each have advantages relative to each other. The LBE has a lower melting point than lead so using LBE simplifies the prevention of primary coolant freezing. On the other hand, LBE costs more than lead, primarily due to the cost of Bi, and has higher radioactivity levels associated with polonium production from bismuth.

6.4.3.1 General corrosion

Stable protective oxides are not formed in sodium, but are possible in lead and LBE. The corrosion response of steels in either lead or lead-bismuth coolant is then a balance between dissolution of the alloying elements and the formation of a protective oxide that prevents dissolution. The key to preventing metal loss due to corrosion is the control of oxygen concentration in the lead-alloy coolant such that sufficient oxygen exists to form and heal protective oxides, but is not so concentrated so as to form lead oxides that could plug the coolant lines.

The Ellingham diagrams shown in Fig. 6.24 provide the basis for understanding corrosion protection in lead and lead-alloy systems [84].

For the operating temperatures chosen, the oxygen concentration in the coolant must be high enough to form protective stable oxides without reaching the Pb-PbO transition that occurs at higher oxygen concentrations.

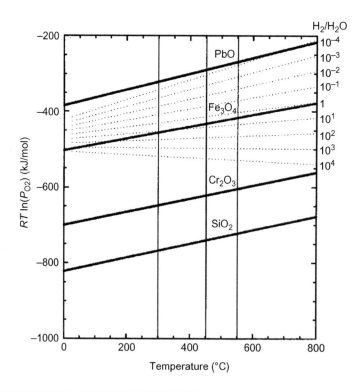

FIG. 6.24

Ellingham diagrams showing the standard Gibbs free energy of formation, ΔG_f°, of oxides of typical steel constituents in comparison to PbO: (A) for liquid Pb; (B) for LBE [83].

6.4 CORROSION IN MOLTEN SALT AND LIQUID METAL-COOLED REACTORS

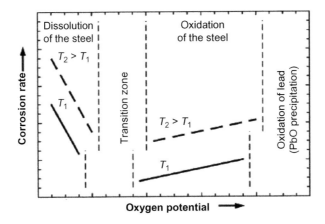

FIG. 6.25

Qualitative performance of steels as a function of oxygen potential in liquid lead alloys [77].

From C. Schroer, O. Wedemeyer, J. Konys, Aspects of minimizing steel corrosion in liquid lead-alloys by addition of oxygen. Nucl. Engin. Des. 241 (2011) 4913–4923.

A second consideration is to keep the oxygen concentration low to slow the oxidation rates of the steels. For austenitic steels, the appropriate oxygen concentration to form a protective oxide in liquid lead is about 10^{-6} mass % at temperatures near 550°C, while the addition of Bi in LBE sets a higher necessary oxygen concentration to prevent dissolution of metal [85]. Fig. 6.25 graphically outlines the balance between dissolution and oxidation as a function of temperature and oxygen concentration.

The goal is to operate in the regime of moderate oxidation. Detailed equations describing the appropriate free energies and solubilities needed to model corrosion and dissolution in lead and LBE alloys can be found in Refs. [84,86].

The oxides that form in lead and LBE are typical for steels in an oxidizing environment. For instance, in an Fe-9Cr-1Mo steel exposed to static LBE, the oxide has a duplex structure composed of an internal Fe-Cr spinel layer and an external magnetite layer [87]. A recent summary by Zhang and Li [88] provides an excellent overview of LBE properties, corrosion mechanisms, corrosion mitigation including oxygen control, and corrosion models. Zhang and Li summarized the knowledge of LBE corrosion as follows:

- At very low oxygen, both austenitic and ferritic-martensitic steels are subject to dissolution, even at low temperature.
- From 300°C to 470°C, with sufficient oxygen ($>10^{-4}$ ppm), protective oxide films can be formed on both austenitic and ferritic-martensitic steels.
- For temperatures above 550°C, austenitic stainless steels undergo heavy dissolution and ferritic-martensitic steels form a very thick and potentially unstable oxide. This thick oxide may be susceptible to erosion at high flow rates. Between 470°C and 550°C, the corrosion behavior in structural steels appears to make transition from oxidation to dissolution. Furukawa et al. determined that at these higher temperatures, the iron oxide form changed from magnetite to wustite which is less adherent and thus more prone to detachment [83].

Mass transport due to temperature differential or dissimilar metal transfer was previously described for molten salts and sodium. This mass transport would also be an issue in lead and lead alloys if it wasn't for the inhibition by formation of oxide layers.

6.4.3.2 Corrosion control techniques in Pb and Pb-Bi

A few studies have pointed toward methods for improving corrosion control in Pb and LBE. Briceno et al. [89] showed that for high oxygen concentration in the coolant, alloys with increased chromium concentration had better corrosion resistance as the chromium promoted the formation of a stable adherent spinel layer. Alternately, at low coolant oxygen concentration, lower bulk chromium (e.g., carbon steels) minimizes Cr dissolution. In addition, changes in oxygen concentration during a test can significantly alter material response. This may pose strict operational concerns on coolant chemistry.

Kurata and Futakawa [90] showed that the high silicon-containing alloy SX (Fe-18Cr-20Ni-5Si) formed a protective oxide at 550°C in tests where Japan Prime Candidate Alloy (JPCA), a material developed for the fusion energy programs, and 316 did not form stable oxides. The result is consistent with the Russian development of alloy EP 823 in that increased silicon allows for the formation of a stable underlying SiO_2 layer that prevents dissolution. Because of concerns on radiation performance and weldability, the silicon level in alloy SX is likely to be too large, but optimizing silicon content is likely to be a critical portion of corrosion control.

At higher temperatures ($\gtrsim 550°C$), corrosion control through the formation of a protective oxide layer is difficult to maintain so protective coatings may be required. Studies led by the Karlsruhe Institute of Technology in Germany have developed a process using intense electron beams to alloy a steel surface with aluminum. This process placed 5–10 μm thick coatings which then was tested on 316, 1.4970, and T91 steels exposed to Pb and Pb-Bi eutectic with considerable improvement in corrosion response [91,92]. As shown in Fig. 6.23, aluminum forms an oxide at very low oxygen concentration providing the basis for this approach.

6.4.3.3 Effect of Pb and Pb-Bi on mechanical properties

Recently, numerous investigations have been reinitiated to understand the effects of Pb or Pb-Bi eutectic on mechanical properties. While much work has been done to understand the effect of liquid sodium on mechanical properties [93], the understanding is less advanced for materials in lead and lead alloys. Ref. [94] summarizes the results from numerous European programs. Significantly:

- Tensile properties appear to be unaffected for both T91 and 316L steels when exposed to oxidizing LBE but does lose some ductility when exposed to an LBE under reducing conditions.
- Low cycle fatigue in 316L contacted with LBE shows only a weak damaging effect but T91 shows a decrease in low cycle fatigue resistance. If the T91 is brought to reducing conditions, the decrease in low cycle fatigue growth is greater, supporting the idea that reducing conditions appear to be detrimental to mechanical properties.
- Creep-rupture tests in flowing LBE at 550°C showed a marked acceleration of creep rate at stresses >180 MPa.

Overall, the mechanical properties of austenitic steels appear to be little affected but those of ferritic-martensitic steels can be affected and must be chosen carefully.

REFERENCES

[1] G.H. Koch, M.P.H. Brongers, N.G. Thompson, Y.P. Virmani, J.H. Payer, Corrosion Costs and Preventative Strategies in the United States, U.S.Federal Highway Administration Publication, 2002. No. FHWA-Rd-01-156.
[2] D.A. Jones, Principles and Prevention of Corrosion, Prentice-Hall, Upper Saddle River, NJ, 1996.
[3] T. Allen, J. Busby, M. Meyer, D. Petti, Materials challenges for nuclear systems, Mater. Today 13 (12) (2010) 14–23.
[4] J.C. Danko, Corrosion in the nuclear power industry, in: Metals Handbook, Ninth Edition, Volume 13 Corrosion, ASM International, Metals Park, OH, 1987, pp. 927–984.
[5] D. Feron, J.-M. Olive (Eds.), Corrosion Issues in Light Water Reactors—Stress Corrosion Cracking, CRC Press, Boca Raton, FL, 2007.
[6] G.S. Was, Y. Ashida, P. Andresen, Irradiation-assisted stress corrosion cracking, Corros. Rev. 29 (2011) 7–49.
[7] G.S. Was, P.L. Andresen, Stress corrosion cracking behavior of alloys in aggressive nuclear reactor Core environments, Corrosion 63 (1) (2007) 19–45.
[8] G.S. Was, J.T. Busby, P.L. Andresen, Effect of irradiation on stress corrosion cracking and corrosion in light water reactors, corrosion in the nuclear industry, in: Corrosion: Environments and Industries, vol. 13c, ASM International, Metals Park, OH, 2006, pp. 386–414. ASM Handbook.

[9] Steam Generator Progress Report, EPRI Steam Generator Database, Electric Power Research Institute, Palo Alto, CA, 2012 Fig. 2–11.
[10] W.J. Shack, T.F. Kassner, P.S. Maiya, J.Y. Park, W.E. Ruther, BWR pipe crack and weld overlay studies, Nucl. Engin. Des. 89 (2–3) (1985) 295–303.
[11] J.R. Strosnider Jr., W.H. Koo, J.A. Davis, M.C. Modes, Pipe Cracking in US BWRs: A Regulatory History, NUREG-1719, U.S. Nuclear Regulatory Commission, 2000.
[12] P.M. Scott, M.L. Calvar, 6th International Symposium on Environmental Degradation of Materials in Nuclear Power Systems—Water Reactors, TMS, Warrendale, PA, 1993, pp. 657–665.
[13] B.M. Capell, G.S. Was, Selective internal oxidation as a mechanism for intergranular stress corrosion cracking of Ni-Ce-Fe alloys, Metall. Mater. Trans. A 39 (2007) 1244.
[14] R. W. Staehle, J. A. Gorman, n.d. Proceedings of the 10th International Conference on Environmental degradation of materials in nuclear power systems: water reactors, NACE international, Houston, TX, Bonus paper.
[15] D.J. Paraventi, W.C. Mosher, Assessment of the interaction of variables in the intergranular stress corrosion crack growth rate behavior of alloys 600, 82 and 182, Proceedings of the 13th International Conference on Environmental Degradation of Materials in Nuclear Power Systems, Canadian Nuclear Society, Toronto, CA, 2007.
[16] R.C. Asher, D. Davies, T.B.A. Kirstein, The corrosion of some zirconium alloys under radiation in moist carbon dioxide-air mixtures, J. Nucl. Mater. 49 (1973/74) 189–196.
[17] D.H. Bradhurst, P.J. Shirvington, P.M. Heuer, The effects of radiation and oxygen on the aqueous oxidation of zirconium and its alloys at 290°C, J. Nucl. Mater. 46 (1973) 53–76.
[18] M.B. Lewis, J.D. Hunn, Investigations of ion radiation effects at metal/liquid interfaces, J. Nucl. Mater. 265 (1999) 325–330.
[19] B. Cox, Effects of irradiation on the oxidation of zirconium alloys in high temperature aqueous environments, J. Nucl. Mater. 28 (1968) 1–47.
[20] G.S. Was, P. Ampronrat, G. Gupta, S. Teysseyre, E.A. West, T.R. Allen, K. Sridharan, L. Tan, Y. Chen, X. Ren, C. Pister, Corrosion and stress corrosion cracking in supercritical water, J. Nucl. Mater. 371 (2007) 176–201.
[21] T. Allen, Y. Chen, X. Ren, K. Sridharan, L. Tan, G. Was, E. West, D. Guzonas, n.d. Materials performance in supercritical water, chapter 100, in Comprehensive Nuclear Materials R. Konings, T. Allen, R. Stoller, S. Yamanaka (Eds.), Elsevier, Netherlands, (in press).
[22] I.G. Wright, P.F. Tortorelli, M. Schütze, Oxide Growth and Exfoliation on Alloys Exposed to Steam, EPRI report No. 1013666. 2007.
[23] P. Kofstad, High Temperature Corrosion, Elsevier, London, 1988.
[24] T. R. Allen, L. Tan, Y. Chen, X. Ren, K. Sridharan, G. S. Was, G. Gupta, P. Ampornrat, Proc. Global 2005, Paper 419 (2005).
[25] S.S. Hwang, B.H. Lee, J.G. Kim, J. Jang, J. Nucl. Mater. 372 (177) (2008).
[26] A. Motta, A. Yilmazbayhan, M. Gomes da Silva, R.J. Comstock, G.S. Was, J.T. Busby, E. Gartner, Q. Peng, Y. Hwan Jeong, J. Yong Park, Zirconium alloys for supercritical water reactor applications: challenges and possibilities, J. Nucl. Mater. 371 (2007) 61–75.
[27] S. Teysseyre, G.S. Was, Stress corrosion cracking of austenitic alloys in supercritical water, Corrosion 62 (2006) 1100.
[28] R. Zhou, E.A. West, Z. Jiao, G.S. Was, Irradiation-assisted stress corrosion cracking of austenitic alloys in supercritical water, J. Nucl. Mater. 395 (2009) 11–22.
[29] S. Teysseyre, Z. Jiao, E. West, G.S. Was, Effect of irradiation on stress corrosion cracking in supercritical water, J. Nucl. Mater. 371 (2007) 107–117.
[30] S. Teysseyre, G.S. Was, T.R. Allen, J. Busby, P.J. King (Eds.), Stress corrosion cracking of neutron irradiated steel in supercritical water, 13th International Conference on Degradation of Materials in Nuclear Power Systems—Water Reactors, Canadian Nuclear Society, Toronto, 2007.
[31] W. Hoffelner, Materials for the very high temperature reactor: a versatile nuclear power station for combined cycle electricity and heat generation, Chimia 59 (12) (2005) 977–982.
[32] K.L. Murty, I. Charit, Structure materials for Gen IV nuclear reactors: challenge and opportunities, J. Nucl. Mater. 38 (2008) 189–195.
[33] G.O. Hayner, R.L. Bratton, R.E. Mizia, W.E. Windes, Next Generation Nuclear Plant Materials Research and Development Program Plan, Idaho Falls, Idaho National Laboratory, 2006.

[34] H. Burlet, J.M. Gentzbittel, C. Cabet, P. Lamagnère, M. Blat, D. Renaud, S. Dubiez-Le Goff, D. Pierron, Structural Materials for Innovative Nuclear Systems, OECD Publishing, Berlin, Germany, 2008.
[35] C. Cabet, A. Terlain, P. Lett, L. Guetaz, J.M. Gentzbittel, High temperature corrosion of structural materials under gas-cooled reactor helium, Mater. Corros. 57 (2) (2006) 147–153.
[36] F. Rouillard, C. Cabet, K. Wolski, M.J. Pijolat, Oxide-layer formation and stability on a nickel-base alloy in impure helium at high temperatures, Oxid. Met. 68 (2007) 133–148.
[37] H.M. Yun, P.J. Ennis, H. Nickel, H. Schuster, The effect of high temperature reactor primary circuit helium on the formation and propagation of surface cracks in alloy 800H and Inconel 617, J. Nucl. Mater. 125 (1984) 258–272.
[38] K. Natesan, P.S. Shankar, Uniaxial creep response of alloy 800H in impure helium and in low oxygen potential environments for nuclear reactor applications, J. Nucl. Mater. 394 (2009) 46–51.
[39] P.S. Shankar, K. Natesan, Effect of trace impurities in helium on the creep behavior of alloy 617 for very high temperature reactor application, J. Nucl. Mater. 366 (1–2) (2007) 28–36.
[40] Y. Kurata, T. Tanabe, I. Mutoh, H. Tsuji, J. Hiraga, M. Shindo, T. Suzuki, J. Nucl. Sci. Technol. 36 (12) (1999) 1160–1166.
[41] B. Huchtemann, The effect of alloy chemistry on creep behavior in a helium environment with low oxygen partial pressure, Mater. Sci. Eng. A121 (1989) 623.
[42] R.H. Cook, Creep properties of Inconel 617 in air and helium at 800 to 1000°C, Nucl. Technol. 66 (1984) 283–288.
[43] D. Kumar, R.R. Adharapurapu, T.M. Pollock, G.S. Was, High-temperature oxidation of alloy 617 in helium containing part-per-mission levels of CO and CO_2 as impurities, Metall. Trans. A. 42A (2011) 1245–1265.
[44] K. G. E. Brenner, Ternary diagrams for characterization of metallic corrosion in high temperature reactors, In Proc. Conf. on Gas-Cooled Reactors Today, Bristol, UK. 1982
[45] K.G.E. Brenner, L.W. Graham, The development and application of a unified corrosion model for high-temperature gas-cooled reactor systems, Nucl. Technol. 66 (1984) 40.
[46] C. Cabet, F. Rouillard, Corrosion of high temperature reactor structural materials, J. Eng. Gas Turbines Power 131 (2009).
[47] F. Rouillard, C. Cabet, K. Wolski, M.J. Pijolat, Oxidation of a chromia-forming nickel base alloy at high temperature in mixed diluted CO/H_2O atmosphere, Corros. Sci. 51 (2009) 752–760.
[48] H.J. Christ, U. Kuneke, K. Meyer, H.G. Sockel, Oxid. Met. 30 (1–2) (1988) 27–51.
[49] H.J. Christ, D. Scwanke, T. Uihlein, H.G. Sockel, Oxid. Met. 30 (1–2) (1988) 1.
[50] L.R. Liu, T. Jin, N.R. Zhao, Z.H. Wang, X.F. Sun, H.R. Guan, Z.Q. Hu, Effect of carbon addition on the creep properties in a Ni-based single crystal superalloy, Mater. Sci. Eng. A 385 (2004) 105–112.
[51] P.J. Ennis, K.P. Mohr, H. Schuster, Nucl. Technol. 66 (1984) 363.
[52] M.W. Rosenthal, P.R. Kasten, R.B. Briggs, Nucl. Technol. 8 (2) (1970) 111.
[53] D.F. Williams, Assessment of Candidate Molten Salt Coolants for the NGNP/NHI Heat-Transfer Loop, ORNL/TM-2006/69, 2006.
[54] C.W. Forsberg, P.F. Peterson, R.A. Kochendarfer, Design options for the advanced high-temperature reactor, International Congress on Advanced Nuclear Power Plants, Anaheim, California, June 8–15, 2008, American Nuclear Society, La Grange Park, Illinois, 2008.
[55] R.W. Moir, J.D. Lee, F.J. Fulton, F. Hiegel, W.S. Neef Jr., A.E. Sherwood, D.H. Berwald, R.H. Whitley, C.P. Wong, J.H. Devan, W.R. Grimes, S.K. Ghose (Eds.), Helium-Cooled Molten-Salt Fusion Breeder, 1984. UCID-20153.
[56] C.E. Till, Y.I. Chang, W.H. Hannum, The integral fast reactor-an overview, Prog. Nucl. Energy 31 (1/2) (1997) 3–11.
[57] S. Delpech, C. Cabet, C. Slim, G.S. Picard, Molten fluorides for nuclear applications, Mater. Today 13 (12) (2010) 34–41.
[58] IAEA-TECDOC-1569, Liquid Metal Cooled Reactors: Experience in Design and Operation, 2007.
[59] E.O. Adamov, White Book of Nuclear Power, N.A. Dollezhal Research Development Institute of Power Engineering, Moscow, Russia, 2001.
[60] W.D. Manly, Fundamentals of Liquid Metal Corrosion, ORNL-2055, 1958.
[61] J. Zhang, P. Hosemann, S. Maloy, Models of liquid metal corrosion, J. Nucl. Mater. 404 (2010) 82–96.
[62] J.R. Keiser, Compatibility Studies of Potential Molten-Salt Breeder Reactor Materials in Molten Fluoride Salts, ORNL/TM-5783, 1977.
[63] J.H. DeVan, R.B. Evans III, Corrosion Behavior of Reactor Materials in Fluoride Salt Mixtures, ORNL-TM-328, 1962.
[64] D. Olander, Fundamental Aspects of Nuclear Reactor Fuel Elements, U.S Dept of Energy, 1976.
[65] D. Olander, Redox condition in molten fluoride salts Definition and Control, J. Nucl. Mater. 300 (2002) 270–272.

[66] P. Calderoni, P. Sharpe, H. Nishimura, T. Terai, Control of molten salt corrosion of fusion structural materials by metallic beryllium, J. Nucl. Mater. 386–388 (2009) 1102–1106.
[67] L. Olson, K. Sridharan, M. Anderson, T. Allen, Nickel-plating for active metal dissolution resistance in molten fluoride salts, J. Nucl. Mater. 411 (2011) 51–59.
[68] G. Muralidharan, D.F. Wilson, L.R. Walker, M.L. Santella, D.E. Holcomb, Cladding Alloys for Fluoride Salt Compatibility Final Report, ORNL/TM-2010/319, 2011.
[69] A.R. Shankar, U.K. Mudali, R. Sole, H.S. Khatak, B. Raj, Plasma-sprayed yttria-stabilized zirconia coatings on type 316L stainless steel for pyrochemical reprocessing plant, J. Nucl. Mater. 372 (2008) 226–232.
[70] J.W. Koger, A.P. Litman, Mass Transfer between Hastelloy N and Haynes Alloy No. 25 in a Molten Fluoroborate Mixture, ORNL-TM-3488, 1971.
[71] J.W. Koger, Effect of FeF_2 Addition on Mass Transfer in a Hastelloy N-LiF-BeF_2-UF_4 Thermal Convection Loop System, ORNL-TM-4188, 1972.
[72] A. Misra, J. Wittenberger, Fluoride salts and container materials for thermal storage applications in temperature range 973 to 1400 K. NASA-Lewis Technical Memorandum 89913, 1987.
[73] J.R. Keiser, D.L. Manning, R.E. Clausing, J.R. Keiser, Corrosion resistance of some Nickel-Base, Alloys to Molten Fluoride Salts Containing UF4, and Tellurium, in: Molten Salts, The Electrochemical Society, New York, 1976, pp. 315–328. Probably better but less accessible, Status of Tellurium–Hastelloy N Studies in Molten Fluoride Salts, (1977).
[74] J. Weeks, H.S. Isaacs, Corrosion and deposition of steels and nickel-based alloys in liquid sodium, in: M.G. Fontana, R.W. Staehle (Eds.), Advances in Corrosion Science and Technology, vol. 3, 1973, p. 45.
[75] T. Furukawa, S. Kato, E. Yoshida, Compatibility of FBR materials with sodium, J. Nucl. Mater. 392 (2009) 249–254.
[76] C.K. Mathews, Liquid sodium—the heat transport medium in fast breeder reactors, Bull. Mater. Sci. 16 (6) (1993) 477–489.
[77] L.F. Epstein, Chem. Eng. Prog. Symp. Ser. 20 (53) (1957) 67.
[78] B. Raj, S. L. Mannan, P. R. Vasudeva Rao, M. D. Mathew, Development of fuels and structural materials for fast breeder reactors. Sadhana Vol. 27, Part 5, 2002, 527–558.
[79] NUREG 1368, 1994 Preapplication Safety Evaluation. Report for the Power Reactor Innovative Small Module (PRISM) Liquid-Metal Reactor.
[80] A.W. Thorley, C. Tyzack, Proceedings of Liquid Alkali, Metals, BNES, London, 1973. 257 p.
[81] A. Moisseytsev, Y. Tang, S. Majumdar, C. Grandy, K. Natesan, Impact from the adoption of advanced materials on a sodium fast reactor design, Nucl. Technol. 175 (2011).
[82] T.R. Allen, D.C. Crawford, Lead-cooled fast reactor systems and the fuels and materials challenges, Sci. Technol. Nucl. Install. (2007), 97486 https://doi.org/10.1155/2007/97486.
[83] T. Furukawa, G. Müller, G. Schumacher, A. Weisenburger, A. Heinzel, F. Zimmermann, K. Aoto, J. Nucl. Sci. Technol. 41 (3) (2004) 265.
[84] C. Schroer, J. Konys, Physical Chemistry of Corrosion and Oxygen Control in Liquid Lead and Lead-Bismuth Eutectic, FZKA Report, Forschungszentrum Karlsruhe, Germany, 2007. FZKA 7364.
[85] C. Schroer, O. Wedemeyer, J. Konys, Aspects of minimizing steel corrosion in liquid lead-alloys by addition of oxygen, Nucl. Engin. Des. 241 (2011) 4913–4923.
[86] H. Steiner, C. Schroer, Z. Voß, O. Wedemeyer, J. Konys, Modeling of oxidation of structural materials in LBE systems, J. Nucl. Mater. 374 (2008) 211–219.
[87] L. Martinelli, F. Balbaud-Célérier, A. Terlain, S. Delpech, G. Santarini, J. Favergeon, G. Moulin, M. Tabarant, G. Picard, Oxidation mechanism of a Fe–9Cr–1Mo steel by liquid Pb–bi eutectic alloy (part I), Corros. Sci. 50 (2008) 2523–2536.
[88] J. Zhang, N. Li, Review of studies on fundamental issues in LBE corrosion, J. Nucl. Mater. 373 (2008) 351–377. LA-UR-0869.
[89] D. Gómez Briceño, F. J. Martı́n Muñoz, L. Soler Crespo, F. Esteban, C. Torres, Behaviour of F82H mod. stainless steel in lead–bismuth under temperature gradient, J. Nucl. Mater. 296 (1–3) (2001) 265–272.
[90] Y. Kurata, M. Futakawa, J. Nucl. Mater. 325 (2004) 217.
[91] V. Engelko, G. Mueller, A. Rusanov, V. Markov, K. Tkachenko, A. Weisenburger, A. Kashtanov, A. Chikiryaka, A. Jianu, Surface modification/alloying using intense pulsed electron beam as a tool for improving the corrosion resistance of steels exposed to heavy liquid metals, J. Nucl. Mater. 415 (2011) 270–275.
[92] A. Weisenburger, C. Schroer, A. Jianu, A. Heinzel, J. Konys, H. Steiner, G. Müller, C. Fazio, A. Gessi, S. Babayan, A. Kobzova, L. Martinelli, K. Ginestar, F. Balbaud-Célérier, F.J. Martin-Muñoz, L. Soler Crespo, Long term

corrosion on T91 and AISI1 316L steel in flowing lead alloy and corrosion protection barrier development: experiments and models, J. Nucl. Mater. 415 (2011) 260–269.
[93] K. Natesan, M. Li, O.K. Chopra, S. Majumdar, J. Nucl. Mater. 392 (2) (2009) 243–248.
[94] D. Gorse, T. Auger, J.-B. Vogt, I. Serre, A. Weisenburger, A. Gessi, P. Agostini, C. Fazio, A. Hojna, F. Di Gabriele, J. Van Den Bosch, G. Coen, A. Almazouzi, M. Serrano, Influence of liquid lead and lead–bismuth eutectic on tensile, fatigue and creep properties of ferritic/martensitic and austenitic steels for transmutation systems, J. Nucl. Mater. 415 (2011) 284–292.

FURTHER READING

[1] C. Cabet, J. Jang, J. Konys, P.F. Tortorelli, Environmental degradation of materials in advanced reactors, in: Y. Geurin, G.S. Was, S.J. Zinkle (Eds.), Materials Challenges for Advanced Nuclear Energy Systems, MRS Bulletin, vol. 34, 2009, p. 35. No. 1.
[2] D. Kumar, G.S. Was, Existence of a critical temperature for corrosion of alloy 617 in He-CO-CO_2 environment, Materials Research Society Proceedings, Volume 1043E, Warrendale PA, 2008. 1043-T01–03.
[3] Y. Kurata, M. Futakawa, K. Kikuchi, S. Saito, T. Osugi, Corrosion studies in liquid Pb–Bi alloy at JAERI: R&D program and first experimental results, J. Nucl. Mater. 301 (1) (2002) 28–34.

ZIRCONIUM ALLOYS FOR LWR FUEL CLADDING AND CORE INTERNALS

Suresh Yagnik*, Anand Garde†

Senior Technical Executive, EPRI, Palo Alto, CA, United States*
Engineer Emeritus, Westinghouse, Columbia, SC, United States†

CHAPTER OUTLINE

- 7.1 Overview of Zr-Alloys 248
- 7.2 Fabrication and Microstructure 251
  - 7.2.1 General Comments 251
  - 7.2.2 Lattice Structure and Second-Phase Particles 252
  - 7.2.3 Basic Zr-Alloy Processing and Fabrication 253
  - 7.2.4 Anisotropy of Zr-Alloys 255
  - 7.2.5 Texture 255
- 7.3 Corrosion and Crud 257
  - 7.3.1 General Comments 257
  - 7.3.2 Corrosion of Zr-Alloys 258
  - 7.3.3 Crud Deposition on Fuel Rods 258
  - 7.3.4 PWR Coolant Chemistry 260
  - 7.3.5 BWR Coolant Chemistry 261
  - 7.3.6 Fuel Failures Due to Severe Corrosion and Crud Deposition 262
- 7.4 Hydriding and Mechanical Integrity 262
  - 7.4.1 General Comments 262
  - 7.4.2 Effect of Hydriding on Unirradiated Mechanical Properties 264
  - 7.4.3 Effect of Hydriding on Irradiated Mechanical Properties 266
  - 7.4.4 Effect of Hydrogen on Post-Accident Transient Mechanical Properties 272
- 7.5 Irradiation Effects 272
  - 7.5.1 General Comments 272
  - 7.5.2 Effect of Irradiation on Corrosion Resistance 274
  - 7.5.3 Irradiation Hardening and Embrittlement 275
  - 7.5.4 Irradiation Growth 275
  - 7.5.5 Irradiation Creep 276
- 7.6 Failure Mechanisms 278
  - 7.6.1 General Comments 278
  - 7.6.2 Debris Fretting 279
  - 7.6.3 Grid To Rod Fretting (GTRF) 280

7.6.4 Pellet-Cladding Mechanical Interaction (PCMI) 280
7.6.5 Pellet-Cladding Interaction-Stress Corrosion Cracking (PCI-SCC) 282
7.6.6 Less Common Failure Mechanisms 283
7.7 Summary/Conclusions 284
References 286

7.1 OVERVIEW OF Zr-ALLOYS

Zirconium alloys (Zr-alloys) are used as fuel cladding (FC) and core internals in light water reactors (LWRs) because of their low neutron absorption cross section, good-in-service corrosion resistance, adequate high-temperature mechanical strength, and dimensional stability under radiation. The preference of Zr-alloys in the LWR cladding application over stainless steels is for the same reasons as listed for FC for the nuclear ship N.S. Savannah [1]: to reduce the required fuel pellet enrichment and increase core life (fuel discharge burnup).

Primarily used in the nuclear industry and to a smaller extent in the chemical industry, current zirconium metal production worldwide exceeds 7000 tons/year. The main difference in Zr-alloys used in the nuclear and nonnuclear chemical applications is the low level of hafnium for the nuclear applications due to the high neutron cross section of hafnium. The first nuclear application of Zr-alloys in the western hemisphere was enhanced by the two early developments in Zr-alloy processing: application of the Kroll reduction process for the economic production of zirconium sponge [2] and effectiveness of tin addition as an alloying element to eliminate the degradation of aqueous corrosion resistance of zirconium due to the presence of the common impurity element nitrogen [3] and to improve mechanical strength. Further improvements in cladding performance have been implemented through new alloys containing niobium in pressurized water reactor (PWR) cladding and by increasing Fe concentrations in boiling water reactor (BWR) cladding.

FC is essentially a thin, long, hollow seamless tube of Zr-alloy. The cladding tube is filled with <5 wt% enriched (in U^{235}) uranium dioxide fuel pellets and an inert gas and sealed by welding Zr-alloy plugs at both ends. This combined metal and ceramic component is referred to as a fuel rod. As shown in Figs. 7.1 and 7.2, a large number of fuel rods are held together by grid spacers. The geometrical arrangements of different fuel assembly components and dimensions of different components vary with the type of nuclear reactor and the fuel supplier/designer. The grid spacers are typically made of Zr-alloy for PWR fuel and alloy 718 for BWR fuel. The fuel assemblies also contain Zr-alloy guide tubes (GTs) and some contain instrument tubes. The BWR fuel assembly is encased in a channel box, also made of Zr-alloy.

Several such fuel assemblies are held in place in the reactor core surrounded by flowing high-temperature, high-pressure coolant water (in PWR) and water and steam (in BWR) to harness the energy generated by the nuclear fission process. The cladding tube for the nuclear fuel rod is an important primary structural component of LWR in that it serves as the first engineered barrier against release of fission products into coolant water during irradiation and into the surroundings in case of an accident during post-irradiation handling and transport of fuel assemblies.

Apart from the square cross-sectional fuel assemblies used in PWRs and BWRs described above, other fuel assembly cross sections are also used. In Canadian (CANDU, Canadian deuterium uranium) reactors, circular cross-sectional short assemblies (fuel bundles) containing natural enrichment fuel pellets are used where online fuel assembly reloading is achieved without shutting the reactor power for refueling outages. In the Russian (VVER, Russian design PWR) reactors, the fuel assembly cross section is hexagonal. Although Zr-alloys are used in LWRs for several non-cladding applications, the current chapter discusses mainly cladding applications as cladding is the critical component, being the first barrier to the release of radioactivity.

Zircaloys were developed under the US naval propulsion program from the early 1950s and subsequently adopted into commercial LWRs [4]. In PWRs, the cladding has been Zircaloy-4 (containing Zr-Sn-Fe-Cr) until the 1990s. Although still in use in some PWRs, Zircaloy-4 is being replaced as the fuel discharge burnups have

7.1 OVERVIEW OF Zr-ALLOYS

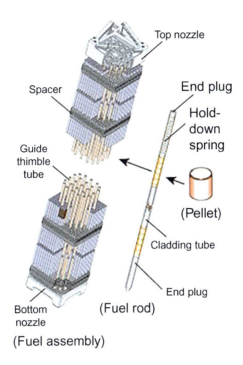

FIG. 7.1

Schematic representation of a PWR fuel assembly (Mitsubishi Nuclear Fuel Co. Ltd.) showing guide thimbles for control rods, a fuel rod, and a fuel pellet.

Source: Internet.

increased and PWR power ratings are increased. The advanced PWR cladding alloys are Nb-containing alloys[1]: ZIRLO™, Optimized ZIRLO, and M5™. In Russia, Zr-1%Nb alloy has been used from the beginning due to a different chemistry used for coolant pH control (potassium addition) compared to the coolant chemistry used in western PWRs (lithium addition). In Canadian heavy water reactor designs, Zircaloy-4 is used as cladding and the pressure tubes are made from Zr-2.5%Nb. In BWRs, the cladding has essentially remained as Zircaloy-2 (containing Zr-Sn-Fe-Cr-Ni). The advanced BWR cladding alloys currently being implemented are Zircaloy-2 with higher levels of iron.

With the expansion of the commercial nuclear power industry in the late 20th century, it was necessary to improve the in-reactor performance of first-generation Zr-alloys for longer in-reactor exposure under more demanding coolant chemistry conditions (higher temperatures due to power uprates, higher pH to control crud[2] deposition on the fuel, new additions to coolant for different reasons such as reduction of cracking of ferrous materials used in LWRs in the noncore applications and reduction of exposure dose for plant servicing personnel, etc.). Higher corrosion resistance and lower irradiation growth were the initial drivers. As a result, the cladding alloy development work in the 1980s culminated in commercialization of two new Nb-containing alloys, ZIRLO and M5. Optimized ZIRLO is a further improvement of ZIRLO [5]. M5 is also being further optimized by minor composition variations such as addition of tin and increase of iron levels [6]. Nominal compositions of several commercial cladding alloys are listed in Table 7.1 [7–9].

---

[1]The trademark symbols for ZIRLO and M5 are not repeated subsequently.
[2]The term 'crud' originated from *C*halk *R*iver *U*nidentified *D*eposits.

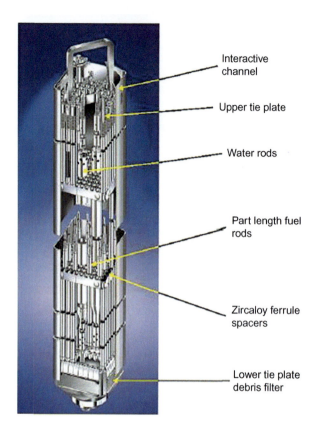

FIG. 7.2

Schematic representation of a BWR fuel assembly (Global Nuclear Fuel).

From http://fukushima.ans.org/report/Fukushima_report.pdf. Copyright 2012 by the American Nuclear Society, LaGrange Park, Illinois.

Table 7.1 Cladding Alloy Compositions (in wt% Unless Specified in Weight Parts per Million, wppm)								
Alloy	Sn	Fe	Cr	Nb	Ni	O	C	Other
Zircaloy-2 (ASTM B811)	1.2–1.7	0.07–0.2	0.05–0.15	–	0.03–0.08	0.09–0.16	0.027 Maximum	
Zircaloy-4 (ASTM B811)	1.2–1.7	0.18–0.24	0.07–0.13	–	–	0.09–0.16	0.027 Maximum	
ZIRLO[7] (Nominal)	1	0.1	–	1	–	1250 wppm		
M5[8] (Nominal)	–	150–600 wppm	–	1	–	900–1800 wppm	25–120 wppm	S: 0–35 wppm
E110 (Nominal)	–	0.006–0.012	–	0.95–1.05	30–60 wppm	500–700 wppm	50–100 wppm	
Optimized ZIRLO[9]	0.67	0.1	–	1	–	1250 wppm		

The cladding alloy development work in general involves melting of small quantities of new alloys with different composition and/or heat treatment steps. These developmental alloys are evaluated for corrosion resistance in autoclave tests; alloys with improved corrosion resistance are fabricated into seamless tubes by pilgering and irradiated in test reactors to evaluate the impact of neutron irradiation damage on the microstructure and the magnitude of irradiation growth. Cladding mechanical properties in as-fabricated condition and changing with irradiation are also evaluated. Cladding alloys with adequate irradiated microstructure stability are further nondestructively tested as fuel rods (loaded with UO_2 pellets) prior to their irradiation in test/commercial reactors. Some of the factors evaluated in such trials are the pellet/cladding mechanical and chemical interactions. Additional fuel performance parameters are measured in poolside and hot-cell examinations to develop material performance models that are used to secure licensing approval from the regulatory bodies prior to the commercial use of the cladding alloy for a full batch reload of nuclear fuel.

Historically, two other developments in the zirconium cladding alloys are worth mentioning: duplex cladding for PWR fuel and liner cladding for BWRs fuel. Duplex cladding refers to a composite tube where the outer thin layer of softer but more corrosion-resistant alloy is metallurgically bonded to the major tube cross section made of higher strength Zircaloy-4. Thus, the duplex cladding has a thin outer layer of corrosion-resistant alloy to improve the coolant side corrosion resistance of the composite cladding while the major thicker interior portion of the cladding tube wall is made of Zircaloy-4 to retain the high strength and mechanical properties for the composite tube [10]. The duplex PWR cladding concept was conceived in an Electric Power Research Institute (EPRI)-led international collaborative program and was first commercially developed by the German fuel fabricator KWU (Kraftwerk union) (which is now a part of the French company Framatome). Duplex cladding has been widely used in Germany, Spain, and Swiss PWRs. The use of duplex cladding is, however, declining due to several factors, including high costs of coextrusion of tubes as well as recycling of rejected tubing during fabrication.

The liner BWR cladding originally started as a thin soft layer of practically pure zirconium on the cladding inside wall [11], while retaining the outer major fraction of tube wall still as Zircaloy-2 cladding. (A similar concept was tried by the Atomic Energy of Canada (AECL) for the Zircaloy-4 cladding used in CANDU reactors by coating graphite on the inside surface of the tube.) This soft inside layer mitigates FC failures due to pellet-cladding interaction (PCI; discussed later). However, as will be discussed later, the zirconium liner fuel had poor secondary degradation resistance due to its poor corrosion resistance and higher hydrogen pickup by the liner after primary failure [12]. The current generation liner materials are dilute alloys of zirconium, rather than pure zirconium, with less than a half percent of alloying element addition, usually a combination of tin, iron, and oxygen.

7.2 FABRICATION AND MICROSTRUCTURE
7.2.1 GENERAL COMMENTS

Zirconium (Zr) is a refractory metal that is found in zircon (zirconium silicate) sands. Owing to the high melting point and high chemical affinity for oxygen and nitrogen, the conventional melting of the metal for fabrication is not economically viable. The ore is processed in the western hemisphere by the Kroll Process to yield zirconium sponge, which is subsequently alloyed with appropriate alloying elements and arc melted at least twice in vacuum. This results in Zr-alloy ingots (typically ~76 cm in diameter and weighing several tons). Each ingot subsequently goes through a number of metal working steps, such as forging, extrusion, and pilgering and heat treatment steps to eventually result in several thousands of Zr-alloy seamless cladding tubes (typically 4 m long, 1 cm in diameter; refer to Figs. 7.1 and 7.2). Pilgering is a heavy wall reduction tube-rolling process using grooved dies and a mandrel supporting the tube interior. The mandrel shape could be either cylindrical or tapered depending on the pilger process vendor. The outer surface of the cladding tube is also given a certain finish (acid etch or belt polish) followed by a final cleaning. The tube inner surface is treated either with acid etch and/or grit blasting. In the past Russian manufacture, the starting zirconium metal was electrolytic powder instead of sponge made by the Kroll process. The outer surface of the as-fabricated Russian cladding has a thin oxide layer to improve the wear resistance.

For Zr-alloy strip fabrication, hot and cold rolling steps are used. The recent trend is to use continuous annealing between rolling steps instead of batch annealing to improve the uniformity of properties of the strip. In batch processing, several separate pieces of the intermediate or final products are placed in an annealing furnace held at a controlled set temperature. This results in different regions within the batch experiencing some variations in time at temperature, which in turn results in nonuniformity of material properties such as corrosion resistance, ductility, and strength. The continuous annealing process rectifies this problem by passing every part of the charge through a small furnace ensuring that time at high temperature is constant for all portions of the annealing charge. This results in a fine uniform grain structure over the entire length of the strip material. The fine grain structure increases strip ductility, thereby reducing cracking at the highly deformed regions of grid strips generated during grid stamping. Ensuring grid integrity this way reduces grid-to-rod fretting (GTRF) fuel leaking failures. During the fabrication of all Zr-alloy components, a beta-quenching step (i.e., heating to the high temperature (>1073 K) into beta-phase followed by quenching in water) is used in the later stages of the fabrication process to homogenize the chemical composition of the alloy and control the microstructure of the finished component (mainly the second-phase particle (SPP), size composition, and size distribution). SPP size control is considered important to achieve high corrosion resistance [13] of Zircaloys and limit irradiation growth [14–16]. The final heat treatment can result in different as-fabricated microstructures (stress-relief annealed, partially recrystallized, or fully recrystallized) that also influence the mechanical properties, irradiation growth, and irradiation creep response of Zr-alloy components.

7.2.2 LATTICE STRUCTURE AND SECOND-PHASE PARTICLES

Zr has a hexagonal close packed (HCP) lattice structure (see Fig. 7.3). The addition of tin improves the strength and aqueous corrosion resistance of Zr with nitrogen impurity. In the early years, about 1.5% tin addition was used for Zircaloys. In the 1980s, with improvements in sponge purity, the nitrogen impurity levels have decreased and lowering the tin levels has improved the corrosion resistance of Zr-alloys, including Zircaloys [17]. Similarly, the reduction of tin level of ZIRLO was implemented to achieve the better corrosion resistance of Optimized ZIRLO. While tin and oxygen remain in solid solution within the Zr metal matrix, the small additions of Fe, Cr, Ni, and Nb

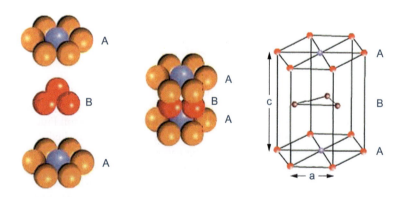

FIG. 7.3

Hexagonal closed-pack lattice structure of Zr in which atomic planes A,B,A show the sequence of successive closed-packed planes of Zr atoms.

From https://math.stackexchange.com/questions/2204090/height-of-hcp-unit-cell.

FIG. 7.4

SEM Micrograph of recrystallized Zircaloy showing SSPs [18].

Courtesy: Sandvik Special Metals (SSM), USA.

precipitate out as intermetallic compounds that appear as SPPs under microscopic examinations (see Fig. 7.4 [18]). These particles in Zircaloys are primarily Zr (Fe Cr)$_2$ or Zr$_2$(Fe, Ni) phases. Zircaloy-2 has both types of SPPs while Zircaloy-4-which contains no Ni (refer to Table 7.1)-has only the Fe-Cr type SPPs. The Zr-Nb alloys have beta-Nb and Zr-Nb-Fe SPPs.

As noted before, the SPP size and size distribution directly impact the corrosion and hydrogen pickup. They are controlled differently during manufacture of Zircaloy-2 cladding for BWRs (with oxidizing liquid and a vapor coolant environment) and Zircaloy-4 for PWRs (in a single-phase coolant with slight hydrogen overpressure resulting in reducing the corrosion environment). Control of SPPs in Zr-Nb alloys uses a totally different approach from that for Zircaloys. The beta-Zr-phase formation is avoided during fabrication of all Zr-alloys (by controlling temperatures used in the fabrication heat treatment steps) as it has a detrimental impact on the alloy corrosion resistance. The eutectoid temperature for Zr-Nb alloys is lower than that for the Zr-Fe, Zr-Cr, and Zr-Ni alloys.

7.2.3 BASIC Zr-ALLOY PROCESSING AND FABRICATION

The as-fabricated cladding tube microstructure depends on the amount of cold work introduced in the last fabrication-deformation step and the time/temperature of the final heat treatment of the tube (e.g., see Ref. [19]). A lower temperature final annealing at ~750K results in a stress-relief-annealed (SRA) elongated grain microstructure as used for the PWR cladding made of Zircaloy-4. A high-temperature final annealing at ~850K results in an equiaxed recrystallized annealed (RXA) grain microstructure as used for BWR fuel (Zircaloy-2). PWR FC (M5) is also used in an as-fabricated RXA microstructure. An intermediate temperature final annealing results in a partially recrystallized annealed (p-RXA) microstructure containing both equiaxed and elongated grains (as used in some PWR cladding). The p-RXA microstructure has been used for Zircaloy-4 by some fuel suppliers and is currently used for Optimized ZIRLO cladding [5].

The microstructure of Nb-containing Zr-alloys varies with the alloy composition, fabrication vendor, and the specific component application. ZIRLO in grid spacers applications and M5 in both cladding and grid spacer applications have fully RXA α-Zr (low-temperature alpha phase of zirconium or its alloy) grains dispersed with β-Nb particles. ZIRLO for cladding applications has an SRA microstructure. Aside from Nb-containing Zr-alloys used in PWRs, the cladding for Russian designed VVER and RBMK (Russian design BWR) water reactors also contains ~1 wt% Nb and pressure tubes in heavy water reactors contain ~2.5% Nb.

The SRA and RXA Zircaloy microstructures, are shown respectively in Fig. 7.5 A and 7.5B. SRA cladding exhibits distorted elongated grains whereas the RXA cladding has relatively uniform equiaxed grains. Similarly, the Zr-Nb PWR cladding types also have distinct microstructural differences depending on the final anneal temperature.

The RXA equiaxed grain structure has lower unirradiated strength, lower initial irradiation growth strain, and lower irradiation creep strain. SRA elongated grain structure results in higher strength, higher irradiation growth strain, and higher irradiation creep strain.

A micrograph of SRA ZIRLO would be very similar to that of an SRA Zircaloy-4 microstructure shown in Fig. 7.5A, although the SPP composition and size distribution in two alloys would be different.

Since numerous heat treatments of different durations are involved from the ingot stage to the final Zircaloy tubing (which also depends on the manufacturer), the final material condition is often represented by a single parameter known as the cumulative A-parameter. The A-parameter concept is developed only for Zircaloys and does not apply to Zr-Nb alloys. This allows a broad comparison of the processing history of Zircaloy products. The parameter is derived by a summation of all fabrication steps where 'i' represents each individual step [13].

$$A = \sum t_i \exp(-Q/RT_i)$$

where A is the cumulative annealing parameter
t_i is the time in hours at temperature T_i
Q is an activation energy
R is the gas constant
T_i is the temperature in Kelvin

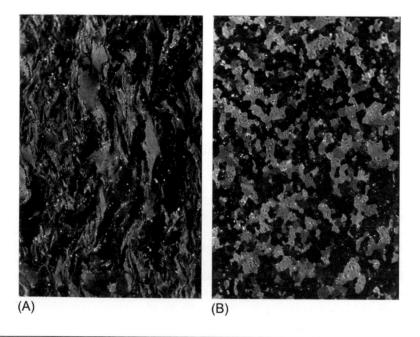

(A) (B)

FIG. 7.5

(A) Zircaloy-4 SRA Microstructure in Polarized Light; approximate magnification ×700 [18]. (B) Zircaloy-4 RXA Microstructure in Polarized Light; approximate magnification ×700 [18].

Courtesy: Sandvik Special Metals (SSM), USA.

Typical values of A range from 10^{-19} to 10^{-17} h with a Q/R value of 40,000 K. Since corrosion performance of the Zircaloy depends in part on heat treatment (e.g., through the average SPP size in alloy microstructure), the cumulative A-parameter was thought to be a single convenient representation of the effects of heat treatments on corrosion. However, the correlation between the A-parameter and corrosion is often not very rigorous and predictable due to the fact that the corrosion phenomenon of Zr-alloys also depends on several other parameters, including coolant chemistry.

7.2.4 ANISOTROPY OF Zr-ALLOYS

Most of the material properties of Zr-alloys are different in different crystallographic directions, that is, the alloys are anisotropic. The reason for the anisotropy is the nonsymmetric HCP crystal structure of zirconium and the marked crystallographic texture [20]. During fabrication processing, the intermediate components deform easily in certain directions and not so in other directions. As a result, the as-fabricated components of Zr-alloys (tubes, strips, rods, etc.) have a preferred orientation of microstructural grains in certain directions described as texture. The texture of Zr-alloys is commonly described by Kearns factors as discussed in the following section.

7.2.5 TEXTURE

All Zr-alloy components are polycrystalline, that is, the component microstructure consists of many individual crystals or grains. The orientation of grains of as-fabricated zirconium components is not random; it develops a preferred orientation during the fabrication processes used to achieve the component shape. Different deformation modes observed in unirradiated alpha-zirconium are listed in Table 7.2 [21]. The deformation modes of HCP metals depend on the c/a ratio for the specific metal where 'c' and 'a' are metal-specific dimensions of the HCP lattice in Fig. 7.3 either normal to the basal plane or within the basal plane.

For a-axis tension or compression, $<a>$ slip on prism plane is observed at low temperature, while $<c>$ slip is observed at high temperatures. For c-axis tension or compression, different twinning modes are observed. The nonrandom preferred texture of zirconium components affects most of the properties of these components including irradiation growth, irradiation creep, mechanical properties, hydride orientation, and corrosion resistance.

Table 7.2 Observed Deformation Modes in Unirradiated Zirconium

	Observed Deformation Modes in α-Zr.		
System	Active Temperature Range	Stress State	Comments
$\{10\bar{1}0\} \langle 11\bar{2}0 \rangle$ slip	<862°C	a-Axis tension or compression	Common, prism slip
$\{0002\} \langle 11\bar{2}0 \rangle$ slip	>500°C	a-Axis tension or compression	Common, basal slip
$\{10\bar{1}1\} \langle 11\bar{2}0 \rangle$ slip	>500°C	a- or c-Axis tension or compression	High stresses required, pyramidal slip
$\langle c+a \rangle$ slip		c-Axis tension or compression	Planes and directions unidentified, pyramidal slip
$\{10\bar{1}2\} \langle \bar{1}011 \rangle$ twin	<862°C	c-Axis tension	Common
$\{11\bar{2}1\} \langle \bar{1}\bar{1}26 \rangle$ twin	<862°C	c-Axis tension	Less frequent
$\{11\bar{2}2\} \langle \bar{1}\bar{1}23 \rangle$ twin	<862°C	c-Axis compression	Common
$\{10\bar{1}1\} \langle \bar{1}012 \rangle$ twin	>400°C	c-Axis compression	Predominant at high temperatures

Source: D.G. Franklin, G.E. Lucas, A.L. Bement, Creep of Zr-alloys in Nuclear Reactors, ASTM STP 815, 1983, p. 229.

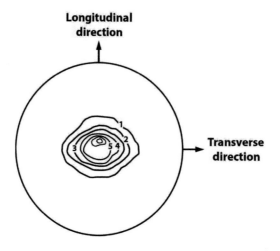

FIG. 7.6

Direct Pole figure for rolled zirconium plate from Ref. [22].

The texture of Zr-alloys can be represented by stereographic projection. The texture is presented either as a direct pole figure or an inverse pole figure. In a direct pole figure, a stereographic projection is presented with respect to some characteristic direction in a polycrystalline specimen on which densities of poles of a particular type are mapped as a function of their orientation with respect to a specific direction. An (0002) pole figure is the most common type generated for a zirconium specimen. An example is given below in Fig. 7.6 for a rolled plate [22].

The reference direction for the pole figure above is normal to the rolling plate. The line marked '3' in this figure represents the locus of points for which the measured basal pole density is three times the density found in a non-textured (random orientation) specimen. A qualitative interpretation of the preceding figure is that the majority of the grains in the specimen are oriented with the c-axis of their respective unit cells lying along a direction close to the normal direction of the plate. In an inverse pole figure, a stereographic projection with relative intensities for all diffracting planes is presented. This is accomplished by plotting contours on a standard (0001) projection. Owing to the symmetry, only one-twelfth of the standard projection needs to be used.

Texture characterization can be measured directly by X-ray diffraction. It is the most widely used method. However, texture characterization can also be achieved indirectly by the following different techniques: optical examination under polarized light, and ultrasonic analysis. Typical textures developed in different geometric-shaped components are shown in Fig. 7.7 excerpted from Ref. [23]. When a cold worked Zr-alloy is annealed, during recrystallization, a 30-degree rotation about the basal pole of the lattice can occur, resulting in an annealed texture.

Many important properties associated with the in-reactor performance of Zr-alloys are related to the texture. These include yield strength, ductility, hydride orientation, creep strength, growth strain, pellet-cladding interaction resistance, thermal expansion, etc. A simplified way to correlate the texture dependence of properties of Zr-alloy is via the Kearns texture parameters [24]. The Kearns parameter f for a specific direction i (f_i) is the effective fraction of basal poles oriented in that specific direction i. The Kearns parameter values for the three principal directions for basal poles are the resolved fraction of basal poles aligned with each of the three directions. The sum of the three Kearns factors in the three principal directions ideally has to be unity [25,26]. For example, the Kearns factors for basal poles for the Zircaloy-2 cladding tube in the tube axial, transverse, and radial directions are typically about 0.05, 0.35, and 0.60.

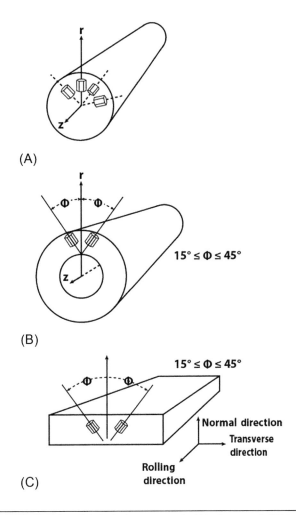

FIG. 7.7

Basal pole orientation for typical zirconium fabricated components (A) wire or rod, (B) tube, and (C) plate or sheet.

After D.G. Franklin, G.E. Lucas, A.L. Bement, Creep of Zr-alloys in Nuclear Reactors, ASTM STP 815 American Society for Testing and Materials, 1983, p. 233.

7.3 CORROSION AND CRUD
7.3.1 GENERAL COMMENTS

Waterside corrosion of cladding during service has received considerable attention not only because it directly relates to cladding thinning—and therefore a reduction of the load-bearing cross section of the tube wall—but also because increased corrosion results in increased hydrogen pickup by the cladding, which is detrimental to its ductility (see Section 7.4). There is a design limit on the maximum oxide layer thickness of $\sim 100\,\mu m$ from which *an estimated* average hydrogen concentration (based on an assumed hydrogen pickup fraction) in the cladding can be calculated.

7.3.2 CORROSION OF Zr-ALLOYS

The in-reactor corrosion rate of Zr-alloys is generally higher than ex-reactor corrosion, measured in either steam or water autoclave tests. The ex-reactor autoclave corrosion resistance is used as an initial screening criterion in the alloy development process. However, in-reactor corrosion performance evaluation by lead fuel irradiation is an essential confirmatory step prior to alloy selection for commercial use. The factors responsible for this difference include the coolant chemistry, radiation damage to both the metal and oxide phases, including SPP evolution, radiolysis of coolant water, and the presence of heat flux across the Zr-alloy component. Each of these factors is briefly discussed below. Gamma radiation also plays an important role in the corrosion behavior (see, e.g., a recent paper by Rishel and Kammenzind [27]).

In PWRs, the most common type of corrosion is known as uniform corrosion. Here, the oxide front moves into the substrate cladding, thus oxidizing Zr metal into ZrO_2. The oxide builds up as an adherent layer of fairly uniform thickness on the cladding outer wall, as shown in Fig. 7.8 [28]. The rate of corrosion depends on the temperature at the oxide/metal interface (typically in the range of 598–628 K). In BWRs, with about half the system pressure of the PWRs, the cladding oxide/metal interface temperature is lower in the range of ~553–568 K. Although BWR coolant is more oxidizing and boiling is present in the coolant, the uniform corrosion kinetics is much slower in BWR cladding than in PWR. However, a localized corrosion called nodular corrosion can affect BWR cladding due to the higher oxygen potential of the coolant, where local lenticular oxide nodules develop on the cladding outer wall. These nodules nucleate in early stages of cladding exposure in a BWR coolant, and subsequently each nodule simply grows in size. Fig. 7.8 [28] and Fig. 7.9 [29] show, respectively, typical examples of uniform and nodular corrosion.

7.3.3 CRUD DEPOSITION ON FUEL RODS

Crud consists of essentially oxides of Fe and/or Ni generated by the corrosion and erosion of the primary circuit non-fuel component surfaces (e.g., alloy 600 steam generators and stainless steel piping in PWRs) that get deposited on hot FC surfaces in the core. Corrosion that occurs on the surfaces of the balance of plant provides the source of the metallic species in the crud, which are predominately nickel ferrite spinels for PWRs and ferrous and ferric oxides in BWRs.

The coolant in PWR represents a reducing environment due to hydrogen overpressure, while that in BWR it is oxidizing. In addition, in PWRs the coolant contains boron, added as boric acid as a chemical shim for reactivity control. To neutralize the acid and to maintain a desired level of coolant pH at operating temperature, lithium hydroxide is added (see Table 7.3 based on Ref. [30]). The pH is maintained to limit the out of core primary circuit corrosion and thereby reduce the crud buildup.

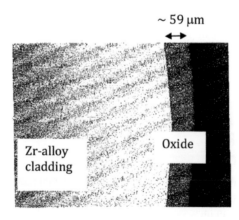

FIG. 7.8

Cross-sectional view of uniform oxide layer ~59 μm (thick) on Zircaloy-4 fuel cladding from a PWR [28].

Reprinted with permission from L. F. P Van Swam, and S. H. Shann, "The Corrosion of Zircaloy-4 Fuel Cladding in Pressurized Water Reactors," Zirconium in the Nuclear Industry: Ninth International Symposium, ASTM STP-1132, C. M. Eucken and A. M. Garde, Eds., American Society for Testing and Materials, Philadelphia 1991, pp. 758–781., Copyright ASTM International.

7.3 CORROSION AND CRUD

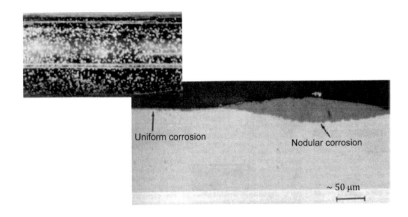

FIG. 7.9

Typical appearance of nodular corrosion in visual inspection of cladding outer surface (top left) and metallographic examination of cladding cross-section (bottom right).

Figures courtesy of Ron Adamson.

Table 7.3 Typical Reactor Environment in PWR and BWR to which Zr-Alloy Cladding is Exposed[a]

	PWR	BWR
Pressure (MPa)	~15	~7
Inlet coolant temp (K)	553–563	545–551
Outlet coolant temp (K)	583–603	553–573
Neutron flux (n cm^{-2} s^{-1}) $E > 1$ MeV	6–9 × 10^{13}	4–7 × 10^{13}
Coolant chemistry[b]		
O_2 (ppb)	<0.05	~200
H_2 (wppm)	2–5	~0.03[c]
pH	6.9 to 7.4	~7
B wppm (as H_3BO_3)	0–2200	–
Li wppm (as LiOH)	Up to 3.5	–

[a] Excerpted from IAEA TECDOC-996; International Atomic Energy Agency, Vienna, Austria Jan 1998 (Ref. 30).
[b] In longer cycle high fuel duty reactors addition of Zn in PWRs and BWRs and addition of noble metals to the coolant in BWRs is also implemented.
[c] Some BWRs use hydrogen water chemistry in which case the hydrogen level in the lower axial elevations of the fuel assembly will be higher.

Local boiling at the cladding surfaces also enhances crud deposition [31,32]. The crud layer generally builds on top of the oxide layer [33], as schematically shown in Fig. 7.10A. The crud deposits can be a mixture of fluffy and loose or tenacious and adherent layers. In the latter category, the crud layer can be an additional impediment to heat transfer [34].

The mechanism of crud deposition involves corrosion products from the balance of plant materials being transported to the core where local heat transfer conditions lead to such deposition on the fuel rod surface. The crud buildup on the cladding surface locally raises the surface temperature of the cladding. It may also lead to the concentration of aggressive species like lithium from the coolant on and within the surface layers of the fuel rod that

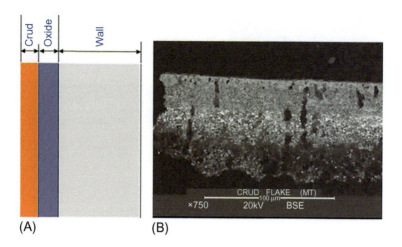

FIG. 7.10
(A) Schematic representation of a crud layer on top of cladding and oxide layers (B) Top-down view of CRUD, showing the boiling chimneys and the porous nature of the corrosion deposits [33].

increase local corrosion rates due the formation of boiling chimneys on and within the porous crud layers [see Fig. 7.10B from Ref. 33]. There have been only four PWR corrosion-related fuel failure events in the US since 1990, and interestingly, local tenacious crud had a key role in all these failures [35], as significant deposits can challenge fuel reliability through a phenomenon known as crud-induced localized corrosion (CILC). In PWRs, since the fuel rod surface temperatures are the highest at the upper spans of the fuel rod, the crud initially has a greater tendency to deposit there. Subsequently, this may result in increased boron deposition in the upper spans. This can lead to operational issues. For example, in a phenomenon known as axial offset anomaly (AOA), or crud-induced power shift (CIPS), boron precipitation in the crud layer at sufficient levels alters the local power as a result of the very large thermal neutron absorption cross section of boron. Thus, the power is lowered in regions with high boron precipitation, forcing power shift to the lower regions of the fuel rod. The primary impact of AOA or CIPS is on operations since the plant has limits on acceptable levels of axial offset in power between the upper and lower halves of the core. In addition to the impact of corrosion product transport and boiling duty on crud deposition, there is also an impact of coolant impurities and additives on crud deposition.

7.3.4 PWR COOLANT CHEMISTRY

In PWRs, the primary system coolant chemistry control is aimed at achieving four objectives:

1. to maintain low radiation fields to minimize worker radiation exposures during refueling and maintenance outages;
2. to minimize primary water stress corrosion cracking of steam generator tubes (alloy 600) to maintain steam generator efficiency;
3. to prevent excessive oxidation of Zircaloy FC (due to deposition of insulating crud layer on FC), which may lead to cladding failure and radioactivity release into the primary coolant; and
4. to reduce the crud deposition on the fuel.

For implementing primary coolant chemistry in PWRs, a balanced approach addressing the four objectives is pursued based on the current knowledge and experience. It is known from out-of-reactor autoclave tests that the corrosion resistance of Zircaloy is greatly diminished when the concentration of lithium, added in the form of lithium hydroxide, is increased to several tens of wppm in the absence of boron [36]. Boron added as boric acid has been shown to abate the negative effect of lithium on corrosion. Thus, an Li/B chemistry protocol is followed throughout the reactor cycle. Under extended burnup, especially in high power density plants, with subcooled boiling conditions and longer cycles, the need for higher lithium is a concern for enhanced FC corrosion, based on the argument that lithium in the coolant might become concentrated in the porous oxide. Similarly, the need for concomitant higher boron concentrations, on the other hand, may cause AOA (or CIPS) during the cycle as discussed earlier in this section.

Zinc is added to PWR [37] primary coolant water to mitigate primary water stress corrosion cracking (PWSCC) of ex-core primary circuit reactor components and to reduce the radiation dose to the plant servicing personnel during outage maintenance and refueling service work.

7.3.5 BWR COOLANT CHEMISTRY

Unlike localized subcooled boiling and temperature peaking in the upper spans of PWR assemblies, the BWR fuel bundles are operated at lower temperatures than PWR and with a relatively flat axial temperature profile (see Fig. 7.11 [38]). Also shown in this figure are several typical flow regimes along axial elevation of the BWR fuel. As mentioned in Table 7.3, the coolant environment is more oxidizing in BWRs than in PWRs. In the temperature and boiling environment in the BWR core, there are several BWR coolant chemistry variations, namely, (i) normal water chemistry (NWC), (ii) hydrogen water chemistry (HWC), (iii) zinc additions, and (iv) noble metal chemical application (NMCA) that are implemented. These coolant additives serve two important purposes in BWRs: mitigation of intergranular stress corrosion cracking (IGSCC) of BWR core internals and radiation dose reduction.

However, not all BWRs implement all these coolant additives and, in fact, coolant chemistry practices have evolved with time in BWRs [39]. For example, when HWC was initially implemented, two disadvantages were

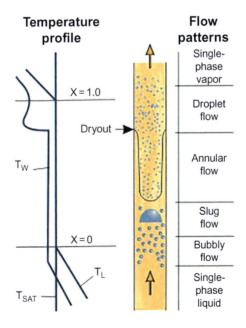

FIG. 7.11

Typical temperature profile representing flow patterns in a BWR fuel channel [38]. Symbols mean as follows: T_w, wall temperature; T_L, liquid temperature, T_{SAT}, saturation temperature, and X, void fraction.

encountered—(a) significant increases in the main steam line dose rate during operation due to ^{16}N partitioning to the steam phase and (b) a significant increase in the shutdown dose rates in the dry well due to increased incorporation of ^{60}Co into stainless steel corrosion films. Together, these effects impeded the adoption of HWC. More recently, online NMCA, albeit with reduced levels of hydrogen than under HWC alone, is commonly applied [40].

The factors affecting crud and corrosion in BWRs are broadly similar to those in PWRs, viz: coolant chemistry upsets (including intrusion of impurities), anomalies in local power, and thermal-hydraulic conditions with cladding metallurgy and composition (including impurities) playing a relatively smaller role. In the following subsection, we provide a couple of recent examples of severe crud and corrosion events that have led to fuel failures.

7.3.6 FUEL FAILURES DUE TO SEVERE CORROSION AND CRUD DEPOSITION

The corrosion performance of today's cladding is generally very good. However, occasionally, certain combinations of coolant chemistry, core design, and reactor operation in some LWRs have resulted in higher temperatures on the FC surface. The higher surface temperature accelerates the corrosion rate and can lead to through-wall corrosion or sufficient corrosion to locally embrittle the cladding by excessive hydrogen pickup [41].

As mentioned before, the four crud and corrosion related fuel failures in the US PWRs since 1990 [35] were caused by localized crud deposition on feed assemblies with higher power peaking. Core designs for all four units had significant changes applied to the reactor fueling for the cycles in which the failures occurred. Thus, these cycles were designed outside of the prior operating experience in terms of fuel power levels and peaking factors, compared to where no fuel failure had occurred previously. Also, the failures occurred at unique locations—at peripheral fuel rods in the assembly or next to control rod GTs—indicating local power peaking or nonuniform thermal-hydraulic conditions. In addition, there have been a few cases of accelerated corrosion of cladding observed over the same time period in the US PWRs.

Fig. 7.12 [42] shows the so-called "distinctive crud pattern" (i.e., localized corrosion, tenacious crud deposition, and spallation) in Cycle 10 of Three Mile Island (TMI)-1 PWR. Although the as-fabricated cladding was SRA, in subsequent hot cell examinations, local regions of recrystallization in the cladding were found, indicating that parts of cladding had reached an estimated ~175°C higher temperatures than expected during some period of time during the reactor cycle. An example of crud and corrosion related fuel failure in BWRs is River Bend Cycle 11. Excerpted here in Fig. 7.13 are photographs of the fuel from poolside inspection from Ref. [43].

Excessive corrosion and deposition of tenacious crud depends primarily on the local heat transfer and thermal-hydraulic conditions, coupled with deviations in coolant chemistry. However, the CILC does not depend on the Zr-alloy composition.

7.4 HYDRIDING AND MECHANICAL INTEGRITY
7.4.1 GENERAL COMMENTS

Zr-alloys are exposed to high-temperature, high-pressure coolant water during service and they are also simultaneously under irradiation flux. As a result, the alloys degrade by corrosion (and resulting hydriding of the alloy) and by radiation induced changes in the alloy itself. With increasing time under these in-service conditions, both degradation effects increase. Radiation damage saturates earlier, but the effects of hydriding (e.g., embrittlement) continue. Thus, for normal operation at higher burnups, the effect of hydriding dominates over the effect of radiation damage. For accident transients, a part of the radiation damage may be eliminated due to the high cladding temperature attained and the hydriding impact on the mechanical properties may start dominating. For the loss of coolant accident (LOCA), the post-quench ductility depends strongly on both the hydrogen and oxygen levels of the cladding [44].

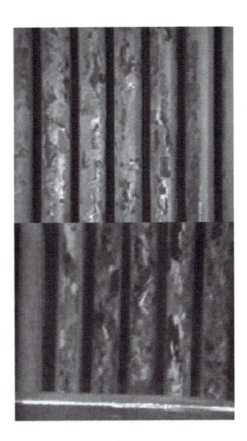

FIG. 7.12

Localized corrosion, crud, and spalling surface of the high-powered TMI-1 Cycle 10 fuel [42].

From R. Tropasso, J. Wilse, and B. Cheng, "Crud-Induced Cladding Corrosion Failures in TMI-1 Cycle 10," Proceedings of the 2004 International Meeting on LWR Fuel Performance, Orlando, Florida, September 19–22, 2004 Paper 1070. Copyright 2004 by the American Nuclear Society, LaGrange Park, Illinois.

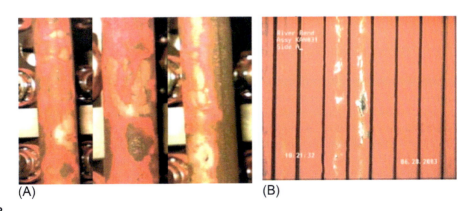

(A) (B)

FIG. 7.13

(A) Examples of tenacious crud appearing as reddish patches on the cladding surface that could not be removed with washing and aggressive brushing; the whitish patches are underlying oxidation of cladding and (B) Fuel Failure location was noted from nearly vertical trails, most likely caused by escaping fission gas bubbles and pellet material erosion streams from the leaking peripheral rod.

From Proc. of 2004 Intl. Mtg. on LWR Fuel Perf., Orlando, FL, Sept 19–22, 2004, paper # 1016.

7.4.2 EFFECT OF HYDRIDING ON UNIRRADIATED MECHANICAL PROPERTIES

The predominant corrosion reaction between the alloy and coolant water is:

$$Zr + 2H_2O \rightarrow ZrO_2 + H_2.$$

A fraction of the total hydrogen that is generated from this reaction is picked up by the alloy and the rest is removed by coolant flow or is dissolved in the coolant water. A part of the hydrogen that is picked up by the alloy remains in solid solution within the alloy lattice and the rest—which is in excess of the solubility limit at the irradiation temperature—is precipitated out as hydrides in the alloy. The hydrides are inherently brittle and can have deleterious effects on the mechanical integrity of the cladding [45].

7.4.2.1 Hydrogen pickup fraction and factors influencing it

The ratio between the hydrogen picked by the alloy and the total hydrogen generated from corrosion reaction is expressed either as a percent or fractional hydrogen pickup (HPU). Despite extensive research on the subject, the mechanism of HPU is not well understood. HPU depends on many factors including alloy type and impurity level, SPP size, and size distributions (depending on the alloy processing and heat treatment), coolant chemistry, and thickness and the nature of the growing oxide layer (e.g., porosity, cracking, and delamination). Typical percent HPUs in Zircaloy-4 PWR FC range between 10% and 25%. In Zircaloy-2 BWR cladding, the percent HPUs are typically slightly lower initially (due to the higher oxygen potential in the BWR coolant) but increase with exposure more rapidly than in Zircaloy-4 (possibly due to the irradiation induced dissolution of SPPs, especially those containing nickel). The HPU for non-heat flux bearing components (such as water rods, GTs, grids, and channels) can be higher than in the cladding [46–48]. There is also a significant scatter in measured percent HPU data. Although HPU percent (or fraction) is not an inherent *material* property of the alloy, as it is often mistakenly interpreted to be, it can serve as a useful guide to relate to the mechanical property of the hydrided alloy. Assuming an estimated HPU fraction, the hydrogen content in the cladding can be estimated. Since cladding is the first engineered barrier against release of fission products into the coolant, the effects of hydrogen pickup and hydriding on the mechanical integrity of the cladding becomes a concern, especially at high burnups [49]. In addition to HPU, the hydrogen impact on the mechanical properties also depends on hydride morphology and hydride distribution. This is discussed further in the following sections.

7.4.2.2 Hydride morphology and redistribution

Precipitated hydrides in Zr-alloys have been predominantly the delta phase (ZrH_x; where $x \approx 1.66$) [50]. The orientation of precipitated hydrides depends on several factors such as the texture of the fabricated component, microstructural grain boundary orientation, and applied stress during the hydride precipitation process. The texture produced by the pilgering process commonly used to fabricate cladding tubes induces hydride precipitation in the circumferential direction. That is, when viewed at the tube wall cross section, the hydride platelets are predominantly oriented perpendicular to the radial direction. In addition, since the grain boundaries are preferred sites of hydride precipitation, the orientation of grain boundaries in the as-fabricated microstructure also influences the hydride orientation. In the case of RXA cladding microstructures, more randomly orientated hydrides are observed due to the corresponding equiaxed grains and random grain boundary orientations. Hydride precipitation under compressive circumferential stress (as present during fuel rod irradiation with external coolant pressure greater than the rod internal pressure) results in circumferentially oriented hydrides in the cladding during the normal operation. In contrast, the hydride precipitation under tensile stress due to rod internal pressure, during the later cooldown phase of dry storage of the fuel, can lead to radial hydride orientation, which is more deleterious orientation for cladding mechanical integrity considerations [51]. Thus, the net orientation of hydride precipitates is a combined effect of texture, microstructural grain shape, and stresses present during the hydride precipitation process.

A significant thermal gradient across the cladding wall is present due to the heat transfer from the pellet to the coolant. Minor axial thermal gradients are present due to the pellet-pellet interfaces and infrequent occurrence of axial pellet-pellet gaps that may develop at high burnups. Since hydrogen in the cladding moves down the temperature gradients, these thermal gradients lead to redistribution of hydrogen in the cladding. As an example, a partial cross section of the cladding wall is shown in Fig. 7.14. Since the temperature at the cladding inside surface is higher than at the outside surface, dissolved hydrogen migrates down the temperature gradient. This typically forms a hydride rim at the oxide/metal interface. Such hydride rims have a high local hydrogen concentration that can be an initiation site for a radial crack. Similarly, when a part of the thick oxide layer spalls and creates a local cold spot in the cladding during service, it can also attract hydrogen to the cold spot in the oxide-spalled region, forming a hydride blister at the oxide/metal interface. The mechanical response of FC to localized hydrides, such as in rims and blisters, decreases the cladding ductility. One possible explanation for it [52] is that the mechanical properties of the hydride/metal composite are a function of the local hydride volume fraction, and that a higher hydride volume fraction leads to lower ductility.

7.4.2.3 Fuel cladding integrity concerns with increasing hydriding

Considerable work has been reported in the open literature that involves measurements of ductility of Zr-alloy test specimens. Such testing is either performed on unirradiated specimens that are intentionally hydrided to different hydrogen contents, or on irradiated specimens with hydrogen distribution in their as-irradiated condition. One example of such ductility tests on unirradiated sheet material at room temperature (RT) (shown in Fig. 7.15) is that from Bai et al. [53] where the data suggest that Zr-alloys can reach zero ductility somewhere in the range of 200–1200 wppm hydrogen, depending on the microstructure (related to the fabrication process) for the alloy. The higher number applies to material with an alpha-phase microstructure, while the lower number is for the transformed-beta (beta-quenched) microstructure. The properties of the transformed beta structure depend strongly on the extent of solute partitioning, which in turn depends on the cooling rate. At higher deformation temperatures and for tubing specimens, the hydrogen concentrations for zero ductility are generally different. Fig. 7.15 shows that for uniformly hydrided test specimens, a clear decrease in ductility is seen with hydrogen content, but this is also a strong function of metallurgical processing of the test specimens.

Hydride localization (rims, blisters) decreases the ductility of FC even more. The dependence of mechanical properties of cladding on hydride distribution in the cladding has been investigated with the size of the hydride blister [54] and the thickness of the hydride rim [55].

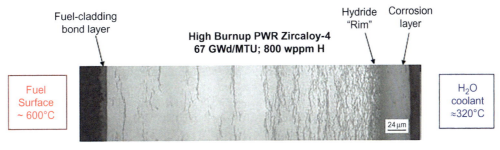

FIG. 7.14

Photomicrograph of a cladding cross section showing circumferential hydrides redistributed under an in-service temperature gradient across the cladding wall. *GWd/MTU*, unit of burnup, Giga Watt Days per metric ton of uranium.

Source: EPRI.

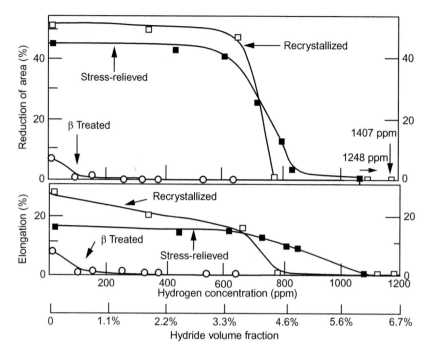

FIG. 7.15

Ductility of unirradiated SRA and RXA sheet specimens at room temperature as a function of hydrogen concentration.

After J.B. Bai, C. Prioul, D. François, Hydride embrittlement in Zircaloy-4 plate: Part I. Influence of microstructure on the hydride embrittlement in Zircaloy-4 at 20°C and 350°C, Metall. Mater. Trans. A 25 (6) (1994) 1185–1197.

7.4.3 EFFECT OF HYDRIDING ON IRRADIATED MECHANICAL PROPERTIES

Zr-alloys exhibit increased strength and reduced ductility with irradiation. For cladding alloys in service, these changes are inherently coupled with the effects of hydrogen due to the hydrogen pickup from the corrosion reaction mentioned previously. Deformation of LWR-irradiated Zr-alloys has been the subject of many investigations over the years. With the advent of newer Nb-containing alloys irradiated to increasingly higher fluences, this subject continues to be very extensively reported. Essentially, most published studies report on mechanical tests performed on certain sample geometry cut out from FC either as tubes, rings, or arcs. These tests are performed under a variety of loading configurations (tensile in the axial or azimuthal direction with respect to the fuel rod, pressurization at temperatures ranging from room temperature to ~625 K, etc.). The hydrogen contents are reported either as estimates, based on an assumed hydrogen pickup fraction, measured average values from the vicinity of the test samples, or as local hydrogen content measured in the gauge section.

Yagnik et al. [56] have reported on the ductility of irradiated Zircaloy-4 cladding and GTs, respectively, in SRA and RXA conditions in the temperature range 298–623 K. These materials had been exposed to an estimated neutron fluence of 0.8 to 10×10^{25} n/m^2 ($E > 1$ MeV) in two different PWRs. The hydrogen content in the irradiated cladding was in the range of ~200–600 wppm, exhibiting the typical through-wall distribution of circumferential hydrides (see Fig. 7.14). In comparison, the hydrogen content in the irradiated GT material was in the range of ~250–1800 wppm, exhibiting a fairly uniform through wall distribution of circumferential hydrides (see Ref. 47). Tensile tests in the axial direction and hydraulic burst tests were conducted on 130–150-mm-long tubular specimens. In addition, smaller specimens machined in the form of 55 mm long curvilinear dog bone and 10 mm

7.4 HYDRIDING AND MECHANICAL INTEGRITY

slotted semicircular arc specimens were tested in plane stress and plane strain configurations. Corresponding unirradiated archive materials in as-received condition and with uniform hydrogen charging up to 1200 wppm were also tested by identical methods.

Fig. 7.16, from Ref. [56], shows uniform elongation (UE) and total elongation (TE) in recrystallized Zircaloy-4 measured at 575 K as a function of local hydrogen content in the gauge section. The data in this figure are for small slotted semicircular arc samples of recrystallized GT and FC materials. The loss of ductility due to irradiation is quite evident. The ratio TE/UE decreases after irradiation. However, both unirradiated and irradiated Zircaloy-4 retained high ductility, especially at high temperatures, even at higher hydrogen concentrations than those anticipated in service. Irradiated GT specimens showed lower ductility compared to unirradiated FC for equivalent hydrogen levels.

The hoop strengths measured by burst tests on large tubular recrystallized GT irradiated specimens showed little or no dependence on hydrogen concentration as shown in Fig. 7.17 [56]. However, the corresponding total and uniform strains decreased slightly as local hydrogen concentration increased, as seen in Fig. 7.16, albeit on smaller arc samples. The main factor affecting the strength is the fast neutron fluence accumulated during irradiation.

As stated before, hydrogen embrittlement of cladding occurs under normal operation. In addition, the following effects of hydrogen on cladding mechanical integrity, which occur under different conditions, are also important considerations and can be itemized under the following four categories:

(a) Incipient cracking
(b) Secondary hydriding

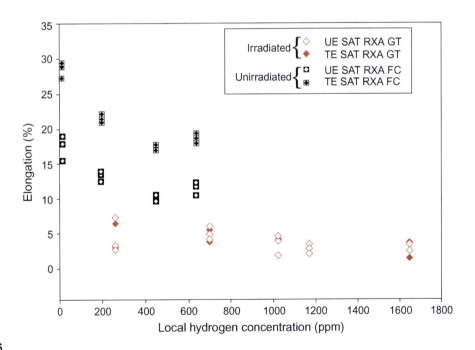

FIG. 7.16

Ductility of recrystallized Zircaloy-4 in irradiated and unirradiated conditions at 575 K.

After S.K. Yagnik, A. Hermann, R-C Kuo, Ductility of Zircaloy-4 Fuel Cladding and Guide Tubes at High Fluences, ASTM STP 1467, in: P. Rudling, B. Kammenzind (Eds.), ASTM International, 2005, pp. 604–631.

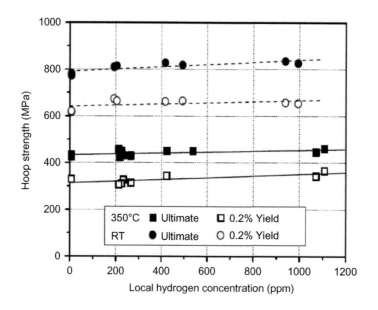

FIG. 7.17

Hoop strength of irradiated recrystallized Zircaloy-4 GT as a function of local hydrogen content at room temperature and 623 K.

After S.K. Yagnik, A. Hermann, R-C Kuo, Ductility of Zircaloy-4 Fuel Cladding and Guide Tubes at High Fluences, ASTM STP 1467, in: P. Rudling, B. Kammenzind (Eds.), ASTM International, 2005, pp. 604–631.

(c) Delayed hydride cracking
(d) Hydride reorientation
   (a) Incipient cracking can develop in the highly brittle hydride rim at the oxide/metal interface. This initial crack can propagate radially under strong pellet cladding mechanical interaction (PCMI) due to fuel swelling during power ramps and/or delayed hydrogen cracking (DHC), eventually becoming through wall, thus failing the fuel rod [57,58].
   (b) Secondary hydriding, schematically shown in Fig. 7.18 [59], occurs in a fuel rod that is already failed by one of the primary failure mechanisms, which will be described later on in Section 7.6. Subsequent to such an initial small breach, several complex chemical reactions (including with radiolytically generated species such as H_2O_2) and transport processes can ensue inside the fuel rod. The steam that enters through a primary hole/crack in the cladding reacts with the Zr-alloy inner wall and produces hydrogen that eventually fills the fuel-cladding gap at locations away from the primary defect. In the immediate vicinity of the primary failure, oxidation of the cladding inside surface is dominated as the steam to hydrogen ratio is high. As the axial distance from the primary failure location increases, the steam to hydrogen ratio in the fuel rod interior decreases.

   The steam to hydrogen ratios are also impacted by significant closure of the pellet/cladding gap. With a much lower availability of steam (steam starvation) away from the location of the primary failure, the extent of hydriding of the cladding interior surface increases [60]. Massive secondary hydriding (as shown in Fig. 7.19 [61]) of the inner cladding surface is of greater consequence in BWR compared to PWR fuel rods. This is largely because the cladding inside surface for BWR fuel is essentially a dilute alloy of zirconium liner, which tends to react more readily with steam/water (i.e., lower corrosion resistance) than Zircaloy. Hydrogen penetrates the oxide scale on the cladding inner wall and forms platelets or blisters of zirconium hydride. These, being quite brittle, can lead to a large secondary failure if the cladding is stressed at the point where they are located. The zirconium corrosion and hydriding

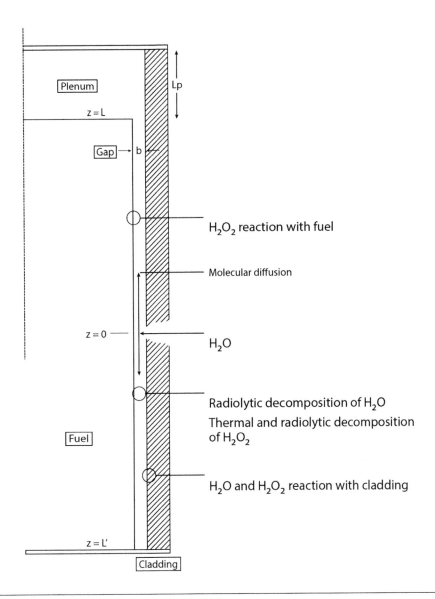

FIG. 7.18

Schematic representation of post-defect secondary hydriding.

After Secondary Hydriding of Defected Zircaloy Clad Fuel Rod, EPRI Technical Report, TR-101773, 1993.

severely embrittles the cladding and renders it susceptible to large-scale splitting when the stress is applied in a power ramp. Such an event is immediately detectable by the presence of washed-out fuel in the coolant and requires immediate reactor shutdown to remove the fuel assembly containing the broken rod.

(c) Delayed hydride cracking (DHC) is a flaw propagation mechanism that does not require significantly higher hydrogen concentrations above the terminal solid solubility (TSS), unlike massive local hydriding in the case of secondary hydriding. With sufficient time and enough hydrogen above TSS accumulating at an existing crack tip, an extension in the crack can occur by DHC under tensile loading. This failure mechanism is implicated in CANDU pressure tubes [62]. The basic mechanism, shown schematically in Fig. 7.20, involves the following steps:

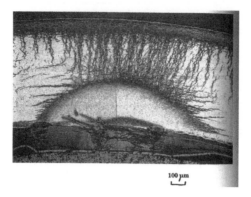

FIG. 7.19

Massive secondary hydriding in a BWR cladding [61].

From K. Edsinger "A Review of Fuel Degradation in BWR," Proc. Intern. Topical Meeting "LWR Fuel Performance" Park City, Utah, USA, April 10–13, 2000. Originally from Jonsson, L. Hallstatius, B. Grapengiesser and G. Lysell, "Failure of a Barrier Rod in Oskarshamn 3", Proc., Int. Topical Meeting on LWR Fuel Performance, ANS, Avignon, France, April 21–24, 1991, pp. 371–377.

- H in solution diffuses to a stress concentrator (i.e., at an existing crack tip)
- TSS is exceeded and hydrides precipitate in front of the crack
- Tensile strain at the crack tip fractures adjacent hydrides
- Crack advances a little and the process repeats itself

(d) Hydride reorientation: A key technical issue for all Zr-alloy LWR FC is the potential for hydride reorientation during the process of loading fuel in dry storage casks. As noted above in subsection 7.4.2.2, in SRA PWR cladding during normal irradiation, the hydrides are typically circumferentially oriented precipitates in r–θ planes of the cladding (see Fig. 7.14). In BWR RXA cladding, on the other hand, the hydrides during normal operation are randomly oriented in the cladding. The hydride precipitates in both cases can change their orientation into the radial direction, depending on the hoop stress in the cladding and amount of dissolved hydrogen available for re-precipitation. Hydride reorientation also occurs in the later stages of dry storage when cooldown under tensile hoop stress is encountered.

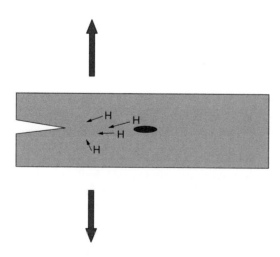

FIG. 7.20

Schematic representation of the DHC mechanism.

The presence of such radially oriented hydrides weakens the cladding's mechanical integrity, especially at low temperatures, leading to operational and regulatory concerns over handling and accident conditions during transportation.

Thus, when the cladding is subjected to high temperatures in certain post-irradiation operation, such as the loading of dry storage casks mentioned above, a fraction (above TSS of hydrogen in Zr-alloy) of its predominantly circumferentially oriented hydrides are dissolved into the alloy matrix. Subsequently, when the cladding cools, that fraction of the dissolved hydrogen may re-precipitate in the form of radially oriented hydrides in response to the prevailing tensile hoop stress in the cladding. There is a critical threshold stress necessary for hydride reorientation to occur [63,64].

The impact of radial hydride precipitation on cladding embrittlement also depends on concurrent recovery of irradiation damage that will occur as a result of the high temperatures [65,66] achieved at the start of dry storage. Such irradiation damage recovery occurs first during the initial dry storage period of heat up of the cladding due to decay heat and could result in an increase in ductility of the cladding, as shown in Fig. 7.21 from Ref. [67]. This recovery phenomenon also tends to reduce the impact of radially reoriented hydrides on cladding ductility during the early heat up stage of dry storage.

The extent of radial hydride reorientation during dry storage also depends on the extent of circumferential hydride precipitation in the cladding occurring during the normal at-power prior operation [68]. If a significant number of circumferential hydrides from the normal operation are present (as is the case with SRA Zircaloy-4 cladding), the precipitation of hydrides during cooldown in dry storage also occurs mainly in the circumferential direction, often right at the locations of the original hydrides that had not dissolved during the heat up portion of the dry storage procedure. This is referred to as the 'memory effect' of hydride nucleation. The orientation of the hydrides is characterized by the radial hydride fraction (RHF) that is used as an acceptance criterion for the hydrogen charged as-fabricated cladding. RHF depends on the as-fabricated cladding microstructure resulting from the cladding fabrication process.

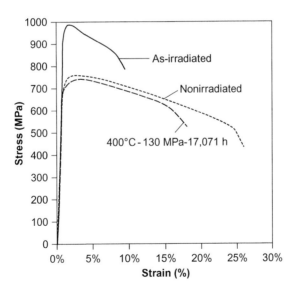

FIG. 7.21

Engineering stress–strain curve from ring tensile tests performed at room temperature on Zircaloy-4 nonirradiated, as irradiated, and following a long duration creep test on as-irradiated specimen at specified condition.

After B. Bourdiliau, F. Onimus, C. Cappelaere, V. Pivetaud, P. Bouffioux, V. Chabretou, A. Miquet, Impact of Irradiation Damage Recovery during Transportation on the Subsequent Room Temperature Tensile Behavior of Irradiated Zirconium Alloys, ASTM STP 1529, in: M. Limbäck, P. Barberis (Eds.), American Society for Testing and Materials, 2011, pp. 929–953.

7.4.4 EFFECT OF HYDROGEN ON POST-ACCIDENT TRANSIENT MECHANICAL PROPERTIES

Commercial use of all Zr-alloys requires licensing approval from the regulatory bodies concerning their anticipated in-LWR performance in normal, off-normal, and postulated accident conditions. This is achieved through performance models (such as corrosion, irradiation growth, irradiation creep, irradiation and hydrogen induced embrittlement, etc.) based on post-irradiation property measurements at pertinent high temperatures. Such models and data analyses are submitted to regulatory bodies for approval.

An important consideration in this regard is the mechanical integrity and maintenance of core coolability following a loss-of-coolant accident (LOCA). This has been the subject of intense interest and numerous publications since the 1970s [69–72]. A detailed discussion of this field, including LOCA properties (such as post-quench ductility, breakaway oxidation, etc.) and modeling approaches, is beyond the scope of this chapter. However, some brief remarks are made below.

The LOCA regulation implemented in the 1970s considered the post-quench ductility impact of oxygen charged into the cladding due to the high-temperature transient in terms of the equivalent cladding reacted (ECR). An ECR of 17% and a limit on the maximum temperature of cladding attained during the LOCA transient of about 1473 K have been included in the current regulation. As a result of the recent testing and concerns with effects of hydrogen, especially in high burnup fuel [73,74], the LOCA regulations are expected to be revised in the USA and elsewhere. The new regulations may include two major considerations: (1) experimental verification that breakaway oxidation time for different cladding lots is beyond the limiting duration of the LOCA transient for the fuel design used in the specific reactor; and, (2) post-quench ductility dependence both on ECR (an effect of oxygen) and cladding hydrogen level at the start of the LOCA transient.

Additional licensing requirements associated with postulated reactivity initiated accident (RIA) are also being implemented. The RIA phenomenon is sometimes also called the reactivity *insertion* accident, or rod ejection accident. This topic is beyond the scope of this chapter. Interested readers can pursue it further in these Refs. [75–82].

7.5 IRRADIATION EFFECTS
7.5.1 GENERAL COMMENTS

When Zr-alloys are exposed to the neutron bombardment, atomic defects are generated in the alloy crystal structure. The energetic neutrons knock the zirconium atoms from their normal equilibrium sites in the HCP crystal lattice to nonequilibrium sites. As a result, vacancies and interstitials are generated in the crystal structure.

Under irradiation, such point defects are created in displacement cascades, and they tend to migrate and form <a> and <c> dislocation-loops (see below). Although some vacancies and interstitials recombine to eliminate each other, most of them condense to form discs creating either vacancy loops or interstitial loops. The loops condense on either a basal or prism plane. The loops formation is a result of anisotropic diffusion of different point defects to different sinks. The loops are designated by the burgers vector of the loop boundary dislocation. Thus, loops condensing on the prism plane are called <a> loops while the loops condensing on the basal plane are called <c> loops. The <a> loops form at low neutron fluence, while <c> loops form at high fluence. The <a> loops are smaller than the <c> loops. The presence of alloying element such as Nb can delay the formation of <c> loops [83]. Dissolution of iron bearing SPPs enhances nucleation of <c> loops.

Both types of loops are obstacles to dislocation motion and thereby increase the strength of the irradiated alloy. Such strengthening with irradiation, which saturates at low fluence, is mainly due to dislocation obstacles generated by <a> loops. In addition, the dislocation loop formation generates a local stacking fault that contributes to irradiation hardening. Typical images of <a> and <c> dislocations, as seen by transmission electron microscopy (TEM) of irradiated Zr-alloys, are shown in Fig. 7.22 [65] and Fig. 7.23, respectively [84].

In addition to the generation of point defects and dislocation loops, radiation damage changes the solubility of different alloying elements in the alloy matrix, resulting in evolution of SPP size, composition, and crystallinity. Some SPPs (especially the iron rich SPPs) dissolve while others become amorphous and some new precipitates form as in Zr-Nb alloys. In Zircaloys, the Laves phases (Zr-Fe-Cr and Zr-Fe-Ni particles) become amorphous and eventually dissolve in the matrix. Such SPP dissolution enhances both the corrosion rate and hydrogen pickup [85]. In Zr-Nb alloys, the SPPs are generally stable and are not eliminated by radiation damage [86]. Beta-Nb particles appear to be stable while the Zr-Fe-Nb particles lose iron [87,88]. Owing to such radiation induced effects on SPP, the in-reactor corrosion rate enhancement of Zr-Nb alloys can be lower than that for the Zr-Sn alloys.

During the deformation of irradiated zirconium alloys, the moving dislocations are generated and interact with the preexisting dislocation loops generated by the radiation damage. Thus, the loop obstacles provide radiation hardening by increasing the flow stress. At a sufficiently high level of flow stress, there is a sudden avalanche of dislocation movement in closely parallel planes leading to bands of dislocations sweeping through regions within the microstructure. This phenomenon is called dislocation channeling. Such localized dislocation motion results in inhomogeneous deformation in the alloy.

The irradiation induced microstructural changes mentioned above (i.e., generation of point defects and dislocation loops, evolution of size, shape, and amorphization of SPP, and formation of dislocation channels during subsequent deformation) affect some important performance characteristics of irradiated zirconium alloy components, such as their corrosion resistance, mechanical properties, and dimensional changes (irradiation creep and irradiation growth). These are discussed below.

Since irradiation effects are very complex phenomena and depend on many variables and factors that are not well understood, test reactor studies are needed to characterize the consequences of neutron exposure on specific property changes. Nonetheless, for a variety of reasons, test reactor irradiations cannot be expected to fully duplicate in-service behavior of Zr alloys in a commercial power reactor environment.

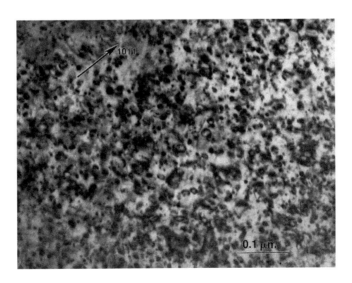

FIG. 7.22

<a> loops in neutron irradiated Zircaloy-2 as observed by TEM [65].

Courtesy: AECL, Canada.

FIG. 7.23

$<c>$ loops as observed by TEM [84].

7.5.2 EFFECT OF IRRADIATION ON CORROSION RESISTANCE

As stated earlier, radiation damage to the microstructure degrades the corrosion resistance of Zr-alloys due to radiation induced dissolution of SPPs. The irradiation induced corrosion resistance degradation is less significant for Zr-Nb alloys due to their SPP stability.

There are two aspects of corrosion resistance, development of the waterside oxide layer (which results in loss of the load bearing metallic cladding wall) and hydrogen pickup by the metal. Integrity of the waterside oxide layer is also an important factor. Spallation of thick waterside oxide (local removal of a part or entire oxide layer thickness) leads to a local cold spot formation during operation, resulting in hydride localization in cladding that can lead to cladding fracture due to hydride embrittlement.

The hydrogen generated by corrosion that is picked up by the cladding is nonuniformly distributed through the wall. Since hydrogen from the inner parts of the cladding (where temperatures are higher) migrates down the temperature gradient to the outer parts due to the heat flux that exists across the cladding wall thickness during operation, coupled with the lower hydrogen solubility in alpha phase at lower temperatures, induces hydride precipitation at the outer oxide-metal interface. Since the zirconium hydride has lower corrosion resistance, and the presence of hydrides at the interface disturbs the oxide coherency, both of these factors increase the in-reactor corrosion rate compared to isothermal corrosion in autoclaves [89]. Although Ref. [89] only evaluated Zircaloy-4, enhancement of corrosion due to hydride precipitation at the metal/oxide interface is observed for all other zirconium cladding alloys used in the nuclear industry.

The presence of the radiation fields also causes radiolysis of coolant water, generating oxidizing radiolytic species, which also increase the corrosion rates under irradiation compared to out-of-reactor autoclave conditions. With hydrogen overpressure in the coolant in PWRs, the extent of radiolysis is suppressed. However, owing to the more oxidizing coolant environment, the radiation induced enhancement of Zr-alloy corrosion is typically higher in BWRs compared to PWRs.

The coolant chemistry used in LWRs is another important factor affecting the in-reactor cladding corrosion. In PWRs, the following primary coolant additives are used to various extents for different specific reasons. Some examples are lithium (to control pH which controls crud deposition on FC), boron (to control reactivity), hydrogen (to control oxygen partial pressure), and zinc (to reduce service personnel exposure), etc. In BWRs, coolant chemistry variations include HWC, Noble Metal Chemical Addition, zinc injection, iron injection, etc. In VVERs used in Russia and other east European countries, KOH is used to control the coolant pH.

7.5.3 IRRADIATION HARDENING AND EMBRITTLEMENT

As stated before, dislocation loops are significant obstacles to dislocation motion which result in irradiation hardening. Further, strain localization and inhomogeneous deformation due to dislocation channeling can result in low macroscopic strain (i.e., low ductility) fracture. The latter phenomenon is generally described as irradiation embrittlement. Crack initiation and propagation at the brittle hydrides located within dislocation channels further increase embrittlement.

Since the critical shear stress value varies for different slip systems, the slip system with the lowest critical resolved shear stress is activated first. In unirradiated Zr-alloys, plastic deformation at a reactor operating temperature of about 623 K is associated with <a> dislocations moving on the prism plane. Interactions between moving dislocation and irradiation induced defects (mainly <a> loops) result in irradiation-induced work hardening. When flow stress is increased due to irradiation hardening, other slip planes (basal and pyramidal) can become active. For plastic deformation, the moving <a> dislocation on the prism plane has to overcome the obstacle of <a> loops. At low stress, this is accomplished by a dislocation climb resulting from the migration of radiation induced point defects to the core of edge dislocation. Such a dislocation climb is an important mechanism for creep of Zr-alloys. Dislocation channeling makes the deformation of irradiated Zr-alloys very inhomogeneous at the grain (or microscopic) level. If the dislocation channel can easily propagate in the neighboring grain (as is likely for the basal plane channel), on a macroscopic level a shear band can be observed on the deformed surface. The number of dislocation channels within a grain, the width of the individual channels, and the extent of their propagation across multiple grains determine the macroscopic ductility of irradiated Zr-alloys. Addition of niobium to Zr-alloys generally improves ductility of alloys in the irradiated state. This observation is postulated [90,91] to be associated with work hardening provided by niobium within the dislocation channels. The extent of interaction between the irradiation damage introduced dislocation loops and the movement of dislocations involved in achieving the deformation strain determines the resolved shear stress to activate a specific slip system. As a result, the activated slip system for irradiated Zr-alloys could be different from those for the unirradiated Zr-alloys. The presence of alloying elements and SPPs can change the resolved shear stress further.

Irradiated Zr-alloys have lower ductility than their unirradiated counterparts due to two factors: irradiation damage and hydrogen pickup. While embrittlement due to irradiation damage can be eliminated by annealing at high-temperature transients, embrittlement due to hydrogen pickup and hydriding cannot be eliminated by a thermal transient.

7.5.4 IRRADIATION GROWTH

The distortion of anisotropic metals during irradiation in the absence of applied stress is called irradiation growth. Irradiation growth is considered a constant volume phenomenon, caused by the anisotropy in point defect diffusion to different sinks on different planes [92,93]. It is a complex kinetic nonequilibrium process. Vacancy loops preferentially condense on the basal planes, while interstitial loops prefer to condense on prism planes. This, in turn, leads to <a> axis expansion due to interstitial loops and <c> axis contraction due to vacancy loops, resulting in asymmetrical-dimensional changes in the zirconium alloy components. Fabrication of anisotropic metal such as zirconium always leads to the development of nonrandom textures, because of the preferred direction of deformation of individual grains during the forming process. Thus, the dimensional changes occur preferentially in one direction, manifested as growth. The irradiation growth anisotropy factor (G), in a given direction, is expressed as $(1 - 3f)$, where f is the resolved fraction of basal poles in that direction. Since beta-quenched material has a random texture with the f value close to 0.33, its irradiation growth is close to zero.

In-reactor dimensional changes in zirconium alloys result from a complex interplay of many factors, such as (1) alloy type and composition, including the addition of elements such as niobium, iron, and tin (e.g., niobium addition generally reduces growth); (2) fabrication process, including cold work, texture, and residual stresses (i.e., cold work generally increases growth); (3) irradiation temperature; (4) hydrogen levels (i.e., higher hydrogen levels increase growth), and (5) SPP evolution (i.e., dissolution of SPP in Zircaloys increases growth). In many cases, the observed dimensional changes in LWR fuel assembly components—especially at high exposures—cannot be fully explained based on current growth and creep models. There is a clear need for fundamental data to better elucidate growth behavior.

Current designs have experienced some unexpected zirconium component growth in LWR cores at high fluences, perhaps exacerbated by accelerated hydrogen pickup or inadvertent lack of control of alloy fabrication parameters or heat treatments. As utilities strive to design even higher-duty burnup cores, along with significant power uprates, to improve cycle economy, the need for reliable irradiation growth data will be more critical than ever.

Alloying element additions and the SPP evolution with irradiation damage (which also depend on the alloy composition) have a strong effect on irradiation growth. For Zircaloys, the SPP dissolution at high burnups leads to accelerated growth. Hydrogen pickup in these alloys at high burnup also accelerates the irradiation growth. Migration of iron from the radiation induced dissolution of SPPs in Zircaloys interacting with c-dislocations in the matrix surrounding the dissolving SPP is also believed to be correlated with the growth acceleration [94]. Addition of niobium to Zr-alloys generally reduces the irradiation growth due to the stability of SPP under neutron irradiation. Minor impurity atoms, such as carbon or sulfur, may lead to unexpected irradiation growth under certain conditions.

Cold worked and stress-relief annealed Zr-alloys exhibit higher irradiation growth than recrystallized materials due to higher dislocation density. Since SPP dissolution due to irradiation exposure has an impact on irradiation growth, the Zircaloy fabrication history, which affects the SPP size in the as-fabricated microstructure, would also have an impact on irradiation growth.

The effect of irradiation temperature and hydrogen pickup on irradiation growth of a Zr-alloy depends on the microstructure and SPP evolution of the specific alloy. Owing to the large size difference between zirconium and hydrogen atoms, both hydrogen in solution and hydride formation in Zr-alloys lead to volume change. In addition, hydrogen in solution may interact with point defects and dislocations, which may have additional effects on irradiation growth that are not yet well understood. Since all the interactions between different parameters are not clearly understood, characterization of irradiation growth of the new alloy is an important part of alloy development activities. Even with such evaluations, occasional unexpected irradiation growth issues have been encountered during commercial implementation of new alloys.

7.5.5 IRRADIATION CREEP

The time-dependent deformation in the presence of neutron flux and applied stress is called irradiation creep. Total creep strains include both irradiation and thermal creep strains. Excessive creep and growth of Zr-alloy structural components during service can lead to the occurrence of operational incidents such as incomplete control rod insertion (IRI) in PWRs [95] or channel bowing in BWRs with concomitant safety implications [96]. Therefore, reliable irradiation creep and growth data are necessary to improve fuel designs and predict dimensional changes during service. Clearly, the large axial distortions in GTs that led to IRI incidents in PWR fuel assemblies were not foreseen. At high fluences, this phenomenon is likely exacerbated by higher than normal hydrogen pickup due to irradiation-induced dissolution of SPPs in Zircaloy microstructures. In addition, design changes associated with higher hold-down spring loads and thinner GT wall thickness were also implicated in IRI events. Immediate remedies to IRI problems included adjustment of the design changes by increasing the GT wall thickness and reducing the axial load reduces axial buckling tendency of the thimble tubes. Nevertheless, additional research that fundamentally elucidates irradiation creep and growth mechanisms must also be a part of the solution to operational problems in LWRs due to dimensional changes in Zr-alloys under irradiation.

The temperature dependence of the thermal creep of Zr-alloys can be divided into three different temperature regions [97], depending on the extent of recovery of dislocation/defect structure of the microstructure. At low temperatures (below 0.25 times beta transformation temperatures, T_β), no recovery takes place and creep strain has a logarithmic relationship with time. At an intermediate temperature range (0.25 to $0.4T_\beta$), recovery becomes increasingly important but is not sufficient to compensate for the strain hardening. In this intermediate temperature region, strain is exponentially related to time. At high temperatures ($>0.4T_\beta$), softening due to recovery and recrystallization compensate for strain hardening and the thermal strain has a close to linear relationship with time. In each of these temperature regions, the creep strain rate depends on stress in an exponential relation with the stress exponent depending on the dominant creep mechanism. The creep of Zr-alloys strongly depends on metallurgical variables such as concentration of alloying elements and impurity elements, cold work, and material texture. Mathematically, the stress, strain rate, temperature (T) dependence of the thermal creep rate ($\dot{\varepsilon}$) in its simplest form is represented by the equation:

$$\dot{\varepsilon} = A\, \sigma^n \, \exp(-Q/RT)$$

where $\dot{\varepsilon}$ is the strain rate, σ is the stress, n is the stress exponent, R is the gas constant, T is temperature in degrees Kelvin, Q is the activation energy, and A is the material dependent constant [98].

For diffusion controlled creep (volume diffusion Nabarro-Herring creep or grain boundary Coble creep), the stress exponent n is unity. For dislocation controlled creep, n is typically about 4. The dislocation controlled creep is typically due to climb-glide mechanisms. The values of Q change according to the controlling creep mechanism (such as activation energy for diffusion of an element if it is pinning dislocations) and may vary for different temperature regions. In reality, all of these factors operate simultaneously; hence, the net creep rate has a complex dependence on the combination of individual factors.

Since radiation damage introduces point defects in the Zr-alloy microstructures which affect dislocation motion, creep rates of irradiated material are different from those in the unirradiated state. Generally, at temperatures lower than 375°C, irradiation creep dominates, while at higher temperatures, thermal creep dominates [99–103]. Irradiation creep for a recrystallized microstructure is lower than that of cold worked alloys due to a higher degree of radiation hardening in the recrystallized condition. Irradiation creep mechanisms for Zr-alloys can be divided into three groups: irradiation retarded creep, irradiation induced creep, and irradiation enhanced creep. Each group may be the dominant mechanism for the specific alloying element, level of stress, temperature, and the extent of radiation damage. Irradiation retarded creep is a manifestation of radiation hardening and is claimed [104] to be the reason for sulfur additions to the advanced alloy M5. Irradiation induced creep mechanisms add to the thermal creep and could be either due to stress induced preferred nucleation (SIPN) of dislocation loops or stress induced preferred absorption (SIPA) of point defects leading to anisotropic growth of the loops. Irradiation enhanced creep also can be associated with mobile dislocation production, solute trapping, enhanced climb and glide, and enhanced jog dragging mechanisms. For a detailed discussion of the different creep mechanisms, the reader is referred to Ref. [105].

Since complete understanding of all possible creep mechanisms operating simultaneously in Zr-alloys is still lacking, all code calculations of the creep depend heavily on empirical relationships based on experimental data. Generally, the creep rate in principal loading directions is the main concern. Creep strain is an important metric for nuclear fuel rod performance. A few examples include: (1) uniform decrease in cladding diameter during the initial irradiation of the fuel rod, (2) impact of cladding creep on pellet cladding interaction (which also takes into account pellet swelling), (3) cladding creep during postulated LOCA, and (4) development of gap between fuel rod outer surface and grid spring supports due to load relaxation that can lead to GTRF fuel failures, as further discussed in the next section.

7.6 FAILURE MECHANISMS
7.6.1 GENERAL COMMENTS

There are several primary failure mechanisms described in this section, where the initial breach of the FC wall during service leads to an increase in concentration of radioactive isotopes in the coolant. The dimensions of the breach area can be small, of the order of 0.1 mm or less. Despite their small size, these flaws permit some volatile fission products that have been released from the fuel (Xe, Kr, I, and perhaps Cs and Te) to escape to the coolant and become dispersed throughout the entire primary circuit. Such primary failures are undesirable because they result in operational and fuel handling difficulties during the next reactor outage. More importantly, continued operation with a primary failure can also lead to secondary degradation of the fuel rod, including severe hydriding (see Section 7.4), which may further weaken the cladding and can lead to larger wall breaches resulting in oxidation and erosion of highly radioactive fuel pellet material into the core and increasing personnel radiation exposures.

With such direct impacts on the safety and economics of nuclear power generation, the fuel industry has become very focused on maintaining a high degree of fuel reliability. The fuel failure rates in recent years have dropped to ppm levels (Fig. 7.24), which is an impressive reliability record for any industry, especially considering that there are several thousand fuel rods operating in each of some ~400 reactors around the world.

After fuel assembly that contains a failed fuel rod has been discharged into the fuel pool at the reactor site, investigation of failure root cause is initiated starting with visual inspection and other nondestructive techniques. If the cause of the failure cannot be identified through such poolside (underwater) examinations, it may be necessary to extract the affected rod from the assembly and ship it to a hot cell facility for further closer inspections

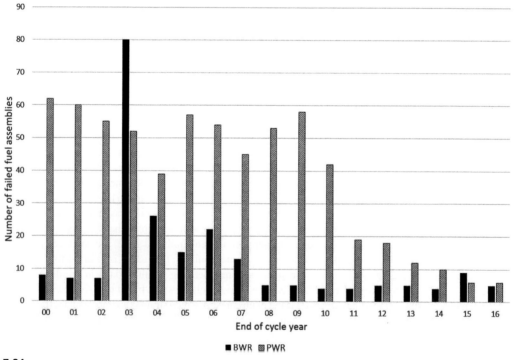

FIG. 7.24

History of fuel failures in the USA 2000–2016.

Courtesy: EPRI Fuel Reliability Program.

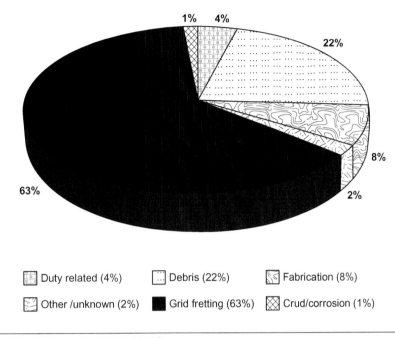

FIG. 7.25

Causes of reported fuel failure in the US 2000–16.

Courtesy: EPRI Fuel Reliability Program.

in dry environment and by destructive investigations as necessary. Several fuel failure mechanisms have been identified in this manner over several decades of industry experience in inspecting failed fuel. Causes of reported fuel failure in the US are shown in Fig. 7.25. In the following paragraphs, several primary fuel failure mechanisms are briefly highlighted.

7.6.2 DEBRIS FRETTING

Small foreign objects (debris) such as pieces of wires and metallic small pieces left over from maintenance work conducted occasionally get introduced in to the core. These objects circulate into the core region by the forceful coolant flow and may often become trapped between the grid spacer and a fuel rod. The trapped debris and fuel rod vibrate relative to one another under high coolant flow, eventually causing a through-wall opening in the cladding wall. This is the reason that even very small debris must be prevented from entering the primary loop, including any that may be trapped in fresh assemblies from fuel fabrication plants. All power plants and fuel fabrication plants have implemented rigorous foreign material exclusion programs to prevent such unintentional introduction of debris during reactor operation and fuel fabrication. An example of debris failure is shown in Fig. 7.26 [106].

Many fuel assembly designs have introduced debris filters in the form of cross-sectional strainers under each individual fuel assembly to prevent debris from reaching the upper elevations where the spacer grids are. Over the years, debris filters have been improved with innovative serpentine strainers that minimize direct flow path from the assembly bottom, thus increasing efficiency in trapping the foreign objects (see, e.g., Ref. [107]). In BWRs, feed water strainers are also being introduced to prevent introduction of external debris into the core region [108]. In some cases, wear resistance coatings on the fuel rod cladding surface (at the bottom end of the fuel rod where debris enters the core) are applied to prevent through-wall clad wear penetration even in the presence of a minor amount of debris.

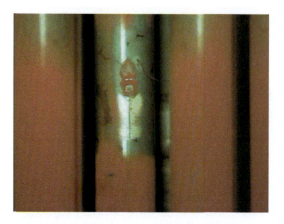

FIG. 7.26

Typical example of debris wear failure (due to a piece of wire).

From J. Siphers, R. Scneider, "Brunswick fuel failure reduction," Proceedings of 2010 LWR Fuel Performance/Top Fuel/WRFPM Orlando, Florida, USA, September 26–29, 2010. Copyright 2010 by the American Nuclear Society, LaGrange Park, Illinois.

7.6.3 GRID-TO-ROD FRETTING (GTRF)

The GTRF failure mechanism arises from vibration of the fuel rod relative to one or more of the assembly spacer grid support features. The relative vibrations are caused by high axial coolant flow, cross flow, development of gaps between fuel rod and grid support feature, and sub-channel turbulence [109,110]. The spacer grids have springs (or metallic vanes), with intricate curved shapes, serving two primary functions: to promote adequate heat transport downstream of the grid and to hold the fuel rod in its place in the fuel assembly lattice structure. Under irradiation, the spring material relaxes and loses its spring force (i.e., its grip) on the fuel rod. This introduces a gap between the fuel rod and grid support and increases the possibility that the fuel rod surface may wear against the part of the grid spacer that holds it. Fig. 7.27 [111] shows an example of such a failure. GTRF can be minimized by increasing the component contact area, modifying the spring mechanical design to reduce vibrations, and providing better spring material selection. Remedies including wear-resistant coatings on the cladding outer surface in the spacer grid regions are also employed.

7.6.4 PELLET-CLADDING MECHANICAL INTERACTION (PCMI)

An illustration of mechanical interaction between the fuel pellet and the cladding is shown schematically in Fig. 7.28, in which the radial scale is intentionally exaggerated. As designed, the as-fabricated fuel rod has a finite gap between the pellet outer diameter and cladding inner diameter (see Fig. 7.28A). Under hot conditions at power, the ceramic UO_2 material develops radial and axial cracking due to thermal stresses. Such cracking causes some degree of fuel relocation and eccentricity, but the overall cylindrical shape of the pellet is retained. With irradiation, the pellet diameter initially begins to contract due to removal of as-fabricated pores (a phenomenon called fuel densification). At the same time, during the first reactor cycle of irradiation, the cladding is creeping inward due to the coolant pressure being higher than the rod internal pressure. The net result is that the pellet cladding gap closes as irradiation proceeds. Subsequently, the pellet begins to swell due to the creation of fission products, both gaseous and solid. Overall, the pellet takes on the shape of an hourglass. The evolution of porosity and density of the pellet material while cladding collapses on to the fuel outer surface results in mechanical interaction between pellet and cladding.

FIG. 7.27

Fretting breach of cladding [111].

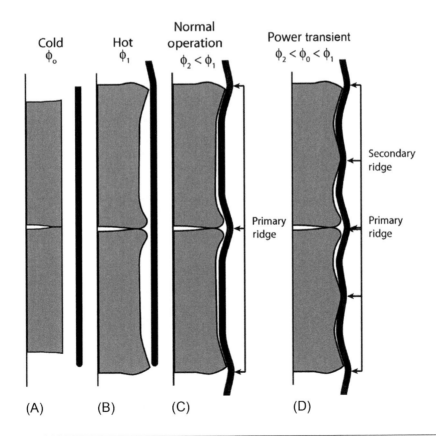

FIG. 7.28

Development of PCMI with increasing burnup. (The radial dimension is intentionally exaggerated for illustrative purposes.)

The mutually imparted stress between pellet and cladding becomes larger as the fuel burnup increases. This phenomenon is called PCMI. Under strong PCMI, and if the cladding is locally weaker due to hydriding (see Section 7.4), or has an initial part-wall fabrication flaw, a cladding breach may result. The small radial clearance that exists between the outside diameter of the ceramic fuel pellets and the inner diameter of the cladding tube facilitates loading pellets in the cladding tube during fuel rod fabrication. However, minor pellet chips can be generated during pellet loading and handling. Such fuel pellet chips can become lodged in the pellet-clad gap and become local stress concentrators. For many BWR designs, damage to the soft liner on the cladding inner surface can also occur due to such chips being lodged in the pellet-clad gap.

Such missing pellet surface (MPS) objects have been identified as a root cause of fuel failures in both PWR [112] and BWR [113]. The MPS can impart local bending stresses as pellet swells against the cladding and also results in a local cold spot where hydrogen can migrate from nearby regions of the cladding. Both of these effects increase the potential for fuel failure. MPS can also create adverse conditions due to the PCI-SCC mechanism, as described below.

7.6.5 PELLET-CLADDING INTERACTION-STRESS CORROSION CRACKING (PCI-SCC)

PCI-SCC fuel failures are normally induced following substantial power changes. The mechanism, shown schematically in Fig. 7.29 [114], involves a synergistic combination of mechanical and chemical interactions between the UO_2 fuel pellets (with iodine and possibly other fission products) and the Zr-alloy cladding. PCI-SCC failures were more prevalent in the early days of BWRs. The advent of soft zirconium liner layer on the clad ID reduced these occurrences (see Ref. [11]). PCI failures are generally low-strain failures and the cladding crack exhibits a zigzag pattern at high magnifications. PCI failures can also propagate axially [115], as in the case of the cladding breach shown in Fig. 7.30, or can lead to secondary degradation [116].

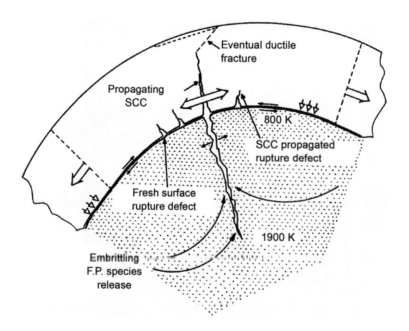

FIG. 7.29

A schematic representation of the PCI-SCC failure mechanism in Zr-alloy cladding [114].

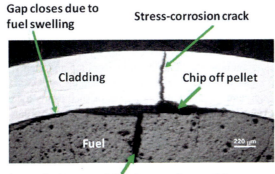

FIG. 7.30

A cross-sectional image of the pellet-clad region where a PCI-SCC crack has propagated through the cladding wall. The presence of MPS and the radial crack in pellet tends to increase the risk of the PCI-SCC crack initiation [115].

From International Atomic Energy Agency, GROESCHEL, F. et al., "Failure root cause of a PCI suspect liner fuel rod", Fuel failure in water reactors: Causes and mitigation, IAEA-TECDOC-1345, IAEA, Vienna (2003). Reproduced with permission from IAEA.

7.6.6 LESS COMMON FAILURE MECHANISMS

The following less common fuel failure mechanisms were encountered in the past but have been essentially eliminated in the modern fuel:

7.6.6.1 Axial splits in BWR cladding: Several BWRs in the 1990s experienced dramatic failures exhibiting long axial splits (see, e.g., Ref. [117]). These were clearly secondary failures following a primary one. The liner BWR cladding with a soft layer of practically pure zirconium tends to more rapidly corrode and pick up more hydrogen in comparison to the outer Zircaloy-2 wall. As schematically shown in Fig. 7.18, following a primary failure of the fuel rod, excessive corrosion and secondary hydriding could occur on the inside of the FC. It is still debatable whether the subsequent axial splits occur over a short period of time, due to power maneuvering of weakened cladding, or more slowly over a relatively longer period of time by the DHC mechanism. As a remedy to such long tube split failures, the current generation of liner materials are dilute alloys of zirconium with less than a half percent of alloying element addition, usually a combination of tin, iron, and oxygen. These current liner alloys tend to corrode less rapidly than the pure zirconium liner and have a lower hydrogen pickup after water ingress into the tube after a primary failure; therefore, they reduce the possibility of long splits of cladding due to the secondary hydriding and degradation.

7.6.6.2 Oxide Spallation: Prior to 1985, when the fabrication process for Zircaloys was not optimized, excessively thick ($\sim$200 μm) growth of waterside oxide layers showed a greater propensity to spall, creating local cold spots during service. The local cold spot can accumulate hydrogen and weaken the cladding. While the thick oxide layers were due to excessive nodular corrosion in BWRs, the uniform oxide itself in localized regions became fairly thick in PWRs. Such failures have now been eliminated in the modern cladding fabrication process by improving both nodular and uniform corrosion resistance and by switching to more corrosion-resistant niobium containing Zr-alloys for PWR fuel applications. However, a different oxide delamination phenomenon was recently reported for high corrosion resistance Zr-alloys with thin oxide layers termed 'oxide peeling' in Ref. [118]. No fuel failures have been reported due to such oxide peeling.

7.6.6.3 Weld failures: In early years, flaws in end plug welding in cladding remained, for example, due to atmospheric nitrogen and oxygen intrusion in the weld from the residual air in weld chambers, improperly cleaned surfaces being welded, or accidental deposition of weld electrode (tungsten) in the weld, etc. After more than five decades of nuclear fuel fabrication experience, the nuclear fuel suppliers have totally eliminated this failure mechanism.

7.7 SUMMARY/CONCLUSIONS

Fig. 7.31 chronologically shows significant milestones in application of Zr-alloys in the nuclear industry. Since the birth of the commercial nuclear power industry about six decades ago, Zr-alloys have retained their unique application as LWR FC and other core components due to neutron economy and their requisite material properties for minimizing component failures under very challenging in-service conditions. Good fuel performance means maintaining adequate margins to design and safety limits. Examples of such limits for fuel rods include allowable internal pressure inside the fuel rod, corrosion layer thickness of the cladding (related to wall thinning), hydrogen pickup in the cladding (related to mechanical ductility), cladding strain (related to ensuring cladding ductility at high burnups), dimensional stability (related to irradiation growth and creep), and acceptable performance under design basis accidents. Clearly, improvements in fuel performance drive overall improvements in plant economics—higher operating capacity factors and increased average fuel burnup at discharge.

Zr-alloy chemical composition, impurities, microstructure, types of SPPs, and texture have a profound impact on the in-service performance of cladding and other Zr-alloy core components. Many of these initial alloy characteristics also evolve with irradiation. Much of the research reported on Zr-alloys has been devoted to understanding, collecting data, and modeling these effects. The considerations for optimum heat treatments used are different in the fabrication of Zircaloys compared to Zr-Nb alloys. For Zircaloys, the processing heat treatments are based on the optimum as-fabricated SPP size for the specific application. Coarse SPP size is preferred for Zircaloy-4 used in PWR fuel applications to ensure adequate uniform corrosion resistance under a low oxygen potential PWR coolant. In the early years, a fine SPP size was preferred for Zircaloy-2 used in BWR fuel applications for adequate nodular corrosion resistance in a high oxygen potential BWR coolant. In recent years, however, as the discharge burnup of the BWR fuel has increased, the optimum as-fabricated SPP size is slightly increased to intermediate size for good nodular corrosion resistance in early life and good uniform corrosion resistance later in life. The presence of beta-zirconium in the as-fabricated microstructure of the nuclear zirconium component is generally undesirable from the corrosion resistance point of view. The intermediate heat treatment temperature and % area reduction in cold work steps after the beta-quenching step are the two processes used to limit the beta-zirconium phase in the microstructure.

Excessive corrosion and crud deposition on cladding and other core components strongly depend on the reactor coolant chemistry and local thermal-hydraulic conditions. These can arise unexpectedly in both BWR and PWR systems if coolant chemistry upsets, coupled with local power anomalies, are encountered as occasionally happens. Zr-alloy cladding and impurities contents that are within their respective specification ranges per se have only minor or indirect roles in CILC failures of Zr-alloys in service.

Hydrogen picked up by FC and other Zr-alloy core components impacts their mechanical integrity. A fuel failure is any breach of the cladding that allows the coolant to enter the fuel rod and contact the fuel pellets and fission products. Fuel failures are not generally a regulatory issue because regulatory limits on fuel failure—from personnel and public exposure to radiation standpoints—greatly exceed typical operational experiences. However, fuel reliability, or the absence of fuel failures, is nonetheless one of the most important aspects of commercial nuclear power. Costs associated with fuel failures include replacement of the fuel and of lost power generation. Fuel failures can cause a plant to shut down earlier in the cycle in some cases. Depending on the specific details, this can result in a cost on the order of one million dollars per day in lost generation and power replacement costs. Fuel failures also impact the radiation dose rates in working areas of a plant which can challenge efforts to keep worker doses low.

Zr-alloy development work is still continuing with objectives that evolve with time, such as lowering hydrogen pickup to accommodate impending new LOCA licensing requirements based on metal hydrogen level limits and requirements for smaller scatter in irradiation growth strains at high burnups. These new alloys have various combinations of lower levels of tin, niobium, iron, and chromium. More recently, so-called "accident tolerant" fuel (ATF) claddings are being evaluated, especially after the Fukushima tsunami/earthquake accident. With the loss of site power to maintain operation of coolant circulation pumps at some of the Fukushima reactor units, the core

7.7 SUMMARY/CONCLUSIONS

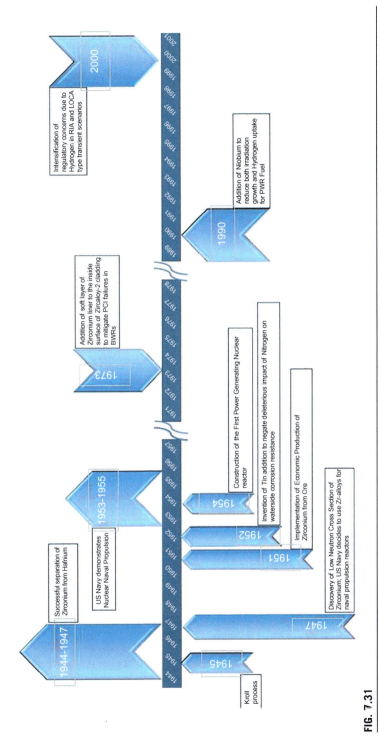

FIG. 7.31

Significant zirconium developments in the nuclear industry.

and used fuel pools encountered high temperatures. As a result, significant quantities of hydrogen were generated due to chemical reaction between zirconium and high-temperature steam, and the explosion of the hydrogen resulted in release of radioactivity to the environment. To reduce the risk of such explosion, it is envisioned that the proposed accident tolerant fuel and cladding designs would replace the Zr-alloy that would be phased out. Zr-alloys would be replaced with materials that limit hydrogen generation during accident scenarios. The proposed cladding materials are Fe-Cr-Al alloys and silicon carbide. However, significant developmental work (such as sealing the silicon carbide tube at both ends, increasing fuel enrichment, or reducing the tube wall thickness to compensate for the high neutron cross section for ferrous alloy cladding, etc.) needs to be completed before such alternate cladding concepts can be implemented in water cooled reactors. Owing to the development effort still needed, commercial implementation of ATF is likely to be at least several decades away.

REFERENCES

[1] C.L. Whitmarsh, Review of Zircaloy-2 and Zircaloy-4 Properties Relevant to N. S. Savannah Reactor Design, ORNL-3281, Oak Ridge National Laboratory, Oak Ridge, TN, 1962.

[2] H.G. Rickover, USN, The Decision to Use Zirconium in Nuclear Reactors, in: R.B. Adamson (Ed.), Zirconium Production and Technology: The Kroll Medal Papers, American Society for Testing and Materials, West Conshohocken, PA, 2010, pp. 9–17. ASTM RPS2.

[3] D.E. Thomas, in: B. Lustman, F. Kerze (Eds.), The Metallurgy of Zirconium, McGraw-Hill, New York, NY, 1955, p. 637.

[4] S. Kass, Development of the Zircaloys, USAEC report WAPD-T-1549, U.S. Atomic Energy Commission, Bettis Atomic Power Laboratory, West Mifflin, PA, September 1963.

[5] G. Pan, A.M. Garde, A.R. Atwood, in: R. Comstock, P. Barberis (Eds.), Performance and Property Evaluation of High-Burnup Optimized ZIRLO™ Cladding, ASTM International, West Conshohocken, PA, 2015, pp. 607–627. ASTM STP 1543.

[6] V. Chabretou, P.B. Hoffman, S. Trapp-Pritsching, G. Garner, P. Barberis, V. Rebeyrolle, J.J. Vermoyal, Ultra Low Tin Quarternary Alloys PWR Performance, in: M. Limback, P. Barberis (Eds.), Irradiation Growth, and Mechanical Properties, ASTM International, West Conshohocken, PA, 2011, pp. 801–826. ASTM STP 1529.

[7] R.J. Comstock, G. Schoenberger, G.P. Sabol, in: E.R. Bradley, G.P. Sabol (Eds.), Influence of Processing Variables and Alloy Chemistry on the Corrosion Behavior of ZIRLO Nuclear Fuel Cladding, American Society for Testing and Materials, West Conshohocken, PA, 1996, pp. 710–725. ASTM STP 1295.

[8] A. Stern, J.-C. Brachet, V. Mailott, D. Hamon, F. Barcelo, S. Poissonnet, A. Pineau, J.-P. Mardon, A. Lesbros, in: B. Kammenzind, M. Limbäck (Eds.), Investigations of the Microstructure and Mechanical Properties of Prior-β Structure as a Function of the Oxygen Content in two Zirconium Alloys, ASTM International, West Conshohocken, PA, 2009, pp. 71–90. ASTM STP 1505.

[9] G. Pan, A.M. Garde, A.R. Atwood, R. Kallstorm, D. Jadernas, High Burnup Optimized ZIRLO™ Cladding Performance, Top Fuel 2013 Conference Proceedings, Paper 8427, Charlotte, North Carolina, USA, 2013.

[10] O.A. Besch, S.K. Yagnik, K.N. Woods, C.M. Eucken, E.R. Bradley, in: E.R. Bradley, G.P. Sabol (Eds.), Corrosion Behavior of Duplex and Reference Cladding in NPP Grohnde, American Society for Testing and Materials, West Conshohocken, PA, 1996, pp. 805–824. ASTM STP 1295.

[11] J.S. Armijo, L.F. Coffin, H.S. Rosenbaum, in: A.M. Garde, E.R. Bradley (Eds.), Development of Zirconium-Barrier Fuel Cladding, American Society for Testing and Materials, West Conshohocken, PA, 1994, pp. 3–18. ASTM STP 1245.

[12] D.H. Locke, Mechanisms of Deterioration of Defected LWR Fuel, Proc. Specialists' Meeting on Defected Zirconium Alloy Clad Ceramic Fuel in Water-Cooled Reactors, Chalk River, Canada, September 17–21, 1979, IWGFPT/6, p101, International Atomic Energy Agency, 1980.

[13] F. Garzarolli, E. Steinberg, H.G. Weidinger, in: L.F.P. Van Swam, C.M. Eucken (Eds.), Microstructure and Corrosion Studies for Optimized PWR and BWR Zircaloy Cladding, American Society for Testing and Materials, West Conshohocken, PA, 1989, pp. 202–212. ASTM STP 1023.

[14] V. Fidleris, R.P. Tucker, R.B. Adamson, in: R.B. Adamson, L.F.P. Van Swam (Eds.), An Overview of Microstructural and Experimental Factors that Affect the Irradiation Growth Behavior of Zirconium Alloys, American Society for Testing and Materials, West Conshohocken, PA, 1987, pp. 49–85. ASTM STP 939.

[15] M. Griffiths, R.W. Gilbert, V. Fidleris, in: L.F.P. Van Swam, C.M. Eucken (Eds.), Accelerated Irradiation Growth of Zirconium Alloys, American Society for Testing and Materials, West Conshohocken, PA, 1989, pp. 658–677. ASTM STP 1023.

[16] S. Valizadeh, G. Ledergerber, S. Abolhassani, D. Jadernas, M. Dahlback, E.V. Mader, J. Wright, L. Hallstadius, in: P. Barberis, M. Limbäck (Eds.), Effects of Secondary Phase Particle Dissolution on the in-Reactor Performance of BWR Cladding, ASTM International, West Conshohocken, PA, 2011, pp. 729–753. ASTM STP 1529.

[17] A.M. Garde, S.R. Pati, M.A. Krammen, G.P. Smith, R.K. Endter, in: A.M. Garde, E.R. Bradley (Eds.), Corrosion Behavior of Zircaloy-4 Cladding with Varying Tin Content in High-Temperature Pressurized Water Reactors, American Society for Testing and Materials, West Conshohocken, PA, 1994, pp. 760–778. ASTM STP 1245.

[18] J.H. Schemel, Zirconium Alloy Fuel Clad Tubing: Engineering Guide, first ed., Sandvik Special Metals, Kennewick (WA), 1989.

[19] D.L. Baty, W.A. Pavinich, M.R. Dietrich, G.S. Clevinger, T.P. Papazoglou, in: D.G. Franklin, R.B. Adamson (Eds.), Deformation Characteristics of Cold-Worked and Recrystallized Zircaloy-4 Cladding, American Society for Testing and Materials, West Conshohocken, PA, 1984, pp. 306–339. ASTM STP 824.

[20] E. Tenckhoff, Operable deformation systems and mechanical behavior of textured Zircaloy tubing, Zirconium in nuclear applications, ASTM STP 551, American Society of Testing of Materials (1974) 179–200.

[21] D.G. Franklin, G.E. Lucas, A.L. Bement, Creep of Zr-alloys in Nuclear Reactors, American Society for Testing and Materials, West Conshohocken, PA, 1983, p. 229. ASTM STP 815.

[22] R.M. Pelloux, R. Ballinger, G. Lucas, Effects of anisotropy and irradiation on the deformation behavior of Zircaloy-2, Electric Power Research Institute, EPRI NP-982 (January 1979).

[23] D.G. Franklin, G.E. Lucas, A.L. Bement, Creep of Zr-alloys in Nuclear Reactors, American Society for Testing and Materials, West Conshohocken, PA, 1983, p. 233. ASTM STP 815.

[24] J.J. Kearns, in: R. Comstock, P. Barberis (Eds.), Reflections on the Development of the "f" Texture Factors for Zirconium Components and the Establishment of Properties of the Zirconium-Hydrogen System, ASTM International, West Conshohocken, PA, 2015, pp. 3–22. ASTM STP 1543.

[25] J.J. Kearns, Thermal Expansion and Preferred Orientation of Zircaloy, WAPD-TM-472, 1965.

[26] M. Kimpara, K. Fujita, T. Kakuma, N. Nagai, in: G. Franklin, R.B. Adamson (Eds.), On the Texture Measurements for Zircaloy Cladding Tube, American Society for Testing and Materials, West Conshohocken, PA, 1984, pp. 244–255. ASTM STP 824.

[27] D.M. Rishel, B.F. Kammenzind, in: R. Comstock, A.T. Motta (Eds.), The Role of Gamma Radiation on Zircaloy-4 Corrosion, American Society for Testing and Materials, West Conshohocken, PA, 2018; pp. 555–595. ASTM STP 1597.

[28] L.F.P. Van Swam, S.H. Shann, in: C.M. Eucken, A.M. Garde (Eds.), The Corrosion of Zircaloy-4 Fuel Cladding in Pressurized Water Reactors, American Society for Testing and Materials, West Conshohocken, PA, Philadelphia, 1991, pp. 758–781. ASTM STP-1132.

[29] T.R. Allen, R.J.M. Konings, A.T. Motta, Comprehensive Nuclear Materials, 5, Elsevier Ltd, Amsterdam, The Netherlands, 2012, pp. 49–68.

[30] IAEA, TECDOC-996, Waterside Corrosion of Zirconium Alloys in Nuclear Power Plants, International Atomic Energy Agency, Vienna, Austria, Jan 1998.

[31] R.V. Macbeth, R. Trenberth, R.W. Wood, An Investigation into the Effect of 'crud' Deposits on Surface Temperature, Dry-out and Pressure Drop, with Forced Convection Boiling of Water at 69 Bar in an Annular Test Section (AEEW-R-705), United Kingdom, 1971.

[32] C. Pan, B. Jones, A.J. Machiels, Concentration level of solutes in porous deposits with chimneys under wick boiling conditions, Nucl. Eng. Des. 99 (1987) 317–327.

[33] PWR Axial Offset Anomaly, (AOA) Guidelines, Revision 1, EPRI Technical Report #1008102, 2004.

[34] J. Deshon, D. Hussey, B. Kendrick, J. McGurk, J. Secker, M. Short, Pressurized water reactor fuel crud and corrosion modeling, JOM 63 (8) (2011) 64–72.

[35] Fuel Reliability Guidelines, PWR Fuel Cladding Corrosion and Crud, Revision 1: Volume 1, Guidance, EPRI Technical Report, 3002002795, 2014.

[36] I.L. Barnwell, P.D. Parsons, D.R. Tice, in: C.M. Eucken, A.M. Garde (Eds.), Corrosion of Zircaloy-4 PWR Fuel Cladding in Lithiated Borated Water Environments, American Society for Testing and Materials, West Conshohocken, PA, 1991, pp. 628–642. ASTM STP-1132.

[37] J.N. Esposito, G. Economy, W. A. Byers, J.B. Esposito, F.W. Pement, R.J. Jacko, C.A. Bergmann, The Addition of Zinc to Primary Reactor Coolant for Enhanced PWSCC Resistance, Proc. Fifth International Conference on Environmental Degradation of Materials in Nuclear Power Systems – Water Reactors, American Nuclear Society, LaGrange Park, IL, 1991.

[38] http://www.thermopedia.com/content 605 (or equivalent websites and text books).

[39] Plant-Specific Optimization of LWR Water Chemistry. EPRI Technical Report TR-107329 (1997).

[40] R.L. Cowan, J. Varela, S.E. Garcia, The effect of on-line noble metal addition on the shutdown dose rates of boiling water reactors, Proc. Fifteenth Environmental Degradation Conference, Colorado Springs, CO (USA), 2011.

[41] K. Edsinger, C.R. Stanek, B.D. Wirth, Light water reactor fuel performance: Current status, challenges, and future high fidelity modeling, JOM 63 (8) (2011) 48–53.

[42] R. Tropasso, J. Wilse, and B. Cheng, Crud-Induced Cladding Corrosion Failures in TMI-1 Cycle 10, Proc. of the 2004 International Meeting on LWR Fuel Performance, Orlando, Florida, September 19–22, 2004 Paper 1070.

[43] E. J. Ruzauskas, D.L. Smith, Fuel failures during Cycle 11 at River Bend, Proc. of the 2004 International Meeting on LWR Fuel Performance Orlando, Florida, September 19–22, 2004 Paper 1016.

[44] M.C. Billone, Argonne National Laboratory, LOCA Embrittlement Criteria, US NRC ADAMS Number ML051010265, 2005.

[45] J.B. Bai, C. Prioul, S. Lanslart, D. Pran-Ois, Brittle fracture induced by hydrides Zircaloy-4, Scr. Metall. 25 (1991) 2559–2563.

[46] Hot cell examination of Dresden-2 fuel rods after 4 cycles of hydrogen water chemistry, EPRI Technical Report, TR-108782, 1997.

[47] S.J. King, R.L. Kesterson, K.H. Yueh, R.J. Comstock, W.M. Herwig, S.D. Ferguson, in: G. Moan, P. Rudling (Eds.), Impact of Hydrogen on Dimensional Stability of ZIRLO Fuel Assemblies, ASTM International, West Conshohocken, PA, 2002, pp. 471–489. ASTM STP 1423.

[48] H. Pettersson, B. Bengtson, T. Andersson, H.-J. Sell, P.-B. Hoffmann, F. Garzarolli, in: Investigation of Increased Hydriding of Guide Tubes in Ringhals 2 during Cycle Startup, Proc. of the LWR Fuel Performance Meeting/Top Fuel, San Francisco, 2007.

[49] P. Rudling, R. Adamson, B. Cox, F. Garzarolli, A. Strasser, High burnup fuel issues, Nuc Engg. & Technol 40 (1) (2008).

[50] R.L. Beck, W.M. Mueller, Zirconium Hydrides and Hafnium Hydrides, Metal Hydrides, Academic Press, New York and London, 1968, p. 246.

[51] R. Choubey, M.P. Puls, Crack initiation at long radial hydrides in Zr-2.5Nb pressure tube material at elevated temperatures, Metall. Mater. Trans. A 25 (5) (1994) 993–1004.

[52] A.M. Garde, G.P. Smith, R.C. Pirek, in: E.R. Bradley, G.P. Sabol (Eds.), Effects of Hydride Precipitate Localization and Neutron Fluence on the Ductility of Irradiated Zircaloy-4, American Society for Testing and Materials, West Conshohocken, PA, 1996, pp. 407–430. ASTM STP 1295.

[53] J.B. Bai, C. Prioul, D. François, Hydride embrittlement in Zircaloy-4 plate: Part I. Influence of microstructure on the hydride embrittlement in Zircaloy-4 at 20°C and 350°C, Metall. Mater. Trans. A 25 (6) (1994) 1185–1197.

[54] A. Hermann, S.K. Yagnik, D. Gavillet, in: B. Kammenzind, M. Limbäck (Eds.), Effect of Local Hydride Accumulations on Zircaloy Cladding Mechanical Properties, ASTM International, West Conshohocken, PA, 2009, pp. 140–162. ASTM STP 1505.

[55] S.K. Yagnik, J.-H. Chen, R.-C. Kuo, in: R. Comstock, P. Barberis (Eds.), Effect of Hydide Distribution on the Mechanical Properties of Zirconium Alloy Fuel Cladding and Guide Tubes, ASTM International, West Conshohocken, PA, 2015, pp. 1077–1106. ASTM STP 1543.

[56] S.K. Yagnik, A. Hermann, R.-C. Kuo, in: P. Rudling, B. Kammenzind (Eds.), Ductility of Zircaloy-4 Fuel Cladding and Guide Tubes at High Fluences, ASTM International, West Conshohocken, PA, 2005, pp. 604–631. ASTM STP 1467.

[57] L.O. Jernkvist, in: Lower threshold LHGR for initiation of outside-radial cracking in liner fuel cladding, *Transactions*, SMiRT-22, San Francisco, California, USA - August 18-23 2013.

[58] M.P. Puls, *The effect of hydrogen and hydrides on the integrity of zirconium alloy components: delayed hydride cracking*, Engineering Materials, Springer-Verlag, London (2012); https://www.springer.com/us/book/9781447141945.

[59] Secondary Hydriding of Defected Zircaloy Clad Fuel Rod, EPRI Technical Report, TR-101773(1993).

[60] R.L. Yang, B.C. Cheng, S.K. Yagnik, B.C. Cheng, H.H. Klepfer, N. Kjaer-Pederson, P. Rank, in: EPRI failed fuel degradation R&D program, Proc of International Topical. Meeting on LWR Fuel, Performance, American Nuclear Society, West Palm Beach, FL (USA), April 17-21, 1994, pp. 435–446.

REFERENCES

[61] Jonsson, L. Hallstatius, B. Grapengiesser and G. Lysell, Failure of a Barrier Rod in Oskarshamn 3, Proc. of Int. Topical Meeting on LWR Fuel Performance, ANS, Avignon, France, April 21-24, 1991, pp. 371–377.

[62] E.C.W. Perryman, Pickering pressure tube cracking experience, Nuclear Energy 17 (1978) 95–105.

[63] J.-J. Won, M.-S. Kim, K.-T. Kim, Heat-up and cool-down temperature dependent hydride reorientation behaviors in zirconium alloy cladding tubes, Nucl. Engg. & Technol 46 (5) (2014).

[64] H. Chung, Understanding hydride- and hydrogen-related processes in high-burnup cladding in spent-fuel storage and accident situations, Proc. of the International Meeting on LWR Fuel Performance, Orlando, FL, U.S.A., pp. 470 (2004).

[65] M. Griffiths, Personal communication, June 2019.

[66] M. Aomi, T. Baba, T. Miyashita, K. Kamimura, T. Yasuda, Y. Shinohara, T. Takeda, in: B. Kammenzind, M. Limbäck (Eds.), Evaluation of Hydride Reorientation Behavior and Mechanical Properties for High-Burnup Fuel-Cladding Tubes in Interim Dry Storage, ASTM International, West Conshohocken, PA, 2009, pp. 651–671. ASTM STP 1505.

[67] B. Bourdiliau, F. Onimus, C. Cappelaere, V. Pivetaud, P. Bouffioux, V. Chabretou, A. Miquet, in: M. Limbäck, P. Barberis (Eds.), Impact of Irradiation Damage Recovery during Transportation on the Subsequent Room Temperature Tensile Behavior of Irradiated Zirconium Alloys, American Society for Testing and Materials, West Conshohocken, PA, 2011, pp. 929–953. ASTM STP 1529.

[68] M. Billone, T. A. Burtseva, Y. Y. Liu, Effects of drying and storage on high-burnup cladding ductility, Proc. International High-level Radioactive Waste Management Conference, Albuquerque, NM, April 28–May 2, 2013.

[69] J. V. Cathcart, R.E. Pawel, R.A. McKee, R.E. Druschel, G.J. Yurek, J.J. Campbell, S.H. Jury, Zirconium metal-water oxidation kinetics IV reaction rate studies, Oak Ridge National Laboratory (ORNL)/NUREG-17, August 1977, ADAMS Accession No. ML052230079.

[70] D. A. Power, R. O. Meyer, Cladding swelling and rupture models for LOCA analysis, NUREG-0630, 1980, NRC ADAMS Accession No. ML053490337.

[71] H.M. Chung, T.F. Kassner, Embrittlement criteria for Zircaloy fuel cladding applicable to accident situations in light-water reactors: Summary report, NUREG/CR-1344, 1980, NRC ADAMS Accession No. ML040090281.

[72] D. F. Ross, Compendium of ECCS research for realistic LOCA analysis, final report, NUREG-1230, R4, 1988, NRC ADAMS Accession No. ML053490333.

[73] M. Billone, Y. Yan, T. Burtseva, R. Daum, Cladding embrittlement during postulated loss-of-coolant accidents, NUREG/CR-6967, 2008, NRC ADAMS Accession No. ML082130389.

[74] Nuclear fuel behavior in loss-of-coolant accident (LOCA) conditions, State-of-the-art Report, Nuclear Energy Agency, NEA/CSNI/R (2009) OECD, 2009. ISBN: 978-92-64-99091-3.

[75] Nuclear fuel behavior under reactivity-initiated accident (RIA) conditions, State-of-the art Report, Nuclear Energy Agency, NEA/CSNI/R (2010), OECD, 2010. ISBN 978–92–64-99113-2.

[76] J. Papin, F. Barré, C. Grandjean, M. Petit, J.-C. Micaelli, Summary and interpretation of the CABRI Rep-Na program, Nucl. Technol. 157 (2007) 230–250.

[77] T. Fuketa, F. Nagase, K. Ishijima, T. Fujishiro, NSRR/RIA experiments with high burnup PWR fuels, Nucl. Saf. 37 (4) (1996) 328–342.

[78] S. Kobayashi, N. Ohnishi, T. Yoshimura, W.G. Lussie, Experimental results of some cluster tests in NSRR, J. Nucl. Sci. Technol. 15 (6) (1978) 448–454.

[79] T. Fuketa, T. Sugiyama, M. Umeda, K. Tomiyasu, H. Sasajima, Behavior of high burnup PWR fuels during simulated reactivity-initiated accident conditions, Top Fuel 2006 Conference, Salamanca, Spain, 2006, pp. 279–283.

[80] P.M. Clifford, in: The US nuclear regulatory Commission's strategy for revising the RIA acceptance criteria, Proc. of the International LWR Fuel Performance Meeting, San Francisco, CA, American Nuclear Society, La Grange, IL, 2007, pp. 543–545.

[81] R. Landry, US NRC technical & regulatory basis for the reactivity-initiated accident interim acceptance criteria & guidance, ADAMS accession number ML070220400, US Nuclear Regulatory Commission, Washington, DC, 2007.

[82] P. M. Clifford, Technical & regulatory basis for the reactivity-initiated accident acceptance criteria and guidance, revision, March 16, 2015, ADAMS Accession No. ML14188C423, US Nuclear Regulatory Commission, Washington, DC, 2007.

[83] G.P. Kobylyansky, A.E. Novoselov, Z.E. Ostrovsky, A.V. Obukhov, V.Y. Shishin, V.N. Shishov, A.V. Nikulina, M. M. Peregud, S.T. Mahmood, D.W. White, Y.P. Lin, M.A. Dubecky, in: B. Kammenzind, M. Limbäck (Eds.), Irradiation-Induced Growth and Microstructure of Recrystallized, Cold Worked and Quenched Zircaloy-2, NSF, and E635 Alloys, ASTM International, West Conshohocken, PA, 2009, pp. 564–582. ASTM STP 1505.

[84] R.A. Holt, R.W. Gilbert, C-component dislocations in neutron irradiated Zircaly-2, J. Nucl. Mater. 116 (1983) 127.

[85] P. Tagtstrom, M. Limback, M. Dahlback, T. Andersson, H. Pettersson, in: G. Moan, P. Rudling (Eds.), Effects of hydrogen pickup and second phase particle dissolution on the in-reactor corrosion performance of BWR claddings, ASTM International, West Conshohocken, PA, 2002, pp. 96–118. ASTM STP 1423.

[86] V.N. Shishov, in: M. Limback, P. Barberis (Eds.), The evolution of microstructure and deformation stability in Zr-Nb-Sn-Fe alloys under neutron irradiation, ASTM International, West Conshohocken, PA, 2011, pp. 37–66. ASTM STP 1529.

[87] V.F. Urbanic, M. Griffiths, in: G.P. Sabol, G.D. Moan (Eds.), Microstructural Aspects of Corrosion and Hydrogen Ingress in Zr-2.5Nb, American Society for Testing and Materials, West Conshohocken, PA, 2000, pp. 641–657. ASTM STP 1354.

[88] P. Bossis, B. Verhaeghe, S. Doriot, D. Gilbon, V. Chabretou, A. Dalmais, J.P. Mardon, M. Blatt, A. Miquet, in: B. Kammenzind, M. Limbäck (Eds.), PWR Comprehensive Study of High Burnup Corrosion and Growth Behavior of M5 and recrystallized Low-Tin Zircaloy-4, ASTM International, West Conshohocken, PA, 2009, pp. 430–456. ASTM STP 1505.

[89] A.M. Garde, in: C.M. Eucken, A.M. Garde (Eds.), Enhancement of Aqueous Corrosion of Zircaloy-4 due to Hydride Precipitation at the Metal-Oxide Interface, ASTM International, West Conshohocken, PA, 1991, pp. 566–594. ASTM STP 1132.

[90] A.M. Garde, in: L.F.P. Van Swam, C.M. Eucken (Eds.), Effects of Irradiation and Hydriding on the Mechanical Properties of Zircaloy-4 at High Fluence, ASTM International, West Conshohocken, PA, 1989, pp. 548–569. ASTM STP 1023.

[91] A. M. Garde, D. Mitchell, Comparison of ductility of irradiated ZIRLO cladding and Zircaloy-4, 2012 Top Fuel Conference, Manchester, England, September 2012, Paper AO149.

[92] R.B. Adamson, R.P. Tucker, V. Fidleris, in: G.D. Franklin (Ed.), High-Temperature Irradiation Growth in Zircaloy, American Society for Testing and Materials, West Conshohocken, PA, 1982, pp. 208–234. ASTM STP 754.

[93] D.O. Northwood, in: J.A. Sprague, D. Kramer, (Eds.), Irradiation Growth in Zirconium and Its Alloys, American Society for Testing and Materials, West Conshohocken, PA, 1979, pp. 62–76. ASTM STP 683.

[94] Y. de Carlan, C. Regnard, M. Griffiths, D. Gilbon, C. Lemaignan, E.R. Bradley, G.P. Sabol, Influence of Iron in the Nucleation of $<c>$ Component Dislocation Loops in Irradiated Zircaloy-4, American Society for Testing and Materials, West Conshohocken, PA, 1996, pp. 638–653. ASTM STP 1295.

[95] NRC Bulletin 96-01, Control Rod Insertion Problems, March 8, 1996, US Nuclear Regulatory Commission, Washington, DC.

[96] F. Garzarolli, R. Adamson, P. Rudling, A. Strasser, BWR Fuel Channel Distortion, ZIRAT16, ANT International, Sweden, 2011.

[97] D.G. Franklin, G.E. Lucas, A.L. Bement, Creep of Zr-Alloys in Nuclear Reactors, American Society for Testing and Materials, West Conshohocken, PA, 1983, pp. 16–17. ASTM STP 815.

[98] K.L. Murty, Creep studies for Zircaloy life predictions in water reactors, Journal of Metals 51 (10) (1999) 32–39.

[99] G.B. Piercy, Mechanisms for the in-reactor creep of zirconium alloys, J. Nucl. Mater. 26 (1) (1968) 18–50.

[100] V. Fidleris, Uniaxial in-reactor creep of zirconium alloys, J. Nucl. Mater. 26 (1) (1968) 51–76.

[101] E.R. Gilbert, In-reactor creep of reactor materials, Reactor Technology 14 (1971) 258–285.

[102] R. Adamson, F. Garzarolli, C. Patterson, In-Reactor Creep of Zirconium Alloys, ZIRAT14, ANT International, Skultuna, Sweden, 2009.

[103] J.H. Moon, P.E. Cantonwine, K.R. Anderson, S. Karthikeyan, M.J. Mills, Characterization and modelling of creep mechanisms in Zircaloy-4, J. Nucl. Mater. 353 (2006) 177–189.

[104] F. Ferrer, A. Barbu, T. Bretheau, J. Crepin, F. Williame, D. Charquet, in: G. Moan, P. Rudling (Eds.), The Effect of Small Concentrations of Sulfur on the Plasticity of Zirconium Alloys at Intermediate Temperatures, ASTM International, West Conshohocken, PA, 2002, pp. 863–887. ASTM STP 1423.

[105] D.G. Franklin, G.E. Lucas, A.L. Bement, Creep of Zr-Alloys in Nuclear Reactors, American Society for Testing and Materials, West Conshohocken, PA, 1983, pp. 72–79. ASTM STP 815.

[106] J. Siphers, R. Schneider, Brunswick fuel failure reduction, Proc. of 2010 LWR Fuel Performance/Top Fuel/WRFPM Orlando, Florida, USA, September 26–29, 2010.

[107] http://www.power-eng.com/articles/npi/print/volume-7/issue-4, [R. Buechel, Fuel Reliability: How it affects the industry, and one fuel vendor's journey to flawless fuel performance, Aug 2014] and other similar online articles.

[108] https://nuclear.gepower.com/service-and-optimize/solutions.

[109] P.R. Rubiolo, M.Y. Young, On the factors affecting the fretting-wear risk of PWR fuel assemblies, Nucl. Eng. Des. 239 (2009) 68.

[110] Y.H. Lee, K.H. Yoon, H.K. Kim, H.S. Kang, Asymmetric clearance effect in grid-rod-fretting, Proc. of Top Fuel 2013, Charlotte, North Carolina, September 15–19, 2013.

[111] R. Buechel, Z. Karoutas, R. Lu, in: Grid To Rod Fretting Performance of Westinghouse Fuel, Proc. of the 2008 Water Reactor Fuel Performance Meeting, Seoul, Korea, Paper 8080, October 19– October 22, 2008.

[112] Y. Aleshin, C. Beard, G. Mangham, D. Mitchell, E. Malek, M. Young, The effect of pellet and local power variations on PCI margin, Proceedings of Top Fuel 2010 September 26–29, Orlando, Florida, USA Paper 041, 2010.

[113] C. Powers, P. Dewes-Erlangen, M. Billaux, R. Perkins, E. Ruzauskas, E. Armstrong, R. Ralph, G. Blomberg, G. Lysell, B. Cheng, Hot Cell Examination Results of Non-classical PCI Failures at La Salle, Proc of Water Reactor Fuel Performance Meeting, Kyoto (Japan), 2005.

[114] J.T.A. Roberts, Structural Materials in Nuclear Power Systems, Plenum Press, New York, 1981 (p. 66).

[115] J.H. Davies, J.S. Armijo, Post-Irradiation Examination of Failed KKL Fuel Rod, in: G. Muhling, K. Kernforschungszentrum (Eds.), Material Development for Fuel Elements in LWRs, Karlsruhe, Germany, 1993.

[116] F. Groeschel, G. Bart, R. Montgomery, S.K. Yagnik, Failure root cause of a PCI suspect liner fuel rod. Fuel Failure in Water Reactors: Causes and Mitigation, IAEA-TECDOC-1345, IAEA, Vienna, 2003.

[117] G. Lysell, V. Grigoriev, P. Efsing, Axial splits in BWR fuel rods, Proc. International Topical Meeting LWR Fuel Performance, Park City, Utah, USA, April 10–13, 2000.

[118] A.M. Garde, G. Pan, A.J. Muller, L. Hallstadius, R. Comstock, P. Barberis (Eds.), Oxide Surface Peeling of Advanced Zirconium Alloy Cladding after High Burnup Irradiation in Pressurized Water Reactors, ASTM International, West Conshohocken, PA, 2015, pp. 673–692. ASTM STP 1543.

AUSTENITIC STAINLESS STEELS

Gary S. Was*, Shigeharu Ukai†
University of Michigan, Ann Arbor, MI, United States* Hokkaido University, Sapporo, Japan†

CHAPTER OUTLINE

- 8.1 Introduction 294
- 8.2 Application in LWRs and GenIV Reactors 294
  - 8.2.1 Light Water Reactors 294
  - 8.2.2 Sodium-Cooled Fast Reactors 295
- 8.3 Radiation-Induced Metallurgical Changes 297
  - 8.3.1 Radiation-Induced Segregation 297
  - 8.3.2 Dislocation Microstructure 302
  - 8.3.3 Phase Stability 303
  - 8.3.4 Transmutation 305
- 8.4 Radiation-Induced Mechanical Property Changes and Degradation Modes 308
  - 8.4.1 Irradiation Hardening 308
  - 8.4.2 Reduction in Fracture Toughness and Embrittlement 309
  - 8.4.3 High-Temperature He Embrittlement 311
  - 8.4.4 Void Swelling 311
  - 8.4.5 Irradiation Creep and Fatigue 315
  - 8.4.6 In-Reactor Creep Rupture Properties 318
- 8.5 PCI/FCCI Effects With Fission Fuels 318
- 8.6 Chemical Compatibility With Coolants 321
- 8.7 Stress-Corrosion Cracking 323
  - 8.7.1 SCC in BWRs 323
  - 8.7.2 IGSCC in PWRs 323
- 8.8 Combined Effects of Water Environment and Radiation 326
  - 8.8.1 Irradiation-Assisted Stress-Corrosion Cracking 326
  - 8.8.2 Irradiation-Accelerated Corrosion 332
  - 8.8.3 Corrosion Fatigue 332
  - 8.8.4 Hydrogen Embrittlement 334
  - 8.8.5 Fracture Toughness 334
- 8.9 Perspectives and Prospects 337
- References 340
- Further Reading 347

8.1 INTRODUCTION

Austenitic alloys used in light water reactor (LWR) service actually span a rather limited range of materials, primarily covering a narrow selection from the 300 series of stainless steels in the United States, European, and Asian service, with primary focus on AISI 304 and AISI 316 stainless steels with some minor emphasis on various stabilized variants of these steels such as AISI 321, 347, and 348, and some developmental alloys such as D9 and high-temperature ultrafine precipitate strengthened (HT-UPS). Various stabilized steels are commonly used in Germany, Russia, and former Soviet states, but the compositional variation is not too different from Western steels. A much wider range of austenitic alloys has been explored in fast reactors but these reactors have different neutron spectra, lower rates of helium and hydrogen production and most importantly, an absence of water and therefore thermal neutrons. More importantly, these reactors typically operate at one to two orders magnitude higher displacement rate and much higher operating temperatures than LWRs. Only in first-generation fast reactors (BN-350, DFR), mostly now decommissioned, did the lower end of the operating temperature range coincide with LWR operation.

The emphasis on fast reactors is important because some of the more worrisome degradation modes associated with radiation express themselves only at very high neutron exposure and were first identified and also extensively studied in fast reactors. Such high exposure levels are only now being reached as LWRs move into extended operation for >40 years. Therefore, our ability to search for more radiation-resistant alloys for LWR service must draw heavily on experience accumulated in fast reactors.

This chapter begins with a section on the application of austenitic stainless steels in both LWRs and Generation IV (GenIV) reactor designs and concepts. Following is a summary of metallurgical changes induced by radiation, such as radiation-induced segregation (RIS), dislocation microstructure evolution, phase stability, and transmutation. Mechanical properties altered by irradiation and key degradation modes is discussed next, followed by sections on the behavior of stainless steels in contact with fuels in both LWR and sodium fast reactor (SFR). This is followed by a short treatment of the chemical compatibility of stainless steels with various coolants including water, sodium, lead, lithium, and molten salt. Stress-corrosion cracking (SCC) is briefly reviewed and then the combined effects of a water environment and radiation are provided. The chapter closes with perspectives and prospects for the future use of stainless steels including accident tolerant fuels.

8.2 APPLICATION IN LWRs AND GenIV REACTORS
8.2.1 LIGHT WATER REACTORS

Stainless steels are widely used in primary circuits of pressurized water reactors (PWRs) and in boiling water reactor (BWR) core components. Types 304 and 316 austenitic stainless steels are the main materials used for the structures supporting the nuclear core in the reactor pressure vessel and for pressurized boundary piping of PWR primary circuit and BWR cores. Stainless steels are found in instrument dry tubes, control rod absorber tubes, fuel bundle cap screws, control blades, handles and sheaths, core shroud and shroud head bolts, top guide baffle former, plate and bolts, jet pump beams, as well as various springs and bolts. The internal surfaces of low alloy steel components (vessel and large pipes mainly) are also clad with type 308/309 stainless steel weld overlays. Higher strength stainless steels (mainly A286 precipitation hardened austenitic steel, A410 and 17-4 pH martensitic stainless steels) are also widely used in PWRs for components such as bolts and nuts, springs and valve elements.

Operating experience with stainless steels over several decades of years has generally been excellent. Nevertheless, SCC failures have been observed. In a recent survey [1], 137 events have been reported regarding SCC occurrence of austenitic stainless steels in PWR primary systems: canopy and omega seals, CEDM housings, RVO-ring leak-off lines, valve drain lines are the locations where the number of cracks events is the highest. Intergranular stress-corrosion cracking (IGSCC) in steam lines of BWRs early in the life of many LWRs resulted from a combination of oxidizing conditions, weld-induced residual stresses, and sensitization of grain boundaries caused by heat treatment of high carbon steels in the temperature range in which chromium carbides formed rapidly on the boundaries, depleting them of chromium and making them susceptible to attack. IGSCC of Alloy 600 occurred in steam generators, on both the primary and secondary sides, and was driven by a susceptible microstructure and the creation of crevices on the secondary side in which crevice chemistry was favorable for intergranular attack. In the early days, stainless steels were used as fuel cladding because of its good resistance to corrosion. However, the high neutron absorption (relative to zirconium) and the emergence of SCC led to its removal from this application. More recently, irradiation-assisted stress-corrosion cracking (IASCC) of core components has posed the greatest threat to core materials integrity. Ref. [2] contains a list of the components susceptible to IASCC in water reactors, the reactor types (4) and alloys (10) in which it has been observed. Note that this list contains several nickel-base austenitic alloys. In fact, the most serious problem is in SCC of nickel base alloy 600 used in steam generator tubes, control rod drive mechanisms housing, and reactor vessel penetrations.

8.2.2 SODIUM-COOLED FAST REACTORS

In the sodium-cooled fast reactors (SFRs) envisioned in the GenIV program, fuel pins are expected to be irradiated at a higher temperature and to a higher burnup than in previously operated reactors, which are, for example, 370°C (inlet)/510°C (outlet) in the coolant temperatures at EBR-II, and the approved maximum burnup and neutron dose are limited in about 100 MWd/kg and 100 dpa. As the target values in SFR, the maximum cladding temperature is required to be 650–700°C to attain 530–550°C in the coolant outlet temperature, and core average burnup reaches 150–200 MWd/kg corresponding to 200–250 dpa in peak neutron dose [3–5]. Integrity of the fuel pins strongly depends on whether the cladding can withstand the irradiation environment. It is preferred that the cladding have a low swelling and high ductility throughout its lifetime. Sufficient creep strength is primarily required at a high temperature.

Austenitic stainless steels have excellent creep strength at high temperature, however, radiation-induced void swelling is a life-limiting factor for these steels when used for SFR fuel cladding. Extensive efforts have been devoted in all the countries developing the SFR, to improve commercially available 316 class austenitic stainless steels. Twenty percent of cold-worked (CW) austenitic steels are currently utilized as the cladding and duct materials in JOYO, MONJU, BOR-60, and BN600. These austenitic stainless steels have experienced the maximum dose in fuel assemblies reaching around 100 dpa. A life-limiting factor of the austenitic stainless steel fuel assemblies is a functional limit due to swelling-induced distortions of a fuel pin bundle in duct tubes.

Improvements in swelling resistance and creep strength in austenitic steels have been made by adding minor alloying elements and adjusting cold-working level to produce fine precipitates during irradiation. Phosphorus and boron are beneficial for the formation of finely dispersed carbides especially in CW austenitic steels [6,7]. Titanium and niobium are well known to provide MC carbide precipitation and to stabilize dislocations. These effects on creep rupture strength have been demonstrated in the development process of Japanese PNC316 and PNC1520 cladding as shown in Fig. 8.1 [7,8].

An increase of Ti up to about the stoichiometric content for MC carbide formation (0.25 wt% Ti) in 20 wt% Ni alloy, PNC1520, has superior strength to PNC316. Similar modifications applied to 20% CW Fe-15Cr-15Ni-Ti alloy, D9, have successfully attained an improved strength in the United States [9]. The alloy HT-UPS was developed as a spin-off based on the earlier developments of alloy D9 for the breeder reactor and alloy PCA (primary candidate alloy) for the US Fusion Materials Program (D9 and PCA are essentially the same composition) [10]. This steel has multiaddition of B, P, Ti, Nb, and V in the base composition of Fe-14Cr-16Ni-2.5Mo-2Mn (mass %) that acts as catalysts. The excess titanium forms TiFeP-type phosphide. These carbide and phosphide precipitates are extremely fine and significantly improve the high-temperature creep strength. These alloying principles were constructed via the fast reactor, fusion reactor, and Basic Energy Sciences programs within the U.S. Department of Energy to develop an austenitic steel with outstanding high-temperature creep resistance for advanced steam cycle coal-fired power generation systems.

The swelling of PNC316 and PNC1520 was also improved by extending the incubation period in accordance with the improvement in creep strength [11]. The trend apparently suggests that the fine dispersion of precipitates and cold-work are beneficial for suppression of radiation-induced void development during irradiation. Specifically, stabilization of dislocations by MC carbide precipitates and dispersion of fine phosphide precipitates during

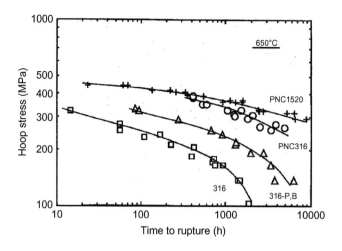

FIG. 8.1

Creep rupture strength of 20% cold-worked modified 316 and advanced austenitic stainless steels developed by JAEA Japan: results of internally pressured creep tests [7].

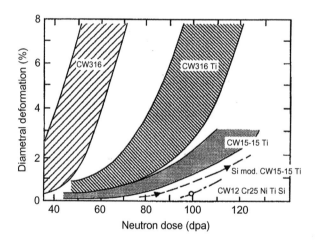

FIG. 8.2

Diametral increase of French SFR fuel pins with modified austenitic stainless steels [13].

From C. Brown, V. Levy, J.L. Seran, K. Ehrlich, R.J.C Roger, H. Bergmann, Cladding and Wrapper Development for Fast Breeder reactor high performance, Int. Conf. on Fast Reactor and Related Fuel Cycles, Atomic Energy Soc. Japan, 1991, Kyoto, 1, 7.5–1.

irradiation are the dominant factors in suppressing swelling in the temperature range between 500°C and 600°C [11,12]. There is a substantial underpinning of mechanistic understanding for the basis of current approaches to the development of swelling and creep resistant behavior in several alloy systems [13,14].

The maximum diametral deformation of French fuel pins with various cladding materials is indicated in Fig. 8.2 [15].

The beneficial effect of titanium has also been demonstrated in French CW316Ti [16]. The French CW15-15Ti, incorporated a fairly high amount of Ti and C and higher Ni/Cr ratio and exhibited improved swelling resistance compared to that of CW316Ti. Further improvements were observed in Si-modified CW15-15Ti (increased Si from 0.43 to 0.95 wt%) and in CW12Cr25NiTiSi. In the high Si alloys, a fine dispersion of γ' [$Ni_3(Si, Ti)$] precipitates was observed [17].

Similar austenitic stainless steels have been applied to SFR core materials. German DIN 1.4970 alloy (15Cr-15Ni-0.5Ti-0.4Si) is the same as French 15-15Ti, and DIN 1.4981 (17Cr-17Ni-0.5Si-0.7Nb) is a Nb-stabilized alloy; both types of commercial alloys have been studied by neutron exposure [18]. UK FV548 (17Cr-12Ni-0.4Si-0.7Nb) is also Nb-stabilized [19]. Russian ChS-68 has nominal composition of 16Cr-15Ni-0.5Si-0.35Ti that has been utilized in driver fuels of BOR-60 and BN600 [20]. All of the austenitic stainless steels utilized in LWR and SFR are listed in Table 8.1.

8.3 RADIATION-INDUCED METALLURGICAL CHANGES

Neutron irradiation causes changes in metallic materials by two primary processes, such as displacements of atoms from their lattice sites and transmutation from one element to another although at high exposures, sometimes the shifts in isotopic ratio (without changes in elemental identity) become important. For LWRs most transmutation processes in austenitic steels are generally but not universally of second order in importance. Atomic displacements, however, have a pronounced effect on the material, so much so that the exposure dose is expressed in terms of displacements per atom (dpa), the calculated number of times on average each atom has been displaced from its lattice site.

The displacement process produces two types of crystalline defects, vacant crystalline positions (vacancies) and displaced atoms in interstitial crystalline positions (interstitials). These two defect types are both mobile, but move by different diffusional modes and at vastly different velocities, with interstitials diffusing much faster than vacancies. Both defect types have the ability to recombine with the opposite type (annihilation) or to form agglomerations of various types and geometries. The developing ensemble of various defect cluster types with increasing dose induces significant changes in physical and mechanical properties, as well as resulting in significant dimensional distortion and modification of phase stability and compositional distribution of many elements. The latter is a consequence not only of highly increased diffusivities but the introduction of new diffusional modes and segregation mechanisms.

When subjected to displacive irradiation, especially at elevated temperatures, an intricate and coordinated coevolution of microstructure and microchemistry commences that is dependent primarily on the alloy starting state, the dpa rate, and the temperature, and secondarily dependent on variables such as He/dpa rate, H/dpa rate and applied or internally generated stresses. In general, the starting microstructure and microchemistry of the alloy determine only the path taken to the radiation-defined quasiequilibrium state, and not the final state itself. If an alloy experiences enough displacements, it effectively forgets its starting state and arrives at a destination determined only by irradiation temperature and dpa rate. This quasiequilibrium or dynamic-equilibrium state does not consist of microstructural components existing at relatively fixed densities and size distributions, but rather individual dislocations, loops, precipitates, or cavities that at any moment may be growing, shrinking, or even disappearing by shrinkage or annihilation.

As a consequence of this microstructural and microchemical evolution a variety of physical and chemical phenomena occur, most of which result in changes that are viewed as a degradation of the properties of the steel that may have been carefully optimized to achieve the alloy starting condition. The various categories of degradation are summarized below.

8.3.1 RADIATION-INDUCED SEGREGATION

RIS describes the redistribution of major alloying elements and impurity elements at point defect sinks [25–36]. Because IASCC is an intergranular process, the sinks of greatest interest are the grain boundaries. RIS is driven by the flux of radiation-produced defects to sinks, and is therefore fundamentally different from thermal segregation or elemental depletion due to grain boundary precipitation. Vacancies and interstitials are the basic defects produced by irradiation and can reach concentrations that are orders of magnitude higher than those at thermal equilibrium, resulting in large increases in diffusion rates, especially since self-interstitials are rare under thermal conditions and solute interstitials are limited to small atoms such as carbon. If the relative participation of alloying elements in the

Table 8.1 Compositions of Austenitic Alloys Considered in This Section

Alloy Designation	Fe	Ni	Cr	Mn	Mo	Al	Ti	Si	C	Other	References
304	Bal.	8.00–10.50	18.00–20.00	<2.00	—	—	—	<1.00	<0.08	—	[19]
304L	Bal.	8.00–12.00	18.00–20.00	<2.00	—	—	—	<1.00	<0.03	—	[19]
309	Bal.	12.00–15.00	22.00–24.00	<2.00	—	—	—	<1.00	<0.08	—	[19]
310	Bal.	19.00–22.00	24.00–26.00	<2.00	—	—	—	<1.50	<0.25	—	[19]
310H	Bal.	19.00–22.00	24.00–26.00	<2.00	—	—	—	<1.50	0.07	—	[20]
316	Bal.	10.00–14.00	16.00–18.00	<2.00	2.00–3.00	—	—	<1.00	<0.08	—	[19]
316L	Bal.	10.00–14.00	16.00–18.00	<2.00	2.00–3.00	—	—	<1.00	<0.03	—	[19]
321	Bal.	9.00–12.00	17.00–19.00	<2.00	—	—	0.60	<1.00	<0.08	—	[19]
347	Bal.	9.00–13.00	17.00–19.00	<2.00	—	—	—	<1.00	<0.08	<0.10 Nb+Ta	[19]
348	Bal.	9.00–13.00	17.00–19.00	<2.00	—	—	—	1.00	0.08	<0.10 Nb+Ta, <0.20 Co	[19]
800	Bal.	32.50	21.00	—	—	0.38	0.38	—	0.05	—	[21]
800H	Bal.	33.00	21.00	—	—	—	—	—	0.08	—	[21]
825	31.00	Bal.	22.00	0.11	3.29	0.12	0.98	0.24	0.01	2.00 Cu, 0.67 Co	[22]
D9	Bal.	15.24	13.88	2.12	2.12	—	0.23	0.64	0.05	—	[23]
HT–JPS	Bal.	16.00	14.00	2.00	2.50	—	0.30	0.15	0.08	0.15 Nb, 0.50 V	[24]
PNC316	Bal.	14	17	1.7	2.5	—	0.1	0.8	0.06	0.1 Nb, 0.025 P, 0.004 B	[12]
PNC1520	Bal.	20.5	14.5	1.7	2.5	—	0.25	0.8	0.06	0.1 Nb, 0.025 P, 0.004 B	[8]
CW316Ti	Bal.	13	17	1.7	2.3	—	0.36	0.65	0.06	—	[14]
CW15-15Ti (DIN 1.4970)	Bal.	15	15	1.3	1.2	—	0.5	0.4	0.1	0.003 B	[15]
DIN 1.4981	Bal.	16.5	16.5	1.3	1.8	—	0.06	0.5	0.06	0.7 Nb	[16]
FV548	Bal.	11.5	16.5	1.0	1.4	—	—	0.3	0.09	0.7 Nb	[17]
ChS-68	Bal.	14.0–15.5	15.0–16.5	1.3–2.0	1.9–2.5	—	0.2–0.5	0.3–0.6	0.05–0.08	0.1–0.3 V, 0.002–0.005 B	[18]

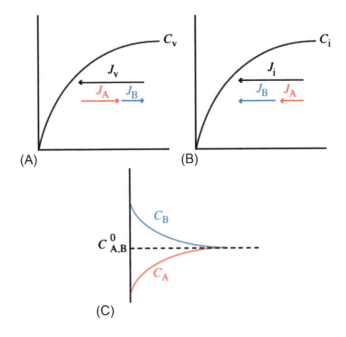

FIG. 8.3

Schematic illustration of radiation-induced segregation in a binary, 50A–50B system showing (A) the development of the vacancy concentration profile by the flow of vacancies to the grain boundary balanced by an equal and opposite flow of A and B atoms, but not necessarily in equal numbers, (B) the development of the interstitial concentration profile by the flow of interstitials to the grain boundary balanced by an equal and flow of A and B atoms migrating as interstitials, but not necessarily in equal numbers, and (C) the resulting concentration profiles for A and B.

From G. S. Was, Fundamentals of Radiation Materials Science: Metals and Alloys, Springer, Berlin, 2007.

defect fluxes is not the same as their relative concentration in the alloy, then a net transport of the constituents to or from the grain boundary will occur, as shown in Fig. 8.3.

This unequal participation of solutes in the vacancy and/or interstitial fluxes to sinks results in either enrichment or depletion of an alloying element at the grain boundary or other sink [31]. The species that diffuse more slowly by the vacancy diffusion mechanism are enriched by default as the faster diffusers become depleted in the vicinity of the sink boundary as they migrate up the vacancy gradient. This process is called the "inverse Kirkendall" effect.

Enrichment and depletion can also occur by association of a solute with the interstitial flux. In general, the undersized species will enrich and the oversized species will deplete by a process called "solute drag" [26]. Some elements can flow in both directions via their interaction with vacancies and interstitial fluxes. For instance, silicon is both a fast diffusing element via vacancy exchange but an undersized element via interstitial interaction. The magnitude of the buildup/depletion is dependent upon several factors such as whether a given constituent migrates more rapidly by one defect mechanism or another at a given temperature and sink type, the binding energy between solutes and defects, and the dose, especially dose rate and the temperature. RIS profiles are also characterized by their relative narrowness, often confined to within 5–10 nm of the grain boundary, as shown in Fig. 8.4 for an irradiated stainless steel.

Segregation in general is a strong function of irradiation temperature, dose, and dose rate (Fig. 8.5).

Segregation peaks at intermediate temperatures since a lack of mobility suppresses the process at low temperatures, and back-diffusion of segregants minimizes segregation at high temperatures where defect concentrations approach their thermal equilibrium values. For a given dose and temperature, a lower dose rate usually results in a greater amount of segregation. At high dose rates, the high defect population results in increased recombination that reduces the numbers of defects that is able to diffuse to the grain boundary. Fig. 8.5 shows the interplay between temperature and dose rate for an austenitic stainless steel. Generally, reactor irradiations occur in the dose rate range 10^{-8}–10^{-7} dpa/s, proton irradiations

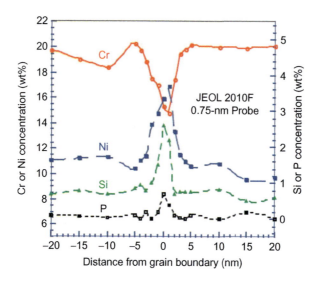

FIG. 8.4

Compositional profiles across grain boundaries obtained by dedicated STEM analysis from a low strain, high-purity 348 stainless steel swelling tube specimen irradiated to 3.4×10^{21} n/cm^2 at 288°C in a BWR. Composition profiles were measured using a field emission gun scanning transmission electron microscope (FEGSTEM).

From A.J. Jacobs, R.E. Clausing, M.K. Miller, C.M. Shepherd, in: D. Cubicciotti (Ed.), Influence of grain boundary composition on the IASCC susceptibility of type 348 stainless steel, Proc. 4th Int. Symp. on Environmental Degradation of Materials in Nuclear Power Systems—Water Reactors, NACE, Houston, TX, 1990, pp. 14–21.

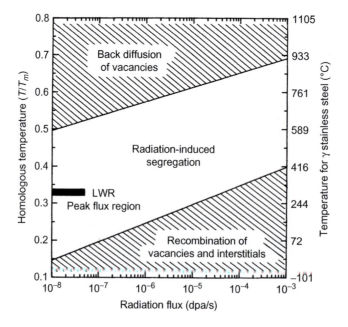

FIG. 8.5

Schematic of dependence of RIS on homologous temperature and dose rate for austenitic stainless steels.

From G.S. Was, Fundamentals of Radiation Materials Science: Metals and Alloys, Springer, Berlin, 2007.

are conducted at 10^{-5} dpa/s and electron and heavy ion irradiations are typically conducted at 10^{-3} dpa/s. RIS occurs in the intermediate temperature range, and this range shifts to higher temperature with increasing dose rate to compensate for the progressively higher recombination rate.

In Fe-Cr-Ni alloys, the vacancy exchange (inverse Kirkendall) mechanism successfully explains the observed major element segregation [32,33]. Studies have shown that nickel segregates to grain boundaries while chromium and iron deplete. The directions of segregation are consistent with an atomic volume effect in which the subsized solute migrates preferentially with the interstitial flux, and the oversized solute participates preferentially with the vacancy flux. The results are also consistent with the diffusivity of the solutes in Fe-Cr-Ni, in which Ni is the slow diffuser, Cr is the fast diffuser, and Fe is intermediate. In commercial austenitic stainless steels, chromium depletes at grain boundaries and nickel enriches, while iron can either deplete or enrich according to the magnitude of the diffusion coefficient relative to those of the other elements [36].

RIS increases with accumulating neutron dose in LWRs but generally saturates after several (~5) dpa in the near −300°C temperature range.

Fig. 8.6 shows grain boundary chromium depletion for austenitic stainless steels as a function of dose [32–44]. As the slowest diffusing element, nickel becomes enriched at the grain boundary. Since iron depletes in 304 and 316 stainless steels, the nickel enrichment makes up for both chromium and iron depletion and can reach very high levels, up to ~30 wt% at LWR-relevant temperatures in some cases.

Minor alloying elements and impurities also segregate to grain boundaries. Mn and Mo strongly deplete at the grain boundary under irradiation [45]. Minor alloying or impurity elements such as Si and P, both undersized fast diffusers, also segregate under irradiation. Silicon strongly enriches at the grain boundary to as much as 10 times the bulk (0.7–2.0 at.%) composition in the alloy [44] and can be important in IASCC. Phosphorus is present at much lower concentrations and is only modestly enriched at the grain boundary due to irradiation [31,45]. Phosphorus tends to segregate to the grain boundary following thermal treatment, which reduces the amount of additional

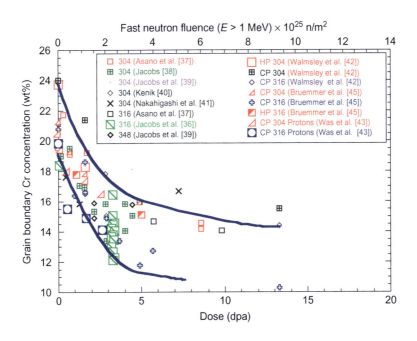

FIG. 8.6

Dose dependence of grain boundary chromium concentration for several 300-series austenitic stainless steels irradiated at temperatures ~300°C [33].

From G.S. Was, Recent Developments in Understanding Irradiation Assisted Stress Corrosion Cracking, Proc. 11th Int'l Conf. Environmental Degradation of Materials in Nuclear Power Systems – Water Reactors, American Nuclear Society, La Grange Park, IL, (2004) 965–985.

segregation to the grain boundary occurring during irradiation, often making the segregation contribution due to irradiation difficult to detect [45]. Undersized solutes such as C, B, and N should also segregate, but there is little evidence of RIS for these elements, due in part to the difficulty of measurement and the fact that C often controls the location and chemical activity of P, B, and N. Another potential segregant is helium, produced by the transmutation of ^{10}B as well as by both fast and especially thermal reactions with nickel isotopes. The mobility of He-vacancy complexes is low at LWR core temperatures, but the opportunity for accumulation at the grain boundary is increased by segregation of B and Ni to the boundary (although the transmutation reaction causes He to distribute within ~5 μm of the grain boundary). Overall, the behavior and consequences of these minor elements on segregation under irradiation is not well understood.

Oversize solutes such as Zr and Hf singly dissolved in the matrix can affect the microchemistry or microstructure of the alloy, thereby altering the IASCC susceptibility. They are believed to affect RIS by binding with vacancies, thereby acting as temporary vacancy traps. This slows down vacancy migration, resulting in an increase of recombination of vacancies and interstitials and thereby reducing RIS. Kato et al. [46] conducted electron irradiations of several stainless steels at temperatures of 400–500°C up to 10 dpa. Results showed that some solutes (Zr and Hf) consistently produced a large suppression of radiation-induced chromium depletion while others resulted in less suppression or suppression only over certain temperature ranges. Fournier et al. [47] conducted irradiation of 316 containing Hf or Pt using 3 MeV protons (400°C) and 5 MeV Ni ions (500°C). Ni irradiations showed little effect of the oversize impurity in reducing grain boundary chromium depletion (Cr depletion increased in the case of Hf), but proton irradiation showed a significant suppression of RIS of chromium at low dose (2.5 dpa) with the effect diminishing at higher (5.0 dpa) dose. Pt was found to have a smaller effect on Cr segregation. Ti and Nb similarly produced little change in the grain boundary chromium concentration after irradiation with 3.2 MeV protons to 5.5 dpa at 360°C. In Zr-doped 304 SS there were no consistent results of suppression of grain boundary chromium after 3.2 MeV proton irradiation to 1.0 dpa at 400°C [48]. Neutron irradiation at very low dose (0.5 dpa) shows a small effect of Ti and Nb on grain boundary Cr [49].

Overall, the data on the effect of oversize solutes on RIS of chromium are rather inconsistent. One possibility for this inconsistency is the tendency of oversize elements to form precipitates such as carbides and intermetallic phases, thereby altering the matrix composition but in a very temperature-dependent and possibly dpa rate-dependent manner.

8.3.2 DISLOCATION MICROSTRUCTURE

The microstructure of austenitic stainless steels under irradiation changes rapidly at LWR service temperatures. Point defect clusters (called "black dot damage" when then-available electron optics could not resolve the details) begin to form at very low dose. These evolve into resolvable dislocation loops that later unfault to form network dislocations that evolve with dose over several dpa. These dislocation ensembles serve as precursors to the formation and growth of He-filled bubbles, voids, and various precipitates in core components in locations exposed to higher dose and temperatures [50–56]. At 300°C and below, the microstructure is dominated by small clusters and small faulted dislocation loops. Above 300°C, the microstructure contains larger faulted loops, unfaulted prismatic loops plus network dislocations, and cavities (bubbles and voids) that appear at higher doses.

The primary defect structures in LWRs are vacancy clusters (both 2D and 3D in nature) and interstitial clusters (2D only), the latter eventually growing into Frank faulted dislocation loops. The nuclei of the clusters are formed either by random diffusional meeting of two or more defects or during the collapse of the damage cascade arising from primary and secondary atom collisions after an interaction of an atom with a high energy neutron. The larger faulted dislocation loops nucleate and grow as a result of the high mobility of interstitials and an inherently larger attraction for interstitials than for vacancies. The loop population grows in size and number density until absorption of vacancies and interstitials equalize, at which point the population of loops, mostly interstitial, has saturated. Fig. 8.7 shows the evolution of loop density and loop size as a function of irradiation dose during LWR irradiation at 280°C.

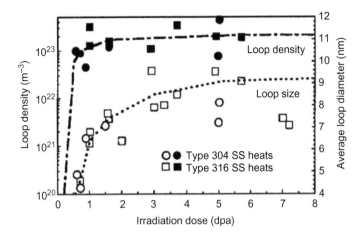

FIG. 8.7

Measured change in density and size of interstitial loops as a function of dose during LWR irradiation of 300-series stainless steels at 275–290°C [56].

From S.M. Bruemmer, E.P. Simonen, P.M. Scott, P.L. Andresen, G.S. Was, J.L. Nelson, J. Nucl. Mater. 274 (1999) 299 Originally from S.M. Bruemmer, B.W. Arey, L.A. Charlot, D.J. Edwards, in: S.M. Bruemmer and F.P. Ford (Ed.), Proceedings of the Ninth International Symposium Environmental Degradation of Materials in Nuclear Power Systems – Water Reactors, The Minerals, Metals and Materials Society, to be published.

Note that saturation of loop number density occurs very quickly, usually by ~1 dpa, while loop size continues to evolve up to ~5 dpa. The specific number density and size are dependent on irradiation conditions and alloying elements, but the loop size rarely exceeds 20 nm and densities are usually of the order of $1 \times 10^{23} \, m^{-3}$.

Oversize solutes can also affect the evolving dislocation microstructure by mechanisms similar to RIS. Proton and nickel ion irradiations show that the addition of Hf to a 316 SS base alloy increased its loop density, decreased its loop size, and eliminated the appearance of voids [47]. Platinum addition to 316 SS resulted in no change in loop density and a small increase in loop size, but increased void size and density. The good agreement between proton and Ni ion irradiation results indicates that the major effect of the oversized solute is not due to the cascade (where there are large differences between proton and nickel ion irradiation), but rather is due to the postcascade defect partitioning during the microstructure evolution. Electron irradiation experiments by Watanabe et al. [57] and proton irradiation experiments by Busby et al. [58] showed that stainless steel with Ti additions had slightly lower dislocation loop densities and larger sizes compared to the base alloy. Nb addition increased loop size only. In contrast to the base alloy, neither the Ti nor the Nb-doped alloys formed voids under the conditions tested. Zirconium addition to 304 SS resulted in reduced hardness, decreased loop density, and no change in loop size in proton irradiation to 1.0 dpa at 400°C and compared to the base alloy [59]. Zirconium-containing samples also had a lower void density with no change in void size as compared to the base alloy.

8.3.3 PHASE STABILITY

Irradiation can have profound effects on the formation or dissolution of phases by altering the composition and stability of thermally stable phases and by creating local changes in microchemistry that favor the formation of nonequilibrium metastable phases or nonequilibrium compositions of thermally stable phases. Because the phase structure of an alloy can significantly affect the physical and mechanical properties of the material, understanding how irradiation affects phase stability is of great importance for design of engineering materials and the prediction of limits to their continued functionality at higher exposures.

The most direct way in which irradiation can alter phase stability is by causing local enrichment or depletion of one or more solutes at a microstructural sink such that the solubility limit is crossed for a given element, or when local compositions are reached that ensure co-precipitation into a new phase. But irradiation can also contribute to the dissolution of otherwise stable phases by either ballistic recoil dissolution, or by causing disordering and or amorphization, or by creating antisite defects (in which constituents in an ordered structure change sublattices). Depending on temperature and dpa rate there are also concurrent healing mechanisms operating, involving reordering and reprecipitation at the surface of the precipitate. Under some conditions, irradiation also leads to the formation not only of metastable phases but new phase sequences as competition for critical elementals shifts as newer phases out-compete earlier-produced phases for critical elements, especially Ni, Si, P, and C. At any given moment, the individual precipitate is responding to all of these processes with its evolution strongly dependent on the irradiation temperature, dpa rate, and current local matrix composition. Thus, one frequently observes a given phase only in a limited range of temperatures, and the balance of phases often shifts with changes in dpa rate.

For stainless steels in LWR service the best example is the precipitation of the γ' (Ni_3Si) phase in austenitic stainless steels, which is a consequence of RIS of nickel and silicon to sinks, primarily small Frank faulted loops. Fig. 8.8 shows the radiation-induced phase often formed in austenitic stainless steels under LWR-relevant irradiation conditions. This phase is sometimes a precursor to formation of other Ni and Si-rich phases such as G-phase or nickel phosphides.

Stainless steels are widely used in the cores of current and advanced reactor systems and have been comprehensively studied. Ten phases have been observed to be affected by irradiation of austenitic stainless steels [50,60,61]. Lee [61] grouped these phases into three categories: radiation-induced, radiation-modified, and radiation-enhanced. Table 8.2 gives the phases belonging to each category along with a crystallographic and morphological description of the phases.

The radiation-induced phases include γ', G, and M_xP phases and appear only under irradiation, not under thermal conditions. The radiation-modified group consists of phases that occur during both irradiation and thermal

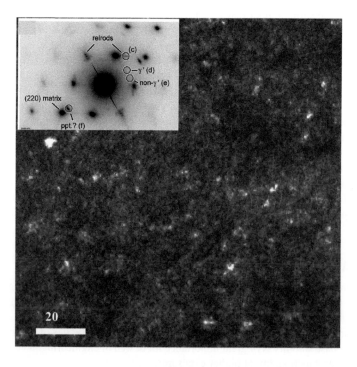

FIG. 8.8

γ' and other phases formed in 316 SS at very small sizes after irradiation in reactor to 7 dpa at 299°C [58]. Other not-yet identifiable phases are also being to form.

8.3 RADIATION-INDUCED METALLURGICAL CHANGES

Table 8.2 Phases Observed in Austenitic Stainless Steel Alloys After Neutron Irradiation [61]

Radiation	Phase	Crystal Structure	Lattice Parameter (nm)	Morphology	Orientation in γ Matrix	Volume Misfit
Induced	γ′ (Ni$_3$Si)	Cubic, A1 Fm3m	0.35	Spherical	Cube on cube	−0.1
	G (M$_6$Ni$_{16}$Si$_7$)	Cubic, A1 Fm3m	1.12	Small rod	Random	0.05
	M$_2$P (FeTiP)	Hexagonal, C22, P321	0.6 (c/a = 0.6)	Thin lath	(1210)$_{ppt}$ ∥ (001)$_\gamma$	−0.4
	Cr$_3$P	Tetragonal, S$_4^2$ 14	0.92 (c/a = 0.5)	–	–	∼0
Modified	η (M$_6$C)	Cubic, E9, Fd3M	1.08	Rhombohedral	Cube on cube or twin	0.1
	Laves (A$_2$B)	Hexagonal, C14 P6$_3$/mmc	0.47 (c/a = 0.77)	Faulted lath	Various	−0.05
	M$_2$P (FeTiP)	Same	Same	Same	Same	Same
Enhanced	MC	Cubic, B1 Fm3m	0.43	Spherical	Spherical	0.7
	η (M$_6$C)	Same	Same	Same	Same	Same
	τ (M$_{23}$C$_6$)	Cubic, D8$_4$ Fm3m	1.06	Rhombohedral platelet	Cube on cube or twin	0.1
	Laves (A$_2$B)	Same	Same	Same	Same	Same
	σ	Tetragonal, D8$_b$ p4/mnm	0.88 (c/a = 0.52)	Various	Various	∼0
	χ	Cubic, A12 143m	0.89	Various	Various	0.05

From E.H. Lee et al. Proc. Conf. on Phase Stability During Irradiation, Met. Soc. AIME, Pittsburgh PA Oct. 5–9 (1980) 191–218. Used with permission of The Minerals, Metals & Materials Society.

aging (but with possibly different compositions under irradiation) and includes the η (M$_6$C), Laves, and M$_2$P (FeTiP) phases. The radiation-enhanced category consists of phases that regularly occur during thermal processing, but are either produced more rapidly, more abundantly at lower temperatures during reactor irradiation. Phases in this category include M$_6$C, M$_{23}$C$_6$, and MC carbides and σ and χ intermetallic phases.

Radiation-enhanced diffusion affects MC precipitation by enhancing its growth compared to that under thermal conditions. While it is expected that oversized phases are more stable under irradiation and undersized phases are less stable, of the phases given in Table 8.2, three have positive misfits of at least 10% and of those, only M$_6$C is stable under a wide range of irradiation conditions. Both MC and M$_{23}$C$_6$ are oversized and will be stabilized by vacancy supersaturation. Conversely, several of the phases that are undersized are induced or enhanced by irradiation; Laves, M$_2$P, and γ′. Clearly, the effect of solute segregation and other material and irradiation parameters play a large role in the phases that are stable under irradiation.

As will be discussed in later sections the phase evolution is important in that it frequently is the precursor to the onset of void swelling, altering the matrix composition such that it becomes more favorable to void nucleation. Additionally, phase evolution can change the elastic moduli of the matrix, introduce or eliminate barriers to dislocation motion, and introduce new failure modes. It is also recognized that the development of unusual phases may in some cases have little effect on the performance of the material in LWR service.

8.3.4 TRANSMUTATION

Most investigations concerning transmutation involve the loss or production of a given element. For instance, Mn has only one naturally occurring isotope, which transmutes to an iron isotope, especially in highly thermalized neutron spectra. Vanadium is also formed from chromium. In general, however, these changes are relatively small and are of minor consequence for most light water applications.

In nickel-containing alloys irradiated in thermalized neutron flux-spectra, however, the formation and reaction of nickel isotopes with thermal neutrons can lead to significant time-varying changes in dpa rate, gas formation, and nuclear heating. Depending on the neutron spectrum and the nickel level of the alloy, these changes can range from insignificant to dominant [62]. For AISI 304 and 316 steels the effects may develop into significant considerations at higher dose levels over a 40–80-year period.

Nickel has five naturally occurring stable isotopes with ^{58}Ni comprising ~67.8% natural abundance, ^{60}Ni comprising 26.2%, and ~6.1% total of combined ^{61}Ni, ^{62}Ni, and ^{64}Ni. During irradiation in a highly thermalized neutron spectrum, all nickel isotopes are transmuted, primarily to the next higher isotopic number of nickel. Before transmutation via thermal neutrons becomes important via the ^{58}Ni → ^{59}Ni → ^{60}Ni sequence, nickel contributes to the majority of transmutant helium and hydrogen, primarily arising from reactions with neutrons above ~6 MeV.

There is no natural ^{59}Ni or ^{63}Ni at the beginning of radiation. However, ^{59}Ni, which has a half-life of 76,000 years, is formed at significant levels from ^{58}Ni via thermal neutron absorption. The recoil of the ^{59}Ni upon emission of the gamma ray produces about five displacements per event and thus can contribute significantly to the overall displacement damage in some instances.

The isotope ^{59}Ni undergoes three strong reactions with thermal and resonance (~0.3 keV) neutrons. These reactions in order of increasing cross-section are (n, γ) to produce ^{60}Ni and (n, p) to produce hydrogen and (n, α) to produce helium. Once significant ^{59}Ni forms, helium/dpa ratios on the order of 10–25 appm/dpa can be experienced along the length of a baffle bolt [63,64] while comparable rates in fast reactors are on the order of 0.1–0.2 appm/dpa. Because H is ubiquitous in LWRs and permeates readily through unirradiated structural materials, it is unlikely that the comparatively small contribution of H from transmutation is important.

In thermalized spectra and as ^{59}Ni builds up, the latter two reactions can quickly overwhelm the gas production produced at high neutron energies. Most importantly, these thermal neutron reactions of ^{59}Ni are quite exothermic in nature and release large amounts of energy, thereby causing increases in the rate of atomic displacements, and concomitant increases in nuclear heating rates. Nuclear heating by elastic collisions with high-energy neutrons is usually too small to be of much significance.

The ^{59}Ni (n, α) reaction releases 5.10 MeV, producing a 4.8 MeV alpha particle which loses most of its energy by electronic losses, depositing significant thermal energy but producing only ~62 atomic displacements per event. However, the recoiling ^{56}Fe carries 340 keV which is very large compared to most primary knock-on energies, and produces ~1701 displacements per event [65,66].

The thermal (n, p) reaction of ^{59}Ni produces about one proton per six helium atoms, reflecting the difference in thermal neutron cross-sections of 2.0 and 12.3 b, and is somewhat less energetic (1.85 MeV), producing a total of ~222 displacements per event [67]. Note that only 4.9 displaced atoms are created by each (n, γ) recoil of ^{60}Ni.

Since ^{59}Ni is progressively transmuted to ^{60}Ni and ^{58}Ni is continuously reduced in concentration, the ^{59}Ni concentration rises to a peak level at 4×10^{22} n/cm^2 where the 59/58 ratio peaks at ~4% and then declines, as shown in Fig. 8.9A.

Given the long half-life of ^{59}Ni, its decay is not a factor and the increased damage rate is determined only to the accumulated thermal neutron fluence and the nickel content of the alloy. An extreme example of this increase in dose is shown for pure nickel in Fig. 8.9B. Note that this calculated increase arises only from the ^{59}Ni (n, α) reaction. An additional but smaller increase will occur as a result of the ^{59}Ni (n, p) reaction.

At the peak ^{59}Ni level the heating rates from the energetic (n, α) and (n, p) reactions are 0.377 and 0.023 W/g of nickel, significantly larger than the neutron heating level of ~0.03 W/g of nickel. Thus, an increase in nuclear heating of ~0.4 W/g of nickel must be added to the gamma heating rate at the peak ^{59}Ni level. Fractions of the peak heating rates that are proportional to the ^{59}Ni level should be added at nonpeak conditions. Gamma heating is the primary cause of temperature increases in the interior of thick plates, and temperature has a major influence on void swelling.

Gamma heating is also a strong function of the thermal-to-fast neutron ratio and the neutron flux, being ~40 W/g in the center of the HFIR test reactor where the thermal to fast neutron (T/F) ratio is ~2.0. In PWR near-core internals, however, the *T/F* ratios are lower by a factor of 2–10, depending on location, and the gamma

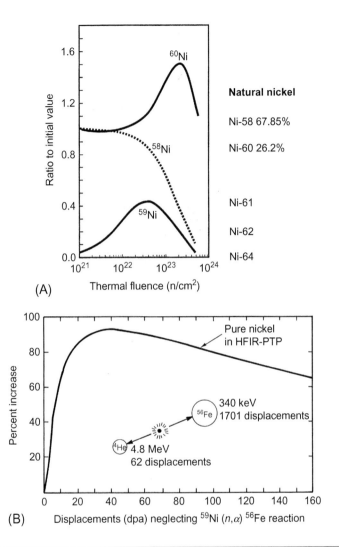

FIG. 8.9

(A) Transmutation-induced evolution of three nickel isotopes during irradiation in thermalized neutron spectra, and (B) increase in dpa arising from the effect of ^{59}Ni to produce helium when pure nickel is irradiated in the HFIR test reactor in the peripheral target position where the thermal/fast ratio is 2.0. The rate of increase will be increased another few percentage if the ^{59}Ni (n, p) reaction is taken into account [65].

heating rates in the baffle-former assembly are ~1–3 W/g. In this case an additional 0.4 W/g of nuclear heating can be a significant addition to total heating, especially for high nickel alloys.

Previously the liquid metal reactor and LWR communities have focused primarily only on the effect of ^{59}Ni reactions on the gas generation rates, but it is now obvious that the displacement and heating effects must also be taken into account. Additionally, another concern may arise in that small nickel-rich phases such as gamma-prime, Ni-phosphides, and G-phase may become less stable due to recoil dissolution as the ^{56}Fe recoils originating in the precipitates, thereby altering the phase evolution in thermalized neutron spectra compared to nonthermalized spectra such as found in fast reactors. These precipitates are known to form as a direct result of irradiation and to contribute to hardening, swelling, and irradiation creep processes [68,69]. The size of these precipitates at PWR-relevant temperatures is often comparable to or smaller than the ~80 nm range of the recoiling ^{56}Fe atom.

8.4 RADIATION-INDUCED MECHANICAL PROPERTY CHANGES AND DEGRADATION MODES

8.4.1 IRRADIATION HARDENING

Under applied stress, the dislocation network interacts elastically with the dislocation loops, precipitates, and cavities produced by irradiation, producing an increase in the yield strength of the alloy, which can be detected in tensile tests or indentation hardness measurements. This is accompanied by a decrease in elongation, which is affected more than reduction in area because necking occurs relatively early in most tensile tests of irradiated stainless steel. The saturation strength level of AISI 316 stainless steel is very dependent on irradiation temperature, as shown in Fig. 8.10, reflecting the strong temperature dependence of the dislocation loop structure [70].

CW alloys can either decrease or increase in yield strength depending on the cold-work level, irradiation temperature, and dpa rate but will quickly approach the same saturation strength reached by the annealed alloy.

Both the source hardening model [71] and the dispersed barrier hardening model [72] provide reasonable correlations between hardening and the dislocation loop microstructure. In the dispersed barrier hardening model, the increase in hardness is proportional to $(N_{loop} \times d_{loop})^{1/2}$, where N_{loop} is the loop number density and d_{loop} is the loop diameter. Similar formulations apply to precipitates and cavities, but it should be noted that precipitation frequently removes elements that strengthen the matrix such as Ti and C, any thereby produce a concurrent softening of the matrix that partially offsets the hardening due to precipitates.

As dislocations move through the irradiated microstructure, they clear many of the barriers associated with loops. This creates a preferential *channel* where other dislocations can readily pass, and gives rise to a series of

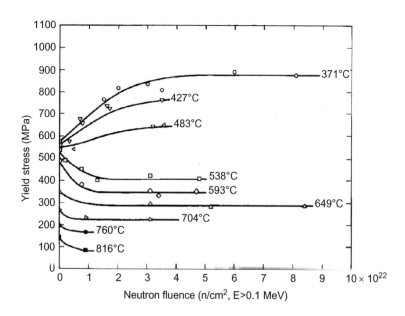

FIG. 8.10

Changes of yield stress in AISI 316 stainless steel irradiated in EBR-II at temperature range between 371°C and 816°C [70].

8.4 RADIATION-INDUCED MECHANICAL PROPERTY CHANGES

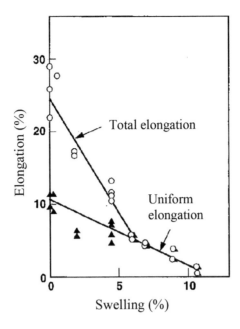

FIG. 8.11

Loss of ductility with increasing swelling for 20% CW316Ti irradiated in PHENIX, tensile test temperature: 180°C [73].
From A. Fissolo, R. Cauvin, J.P. Hugot and V. Levy, Effects of Radiation on Materials: 14th Int. Symp., Vol. II, STP 1046, Philadelphia: ASTM, (1990) 700–713.

channels whose width and spacing vary. In general, the concept of *brittleness* of irradiated stainless steels is misleading, because the local ductility is almost always very high, as evidenced by extensive reduction-in-area.

As swelling increases, its influence on mechanical properties becomes more pronounced. Fig. 8.11 shows that the ductility of CW316Ti fell continuously with increasing swelling, tensile rupture occurring without reduction of area at higher than 6% swelling. In the swelling range above 10%, the steel exhibited no ductility at all [73].

8.4.2 REDUCTION IN FRACTURE TOUGHNESS AND EMBRITTLEMENT

The effect of neutron irradiation on the fracture toughness at 250–320°C of austenitic stainless steels irradiated in LWRs (288–316°C) up to about 17 dpa was reviewed by Chopra and Rao [74] and is shown in Fig. 8.12. The data show a rapid decrease in fracture toughness over a neutron dose range of 1–5 dpa, Note that the neutron dose at the onset of the rapid decrease varies only somewhat with the alloy. In fact, the value of the fracture toughness, J_{IC} drops from values between 600 and 823 kJ/cm^2 in the unirradiated condition to values as low as 20 kJ/cm^2 at ∼6 dpa for 304 stainless steel irradiated and tested at 288°C [75].

Some of the materials irradiated above 4 dpa at LWR temperatures show very low fracture toughness with J_{IC} values near zero. For type 304 SS irradiated to 4.5–5.3 dpa, 9 out of 10 CT specimens showed no ductile crack extension, and values of the plane strain fracture toughness, K_{IC}, were 52.5–67.5 MPa m$^{1/2}$ (47.7–61.4 ksi in$^{1/2}$) [76]. The lowest fracture toughness, with K_{JC} values in the range of 36.8–40.3 MPa m$^{1/2}$ (33.5–36.6 ksi in$^{1/2}$), was for a type 347 SS irradiated to 16.5 dpa in a PWR [62] and for a type 304 steel irradiated to 7.4–8.4 dpa in a BWR [77].

Chopra and Rao [74] note that the fracture toughness has also been observed to be orientation dependent. Fracture toughness J-R tests have been conducted on type 304 control-rod and type 304L top guide materials irradiated to 4.7–12 dpa, and on type 304 control-rod material irradiated to 7.4 and 8.4 dpa. The results show lower fracture toughness in the T-L orientation than in the L-T orientation [78]. The lower fracture toughness along the T-L

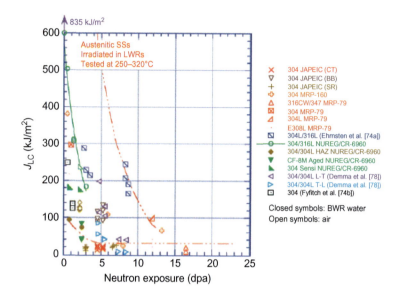

FIG. 8.12

Fracture toughness as a function of neutron dose for austenitic alloys irradiated in LWRs (288–316°C) and tested in the temperature range 250–320°C [74].

From O. K. Chopra and A. S. Rao, A review of irradiation effects on LWR core internal materials – neutron embrittlement, J. Nucl. Mater. 412 (2011)195–208.

orientation has been attributed to the presence of stringers consisting of long, narrow particles oriented in the rolling direction, which result in a long and narrow quasicleavage structure parallel to the crack advance, thereby accelerating the crack advance [74].

Microstructural characterization of type 304 control-rod material has revealed a fine distribution of γ' phase with particle size in the range of 2–10 nm and an average size of 4.4 nm [60]. The γ' phase has also been observed at dose levels above 4 dpa in CW type 316 alloy irradiated under PWR conditions [79]. The presence of precipitates can change the strain hardening behavior, and changes in material microchemistry due to RIS can change the deformation behavior, both of which may affect the fracture toughness of the material. The contribution of additional precipitate phases, voids, and cavities on fracture toughness needs to be investigated.

There is also evidence that void swelling may be correlated with reduced fracture toughness. A classic example is a type 316 stainless steel irradiated in EBR-II to 130 dpa at 400°C, resulting in 14% void swelling that fractured during handling at room temperature [80]. It has been proposed that the combination of RIS of Ni and Si to grain boundaries and void surfaces, and the formation of γ' reduces the stacking fault energy, resulting in greater slip planarity and an increase in the propensity for flow localization and fracture in the channels [74,80]. Hamilton and coworkers [81] expand on this possibility, pointing out that while the fracture toughness of 316 stainless steel declines with void swelling, it is actually the tearing modulus that plunges to zero. They show that nickel segregation at void surfaces drives the matrix between voids toward a martensite instability associated with a reduced stacking fault energy, which when compounded by stress concentration between voids, causes a hardening of the matrix with ε-martensite which converts to α-martensite at the crack tip. It is also shown that other 300 series steels exhibit the same behavior once ~10% swelling is reached.

While all of the processes by which irradiation induce embrittlement are not completely known, it is understood that restriction or localization of plastic flow, formation of a distribution of hard precipitates, and swelling all correlate with a reduction in fracture toughness. Localized deformation (LD) occurs after relatively low dose and, while the restriction of plastic flow becomes greater at higher dose, the degree of LD should not change dramatically over the fluence interval between 60 and 80 years of life. However, second-phase formation is much less well known and it is possible that high doses could lead to the formation of precipitates that have not yet been observed. With swelling, it is more certain that void swelling will become more severe with extended operation. So if swelling has an impact on fracture toughness, then it will only become worse with increasing age.

8.4.3 HIGH-TEMPERATURE He EMBRITTLEMENT

The D-T fusion neutrons with energy 14 MeV produce He through the direct (n, α) reaction in nearly all common structural materials such as Fe, Cr, and Ni, since this neutron energy is far above the threshold energy for the (n, α) reaction (6 MeV). The thermal neutrons produce He in Ni-bearing austenitic alloys by the two-step reaction sequence of ^{58}Ni(n, γ)^{59}Ni(n, α). In typical operating conditions, He to dpa ratio (He/dpa) is <1 for fast reactors and about 10 for fusion reactors. The spallation proton-neutron environment creates a much higher He/dpa ratio of up to 100, which is likely to have significant effects on the microstructural and mechanical evolution. For instance, the specimens irradiated in the mixed-spectrum reactor (500–1000 appm He) exhibit consistently lower tensile ductility than the fast spectrum reactor specimens (<5 appm He) even at low dpa, especially above 600°C, where He embrittlement becomes significant [82].

Concerning the creep rupture property of austenitic stainless steels, deterioration of the rupture time is induced by He embrittlement that coincides with an increasing transition from transgranular to intergranular fracture [83]. Helium embrittlement is attributed to the nucleation of helium bubbles on grain boundaries, their subsequent growth by helium absorption or coarsening, followed by their stress-induced transformation to unstably growing cavities. One of the extreme results in He embrittlement on creep rupture time was shown by Bloom for the Ni and B bearing 20% CW316 stainless steel tested at 550°C and 310 MPa following irradiation between 535°C and 605°C in the mixed spectrum HFIR that produced up to 3190 appm He and 85 dpa [84].

These fission reactor irradiation tests for He embrittlement research were complemented by accelerator-based He ion implantation experiments [85,86]. It was shown that He embrittlement led to a large reduction in the rupture time especially at lower stress for SA 316 stainless steel at 750°C for in-beam creep at an implantation rate of 100 appm He/h [85]. He embrittlement was also observed for a Ti-modified austenitic stainless steel (DIN 1.4970) at the same test conditions [85]. Yamamoto et al. have extensively conducted a series of posthelium implantation creep testing at 650°C for Fe-15% Cr-25% Ni austenite stainless steels containing the stabilized elements [87–90]. 60–70 appm He was implanted using a 26 MeV helium-3 ion beam from a cyclotron. The various MX type of precipitates (M:V, Ti, Nb, Zr; X:C, N) were stabilized after helium implantation and creep testing at 650°C. Fig. 8.13 illustrates the typical failure morphology of grain boundary decohesion in helium-implanted samples.

The faces of this fracture mode were often associated with small dimples, and sponge-like structure, which is very common for helium embrittled materials, indicating that the fracture process is related to grain boundary helium bubbles. It was observed that He implantation caused deterioration in terms of both creep rupture time and elongation, however, suppression of helium embrittlement can be achieved through a higher dispersion density of precipitates, because of their high capability of bubble entrapment within grains, as shown in Fig. 8.14.

It was also suggested that the enhancement of grain boundary decohesion by helium is a result of unstable growth of supercritical helium bubbles. The models of He embrittlement have been developed by researchers [91–93] in conjunction with such kind of He-ion implantation studies.

8.4.4 VOID SWELLING

Stainless steels are susceptible to a phenomenon called void swelling in which crystallographically faceted, vacuum-filled cavities (voids) form at high densities with the material lost from the cavities redistributed to grain boundaries, thereby increasing the volume of the steel. The process, once initiated, is unsaturable and levels of volume increase (swelling) approaching 100% have been observed several times at intermediate temperatures where void swelling is most important. The process can be broken roughly into two regimes, the incubation or transient regime and the posttransient steady-state regime. The latter regime is very predictable with a rate of

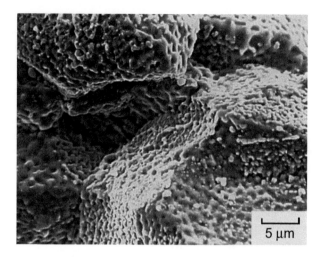

FIG. 8.13

SEM fractograph shows typical "sponge" like structure on intergranularly fractured facets after postimplantation creep rupture; 650°C, 50 appm He, stress 120 MPa, rupture time 839 ksi [88].

From N. Yamamoto, J. Nagakawa, Y. Murase, H. Shiraishi, J. Nucl. Mater. 258–263 (1998) 1628–1633.

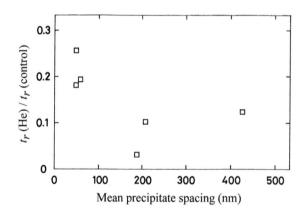

FIG. 8.14

Creep life ratio of helium implanted (50 appm He) specimens to unimplanted control specimen at 650°C, as a function of mean spacing of intragranular precipitates [88].

Adapted from N Yamamoto, J Nagakawa, H Shiraishi, J. Nucl. Mater., 226 (1995), p. 185.

~1%/dpa independent of composition, starting microstructure, thermal-mechanical processing, and environmental factors such as temperature, dpa rate, and stress level or state. However, the duration of the transient regime is highly sensitive to all of these variables as demonstrated in many fast reactor experiments. The transient regime of various austenitic steels can vary from 10 to 100 dpa depending on the steel, irradiation temperature, and dpa rate. Reviews of this important topic can be found in Refs. [65,94,95], the latter specifically focusing on LWR conditions.

As described in Section 8.1, an addition of an appropriate amount of Ti(MC), P, and Si can significantly improve swelling resistance in terms of prolonged incubation dose. A very fine distribution of MC carbides can be formed by co-addition of P, B, and cold-work, and they have a coherent interface with matrix. The fine MC carbide can trap tiny cavities at the interface with the matrix during the swelling incubation period. However, MC carbides

8.4 RADIATION-INDUCED MECHANICAL PROPERTY CHANGES

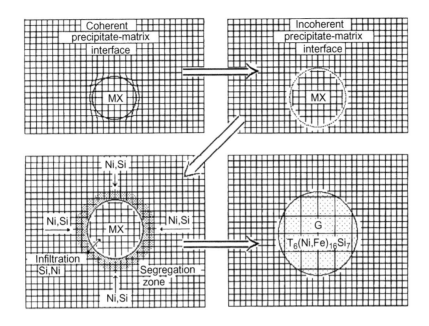

FIG. 8.15

Scheme of transformation in MX-type precipitate during irradiations [97].

From V.N. Voyevodin, I.M. Neklyudov, V.V. Bryk, O.V. Borodin, J. Nucl. Mater. 271–272 (1999) 290–295.

incorporate the segregation of the undersize elements such as Ni and Si at the carbide-matrix interfaces that is caused by RIS. This type of RIS cannot be avoided during irradiation, and MC carbides are often replaced with radiation-induced G phase having an incoherent interface with the matrix [96]. The schematic figure of structural change from MC carbide to G phase during irradiation is shown in Fig. 8.15 [97].

When the coherent-incoherent structural change in MC carbides proceeds, swelling will be accelerated toward the steady-state regime. The phosphide phase developed during irradiation in Ti-bearing austenitic steels was reported to be FeTiP [16]. The FeTiP phosphide, once precipitated during irradiation, possibly dissolves or changes to G phase and M_6C carbide in similar way to MC carbide at higher doses. Such a compositional change was observed during ion irradiation [98]. The dissolution of phosphides, coinciding with void evolution, has been observed at higher doses in a Fe-15Cr-20Ni-P model alloy, 316 class stainless steels [12,99] by neutron irradiation, and in a modified 316 stainless steel irradiated by Ni ion with simultaneous He injection [100]. Therefore, it is noted that an important process for swelling onset is closely associated with the instability of phosphides and MC carbides caused by RIS.

The role of Ti-containing γ' [$Ni_3(Si,Ti)$] in the French Si-mod.CW15-15Ti steel is worth noting, because it exhibited further improvement in swelling resistance than that of CW15-15Ti. The γ' precipitate with coherent interface is responsible for the swelling suppression in the Si-mod.CW15-15Ti steel, since its interface with matrix effectively act as the trapping of tiny cavities and strong sinks for point defect [101]. That is a reason why the incubation dose is so prolonged in Si-mod.CW15-15Ti steel, as shown in Fig. 8.2.

In addition to dose and temperature, swelling behavior is sensitive to dose rate, temperature change, and stress, as shown schematically in Fig. 8.16.

With regard to effects of dpa rate on void swelling, a relatively lower dpa rate causes a shorter incubation dose as compared with higher dpa rate [102,103]. Thus, the irradiation with lower dpa rate accelerates the void swelling. This is thought to be a consequence of several rate-dependent processes such as dissolution of phosphide, microchemical evolution of the alloy matrix, and dislocation loop evolution. Change in temperature during irradiation

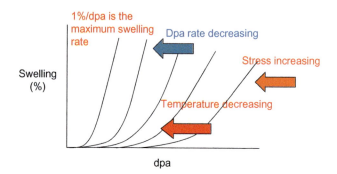

FIG. 8.16

Schematic representation of the parametric dependence of swelling of annealed AISI 304 in the temperature and dpa rate ranges of PWR interest [94].

From F. A. Garner, Radiation Damage in Austenitic Steels, Chapter 65 in Comprehensive Nuclear Materials, R. Konings, Ed., Elsevier, Volume 4, 2012.

can sometimes have a pronounced effect on microchemical evolution and thereby on void swelling [104]. From a JOYO irradiation experiment of PNC316, void formation was accelerated by a step-wise decrease in temperature, while void formation was suppressed as temperature increased, as compared with constant temperature irradiation [105]. Various similar results are reported in EBR-II irradiation tests [106]. Concerning stress effect, an enhancement of the swelling rate was modeled by the hydrostatic stress; this implied that compressive stresses decrease the swelling [107]. The other study suggested that any stress state, such as tensile, compressive, shear or mixed, would accelerate the onset of swelling [108]. In addition, the effect of a hoop stress induced by temperature resulted in a swelling gradient across the cladding wall and was investigated using fuel pins irradiated in JOYO and FFTF. The cladding was made of PNC316. The void distribution is nonmonotonic through the cladding wall thickness; swelling at the mid-wall part is obviously lower than at the outer and inner parts [109]. In parallel, the secondary stress induced by the swelling difference across the cladding wall thickness was analyzed by means of finite element method (FEM), and effect of the stress-enhanced swelling was evaluated [110].

Voids have been observed in BWR components but never at a high enough level to have any significant impact on important properties, primarily due to the low dpa levels reached in the shroud. Voids have been observed in various PWR components (baffle bolts and flux thimbles at <15 dpa) at larger swelling levels but still well below the levels characteristic of steady-state swelling. At larger dpa levels, especially at extended PWR lifetimes, the possibility of swelling at 1%/dpa cannot be discounted. Swelling is known to be accelerated by helium and hydrogen generation, both of which are much larger per dpa in LWRs than in fast reactors. Lower dpa rates in PWRs compared to fast reactors are also known to accelerate the onset of swelling, but other factors may work to retard the onset of swelling, especially the lower temperature range of PWRs and their very different temperature history.

Swelling can introduce changes in elastic moduli and thermal resistivity but these are of second-order significance. Most importantly, swelling can lead to significant increases in volume and when compounded with irradiation creep (see next section) can induce significant distortion with consequences involving interaction with other components, high withdrawal loads, coolant blockage and generation of stress on bolts, maintaining loads contributing to SCC, and other problems.

Unfortunately, as described earlier, swelling levels in the 5%–10% range introduce a new severe failure mode in austenitic steels that involves a martensite instability arising from the consequences of nickel segregation on void surfaces, decreasing austenite stability in the volume between voids, and stress concentration between voids. The result is a tearing modulus that plunges to zero, with severity increasing with decreasing deformation temperature and increasing with increasing swelling. This failure mode is well understood and relatively independent of everything except void swelling.

8.4.5 IRRADIATION CREEP AND FATIGUE

Under irradiation, creep deformation at lower temperature is accelerated by orders of magnitude higher than thermally activated creep. The irradiation creep is not destructive on the microstructural level, reducing stress concentrations and healing internal flaws. It was consistently shown that the irradiation creep rate is the sum of two terms; one is independent of the swelling and the other is strongly coupled with the swelling rate. The major components of the instantaneous creep rate can be expressed by the following empirical Equation [111,112]:

$$\frac{\dot{\varepsilon}}{\bar{\sigma}} = B_0 + D\dot{S}$$

where $\dot{\varepsilon}$ is an instantaneous irradiation creep rate, $\bar{\sigma}$ is an effective stress, B_0 is the creep compliance, D is the irradiation creep-swelling coupling coefficient, and $\dot{S}$ is the instantaneous volumetric swelling rate per dpa.

Irradiation creep has been studied by the irradiation of pressurized tubes. On the basis of the most extensive review by Garner [113], B_0 of austenitic stainless steels is typically in the range from 0.5 to 4×10^{-6} MPa^{-1} dpa^{-1}. Creep compliance B_0 estimated for the various kinds of steels including austenitic and ferritic stainless steels and Ni-base alloys is shown in Fig. 8.17 [114], where B_0 is insensitive to alloy composition and structure.

This swelling independent term has been well interpreted in terms of the stress-induced preferential absorption (SIPA) mechanism under active investigations for many years. The SIPA mechanism proposed by Heald and Speight [115], Wolfer and Ashkin [116], and Bullough and Willis [117] depends solely on differences in the climb rate of dislocations whose orientation relative to the applied stress influences their absorption of point defects. Dislocations with their Burgers vector parallel to the external stress direction can more easily absorb interstitial atoms and can take climb motion than those with their Burgers vector perpendicular to the external stress. Thus, SIPA produces a stress-induced bias differential among dislocations, causing anisotropic dislocation climb and loop growth, which leads to the total deviatoric strain along the external stress. This creep mechanism provides the best opportunity to explain many of the irradiation creep observations under condition where swelling is not significant.

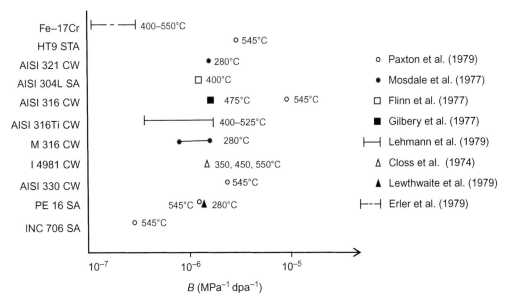

FIG. 8.17

Creep compliance (B_0) for the various steels including austenitic and ferritic stainless steels and Ni-base alloys [114].

From K. Ehrlich, J. Nucl. Mater. 133–134 (1985) 119–126.

Concerning irradiation creep driven by swelling, Gittus proposed the I-creep model to account for coupling between irradiation creep and swelling [118]. In this model, irradiation creep is induced by sequential climb-glide processes of dislocations due to absorption of interstitial atoms under dislocation bias, while excess vacancies produce voids. Therefore, this mechanism predicts an irradiation creep rate proportional to swelling rate [119], which means that the D value turns out to be constant. The value of D derived from irradiation tests using the pressurized tubes of austenitic stainless steels is very limited, and values of around 2×10^{-3} MPa^{-1} were reported from CEA 316Ti and 15–15Ti [113,120] and PNC316 [121]. These values of D were derived from the integral form of the instantaneous creep rate equation, assuming constant stress and constant temperature with respect to neutron dose. On the other hand, Garner suggested by taking into account the instantaneous swelling rate, the D value approached 6×10^{-3} MPa at low swelling rates and it declined at higher average swelling rate, with approaching an asymptotic value of 2×10^{-3} MPa^{-1}. This is shown in Fig. 8.18A.

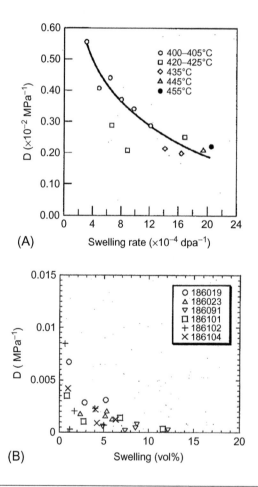

FIG. 8.18

Irradiation creep-swelling coupling coefficient (D): (A) as a function of swelling rate for CW316Ti assuming $B_0 = 1.25 \times 10^{-6}$ MPa^{-1} dpa^{-1} (Refs [113,120]), (B) as a function of swelling for PNC316 irradiated in fuel pin [122].

(A) From F. A. Garner, Chapter 6. Irradiation Performance of Cladding and Structural Steels in Liquid Metal Reactors, Materials Science and Technology, Vol. 10A: A Comprehensive Treatment, VCH Publishers, Weinheim, 1994, 483–514; A. Maillard, H. Touron, J. L. Seran et al., Effects of radiation on materials, 16th Int. Symp., STP 1175, Philadelphia, ASTM (1993). (B) From S. Ukai, S. Ohtsuka, J. Nucl. Science and Technol. Vol. 44 (2007) 743–757.

8.4 RADIATION-INDUCED MECHANICAL PROPERTY CHANGES

Similar trend of D values against swelling was reported for MONJU fuel pin cladding PNC316 irradiated in FFTF. Fig. 8.18B shows that the D value tends to decline and almost disappear as swelling develops for PNC316-based fuel pins, where the results of the six fuel pins are plotted [122]. Thus, D value is not constant, and depends on swelling.

Failure of materials caused by cyclic loads is called fatigue. Two types of fatigue are distinguished by the number of cycles to failure. High-cycle fatigue, HCF, with a number of cycles to failure over 10^4, is characterized by low stress amplitudes and large elastic portions in the strain cycle. Crack initiation consumes most of the fatigue cycles, while crack propagation, usually transgranular in character, and amounts to <10% of the number of cycles. Low-cycle fatigue, LCF, has a large plastic portion in the strain cycle and a large fraction of fatigue life, >50%, consists of crack propagation, consequently the number of fatigue cycles is below 10^4. LCF crack propagation and final fracture have a transgranular character at high strain rates and tend to be intergranular at low cyclic rates in the creep range [123].

Information about the effects of irradiation on the fatigue behavior of austenitic stainless steels is very scarce. The results of type 316 are plotted in Fig. 8.19 for an irradiation and test temperature of 430°C [124,125].

A moderate reduction of the cycles to fracture including a small decrease of the endurance limit has been shown. This degradation in the fatigue life has been attributed to a matrix hardening caused by a radiation-induced formation of precipitates. At higher irradiation temperatures (550°C) no influence of irradiation has been observed. However, another important result is reported for the type 304 SS, if combined creep-fatigue tests are undertaken on postirradiated samples, where two parameters influence the fatigue life: the strain rate and additional hold-times in tension [126]. Since the strain rate effect is reduced with increasing hold times, it is mainly the creep in tension that reduces the cycles to fracture, even for the unirradiated material. It is, however, of most practical importance that small amounts of helium generated during irradiation (5 ppm) can reduce the cycles to fracture by more than one order of magnitude, especially if one takes into account demonstration operation with pulse length of the order of 1000 s and longer. One of the key parameters contributing to combined creep-fatigue degradation is therefore the helium-to-displacement rate.

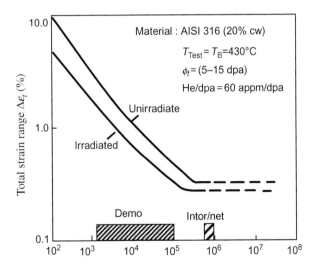

FIG. 8.19

Effect of preirradiation on the cycle to fracture in 20% CW316SS: $T_{irra} = T_{test} = 430°C$, He/dpa = 60 appm/dpa, dose: 5–15 dpa [124,125].

Based on M.L. Grossbeck, K.C. Liu, Nucl. Technol. 58 (1982) 53; M.L. Grossbeck, K.C. Liu. J. Nucl. Mater. 103–104 (1981) 853.

8.4.6 IN-REACTOR CREEP RUPTURE PROPERTIES

Thermal creep rupture under irradiation is a critical issue for fuel pin design. Lovell et al. [127] found that in-reactor rupture lives were equal to or greater than in air for 20% cold worked (CW) AISI 316SS, based on the irradiation tests of instrumented subassembly in EBR-II. Puigh and Schenter [128] also reported as the first experiment of in-reactor creep rupture test using MOTA in FFTF that in-reactor creep rupture lives agreed well with the in-air creep rupture data. However, Puigh and Hamilton [129] pointed out that neutron irradiation eventually led to a decrease in rupture lives for 20% CW316SS and 10% CWD9 steel based on the FFTF-MOTA experiment, when compared with the in-air creep rupture data, which is contrary to his earlier indications [128].

Ukai et al. [130] directly compared the in-air, in-sodium, and in-reactor creep rupture data for the pressurized tubes made of PNC316 in a hoop stress vs Larson-Miller parameter (LMP) diagram, which are summarized in Fig. 8.20A.

In-reactor data were obtained from the FFTF-MOTA experiment. The in-sodium creep rupture strength is similar to in-air data, but beyond approximately 16.5×10^3 for LMP, the rupture lives in-sodium become relatively shorter than in-air. On the other hand, neutron irradiation caused a remarkable reduction in rupture strength at low-to-high LMP region. Fig. 8.20B shows 0.2% proof strength of PNC316 after aging, sodium and neutron exposures at three temperature levels of 600°C, 650°C, and 700°C as a function of LMP. At 600°C, there is little difference in the 0.2% proof strength between the aged and sodium-exposed samples, whereas it was remarkably reduced after neutron exposure. This behavior surprisingly coincided with reduction of in-reactor creep rupture life, when compared with Fig. 8.20A at low LMP. Such strength reduction during and after neutron irradiation could be mainly attributed to the earlier recovery of dislocation structure introduced by cold-working. At 650°C and 700°C (above LMP of 16.5×10^3), 0.2% proof strength after sodium exposure and in-sodium creep rupture lives showed a tendency toward more decrease than after aging and in-air. Flowing sodium enhanced the dissolution of phosphorus and boron into the sodium, which caused earlier coarsening of MC precipitates [131]. This behavior gave rise to degradation of 0.2% proof strength and creep rupture strength as well. Irradiation at 650°C and 700°C and corresponding LMP above approximately 16.5×10^3 tended to decrease more than sodium exposure due to enhanced of dissolution of phosphorus and boron into the sodium.

8.5 PCI/FCCI EFFECTS WITH FISSION FUELS

In the 1960s and 70s, stainless steel cladding was used in both PWR and BWR fuel. The cladding generally exhibited better performance than Zr alloy cladding in first-generation PWRs (fuel pin defect rate of 0.01% vs 0.1%–0.3% for Zr alloy cladding). However, steady improvement of the Zr alloy cladding performance during the 1960s and early 1970s along with the superior (low) parasitic absorption of neutrons by Zr led to nearly universal adoption of Zr alloy cladding by the early 1970s [132,133]. However, due to the combination of stress and fuel-clad interaction, an oxidizing environment (in BWRs) and the effect of irradiation, stainless steel fuel cladding experienced SCC, later identified as IASCC. Stainless steel fuel cladding failures were first reported in the Connecticut Yankee PWR in 1977 (304 stainless steel) and in PWR test reactors [134–142]. At the West Milton PWR test loop, intergranular failure of vacuum annealed type 304 stainless steel fuel cladding was observed [143] in 316°C ammoniated water (pH 10) when the cladding was stressed above yield. Similarly, IASCC was observed in creviced stainless steel fuel element ferrules in the Winfrith SGHWR [144,145], a 100MWe plant in which light water is boiled in pressure tubes, where the coolant chemistry is similar to other BWR designs. The 20%Cr/25%Ni/Nb stainless steel differs from type 304 primarily in Ni and Nb content, and compared to most heats of 304 of that time, it had lower sulfur ($\approx 0.006\%$) and phosphorus ($\approx 0.005\%$) contents. The ferrules were designed for a 5-year exposure during which the peak fast neutron fluence is $3–5 \times 10^{21}$ n/cm^2. The emergence of IASCC along with the significant enrichment penalty due to their relatively high neutron absorption, caused them to be phased out of service.

8.5 PCI/FCCI EFFECTS WITH FISSION FUELS

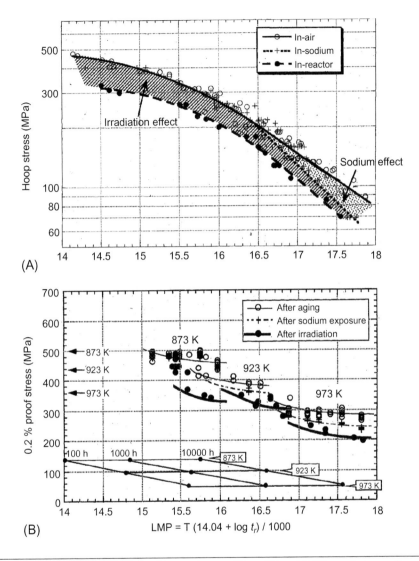

FIG. 8.20

Creep rupture curves for in-air, in-sodium, and in-reactor conditions (A) and 0.2% proof stress after aging, sodium and irradiation exposures (B), as a function of Larson-Miller parameter [130].

From S. Ukai, S. Mizuta, T. Kaito, H. Okada, J. Nucl. Mater. 278 (2000) 320–327.

Internal wastage of the fuel pin cladding in SFR is caused by fuel-cladding chemical interaction (FCCI) during reactor operation. FCCI is caused by fission product-accelerated oxidation and corrosion of the cladding that is frequently observed in SFR fuel pins involving reactive and volatile FPs, such as Cs, Te, I, and Mo. Specifically, Cs and Te contribute to the most aggressive intergranular attack modes [146]. FCCI was recognized as one of the major factors determining integrity and lifetime of the oxide fuel pins. Among the three major constituents of austenitic stainless steel cladding, Fe, Cr, and Ni, chromium has the greatest affinity for oxygen and forms the most stable and protective oxide layer, Cr_2O_3 at the internal surface of the cladding. However, inside fuel pins, this protective Cr_2O_3 layer is destroyed by chemical attack of the volatile and reactive fission products (FPs). The generated volatile FPs are released and accumulated at the fuel-cladding gap with increasing burnup. When fuel surface oxygen potential exceeds the threshold necessary for cladding oxidation, excess oxygen, and corrosive FPs can interact with the cladding at the inner surface.

Abundant in-pile and out-of-pile tests for austenitic stainless steel cladding have been conducted since the 1970s using a large number of type 316 SS class fuel pins from an initial stage of SFR development. Fig. 8.21 shows an example of the measured cladding wastage vs. cladding inner surface temperature at different fuel O/M ratios and local burnup [147].

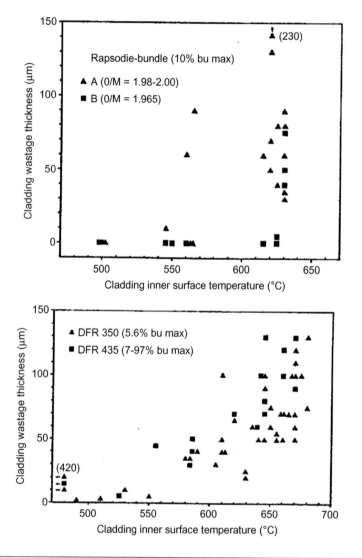

FIG. 8.21

Cladding inner wastage of fuel pins irradiated to a burnup of 5–10 at.% and with initial O/M ratios of 1.955–2.00 as a function of cladding inner surface temperature [147].

From M. Coquerelle, O. Gotzmann and S. Pickering, Inner Cladding Attack in Mixed-Oxide Fuel Pins. European Nuclear Conference, Paris, April 21–25, 1975. Transactions of the ANS, 20 (1975) S. 287–89. Copyright 1975 by the American Nuclear Society, LaGrange Park, Illinois.

Two types of cladding were used in this test: solution annealed austenitic stainless steel 1.4988 (17Cr-13Ni-1.3Mo-1Nb-0.7 V-0.1 N-0.08C) and 10%–15% CW and annealed austenitic stainless steel 1.4970 (15Cr-15Ni-1.2Mo-0.45Ti-0.1C). The cladding wastage due to FCCI was affected by the fuel stoichiometry; higher O/M fuel with 1.98–2.00 increased cladding wastage as compared to lower O/M (1.965) fuel. Similar trends have been acquired in another irradiation test [148]. Fig. 8.21 also shows a remarkable increase in the cladding wastage with increasing temperature above a threshold of 520°C. However, the burnup dependence is not obvious. Even at a burnup as low as 0.5 at.%, intergranular penetration may extend occasionally to 70 μm. As burnup increases, the maximum depth of wastage does not normally increase greatly, usually <130 μm. The corrosion spreads and becomes more uniform over the inner cladding surface. On the basis of FCCI data conducted in the irradiation experiments so far, the modification of austenitic stainless steels by addition of minor elements such as Ti, P, B, and Si cannot reduce the corrosion thickness although it was demonstrated that high-temperature strength and void swelling have been considerably improved as shown in the previous sections.

8.6 CHEMICAL COMPATIBILITY WITH COOLANTS

Stainless steels exhibit excellent corrosion resistance in high-temperature water due to formation of protective layers of chromium-rich oxides that act as a diffusion barrier. The morphology of the oxide scale includes an outer layer mainly composed of magnetite (Fe_3O_4) and spinel iron-chromium oxide ($FeCr_2O_4$ with Fe/Cr ratio around 3) crystallites and an inner, compact layer composed of a chromium-rich oxide. The outer oxide layer is assumed to be formed by re-deposition of iron and chromium dissolved in solution due to general corrosion of stainless steels. The inner oxide layer is richer in chromium and is believed to form by the inward solid-state diffusion of oxygen. The main difference between inner and outer oxide is that the inner oxide layer is composed of a Cr-rich oxide. The depth of the oxide layer is usually in the range 200–300 nm after 1000 h of exposure to PWR primary water conditions. Between CW materials and non-CW stainless steels (304 and 316 L), the main difference appears between the inner oxide scale microstructure after 1000 h of exposure in PWR nominal primary conditions: the inner Cr-rich oxide is continuous on annealed materials while the inner Cr-rich oxide is discontinuous on CW materials [149]. While there are differences in the morphology and composition of the oxide film in a BWR (NWC or HWC) and a PWR, the basic two-layer structure is the same, with the inner layer providing oxidation resistance to the underlying metal.

Besides general corrosion, stainless steels undergo other modes of degradation in high-temperature water, most all in the form of localized corrosion. These include pitting, crevice corrosion, intergranular corrosion (IGC), SCC, and IASCC. Pitting and crevice corrosion are induced and exacerbate by water impurities (chlorides, sulfates) and oxidizing species such as Cu from the condenser. These problems are treatable by sound water chemistry practice. SCC is discussed next (Section 8.7) and IASCC is treated in Section 8.8.

Coolant sodium with excellent thermal conductivity (0.648 W/cm °C at 550°C) can efficiently transfer the heat of the reactor core to the power generation system in SFR. Corrosion and mechanical property changes of the austenitic stainless steels exposed in sodium have been studied for more than five decades. Solubility of nickel, iron, and chromium into sodium is low, around 1 ppm at 650°C. The dissolution process of the major elements into sodium dominates the corrosion rate of austenitic stainless steel, and this process is expressed according to the solid-state diffusion of the elements in the steel. The dissolved oxygen (DO) also affects the sodium corrosion. Fig. 8.22 shows a temperature dependence of the corrosion rate of 304, 316, and 321 SS measured at 2.5 and 9 ppm DO [150].

This figure covers a typical SFR condition: temperature of 450–650°C, sodium velocity of 2–4.8 ms^{-1}, and DO of 2.5–5.0 wppm. Concerning mechanical property changes of the sodium-exposed steels, as described in Fig. 8.20 of the previous section, in-sodium creep rupture strength is reduced from the in-air at LMP over 16.5×10^3, which corresponds to 10^3 h at 700°C. The sodium exposure slightly affects the strength reduction at higher temperature and at longer time through the carbide coarsening induced by dissolution of phosphorus and boron. The low-cycle fatigue lives of 304 SS in sodium are reported to be greater than those in the in-air condition [150]. This is related to

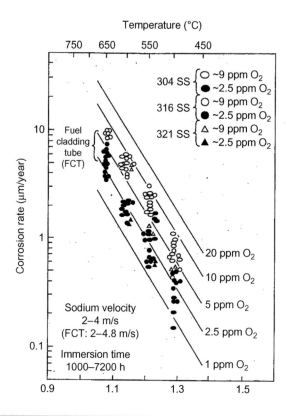

FIG. 8.22

Corrosion rate of various kinds of austenitic stainless steels vs temperature, sodium velocity: 2–4 ms^{-1}, exposed time: 1000–7200 h [150].

From T. Furukawa, S. Kato, E. Yoshida, "Compatibility of FBR materials with sodium," Journal of Nuclear Materials 392 (2009) 249–254.

a surface condition. Many micro-cracks are observed on the oxide scale for in-air specimens, while micro-cracks are hardly observed on the in-sodium specimens because of reduced atmosphere under sodium.

In the last 10 years, the study of the utilization of Pb and Pb-Bi eutectics (LBE) has been the focus for application to nuclear waste transmutation systems and Pb-Bi cooled reactors. For instance, LBE is a candidate for both the subcritical-reactor coolant and the spallation neutron source target for the accelerator-driven nuclear transmutation system (ADS). In addition, Pb or Pb-Bi-cooled fast reactor (LFB) is one of the GenIV systems. Concerning the LBE corrosion behavior, solubility of major elements of austenitic steels plays an important role. On the basis of correlations of various metal solubilities published by IPPE [151], the solubility of Ni in LBE is as high as 37,000 wppm at 600°C, indicating that austenitic stainless steels containing high-Ni content could be at a disadvantage for application to LBE structural materials. The compatibility of austenitic steels with Pb and LBE has been studied [152]. Generally, in the low-temperature range, for example, below 450°C, and with an adequate oxygen activity in the liquid metal, austenitic steels form an oxide layer which behaves as a corrosion barrier [153]. However, in the higher temperature range, that is, above 500°C, corrosion protection through the oxide scales seems to fail. Indeed, a mixed corrosion mechanism has been observed, where dissolution of both oxide scales occurred [153]. However, in this high-temperature range, it has been demonstrated that the corrosion resistance of the austenitic steels can be enhanced by coating the steel with FeAl alloys. Several mechanical tests (e.g., tensile, fatigue, slow strain rate, etc.) in Pb and LBE have also been conducted [152].

The molten salt reactor (MSR) is one of the GenIV systems. The fuel of the MSR is based on the dissolution of the fissile material (^{235}U, ^{233}U, or ^{239}Pu) in an inorganic liquid that is pumped at a low pressure through the reactor vessel and the primary circuit, and thus also serves as the primary coolant. One of the candidate fuel types is ^{7}LiF-BeF$_2$-^{232}ThF$_4$-UF$_4$ system [154], where U and Th are dissolved in the molten fluorides and ^{232}ThF$_4$ is a fertile material that is used to produce fissile ^{233}UF$_4$ by neutron capture. This type of fuel is a candidate for the graphite-moderated thermal breeder reactor. For actinide burner by the nonmoderated MSR concept, ^{7}LiF-(NaF)-BeF$_2$-AnF$_3$ system is proposed, where An means actinide elements. The molten fluorides, called FLIBE, are used in these systems. A compatibility and corrosion of the structural material with the molten fluorides are significantly concern and critical issue for the MSR development. ORNL has conducted a series of corrosion tests using either thermal convection or forced convection flow loops, to select the materials best suited to this application. The available commercial Ni-base alloys and austenitic stainless steel have been tested. Recent corrosion tests of Ni-base alloys (Hastelloy-N, Hastelloy-X, Haynes-230, Inconel-617) and austenitic Incoloy-800H (Fe-20Cr-32Ni-0.8Mn-0.6Ti-0.5Al) were performed in molten fluoride salt at 850°C. Corrosion was noted to occur predominantly by dealloying of Cr from the alloys, and weight-loss due to corrosion was generally correlated with the initial Cr-content of the alloys [155]. Details of the material corrosion by the molten fluorides are reviewed in Ref. [156].

8.7 STRESS-CORROSION CRACKING

8.7.1 SCC IN BWRs

Austenitic stainless steels are susceptible to SCC in LWR environments. The metallurgical state of the alloy, the environment and the application of stress all play roles in SCC. The prime environmental factor is the electrochemical potential (ECP). Extensive work by Ford and Andresen [157,158] has shown that crack propagation is extremely sensitive both to the corrosion potential and solution conductivity. Relatively modest increases of either environmental component cause large increases in the crack propagation rate in °C water [159]. The corrosion potential is a function of the oxygen concentration such that an increase in DO produces an increase in the corrosion potential. These observations are most applicable to relatively oxidizing conditions typical of a BWR using normal water chemistry (NWC) which is generally characterized by 0.2 ppm DO and a conductivity below 0.2 μS/cm for a corrosion potential above 50 mV$_{SHE}$. Cracking is significantly reduced by suppressing the corrosion potential, which is achieved by adding hydrogen (using hydrogen water chemistry, HWC). Corrosion is also a strong function of water purity as high cation or anion concentrations and metal salt/oxide precipitation can result in crevices in oxidizing environments or boiling on heat transfer surfaces [160]. This problem was addressed by improving water purity in BWRs such that in today's units, the coolant purity, as measured by water conductivity, approaches that of theoretical-purity water. IGSCC is also affected by material condition, most importantly by grain boundary chromium depletion. Heat treatments applied to stainless steels, either during fabrication or locally during welding, can induce the formation of chromium carbides at grain boundaries, and depletion of chromium in the region near the grain boundary. The loss of chromium means that passivity cannot be achieved locally and grain boundaries become susceptible to IGC or IGSCC. The latter occurs either due to service stresses or weld residual stresses. The use of low carbon stainless steels and better control over weld residual stresses can mitigate cracking.

8.7.2 IGSCC IN PWRs

In PWRs [160], the major SCC issues are in the nickel-base alloys used in steam generators and in vessel penetrations and stainless steels used in the reactor core. Since steam generator tubes comprise some 75% of the surface area of the primary circuit in contact with the coolant, their performance is critical to that of the reactor. While there exist many degradation modes of alloy 600 steam generator tubes, SCC has dominated its performance over the past

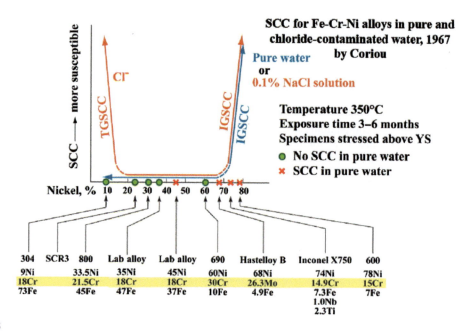

FIG. 8.23

SCC severity of austenitic alloys as a function of nickel content in pure water and 0.1% sodium chloride.

Courtesy R. W. Staehle; Data from Coriou, Montage by Staehle.

25 years on both the primary and secondary sides. SCC of nickel-base alloys is mostly absent of irradiation but is very sensitive to composition and water chemistry. Fig. 8.23 shows the propensity for cracking of austenitic alloys as a function of nickel content in both pure water and 0.1% NaCl.

Note that SCC in chloride-containing water is at a maximum at the extremes in nickel content, but SCC in pure water occurs only at high nickel concentrations. Unfortunately, alloy 600 used originally in steam generators and vessel penetrations contains approximately 72% Ni and so is highly susceptible to SCC in pure water. Neither microstructure modification (thermal treatment) nor water chemistry control has been unsuccessful in eliminating SCC in alloy 600.

On the primary side [161], susceptibility of alloy 600 to IGSCC first became apparent in steam generator tubes in the U-bends and in the rolled transitions at the top of the tubesheets. This 1970s observation was followed by IGSCC of pressurizer nozzles and control rod drive mechanism penetrations in the upper heads of PWR reactor pressure vessels in the 1980s. IGSCC in alloys 182 and 82 weld metal has also begun to crack in major primary circuit welds after extended periods of service between 17 and 27 years [161]. In addition to alloy 600, the gamma-prime strengthened analogue, X-750, used in pins that attach the CRDM guide tubes to the upper core plates have also undergone IGSCC. Staehle [162] has identified some seven different SCC modes defined by their potential-pH combinations. In fact, SCC in alloy 600 appears to track the Ni-NiO stability line in which cracking is at a maximum at the line and drops off both above and below [163]. Since this range of potential and pH span those typically achievable in service, control of SCC in alloy 600 has proved very difficult. The eventual remedy to the alloy 600 cracking problem has been replacement with alloy 690 (and weld metal counterparts 152 and 52), which has a nominal composition of 60Ni-30Cr-10Fe and exhibits much higher resistance to SCC in pure water, chlorides, and alkaline solutions. However, the history with Alloy 600 is that it was difficult to crack in pure high-temperature water in the laboratory, prompting the belief that it would be resistant in service. In fact, even after years in operation, few incidents of cracking were reported. However, the incubation period was found to be of the order of

8.7 STRESS-CORROSION CRACKING

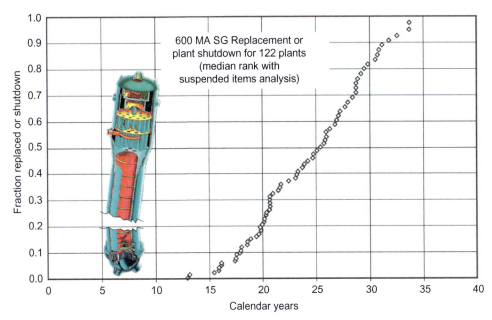

FIG. 8.24

Fraction of steam generators using 600 MA steam generator tubing that were replaced or shut down over time for 122 plants.

Courtesy R. W. Staehle and J. Gorman.

11–12 years, Fig. 8.24, and eventually, all plants began to experience cracking that has eventually led to widespread replacement with steam generators tubed with Alloy 690.

The concern today is that the superior resistance to SCC may be due to a longer incubation period and that in fact, it will eventually begin to crack. Current efforts are focused on trying to determine if there are microstructures, processing routes or water chemistries for which the alloy is susceptible to SCC. Recently, it was found that a single cold-rolling operation to 20%–30% reduction in thickness increases the crack growth rate (CGR) in pure water by over a factor of 100 [164]. Similar efforts are growing to identify susceptibility to crack initiation.

IGSCC also occurs on steam generator tubes on the secondary side (ODSCC) and has been responsible for tube plugging and even steam generator replacement [161]. It was first observed in the tube/tube sheet crevice before full depth rolling was adopted. It also occurs more commonly in the crevices between the tubes and carbon steel tube support plates, and has been observed under the sludge piles that accumulate on tube sheets. Cracking has been observed to be more severe at the bottom of steam generators where the temperature difference between the primary and secondary water is greatest, providing evidence that impurity hideout that increases as a function of superheat. IGSCC has also been observed on the free-span of recirculating steam generators. It has been attributed to one of two causes. The first is the formation of silica-rich deposits that formed bridges between tubes in regions of high heat flux, and resulting in degradation that is probably the same as that due to deposits at tube/tube support plate crevices. The second is lead-induced cracking due to maintenance activities that accidentally left lead blankets in the steam generator after a service outage. Finally, once-through steam generators (OSTGs) exhibit free span IGSCC in the superheated steam zone, which has been attributed to damage on the outer tube surfaces during tube installation. In summary, secondary side cracking is a complicated process involving many different modes. Measures taken to mitigate ODSCC include application of thermal treatment (715°C for 16h), that has proven to be beneficial, but does not provide complete immunity. The elimination of crevices in the tube support plate by adopting quatrefoil broached hole designs has also helped to reduce impurity accumulation that is so important in the IGSCC process. Finally, replacement of tubes with alloy 690 has provided the greatest increment of resistance to IGSCC as the increased chromium content provides protection on both the primary and secondary sides.

8.8 COMBINED EFFECTS OF WATER ENVIRONMENT AND RADIATION
8.8.1 IRRADIATION-ASSISTED STRESS-CORROSION CRACKING

Perhaps the single most important way in which irradiation can affect material performance in LWR core materials is in the inducement of SCC. Early plant and laboratory observations showed that the same basic dependencies existed for unirradiated and irradiated stainless steels, and that increasing fluence produces a well-behaved increase in SCC susceptibility. Intergranular (IG) SCC is promoted in austenitic stainless steels above a "threshold" fluence. This occurs in oxygenated (e.g., BWR) water above 2 to $5 \times 10^{20}\,\text{n/cm}^2$ ($E > 1\,\text{MeV}$), which corresponds to about 0.3–0.7 dpa, and the "threshold" fluence depends on the stress, water chemistry (esp. sulfate and chloride), operating time, and other factors. (Therefore, it is not a true threshold, but rather indicates that a minimum fluence is required before irradiation consequentially "assists" SCC.) Attempts to reproduce the same level of IG cracking in inert environments have been unsuccessful, confirming that it is an environmental cracking phenomenon, not simply a change in the microstructure, mechanical properties and the overall response of the irradiated material in an inert environment. IASCC field experience is summarized by the trends and correlations in the following section [2].

8.8.1.1 Observations and correlations from field experience
- Water impurities, especially chloride and sulfate, strongly and similarly affect IASCC in BWR water. This correlation applies equally to low and high flux regions and to stainless steels and nickel-base alloys, and is closely paralleled in out-of-core components. At higher levels, the same impurities can affect SCC in PWRs. If high corrosion potential conditions form in the PWR primary (where B and Li are present), the crack chemistry is dramatically altered, and high growth rates can result.
- Field and laboratory data demonstrate that corrosion potential is a very important parameter, with its effect being consistent from zero to low to high fluence, except in some high-fluence materials and/or under high stress intensity factor conditions where high growth rates are always observed (see Section 2 of this report). Materials prone to high radiation-induced changes in grain boundary Si level may exhibit a very limited effect of corrosion potential. While corrosion potential usually affects crack initiation and growth, there is no evidence of *threshold* potential.
- Cracking is enhanced by crevices, primarily because of their ability to create a more aggressive crevice chemistry from the gradient in corrosion potential (in BWRs) or in temperature (most relevant to PWRs), which in turn can accelerate crack initiation. Stress and strain concentration can also occur in crevices.
- Temperature increases IASCC severity for both initiation and growth rate.
- IASCC of annealed stainless steel was once thought to occur only at fluences above $\approx 0.3 \times 10^{21}\,\text{n/cm}^2$. But significant intergranular cracking in BWR core shrouds (which do not have thermal sensitization) occurs over a broad range of fluences, showing that a true fluence threshold does not exist. The observations of SCC in unirradiated, unsensitized stainless steel (with or without cold work) also undermine the concept of a *threshold* fluence below which no SCC occurs. Thresholds in corrosion potential, water impurities, temperature, etc. have also been disproven.
- Irradiation has a complex effect on SCC susceptibility. While radiation-induced segregation and radiation hardening increase susceptibility, radiation-induced creep relaxation of constant displacement loads (e.g., bolts and welds) tends to reduce susceptibility. For these reasons, SCC in BWR shrouds and PWR baffle bolts does

8.8 COMBINED EFFECTS OF WATER ENVIRONMENT AND RADIATION

not always correlate strongly simply with fluence. SCC can only be interpreted and predicted by accounting for the conjoint effects of multiple factors.
- Hardening has been cited as a key factor in IASCC susceptibility. Fig. 8.25A shows that increasing yield strength increases the CGR in nonsensitized austenitic stainless steel tested in 288°C BWR NWC.
- This dependence carries over to irradiated material. Fig. 8.25B shows a correlation between yield strength and susceptibility to IASCC in SSRT tests although the correlation is complicated by other radiation-induced

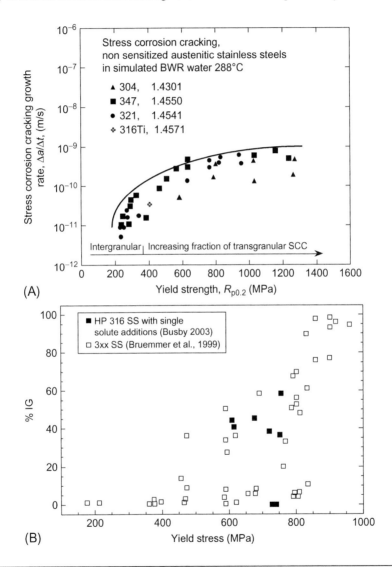

FIG. 8.25

Effect of yield strength on IGSCC. (A) Crack growth rate of cold worked, unirradiated 300-series stainless steels tested in 288°C simulated BWR water, and (B) %IGSCC in SSRT tests on 300-series stainless steels where hardening is by irradiation [69].

From G.S. Was, Recent Developments in Understanding Irradiation Assisted Stress Corrosion Cracking, Proc. 11th Int'l Conf. Environmental Degradation of Materials in Nuclear Power Systems – Water Reactors, American Nuclear Society, La Grange Park, IL, (2004) 965–985.

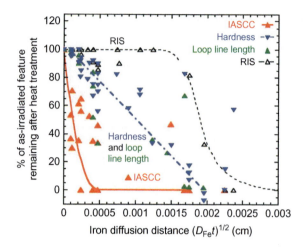

FIG. 8.26

Removal of RIS and dislocation microstructure as measured by loop line length and hardness with extent of annealing as measured by $\sqrt{(Dt)}$ for iron, where D is the diffusivity of Fe and t is time. This relationship accounts for annealing at different times and temperatures. The effect on IASCC in slow strain rates is also shown [69].

From G.S. Was, Recent Developments in Understanding Irradiation Assisted Stress Corrosion Cracking, Proc. 11th Int'l Conf. Environmental Degradation of Materials in Nuclear Power Systems – Water Reactors, American Nuclear Society, La Grange Park, IL, (2004) 965–985.

changes, especially RIS. The correlation is better at high values of yield strength (>800 MPa) and at very low values (<400 MPa), with more scatter at intermediate values (400–800 MPa).
- Hardening alone is not sufficient to explain IASCC. Annealing experiments tracked the change in hardness and dislocation loop microstructure vs. annealing condition [69]. Fig. 8.26 shows the change in hardness and the change in the dislocation loop line length ($N_{avg} \times d_{avg}$) along with the change in IASCC susceptibility in SSR tests as annealing progresses.
- Except for very short annealing times, the loop line length and the hardness closely track each other, as expected if the loop structure is controlling the hardening. At very small values of $(Dt)^{1/2}$, the hardening remains flat before decreased with annealing time. While both hardening and cracking are reduced with increased annealing, the behavior of the hardness does not fully explain the response.
- High stress and dynamic strains were responsible for the earliest incidents of IASCC in plants, but cracking has since been observed at quite low stresses at moderate to high fluences during longer operating exposure. Laboratory and field data indicate that IASCC growth can occur at stress intensity factors well below 10 MPa $\sqrt{m}$, and initiation can occur at <20% of the irradiated yield stress in some cases.
- Bulk cold work, surface cold work (especially abrasive surface grinding), and weld residual strain in the heat affected zone tend to exacerbate all forms of SCC although it can also delay the onset of some radiation effects (especially swelling) by creating more sinks for interstitials and vacancies. The effect of cold work is pronounced on the CGR, but cold work (especially abusive surface cold work) can strongly affect crack initiation.
- RIS is understandably implicated in IASCC of stainless steels in oxidizing environments, in part because of the deleterious effect of Cr depletion from thermal sensitization in extensive data from lab and plant operational experience [45,165,166]. As shown in Fig. 8.6, grain boundary chromium depletion during irradiation can be severe [37–44].

Fig. 8.27A shows a correlation between grain boundary chromium level and IGSCC susceptibility in stainless steels where the grain boundary depletion is due to thermal sensitization [80].

8.8 COMBINED EFFECTS OF WATER ENVIRONMENT AND RADIATION

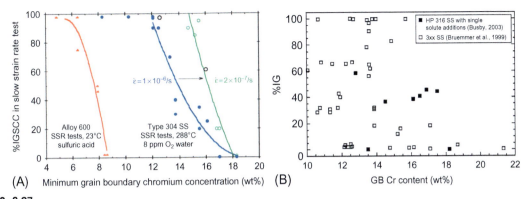

FIG. 8.27

Effect of grain boundary Cr content on IGSCC for: (A) sensitized stainless steel and Alloy 600 and (B) irradiated stainless steels [166a]. SSR tests are slow strain rate tests in which the specimen is monotonically strained vs time [69].

From G.S. Was, Recent Developments in Understanding Irradiation Assisted Stress Corrosion Cracking, Proc. 11th Int'l Conf. Environmental Degradation of Materials in Nuclear Power Systems – Water Reactors, American Nuclear Society, La Grange Park, IL, (2004) 965–985.

Much data have been accumulated to support the role of chromium depletion as an agent in IGSCC of austenitic alloys in oxidizing conditions. Numerous studies show that as the grain boundary chromium level decreases, intergranular SCC increases. Typical chromium depleted zone widths are of the order 100–300 nm full-width at half-maximum (FWHM), providing a significant volume of depleted material adjacent to the grain boundary.

Fig. 8.27B shows a similar correlation between grain boundary chromium level and IASCC susceptibility as measured by the %IG cracking on the fracture surface during slow strain rate experiments. A major difference between Cr depletion profiles resulting from RIS and those due to precipitation reactions during heat treatment is that the width of the RIS profiles can be as much as two orders of magnitude smaller, typically 5–10 nm. There is large scatter in the data that make a direct correlation difficult to support, and differences in testing conditions undoubtedly contribute. Special, multistep thermal treatments can create ~10 nm Cr depletion profiles, and SCC is controlled by the minimum Cr content of the profile even though it is very narrow.

Among the minor alloy elements, only Si is known to segregate to high levels and Si segregation is correlated with IASCC. Experiments by Busby et al. [167,168] on a high-purity 316-base alloy doped with 1 wt% Si showed severe IASCC in NWC and in primary water after irradiation to 5.5 dpa at 360°C. Scanning transmission electron microscopy (STEM) measurements of grain boundary Si confirm levels up to 6 wt%. Past studies comparing Auger electron spectroscopy (AES) and STEM results have shown that the actual concentration of Si at the grain boundary plane may be as high as 15–20 wt%. Though the electron beam probe in STEM is very small, the measurement underpredicts the concentration at the grain boundary by a factor of 2–3. Yonezawa et al. [169–172] and Li et al. [173] have provided extensive evidence to show that increased Si in stainless steel results in increased IGSCC in alloys tailored to imitate the composition of grain boundaries under irradiation. However, it should be noted that these alloys were not irradiated and this difference may be important in the relevance of such experiments to IASCC. Using 1.5%–5% Si stainless steels of both standard (e.g., 304L) base composition and synthetic irradiated grain boundary composition, Andresen has observed significantly increased growth rates, reduction in the benefit of lowering corrosion potential, and very little effect of stress intensity factor between 27 and 13 ksi $\sqrt{\text{in.}}$ (30 and 14 MPa-m$^{1/2}$).

The fluence at which IASCC is observed depends on applied stress and strain, corrosion potential, solution conductivity, crevice geometry, cold work, prior sensitization, etc. At sufficiently high conductivities, cracking has been observed in solution annealed stainless steel in the field and in the laboratory. Thus, while potentially useful in an engineering context, the concept of a *threshold* fluence (or stress, corrosion potential, etc.) is scientifically misleading. IASCC initiation and growth must be understood in terms of the interdependent effects of many parameters.

8.8.1.2 What is known about the mechanism of IASCC

Our understanding of the effect of irradiation on SCC is that it can impact SCC in four principal ways: segregation, hardening and/or deformation, dynamic strain, and radiolysis. At high fluence, differential swelling between baffle plates and bolts can produce re-loading, and He generation or the precipitation/dissolution of phases can also play a role. The processes believed to be important in the IASCC mechanism are summarized below:

- RIS is a process that produces enrichment in some species (e.g., Ni and Si) at grain boundaries and other defect sinks, and depletion in other species (e.g., Cr). Even though the distance over which RIS occurs is very limited (a few nm), studies of unirradiated materials have shown that the narrow profiles can affect SCC. While chromium depletion is believed to be important only in the case of an oxidizing environment (or pH shifted crack chemistry), segregation of other species, such as Si, could create deleterious results in low potential and perhaps all environments.
- Radiation hardening (RH) results from radiation damage and the creation of vacancy and interstitial loops, which impede dislocation motion. Once a few dislocations move along a given slip plane, they clear a "channel" of most of these barriers, and subsequent dislocation motion occurs primarily in these channels. The localization of deformation into these channels creates very high local stresses and strains at their intersection with grain boundaries. LD correlates strongly with IASCC susceptibility and may play a key role in the process.
- Dynamic strain is recognized as being important in SCC in hot water, and it is widely considered to be a fundamental parameter, both by experimental observation and conceptually because the slip offsets caused by plastic strain disrupt the surface passive films that in turn impart corrosion resistance. Experimentally, laboratory and in-reactor data show that there is a strong effect of very slow straining or test perturbation on crack nucleation, short crack coalescence, and crack growth. Analysis of field data on PWR baffle former bolts showed a major influence of load following (daily reactor power changes) on incidence of SCC in bolts made from the same materials and used in multiple plants. If cyclic loading is present, the dynamic strain at the crack tip, or crack tip strain rate, results from the reversed slip processes that cause (inert) fatigue crack advance. In slow strain rate tests, the crack tip strain rate results from the applied strain rate, which is partitioned among the number of growing cracks. At constant load (or constant K), the dynamic strain results from the crack advance process itself although thermal and irradiation creep can play a role. As the crack advances, the stress/strain field at the crack tip must be redistributed, which requires dislocation motion. This creates a circularity: dynamic strain causes crack advances which sustains the dynamic strain. It is this interplay that makes SCC much more difficult to study than, for example, corrosion fatigue.

    Constant load tests have not been as effective in either initiating or sustaining cracking as tests involving dynamic strain. Yet when cracking occurs, it can be at applied loads that are surprisingly small. The database indicates that stresses at or above 0.4 of the yield strength can be sufficient to induce IGSCC in irradiated stainless steels. (Some observations of cracking at stresses lower than this have been made.) This is surprising because it is well below the stress for general yielding. However, localized yielding can occur at stresses at least as low as 70% of the nominal yield strength. Since plastic deformation in irradiated stainless steels is in the form of heterogeneous dislocation channels, the low value of stress at which crack initiation occurs may be an indication that LD occurs and is a necessary condition.
- The radiolysis of water results in its decomposition into various primary species including both radicals (e.g., e^-_{aq}, H, OH, HO_2) and molecules (e.g., H_2O_2, H_2, O_2), which can be oxidizing (e.g., O_2, H_2O_2, HO_2) or reducing (e.g., e^-_{aq}, H, H_2). The concentrations of radiolytic species are roughly proportional to the square root of the radiation flux in pure water. The integrated effect of various oxidants and reductants on environmental cracking is best described by changes in corrosion potential, which controls the thermodynamics and influences the kinetics of most reactions. There is no evidence that the specific species formed are important, and indeed their effect on cracking appears to be fully captured by their overall effect on the corrosion potential of the material. Further, there is sufficient evidence to show that IASCC readily occurs in materials that are preirradiated and then tested in the lab where radiation and radiolysis do not occur.

8.8 COMBINED EFFECTS OF WATER ENVIRONMENT AND RADIATION

Other processes that are thought to be important in at least some instances include radiation-induced creep, swelling, and microstructure processes such as precipitation and/or dissolution of phases.

Radiation creep relaxation reduces preload stresses such as in bolts as well as stresses associated with welding. During active irradiation, however, radiation creep can promote dynamic strain, and thereby SCC.

Swelling, especially differential swelling between components, occur to a limited extent at temperatures above ~300°C and can be sufficient to produce re-loading of components such as PWR baffle former bolts. Swelling occurs differently in different materials, and is delayed in CW materials such as bolts, which are embedded into more swelling prone annealed plates. Stresses due to swelling are balanced by radiation creep relaxation, but the resulting stress, while low, can be sufficient to drive IASCC.

Irradiation can induce precipitation of second phases in some cases, and in others, it can cause disordering/dissolution of existing phases. Changes in phase stability can result in local composition changes and changes in the mechanical behavior of the alloy. While there is no clear evidence that such changes affect IASCC response, this may only reflect the limited characterization and IASCC studies that have been performed on high-fluence materials.

8.8.1.3 What is not known about the mechanism of IASCC

Uncertainty in the mechanism for IASCC creates considerable difficulty in identifying long-term issues for IASCC in existing plants. However, given the processes that are believed to contribute to IASCC, it is possible to identify those that are likely to become more important. For example, the data suggest that, with the possible exception of Si, RIS at grain boundaries saturates at relatively modest doses (~5 dpa) and does not change much with increasing dose. So an increase in severity of IASCC driven by RIS (apart from Si) does not seem likely. The same is true for the dislocation microstructure and the resulting hardening. However, postirradiation annealing (PIA) slow strain rate experiments have shown that IGSCC can be nearly completely ameliorated by heat treatments that leave RIS unaffected and the dislocation loop microstructure only slightly changed. Precipitation of second phases remains a significant unknown at high doses. G-phase and γ' are known to form at lower doses. Their behavior at high dose, however, is unknown. The observation of Cu precipitation in 304 SS [174] suggests that other phases may form and become stable at higher doses, which could then affect IASCC susceptibility.

Swelling is one process that is very likely to become a bigger problem with increasing operating time. The temperature range of most core internals is at the lower edge of the swelling range, but gamma heating in some areas is predicted to increase the temperature well above 400°C. With increasing dose, significant swelling could occur. Increased swelling of certain components, such as baffle plates, means increased stresses on the baffle bolts and distortion to the baffle structure. The extent to which irradiation creep can mitigate swelling is also relatively unknown, and swelling does not simply produce tensile loading, but also shear loading.

Additional unknowns include as follows:

- While dynamic straining is clearly important in IASCC, the reason it impacts IASCC is poorly understood. Constant load experiments have shown that crack initiation is dependent on the load and especially load fluctuations. The role of loading history is poorly understood. Recent studies have shown that load history (and the strain path prior to loading) can have a marked effect on the susceptibility to crack initiation. A better understanding of how load history affects IASCC is greatly needed.
- The role of irradiation, including radiation creep and the development of microstructure where dislocations occur in channels, in the initiation of cracks is poorly understood. Irradiation often shortens the incubation time for crack nucleation.
- Little is known whether irradiation affects crack initiation in the same way as crack growth. Some of the dependencies on initiation and growth are similar, but others are not, and this makes it hard to know how similar the effect of irradiation is on initiation and growth.
- The influence of cold work on crack initiation is poorly understood. In unirradiated materials, CGRs are quite sensitive to cold work. But in initiation tests, the correlation is much less clear, apart from severe surface cold work/grinding, which accelerates initiation.

- Postirradiation SCC tests show that the irradiated microstructure responds to stress by forming intense, LD bands that could contribute to crack initiation or crack growth. However, less is known about the more practical situation in which small amounts of plasticity are accumulated during irradiation and whether the simultaneity of irradiation and deformation lessens the degree of LD compared to the case in which the two processes occur sequentially.

8.8.2 IRRADIATION-ACCELERATED CORROSION

The corrosion resistance of austenitic stainless steels to high-temperature water is generally excellent. Corrosion rates are quite low at temperatures characteristic of both BWR and PWR operation although the oxide layer is much thicker (~0.2–1 μm) than at room temperature. The resistance is provided by the formation of a dense, adherent oxide that is both slow growing and protective. The film morphology of the oxide scale includes an outer layer composed of a mixture of magnetite (Fe_3O_4) and iron-chromium spinel ($FeCr_2O_4$) and an inner, more compact layer of chromium rich oxide. Crystallites often form on the surface and are associated with redeposition of iron and chromium from the solution. The inner layer is the most protective and is formed by the diffusion of anions to the oxide-metal interface and/or the diffusion of oxygen to through the oxide to the interface.

Irradiation might affect corrosion or oxidation in at least three different ways. Radiation interaction with water can result in the decomposition of water into radicals and more stable oxidizing and reducing species that can increase the corrosion potential. Second, irradiation of the solid surface can produce excited states that can alter corrosion, such as in the case of photo-induced corrosion. Last, displacement damage in the solid will result in a high flux of defects to the solid-solution interface that can alter and perhaps accelerate interface reactions. Defect fluxes to the oxide metal interface will also be increased by damage to the alloy. Radiation can affect oxide properties such as density, thermal conductivity, and crystal structure or phase fraction. Irradiation may also result in higher hydrogen pick-up fractions leading to additional issues in core components. Largely because of the difficulty in conducting controlled experiments, little data exist to provide an understanding of the role of irradiation of the solid in the corrosion process. However, experiments that *include* damage to the substrate show larger increases in oxidation rate than those in which the effect is confined to the water. While irradiation is believed to affect corrosion of stainless steels, no data exist.

8.8.3 CORROSION FATIGUE

Very little data exist on the effect of neutron irradiation on fatigue in current LWRs. Chopra and Rao [175] have reviewed the literature on the effect of neutron irradiation on fatigue crack growth and noted that most all of the data have been generated in the fast breeder reactor program. In air, irradiation does not appear to enhance fatigue CGRs [175]. However, recent CGR data have been collected on solution annealed types 304 and 316 SS irradiated up to 3 dpa and tested in high and low DO environments. Fig. 8.28 plots the CGR in the environment against that in air for the same loading conditions.

The data show that at low dose, there is little departure from the 45 degrees line at low DO—that is, there is no environmental enhancement. However, increasing DO results in an increase in the CGR, and increasing dose causes an increasing departure. Data on CGR of irradiated HAZ weld material show very similar CGRs and increases in CGR relative to air.

The role of irradiation on fatigue poorly understood and the amount of data are minimal. Due to the reduced uniform strain and increasing localization of plastic deformation resulting from irradiation, it may be expected that fatigue crack growth should respond accordingly. In particular, in stage III fatigue crack growth, where the growth rate is limited by the fracture toughness, reduction in fracture toughness due to irradiation will result in an increase in CGR. In stage I, the threshold stress intensity range ΔK_{th} is sensitive to the chemical environment, the load ratio (R), grain boundary impurity segregation and to the tendency for high-strength materials to undergo flow localization. This latter sensitivity is supported by empirical data that show that ΔK_{th} decreases with increasing yield strength in unirradiated 316 SS [176]. Consequently, the severe localization of plasticity caused by irradiation may be expected to lead to decreases in the threshold stress intensity.

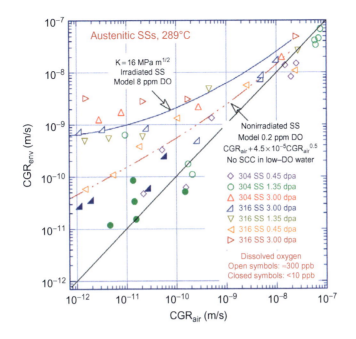

FIG. 8.28

Fatigue crack growth rate for irradiated austenitic stainless steels tested in 289°C water containing varying amounts of dissolved oxygen [175].

From O. K. Chopra, A. S. Rao, A review of irradiation effects on LWR core internal materials – IASCCsusceptibilty and crack growth rates of austenitic stainless steels, J. Nucl. Mater. 409 (2011) 235–256.

However, the data on the effect of irradiation on fatigue in austenitic stainless steels in the low-to-intermediate temperature range is mostly in stage II that is described by the Paris equation. In this regime, crack propagation is primarily dependent on the elastic properties of the solid, and less so on the microstructure and on plastic deformation processes. Limited data show that the CGR is, in fact, relatively insensitive to irradiation to doses of up to about 30 dpa. The CGR of 316 stainless steel irradiated to 2.03×10^{21} n/cm² at 380°C is bounded on the high side by crack growth in mill-annealed plate and on the low side by 20% CW plate, both in the unirradiated condition [177]. As such, irradiation of austenitic stainless steels to low or intermediate doses does not result in measurable increases in fatigue crack growth. However, with increasing temperature, the generation and accumulation of helium into bubbles can affect the fatigue crack propagation.

Nevertheless, it may be expected that under irradiation, LCF life should decrease due to decreased ductility, and HCF life should increase due to increased strength. This behavior was indeed measured for 304 stainless steel at both room temperature and 325°C following irradiation to 8×10^{22} n/cm² ($E > 0.1$ MeV) at a temperature of 400°C [2]. The beneficial effect of irradiation in HCF is likely to be because despite significant hardening and a reduction in work hardening coefficient by a factor of 2.7, the alloy retained ductility to 4%–5% elongation.

Irradiation is important in HCF, LCF, and fatigue crack growth. The existing database for the effect of irradiation in LWR environments on fatigue crack growth is limited to very low doses, on the order of 3 dpa. Hence, there are no data on the role of irradiation at high doses encountered in core internal components. Similarly, there is very little existing data on irradiation effects on either LCF or HCF in hot water. In all cases, hardening and flow localization are expected to play a role. While both tend to saturate at relatively modest doses (~5 dpa), hardening and localization can also be affected by precipitation reactions or the growth of voids or bubbles. Since these all tend to be higher dose phenomena in LWRs, the response of fatigue to such high-dose processes is essentially unknown and represents major areas for future study.

8.8.4 HYDROGEN EMBRITTLEMENT

Substantial amounts of hydrogen are typically found in irradiated stainless steels removed from core internal structures of both BWRs and PWRs. Whereas only a few ppm by weight are found in unirradiated stainless steels exposed to these environments, irradiation damage significantly increases the number of trapping sites for hydrogen so that 30 ppm is typical with extreme cases up to 100 ppm [178]. Since the yield strength of austenitic stainless steels increases substantially with neutron irradiation damage and ductility is reduced, it is important to determine whether hydrogen embrittlement cracking is possible both at normal operating temperatures as well as at lower temperatures encountered during plant shut down.

To address this issue, a project was conducted at NRI on hydrogen charged and irradiated tensile specimens tested at a slow strain rate both in argon and in PWR primary water at 100 and 320°C to evaluate the effect of hydrogen on cracking [179]. Hydrogen embrittlement studies of austenitic stainless steel after irradiation (in the BOR-60 fast reactor) to ~10 dpa and charging with gaseous hydrogen to concentrations typical of BWR or PWR core internals found no evidence of hydrogen-induced cracking at either 100°C or 320°C. The higher temperature result was not unexpected but the lower temperature result was surprising given the extent of irradiation hardening of the tested materials and merits reexamination.

With one exception, the fracture surfaces were ductile although occasionally a brittle transgranular facet was found in the irradiated specimens with higher hydrogen levels (whether precharged or acquired by exposure to PWR primary water). The occurrence of transgranular facets in the irradiated specimens, sometimes apparently initiated internally since they apparently unconnected to the surface, seemed to increase with increased hydrogen content, lower strain rate, and particularly lower test temperature. It was clear from the tests carried out on irradiated material at 320°C that there was tendency for hydrogen to enter the specimens that were not precharged and to leave those that were precharged so that all tended to equilibrate at ~20–30 ppm at the end of test in PWR primary water. At 100°C, the hydrogen loss rate was obviously lower from the precharged specimens so that higher end-of-test hydrogen concentrations were observed in irradiated material. There was no significant effect of hydrogen on the macroscopic tensile properties or fracture surfaces of the unirradiated type 316 stainless steel.

Overall, these experiments did not demonstrate any particular sensitivity to hydrogen embrittlement or intergranular cracking of 15% CW type 316 stainless steel irradiated to 9.4 dpa, with one exception of a test in PWR primary water at 320°C. Some small tendency to transgranular cracking was observed, enhanced at lower temperature, and possibly caused by hydrogen embrittlement. Such observations at high strains and high hydrogen contents have been made by others in unirradiated stainless steels. Thus, these tests did not demonstrate any significant effect of hydrogen on intergranular cracking of irradiated steels [178].

These results are consistent with two other studies [178]. In one study, tensile specimens were preirradiated to ~10 dpa in the BOR-60 fast reactor to examine the role of in-pile radiolysis on IASCC in hydrogenated PWR primary water. No evidence from SSRT experiments was obtained of any different behavior between in-pile and out-of-pile tests. This conclusion was expected from fundamental considerations of the role of excess dissolved hydrogen in water radiolysis. A second project was undertaken to compare IASCC susceptibility on the same irradiated heat of austenitic stainless steel (harvested from a BWR in this case) in BWR and PWR environments. The results of the project demonstrated the clear benefit of decreased corrosion potential caused by the presence of hydrogen in solution rather than oxygen, as measured by SSRT.

However, as covered in Section 8.4.4, it now appears that as void swelling develops there is an increasing tendency to store much larger amounts of hydrogen, especially in helium-nucleated bubbles and voids. Therefore, the issue of hydrogen-related embrittlement requires further study.

8.8.5 FRACTURE TOUGHNESS

Structural integrity margins, including leak-before-break analyses, are central to the avoidance of excessive loss of coolant and severe accidents. They are currently defined from measurements of the fracture properties of the structure in air in accordance with ASTM standards but may not be adequately conservative given the recent

8.8 COMBINED EFFECTS OF WATER ENVIRONMENT AND RADIATION

observations of significant decreases in fracture toughness of structural materials when the tests are conducted in water. While historical plant operation provides confirmation of our knowledge of SCC and corrosion fatigue, there is little if any plant experience that can characterize the nature of the environmental fracture issue.

Environmental effects on mechanical properties [180]—with a focus on constant and monotonically increasing loading—can be categorized into high temperature (e.g., >150–200°C) and low temperature (e.g., <150–200°C) response, with the latter often dominated by hydrogen effects—and into SCC (e.g., growth at constant stress intensity factor), J-R tearing resistance, and fracture toughness (K_{IC}). While the traditional view is that fracture behavior (e.g., J-R or toughness) is affected only, for example, by microstructure and test conditions (e.g., temperature and strain rate), there is every reason to consider environmental effects in connection with these properties where such exposure is inherent to the application.

SCC in high-temperature water has been extensively investigated, and some observations of high SCC growth rates have been reported below 150°C although only in a limited number of Ni alloys [181]. Unusually high growth rates have been observed in high-temperature water under +dK/da conditions [182], that is, when the change in stress intensity factor (K) vs crack depth is large, which is most likely when cracks are small or large (note that most cracks in plant components grow under dK/da control, that is, K only changes because the crack grows). Apart from the Ni alloy SCC studies [181], neither SCC nor +dK/da effects have been consequentially investigated at lower temperature, and there are particular concerns for the higher yield strength materials, such as CW stainless steels (including the weld heat affected zone), irradiated stainless steel, and low alloy steels.

J-R tests measure resistance to tearing, and various investigators have observed very large reductions in tearing resistance when tests are performed in water [183]. These tests have not been comprehensive in temperature, environment, material, displacement/strain rate, and high-temperature water preexposure effects (to saturate the metal with hydrogen). There are limited environmental J-R data on CW or irradiated stainless steel that is emerging, but none on pressure vessel steels (especially steels that have undergone some loss of properties from irradiation).

Fracture toughness (K_{IC}) represents the resistance to sudden, unstable fracture, and is commonly characterized using Charpy or CT specimens. Like other mechanical properties, there is no reason to expect that K_{IC} is a function only of material and temperature, and not of the environment and/or hydrogen content of the material. It is known that hydrogen readily diffuses through Fe and Ni alloys in ~300°C water [184], and it is certainly reasonable to be concerned for the presence of hydrogen throughout a metal on fracture properties. Only isolated and unintentional observations have been reported to date, but they involve ~two dozen specimens in several experienced laboratories. An example is shown in Fig. 8.29 for stainless steel and Alloy 182 weld metal exposed in 288°C water.

All observations to date have been in ~300°C water, and the possibility exists that larger effects will be observed if the materials are cooled to <150°C, where hydrogen effects are known to be larger. It is possible without an environmental effect, but plasticity (suck-in) is very small in most specimens, and similar behavior is seen during air fatigue crack growth only at very high "K" values (e.g., >250 MPa $\sqrt{m}$).

This decrease in fracture toughness due to environmental effects is expected to be greatest in higher yield strength materials and in materials that are inhomogeneous (e.g., Ni-alloy weld metals). Materials such as Alloy X-750, Alloy 182/82 weld metals, and CW stainless steel are of concern, and the effects of irradiation on stainless steel and low alloy steel may increase their susceptibility, because their K_{IC} value is already decreased due to irradiation fluence. A significant reduction in K_{IC} for the low alloy RPV steel would impact not only on license renewal activities but also on the risk-informed reformulations of the pressurized thermal shock (PTS) rule. The conjoint effect of radiation embrittlement and environment effects should be carefully examined. Unfortunately, there is limited understanding and no quantitative prediction approach for these phenomena. There is a basis for expecting effects of strain rate, temperature, hydrogen content, etc., but not for the extent of the phenomenon or the impact of different environments, microstructures, irradiation, or loading modes (e.g., normal operation, seismic loads, etc.).

In summary, there is growing evidence that the fracture behavior (e.g., J-R and K_{IC} values) of structural materials can be reduced significantly by stressing in water. Some J-R data exist to characterize elements of the problem, but the broad nature and extent of the concerns as a function of loading, temperature, material/microstructure,

336 CHAPTER 8 AUSTENITIC STAINLESS STEELS

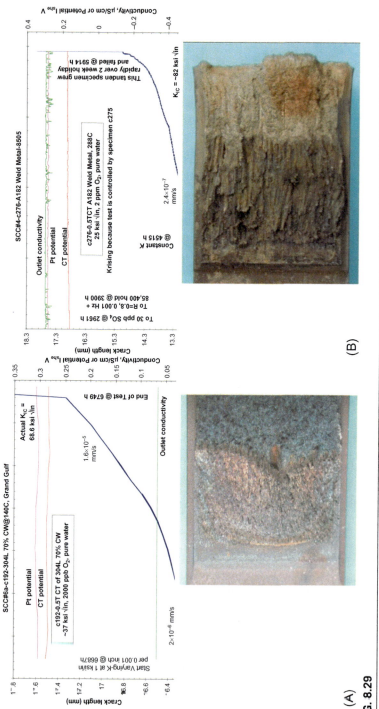

FIG. 8.29

Examples of sudden failure of (A) cold-worked stainless steel and (B) alloy 182 weld metal tested in 288°C water at increasing K until failure occurred. The load and crack depth at failure are well defined, and the resulting "K_{IC}" (not necessarily obtained valid conditions) is relatively low [184].

From an Unpublished internal report, An Interim Review of the Cooperative Irradiation-Assisted Stress Corrosion Cracking Research (CIR) Program – Revision 1, Electric Power Research Institute, Palo Alto, CA, EPRI report 1015493, November 2007.

and environment are undefined. Importantly, we can speculate as to the mechanism, but we cannot quantitatively predict the effects. While historical plant operation provides confirmation of our knowledge of SCC and corrosion fatigue, there is little if any plant experience that can characterize the nature of the environmental fracture issue. The consequence of the phenomenon is that the structural integrity margins would be reduced and the risk of loss-of-coolant accidents increased to an extent that depends on specific material-component operating conditions.

8.9 PERSPECTIVES AND PROSPECTS

The principle challenge with the use of austenitic stainless steels in LWRs is their susceptibility to IASCC. Several alloy modifications have been proposed to take advantage of empirical data that indicate increased resistance to cracking. An experimental program to examine the susceptibility of several austenitic stainless steels and nickel-base alloys to IASCC was conducted by Garzarolli et al. [185–189] over a period of >10 years. Included in this study were 304, 316, 348, 1.4541, and 1.4981 stainless steels, and Alloys 800, 718, and X-750. The focus was on the high C and Nb-containing 348 SS as the formation of NbC and a fine grain microstructure was believed to provide better in-reactor behavior. Swelling tube and expanding mandrel experiments were conducted in both BWR and PWR environments. While results were not definitive, a combination of high Nb and high C along with low Si + P showed better (though not complete) resistance to IASCC. Fig. 8.30A and B shows summaries of various 348 alloy variations as well as other alloys and how they performed in a PWR environment.

Alloy D9 is an advanced austenitic alloy of composition Fe-15Cr-15Ni alloy containing TiC particles that was developed as part of the United States. National Cladding and Duct Development Program in the 1970s and 1980s [189]. The alloy was designed to have greater tensile strength than 316 SS at high temperatures below 600°C and superior swelling resistance. Swelling resistance is provided by the high density of very fine TiC particles that act as traps for vacancies and interstitials and also for He produced by transmutation. As shown in Fig. 8.31, the swelling is well below that for 316 SS.

While it appears promising, the technical database for the alloy is considerably less extensive than that for 316 SS on, for example, creep, fabrication, joining, etc. Further there is no information on the IASCC susceptibility of this alloy and how it compares to that for 304 or 316 SS [190]. There are no known applications of alloy D9 in LWRs.

High Ni + Cr alloys such as alloy 800 or alloy 800H (33Ni-21Cr) are attractive because they should exhibit good corrosion resistance and better swelling resistance than the 300 series stainless steels. In fact, Alloy 800 is indeed quite corrosion resistant in high-temperature water in the range of 300°C. Oxides on 800 exhibit the typical duplex structure reported for stainless steels and carbon steels in high-temperature aqueous media and growth rates are similar to those in stainless steels [191]. Similarly, while inferior to higher Ni-containing alloys, the swelling resistance of Alloy 800 is superior to austenitic stainless steels. The challenge for this alloy class is the susceptibility to IASCC. Conflicting reports show that commercial Alloy 800 behaves similar to 304 SS in its susceptibility to IASCC [186]. However, experiments on high-purity austenitic alloys containing high levels of both Cr and Ni show very high resistance to IASCC. High-purity alloys of Fe-18Cr-25Ni, Fe-25Cr-25Ni show immunity to IASCC following irradiation with either neutrons or protons, and alloy Fe-21Cr-32Ni shows no evidence of IASCC after proton irradiation to 5 dpa and following straining to 7% in 288°C simulated BWR NWC, Fig. 8.32 [192].

The resistance of these alloys to IASCC is coincident with their lesser tendency to exhibit pronounced dislocation channeling during straining. While these results are preliminary, they are encouraging and provide the impetus for further study.

HT-UPS (Fe-14Cr-16Ni base) austenitic stainless steel is one of the candidate structural materials for next generation fast reactors because it exhibits creep properties superior to any other advanced austenitic stainless steels or even comparable to a Ni-base alloy [193–195], as shown in the LMP plot in Fig. 8.33.

Dislocations introduced by cold working act as nucleation sites for nanoscale MC carbides, which increase the resistance of creep deformation. Aging of HT UPS steels for 1000 h at 650°C causes the formation of precipitates of nanoscale MC carbides on the dislocations. No recovery of dislocations or recrystallization took place, indicating that the formation of nanoscale MC carbides pinned dislocations and maintained the dense dislocation network

338 CHAPTER 8 AUSTENITIC STAINLESS STEELS

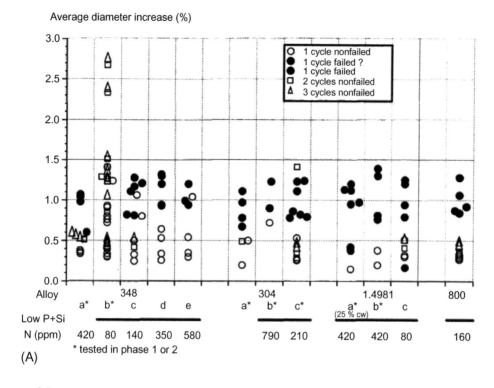

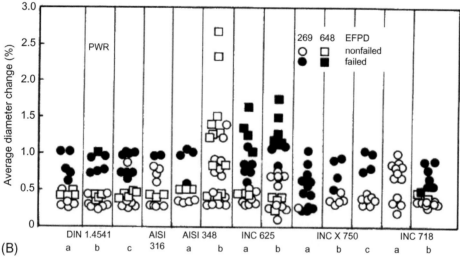

FIG. 8.30

(A) Failure results for swelling mandrel experiments in PWR environments [187] and (B) strain to failure of various alloys in the core of a PWR [187].

From "Deformability of High-purity Stainless Steels and Ni-Base Alloys in the Core of a PWR, F. Garzarolli, P. Dewes, R. Hahn, J. L. Nelson, Sixth International Symposium on Environmental Degradation of Materials in Nuclear Power Systems – Water Reactors, TMS, Warrendale, PA, 1993, p. 607. Copyright © 1993 by The Minerals, Metals & Materials Society. Used with permission.

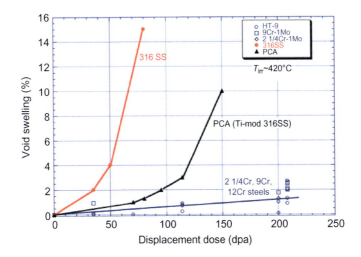

FIG. 8.31

Comparison of void swelling as a function of dose for alloys 316SS, D9 and several ferritic steels.

Courtesy S. Zinkle.

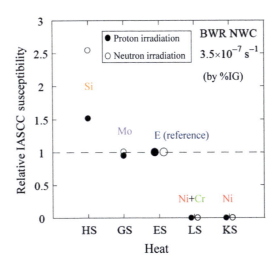

FIG. 8.32

IASCC susceptibility (by %IG on the fracture surface) relative to reference heat E following neutron or proton irradiation and CERT testing in 288°C simulated BWR NWC.

even after 1000h aging. Additions of N and Nb were made to stabilize the nanoscale M(C,N) dispersion and Ti and V were removed to maintain ductility [196]. In addition to mechanical properties, the high density of fine precipitates is also expected to help reduce void swelling and increase creep strength during irradiation.

As a closing note, stainless steels could contribute to accident tolerant fuel in LWRs. The shortcomings of the Zr alloy/ UO_2 fuel system to severe accident conditions such as what occurred at the Fukushima Dai-ichi nuclear power plant in Japan following the 2011 earthquake and tsunami has led to increased interest in improving the

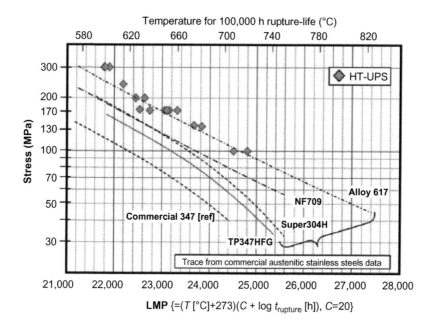

FIG. 8.33

Larson-Miller parameter plot of the HT-UPS steel and commercially available heat-resistant steel alloys [193].

From P.J. Maziasz, "Developing an austenitic stainless steel for improved performance in advanced fossil power facilities" *Journal of Metals, 41 (1989) (7) 14–20.*

safety of nuclear reactors to rare but credible accident scenarios [197,198]. For loss of coolant scenarios, the rapid increase in Zr-alloy oxidation rate with increasing temperature will degrade the mechanical properties of the cladding and could lead to cladding breach and/or fracture, which in turn could lead to reduced flow of primary coolant within the fuel assemblies in the core due to coolant channel blockage by cladding fragments. The high heat of oxidation for Zr can also make a large contribution to the core heating; considering that a typical 1000 MWe LWR core contains ~30,000 kg of cladding, ~55 MW-h of heat would be released if it were completely oxidized (note: BWR cores also contain nearly a comparable amount of Zr in the channel boxes). This could set up an autocatalytic reaction where the heat from Zr oxidation drives a temperature increase in the core, leading to more rapid oxidation and heat generation. Finally, the oxidation of Zr by steam leads to the production of potentially explosive hydrogen gas (~1200 kg if all of the cladding was oxidized by steam). The high value of the heat of oxidation is at the root of the problem and various solutions have been proposed for alternate fuel cladding. A potential solution is the use of a co-extruded tube consisting of an inner layer of ferritic-martensitic alloy to provide strength and SCC resistance, and an outer layer of austenitic stainless steel to provide optimum corrosion resistance. The bilayer structure provides high strength, resistance to general corrosion, resistance to SCC penetration, and a lower heat of oxidation than for Zr alloys.

REFERENCES

[1] G.O. Ilevbare, F. Cattant, N. K. Peat, SCC of stainless steels under POW service conditions, in: Proceedings of the 7th International Conference on Contribution of Materials Investigations to Improve the Safety and Performance of LWRs, Pontevraud 7, 26 30, September 2010.
[2] G.S. Was, Y. Ashida, P. Andresen, Irradiation-assisted stress corrosion cracking, Corrosion Rev. 29 (2011) 7–49.
[3] U.S. DOE and Generation IV International Forum, A Technology Roadmap for Generation IV Nuclear Energy System, GIF-002-00 (2002).

[4] Status of liquid metal cooled fast reactor technology, IAEA-TECDOC-1083, IAEA (1999).
[5] M. Hori, A. Takeda, Int. Conf. on Fast Reactor and Related Fuel Cycles, Atomic Energy Soc. Japan, Kyoto, 1.5–1, 1991.
[6] Y. Kondo, T. Yukitoshi, K. Yoshikawa, N. Nagai, S. Ohta, M. Fujiwara, S. Yoshida, C. Tanaka, K. Uematsu, K. Suzuki, Radiation Effects of Breeder Reactor Structural Materials, TMS-American Institute of Mining, New York, 1977, p. 253.
[7] M. Fujiwara, H. Uchida, S. Ohta, S. Yuhara, S. Tani, Y. Sato, Radiation-induced changes in microstructure, 13th Symp. (Part 1), ASTM STP 955, 1987. p. 127.
[8] S. Nomura, S. Shikakura, S. Ukai, I. Seshimo, M. Harada, I. Shibahara and M. Katsuragawa, Int. Conf. on Fast Reactor and Related Fuel Cycles, Atomic Energy Soc. Japan, Kyoto, 1, 7.4–1, 1991.
[9] M.L. Hamilton, G.D. Johnson, R.J. Puigh, F.A. Garner, P.J. Maziasz, W.J.S. Yang, N. Abraham, Residual and Unspecified Elements in Steel, ASTM STP 1042, 1989, p. 124.
[10] P.J. Maziasz, JOM 41 (7) (1989) 14–20.
[11] I. Shibahara, N. Akasaka, S. Onose, H. Okada, S. Ukai, J. Nucl. Mater. 212–215 (1994) 487.
[12] I. Shibahara, S. Ukai, S. Onose, S. Shikakura, J. Nucl. Mater. 204 (1993) 131.
[13] E.H. Lee, L.K. Mansur, Met. Trans. A 23A (1992) 1977–1986.
[14] E.H. Lee, L.K. Mansur, Met.Trans. A 21A (1992) 1021–1034.
[15] C. Brown, V. Levy, J.L. Seran, K. Ehrlich, R.J.C Roger, H. Bergmann, Cladding and wrapper development for fast breeder reactor high performance, in: Int. Conf. on Fast Reactor and Related Fuel Cycles, Atomic Energy Soc. Japan, Kyoto, 1, 7.5–1, 1991.
[16] P. Dubuisson, A. Mailard, C. Delalande, D. Gilbon, J.L. Seran, Effects of radiation on materials, 15th Int. Symp. ASTM STP 1125, 1992, p. 995.
[17] J.L. Seran, V. Levy, P. Dubuisson, D. Gilbon, A. Millard, A. Fissolo, H. Touron, R. Cauvin, A. Chalony, E.L. Boulbin, Effects of radiation on materials, 16th Int. Symp. ASTM STP 1175, 1993, p. 1209.
[18] K. Herschbach, W. Schneider, K. Ehrlich, J. Nucl. Mater. 203 (1993) 233–248.
[19] C. Brown, R.M. Sharpe, E.J. Fulton, C. Cawthorne, Dimensional Stability and Mechanical Behavior of Irradiated Metals and Alloys, British Nuclear Energy Society, London, 1983, pp. 63–67.
[20] S.I. Potollo, Y.V. Konobeev, F.A. Garner, J. Nucl. Mater. 393 (2009) 61–66.
[21] D. Peckner, I.M. Bernstein (Eds.), Handbook of Stainless Steels, McGraw-Hill, New York, 1977.
[22] C.C. Arteaga, J.P. Calderon, et al., Comparison of corrosion resistance of carbon steel and some stainless steels exposed to LiBr-H_2O solution at low temperatures, Int. J. Electrochem. Sci. 7 (2012) 445–470.
[23] M.J. Donachie, S.J. Donachie, Superalloys: A Technical Guide, ASM International, Metals Park, OH, 2002.
[24] N. Hussain, G. Schanz, et al., High-temperature oxidation and spalling behavior of incoloy 825, Oxid. Met. 32 (1989) 405–431.
[25] G.S. Was, Fundamentals of Radiation Materials Science: Metals and Alloys, Springer, Berlin, 2007.
[26] H. Wiedersich, P.R. Okamoto, N.Q. Lam, J. Nucl. Mater. 83 (1979) 98.
[27] P.R. Okamoto, L.E. Rehn, J. Nucl. Mater. 83 (1979) 2.
[28] P.R. Okamoto, H. Wiedersich, J. Nucl. Mater. 53 (1974) 336.
[29] N.Q. Lam, A. Kumar, H. Wiedersich, in: H.R. Brager, J.S. Perrin (Eds.), Kinetics of radiation induced segregation in ternary alloys, Proc. Effects of Radiation on Materials, 11th Conf. ASTM STP 782, ASTM International, West Conshohocken, PA, 1982, p. 985.
[30] J.M. Perks, A.D. Marwick, C.A. English, in: D.I. Norris (Ed.), Fundamental aspects of radiation induced segregation in Fe-Cr-Ni alloys, Proc. Radiation-Induced Sensitisation of Stainless Steels, Central Electricity Generating Board, Berkeley Nuclear Labs, Berkeley, 1987, p. 15.
[31] A. Jenssen, L.G. Ljungberg, J. Walmsley, S. Fisher, Corrosion 54 (1) (1998) 48.
[32] T. Allen, J.T. Busby, G.S. Was, E.A. Kenik, J. Nucl. Mater. 255 (1998) 44.
[33] G.S. Was, T.R. Allen, J.T. Busby, J. Gan, D. Damcott, D. Carter, M. Atzmon, Microchemistry of proton irradiated austenitic alloys under conditions relevant to LWR core components, Proc. Mater. Res. Soc., vol. 540 (Warrendale, PA: Materials Research Society, 1999), 421–432.
[34] A.C. Hindmarsh, GEAR: Ordinary Differential Equation System Solver, Lawrence Livermore Laboratory Report UCID-30001, 1972.
[35] E. Simonen, Pacific Northwest National Laboratory, Private Communication (2001).
[36] A.J. Jacobs, R.E. Clausing, M.K. Miller, C.M. Shepherd, in: D. Cubicciotti (Ed.), Influence of grain boundary composition on the IASCC susceptibility of type 348 stainless steel, Proc. 4th Int. Symp. on Environmental Degradation of Materials in Nuclear Power Systems—Water Reactors, NACE, Houston, TX, 1990, pp. 14–21.

[37] K. Asano, K. Fukuya, K. Nakata, M. Kodama, Changes in grain boundary composition induced by neutron irradiation on austenitic stainless steels, Proc. 5th Int. Symp. on Environmental Degradation of Materials in Nuclear Power Systems—Water Reactors, (Eds.) D. Cubicciotti, E.P. Simonen, R.E. Gold (LaGrange, IL: American Nuclear Society, 1992), 838.
[38] A. Jacobs, in: Effects of low temperature annealing on the microstructure and grain boundary chemistry of irradiated type 304SS and correlations with IASCC resistance, Proc. 7th Int. Conf. on Environmental Degradation of Materials in Nuclear Power Systems—Water Reactors, NACE, Houston, TX, 1995, p. 1021.
[39] A.J. Jacobs, G.P. Wozadlo, K. Nakata, S. Kasahara, T. Okada, T.S. Kawano, S. Suzuki, R.E. Gold, E.P. Simonen (Eds.), The correlation of grain boundary composition in irradiated stainless steel with IASCC resistance, Proc. 6th Int. Symp. on Environmental Degradation of Materials in Nuclear Power Systems – Water Reactors, The Minerals, Metals, and Materials Society, Warrendale, PA, 1993, p. 597.
[40] E.A. Kenik, J. Nucl. Mater. 187 (1992) 239.
[41] S. Nakahigashi, M. Kodama, K. Fukuya, S. Nishimura, S. Yamamoto, K. Saito, T. Saito, Effects of neutron irradiation on corrosion and segregation behavior in austenitic stainless steels, J. Nucl. Mater. 179–181 (1991) 1061.
[42] J. Walmsley, P. Spellward, S. Fisher, A. Jenssen, in: Microchemical characterization of grain boundaries in irradiated steels, Proc. 7th Int. Symp. on Environment Degradation of Materials in Nuclear Power System—Water Reactors, NACE, Houston, TX, 1997, p. 985.
[43] G.S. Was, J.T. Busby, J. Gan, E.A. Kenik, A. Jenssen, S.M. Bruemmer, P.M. Scott, P.L. Andresen, J. Nucl. Mater. 300 (2002) 198.
[44] R. Carter, D. Damcott, M. Atzmon, G.S. Was, S.M. Bruemmer, E.A. Kenik, J. Nucl. Mater. 211 (1994) 70.
[45] S.M. Bruemmer, E.P. Simonen, P.M. Scott, P.L. Andresen, G.S. Was, J.L. Nelson, J. Nucl. Mater. 274 (1999) 299.
[46] T. Kato, H. Takahashi, M. Izumiya, J. Nucl. Mater. 189 (1992) 167.
[47] L. Fournier, B.H. Sencer, G.S. Was, E.P. Simonen, S.M. Bruemmer, J. Nucl. Mater. 321 (2–3) (2003) 192–209.
[48] J. Gan, E.P. Simonen, S.M. Bruemmer, L. Fournier, B.H. Sencer, G.S. Was, J. Nucl. Mater. 325 (2004) 94–106.
[49] S. Dumbill, M. Hanks, in: R.E. Gold, E.P. Simonen (Eds.), Strategies for the moderation of chromium depletion at grain boundaries in irradiated steels, Proc. 6th Int. Symp. Environmental Degradation of Materials in Nuclear Power Systems—Water Reactors, The Minerals, Metals, and Materials Society, Warrendale, PA, 1993, p. 521.
[50] P.J. Maziasz, C.J. McHargue, Int. Met. Rev. 32 (1987) 190.
[51] P.J. Maziasz, J. Nucl. Mater. 205 (1993) 118.
[52] F.A. Garner, J. Nucl. Mater. 205 (1993) 98.
[53] S.J. Zinkle, P.J. Maziasz, R.E. Stoller, J. Nucl. Mater. 206 (1993) 266.
[54] D.J. Edwards, E.P. Simonen, S.M. Bruemmer, J. Nucl. Mater. 317 (2003) 13.
[55] D.J. Edwards, B.A. Oliver, F.A. Garner, S.M. Bruemmer, in: Microstructural evaluation of a cold worked 316SS Baffle Bolt irradiated in a commercial PWR, Proc. 10th Int. Conf. Environmental Degradation of Materials in Nuclear Power Systems – Water Reactors, NACE International, Houston, TX, 2002.
[56] D.J. Edwards, E.P. Simonen, F.A. Garner, L.R. Greenwood, B.M. Oliver, S.M. Bruemmer, J. Nucl. Mater. 317 (2000) 32.
[57] H. Watanabe, T. Muroga, N. Yoshida, J. Nucl. Mater. 239 (1996) 95.
[58] J.T. Busby, M.C. Hash, E.A. Kenik, G.S. Was, in: Precipitation of Second-Phase Particles in a Proton-Irradiation Model Alloy, Proceedings of Microscopy and Microanalysis Society 2003 Meeting, Vol 9, Supplement 2, 604, 2003.
[59] M.J. Hackett, J.T. Busby, M. Miller, G.S. Was, J. Nucl. Mater, 389 (2009) 265–278.
[60] L.K. Mansur, in: G.R. Freeman (Ed.), Kinetics of Nonhomogeneous Processes, John Wiley & Sons, Inc., Hoboken, NJ, 1987, pp. 377–463.
[61] E.H. Lee, et al., Proc. Conf. on Phase Stability During Irradiation, Met. Soc. AIME, Pittsburgh, PA, Oct. 5–9, 1980, pp. 191–218.
[62] M.N. Gusev, O.P. Maksimkin, I.S. Osipov, N.S. Silniagina, F.A. Garner, Unusual enhancement of ductility observed during evolution of a 'Deformation Wave' in 12Cr18Ni10Ti stainless steel irradiated in BN-350, J. ASTM Int. 6, 7 (2009). paper ID JAI102062.
[63] F.A. Garner, L.R. Greenwood, D.L. Harrod, in: Potential high fluence response of pressure vessel internals constructed from austenitic stainless steels, Proc. Sixth Intern. Symp. on Environmental Degradation of Materials in Nuclear Power Systems—Water Reactors, San Diego, CA, August 1–5, 1993, pp. 783–790.
[64] F.A. Garner, L.R. Greenwood, in: Survey of recent developments concerning the understanding of radiation effects on stainless steels used in the LWR power industry, 11th International Conference on Environmental Degradation of Materials in Nuclear Power Systems—Water Reactors, 2003, pp. 887–909.

[65] F.A. Garner, M. Griffiths, L.R. Greenwood, E.R. Gilbert, in: Impact of Ni-59 (n, α) and (n, p) reactions on dpa rate, heating rate, gas generation and stress relaxation in LMR, LWR and CANDU reactors, Proc. 14th International Conference on Environmental Degradation of Materials in Nuclear Power Systems—Water Reactors, Virginia Beach, VA, 2009, pp. 1344–1354.

[66] L.R. Greenwood, A new calculation of thermal neutron damage and helium production in nickel, J. Nucl. Mater. 116 (1983) 137–142.

[67] L.R. Greenwood, F.A. Garner, Hydrogen generation arising from the 59Ni (n, p) reaction and its impact on fission-fusion correlations, J. Nucl. Mater. 233–237 (1996) 1530–1534.

[68] F.A. Garner, Irradiation performance of cladding and structural steels in liquid metal reactors, in: Materials Science and Technology, Vol. 10A: A Comprehensive Treatment, VCH Publishers, Weinheim, 1994, pp. 483–514 (Chapter 6).

[69] G.S. Was, Recent developments in understanding irradiation assisted stress corrosion cracking, in: Proc. 11th Int'l Conf. Environmental Degradation of Materials in Nuclear Power Systems—Water Reactors, American Nuclear Society, La Grange Park, IL, 2004, pp. 965–985.

[70] F.A. Garner, M.L. Hamilton, N.F. Panayotou, G.D. Johnson, J. Nucl. Mater. 103–104 (1981) 803–808.

[71] B.N. Singh, A.J.E. Foreman, H. Trinkaus, J. Nucl. Mater. 249 (1997) 103.

[72] A. Seeger, in: On the theory of radiation damage and radiation hardening, Proc. 2nd United Nations Int. Conf. on Peaceful Uses of Atomic Energy, vol. 6 (New York, NY: United Nations), 1958, p. 250.

[73] A. Fissolo, R. Cauvin, J.P. Hugot, V. Levy, in: Effects of Radiation on Materials, 14th Int. Symp., Vol. II, STP 1046, ASTM, Philadelphia, 1990, pp. 700–713.

[74] O.K. Chopra, A.S. Rao, A review of irradiation effects on LWR core internal materials—neutron embrittlement, J. Nucl. Mater. 412 (2011) 195–208.

[74a] U. Ehrnsten, K. Wallin, P. Karjalainen-Roikonen, S. van Dyck, P. Ould, In: Proceedings. Sixth International Symposium on Contribution of Materials Investigations to Improve the Safety and Performance of LWRs, vol. 1, Fontevraud 6, French Nuclear Energy Society, SFEN, Fontevraud Royal Abbey, France, September 18–22, 2006, pp. 661–670.

[74b] S. Fyfitch, H. Xu, A. Demma, R. Carter, R. Gamble, P. Scott, In: Proceedings of Fourteenth International Conference on Environmental Degradation of Materials in Nuclear Power Systems—Water Reactors, American Nuclear Society, Lagrange Park, IL, 2009.

[75] U. S. Nuclear Regulatory Commission, Fracture Toughness and Crack Growth Rates of Irradiated Austenitic Stainless Steels, NUREG/CR-6826, U.S. Nuclear Regulatory Commission, Washington, DC, 2003, p. 21.

[76] A review of radiation embrittlement for stainless steels, Materials Reliability Program (MRP-79), Rev. 1, EPRI Report 1008204, September 2004.

[77] Fracture toughness testing of decommissioned PWR core internals material samples, Materials Reliability Program (MRP-160), EPRI Report 1012079, September 2005.

[78] A. Demma, R. Carter, A. Jenssen, T. Torimaru, R. Gamble, in Proc. Thirteenth Int'l Conf. Environmental Degradation of Materials in Nuclear Power Systems—Water Reactors, T. R. Allen, P. J. Kin, L. Nelson, Eds., Canadian Nuclear Society, Toronto, 2007, Paper No. 114.

[79] K. Fukuya, K. Fuji, H. Nishioka, Y. Kitsunai, J. Nucl. Sci. Etchnol. 43 (2) (2006) 159–173.

[80] D.L. Porter, F.A. Garner, J. Nucl. Mater. 159 (1988) 114–121.

[81] M. L. Hamilton, F. H. Huang, W. J. S. Yang and F. A. Garner, Mechanical Properties and Fracture Behavior of 20% Cold-Worked 316 Stainless Steel Irradiated to Very High Exposures, Effects of Radiation on Materials: Thirteenth International Symposium (Part II) Influence of Radiation on Material Properties, ASTM STP 956, F. A. Garner, N. Igata and C. H. Henager Jr., Eds., ASTM Philadelphia, PA, 1987, 245–270.

[82] L.K. Mansur, M.L. Grossbeck, J. Nucl. Nater. 155–157 (1988) 130.

[83] E.E. Bloom, J. Nucl. Mater. 85-86 (1979) 795.

[84] E.E. Bloom, F.W. Wiffen, J. Nucl. Mater. 58 (1975) 171.

[85] H. Schroeder, P. Batfaisky, J. Nucl. Mater. 117 (1983) 287.

[86] H. Schroeder, W. Kestemich, H. Ullmaier, J. Fusion Eng. Des. 2 (1985) 65.

[87] N. Yamamoto, J. Nagakawa, H. Shiraishi, J. Nucl. Mater. 226 (1995) 185–196.

[88] N. Yamamoto, J. Nagakawa, Y. Murase, H. Shiraishi, J. Nucl. Mater. 258–263 (1998) 1628–1633.

[89] N. Yamamoto, J. Nagakawa, Y. Murase, H. Shiraishi, J. Nucl. Mater. 258–263 (1998) 1634–1638.

[90] N. Yamamoto, J. Nagakawa, Y. Murase, H. Shiraishi, Fusion Eng. Des. 41 (1998) 111–117.

[91] H. Trinkaus, J. Nucl. Mater. 133 (1985) 105.

[92] H. Ullmaier, J. Nucl. Mater. 133-134 (1985) 100.
[93] H. Trinkaus, B.N. Singh, J. Nucl. Mater. 323 (2003) 229.
[94] F.A. Garner, Radiation damage in austenitic steels, in: R. Konings (Ed.), Comprehensive Nuclear Materials, vol. 4, Elsevier, 2012 (Chapter 65).
[95] F.A. Garner, Void swelling and irradiation creep in light water reactor environments, in: P.G. Tipping (Ed.), Understanding and Mitigating Ageing in Nuclear Power Plants, Woodhead Publishing, Cambridge, 2010, pp. 308–356 (Chapter 10).
[96] P.J. Maziasz, J. Nucl. Mater. 122&123 (1984) 472.
[97] V.N. Voyevodin, I.M. Neklyudov, V.V. Bryk, O.V. Borodin, J. Nucl. Mater. 271-272 (1999) 290–295.
[98] P.J. Maziasz, J. Nucl. Mater. 200 (1993) 90.
[99] P.J. Maziasz, ORNL-6121, 1985.
[100] E.H. Lee, L.K. Mansur, A.F. Rowcliffe, J. Nucl. Mater. 122&123 (1984) 299.
[101] M.P. Shaw, B. Raph, W.M. Stobbs, J. Nucl. Mater. 115 (1983) 1.
[102] J.L. Seran, J.M. Dupouy, in: Effects of radiation on materials, 11th Int. Symp., STP 782, ASTM, Philadelphia, 1982, pp. 5–16.
[103] J.L. Seran, J.M. Dupouy, Proc. Conf. Dimensional stability and Mechanical Behavior of Irradiated Metals and Alloys, vol. 1., Brighton British Nuclear Energy Society, 1983, 22–28.
[104] F.A. Garner, N. Sekimura, M.L. Grossbeck, A.M. Ermi, J.W. Newkirk, H. Watanabe, M. Kiritani, J. Nucl. Mater. 205 (1993) 206–218.
[105] N. Akasaka, K. Hattori, S. Onose, S. Ukai, J. Nucl. Mater. 271&272 (1999) 370–375.
[106] J.P. Foster, A. Boltax, Nucl. Technol. 47 (1980) 181.
[107] J.F. Bates, E.R. Gilbert, J. Nucl. Mater. 71 (1978) 286.
[108] M.M. Hall, Trans. ANS 28 (1978) 146–147.
[109] N. Akasaka, I. Yamagata, S. Ukai, J. Nucl. Mater. 283–287 (2000) 169–173.
[110] T. Uwaba, S. Ukai, J. Nucl. Mater. 305 (2002) 21–28.
[111] K. Ehrlich, J. Nucl. Mater. 100 (1981) 149–166.
[112] J.P. Foster, W.G. Wolfer, A. Biancheria et al., Proc. Conf. Irradiation Embrittlement and Creep in Fuel Cladding and Core Components, BNES, London, (1973) 273.
[113] F.A. Garner, Irradiation performance of cladding and structural steels in liquid metal reactors, materials science and technology, in: A Comprehensive Treatment, vol. 10A, VCH Publishers, Weinheim, 1994, pp. 483–514 (Chapter 6).
[114] K. Ehrlich, J. Nucl. Mater. 133–134 (1985) 119–126.
[115] P.T. Heald, M.V. Speight, Acta Metall. 23 (1975) 1389–1399.
[116] W.G. Wolfer, M. Ashkin, J. Appl. Phys. 47 (1976) 791–800.
[117] R. Bullough, J.R. Willis, Philos. Mag. 31 (1975) 855–861.
[118] J.H. Gittus, Philos. Mag. 25 (1972) 345–354.
[119] P.T. Heald, J.E. Harbottle, J. Nucl. Mater. 67 (1977) 229–233.
[120] A. Maillard, H. Touron, J.L. Seran, et al., in: Effects of radiation on materials, 16th Int. Symp., STP 1175, ASTM, Philadelphia, 1993.
[121] A. Uehira, S. Mizuta, S. Ukai, et al., J. Nucl. Mater. 283–287 (2000) 396–399.
[122] S. Ukai, S. Ohtsuka, J. Nucl, Sci. Technol. 44 (2007) 743–757.
[123] B.V.D. Schaaf, J. Nucl. Mater. 155–157 (1988) 156–163.
[124] M.L. Grossbeck, K.C. Liu, Nucl. Technol. 58 (1982) 53.
[125] M.L. Grossbeck, K.C. Liu, J. Nucl. Mater. 103–104 (1981) 853.
[126] R. Schmitt and W. Scheibe. Proc. Seventh Int. Conf. on Structural Mechanics in Reactor Technology. Chicago, USA. vol. L (1983) 377.
[127] A.J. Lovell, B.A. Chin, E.R. Gilbert, J. Mater. Sci. 16 (1981) 870.
[128] R.J. Puigh, R.E. Schenter, in: Effects of radiation on materials, 12th international symposium, ASTM STP 870, 1985, p. 795
[129] R.J. Puigh, M.L. Hamilton, in: Influence of radiation on materials properties, 13th international symposium (part II), ASTM STP 956, 1987, p. 22.
[130] S. Ukai, S. Mizuta, T. Kaito, H. Okada, J. Nucl. Mater. 278 (2000) 320–327.
[131] Y. Tateishi, J. Nucl. Sci. Technol. 26 (1) (1989) 132.
[132] D.H. Locke, Review of experience with water reactor fuels 1968–1973, Nucl. Eng. Des. 33 (1975) 94.

[133] M.T. Simnad, A brief history of power reactor fuels, J. Nucl. Mater. 100 (1981) 93.
[134] P.L. Andresen, F.P. Ford, S.M. Murphy, J.M. Perks, in: State of knowledge of radiation effects on environmental cracking in light water reactor core materials, Proc 4th Int. Symp. on Environmental Degradation of Materials in Nuclear Power Systems—Water Reactors, NACE International, Houston, TX, 1990, pp. 1–83.
[135] J.B. Brown Jr., B.W. Storhok, J.E. Gates, Corrosion aspects of Iron austenitic stainless steel + inconel in high temperature water, Trans. Am. Nucl. Soc. 10 (1967) 668–669.
[136] C.F. Cheng, Corrosion 20 (11) (1964) 341.
[137] F. Garzaroll, H. Rubel, E. Steinberg, in: Behavior of water reactor core materials with respect to corrosion attack, Proc. Int. Conf. on Environmental Degradation of Materials in Nuclear Power Systems—Water Reactors, NACE, Houston, TX, 1984, pp. 1–24.
[138] H. Hanninen, I. Aho-Mantila, in: Environment sensitive cracking of reactor internals, Proc. 3rd Environmental Degradation of Materials in Nuclear Power Systems—Water Reactors, The American Institute of Mining, Metallurgical, and Petroleum Engineers (AIME), New York, NY, 1988, pp. 77–92.
[139] I. Multer, in: European operating experience, Proc. Joint Topical Meeting on Commercial Nuclear Fuel Technology Today, CNA/ANS, Toronto, 1975.
[140] V. Pasupathi, R.W. Klingensmith, Investigation of Stress Corrosion in Clad Fuel Elements and Fuel Performance in the Connecticut Yankee Reactor. NP-2119, EPRI, Palo Alto, CA, 1981.
[141] L.D. Schaffer, Army Reactor Program Progress Report ORNL 3231, 1962.
[142] J.T. Storrer, D.H. Locke, High burnup irradiation experience in Vulcain, Nucl. Eng. Int. 2 (1970) 93–99.
[143] C.F. Cheng, Intergranular corrosion cracking of type-304 stainless steel in water-cooled reactors, React. Technol. 13 (3) (1970) 310.
[144] F. Garzarolli, D. Alter, P. Dewes, J.L. Nelson, in: Deformability of austenitic stainless steels and nickel-based alloys in the core of a boiling and a pressurized water reactor, Proc. 3rd Int. Symp. on Environmental Degradation of Materials in Nuclear Power Systems—Water Reactors, The American Institute of Mining, Metallurgical, and Petroleum Engineers (AIME), New York, NY, 1988, pp. 657–664.
[145] D.I.R. Norris, C. Baker, J.M. Titchmarsh, in: Compositional Profiles at Grain Boundaries in 20%Cr/25%Ni/Nb Stainless Steel, Proc. Radiation-Induced Sensitization of Stainless Steels, Berkeley Nuclear Labs, Central Electricity Generating Board, London, 1987, pp. 86–98.
[146] M.G. Adamson, E.A. Eitken, R.W. Caputi, P.E. Potter, M.A. Mignanelli, in: Thermodynamics of nuclear materials, Proceedings of an International Symposium, Julich, Jan 29 to Feb. 2, 1979, vol. 1, IAEA, Vienna, 1980, pp. 503–538.
[147] M. Coquerelle, O. Gotzmann, S. Pickering, Trans. Am. Nucl. Soc. 20 (1975) 287–289.
[148] J.W. Weber, E.D. Jensen, Trans. Am. Nucl. Soc. 14 (1971) 175–176.
[149] D. Feron, E. Herms, B. Tanguy, Behavior of stainless steels in pressurized water reactor primary circuits, J. Nucl. Mater. 427 (2012) 364–377.
[150] T. Furukawa, E. Yishida, Material performance in sodium, in: R. Konings (Ed.), Comprehensive Nuclear Materials, vol. 5, Elsevier, 2012 (Chapter 5.13).
[151] P.N. Martynov, K.D. Ivanov, Properties of Lead–Bismuth Coolant and perspectives of Non-electric Applications of Lead-Bismuth Reactor, IAEA-TECDOC-1056, 1997, pp. 177–184.
[152] Handbook on Lead–Bismuth Eutectic Alloy and Lead Properties, Materials Compatibility, Thermal-hydraulics and Technologies, OECD NEA 6195, 2007 Ed.
[153] C. Fazio, A. Alamo, A. Almazouzi, D. Gomez-Briceno, F. Groeschel, F. Roelofs, P. Turroni, J.U. Knebel: GLOBAL 2005 Proceedings, Tsukuba, Japan, 9–13, October 2005.
[154] O. Benes, R.J.M. Konings, Molten salt reactor fuel and coolant, in: R. Konings (Ed.), Comprehensive Nuclear Materials, vol. 3, Elsevier, 2012 (Chapter 3.13).
[155] L.C. Olson, J.W. Ambrosek, K. Sridharan, M.H. Anderson, T.R. Allen, J. Fluorine, Chemistry 130 (2009) 67–73.
[156] V. Ignatiev, A. Surenkov, Material performance in molten salts, in: R. Konings (Ed.), Comprehensive Nuclear Materials, vol. 5, Elsevier, Amsterdam, 2012 (Chapter 5.10).
[157] F.P. Ford, P.L. Andresen, Corrosion in nuclear systems: environmentally assisted cracking in light water reactors, in: Corrosion Mechanisms, Marcel Dekker, New York, NY, 1994, pp. 501–546.
[158] F.P. Ford, P.L. Andresen, in: Development and use of a predictive model of crack propagation in 304/316L, A533B/A508 and Inconel 600/182 in 288C water, Proc. 3rd Int'l Symp. on Environmental Degradation of Materials in Nuclear Power Systems—Water Reactors, The Metallurgical Society of the American Institute of Mining, Metallurgical and Petroleum Engineers (AIME), New York, NY, 1988, p. 789.

[159] F.P. Ford, B. Gordon, R.M. Horn, Corrosion in boiling water reactors, ASM handbook, in: Corrosion: Environments and Industries, vol. 13C, ASM International, Materials Park, OH, 2006, p. 341.
[160] S.J. Zinkle, G.S. Was, Materials challenges in nuclear energy, Acta Mater. 61 (2013) 735–758.
[161] P.M. Scott, P. Combrade, Corrosion in pressurized water reactors, ASM handbook, in: Corrosion: Environments and Industries, vol. 13C, ASM International, Materials Park, OH, 2006, p. 362.
[162] R.W. Staehle, J.A. Gorman, in: Progress in understanding and mitigating corrosion on the secondary side in PWR steam generators, 10th International Conference on Environmental Degradation of Materials in Nuclear Power Systems-Water Reactors, NACE, Lake Tahoe, CA, 2001. special bonus paper.
[163] Staehle R. personal communication. 2012.
[164] D.J. Paraventi, W.C. Moshier, in: Assessment of the Interaction of Variables in the Intergranular Stress Corrosion Crack Growth Rate Behavior of Alloys 600, 82 and 182, 13th Int'l Conf. Environmental Degradation of Materials in Nuclear Power Systems. Whistler, British Columbia, Canada, Canadian Nuclear Society, Toronto, 2007.
[165] G.M. Gordon, K.S. Brown, in: Dependence of creviced BWR component IGSCC behavior on coolant chemistry, Proc. 4th Int. Conf. on Environmental Degradation of Materials in Nuclear Power Systems—Water Reactors, NACE, Houston, TX, 1990, pp. 14–62.
[166] K.S. Brown, G.M. Gordon, Effects of BWR coolant chemistry on the propensity for IGSCC initiation and growth in creviced reactor internals components, Proc. 3rd Environmental Degradation of Materials in Nuclear Power Systems—Water Reactors (New York, NY: The American Institute of Mining, Metallurgical, and Petroleum Engineers (AIME), 1988), 243–248.
[166a] S.M. Bruemmer, G.S. Was, J. Nucl. Mater, 216 (1994) 348.
[167] J.T. Busby, G.S. Was, E.A. Kenik, J. Nucl. Mater. 302 (2002) 20.
[168] J.T. Busby, G.S. Was, in: Irradiation assisted stress corrosion cracking in model austenitic alloys with solute additions, Proc. 11th Int. Conf. Environmental Degradation of Materials in Nuclear Power Systems—Water Reactors, American Nuclear Society, La Grange Park, IL, 2003, p. 995.
[169] T. Yonezawa, K. Fujimoto, H. Kanasaki, T. Iwamura, S. Nakada, K. Ajiki, S. Urata, Corros. Eng. 49 (2000) 655.
[170] K. Fujimoto, T. Yonezawa, T. Iwamura, K. Ajiki, S. Urata, Corros. Eng. 49 (2000) 701.
[171] T. Yonezawa, K. Fujimoto, T. Iwamura, S. Nishida, in: R.D. Kane (Ed.), ASTM 1401, Improvement of IASCC Resistance for Austenitic Stainless Steels in PWR Environment, Environmentally Assisted Cracking: Predictive Methods for Risk Assessment and Evaluation of Materials, Equipment, and Structures, ASTM International, West Conshohocken, PA, 2000, p. 224.
[172] T. Yonezawa, T. Iwamura, K. Fujimoto, K. Ajiki, in: Optimized chemical composition and heat treatment conditions of 316 CW and high chromium austenitic stainless steels for PWR Baffle Bolts, Proc. 9th Int. Conf. Environmental Degradation of Materials in Nuclear Power Systems—Water Reactors, The Minerals, Metals, and Materials Society, Warrendale, PA, 1999, p. 1015.
[173] G.F. Li, Y. Kaneshima, T. Shoji, Corrosion 56 (5) (2000) 540.
[174] Z. Jiao, G.S. Was, Novel features of radiation-induced segregation and radiation-induced precipitation in austenitic stainless steels, Acta Mater. 59 (2011) 1220–1238.
[175] O.K. Chopra, A.S. Rao, A review of irradiation effects on LWR core internal materials – IASCCsusceptibilty and crack growth rates of austenitic stainless steels, J. Nucl. Mater. 409 (2011) 235–256.
[176] W.G. Wolfer, R.H. Jones, J. Nucl. Mater. 103–104 (1981) 1305–1314.
[177] G. Lloyd, J. Nucl. Mater. 110 (1982) 20–27.
[178] An Interim Review of the Cooperative Irradiation-Assisted Stress Corrosion Cracking Research (CIR) Program—Revision 1, Electric Power Research Institute, Palo Alto, CA, EPRI report 1015493, November 2007.
[179] M. Ruscak, P. Chvatal, M. Zamboch, K. Splichal, M. Postler, Influence of Radiolysis and Hydrogen Embrittlement on In-Service Cracking of PWR Internal Structures, Electric Power Research Institute, Palo Alto, CA, 1999. EPRI report TR-112593, November.
[180] P.L. Andresen, in: Emerging issues and fundamental processes in environmental cracking in hot water, Proc. Research Topical Symposium on Environmental Cracking, Corrosion/07, NACE, 2007.
[181] C.A. Grove, L.D. Petzold, Mechanism of SCC of Alloy X750 in High Purity Water, J. Mater. Energy Sys., Vol. 7, No. 2, Sept 1985, 147–162. Also. L.D. Petzold, C.A. Grove, in: R.C. Scarberry (Ed.), Mechanism of SCC of Alloy X750 in High Purity Water, Corrosion of Nickel Alloys, ASM, Metals Park, OH, 1985, p. 165.
[182] P.L. Andresen, M.M. Morra, in: Effects of positive and negative dK/da on SCC growth rates, Proc. 12th Int. Symp. on Environmental Degradation of Materials in Nuclear Power Systems—Water Reactors, TMS, Snowbird, August, 2005.

[183] C.M. Brown, W.J. Mills, in: Load path effects on the fracture toughness of alloy 82H and 52 welds in low temperature water, Proc. 12th Int. Symp. on Environmental Degradation of Materials in Nuclear Power Systems—Water Reactors, TMS, Snowbird, August, 2005.

[184] P.L. Andresen, P.W. Emigh, M.M. Morra, R.M. Horn, in: Effects of yield strength, corrosion potential, stress intensity factor, silicon and grain boundary character on the SCC of stainless steels, Proc. of 11th Int. Symp. on Environmental Degradation of Materials in Nuclear Power Systems—Water Reactors, ANS, 2003.

[185] F. Garzarolli, D. Alter, P. Dewes, J.L. Nelson, in: Deformability of austenitic stainless steels and Ni-base alloys in the core of a boiling and a pressurized water reactor, Proceedings of the Third International Symposium on Environmental Degradation of Materials in Nuclear Power Systems, The Metallurgical Society, 1988, pp. 657–664.

[186] F. Garzarolli, P. Dewes, R. Hahn, J.L. Nelson, in: Deformability of high-purity stainless steels and Ni-base alloys in the core of a PWR, Sixth International Symposium on Environmental Degradation of Materials in Nuclear Power Systems—Water Reactors, The Minerals, Metals and Materials Society, 1993, pp. 607–613.

[187] F. Garzarolli, P. Dewes, R. Hahn, J.L. Nelson, in: In-reactor testing of IASCC resistant stainless steels, Seventh International Symposium on Environmental Degradation of Materials in Nuclear Power Systems—Water Reactors, American Nuclear Society, 1995, pp. 1055–1065.

[188] F. Garzarolli, P. Dewes, S.T. Pritsching, J.L. Nelson, in: Irradiation creep behavior of high-purity stainless steels and n-base alloys, Ninth International Symposium on Environmental Degradation of Materials in Nuclear Power Systems—Water Reactors, The Minerals, Metals & Materials Society (TMS), 1999, pp. 1027–1034.

[189] J.J. Laidler, J.W. Bennet, Core materials studies improve fast breeder performance, Nucl. Eng. Int. 25 (301) (1980) 31–36.

[190] J.T. Busby, C.E. Duty, K.J. Leonard, R.K. Nanstad, S.J. Pawal, M.A. Sokolov, Assessment of Structural and Clad Materials for Fission Surface Power Systems, Oak Ridge National Laboratory Report # ORNL/LTR/FSP/2007–01, September, 2007.

[191] M.G. Alvarez, A.M. Olmedo, M. Villegas, JNM 229 (1996) 93–101.

[192] M.D. McMurtrey, G.S. Was, in: J.T. Busby, G. Ilevbare, P.L. Andresen (Eds.), Role of slip behavior in the irradiation assisted stress corrosion cracking in austenitic steels, 15th Int'l. Conf. Environmental Degradation of Materials in Nuclear Power Systems—Water Reactors, John Wiley & Sons, Hoboken, NJ, 2011, pp. 1383–1394.

[193] R.W. Swindeman, et al., Evaluation of advanced austenitic alloys relative to alloy design criteria for steam service: Part 1—lean stainless steels, ORNL-report, ORNL-6299/P1, 1990.

[194] P.J. Maziasz, Developing an austenitic stainless steel for improved performance in advanced fossil power facilities, J. Metals 41 (7) (1989) 14–20.

[195] J.A. Francis, W. Mazur, H.K.D.H. Bhadeshia, Type IV cracking in ferritic power plant steels, Mater. Sci. Technol. 22 (2006) 1387–1395.

[196] L. Tan, Y. Yamamoto, Summary Report of FY11 Testing and Optimization Progress of Advanced Alloy Development, Oak Ridge National Laboratory, Oak Ridge, TN, 2011. ORNL/TM-2011/327, September 15.

[197] S.J. Zinkle, K.A. Terrani, J.C. Gehin, L.L. Snead, Accident tolerant fuels for LWRs: a perspective, J. Nucl. Mater. 448 (2014) 374–379.

[198] K.A. Terrani, Accident tolerant fuel cladding development: promise, status, and challenges, J. Nucl. Mater. 501 (2018) 13–30.

FURTHER READING

[199] K.G. Samuel, S.K. Ray, G. Sasikala, Dynamic strain ageing in prior cold worked 15Cr–15Ni titanium modified stainless steel (Alloy D9), J. Nuclear Mater. 355 (2006).

[200] Y. Yamamoto, et al., Creep-resistant, Al_2O_3-forming austenitic stainless steels, Science 316 (2007) 433–436.

[201] G.S. Was, P.L. Andresen, Irradiation assisted corrosion, in: D. Feron (Ed.), Nuclear Corrosion Science & Engineering, Woodhead Publishing, Cambridge, UK, 2012 (Chapter 6).

[202] ATEM Characterization of stress-corrosion cracks in LWR-irradiated austenitic stainless steel core components, PNNL EPRI report, 11/2001.

CHAPTER 9

NI-BASED ALLOYS FOR REACTOR INTERNALS AND STEAM GENERATOR APPLICATIONS

Malcolm Griffiths
Department of Mechanical and Materials Engineering, Queen's University, Kingston, ON, Canada ANT International, Mölnlycke, Sweden

CHAPTER OUTLINE

9.1 Introduction	350
9.2 Physical Metallurgy	351
9.3 Thermomechanical Processing	354
9.4 Joining	356
9.5 Mechanical Properties	358
9.6 Fracture Modes	360
9.7 Deformation Mechanisms (Yield Stress and Creep Strength)	362
9.8 Stress Corrosion Cracking	364
9.9 Ni Alloys for Generation-IV Reactors	365
9.10 Chemical Compatibility With Coolants	368
9.11 Radiation Damage and Gas Production in Ni Alloys	369
9.12 Radiation Hardening/Softening and Loss of Ductility in Ni Alloys	376
9.12.1 CANDU Reactors	377
9.12.2 LWR Reactors	380
9.12.3 Fast Reactors	383
9.12.4 Proton Irradiation Facilities	385
9.12.5 Ion Irradiation Facilities	390
9.13 Hydrogen Embrittlement	390
9.14 Helium Embrittlement	391
9.15 Point Defects	395
9.16 Irradiation Creep and Stress Relaxation	396
9.17 Fatigue and Creep Fatigue	399
9.17.1 Fatigue	399
9.17.2 Creep-Fatigue Deformation	400
9.18 Conclusions	401
Acknowledgments	402
References	402

9.1 INTRODUCTION

Nickel alloys are used extensively in nuclear reactor applications because they have high strength and good corrosion properties. However, nickel has some unique nuclear properties (including relatively high thermal neutron absorption cross sections) that limit its use in nuclear reactor cores to specialized components, such as springs and high strength fasteners, which are subject to low neutron doses over their service life. Table 9.1, adapted from Yonezawa's review [1], lists the more common Ni alloys that are currently in use, or have been used, in nuclear power plants. Fyfitch [2] has reviewed corrosion and stress-corrosion cracking (SCC) of Ni alloys in reactor systems. Fyfitch's review also provides a detailed description of where Ni alloy components are used in existing nuclear power plants. Table 9.2 lists the nominal compositions for the commonly used alloys and Table 9.3 lists the

Table 9.1 Common Ni-Based Engineering Alloys Used in Nuclear Reactor Cores and Primary Coolant Circuits

Reactor Type	Alloy						
BWR	600	718	X-750		800	625	
PWR	600	718	X-750		800		690
CANDU	600	718	X-750		800		690
LMFBR[a]		718	X-750	PE16, 706	800		
HTGR	600	718	X-750			625	

[a]Liquid metal fast breeder reactor.

Table 9.2 Nominal Compositions (wt%) of Commercial Ni-Based Alloys

Alloy	Ni	Cr	Fe	Ti	Al	Nb	Mo
Nimonic PE16	43	17	33	1.2	1.3		3.7
Inconel 750	72	15.5	7	2.5	0.7	1	
Inconel 718	53	18	19	0.9	0.6	5	2.5
Inconel 706	42	16	37	1.7	0.3	2.9	0.1
Inconel 690	61	29	9	0.5	0.5		
Inconel 625	61	22	<5	0.3	0.3	3.5	9
Inconel 600	75	16	8	0.3	0.2		
Incoloy 800	33	21	>39.5	0.4	0.4		

Table 9.3 Maximum Allowed Minor Alloying/Impurity Elements (wt%) of Commercial and Developmental Ni-Based Alloys

Alloy	Mn Max	Si Max	C Max	Cu Max	P Max	S Max	B Max	Co Max
Nimonic PE16	0.1	0.2	0.05					0.05
Inconel 750		0.5	0.08	0.5	0.008	0.01		0.05
Inconel 718	0.2	0.2	0.04	0.3	0.015	0.015	0.006	0.05
Inconel 706	0.2	0.2	0.03			0.015		0.05
Inconel 690	0.5	0.5	0.05	0.5	0.025	0.015		0.05
Inconel 625	0.2	0.2	0.05	0.3	0.015	0.015	0.02	0.05
Inconel 600	0.2	0.2	0.08	0.3	0.1	0.015		0.05
Incoloy 800	0.9	0.5	0.08	0.5	0.03	0.03		0.05

minor elements/impurities, which can also have a significant effect on the properties of the alloy. Chapters 1 and 2 in this book provide overviews of structural materials applications in water-cooled and proposed Gen-IV fission reactors, Chapter 5 provides a general overview of radiation and thermomechanical degradation effects in reactor structural materials, and Chapter 6 summarizes corrosion and SCC phenomena. This chapter focuses specifically on Ni-alloy performance in current and future nuclear reactor applications.

9.2 PHYSICAL METALLURGY

Nickel (Ni) has a face-centered cubic (FCC) crystal structure and melts at 1455°C. It is an element that forms the base for a group of alloys (that also includes those based on iron and cobalt) known as superalloys; these are high-temperature materials that are resistant to mechanical and chemical degradation at temperatures close to their melting points [3]. The alloys listed in Table 9.1 are all FCC-based alloys (the so-called γ-phase) because of the high Ni content. The ternary-phase diagram showing the approximate compositions of the three main alloying elements (Ni, Cr, and Fe) for the alloys in Table 9.1 is shown in Fig. 9.1 [4]. Apart from Fe and Cr there are a number of elements that are soluble in Ni (e.g., Al, Ti, Nb, Mo), thus providing a large range of possibilities for solid solution strengthening and also for precipitation hardening.

Some Ni-based superalloys, containing precipitates of the ordered γ′-phase with sufficient size and density, exhibit an unusual property in that the yield strength (YS) and creep resistance either increase or remain relatively constant with increasing temperature up to about 800°C [3]. The γ′-phase imparts a strengthening effect in the alloy

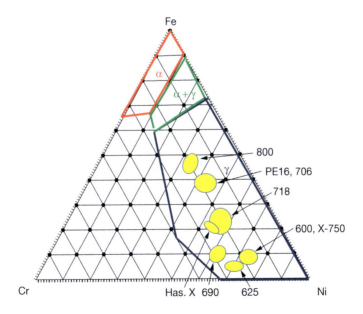

FIG. 9.1

Ternary diagram for the Fe-Cr-Ni system at 400°C illustrating the approximate compositions of the main Ni-rich structural alloys in water-cooled reactors and their thermodynamically stable FCC phase (γ) compositions outlined in *blue*. The austenite phase in the *green* region is metastable leading to a wide range of austenitic, ferritic, martensitic, and duplex stainless steels.

Modified from A. Strasser, F.P. Ford, High Strength Ni alloys in Fuel Assemblies, ANT International, Molnlycke, 2012 and reproduced with kind permission from ANT International.

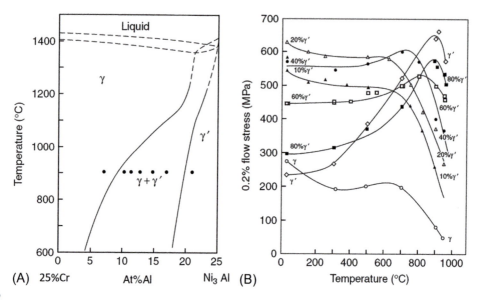

FIG. 9.2

Temperature dependence of the yield stress for a number of Ni-Cr-Al alloys with 75 at.% Ni and varying amounts of Cr and Al giving varying amounts of γ′: (A) alloy compositions in the section of the ternary phase diagram at 75 at.% Ni; (B) 0.2% yield stress as a function of volume fraction of γ′-phase.

Modified from P. Beardmore, R.G. Davies, T.L. Johnston, On the temperature dependence of the flow stress of nickel-base alloys, Trans. Metal. Soc. AIME 245 (1969) 1537–1545, with permission.

that is a function of the γ′ volume fraction, Fig. 9.2 [5]. The γ′-phase itself increases in yield strength with increasing temperature up to about 800°C, whereas the γ-phase exhibits a temperature dependence that is typical of most metals, that is, metals tend to be softer (decreasing YS) with increasing temperature.

The γ′-phase has a composition of Ni_3X; it has an ordered FCC structure with the Ni atoms occupying the cube face positions and element X occupying the cube corners (Fig. 9.3). For most of the γ′-containing Ni alloys used in nuclear applications X is typically Al, but other elements (such as Ti, Nb, and Mo) can also partially substitute for Al. The γ′-phase is often coherent with the base Ni matrix and the compositional difference leads to a small misfit strain (typically <0.5% [3]), thus producing strain contrast when imaged in an electron microscope.

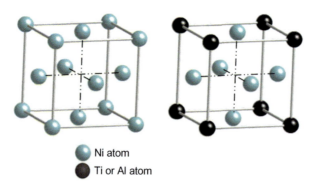

● Ni atom
● Ti or Al atom

FIG. 9.3

Crystal structure of FCC γ-phase (Ni) and γ′-phase [$Ni_3(Ti,Al)$].

With Nb present, a related ordered phase, γ'', which has a body-centered tetragonal (BCT) structure and composition of Ni_3Nb, can be produced in addition to γ'. Carbon, present at concentrations up to 0.1 wt%, combines with reactive elements such as Cr to form carbides. Carbides often nucleate on grain boundaries during high-temperature processing. They are favored for very high-temperature engineering applications in turbine blades because they restrict grain growth and grain boundary sliding during very high-temperature operation [3]. Carbides can also be detrimental to SCC, thus they are not required in Ni alloys used at the relatively low temperatures of nuclear applications [4]. Boron acts in a manner similar to carbon and forms borides on grain boundaries if present in sufficient concentrations. Boron is a beneficial trace element for nonnuclear applications but has an adverse impact on high-temperature stress corrosion cracking (HTSCC), and low-temperature crack propagation (LTCP), of irradiated material for B contents >20 wtppm (parts per million by weight) [6]. Mills et al. [6] attributed the effect of the boron, in lowering fracture toughness and increasing SCC rates in irradiated specimens, to transmutation effects, either through the production of He or through the production of Li. Boron is normally a trace element (<0.01 wt%, i.e., 100 wtppm) for alloys used in nuclear reactor cores. ^{10}B, which constitutes ~20% of the B isotopes, has a very high thermal neutron (n,α) cross section producing He. The maximum He concentration that can be produced is limited by the concentration of ^{10}B and is typically <100 appm (atomic parts per million) for most engineering alloys.

The main elements that are important to the constitution of Ni-based superalloys are shown in Fig. 9.4 [3]. The relative atomic sizes and electron vacancy numbers (N_v), that is, the average numbers of electron vacancies per atom, give an indication of the likelihood that the alloying elements are either in solid solution (similar atomic size) or form the γ'-phase (large mean N_v), [7]. The common precipitate phases are listed in Table 9.4 [4].

The γ'- and γ''-phases are metastable and transform to the hexagonal close-packed η (Ni_3Ti) and orthorhombic δ (Ni_3Nb) phases, respectively, with sufficient aging time. These η- and δ-phases do not contribute to precipitation hardening because of their size, number density, and lack of coherency with the γ matrix. They can be beneficial to mechanical properties when segregated to grain boundaries in restricted amounts since they can inhibit excessive grain growth in the same way as carbides. However, they can also be deleterious to mechanical properties due to their brittle nature and acicular morphology.

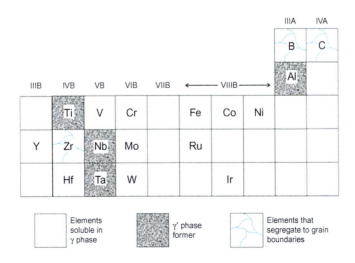

FIG. 9.4

Summary of elements important to the constitution of the Ni-based superalloys, and their relative positions in the periodic table.

Table 9.4 Phases in High Strength Nickel Alloys

γ (Gamma)	Face Centred Cubic (FCC)	Ni-Cr-Fe Solid Solution	Same Structure as Matrix for Alloys 625, 718, and X-750
γ' (gamma prime)	Ordered FCC	Ni(Cr,Fe)$_3$Al(Ti, Nb,Ta)[a]	Precipitation hardening phase for alloys 718 and X-750, associated with coherency between the γ matrix and the precipitate. The γ' shape changes with aging time and temperature, varying between spherical and cubic.
γ" (gamma double prime)	Body-centered-tetragonal (BCT)	Ni(Cr, Fe)$_3$Nb(Ti, Al, Ta)[a]	Additional precipitation hardening phase for alloy 718 with a disc-shaped morphology 60 nm diameter and 5–9 nm thickness.
δ (delta)	Orthorhombic	Ni$_3$Nb	Observed in overaged alloy 718 as a transformation from the coherent γ"-phase. Morphology is acicular and it is incoherent with the γ-phase. Intergranular formation acts as an inhibitor to grain growth
η (eta)	Hexagonal-close packed (HCP)	Ni$_3$Ti	Observed in overaged alloy X-750 as a transformation from the coherent γ'-phase. Morphology is acicular and it is incoherent with the γ-phase. Intergranular formation acts as an inhibitor to grain growth.
σ (sigma)	Tetragonal	FeCr, FeCrMo	Irregular-shaped globules and plate morphologies formed after extended times at high temperatures (540–980°C)
μ (mu)	Rhombohedral	(Fe,Co)$_7$(Mo,W)$_6$[a]	Coarse irregular Widmanstatten platelets formed at high temperatures in alloys with high Mo or W contents.
Laves	Hexagonal	Fe$_2$Nb, Fe$_2$Ti, Fe$_2$Mo	Irregular elongated globules or platelets after extended high-temperature exposure.
MC	Cubic	TiC, NbC, HfC	Primary carbide formed during casting in interdendritic regions
M$_6$C	FCC	Fe$_3$Mo$_3$C	Primary carbide formed during casting in interdendritic regions.
M$_7$C$_3$	Hexagonal	Cr$_7$C$_3$	Intergranular secondary carbides formed at intermediate temperatures with an accompanying Cr denuded zone.
M$_{23}$C$_6$	FCC	(Cr, Fe, W, Mo)$_{23}$C$_6$[a]	Intergranular secondary carbides formed at intermediate temperatures, with an accompanying Cr denuded zone.
M$_3$B$_2$	Tetragonal	Ta$_3$B$_2$, Nb$_3$B$_2$	Borides observed in alloys containing >0.03 wt% B
MN	Cubic	TiN, ZrN	Nitrides observed in cast structures with square to rectangular shapes.

[a]Brackets denote optional substitution.
Modified from A. Strasser, F.P Ford, High Strength Ni-alloys in Fuel Assemblies, ANT International, Molnlycke, Sweden, 2012 and reproduced with kind permission from ANT International.

9.3 THERMOMECHANICAL PROCESSING

There are five generic types of heat treatment that have been developed for Ni-rich engineering alloys and each have a role in defining the temperature–time combinations required for specific phase formations [4]:

> "*Homogenization*" that is sometimes used, but rarely specified, at temperatures approaching the melting temperature. The objective is to diffuse out the gross solidification microchemical gradients, and to take into solution high-melting point phases, such as primary carbides.

"*Solution anneal*" (SA) applied in the range 890–1100°C to take into solid solution those phases that impact on strength and structural integrity while precipitating some carbides on grain boundaries.

"*Direct age*" (DA) refers to hot working followed immediately by aging at temperatures around 704°C to precipitate out the hardening γ'- and γ''-phases.

"*Single step aging*" refers to a single aging treatment around 704°C following the solution anneal.

"*Double-step aging,*" or "*double aging,*" refers to a double-step aging treatment following the solution anneal, with the first aging treatment at about 885°C, followed by the second aging treatment at around 704°C to precipitate out the hardening γ'- and γ''-phases. The first age improves the stress rupture properties of alloys such as X-750 and 718 at temperatures >650°C by promoting grain boundary precipitation (primarily carbides) that limits grain boundary sliding at high temperatures.

Each alloy has its own characteristic phase stability and transformation kinetics, see, for example, Figs. 9.5 and 9.6 [8]. These characteristics are of importance in specifying the thermomechanical treatments and the cooling rates. The size, shape, and spacing of the γ' and γ'' precipitates, determined by the heat treatment [9], are important in controlling the high-temperature creep and yield strength (see Sections 9.5 and 9.7). Various thermomechanical treatments are designed to produce specific metallurgical properties for different Ni-alloy components depending on their applications and are described in the appropriate manufacturers' alloy specifications [10–17].

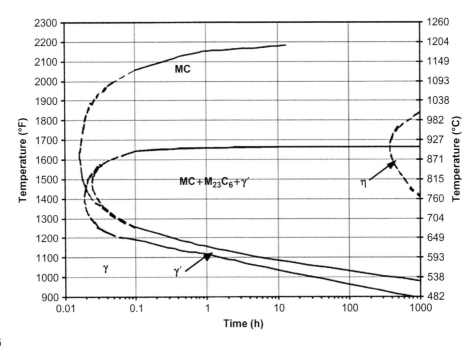

FIG. 9.5

Time-temperature-transformation diagram for Inconel X-750 showing regions of carbide γ' and η precipitation.

Reproduced with permission from G.D. Smith, S.J. Patel, Superalloys 718, 625, 706 and Derivatives, in: E.A. Loria TMS, The Minerals, Metals & Materials Society, 2005. Copyright 2005 by The Minerals, Metals & Materials Society.

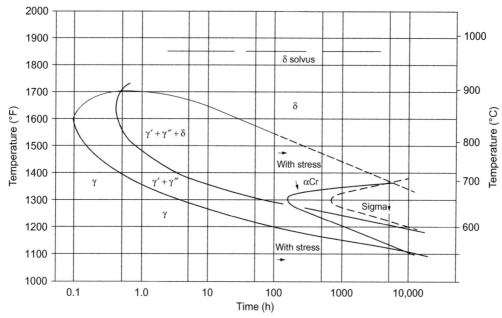

FIG. 9.6

Temperature-time-transformation diagram for Inconel 718 showing variation of nucleation times for γ′- and γ″-phases.
Reproduced with permission from G.D. Smith, S.J. Patel, Superalloys 718, 625, 706 and Derivatives, in: E.A. Loria TMS, The Minerals, Metals & Materials Society, 2005. Copyright 2005 by The Minerals, Metals & Materials Society.

9.4 JOINING

Successful joining processes that have been used on high strength Ni-alloy components for fuel assemblies (FAs) include resistance spot welding, electron beam (EB) welding, laser welding (LW), and brazing [18]. Welds requiring filler metals (FMs), such as alloys 52 and 82, have not been used for fuel assembly components. Welding processes are applied to the completed, annealed assemblies that are subsequently heat treated according to the recommended solution and precipitation-hardening process.

The most important structural welds in nuclear systems (that cannot be isolated from the reactor coolant system) are those between the main coolant line piping and ferritic nozzles on the reactor pressure vessel and other pressure vessel penetrations such as control rod drive mechanisms (CRDMs) and in the SGs. The most common Ni alloys in use in pressurized water reactors (PWRs) and boiling water reactors (BWRs) that are part of welded pressure boundary penetrations are Inconel 600 and 690. Welding electrode (WE), Inconel 182, and FM, Inconel 82, were the first NiCrFe-type welding products capable of depositing crack-free, porosity-free weldments in alloy 600. Inconel 690 is now the preferred alloy for SG piping. Inconel 690 and the corresponding weld metals (Inconel 152 and 52, and variants) have a higher chromium content, see Table 9.5 [18].

The SCC is a significant aging degradation mechanism for major components of both PWRs and BWRs, and is particularly prevalent at welds (see Chapter 6 for a general discussion on corrosion and SCC phenomena). The main causes of concern are SCC in the weld itself or in the component piping being welded. The SCC initiates as a result of residual stresses induced by the welding procedure. An excellent review of SCC issues in various reactor core structures that describes the welds and the problems associated with weld cracking in various reactor designs is provided in one of the IAEA (International Atomic Energy Agency) Nuclear Energy series of documents [19].

Table 9.5 Nominal Compositions (wt%) of Ni-Based Alloys and Welding Consumables for Nuclear Applications

Alloy	Ni	C	Mn	Fe	S	Cu	Si	Cr	Ti	Nb	P	Mo	Al	Other
600	72	0.15	10	6	0.015	0.5	0.5	14–17	–	–	–	–	–	–
FM 82	67	0.1	2.5–3.5	3	0.015	0.5	0.5	18–22	0.75	2–3	0.03	–	–	0.5
WE 182	59	0.1	5–9.5	10	0.015	0.5	1	13–17	1	1–2.5	0.03	–	–	0.5
690	58	0.05	0.5	7–11	0.015	0.5	0.5	27–31	–	–	–	–	–	–
FM 52	Balance	0.04	1	7–11	0.015	0.3	0.5	28–31.5	1	0.1	0.02	0.5	1.1	0.5
WE 152	Balance	0.05	5	7–12	0.015	0.5	0.75	28–31.5	0.5	1–2.5	0.03	0.5	0.5	0.5
FM 52M[a]	Balance	0.04	1	7–11	0.015	0.3	0.5	28–31.5	1	0.5–1	0.02	0.5	1.1	0.5
WE 152M[a]	Balance	0.05	5	7–12	0.015	0.5	0.75	28–31.5	0.5	1–2.5	0.03	0.5	0.5	0.5

[a]Minor additions of boron and zirconium.
Modified from S.D. Kiser, E.B. Hinshaw, J.R. Crum, L.E. Shoemaker, Nickel alloy welding requirements for nuclear service, Focus On Nuclear Power Generation, 2005, pp. 21–26, www.specialmetals.com/assets/smw_docs/Special-Metals.pdf.

Residual stresses in the nozzles and welds have been a particular issue in crack initiation [20]. The SCC of alloy 600 material occurs at locations with high residual stresses resulting from joining. The high residual stresses are mainly the result of weld-induced deformation being imposed on nozzles with cold-worked (CW) machined surfaces. Weld shrinkage, which occurs when welding the nozzle into the high restraint vessel shell, pulls the nozzle wall outward, thereby creating residual hoop stresses comparable to the yield strength in the nozzle base metal and the higher strength cold-worked surface layers. These high residual hoop stresses contribute to the initiation of axial cracks in the cold-worked surface layer and to the subsequent growth of the axial cracks in the lower strength nozzle base material. The lower frequency of cracking in the weld metal relative to the base metal may result from the fact that welds tend not to be cold-worked once they have been created.

Cr depletion adjacent to grain boundaries caused by Cr carbide precipitation can lead to intergranular stress corrosion cracking (IGSCC) of alloy 182 welds. A decrease in carbon content is not effective for improving SCC resistance because Cr carbides are formed even at small carbon contents due to negligible carbon solubility in this alloy. Therefore, carbon stabilization by Nb addition was adopted for the SCC-resistant alloy 182 in the same way as for stabilized stainless steel. The Nb content and C/Nb ratio are controlled in the improved alloy. Alloy 82 contains higher Cr and Nb than alloy 182 and has proved to be SCC resistant in BWRs. The newer Japanese BWRs use alloy 82 or Nb-controlled alloy 182. Alloy 182 is an Ni-based alloy used for dissimilar metal welds, for example, between stainless steel piping and a carbon steel reactor pressure vessel. This weld metal can be susceptible to SCC when sufficiently stressed in the primary coolant. Therefore, there has been a significant effort to determine the conditions under which alloy 182 is most susceptible to SCC. Mitigation of welding issues by material changes have involved migrating from use of alloy 182 to Nb-controlled 182 for BWRs and migrating from the use of alloy 82/182 to 52/152 for PWRs [19].

The Ni-based solid solution strengthened alloy, alloy 617, is the primary material candidate for the intermediate heat exchanger (IHX) in very high-temperature reactor (VHTR), and it is expected that Alloy 617 FM will be used for welds [21].

Welding is a concern for repairs to reactor core components because of the pronounced effect of He [generated via (n,α) reactions] that promotes cavity growth when the material is heated [22–26]. Although temperatures near the melting point of the metal in a heat affected zone during welding are extreme, high temperatures compensate for the very low concentration of He. One assessment carried out for BWR repairs states that the recommended maximum matrix He concentration is 1 appm [27].

9.5 MECHANICAL PROPERTIES

There have been a number of reviews describing the mechanical properties and characteristics of Ni alloys relevant to nuclear reactor applications, the most recent being that of Yonezawa [1] for unirradiated alloys, and Boothby [28], Rowcliffe et al. [29], and Angeliu [30] for irradiated alloys. There will be no attempt to reproduce the work of previous reviews here. However, certain aspects of Ni-alloy usage in the nuclear industry have come to light based on recent experience with CANDU (a trademark of Atomic Energy of Canada Limited) reactors and will be addressed as part of this review.

The needs of the nuclear industry differ from those of the aerospace industry primarily because of the operation in a water, He, molten salt, or liquid metal coolant environment, but both industries share common requirements for high strength and corrosion resistance. For the aerospace industry, high-temperature strength and creep resistance are important properties for Ni alloys. Operation in an aqueous or moist environment is a requirement for most nuclear applications and therefore resistance to corrosion, SCC, and irradiation-assisted SCC (IASCC) is also important [2].

With respect to the alloys listed in Tables 9.1 and 9.2 there are three common strengthening mechanisms: (i) solid solution hardening (Cr, Fe, Al, Nb, Ti, Mo); (ii) coherent precipitate hardening (mostly from Al, Ti, or Nb additions) forming γ' and γ''; (iii) incoherent precipitate strengthening (carbide phases $M_{23}C_6$, M_6C and

9.5 MECHANICAL PROPERTIES

MC (a carbide with M being a metal atom, often Cr), where M is Cr, Fe, Nb, Ti, and Mo). The contribution of each mechanism to the mechanical and creep strength of the Ni alloys used in nuclear reactors is dependent on the various thermomechanical treatments that have been developed to suit different applications [4,9–17].

For the precipitation-hardened (PH) alloys (like X-750 and 718) there is a wide range of properties that are sensitive to the size, number density, and spatial distribution of the precipitates. Carbides typically form during high-temperature heat treatments and are often segregated to grain boundaries [3,4]. The γ' and γ'' coherent precipitates form within grains and also at grain boundaries (Fig. 9.7). The intragranular precipitates, in particular,

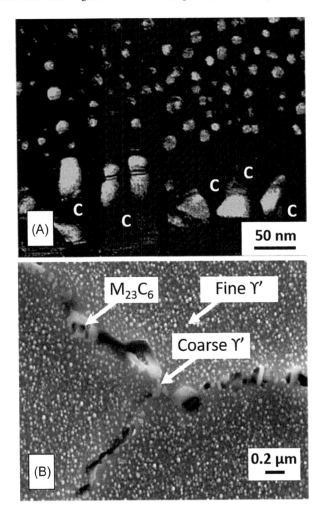

FIG. 9.7

Micrographs showing the fine intragranular γ' precipitates and coarser γ' at the grain boundary together with intergranular carbides [labeled C in (A) and the darker particles in (B)] in Inconel X-750 after precipitation hardening at about 715°C after solution annealing.

(A) Dark-field TEM micrograph modified from R. Bajaj, W.J. Mills, B.F. Kammenzind, M.G. Burke, Effects of neutron irradiation on deformation behaviour of Ni base fastener alloys, in Ninth International Symposium on Environmental Degradation of Materials in Nuclear Power Systems-Water Reactors, TMMMS, 1999, 1069; (B) SEM micrograph supplied by P. Morra. Modified from A. Strasser, F.P. Ford, High Strength Ni-alloys in Fuel Assemblies, ANT International, Molnlycke, 2012.

impart an important strengthening component; their size and density being controlled by aging treatments to give specific properties [1,3,4,9]. In the overaged condition, these precipitates lose their ordered structure and evolve into the larger η- and δ-phases. The γ′-phase, in particular, is responsible for an important property of Ni superalloys, that is, high yield and creep strength up to high temperatures. For some alloys, the yield strength actually increases with increasing temperature [3,31], see Fig. 9.2.

9.6 FRACTURE MODES

A common fracture mode in some Ni alloys is intergranular. This comes about partly because of carbide precipitation at the grain boundaries and also because zones free of γ′ and γ″ precipitates adjacent to grain boundaries are often produced as a result of different heat treatments [4] thereby creating a localized softer layer of material adjacent to the grain boundaries. Localized plasticity can then occur leading to a potential increase in the tendency for intergranular fracture [32,33]. One possible cause for γ′ and γ″ precipitate-free zones (PFZs) is the depletion of Ti, Al, and Nb in the vicinity of η- and δ-phases [4]. In the absence of η- and δ-phases, however, the γ′ and γ″ PFZs adjacent to grain boundaries can also arise because grain boundaries act as vacancy sinks thus lowering self-diffusion and the rate of precipitate formation immediately adjacent to the grain boundary sink [34,35].

The microstructure near grain boundaries is important for SCC and IASCC, either because of the variation in chemistry in the vicinity of the boundary or as a direct consequence of precipitate formation on the grain boundaries. Intergranular fracture in inert environments without embrittling species at grain boundaries takes the form of void nucleation caused by the elastic and plastic incompatibility between the hard grain boundary precipitate and the softer deforming matrix. During deformation the grain boundary precipitates promote micro-void formation and coalescence leading to eventual intergranular failure [35], Fig. 9.8. For many Ni-alloys, the grain boundary precipitates are often carbides that have been formed during high-temperature heat treatment [4]. Softer PFZs

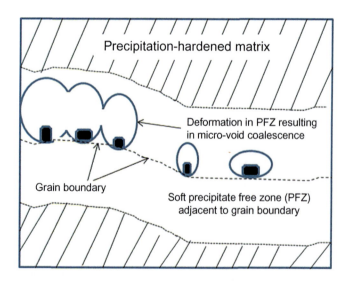

FIG. 9.8

Schematic diagram illustrating the mechanism for micro-void formation and coalescence involving precipitates segregated at boundaries with a precipitate-free zone (PFZ).

Based on the mechanism proposed by S.P. Lynch, Mechanisms of inter-granular fracture, Mater. Sci. Forum 46 (1989) 1–24, https://doi.org/10.4028/www.scientific.net/MSF.46.1.

adjacent to the grain boundaries can be produced when γ' and γ'' are formed at intermediate aging treatments. These precipitates harden the matrix while retaining a softer PFZ near the grain boundary [4,32,33].

Primary carbides, such as NbC and TiC, can be effective grain size refiners, but there is the possibility of promoting uneven grain sizes if the carbides are unevenly distributed during casting and rolling operations. This macroscopic heterogeneity can lead to orientation effects ("banding") on stress corrosion crack initiation and propagation [4,36]. Carbides may also occur in the form of $M_{23}C_6$ (a complex carbide with M often being Cr) particles, often Cr-rich, precipitated on the grain boundary during solution annealing or γ' and γ'' aging treatments [4]. Such carbide precipitation can lead to Cr depletion in the γ matrix adjacent to the $M_{23}C_6$ carbides and this may have a deleterious effect on the SCC resistance in some LWR environments [37–39]. Cr depletion also promotes localized intergranular attack (IGA). In order to minimize IGA and IGSCC of Ni-based alloys it is sometimes necessary to minimize Cr-depletion at grain boundaries. Thus, the strategy has been to extend the time of the heat treatment such that "healing" of the Cr-depleted zones adjacent to grain boundaries can occur. This "healing time" is a function of the temperature and degree of cold work, as well as the residual dissolved carbon content of the Ni-based alloy [37,38]. The effect of heat treatment on Cr depletion at grain boundaries is illustrated in Fig. 9.9 for solution-strengthened Inconel 600 [40].

Cr depletion adjacent to grain boundaries is often referred to as sensitization in that it renders a material susceptible to IGA. The increased susceptibility to IGA does not automatically lead to SCC susceptibility [41]. The IGA can be distinguished from SCC in that it does not depend on stress, although stress, by opening cracks, will naturally enhance IGA. The IGA is particularly important in moist environments exposed to ionizing radiation and made acidic by the radiolysis of gases such as nitrogen (air). The IGA failure due to nitric acid attack was a common failure mode in early designs of Inconel-600 flux detectors exposed to moist air above the moderator in early CANDU reactors [42], and also for other components operating in a moist air environment and in a high gamma radiation field in the CANDU moderator vaults. To mitigate the effects of IGA in the early CANDU reactors the air in the moderator vaults was dried using dehumidifiers. For flux detectors exposed to the moderator, mitigation involved introducing a He cover gas. Later designs of CANDU reactors employed encapsulated flux detector

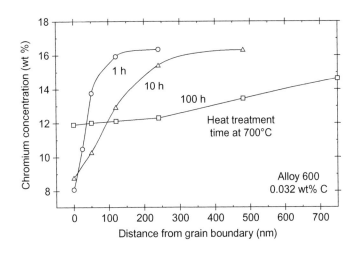

FIG. 9.9

Chromium depletion at grain boundaries of alloy 600 associated with $Cr_{23}C_6$ precipitation following various aging times at 700°C after solution annealing at 1100°C (2010°F).

From S.M. Bruemmer, G.S. Was, Microstructural and microchemical mechanisms controlling inter-granular stress corrosion cracking in light-water-reactor systems, J. Nucl. Mater. 457 (2015) 165–172.

assemblies; the Inconel 600 detectors were encapsulated within a guide tube that was back-filled with He to keep them dry and free from acid attack [43].

In the CANDU reactor, encapsulation only served to delay eventual failure. During service the He cover gas was gradually lost and replaced by air. This loss of He led to an increase in the operating temperature of the assemblies because of the reduction in nuclear heat removal rate as a result of the lower thermal conductivity of air compared with He. Embrittlement of the Inconel 600 sheathed lead wires and detector coils then occurred because of the combined effects of He generation and increased operating temperature that also promoted corrosion in the form of nitridation [43]. The coiled design of Inconel 600 flux detectors typically failed after 5–6 years of service in the CANDU reactor core even though they were operating in a stress-free, dry environment. Failure has been deemed to be the result of He embrittlement [43,44]. Helium embrittlement of Ni-alloys will be discussed in Section 9.14.

9.7 DEFORMATION MECHANISMS (YIELD STRESS AND CREEP STRENGTH)

One popular explanation for the positive temperature dependence of the YS in some superalloys is related to dislocation reactions that occur within the ordered γ'-phase. Gliding dislocations of the type $a/2 \langle 101 \rangle \{11\bar{1}\}$ sometimes act in pairs when cutting through γ' precipitates. The antiphase boundary (APB) created by the passage of a single $a/2 < 101 >$ dislocation on $\{11\bar{1}\}$ can be removed by the passage of a second "superpartial" dislocation. The creation of the APB increases the critical resolved shear stress for that slip system by about 400 MPa [3]. Passage of a second dislocation with the same Burgers vector and slip plane removes the APB thus allowing a slip band to cut through the precipitates with reduced energy. The paired superdislocations are either coupled strongly or weakly depending on the size and distribution of the γ' precipitates [3]. For $a/2 < 101 > \{11\bar{1}\}$ dislocations cutting through an ordered γ' precipitate, there is an increased propensity for cross-slip onto $\{010\}$ planes, forming Kear-Wilsdorf locks [3,45–48], as the temperature is increased. At temperatures above that corresponding to the peak in the YS (about 800°C) there is a preference for $a/2 < 101 >$ slip on the $\{010\}$ plane and the γ'-phase then acts like most other metal phases where softening occurs with increasing temperature [3], see Fig. 9.2.

The Kear–Wilsdorf mechanism is only applicable to the mechanical properties of the alloy in cases where the γ' precipitates are cut by dislocations; this is dependent on the precipitate size, spatial distribution, and volume fraction [3,49]. Read [3] has shown that for a given volume fraction of γ'-phase, the optimum hardening in the Ni-based superalloys occurs for a particle size which lies at the transition from weak to strong coupling between the paired $a/2 < 101 > \{11\bar{1}\}$ superdislocations that traverse the precipitates. For PE16, an optimum precipitate radius (r) was reported to be in the range 26–30 nm [3], although lower values are found for other alloys [49]. Precipitates larger than this optimum value are more likely to be bypassed by dislocations via the Orowan mechanism [3]. Other theories of dislocation interactions with precipitates consider single dislocations only. For single dislocations the tendency to cut or bypass a precipitate is dependent on the dislocation line tension, which is a function of the radius of curvature of the dislocation line. By considering the line tension, Brown has shown that cutting tends to be observed when precipitates sizes are small and Orowan looping tends to be observed when precipitates are large [50].

Although Del Valle et al. [49] concluded that the optimum radius for precipitate strengthening in Inconel X-750, and therefore the transition from precipitate cutting to Orowan looping, was about 15 nm, they also reported instances where this was not the case [49,51]. The two modes of dislocation-precipitate interaction are illustrated in Fig. 9.10A, [52]. Fig. 9.10B shows shearing of γ' precipitates in Inconel X-750 that has a radius of about 50 nm, while Fig. 9.10C shows Orowan looping for precipitates with radius of about 24 nm [49]. Del Valle et al. [49] attributed this departure from normal behavior as an anomalous observation in part because the plastic deformation was low (1.5%) and insufficient deformation had occurred to be representative of the most likely behavior. At higher strains (5%), Del Valle et al. [51] showed that dislocation tangles were normally found around γ' precipitates having an average radius of about 56 nm in Inconel X-750, Fig. 9.10D.

9.7 DEFORMATION MECHANISMS (YIELD STRESS AND CREEP STRENGTH)

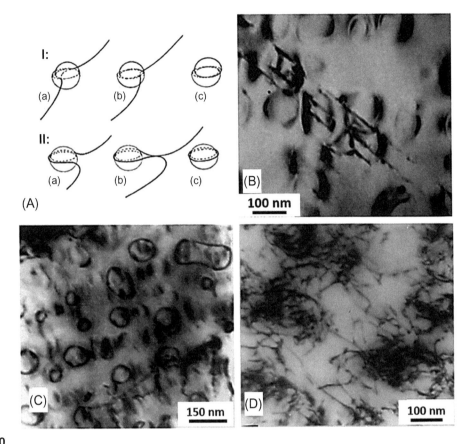

FIG. 9.10

Examples of dislocation-precipitate interactions for γ' precipitates in Inconel X-750: (A) schematic diagram showing two different modes of dislocation–precipitate interaction, I—cutting, II—looping; (B) TEM micrograph showing sheared precipitates (average radius = 50 nm) after 1.5% deformation; (C) TEM micrograph, illustrating Orowan loops around precipitates (average radius = 24 nm) after deformation; (D) TEM micrograph, illustrating dislocation tangles around precipitates (average radius = 56 nm) after 5% deformation.

Modified from J.A. Del Valle, A.C. Picasso, R. Romero, Work-hardening in Inconel X-750: study of stage II. Acta. Mater. 46(6) (1998) 1981–1988; J.A. Del Valle, R. Romero, A.C. Picasso, Bauschinger effect in age-hardened Inconel X-750 alloy, Mater. Sci. Eng. A311 (2001) 100–107; V. Gerold, Precipitation hardening, in: Dislocations in Solids, vol. 4, North-Holland Publishing Company, Amsterdam, NY, 1979.

Del Valle et al. [49] defined the critical particle radius for which the force required to shear the precipitate, by a single dislocation, is equal to the force required to loop it by the Orowan mechanism. They estimated that for Inconel X-750 this radius is 15 nm, that is, precipitates smaller than 15 nm radius are more likely to become sheared by the passage of a single dislocation while those with larger radius are more likely to be bypassed by the Orowan mechanism.

The line tension depends on the radius of curvature of the dislocation and therefore the force imposed on the precipitate by the dislocation will be higher when the precipitate is small and the radius of curvature of the dislocation looping around the precipitate is small. The transition from precipitate cutting to Orowan looping is also a function of the higher energy needed to cut through larger precipitates. For a given volume fraction of the precipitate phase, larger precipitates coincide with larger precipitate spacing that, combined with the larger radius of curvature at the precipitate, favors Orowan looping over cutting [50]. For the condition where Orowan looping

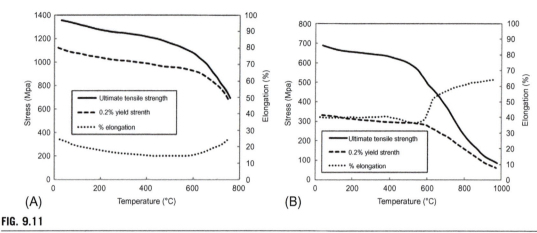

FIG. 9.11

Tensile properties as a function of temperature for: (A) Inconel 706. Solution treatment at 1700–1850°F (925–1010°C) for a time commensurate with sample size, air cool. Precipitation treatment at 1350°F (730°C) for 8h, furnace cool at 100°F (55°C) per hour to 1150°F (620°C) for 8h and the air-cooled [14]; (B) Inconel 600. Annealed at 1600°F (870°C) for 1h and then hot-rolled [10].

Data reproduced with permission from Special Metals Corporation.

dominates, the strengthening is determined by both precipitate size and spacing considerations as outlined by Bacon et al. [53].

For engineering alloys such as Inconel X-750 used in nuclear applications the Orowan mechanism appears to be prevalent, as evidenced by the dislocation tangles observed around γ' precipitates [33,49]. This is probably because the volume fraction of γ' in Inconel X-750 has been reported to be <10% [54] and such a low value would not favor cutting as the dominant mechanism. A low-volume fraction of γ' in Inconel X-750 precludes any anomalous temperature dependence (Fig. 9.2).

The mechanical properties of Ni alloys are largely dependent on solution strengthening, grain size refinement (controlled to some extent by heat treatments to form grain boundary carbides), and precipitation strengthening (primarily γ' and γ''). There are many thermomechanical process routes to tailor the properties for different applications and these are described in each of the pertinent technical specifications [10–17]. The alloying elements responsible for the γ and γ'-phase formation are Ti, Al, Nb, and Ta (Fig. 9.4). Alloys with appreciable concentrations of these elements are termed precipitation-hardenable alloys, for example, Inconel X-750, 718, 706. Alloys with low concentrations of the γ and γ'-phase forming elements, for example, Inconel 600 and 690 are solution-strengthened only. The tensile properties of typical precipitation-hardened alloy (alloy 706) are compared with a solution-treated alloy (alloy 600) in Fig. 9.11. It is clear that the precipitation-hardened alloy (706) has a higher strength but less ductility than the solution-annealed (SA) alloy (600).

9.8 STRESS CORROSION CRACKING

The SCC is the result of a combination of three factors: (i) a susceptible material, (ii) exposure to a corrosive environment, and (iii) tensile stresses above a threshold [2,4]. For Ni-based alloys in PWRs and BWRs, initiation of SCC can involve an incubation period that depends primarily on temperature, residual fabrication stress, and material susceptibility [55]. Concerning the environment, there is a maximum SCC susceptibility at a certain temperature and coolant hydrogen concentration [56,57]. The SCC and IASCC have been the subject of many prior reviews. For reference, comprehensive assessments and reviews on SCC and IASCC can be found in Refs. [1,2,4,36,58–70], and further overview information is also provided in Chapter 6.

Inconel 600 and 690 are the main alloys that have been used in SGs [70]. Initially the Inconel 600 tubes were supplied in mill-annealed (MA) condition but because of residual stresses induced by tube straightening that promoted SCC, a final heat treatment at 715°C for 5–12 h was carried out [55]. Although the heat treatment was initially designed to suppress residual stresses, it also promoted chromium carbide precipitation on grain boundaries with a corresponding Cr depletion adjacent to the boundaries. Inconel 690 has better high-temperature corrosion resistance than Inconel 600 because of the higher Cr content (Table 9.2), and will also be less susceptible to grain boundary sensitization for the same reason. Some studies indicate that Cr depletion does not have a strong effect on SCC in high-temperature oxygenated water for alloys such as Inconel X-750 [71] and this has been attributed to the balancing effect of Ti that tends to segregate to grain boundaries in these alloys [72], presumably because Ti competes with Cr to form carbides with the available carbon. Ti is not present in Inconel 600 and 690 and the effects of Cr depletion in promoting IGA and IGSCC may be more critical. Heat treatments to homogenize the Cr concentration in the vicinity of grain boundaries for alloys such as Inconel 600 and 690 are desirable because, for these alloys, Cr depletion at grain boundaries is known to be an issue for sensitization to IGA [37,40,73]. Alloy 800 is the preferred SG tube material for CANDU reactors and is also used extensively in the SGs of German-designed PWR systems. Degradation of alloy 800 SG tubing has only been found in a relatively small number of tubes at a limited number of stations despite the large number of SG tube operating years accumulated to date [73].

Although heat-to-heat variability overrides certain generic treatments to improve SCC susceptibility [72], it is widely accepted that, for Inconel X-750, a heat treatment that involves a two-stage high-temperature solution anneal (1093°C/1 h) followed by a lower temperature anneal (704°C/20 h) is good for promoting resistance to SCC. Using a lower temperature (885°C/24 h) for the high-temperature anneal condition results in poor SCC resistance [71].

Environments such as high-temperature water are very aggressive and will cause SCC of most materials. High yield strengths tend to promote SCC so radiation hardening is likely to enhance SCC and is a factor that is important in the susceptibility of a component to SCC in the reactor core [6,74,75]. There is some evidence for radiation-enhanced Ni segregation at grain boundaries, which also constitutes a depletion of Cr and other alloying elements, and this could therefore also enhance SCC [64]. Precipitation-hardened Ni alloys tend to become softer with irradiation dose after an initial transient increase caused by point defect clustering (see Section 9.12). It is difficult to reconcile the enhancement of SCC by irradiation due to irradiation hardening for precipitation-hardened Ni alloys for this reason.

The effect of high-temperature water on IGSCC in high Ni-based alloys has been characterized by Coriou et al. [76], Fig. 9.12. The IGSCC in high-temperature water is often referred to as primary water stress corrosion cracking (PWSCC). The SCC of Inconel 600 can be mitigated to some extent by a thermal treatment at about 700°C for >10 h after a high-temperature mill anneal (MA) solution treatment. An MA treatment typically occurs below the secondary carbon solvus (see Fig. 9.5), that is, between about 925°C and 1050°C and may therefore produce some secondary carbide precipitation. Inconel 690 was developed for SG tubes in the early 1980s as a material with excellent IGSCC resistance in PWR primary water. The thermal treatment chosen for Inconel 690 also consisted of heating at about 700°C for >10 h after mill annealing [1].

For Inconel X-750, the effects of heat-treatment conditions on PWSCC resistance are shown in Fig. 9.13. Precipitation hardening at about 715°C after solution annealing at a high temperature near 1075°C, the so-called high-temperature heat (HTH)-treatment condition, was selected for fabricating the most PWSCC resistant alloy [77].

9.9 NI ALLOYS FOR GENERATION-IV REACTORS

Material challenges for future Generation-IV reactor concepts have been described by Zinkle and Was [67], and a general overview is also provided in Chapter 2. A perspective on Ni-alloy usage in advanced reactors has been given by Rowcliffe et al. [29].

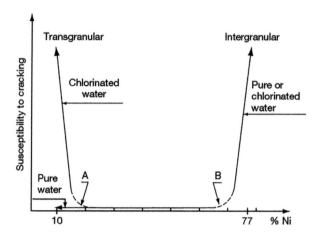

FIG. 9.12

Schematic diagram showing the influence of nickel content on the cracking processes occurring in 18% chromium austenitic alloys, when stressed slightly above the yield point in 350°C water (demineralized or containing 1 g/L of chloride ions) [76].

From T. Yonezawa, Nickel alloys: properties and characteristics, Comprehensive Nuclear Materials, Elsevier, Amsterdam, vol. 1 (2012) 233–266.

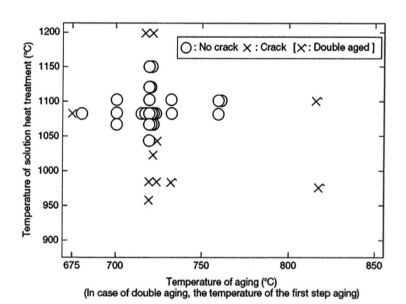

FIG. 9.13

Effect of heat-treatment condition on the stress corrosion cracking susceptibility of Inconel X-750 in high-temperature water [77].

From T. Yonezawa, Nickel alloys: properties and characteristics, Comprehensive Nuclear Materials, Elsevier, Amsterdam, vol. 1 (2012) 233–266.

9.9 NI ALLOYS FOR GENERATION-IV REACTORS

The research being conducted on materials for Generation-IV reactors has mostly focused on corrosion, SCC, and void swelling at high temperatures. Ni alloys are prime candidates for use in Generation-IV applications because of their resistance to corrosion at high temperatures and to swelling at intermediate Ni contents, Fig. 9.14 [78]. He production is not considered as important as corrosion and swelling resistance when choosing alloys for the supercritical water Generation-IV reactor because the thermal neutron flux and/or fluence is deemed low enough that He generation is not a major issue. However, there is a knowledge gap concerning He embrittlement of materials in the lower part of the temperature range that is important to Generation-IV reactor operation (300–550°C). Most data on materials properties at elevated temperatures (>500°C) has been obtained from fast reactors and is therefore not representative of Ni-alloy operation in a high thermal neutron flux, that is, high He generation rates.

Some water-cooled Generation-IV reactors span the range between 300°C and 600°C [67], where increased corrosion and swelling are the biggest concerns. Therefore, apart from having good high-temperature mechanical properties, having a high Cr content for improved corrosion resistance in water environments and an intermediate Ni content for reduced swelling are deemed to be important factors at this time (Table 9.6) [29]. There has not been much data on mechanical properties applicable to Generation-IV temperatures. One notable study targeting high-temperature Generation-IV applications is the work of Nanstad et al. [79]. They showed that alloys 617 and 800H irradiated in high flux isotope reactor (HFIR) to a dose of about 1.5 dpa exhibited significant increases in yield and ultimate tensile strength (UTS) at irradiation temperatures between 580°C and 700°C and test temperatures between 25°C and 700°C. The ductility for alloy 617 remained high (10%) for all conditions, but alloy 800H had very low (0.35%) ductility when irradiated and tested at about 700°C.

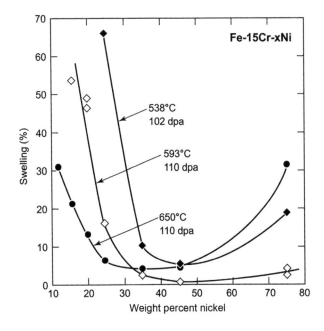

FIG. 9.14

Swelling behavior of Fe-Cr-Ni alloys after neutron irradiation to 100 dpa.

From F. A. Garner, H. R. Brager, Dependence of neutron induced swelling on composition in iron-based austenitic alloys, in: F.A. Garner, D.S. Gelles, F.W. Wiffen (Eds.), Optimizing Materials for Nuclear Applications, TMS-AIME, Warrendale, PA, 1985, 87–109.

Table 9.6 Nominal Compositions (wt%) of Commercial Ni-Based Alloys for Use in Generation-IV Reactors

Alloy	Ni	Fe	Cr	Mo	Nb	Co	W	Ti	Al	C
Incoloy 800H	33	45	21	–	–	–	–	0.4	0.4	0.08
Hastelloy HR120	37	33	25	–	0.7	3	–	–	0.1	0.05
Inconel 706	42	37	16	0.1	2.9	–	–	1.7	0.3	0.03
Nimonic PE16	43	33	17	3.7	–	–	–	1.2	1.3	0.05
Inconel 617	45	23	20	–	–	10	–	0.6	0.8	0.05
Hastelloy X	50	17	22	9	–	1	0.6	–	–	0.15
Inconel 718	53	19	18	2.5	5	–	–	0.9	0.6	0.04
Haynes 230	55	3	22	2	–	4	14	–	0.3	0.10

From A.F. Rowcliffe, L.K. Mansur, D.T. Hoelzer, R.K. Nanstad, Perspectives on radiation effects in nickel-base alloys for applications in advanced reactors, J. Nucl. Mater. 392 (2009) 341–352.

9.10 CHEMICAL COMPATIBILITY WITH COOLANTS

Ni alloys are relatively resistant to corrosion in hot water (~300°C) compared with other structural alloys, but water chemistry can be an issue if Cl is present [4]. They are used in light-water reactors (LWRs) because of their mechanical and corrosion properties [1,4], but they are mostly used in fuel assemblies where neutron exposure is short (up to 4–6 years), or in SGs outside of the reactor core. For heavy-water reactors, such as the CANDU reactor, Ni alloys are used as specialized components (springs primarily) that operate either in a dry gas environment or in colder (60–70°C) moderator. Under abnormal conditions, for example, operation in an irradiation field in a damp environment, Ni alloys such as Inconel X-750 are susceptible to nitric acid attack, the acid forming from radiolysis of moist air [80]. Inconel 600 and 690 are frequently used in SGs in LWR and heavy water reactor [70].

For candidate materials to be used in liquid metal reactors (LMRs) with various liquid metal coolants (Na, Pb, Bi, and Sn) the reader is referred to the work of Miller [81]. Can [82] has identified three main categories that require attention when it comes to selecting structural alloys for use in LMRs: (i) solution attack, (ii) intergranular penetration, and (iii) direct alloying reactions. Ni alloys are deemed to be poor performers in the presence of lead, bismuth, lead–bismuth eutectic (LBE), and tin. Precipitation-hardened Ni alloys such as Nimonic PE-16 and Inconel 706 have been used in sodium-cooled fast reactors in the United Kingdom and United States [83]. These alloys were chosen because they exhibited superior strength and creep resistance, as well as lower swelling, compared with other structural alloys such as stainless steels. However, even in a fast neutron environment, large quantities of He are generated from the high-energy (n,α) reactions [84], see Section 9.11, and these alloys suffer strongly from both helium embrittlement and phase instability [28,29].

The structural materials for the primary loop of molten salt reactors (MSRs) will be subject to extremely corrosive environments and high temperatures. The primary coolant operates between about 550°C (inlet) and 700°C (outlet) and historically the most suitable material for containment is the high nickel alloy Hastelloy N [85,86,86a]. The Ni-based Hastelloy N alloy is considered to be the primary option of metallic structural materials in MSR because it is chemically compatible with molten fluoride salts.

For high-temperature gas-cooled reactors (HTR or HTGR), two alloys proposed for internals, piping, and heat exchangers are Incoloy 800H and Inconel 617 [87]. Inconel 617 is better for use at the highest operation temperatures of 950°C and Incoloy 800H is better for use at temperatures up to 850°C. Incoloy 800H is code certified for temperatures up to 760°C for use in nuclear systems. A draft code case for alloy 617 has been developed [88] but alloy 800H remains the primary candidate for HTR applications. A substantial database has been developed for both alloy 800H and alloy 617. Hastelloy X was also considered for HTRs, but since the high-temperature scaling in Hastelloy X has not been acceptable, a modified version, Hastelloy XR, is being developed. Information about the properties of alloy 617 and alloy 800H can be found in Refs. [87,89]. In all of the studies on corrosion in high-temperature He the effect of impurities in the He has been the major concern. Other alloy component properties are also affected by thermal aging [90–92].

9.11 RADIATION DAMAGE AND GAS PRODUCTION IN NI ALLOYS

The effect of irradiation on Ni alloys in LWR and fast reactor environments has been extensively reviewed [28,29] and no attempt will be made to reproduce the same historic data. Rather, the focus will be on issues that have come to light based on recent experience with CANDU reactors. Unlike LWRs, the heavy-water moderated CANDU reactor operates with a neutron spectrum that is ideal for promoting transmutation effects with Ni [84,93–100]. The unique nuclear properties of Ni coupled with operation in a CANDU reactor environment have severely impacted Ni-alloy component integrity and creep performance in CANDU reactor cores [43,44,96,100].

The most important aspect of Ni-alloy usage in thermal nuclear reactor cores is the enhanced damage rates and gas production (He and H) that occurs because of the transmutation reactions in Ni, depending on the neutron spectrum and exposure time.

The most common isotope of nickel, ^{58}Ni, has a large (n,γ) reaction cross section with thermal neutrons. The ^{59}Ni that is produced, in turn, has large (n,γ), (n,p), and (n,α) reaction cross sections with thermal neutrons leading to enhanced atomic displacement damage and He and H gas production in the alloy. The exposure of an Ni-rich alloy to thermal neutrons will, in time, result in a substantial buildup of ^{59}Ni (a few atomic percent) that changes the overall radiation damage characteristics of the alloy.

Naturally occurring ^{59}Ni is a long-lived cosmogenic radionuclide with a half-life of 76,000 years. It is similar to ^{22}Na, which is also cosmogenic and also has a large (n,α) reaction cross section with thermal neutrons. However, it is dissimilar to ^{22}Na in that the precursor for the production by the (n,γ) reaction, ^{21}Na, is not a stable isotope. Unlike the precursor to ^{59}Ni (i.e., stable ^{58}Ni), there is no substantial reaction path for the production of ^{22}Na. Nickel can therefore be said to be unique among engineering alloying elements because the reactive ^{59}Ni is readily produced from the stable ^{58}Ni isotope (68% abundance) and this can result in the generation of large quantities of He and H in any Ni-rich component after only a few years of service in a power reactor core. The other main source of He in reactor core materials is boron. The ^{10}B isotope has one of the highest (n,α) reaction cross sections of all the elements and, in the absence of Ni, is the main source of He production in reactor core components. For most engineering alloys, boron is generally only present in concentrations ranging up to about 200 appm. Therefore, as the abundance of ^{10}B is ~20%, the maximum amount of He that can be produced in such cases is only about 40 appm. Nevertheless, such small quantities of boron can have deleterious effects on mechanical properties at high temperatures [74,101].

For high Ni-containing alloys in the core of LWRs, the generation of He will be dominated by ^{10}B for thermal neutron fluences up to about 10^{21} n.cm^{-2} [93], Fig. 9.15. So long as Ni-alloy components operate in a low thermal neutron flux, or for short residency times in the core, it is unlikely that large concentrations of He will be generated. Such reasoning was applied to the assessment of radiation damage for the Prometheus space reactor project. Angeliu et al. [30] showed that for the Prometheus fast reactor spectrum and expected neutron exposure, ^{59}Ni effects would be insignificant for the expected life of the reactor. As components are expected to operate at high temperatures, low B-Ni alloys (as well as Co alloys) were recommended for fabrication of the Prometheus pressure vessel [30]. For LWR reactor applications, Ni-alloy components in fuel assemblies [4] are only exposed to the reactor environment for relatively short periods of time (up to 4–6 years). In addition, because of the low thermal flux compared with a CANDU reactor, the He generation in the core of an LWR is relatively low during the first 4 years of exposure in the reactor (<300 appm He). By comparison, for Inconel X-750 spacers in the core of CANDU reactors, the amount of He generated after 20 years of service is >20,000 appm [43,44].

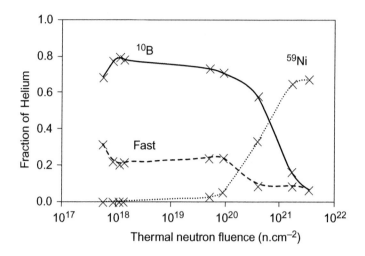

FIG. 9.15

Relative fraction of He produced by ^{10}B, fast neutrons, and ^{59}Ni as a function of thermal neutron fluence ($E<0.5\,eV$) for stainless steel and neutron spectra in the core of BWR reactors. The B concentrations in the samples examined ranged from about 1 to 4 wtppm.

From L.R. Greenwood, B.M. Oliver, Comparison of predicted and measured helium production in US BWR reactors, J. ASTM Int. (2005) JAI 3 (3), Art. no. JAI13490.

In order to understand the impact of the neutron spectrum on material degradation (in particular with respect to He generation) at relatively high doses one needs to consider how Ni transmutes. The two-stage transmutations of ^{58}Ni to ^{59}Ni with subsequent (n,γ), (n,p), and (n,α) reactions are illustrated below [84,93,94]:

$$^{58}Ni + n \rightarrow {}^{59}Ni + \gamma$$

$$^{59}Ni + n \rightarrow {}^{56}Fe + {}^{4}He$$

$$^{59}Ni + n \rightarrow {}^{59}Co + H$$

$$^{59}Ni + n \rightarrow {}^{60}Ni + \gamma$$

Naturally, the rate of gas atom production and the displacement damage created by the nuclear reactions with ^{59}Ni are dependent on the concentration of ^{59}Ni, which evolves during irradiation subject to gain from the ^{58}Ni(n, γ) reaction and loss from the ^{59}Ni(n, γ), ^{59}Ni(n, γ), and ^{59}Ni(n, γ) reactions. The time-dependent equations giving the concentration of ^{59}Ni and the concentration of H and He produced from the two-step reactions ^{58}Ni (n,γ)^{59}Ni(n,p)^{59}Co and ^{58}Ni(n,γ)^{59}Ni(n,α)^{56}Fe are given by [84,94]

$$\frac{N({}^{59}Ni)}{N_o({}^{58}Ni)} = \frac{\sigma_\gamma}{(\sigma_T - \sigma_\gamma)} \cdot (e^{-\sigma_\gamma \varphi t} - e^{-\sigma_T \varphi t})$$

$$\frac{N(H)}{N_o({}^{58}Ni)} = \frac{\sigma_p}{\sigma_T} + \frac{\sigma_p e^{-\sigma_\gamma \varphi t}}{(\sigma_\gamma - \sigma_T)} - \frac{\sigma_\gamma \sigma_p e^{-\sigma_\gamma \varphi t}}{(\sigma_\gamma - \sigma_T)\sigma_T}$$

$$\frac{N(He)}{N_o({}^{58}Ni)} = \frac{\sigma_\alpha}{\sigma_T} + \frac{\sigma_\alpha e^{-\sigma_\gamma \varphi t}}{(\sigma_\gamma - \sigma_T)} - \frac{\sigma_\gamma \sigma_\alpha e^{-\sigma_\gamma \varphi t}}{(\sigma_\gamma - \sigma_T)\sigma_T}$$

where:
$N(H)$ = hydrogen atoms produced.
$N(He)$ = helium atoms produced.

$N_o(^{58}Ni)$ = initial number of ^{58}Ni atoms.
σ_α = spectral-averaged $^{59}Ni(n,\alpha)$ cross section.
σ_p = spectral-averaged $^{59}Ni(n,p)$ cross section.
σ_T = spectral-averaged total absorption cross section of ^{59}Ni.
σ_γ = spectral-averaged $^{58}Ni(n,\gamma)$ cross section.
φ = total flux.
t = irradiation time.

Codes such as SPECTER [95] take neutron flux spectra as input and provide displacement damage rates for most elements. SPECTER also provides H and He gas atom production for most elements and spectral averaged cross sections for ^{58}Ni and ^{59}Ni that are used to calculate H and He gas atom production arising from the production of ^{59}Ni as a function of dose.

^{59}Ni is not a naturally occurring isotope and has to be produced. In a CANDU reactor, the concentration of ^{59}Ni reaches a maximum concentration of about 4% of the ^{58}Ni concentration after about 5–10 years of service depending on the location in the core [43,44,96]. The thermal neutron fluence needed to reach the peak concentration of ^{59}Ni is about 4×10^{22} n.cm^{-2} ($E < 0.5$ eV) and occurs between 5 and 10 years of full-power operation [43,44]. The ^{59}Ni (n,p) and ^{59}Ni (n,α) reactions are important because they extend across a wide range of energies. All naturally occurring isotopes of the main engineering alloying elements (Fe, Cr, and Ni) also have moderately high (n,p) and (n,α) reaction cross sections, but they are only significant at high neutron energies ($E > 1$ MeV); such high energy neutrons are infrequent in the heavily moderated CANDU reactor neutron spectrum. In contrast, the (n,p) and (n,α) reaction cross sections for ^{59}Ni are relatively high down to thermal energies ($E < 0.5$ eV). The (n,γ), (n,p), and (n,α) reaction cross sections for the peak ^{59}Ni concentration in a CANDU reactor core (about 4 at.%) are compared with those for the main naturally occurring isotope in Ni, ^{58}Ni in Fig. 9.16. The cross sections have been scaled based on atomic abundance in each case. For ^{59}Ni, the cross sections have been scaled by a factor of 0.04, that is, corresponding to the peak concentration of ^{59}Ni relative to ^{58}Ni occurring after 5–10 years of full-power service in a CANDU reactor [43,44].

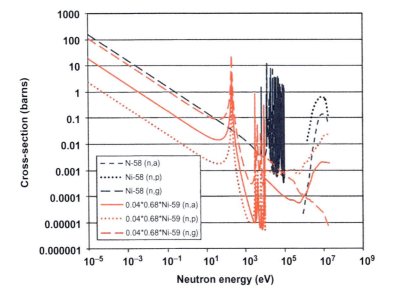

FIG. 9.16

^{59}Ni and ^{58}Ni (n,γ), (n,p), and (n,α) reaction cross sections as a function of neutron energy. The ^{59}Ni cross sections have been scaled by a factor of 0.04, which is the ^{59}Ni content relative to the parent ^{58}Ni after about 5 years of operation in a CANDU reactor.

The ^{59}Ni (n,γ), (n,p), and (n,α) reactions produce about 5, 222, and 1762 atomic displacements, respectively, for each reaction. The largest effect comes from the recoil induced in the emitting atom but these numbers also include the additional atomic displacements arising directly from the ejected particle [84,94]. Codes that calculate the atomic displacement damage and gas production use cross-section data for all the naturally occurring isotopes. The resultant damage cross section can be obtained from cross section libraries such as the one at the Los Alamos National laboratory (LANL) [102]. The LANL displacement cross sections therefore represent the probability of creating an atomic displacement for each Ni atom. Once the ^{59}Ni concentration has reached a maximum value the damage produced per thermal neutrons ($E < 0.5 \text{ eV}$) from the ^{59}Ni (n,α) reaction can exceed the damage production by a fast neutron ($E > 1 \text{ MeV}$), see Fig. 9.17. The damage cross section for ^{59}Ni (n,α) in Fig. 9.17 is represented on a "per Ni atom basis" and is therefore scaled to account for the abundance of ^{59}Ni and ^{58}Ni (0.04 × 0.68) and then multiplied by 1762 to give a damage cross section per Ni atom at the peak of the ^{59}Ni production.

The spectra for fast and thermal reactors are also illustrated in Fig. 9.17 to show the relative importance of the different cross sections when weighted by the neutron spectrum. When comparing with various spectra it is clear that the ^{59}Ni (n,α) reaction gives a substantial contribution to atomic displacement damage in the CANDU core, more so than for the BWR and LWR core. In terms of damage production, the ^{59}Ni reactions are important in a thermalized neutron flux, both because of the large number of displacements per event, and because the ^{59}Ni (n,α) and (n,p) cross sections are large at low neutron energies (Fig. 9.18).

It is clear from Fig. 9.17 that the ^{59}Ni effect will be the most significant for the CANDU reactor because of the high thermal neutron flux. This, coupled with the fact that most Ni-alloy components within the core of LWR reactors are in fuel assemblies that only reside in the core for up to about 4–6 years, means that the ^{59}Ni effect is generally not seen as large a concern for LWR reactors as it is for CANDU reactors. For different spectra one can generate the dpa, He

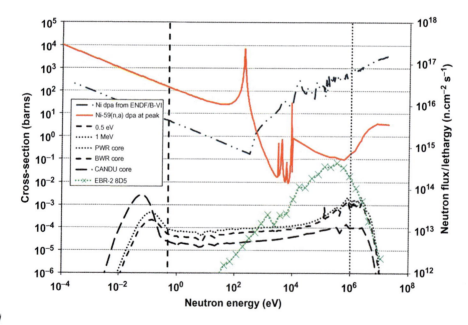

FIG. 9.17

Atomic displacement cross sections for Ni (left hand scale) and neutron spectra for PWR, BWR, CANDU, and EBR-II reactors (right hand scale). The displacement cross sections from LANL includes all naturally occurring isotopes and are compared with displacement cross sections due to the ^{59}Ni (n,α) reaction as a function of neutron energy. The ^{59}Ni values are scaled by a factor of 0.04 × 0.68 to take into account the ^{59}Ni abundance relative to all the Ni isotopes (100%).

9.11 RADIATION DAMAGE AND GAS PRODUCTION IN NI ALLOYS

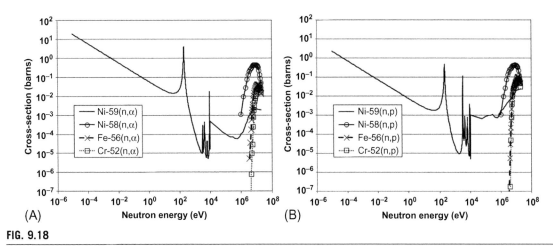

FIG. 9.18

(A) (n,α), and (B) (n,p) reaction cross sections as a function of neutron energy for the most abundant isotopes in Ni alloys containing Ni, Cr, and Fe (^{58}Ni, ^{56}Fe, and ^{52}Cr), scaled by atomic abundance (^{58}Ni = 0.68, ^{56}Fe = 0.92, ^{52}Cr = 0.84). The (n,α) and (n,p) cross sections for ^{59}Ni are also shown when the ^{59}Ni content relative to ^{58}Ni is about 4%, that is, the cross sections have been scaled by an abundance factor of 0.04 × 0.68 for a thermal neutron dose ~4 × 10^{22} n.cm^{-2} ($E < 0.5$ eV) in a CANDU reactor.

and H production as a function of time using the SPECTER code developed by Greenwood and Smither [95]. The spectra differ not only from reactor to reactor but also within a given reactor. Ni-alloy components are expected to reside in different locations within a CANDU reactor core and reflector regions over the life of the reactor. The He (and possibly H) production and the enhanced damage rates have therefore been problematic for lifetime operation in CANDU reactors. The CANDU reactor is especially prone to the effect of ^{59}Ni also because the thermal neutron flux is high everywhere in the reactor. In fact the thermal neutron flux can be higher in some peripheral regions, where Ni-alloy components operate, when compared with the core [44]. For fast reactors, with low thermal neutron fluxes, or when the neutron exposure is low so that there is little ^{59}Ni buildup, the main source of He and H production still comes from Ni, that is, ^{58}Ni, due to the high energy ^{58}Ni (n,α) and ^{58}Ni (n,p) reactions. Fig. 9.18 shows the (n,p) and (n,α) cross sections for the most abundant isotopes and scaled by the atomic abundance. This shows that even in fast neutron spectra high Ni alloys are still susceptible to the effects of high levels of H and He production.

The effect of neutron spectrum in CANDU reactors has been the focus of intense study in recent years because of the effect on stress relaxation of Inconel X-750 tensioning springs for reactivity mechanism guide tubes [96] and also because of the effect on mechanical properties of fuel channel spacers [44,100]. Fig. 9.19A illustrates the location and function of Inconel X-750 spring components in CANDU reactor cores. The corresponding neutron spectra are shown in Fig. 9.19B. Annulus spacers are used in CANDU reactor fuel channels to separate a hot pressure tube from a cold calandria tube and thus maintain an insulating gas gap [43,100], see Fig. 9.19A. The spacers have a dual role: (i) they must remain in position on the pressure tube; (ii) they must support the pressure tube and prevent contact between the pressure tube and the calandria tube. Irradiation can affect the spacer performance through stress relaxation, swelling, and embrittlement [100]. Radiation-enhanced stress relaxation has been a particular issue for tensioning spring components in the reflector region of CANDU reactors. For LWR reactors the thermal flux in the periphery is low and the ^{59}Ni effect is not as much of a concern in that region of the reactor. Neutron spectra for CANDU and PWR reactors are compared in Fig. 9.20. The effects on He production in the core and periphery regions of PWR and CANDU reactors have been determined using the SPECTER code provided by Greenwood and Smither [95] and the results are plotted in Fig. 9.21.

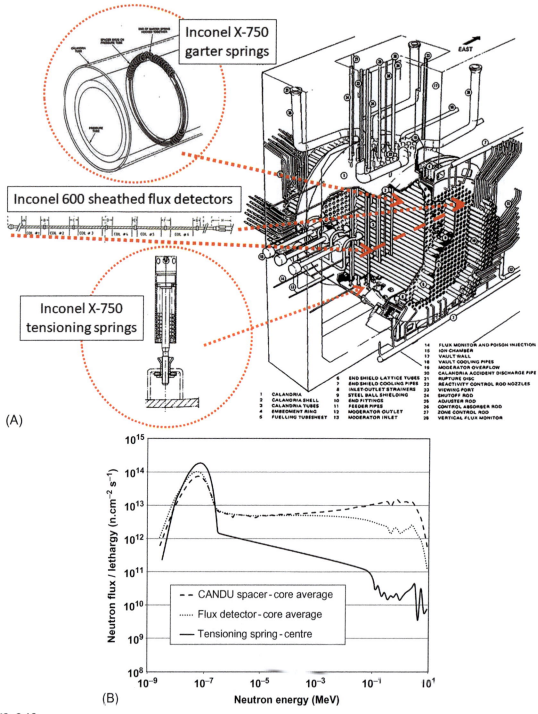

FIG. 9.19

(A) Schematic diagram illustrating location of Ni-alloy components in a CANDU reactor. (B) Neutron spectra corresponding to the three Ni-alloy components identified in (A).

9.11 RADIATION DAMAGE AND GAS PRODUCTION IN NI ALLOYS

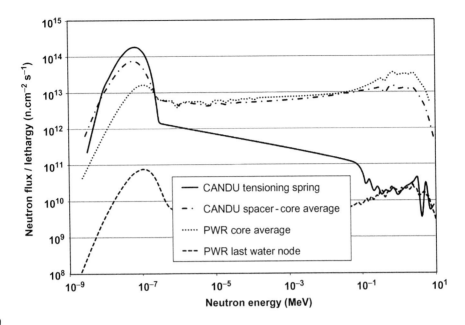

FIG. 9.20

Comparison of neutron spectra in a CANDU reactor compared with a PWR reactor. Core-average values are shown and are compared with equivalent places near the boundary of the containment vessel. Note that for CANDU case moving from the core to the periphery reduces the fast, but not thermal, flux.

Ni-alloy components in the periphery regions of the core of BWR and PWR reactors are expected to maintain their integrity for the lifetime of the reactor and their performance is determined to a large extent by the neutron spectrum. The LWRs tend to have lower thermal neutron fluxes compared with CANDU reactors in the periphery of the core because of the shorter path length through water and increased thermal neutron absorption for light water compared with heavy water.

Ni-alloy components in CANDU reactors are subject to very high He production and displacement damage rates even in the periphery region because of the effect of ^{59}Ni. As the specialized Ni-alloy core components in a CANDU reactor are meant to reside in the CANDU core for longer periods than Ni-alloy components in fuel assemblies in LWRs, the effects of high irradiation doses are turning out to be significant. However, some Ni-alloy components in LWRs (e.g., jet pump beams and tie rods in BWRs) occasionally require weld repairs and the effect of even a small amount of He production can be very important.

One concern with maintenance of BWR and PWR reactors is the effect of He on weldability. Metals are considered to be weldable if they have He contents <1 appm [18]. In order to assess the likelihood of core components being susceptible to He-induced weld cracking [25–27], studies have considered the contribution to He production from different sources. One such study by Greenwood and Oliver [93] showed that He is produced primarily not only from the ^{10}B(n,α) reaction at low thermal neutron doses but also from high neutron energy M(n,α) reactions (where M is a major alloying element, primarily Fe, Cr, and Ni), see Fig. 9.18. For stainless steels containing 11.2 wt% Ni, Greenwood and Oliver showed that the production of He from Ni exceeds that from other sources when the thermal neutron fluence ($E < 0.5$ eV) exceeds 10^{21} n.cm^{-2}, Fig. 9.15. Of course this threshold is reduced proportionally, that is, to about 10^{20} n.cm^{-2} for Inconel X-750, for example, for Ni alloys with higher Ni contents.

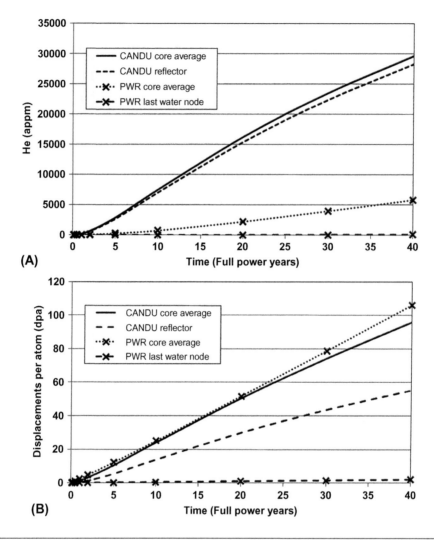

FIG. 9.21

Comparison of (A) He production, and (B) displacements per atom (dpa) production, for Inconel X-750 in a CANDU compared with a PWR reactor core and periphery regions. The He production in a PWR reactor core is relatively small compared with a CANDU core and negligible in the region adjacent to the pressure vessel wall (last water node).

9.12 RADIATION HARDENING/SOFTENING AND LOSS OF DUCTILITY IN NI ALLOYS

Embrittlement of reactor alloys has been attributed to a number of different mechanisms. For most materials hardening of the matrix by radiation damage acts in a manner similar to work hardening and the reduction in ductility (embrittlement) is generally associated with an increase in yield strength of the material [28,33,67]. In cases where embrittlement is characterized by intergranular failure a number of possible mechanisms have been proposed. One involves strain localization leading to dislocation pileup at precipitate grain boundary interfaces promoting localized strain-induced void formation and eventual cracking following the Stroh mechanism, for example, [33]. Strain localization in itself is not sufficient to promote intergranular failure; there are many examples where strain localization occurs in irradiated alloys (e.g., Zr alloys) without intergranular failure [103]. The main difference between Zr alloys and Ni alloys, with carbides decorating the grain boundaries, is that slip bands intersecting carbides cannot propagate into adjacent grains as in Zr alloys. Hard carbides may therefore promote pileup and Stroh-type cracking. Softer PFZs have been linked to intergranular failure in unirradiated Inconel X-750 [32,33] and there may be a similar effect when defect-free zones adjacent to boundaries are found in irradiated samples [104,105]. When it comes to identifying radiation-induced changes that promote intergranular embrittlement, however, there are clear cases where failure not only occurs at reduced strain but also at reduced stress. In conventional radiation hardening, strain localization results in softening after yielding but the YS and ultimate tensile stress are still generally higher than that in the unirradiated state [28,33]. When failure occurs without hardening other mechanisms need to be considered. Apart from strain localization the explanations for intergranular embrittlement fall into three main categories: (i) He segregation at grain boundaries; (ii) precipitate formation at grain boundaries; and (iii) denudation of precipitates and point defect clusters adjacent to grain boundaries.

The brittle failure of a component in a radiation field is often referred to as irradiation assisted stress corrosion cracking (IASCC). However, in many cases the irradiation-induced degradation of a core component results in reduced strength and ductility even when tested in dry conditions and at low stresses, that is, not SCC. In such circumstances the material degradation is better described simply as irradiation embrittlement. By far the most significant factor leading to loss of strength and embrittlement of irradiated Ni-containing reactor core components is the accumulation of He with increasing dose. Because of the unique characteristics of the nuclear reactions of Ni, producing He and H, Ni alloys are most prone to He embrittlement and for this reason He embrittlement will be described and discussed in detail (see Section 9.14).

Ni-rich alloys are used in fuel assemblies in PWR and BWR reactors but as they are in the core for relatively short periods of time (up to 4–6 years) the effect of irradiation is not as severe as in components that are meant to reside in the core for the life of the reactor. Some components cannot be replaced easily but others, for example, flux detectors in CANDU reactors and flux thimbles in PWR and BWR reactors, are expected to remain in the core as long as they can function. Although these components may operate in a relatively stress-free environment they can fail if subject to thermal stresses during reactor power transients or abnormal loading [43,44,106]. Flux detector assemblies made from Inconel 625 (in PWR reactors) or Inconel 600 (in CANDU reactors) containing approximately 70 wt% Ni are severely embrittled after only a few years of service [43]. In the absence of evidence for H retention in CANDU reactor components [100] it is highly likely that the observed embrittlement is the result of operating in a high thermal neutron flux, thus inducing high concentrations of He [43,44]. The embrittling effects of He are much reduced in LWRs compared with PHWRs (CANDU) because the thermal flux is substantially lower (Figs. 9.17, 9.20, and 9.21). Radiation affects the properties of a material in different ways than thermomechanical treatments do. Irradiation effects on the strength and ductility of Ni-alloys irradiated in different reactors or irradiation facilities will now be described and discussed. The description will mostly focus on the effects of irradiation on Inconel X-750 and Inconel 718 (two very similar precipitation-hardened alloys, except that 718 contains precipitates of γ'' in addition to γ') as data for these alloys spans a range of irradiation conditions and reactor types.

9.12.1 CANDU REACTORS

In recent years, evidence has been accumulating that Inconel X-750 spacers operating in CANDU reactors are degrading with operating time [43,44,100,104,105]. The characteristics of the degradation are reduced strength and ductility. When failure occurs in crush tests at slow strain rates (about 10^{-4} s^{-1}), it is intergranular in nature (Fig. 9.22).

For Inconel X-750 spacers and Inconel 600 flux detectors in CANDU reactors, perhaps one of the most important features of the degradation is that the reduction in failure load and ductility in room temperature tests

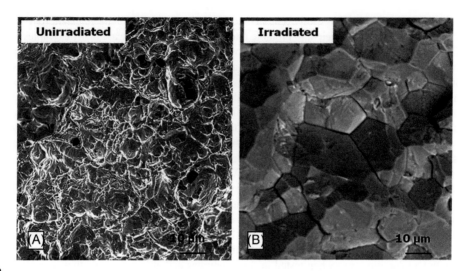

FIG. 9.22

Comparison of fracture surfaces for unirradiated and irradiated (to a dose of 23 dpa at >300°C) spacers after mechanical testing to breakage at room temperature. The unirradiated sample (A) exhibits ductile failure compared with the irradiated sample (B) that exhibits brittle intergranular failure when tested at room temperature after irradiation.

From M. Griffiths, The effect of irradiation on Ni-containing components in CANDU reactor cores: a review, AECL Nucl. Rev. 2(1) (2013) 1–16, erratum AECL Nucl. Rev. 3(2) (2014) 89, reproduced with permission.

are not as severe as for material irradiated at low temperatures. When tested after service, parts operating at temperatures <200°C exhibit higher failure loads and ductility compared with the parts operating at higher temperatures (>300°C) [100]. Concerning the effect of cavities and other point defect clusters on yield strength, one would anticipate a different yield strength at the lower temperature because the cavities and other defect clusters (dislocation loops and small precipitates) are smaller compared with those in the higher irradiation temperature material [100,104,105,107]. In addition, with the exception of cavities, the defect cluster density appears to be lower at the lower temperature [107], Fig. 9.23. Although He generation is the main driver for the cavity formation, having smaller point defect clusters with lower densities at the lower temperature can be understood simply based on the steady-state point defect concentrations assuming that the damage evolution is dictated by the net accumulation of freely migrating point defects and any in-cascade defect cluster production is negligible for the range of neutron energies applicable to the CANDU reactor. For a vacancy migration energy of 1.38 eV [108], the transition from recombination-dominated to sink-dominated regimes occurs between 200°C and 300°C for the applicable damage rates and sink densities (see Section 9.15). The lower freely migrating point defect flux at the lower irradiation temperature reduces the rate of radiation damage evolution and the damage density. Transmission electron microscope (TEM) observations comparing the two irradiation temperature conditions show that the network dislocation structure is relatively clean at the lower temperature but is decorated by point defect clusters at the higher temperature, Fig. 9.23 [107]. The cavity size in the matrix is smaller and has a higher density at the lower temperature, Fig. 9.24, [44]. At higher temperatures an inhomogeneous distribution of larger cavities is apparent within the matrix, Fig. 9.24, and at grain boundaries, Fig. 9.25 [44,105,107].

The insets in Fig. 9.23 show that the γ' precipitates, which are dissolved or dispersed during irradiation, are more diffuse after irradiation at the lower temperature, consistent with the results of Nelson et al. [109]. There is also a temperature dependence on the disordering of the γ' during irradiation with a higher degree of disorder at lower temperatures [105,110]. The apparent increased dissolution at lower temperatures shown in Fig. 9.23 [107] is consistent with the work of Nelson et al. [109]. Using parameters applicable to Ni alloys, such as PE16, Nelson

9.12 RADIATION HARDENING, SOFTENING

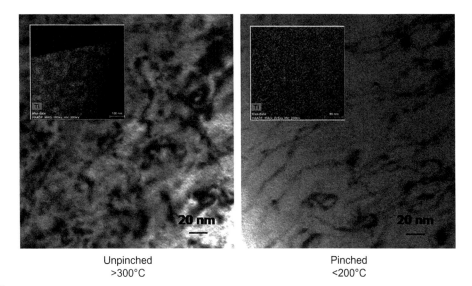

FIG. 9.23

TEM micrographs showing dislocation structure in Inconel X-750 spacer material from pinched (<200°C), and unpinched (>300°C), sections after irradiation to about 55 dpa and 18,000 appm He. The insets are lower magnification EELS maps showing distribution of Ti in each sample. Diffracting vector = (220).

Modified from H.K. Zhang, Z. Yao, G. Morin, M. Griffiths, TEM characterisation of neutron irradiated CANDU spacer material, Inconel X-750, J. Nucl. Mater. 451 (2014) 88–96. Insets provided by Z. Yao, Queens University.

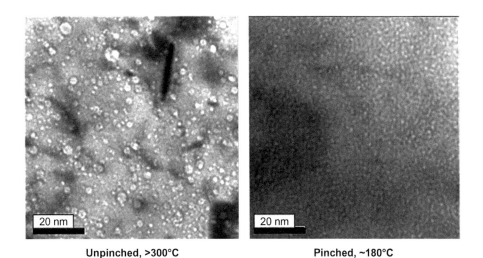

FIG. 9.24

TEM micrographs showing cavity structure in Inconel X-750 spacer material from unpinched (about 300°C) and pinched (about 200°C) sections after irradiation to about 55 dpa/18,000 appm He and 45 dpa/15,000 appm He, respectively.

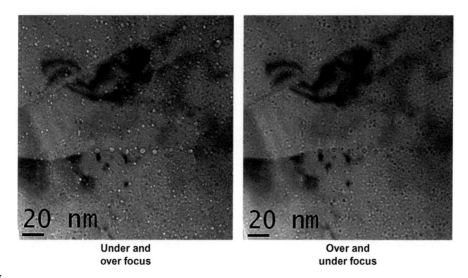

FIG. 9.25

TEM micrographs from a through focus series illustrating the cavity structure near grain boundaries in Inconel X-750 spacer material from unpinched material (irradiated at about 300°C) after irradiation to about 55 dpa/ 18,000 appm He.

et al. [109] showed that, for a displacement damage rate of 10^{-6} dpa s^{-1}, the steady-state radius of Υ' precipitates was vanishingly small at 200°C, but was between 20 and 120 nm at 300°C, depending on the initial precipitate density [109]. They argued that the dissolution and dispersion of the Υ' precipitates was a balance between radiation-induced mixing and a diffusional-driven transformation back to the quasiequilibrium state. Nelson et al. [109] deduced that there was a critical temperature for a given damage rate above which diffusion would start to dominate and reform the dispersed precipitates. At this critical temperature the radiation-enhanced diffusion coefficient would be just sufficient to allow the diffusion of solute atoms to reform the precipitate to balance the loss by radiation-induced dissolution. For a damage rate of 10^{-6} dpa s^{-1} their calculations showed that this critical temperature was somewhere between 200°C and 300°C [109] and is consistent with observations after ion irradiation (see Section 9.12.5).

Mechanical test results on spacers show that there is no positive correlation between the matrix yield strength determined from nanoscale three-point-bend tests (lower for the lower temperature), Fig. 9.26 [111], and the failure load in crush tests of the bulk spring sections (higher for the lower temperature), Fig. 9.27 [111]. Therefore, one needs to consider that the yield behavior is a function of the matrix properties and the ultimate intergranular failure load is dependent on the grain boundary strength. In the absence of any other evidence to the contrary it is likely that the main reason for the loss of UTS and ductility in Ni-alloy components operating in CANDU reactors is because of the accumulation of cavities at grain boundaries. This assertion is supported by observations showing that cracking proceeds along the boundary of the grains, whether this is an α/α boundary or the boundary with a second-phase particle; the cracks propagate along the boundary progressing between the cavities (Fig. 9.28 [105]).

9.12.2 LWR REACTORS

The combination of high strength and corrosion resistance has led to Ni alloys being widely used in BWRs and PWRs for reactor internal structural applications including fuel assembly hold-down springs, control rod guide tube support pins, jet pump beams, tie-rods, and core internal bolting. Some of the Ni-alloy reactor

9.12 RADIATION HARDENING, SOFTENING

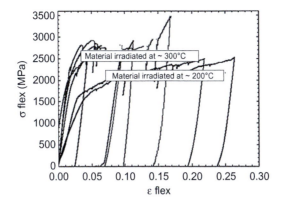

FIG. 9.26

Flexural stress versus strain curves at room temperature for Inconel X-750 spacer material after irradiation to about 55 dpa and 18,000 appm He as a function of the irradiation temperature. Data for specimens operating at higher temperatures (about 300°C) are shown in the top set of curves. Data for specimens operating at lower temperatures (about 200°C) are shown in the bottom set of curves.

Modified from C. Howard, S. Parker, D. Poff, C. Judge, M. Griffiths, P. Hosemann, Characterization of neutron irradiated CANDU-6 Inconel X-750 garter springs via lift out three point bend tests, in: Proceedings of HOTLAB conference, Leuven, Belgium, 2015.

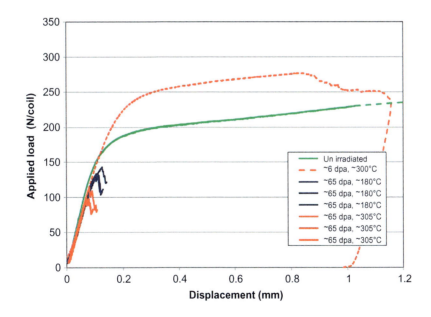

FIG. 9.27

Load-displacement curves at room temperature for Inconel X-750 spacers as a function of irradiation temperature and displacement damage dose (dpa). The He/dpa ratio is approximately 300 appm He/dpa.

Modified from C. Howard, S. Parker, D. Poff, C. Judge, M. Griffiths, P. Hosemann, Characterization of neutron irradiated CANDU-6 Inconel X-750 garter springs via lift out three point bend tests, in: Proceedings of HOTLAB conference, Leuven, Belgium, 2015.

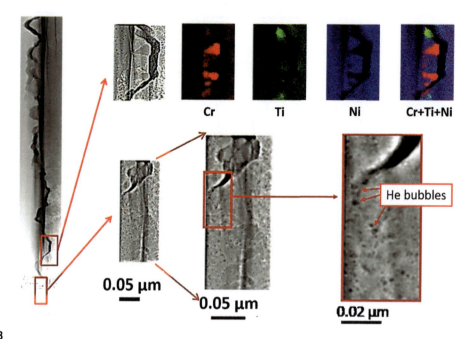

FIG. 9.28

HAADF and EELS spectra showing the passage of a crack along a grain boundary in Inconel X-750 spacer material irradiated to ~55 dpa and ~1800 appm He at about 300°C. The crack follows the α/α interface and the interface with precipitates (mostly carbides) residing on the boundary.

Modified from C.D. Judge, N. Gauquelin, L. Walters, M. Wright, J.I. Cole, J. Madden, G.A. Botton, M. Griffiths, Inter-granular fracture in irradiated Inconel X-750 containing very high concentrations of helium and hydrogen, J. Nucl. Mater. 457 (2015) 165–172, reproduced with permission.

components subjected to high stresses during service have experienced premature failures, mostly linked to SCC (see Section 9.8).

Work on the performance of Ni alloys in BWR and PWR environments has primarily focused on SCC and IASCC. One such study involved testing of internally stressed tubes made from Inconel 625, X-750, and 718 in test assemblies inserted in operating PWR and BWR reactors [112–114]. The results of these tests demonstrated the propensity for SCC failure of these alloys with different metallurgical states compared with some stainless steels. Garzarolli et al. [113] concluded that "Inconel 718 with the proper heat treatment and high purity AISI (American Iron and Steel Institute) 348 behaved well in the environments studied."

The TEM examinations and hardness testing on the Inconel 718 irradiated with neutrons at 288°C from the Garzarolli study [113] have provided information on the changes on the γ' and γ'' precipitate structure related to the changes in yield strength [110,115]. Thomas and Bruemmer [115] showed that the hardness of Inconel 718 after 3.5 dpa at 288°C was similar to the unirradiated state and there was marked softening for a dose of 20 dpa (Fig. 9.29). They showed that the γ' precipitates disordered at low doses and both the γ' and γ'' precipitates dissolved and dispersed with increasing dose (Fig. 9.30). As disordering was mostly complete at doses <3.5 dpa the softening with increasing dose (up to 20 dpa) can be mostly attributed to the dissolution and dispersion of the precipitates rather than any effect of disordering. Qualitatively similar results were obtained at lower neutron irradiation temperatures by Byun and Farrell [116]. Byun and Farrell observed progressive radiation hardening in solution annealed Inconel 718 over a dose range up to 1 dpa. Precipitation-hardened Inconel 718, on the other hand, showed little change in yield strength and some reduction in UTS up to 1 dpa.

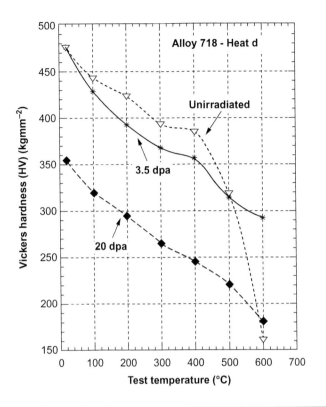

FIG. 9.29

Variation with temperature of the hardness measured in precipitation hardened 718 at different doses (0, 3.5, and 20 dpa). Note that at a given temperature the 20-dpa irradiated material is softer than the 3.5-dpa irradiated material.

From L.E. Thomas, S.M. Bruemmer, Radiation-induced microstructural evolution and phase stability in Ni-base alloy 718, in: Eighth International Symposium on Environmental Degradation of Materials in Nuclear Power Systems, ANS, 1997, 772.

9.12.3 FAST REACTORS

A significant amount of work has been reported on fast reactor irradiations of Ni alloys. Much of this has been described in the recent review by Boothby [28]. Test data exist for various irradiation and test temperatures confounding a comparison of the effects of irradiation on material properties. Most of the published data are for room temperature tests. Therefore, for simplicity, the effect of irradiation on mechanical properties in a fast reactor environment will be limited to room temperature tests.

Ward [117] carried out tensile tests on alloy 718 before and after irradiation in experimental breeder reactor 2 (EBR-II). The samples were irradiated at temperatures between 391°C and 649°C at different locations to doses between 5.5 and 6.49×10^{25} n m^{-2} ($E > 1$ MeV) and the tensile tests were carried out at different temperatures between 22°C and 649°C. Not surprisingly the test results showed an increase in strength with irradiation but there was insufficient information to make any assessment of trends in mechanical properties with increasing dose. Mills [118] examined the effect of irradiation on the tensile properties and fracture toughness of 718 plate. The material was examined in two heat-treated conditions and was irradiated in EBR II at temperatures between 400°C and 430°C to doses between 1.5 and 28 dpa. Mills concluded that the fracture toughness for the specimens at doses between 8 and 28 dpa was independent of dose. Any embrittlement caused by the microstructural changes from the irradiation was considered to have saturated by 8 dpa. The tensile data showed that the YS increases to a maximum value at

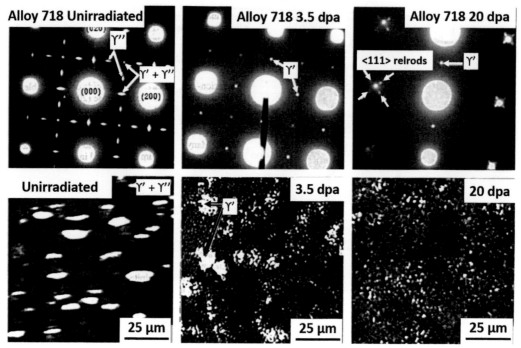

FIG. 9.30

Diffraction patterns and (100) dark-field micrographs showing irradiation-induced dissolution of γ' and γ'' and re-precipitation of γ'-phase in Inconel 718 after neutron irradiation at 288°C. The original γ' particles began breaking up at 3.5 dpa and were not discernible at 20 dpa.

From L.E. Thomas, S.M. Bruemmer, Radiation-induced microstructural evolution and phase stability in Ni-base alloy 718, in: Eighth International Symposium on Environmental Degradation of Materials in Nuclear Power Systems, ANS, 1997, 772.

doses between 1.5 and 7.0 dpa after which it decreases. The UTS showed a maximum at ~1.5 dpa after which it decreased with increasing dose.

Mills and Mastel [33], working on Inconel X-750 irradiated in EBR-II at 400–427°C, showed that after an initial increase in room temperature yield and UTS at low doses (<5 dpa), further irradiation resulted in a gradual decrease in strength with increasing dose (Fig. 9.31). They did not include any detailed results describing the changes in the precipitate structure, and the softening was attributed to coarsening of the dislocation loop structure due to loop unfaulting with increasing dose. Alternatively, or in addition, Υ' precipitate dissolution may lead to lower yield strength with increasing dose and, at high temperatures, >400°C, recovery of the cold-worked network dislocation structure may also be occurring. In the review by Garner [119] it was shown that for 20% cold-worked austenitic alloys the dislocation density, and also yield strength, decreases with increased irradiation dose at high temperatures (>500°C), and increases with dose at low temperatures (due to the contribution from dislocation loop formation). Garner showed that the change in yield strength was due to recovery of the original cold-worked dislocation structure and this recovery tended to occur gradually, being complete after a fast neutron dose of about 1×10^{22} n m^{-2} in EBR-II. Garner also showed that the same decrease in YS with increasing fluence was not exhibited by the same alloy in the solution-annealed condition, indicating that recovery of the cold-worked dislocation structure was responsible for the lower yield strength. Recovery of a cold-worked dislocation structure will occur in any material at elevated temperatures. However, during irradiation this recovery may be enhanced by the higher

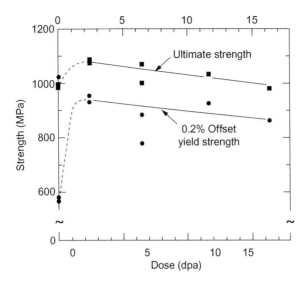

FIG. 9.31

Effect of irradiation on the tensile properties of Inconel X-750 irradiated in EBR-II at 400–427°C. Note that the strength decreases after a maximum at 2–3 dpa.

Modified from W.J. Mills, B. Mastel, Deformation and fracture characteristics for irradiated Inconel X-750, Nucl. Technol. (1986) 102.

point defect concentration, but will also be offset by the point defect clustering (loop formation). Softening due to recovery of network dislocations, but with little hardening from low damage densities, may occur at higher irradiation temperatures (as in the Mills and Mastel case [33]) but is unlikely to be a factor at low temperatures where precipitate dissolution appears to be the more important factor.

9.12.4 PROTON IRRADIATION FACILITIES

Neutron irradiation effects on Ni alloys have been simulated using high-energy (570 MeV) proton irradiation at the Paul Scherrer Institute as part of the SINQ target irradiation program (STIP). The results on Inconel 718 were consistent with those from neutron irradiation in that irradiation increased yield and UTS and also reduced ductility [120]. The most comprehensive set of data on Inconel 718 come from material irradiated at the accelerator production of tritium (APT) facility at the Los Alamos Neutron Science Centre (LANSCE) [121–126]. The mechanical test data cited in the remainder of this section comes from the postirradiation examination of Inconel 718 irradiated at LANSCE. Some material was irradiated in a mixture of protons and spallation neutrons as mechanical test specimens at temperatures between 20°C and 164°C for doses up to about 10 dpa [122]. Other Inconel 718 material came from components exposed to the proton beam: (i) the "water degrader" at temperatures <250°C for doses up to about 10 dpa [123], and (ii) the beam window at temperatures between 367°C and 400°C for doses up to about 20 dpa [124].

The Inconel 718 material from LANSCE was examined by different methods at different laboratories. The results were consistent with fast and thermal reactor data on Inconel X-750 and 718 [33,115,124]. For the most part the results of the different measurements from the LANSCE precipitation-hardened Inconel 718 showed that the material softened with increasing dose for irradiation doses up to about 20 dpa. Results from shear punch tests on samples made from Inconel 718 are shown in Fig. 9.32 [122]. The results are similar to that observed from microhardness tests on strips cut from the Inconel 718 water degrader and these are shown in Fig. 9.33 [123]. The fracture surfaces following three-point bend testing are shown in Fig. 9.34 [123]. The temperature of the testing was not

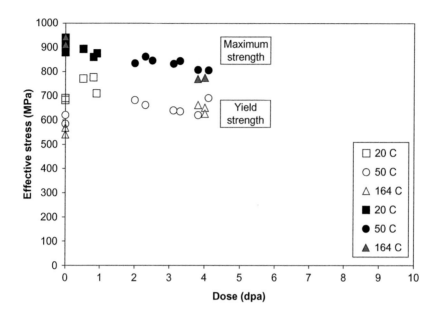

FIG. 9.32

Shear punch test data. Effective shear strength of Inconel 718 irradiated at low temperatures and tested at approximately the same temperature as the irradiation (20–64°C). Open symbols represent data for shear yield strength and solid symbols represent data for shear maximum strength.

From M.L. Hamilton, F.A Garner, M.B. Toloczko, S.A Maloy, W.F Sommer, M.R. James, P.D Ferguson, M,R. Louthan Jr, Shear punch and tensile measurements of mechanical property changes induced in various austenitic alloys by high energy mixed proton and neutron irradiation at low temperatures, J. Nucl. Mater. 283–287 (2000) 418.

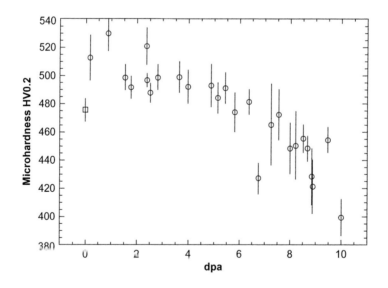

FIG 9.33

Microhardness of irradiated Inconel 718 as a function of irradiation dose at a temperature <250°C. The *square* marker at zero dpa gives the value of the nonirradiated reference material.

From F. Carsughi, H. Derz, G. Pott, W. Sommer, H. Ullmaier, Investigations on Inconel 718 Irradiated with 800 MeV Protons, J. Nucl. Mater. 264 (1999) 78.

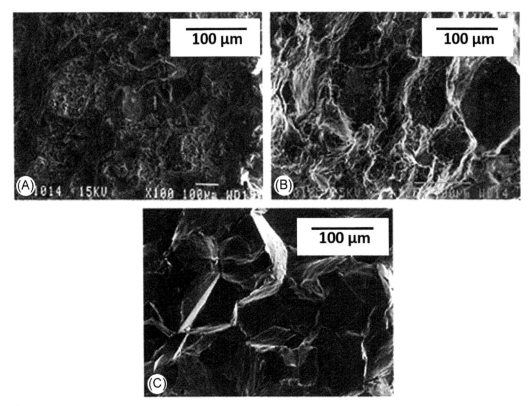

FIG. 9.34

Scanning electron micrographs (SEM) of the fractured area of bent-beam samples exposed to: (A) 0.8×10^{25} pm^{-2}, ~2.5 dpa; (B) 2.6×10^{25} pm^{-2}, ~8.5 dpa; (C) 3×10^{25} pm.$^{-2}$, ~ 10 dpa. Irradiation temperature is <250°C.

From F. Carsughi, H. Derz, G. Pott, W. Sommer, H. Ullmaier, Investigations on Inconel 718 Irradiated with 800 MeV Protons, J. Nucl. Mater. 264 (1999) 78.

specified in Ref. [123] so it is assumed to be ambient (cell temperature). A similar evolution in fracture behavior over the same dose range, with a transition from intragranular to intergranular, was observed for Inconel X-750 spacers irradiated in a CANDU reactor, see Fig. 9.22.

Coincident with softening in the materials from LANSCE was irradiation-induced embrittlement and an increased tendency to intergranular failure at high doses. For most materials irradiation embrittlement is coincident with hardening of the matrix [67,116]; softening and embrittlement do not normally go together. Carsughi et al. [123] showed that there was a change from ductile intragranular to brittle intergranular failure with increasing dose up to 10 dpa at temperatures <250°C (Fig. 9.34) coincident with softening of the material (Fig. 9.33). Other work on the window material (that was also precipitation-hardened Inconel 718) showed that there was severe embrittlement with little evidence of yielding at higher temperatures (between 367°C and 400°C) at higher doses (~20 dpa) [124], Fig. 9.35. This transition to almost zero ductility is similar to that observed for V-15Cr-5Ti irradiated to similar doses (17 dpa) and subject to high He levels (300 appm He) at 520°C [127]. Although no details of the He generation in the window material have been provided it is likely that the He generation per dpa would be comparable with the window degrader (about 150 appm/dpa [126]) and therefore He levels are likely to be about 3000 appm He for the window at 20 dpa. The (p,α) reactions at the center of the beam striking the water degrader

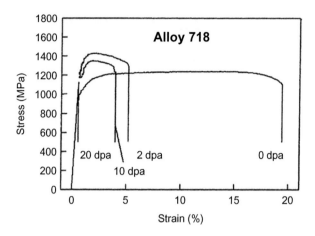

FIG. 9.35

Room temperature stress-strain curves for Inconel 718 material extracted from the LANSCE window and irradiated at a temperature between 367°C and 400°C.

From M.R. James, S.A Maloy, F.D Gac, W.F Sommer, J Chen, H Ullmaier, The mechanical properties of an alloy 718 window after irradiation in a spallation environment, J. Nucl. Mater. 296 (2001) 139.

produced He at a concentration of ~1400 appm, corresponding to ~10 dpa [122,123]. Hydrogen generation levels were about twice that of the helium [126]. For the LANSCE Inconel 718 window material operating between 367°C and 400°C there is an initial increase in the yield strength and UTS for doses up to 2 dpa; both the yield and UTS decrease with increasing dose thereafter [124], consistent with that was observed at lower temperatures [121–123], Fig. 9.32. Comparing room temperature yield strengths for material tested after irradiation at different temperatures, the data in Fig. 9.36 show that the yield strength is low after the irradiation at 20–164°C compared with the irradiation at 367–400°C, consistent with similar results on Inconel X-750 irradiated in CANDU reactors (see Section 9.12.1). The rate of change in yield strength with increasing dose flattens out for doses >10 dpa and can be understood in terms of the competing effects of: (i) precipitate disordering and dissolution leading to a decrease in yield strength; and (ii) cavity evolution (size and density) contributing to an increase in yield strength.

For steels in the solution-annealed condition, where matrix precipitate dissolution during irradiation is not a factor, the yield strength during postirradiation testing (at the irradiation temperature) is dependent on the irradiation temperature as shown in Fig. 9.37 [128]. The peak hardening at 300–330°C has been attributed to a high density of small defect clusters, including dislocation loops and small He bubbles [128]. The lower yield strength at irradiation temperatures <300°C may be related to the effect of point defect recombination lowering the radiation damage density. The lower yield strength at temperatures >300°C has been attributed to a lower density of dislocation loops and network dislocations at the higher irradiation temperatures. The irradiation temperature affects the point defect cluster density and the state of the matrix precipitates and that can account for the differences in yield strength observed in postirradiation testing [129]. The trend for decreasing yield strength with increasing dose in Ni-based alloys has to be related to some change in the microstructure that has been occurring over long periods of irradiation. This change can be attributed to the dissolution of the γ' and γ'' matrix precipitates [110,111,115,122–126].

Disordering of the γ' and γ'' precipitates in Inconel 718 occurs at irradiation temperatures <55°C for doses <1 dpa [125,126]. Hardening is also observed over the same dose range [122] indicating that any effect on the mechanical properties due to disordering (i.e., softening) does not dominate over other radiation-induced hardening or softening mechanisms. These results [122,125,126] are qualitatively similar to those from the neutron-irradiated

9.12 RADIATION HARDENING, SOFTENING

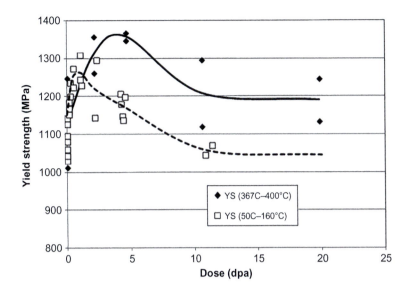

FIG. 9.36

Yield strength of Inconel 718 as a function of dose after irradiation at high temperatures (367–400°C) and low temperatures (20–164°C).

Modified from M.L. Hamilton, F.A Garner, M.B. Toloczko, S.A Maloy, W.F Sommer, M.R. James, P.D Ferguson, M,R. Louthan Jr., Shear punch and tensile measurements of mechanical property changes induced in various austenitic alloys by high energy mixed proton and neutron irradiation at low temperatures, J. Nucl. Mater. 283–287 (2000) 418; M.R. James, S.A Maloy, F.D Gac, W.F Sommer, J Chen, H Ullmaier, The mechanical properties of an alloy 718 window after irradiation in a spallation environment, J. Nucl. Mater. 296 (2001) 139.

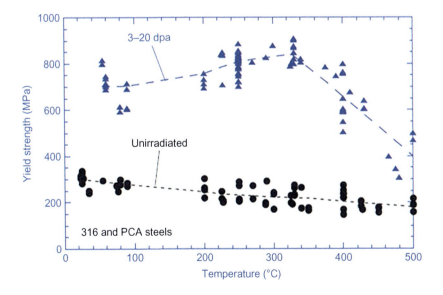

FIG. 9.37

Yield strength as a function of irradiation temperature for neutron-irradiated, solution-annealed austenitic stainless steels in various reactors.

From J.E. Pawel, A.F. Rowcliffe, G.E. Lucas, S.J. Zinkle, Irradiation performance of stainless steels for ITER application, J. Nucl. Mater. 239 (1996) 126–131.

case over the same dose range [116]. Compared with those from neutron irradiation at higher temperatures (288°C), the γ′ and γ″ become disordered at low doses (<3.5 dpa) although some ordered γ′ is retained, or reprecipitated [109], at high doses (up to 20 dpa). Precipitate dissolution, on the other hand, evolves continuously with increasing dose up to 20 dpa [115].

The fact that microchemical redistribution and precipitate formation occurs in irradiated Ni alloys is unequivocal. How much of the chemical changes that occur are the result of irradiation and how much are simply a function of time at temperature is a little more ambiguous. There are few studies in which controlled experiments have been conducted to distinguish between the effects of time at temperature and the effects of irradiation (although two such notable pieces of work, where out-of-flux controls were used, are by Mills et al. [6] and Ward et al. [117]). Because of the variability in microstructures of many engineering alloys very little information exists concerning true radiation effects as opposed to time-temperature-transition effects, which may be enhanced by irradiation. Some irradiation-induced features such as grain boundary solute segregation and depletion are well documented [37,38,64,115]. In cases where equilibrium precipitates dissolve or are dispersed [109] it is often clear that one is observing a radiation-induced phenomenon. One common feature of precipitate-hardened Ni alloys is that, after an initial increase in hardness/strength due to point defect cluster formation at low doses, the material becomes softer (lower YS) with increased irradiation [33,110,111,115,117,122–124]. This softening of precipitation-hardened material has been attributed to radiation-induced dissolution and dispersion of the original γ′ and γ″ precipitates [112, 117, 124–128].

9.12.5 ION IRRADIATION FACILITIES

Work by Hunn et al. [129] and Hashimoto et al. [130] on H, He, and Fe-ion irradiation of Inconel 718 showed that lowering of the yield strength (softening of the material) with irradiation was a phenomenon that was exhibited by precipitation-hardened materials after ion irradiation. Irradiation-induced softening was not exhibited by the same materials in a solution-annealed state. They showed that although systematic hardening of the solution-annealed material occurred for doses up to about 10 dpa, little hardening was observed for precipitation-hardened material for doses up to about 1 dpa, and significant softening was observed between 1 and 50 dpa. Hunn et al. [129] attributed this behavior to the dissolution of γ′ and γ″ precipitates in Inconel 718.

Other ion irradiation studies on Inconel X-750 by Zhang [131] showed that the threshold dose for disordering of the γ′-phase is a function of temperature. Zhang also showed that He stabilizes γ′ precipitates against disordering [131]. Therefore, apart from the effect on cavity stability, the lower He production in fast reactor irradiations may affect other aspects of the microstructure evolution such as precipitate disordering.

Even though γ′ is an ordered phase there is very little indication that disordering, which occurs at low doses, has any significant effect on the softening of Inconel X-750 or 718 that is observed at high doses. In addition, observations of dislocation tangles (Orowan loops) around γ′ precipitates in Inconel X-750 [31–33,51] indicates that dislocation cutting, which would be affected by ordering of the γ′ phase, is not a major deformation mechanism in engineering alloys such as Inconel X-750, and may be related to the low volume fraction and size of the γ′ precipitates.

9.13 HYDROGEN EMBRITTLEMENT

Hydrogen embrittlement is an enigmatic phenomenon in a nuclear reactor environment. Hydrogen is known to embrittle ferritic steels and is a particular problem for the petroleum industry [35,132,133]. Hydrogen is a rapid diffuser and Ni alloys take up hydrogen from the environment. Hydrogen is also generated in a reactor core by high-energy (n,p) reactions with Fe, Cr, and Ni in addition to the hydrogen generated from thermal neutron reactions involving ^{59}Ni [28,29,84,93–96]. There have been instances where surprisingly large quantities of retained hydrogen have been measured in pure Ni and Inconel 718 after neutron exposure [97–99]. There is some debate concerning the conditions for hydrogen retention in engineering alloys given that hydrogen is easily picked up from, and lost to, the environment. It is not known whether hydrogen has any significant effect on material properties in a reactor environment, even if it is retained in some circumstances. For Inconel X-750 spacers in the core of CANDU reactors, hydrogen levels greater than those of unirradiated baseline material were not detected in tests conducted at AECL's Chalk River Laboratories after reactor operation for about 14 equivalent full power years (EFPYs) in a CANDU reactor fuel channel gas annulus (a dry environment) even though up to about 4000 appm H was produced from transmutation reactions [100]. However, it should be noted that the background levels for gases such as hydrogen were high in the Chalk River tests and retention of a small fraction of hydrogen would likely not be detected. Retained H is an additional factor that needs to be taken into account as a potential contributor to swelling and embrittlement. Garner et al. [134] postulated that for a high cavity density, the hydrogen may be directly stored in cavities, most likely in the form of molecular H_2.

Enhanced LTCP, giving rise to intergranular cracking at temperatures <150°C in both air and water, has been observed in Inconel X-750 charged to low levels of hydrogen ($\sim$25–60 wtppm H, i.e., $\sim$1500–3600 appm H) [135]. After gaseous charging the Inconel X-750 material was stored at a low temperature to minimize hydrogen loss, and then tested using a variety of techniques [135]. Uncharged material that had been hot-worked and aged for 24 h at 704°C and then air-cooled, exhibited high toughness and a ductile, intragranular, failure mode when tested in air at 93°C, but low toughness and a brittle, intergranular, failure when tested in water at 93°C. Precharging with hydrogen reduced the fracture toughness for material tested in air and water at low temperatures, and promoted intergranular failure for the material tested in air. Mills et al. [135] attributed the reduced ductility and intergranular failure during low-temperature cracking to the presence of hydrogen. They also concluded that, although hydrogen-assisted cracking was believed to be active during high-temperature SCC, the rate-controlling process for HTSCC involved thermally activated slip. For Ni alloys in reactor coolant chemistries, hydrogen embrittlement is believed to be important at low temperatures while SCC is important at reactor operating temperatures [136], albeit enhanced by the presence of hydrogen. In the presence of high-density cavities, the hydrogen could also affect embrittlement by contributing to cavity stability in the same way He does [134].

9.14 HELIUM EMBRITTLEMENT

James et al. [124] and Carsughi et al. [123] reported a transition to brittle intergranular failure of Inconel 718 with increased dose and stated that He embrittlement would be a possible explanation for their observations. But they also considered that the temperatures were too low for He-bubble segregation at grain boundaries and therefore another explanation was wanting, possibly the formation of a brittle phase at grain boundaries. However, even precipitate formation at the boundaries would require that diffusion of the appropriate alloying elements would occur at a rate commensurate with elevated temperatures. Such enhanced diffusion would also apply to the transport of He atoms.

Although enhanced self-diffusion is the result of higher vacancy concentrations, the He-vacancy complex has a lower migration energy compared with a single vacancy except in the temperature range of $125°C < T < 525°C$ [137], that is, in the temperature range typical of conventional power reactors. One cannot assume that intergranular He embrittlement is solely a high-temperature phenomenon given that cavities are observed accumulating on grain boundaries at temperatures that may be considered low (Figs. 9.25 and 9.28).

Mass transport is a product of concentration and jump rate. For Ni alloys in a neutron spectrum in the core of a CANDU reactor the He generation can exceed 300 appm/dpa. This He generation rate is sufficient to result in enough mass transport for a substantial accumulation of cavities on boundaries, thus weakening them. It is well known that grain boundary cavities are stabilized by He, grain boundary cavities being absent even when voids are produced in the material such as that occurs after irradiation with ions [138], Fig. 9.38. The difference between the grain boundary microstructure evolution in materials with low B or Ni content, and also for materials irradiated in fast reactors where the contribution to He production from thermal neutron (n,α) reactions is negligible, is He

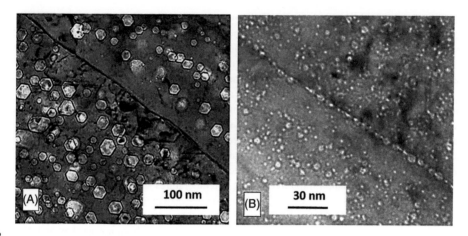

FIG. 9.38

(A) Voids in an Fe-Cr-Ni alloy (MA957) after self-ion (Fe) irradiation to 500 dpa at 450°C. (B) Cavity distribution on and near a grain boundary in an Inconel X-750 spacer after about 14 EFPY, 55 dpa, 1800 appm He at about 300°C. There is a zone denuded of cavities adjacent to the grain boundary thus indicating that the boundry is a sink for vacancies. The boundary is also a sink for He that stabilizes vacancy clusters against absorption on the boundary.

Modified from M.B. Toloczko, F.A. Garner, V.N. Voyevodin, V.V. Bryk, O.V. Borodin, V.V. Melnychenko and A.S. Kalchenko, Ion-induced swelling of ODS ferritic alloy MA957 tubing to 500 dpa, J. Nucl. Mater. 453 (2014) 323; M. Griffiths, G.A. Bickel, S.A. Donohue, P. Feenstra, C.D. Judge, D. Poff, L. Walters, M.D. Wright, L.R. Greenwood, F.A. Garner, Degradation of Ni-alloy components in a CANDU reactor core, in: 16th Int. Symposium on Environmental Degradation in Materials, Asheville, NC, 2013.

generation. Given the information on changes in yield strength and failure modes presented in Section 9.12 it is reasonable to postulate that the grain boundaries weaken with increased dose due to He accumulation. Although the rate of accumulation of grain boundary cavities will likely decrease as the grain interior matrix sink strength increases with increasing irradiation dose, if one assumes that the accumulation of cavities is monotonically increasing with increasing dose, one can hypothesize that the strength of the boundary will decrease with increasing dose to a point that the grain boundary strength will eventually become lower than the yield strength in the matrix, which is itself decreasing with increasing dose, as shown in Fig. 9.39. There is some ambiguity concerning what aspect of the cavity structure is important (size, density, or spacing) [139,140] but many researchers consider that intergranular failure is related to cavity accumulation on boundaries in cases where significant concentrations of He (100's or 1000's appm) are generated [28,29].

Many examples exist showing high densities of He-stabilized grain boundary cavities at high irradiation temperatures [28]. In some high-temperature irradiations so much He accumulates on the grain boundaries that only thin webs of metal remain to maintain bonding between grains as shown by the white filaments on the fracture surface in Fig. 9.40 [141]. There is a dearth of observations on cavities at power reactor operating temperatures (about 300°C) as noted by Rowcliffe et al. [29]. This could be because of the high vacancy migration energy of Ni (about 1.4 eV [108]); the low-temperature behavior (<300°C) is therefore dominated by point defect recombination. Because there are few observations of cavities at low temperatures, void swelling and He embrittlement were generally not considered a concern in the past [29,61,122,123]. However, given that cavities with diameters ~1 nm are at the limit of resolution for many older electron microscopes it is likely that they were simply not observed in many of the early studies. It is only within the past 10 years that observations of cavities at grain boundaries have been made for stainless steel components [142] and Ni alloys [100] removed from power reactors.

In all of the studies on precipitation hardened Inconel X-750 and 718, the main microstructural changes that have been reported concerned the evolution of dislocation loops and the disordering and dissolution of matrix

9.14 HELIUM EMBRITTLEMENT

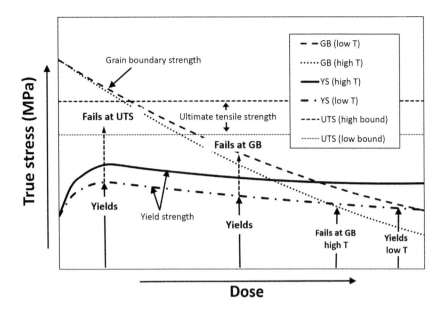

FIG. 9.39

Hypothetical plot showing yield strength (YS), ultimate tensile strength (UTS), and grain boundary strength (GBS) as a function of dose for irradiated precipitation-hardened Ni alloys such as Inconel X-750 and 718.

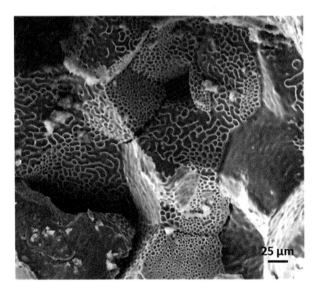

FIG. 9.40

Fracture surface of alloy 600 irradiated in the HFIR at 650°C to 8.5 dpa and 1780 appm He and fractured at 35°C. Micrograph supplied by F.W. Wiffen, Emeritus Scientist at ORNL (Oak Ridge Nuclear Laboratories).

precipitates [109,115,125,126]. There is very little information about microstructural or microchemical changes at grain boundaries, except for Thomas et al. [110] describing enrichment in Ni and a corresponding decrease in Cr, Fe, and Mo that is consistent with observations on other Ni alloys after irradiation [64,109].

One controversial topic concerns radiation-induced embrittlement leading to intergranular failure. Clearly, the strengthening of the matrix will have a tendency to shift the point of failure to weaker locations. Under normal circumstances, the grain boundaries are stronger than the matrix but grain boundary failure can occur if deformation becomes localized at or close to the grain boundary [35]. In many cases, there is yielding in the matrix, or the softer zone adjacent to the grain boundary, before failure at the boundary can occur [32,33,35]. One would not therefore expect a material to fail at stresses lower than the unirradiated material or lower than the YS of the unirradiated matrix. However, there are many cases of irradiated Ni alloys where the material is both weaker and embrittled compared with unirradiated material.

Although He is deemed to be an important component in the effect of irradiation on mechanical properties [143], some studies have concluded that He is not an important factor in controlling mechanical properties relative to other microstructural changes such as precipitation. Based on tensile results reported by Yang et al. [144], it has been proposed that radiation damage and precipitation in the matrix hardens the material against intragranular yielding and segregation of brittle phases at grain boundaries then promotes intergranular failure [144–146]. For this hypothesis to be robust, however, it is not just the brittle nature of the failure that needs to be satisfied but other mechanical properties that need to be consistent with this mode of failure. In contrast to Yang et al.'s observations on precipitate formation at grain boundaries other observations on irradiated precipitation-hardened Ni alloys tend to show dissolution of grain boundary precipitates [110]. Yang's hypothesis depends on following arguments:

(i) A continuous phase forms along grain boundaries. A continuous phase can promote cleavage that propagates within the brittle-phase layer, that is, there are no interruptions of the single phase to allow for crack blunting. Given that γ' is an ordered phase, using a superlattice reflection to image the γ'-phase on the boundaries may well give the impression of a single phase. An examination of the micrographs presented by Yang et al. [144–146] shows that discrete γ' precipitates are visible on the grain boundaries under the same diffraction conditions (presumably using a superlattice reflection, not defined) but they are not continuous. Close examination shows that the images presented by Yang et al. do not show a continuous phase as postulated in Ref. [144]. The fact that the precipitates are ordered is consistent with the temperature of the irradiation (500–650°C), [8]. In examining a failed fuel pin made from Inconel 706 irradiated at 447–526°C, Yang [146] attributes embrittlement to η-phase formation at the grain boundaries, similar to γ'-phase formation at grain boundaries observed for PE16 irradiated at 510°C [145]. At the same time Yang noted that η-phase formation at the grain boundaries is observed after thermal aging but appears to be enhanced by irradiation. Other researchers have also reported a continuous thin layer of γ' along grain boundaries in IN706 and PE16 after fast reactor irradiation to ~15–35 dpa at temperatures up to 575°C [147]. The argument that precipitates segregated to grain boundaries is the reason for the radiation-induced embrittlement is difficult to understand given that it is not exclusively a radiation effect. Also, in other cases of irradiation of precipitation-hardened alloys (Inconel 718) at lower temperatures (288°C), γ' and γ'' precipitate disordering and dissolution is the main irradiation effect that is observed [109].

(ii) The mechanical properties are similar to those exhibited by single crystal γ', $Ni_3(Ti,Al)$. Yang et al. [144–146] compare their results with those of Aoki and Izumi [148] on single crystal $Ni_3(Ti,Al)$ to support the argument that the failure of the alloys in question is governed by the properties of the $Ni_3(Ti,Al)$ phase. However, although a ductility minimum as a function of temperature is cited (albeit at a different temperature than that observed for γ' [3,148]), Yang et al. fail to reconcile this minimum in ductility with a maximum in yield strength for the γ'-phase [3,148] when their results [144] show the opposite, that is, a reduction in strength leading up to the minimum in ductility.

Yang et al. [144–146] are predisposed to argue against He embrittlement as a mechanism for intergranular failure, yet the alternate hypothesis that they present depends on the assumption that the γ'-phase is both brittle and weak. There is no supporting evidence for this hypothesis. Conventional embrittlement (whether intra- or intergranular in nature) occurs concurrently with matrix hardening. In cases where embrittlement is not concurrent with increased strength of the matrix one needs to consider the weakening effect of cavities segregated at the grain or twin boundaries. The perforation of the boundaries is an obvious factor that needs to be considered, irrespective of whether it is the main reason for the intergranular failure of irradiated components or not. However, given that He generation and embrittlement are often associated, the parsimony principle dictates that in many cases intergranular embrittlement is related to the segregation of He at grain boundaries. Although H is also associated with enhanced intergranular failure by SCC and LTCP [135], it is likely that in the presence of a high cavity density the H will be trapped and retained as molecular H_2 within the cavities [134] (see Section 9.13). When it comes to the phase stability of precipitation-hardened Ni alloys the effect of irradiation is to soften the matrix by dissolving and dispersing the hardening phases. The most recent reviews on this topic have concluded that cavity clustering at grain boundaries is the most likely explanation for intergranular failure [28,29]. Boothby [28], citing the same Nimonic PE16 alloy irradiated in EBR-II as Yang et al. [145], concluded that "helium embrittlement, rather than the formation of γ' layers, is primarily responsible for the low ductility failures in postirradiation tests."

9.15 POINT DEFECTS

The irradiation behavior of Ni alloys in power reactors is largely dictated by the microstructure and transmutations producing He and H. The irradiation response is very sensitive to the interstitial and vacancy mobility and the temperature. The point defect effects can be illustrated using observations of the behavior of CANDU reactor components.

CANDU spacers made from Inconel X-750 have large grain sizes (>10 μm diameter) and a dislocation density of about $2-4 \times 10^{14}$ m^{-2} [100]. Given this initial microstructure, and assuming that the microstructure evolution is primarily dictated by the diffusion of freely migrating point defects, rate-theory calculations at the onset of irradiation show that the net interstitial flow to biased sinks will be significantly reduced at temperatures <200°C if the vacancy migration energy in Ni is at the high end of the ranges that exist in the literature. Different values have been derived for vacancy migration energies in Ni. In a recent review by Wolfer, the value for the vacancy migration energy in Ni was quoted to be about 1 eV [149]. However, other sources give values up to about 1.4 eV [108]. As the observation of a lower damage density at 200°C compared with 300°C (Fig. 9.23) could be consistent with recominbination dominating the freely migrating point defect density, a value at the high end of the range is favored for the purposes of the calculations in this section. It should be noted, however, that some point defect clusters will form within the collision cascades and some damage is expected even if recombination of freely migrating point defects is dominant. The transition from a sink-dominated to recombinbation-dominated mass transport regime is illustrated in Fig. 9.41 for the two extremes of vacancy migration energy that exist in the literature for Ni [108,137,149–151]. The approximate displacement damage rates are determined using the SPECTER code [95] and the net freely migrating point defect generation rates are determined arbitrarily assuming a high value for cascade efficiency (10%) for the sake of this calculation. Casacade efficiency values between 1% and 10% may be likely [152], although 1% may be more realistic. It should be emphasized that Fig. 9.41 is valid only if the irradiation produces point defects only and no defect clusters. Single point defects may be dominant at low PKA energies but may not be relevant for high PKA energies, particularly for the HFIR case. Choosing 1.38 eV as an appropriate value to use for vacancy migration energy in Ni [108], calculations also show how self-diffusion in a power reactor (displacement damage rate $\sim 10^{-7}$ dpa s^{-1}) varies as a function of temperature and sink (cavity) density (Fig. 9.42). Using the applicable as-fabricated microstructure parameters at the onset of irradiation, the radiation-enhanced self-diffusion coefficents at 180°C and 305°C are equivalent to the thermal diffusion coefficients at about 500°C and 550°C, respectively.

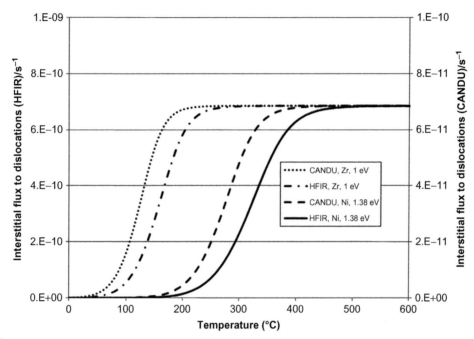

FIG. 9.41

Calculated point defect flux to dislocations at the onset of irradiation. Sink strength (dislocations) = 4×10^{14}, sink strength (grain boundary) = 2.4×10^{13}, 30% interstitial bias for dislocations, 10^{-7} dpa s^{-1} (CANDU) and 10^{-6} dpa s^{-1} (HFIR), 10% damage efficiency, vacancy migration energies of 1 and 1.4 eV.

Once the cavity structure has evolved the radiation-enhanced self-diffusion coefficients change, depending on the temperature, because of the the increasing density of the cavities that act as recombination sites for vacancy and interstitial point defects. The cavity density is also a function of the irradiation temperature. The calculated change in self-diffusion coefficents for the sink strengths corresponds with Fig. 9.24, 1.6×10^{16} at 305°C and 4×10^{16} at 180°C are shown in Fig. 9.42. The decrease in the radiation-enhanced self-diffusion coefficient due to an increase in cavity sink strength is smaller at 180°C compared with 305°C; at 305°C the self-diffusion coefficient is reduced by a factor of about 20 when one includes the effect of the higher sink strength from the cavity evolution, whereas at 180°C it is reduced by a factor of about 4. The self-diffusion coefficients for irradiation at 180°C and 305°C converge as the microstructure evolves and correspond with the thermal self-diffusion values at about 485°C and 495°C, respectively.

9.16 IRRADIATION CREEP AND STRESS RELAXATION

As with PWR and BWR reactors, CANDU reactor guide tubes surround and position the reactivity control mechanisms. In the case of the liquid zone control assemblies in CANDU reactors, to reduce possible flow-induced vibration of the guide tube, Inconel X-750 tensioning springs attach the vertical guide tubes to the base of the calandria vessel [see Fig. 9.19A]. The tensioning springs are located in the reflector region of the reactor (residing between 20 and 40 cm from the bottom of the core for a CANDU-6 reactor) where the fast neutron flux is very low and the temperature is about 60–80°C. On installation the springs are adjusted to produce an initial axial load of

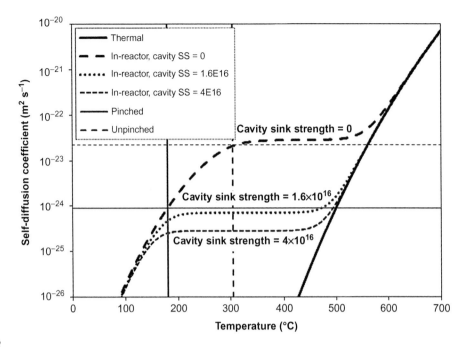

FIG. 9.42

Calculated self-diffusion coefficients at the onset of irradiation, sink strength (SS) for dislocations $= 4 \times 10^{14}$, sink strength (ss) for grain boundaries $= 2.4 \times 10^{13}$, and after the cavity structure has evolved to match a cavity sink strength corresponding to a high irradiation dose: ss $= 1.6 \times 10^{16}$ at 305°C and ss $= 4 \times 10^{16}$ at 180°C corresponding to the microstructures shown in Fig. 9.24. Damage rate $= 10^{-7}$ dpa s^{-1}, 10% damage efficiency, 30% interstitial bias for dislocations, vacancy migration energy $= 1.38$ eV.

1400 pounds in compression [43,96]. Inspection after 18.5 EFPY of service in one CANDU reactor showed that the springs lost most, if not all, of their installed tension. The extent to which the springs had relaxed was unexpected based on assessments of relaxation rates as a function of fast neutron flux from tests of other Inconel X-750 material irradiated in the NRU reactor [96]. However, in the peripheral regions of the CANDU reactor core, the fast neutron flux drops much more rapidly than the thermal neutron flux. Due to the large heavy-water gap, the thermal neutron flux in the periphery can exceed the thermal neutron flux in the center of the core region [see Fig. 9.19B]. In the reflector region, the contribution to displacement damage by processes other than direct collisions is important and spectral effects dominate the damage production. If the alloy has a high nickel concentration then the spectral effects are very much enhanced due to the two-stage reactions that occur in nickel (see Section 9.11). The ^{59}Ni generated from transmutation of ^{58}Ni not only leads to an enhancement in damage production (by 2–3 orders of magnitude in the periphery regions and by a factor of about 2 in the center of the core) but also enhances He production. The He produced could, in principle, result in He embrittlement and failure of the guide tube tensioning springs after long exposures (10–20 years), however, stress relaxation reduces the load to negligible levels after a few years of service so failure by this mechanism is unlikely. In addition, the springs operate in the moderator that is at a temperature where He embrittlement is less severe than in the center of the core (see Section 9.12.1). It is noteworthy that the thermal neutron flux at the pressure vessel boundary in PWRs and BWRs is 3–4 orders of magnitude lower than that for a similar location in a CANDU reactor. Spectral effects for CANDU systems are therefore more significant than for other reactors.

Spectral effects are important for CANDU reactor core components made from Ni alloys because the thermal neutrons make up a larger fraction of the neutron spectrum compared with LWRs and because the Ni-alloy core components reside in the core for the life of the reactor. Tight-fitting Inconel X-750 spacers (garter springs) are used in the fuel channels of CANDU reactors to separate each pressure tube from an outer coaxial and concentric calandria tube thus maintaining an insulating gas gap between the cold moderator and the hot pressure tube [100], Fig. 9.19A. The tight-fitting spacers, as the name implies, are required to remain tight on the pressure tubes at least until the tubes have sagged sufficiently such that the springs become pinched between the hot pressure tube (operating at 260–310°C) and relatively cold calandria tube (in contact with the moderator at 60–80°C). The tight-fitting spacers are expected to remain tight on the pressure tube until such time that they are pinched between the pressure tube and the calandria tube as the pressure tube sags after installation and during the commencement of reactor operation. Stress relaxation leading to de-tensioning is therefore an important issue.

The rate of stress relaxation as a function of displacement damage (dpa) has been determined from stress relaxation tests in NRU at 300°C and 60°C [44,96,100], which are applicable to the operating temperature for the Inconel X-750 spacer and tensioning springs, respectively. When the in-reactor creep tests were originally carried out, the creep rate was determined as a function of fast neutron fluence, or displacement damage due to fast neutrons only [153]. With the advent of more rigorous damage calculations, involving Ni transmutation effects, a reassessment of the creep rate has been made. The new creep rates as a function of fast neutron dose and dpa at two different temperatures are shown in Fig. 9.43. The bent beams have similar microstructures compared with Inconel X-750 spacers used in CANDU reactor cores. Assuming that the irradiation-induced stress-relaxation rates are the same for springs and bent beams, even though the stress states are different, the bent beam relaxation data have been used to account for the observed relaxation behavior of Inconel X-750 spring components in CANDU reactors [96,100]. From these data it is clear that the relaxation rate is faster at lower irradiation temperatures. This, apparently anomalous, creep behavior has been observed in other alloys [154,155].

One model developed to account for the temperature dependence of creep in Ni alloys accounts for the anomalous temperature effect based on a longer transient time for vacancy point defects to reach steady state at lower temperatures [156]. Radiation-enhanced climb and glide is assumed to be the predominant creep mechanism. An

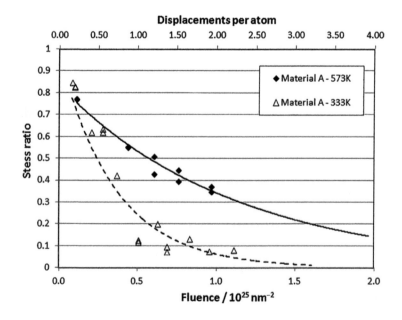

FIG. 9.43

Stress relaxation data for Inconel X-750 bent beams irradiated in NRU.

additional, or alternative, consideration that may be important is the effect of increased recombination rates at the lower temperatures being considered (<100°C). The lower flux of point defects to dislocations at low temperatures (<100°C), Fig. 9.41, is likely to produce less climb. The evidence from TEM (Fig. 9.23) is that the cold-worked dislocation structure is relatively unaffected by point defect clustering at low temperatures, corresponding with the recombination-dominated regime. It is conceivable that conventional thermal creep due to dislocation glide (albeit at a lower temperature) could be dominant at low temperatures but is suppressed at higher temperatures because of the increased defect cluster density.

Radiation-induced thermal creep suppression, commonly observed in Zr alloys [157], could be important in lowering the creep due to dislocation glide at the higher temperatures where point defect clustering locks up the network dislocation structure (Fig. 9.23). It is noteworthy that recent work on 304 stainless steel by Briceno et al. [158], examining the in situ mobility of dislocations in a tensile specimen in an electron microscope, with and without irradiation in a heavy ion beam, showed that radiation damage suppresses, rather than enhances, dislocation motion. Although discussed in terms of the yield strength of the preirradiated material, Briceno et al. [158] effectively showed that dislocation motion during creep is suppressed by irradiation. Although radiation-enhanced climb and glide has been proposed as a mechanism to explain irradiation creep it only applies to edge dislocations. Screw dislocation motion, on the other hand, will always be inhibited by helical climb or jog formation from the absorption of point defects. Although they do not give the Burgers' vector or the character (screw or edge) of the dislocations in their study, the results of Briceno et al. [158] indicate that radiation-enhanced climb and glide should be reconsidered as a mechanism for irradiation creep suppression; radiation damage suppressing rather than enhancing the mobility of all dislocations except pure edge dislocations. The higher creep rate of cold-worked Inconel X-750, at 60°C compared with 300°C, may be understood in terms of the suppression of network dislocation mobility at the higher temperature due to the higher cluster density and point defect absorption on nonedge network dislocations (Fig. 9.23), and the absence of barriers and locking mechanisms allowing free mobility of dislocations at the lower temperature. Although conventional thermal creep may be slow at low temperatures, it is possible that creep due to dislocation glide could be enhanced at neutron fluxes low enough to limit point defect clustering but high enough to increase the point defect concentration in the matrix sufficient to enhance creep due to dislocation glide. Apart from the direct effect on dislocation mobility caused by vacancy or interstitial loop clustering, and point defect absorption on nonedge dislocations, one must also consider the effect of the irradiation temperature on the He-stabilized cavity microstructure (Fig. 9.24).

Inconel X-750 springs are also used in PWR and BWR reactors but their residency time is substantially less than that operating in CANDU reactors. Stress relaxation has also been studied using fast reactor data at elevated temperatures [159]. Other applications of creep in LWR reactors have been reviewed by Garner [160].

9.17 FATIGUE AND CREEP FATIGUE

There are limited data concerning fatigue of Ni alloys under power reactor operating conditions. Fatigue, like SCC, is a high-stress yielding phenomenon, albeit at a stress concentration such as a crack. Creep, on the other hand, tends to be regarded as a low stress phenomenon that, at elevated temperatures or in an irradiation environment, is controlled by diffusion. Synergistic effects are expected during irradiation, especially at high temperatures.

9.17.1 FATIGUE

The fatigue life of structural components and welds is decreased in LWR environments. The fatigue data in simulated BWR and PWR coolant for Ni-Cr-Fe alloys shows that the fatigue life of Inconel 690 is comparable with that of Inconel 600 (Fig. 9.44). Also, the fatigue life of the Ni-alloy welds are comparable to those of the wrought alloys 600 and 690 in the low-cycle regime, that is, $<10^5$ cycles, and are slightly superior to the lives of wrought materials in the high-cycle regime [161].

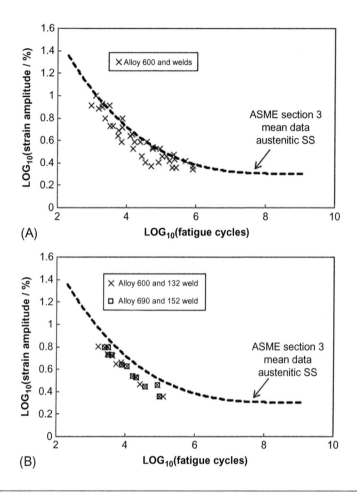

FIG. 9.44

Fatigue curves for Ni alloys 600 and 690, and Ni-alloy welds, in (A) simulated BWR coolant at 280°C, and (B) simulated PWR coolant at 315°C–325°C.

Modified from O.K. Chopra, W.J. Shack, Effect of LWR Coolant Environments on the Fatigue Life of Reactor Materials, NUREG/CR-6909, February 2007.

The fatigue lives of Ni alloys are decreased in LWR environments. The reduction depends on some key material, loading, and environmental parameters. The key parameters that influence fatigue life in these water environments are temperature, dissolved oxygen (DO), strain rate, and strain (or stress) amplitude. The effects of these parameters on fatigue life are consistent with the much larger database on enhancement of crack growth rates in LWR environments. The range of values of the fatigue parameters within which environmental effects are significant has been clearly defined. If these critical loading and environmental conditions exist during reactor operation then fatigue failures are possible [161].

9.17.2 CREEP-FATIGUE DEFORMATION

The VHTR IHX may be joined to piping or other components by welding. Creep-fatigue deformation is expected to be a predominant failure mechanism of the IHX. Weldments used in its fabrication will experience varying cyclic stresses interrupted by periods of elevated temperature deformation. These periods of elevated temperature

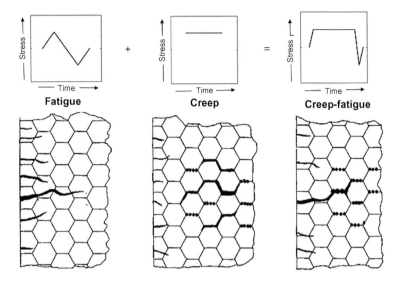

FIG. 9.45

Illustration of creep-fatigue mechanism.

Modified from J.K. Wright, L.J. Carroll, R.N. Wright, Creep and Creep-Fatigue of Alloy 617 Weldments, INL/EXT-14-32966, August 2014.

deformation are greatly influenced by creep behavior of a material. An Ni-based solid solution strengthened alloy, alloy 617, is the primary material candidate for a VHTR-type IHX, and alloy 617 FM will be used for welds [21]. The creep fatigue process is illustrated in Fig. 9.45.

9.18 CONCLUSIONS

1. For in-core nuclear applications of Ni alloys, it is extremely important to take into account the neutron spectrum, especially when considering radiation effects for reactor core components at the periphery of the core where the thermal neutron flux tends to be higher than the fast neutron flux in all reactors. Certain locations in the reflector region of CANDU reactors can have thermal neutron fluxes higher than that in the reactor core. The thermal neutron flux in PWRs and BWRs is 3–4 orders of magnitude lower at the pressure vessel boundary, and an order of magnitude lower at the center of the core, compared with a CANDU reactor. Spectral effects for CANDU systems are therefore more significant than for LWRs.

2. The effect of neutron spectrum must be considered when conducting material tests in test reactors to simulate behavior in a power reactor. For Ni alloys in particular, it is advisable to conduct simulations in any test reactor site that has a spectrum similar to the location of interest in the power reactor, especially when thermal neutron doses are $>1 \times 10^{21}$ n.cm^{-2}.

3. One of the most important properties that defines superalloys is the retention of high yield strength, or even increasing strength, at high temperatures. The main contributory factor is the effect of precipitation hardening, primarily due to the formation of ordered phases, for example, γ' and γ''. These ordered phases are disordered and dispersed during irradiation and this can lead to the unusual observation of decreasing YS with increasing irradiation dose. This decrease in YS with increasing dose is a phenomenon that is common to many precipitation-hardened materials.

4. The SCC is the result of a combination of three factors: (i) a susceptible material, (ii) exposure to a corrosive environment, and (iii) tensile stresses above a threshold. The IASCC refers to cases where irradiation assists in

this cracking process. Irradiation can enhance cracking by additional hardening from radiation damage. However, for precipitation-hardened Ni alloys, radiation-induced softening appears to be the norm. Other factors that promote IASCC and IGSCC are Cr depletion, precipitate formation, and He accumulation at grain boundaries. Although He-induced cavities can be very small at irradiation temperatures <350°C, and have therefore often been overlooked in the past, recent studies have shown that He accumulation on grain boundaries is prevalent at power reactor operating temperatures and is therefore likely to be an important factor that controls the intergranular failure and loss of ultimate strength in Ni-alloy components. He embrittlement has generally not previously been considered an important degradation mechanism at irradiation temperatures <350°C. Recent studies show that this presumption may not be valid.

5. For a given He content, and radiation damage dose, the degree of intergranular embrittlement increases with increasing irradiation temperature [28,29,43,44]. The amount of He needed to cause a given degree of intergranular embrittlement is lower at higher irradiation temperatures. This may be understood based on the mobility of He and vacancies and the lower sink density in the matrix at high temperatures. The lower matrix sink density at higher temperatures would trap less He before it could migrate to grain boundaries. The effect is so severe at high irradiation temperatures that Ni alloys exhibit almost zero ductility during room temperature testing of material that had been irradiated at about 650°C [141]; there is extensive grain boundary coverage by He-stabilized cavities, leaving only thin filaments of material to hold the boundaries together (Fig. 9.40). Most data concerning the effect of temperature on irradiation embrittlement is derived from tests on stainless steels. The effect of high temperature combined with He results in severe embrittlement in stainless steels [23–26,162,163]. Ni alloys, with a propensity for high He production levels in the right spectral conditions, can be expected to exhibit more severe embrittlement effects compared with stainless steels at the same irradiation temperatures because of the higher Ni content.

ACKNOWLEDGMENTS

The author would like to thank Steve Zinkle for his help and useful technical advice in preparing this chapter. The author would also like to thank C.E. Coleman, F.A. Garner, D. Graham, L.R. Greenwood, M.L. Grossbeck, D. Jobba, G.D. Moan, M. Short, S. Xu, Z. Yao, and former colleagues at the Canadian Nuclear Laboratories for useful technical discussions.

REFERENCES

[1] T. Yonezawa, Nickel alloys: properties and characteristics, in: Comprehensive Nuclear Materials, vol. 1, Elsevier, Amsterdam, 2012, pp. 233–266.
[2] S. Fyfitch, Corrosion and stress corrosion cracking of Ni-Base alloys, in: Comprehensive Nuclear Materials, vol. 5, Elsevier, Amsterdam, 2012, pp. 69–91.
[3] R.C. Read, The Superalloys, Fundamentals and Applications, Cambridge University Press, 2006. ISBN-13 978-0-511-24546-6.
[4] A. Strasser, F.P. Ford, High Strength Ni-Alloys in Fuel Assemblies, ANT International, Molnlycke, 2012.
[5] P. Beardmore, R.G. Davies, T.L. Johnston, On the temperature dependence of the flow stress of Nickel-Base alloys, Trans. Metall. Soc. AIME 245 (1969) 1537–1545.
[6] W.J. Mills, M.R. Iebo, J.J. Kearns, R.C. Hoffman, J.J. Kuriuko, R.F. Iuther, G.B. Sykes, in: Effect of irradiation on the stress corrosion cracking behaviour of alloy X-750 and alloy 625, Sixth International Symposium on Environmental Degradation of Materials in Nuclear Power Systems-Water Reactors, TMMMS, 1993, pp. 633–643.
[7] C.T. Sims, N.S. Stoloff, W.C. Hagel (Eds.), Superalloys II; High Temperature Materials for Aerospace and Industrial Power, John Wiley and Sons, New York, 1987.

[8] G.D. Smith, S.J. Patel, Superalloys 718, 625, 706 and derivatives, in: E.A. Loria (Ed.), TMS, The Minerals, Metals & Materials Society, 2005.

[9] R. Cozar, A. Pineau, Morphology of γ' and γ'' precipitates and thermal stability of Inconel 718 type alloys, Met. Trans. 4 (1973) 47–59.

[10] "INCONEL® Alloy 600," Special Metals Corporation, SMC-027, September 2008, Available at: http://www.specialmetals.com/tech-center/alloys.html.

[11] "INCONEL® Alloy 625," Special Metals Corporation, SMC-063, January 2006, Available at: http://www.specialmetals.com/tech-center/alloys.html.

[12] "INCONEL® Alloy 718," Special Metals Corporation, SMC-045, September 2007, Available at: http://www.specialmetals.com/tech-center/alloys.html.

[13] "INCONEL® Alloy X-750," Special Metals Corporation, SMC-067, September 2004, Available at: http://www.specialmetals.com/tech-center/alloys.html.

[14] "INCONEL® Alloy 706", Special Metals Corporation, SMC-067, September 2004, Available at: http://www.specialmetals.com/tech-center/alloys.html.

[15] "INCONEL® Alloy 690," Special Metals Corporation, SMC-067, September 2004, Available at: http://www.specialmetals.com/tech-center/alloys.html.

[16] "INCOloy® Alloy 800", Special Metals Corporation, SMC-067, September 2004, Available at: http://www.specialmetals.com/tech-center/alloys.html.

[17] "NIMONIC® Alloy PE16", Special Metals Corporation, SMC-067, September 2004, Available at: http://www.specialmetals.com/tech-center/alloys.html.

[18] S.D. Kiser, E.B. Hinshaw, J.R. Crum, L.E. Shoemaker, Nickel alloy welding requirements for nuclear service, in: Focus On Nuclear Power Generation, 2005, pp. 21–26. www.specialmetals.com/assets/smw_docs/Special-Metals.pdf.

[19] IAEA Nuclear Energy Series Publications, Stress Corrosion Cracking In Light Water Reactors: Good Practices and Lessons Learned, No. NP-T-3.13 (2011), IAEA nuclear energy series, ISSN 1995–7807, STI/PUB/1522, ISBN 978–92–0–117210–5.

[20] Crack Growth Rates of Nickel Alloy Welds in a PWR Environment, NUREG/CR-6907, ANL-04/3.

[21] J. Wright, L. Carroll, R. Wright, Creep and Creep-Fatigue of Alloy 617 Weldments. INL/EXT-14-32966, 2014. https://doi.org/10.2172/1168621.

[22] H. Trinkaus, H. Ullmaier, High temperature embrittlement of metals due to helium: is the lifetime dominated by cavity growth or crack growth? J. Nucl. Mater. 212–213 (1994) 303–309.

[23] H. Trinkaus, B.N. Singh, Helium accumulation in metals during irradiation—where do we stand? J. Nucl. Mater. 323 (2003) 229–242.

[24] H. Trinkaus, On the modeling of the high temperature embrittlement of metals containing helium, J. Nucl. Mater. 118 (1983) 39–49.

[25] C.A. Wang, M.L. Grossbeck, B.A. Chin, Threshold helium concentration required to initiate cracking during welding of irradiated stainless steel, J. Nucl. Mater. 225 (1995) 59–68.

[26] C.A. Wang, H.T. Lin, M.L. Grossbeck, B.A. Chin, Suppression of HAZ cracking during welding of helium-containing materials, J. Nucl. Mater. 191–194 (1992) 696–700.

[27] BWR Vessel Internals Project, Weldability of Irradiated LWR Structural Components, TR-108707 (1997) BWRVIP-45, www.epri.com.

[28] R.M. Boothby, Radiation effects in nickel-based alloys, in: Comprehensive Nuclear Materials, vol. 4, Elsevier, Amsterdam, 2012, pp. 123–150.

[29] A.F. Rowcliffe, L.K. Mansur, D.T. Hoelzer, R.K. Nanstad, Perspectives on radiation effects in nickel-base alloys for applications in advanced reactors, J. Nucl. Mater. 392 (2009) 341–352.

[30] T.M. Angeliu, J.T. Ward, J.K. Witter, Assessing the effects of radiation damage on Ni-base alloys for the prometheus space reactor system, J. Nucl. Mater. 366 (2007) 223–237.

[31] R.G. Davies, N.S. Stoloff, On the yield stress of aged Ni-Al alloys, Trans. Metall. Soc. AIME 233 (1965) 714–719.

[32] W.J. Mills, in: Effect of temperature on the fracture toughness behavior of Inconel X-750, Fractography and Materials Science, ASTM STP-733, 1981, p. 98.

[33] W.J. Mills, B. Mastel, Deformation and fracture characteristics for irradiated Inconel X-750, Nucl. Technol. 73 (1986) 102.

[34] M.J. Donachie, S.J. Donachie, Superalloys: A Technical Guide, ASM International, Materials Park, OH, 2002.

[35] S.P. Lynch, Mechanisms of inter-granular fracture. Mater. Sci. Forum 46 (1989) 1–24, https://doi.org/10.4028/www.scientific.net/MSF.46.1.
[36] P.L. Andresen, R. Reid, J. Wilson, SCC mitigation of Ni alloys and weld metals by optimizing dissolved hydrogen, Proceedings of 14th International Symposium on Environmental Degradation in Nuclear Power Systems—Water Reactors, Virginia Beach, VA, ANS, 2009, pp. 345–372.
[37] G.S. Was, H.H. Tischner, R.M. Latanision, The influence of thermal treatment on the chemistry and structure of grain boundaries in Inconel 600, Met. Trans. A. 12 (1981) 1397.
[38] T.R. Allen, L. Tan, G.S. Was, E.A. Kenik, Thermal and radiation-induced segregation in model Ni-base alloys, J. Nucl. Mater. 361 (2007) 174–183.
[39] I.J. Kai, G.P. Yu, C.H. Tsai, M.N. Liu, S.C. Yao, The effects of heat treatment on chromium depletion, precipitate evaluation and corrosion resistance of alloy 690, Met. Trans. A 20A (1989) 2057–2067.
[40] S.M. Bruemmer, G.S. Was, Microstructural and microchemical mechanisms controlling inter-granular stress corrosion cracking in light-water-reactor systems, J. Nucl. Mater. 457 (2015) 165–172.
[41] T. Kekkonen, H. Hanninen, The effect of heat treatment on the microstructure and corrosion resistance of Inconel X-750, Corros. Sci. 25 (1985) 789–803.
[42] G.P. Quirk, M.H. Mirzai, P.E. Doherty, W.W. Bek, in: Corrosion surveillance for reactor materials in the calandria vault of pickering NGS A Unit 1, Proceedings of the Sixth International Symposium on Environmental Degradation of Materials in Nuclear Power Systems—Water Reactors, The Minerals, Metals and Materials Society, 1993.
[43] M. Griffiths, G.A. Bickel, S.R. Douglas, in: Irradiation-induced embrittlement of Inconel 600 Flux Detectors in CANDU reactors, Proceedings of the 18th International Conference on Nuclear Engineering (ICONE18), Xi'an, China, May 17–21 (2010) ICONE18–30121, 2010.
[44] M. Griffiths, The effect of irradiation on Ni-containing components in CANDU reactor cores: a review, AECL Nucl. Rev. 2 (1) (2013) 1–16. Erratum AECL Nucl. Rev. 3(2) (2014) 89.
[45] B.H. Kear, H.G.F. Wilsdorf, Dislocation configurations in plastically deformed polycrystalline Cu3Au alloys, Trans. Met. Soc. AIME 224 (1962) 382–386.
[46] H.P. Karnthaler, E.T. Mtjhlbacher, C. Rentenberger, The influence of the fault energies on the anomalous mechanical behaviour of Ni and Al alloys, Acta Mater. 44 (2) (1996) 547–560.
[47] P.B. Hirsch, A-model-of-the-anomalous-yield-stress-for-111-slip-in-L12-alloys, Prog. Mater. Sci. 36 (1992) 63.
[48] D. Caillard, G. Moknat, V. Paidar, On-the-role-of-incomplete-Kear-Wilsdorf-locks-in-the-yield-stress-anomaly-of-Ni$_3$Al, Mater. Sci. Eng. A A234–236 (1997) 695–698.
[49] J.A. Del Valle, A.C. Picasso, R. Romero, Work-hardening in Inconel X-750: Study of stage II, Acta Mater. 46 (6) (1998) 1981–1988.
[50] L.M. Brown, Precipitation and dispersion hardening, in: P. Haasen, V. Gerold, G. Kostorz (Eds.), Strength of Metals and Alloys, Proceedings of the 5th International Conference, Aachen, Federal Republic of Germany, August 27–31, 1979, pp. 1551–1570.
[51] J.A. Del Valle, R. Romero, A.C. Picasso, Bauschinger effect in age-hardened Inconel X-750 alloy, Mater. Sci. Eng. A A311 (2001) 100–107.
[52] V. Gerold, Precipitation hardening, in: Dislocations in Solids, Vol. 4, North-Holland Publishing Company, Amsterdam/New York/Oxford, 1979.
[53] D.J. Bacon, U.F. Kocks, R.O. Scattergood, The effect of dislocation self-interaction on the orowan stress, Phil. Mag. 28 (6) (1973) 1241–1263.
[54] J.W. Ha, B.S. Seong, H.W. Jeong, Y.S. Yoo, Y.S. Choi, N. Kang, Effect of cold-drawing ratio on γ' precipitation in Inconel X-750, Mater Charact 96 (2014) 1–5.
[55] M. Le Calvar, I. de Curières, Corrosion issues in pressurized reactor (PWR) systems, in: Nuclear Corrosion Science and Engineering, 2012, pp. 473–547.
[56] J. Panter, B. Viguier, J.-M. Cloue, M. Foucault, P. Combrade, E. Andrieu, Influence of oxide films on primary water stress corrosion cracking initiation of alloy 600, J. Nucl. Mater. 348 (2006) 213–221.
[57] S. Morton, S.A. Attanasio, J.S. Fish, M.K. Schurman, in: Influence of dissolved hydrogen on nickel alloy SCC in high temperature water, CORROSION 99, 1999.
[58] P.M. Scott, A review of environment-sensitive fracture in water reactor materials, Corros. Sci. 25 (1985) 583–606.

REFERENCES

[59] P.M. Scott, Stress corrosion cracking in pressurized water reactors—interpretation, modeling, and remedies, Corrosion 56 (2000) 771–782.

[60] P.M. Scott, An overview of materials degradation by stress corrosion cracking in PWRs, in: D. Féron, J.M. Olive (Eds.), Corrosion Issues in Light Water Reactors—Stress Corrosion Cracking, European Federation of Corrosion Publications, Woodhead Publishing Limited, Cambridge, Number 51, 2007.

[61] P.M. Scott, A review of irradiation-assisted steress corrosion cracking, J. Nucle. Mater. 211 (1994) 101–122.

[62] J. Gorman, S. Hunt, P. Riccardella, G.A. White, PWR reactor vessel—alloy 600 issues, in: Companion Guide to the ASME Boiler & Pressure Vessel Code, third ed., vol. 3, 2007.

[63] PWSCC of Alloy 600 materials in PWR primary system penetrations, EPRI TR-103696, Research Project 3223-01, Final Report, Palo Alto, CA, July 1994.

[64] G.S. Was, P.L. Andresen, Stress corrosion cracking behaviour of alloys in aggressive nuclear reactor core environments, Corrosion 63 (1) (2007) 19–45.

[65] S.M. Bruemmer, E.P. Simonen, P.M. Scott, P.L. Andresen, G.S. Was, J.L. Nelson, Radiation-induced material changes and susceptibility to inter-granular failure of light-water-reactor core internals, J. Nucl. Mater. 274 (1999) 299.

[66] P.L. Andresen, F.P. Ford, Life prediction by mechanistic modeling and system monitoring of environmental cracking of iron and nickel alloys in aqueous systems, Mat. Sci. Eng. A103 (1988) 167–184.

[67] S.J. Zinkle, G.S. Was, Materials challenges in nuclear energy, Acta Mater. 61 (2013) 735–758.

[68] NACE Resource Center, Stress Corrosion Cracking, http://events.nace.org/library/corrosion/Forms/scc.asp, 2010.

[69] ASM: American Society of Metals, Stress-corrosion cracking, in: ASM Metals Handbook of Corrosion, vol. 13, ASM International, Materials Park, OH, 2002, pp. 828–860.

[70] J. Harris, EPRI report 1003589, PWR Generic Tube Degradation Predictions—US Recirculating Steam Generators with Alloy 600 TT and Alloy 690TT (2003).

[71] C.A. Grove, L.D. Petzold, Mechanisms of stress-corrosion cracking of alloy X-750 in high-purity water, J. Mater. Energy Sys. 7 (2) (1985) 147–162.

[72] L.G. Ljungberg, B. Bengtsson, Grain boundary segregation in alloy X-750 of varying heat treatment and stress corrosion susceptibility, in: J.L. Nelson (Ed.), Proceedings 1986 Workshop on Advanced High-Strength Materials, 1989. EPRI NP-6363 Paper 19.

[73] Y. Lua, S. Ramamurthy, G. Goszczynskic, An aging assessment on ex-service alloy 800 steam generator tubing, Nucl. Eng. Des. 242 (2012) 91–99.

[74] R. Bajaj, W.J. Mills, M.R. Lebo, B.Z. Hyatt, M.G. Burke, Irradiation assisted stress corrosion cracking of HTH alloy X-750 and alloy 625, Seventh International Symposium on Environmental Degradation of Materials in Nuclear Power Systems-Water Reactors, TMMMS, 1995, p. 1093.

[75] H. Coriou, L. Grall, C. Mahieu, M. Pelas, Sensitivity to stress corrosion cracking and inter-granular attack of high Ni austenitic alloys, Corrosion 22 (1966) 280.

[76] H. Coriou, L. Grall, Y. Le Gall, S. Vettier, Corrosion Fissurante sous Contrainte de l'Inconel dans l'eau à Haute temperature, Proceedings of 3eme Colloque de Metallurgie sur la Corrosion, Saclay, North Holland, Amsterdam, 1959, pp. 161–169.

[77] T. Yonezawa, K. Onimura, N. Sakamoto, N. Sasaguri, H. Nakata, H. Susukida, in: Environmental degradation of materials in nuclear power systems—water reactors, Proceedings of the International Symposium on Environmental Degradation of Materials in Nuclear Power Systems—Water Reactors, Aug 22–25, 1983, American Nuclear Society, Myrtle Beach, SC, 1983, pp. 345–366.

[78] F.A. Garner, H.R. Brager, Dependence of neutron induced swelling on composition in iron-based austenitic alloys, in: F.A. Garner, D.S. Gelles, F.W. Wiffen (Eds.), Optimizing Materials for Nuclear Applications, TMS-AIME, Warrendale, PA, 1985, pp. 87–109.

[79] R.K. Nanstad, D.A. McClintock, D.T. Hoelzer, L. Tan, T.R. Allen, High temperature irradiation effects in selected generation IV structural alloys, J. Nucl. Mater. 392 (2009) 331–340.

[80] B.A. Cheadle, A. Celovsky, M. Ghafoor, W. Butt, in: Assessment of the Integrity of Kanupp Fuel Channels, 96 CNA/CNS Conference, Fredericton, New Brunswick, 1996. Issued as AECL-11710.

[81] E.C. Miller, Corrosion of materials by liquid metals, in: Liquid Metals Handbook, 1952 (Chapter 4).

[82] L. Can, Analysis of Coolant Options for Advanced Metal Cooled Nuclear Reactors, (M.Sc. thesis), Naval Post-Graduate School, Monterey, CA, 2006.

[83] IAEA Nuclear Energy Series, Structural Materials for Liquid Metal Cooled Fast Reactor Fuel Assemblies—Operational Behaviour, No. NF-T-4.3, http://www-pub.iaea.org/MTCD/Publications/PDF/Pub1548_web.pdf, 2012.
[84] L.R. Greenwood, A new calculation of thermal neutron damage and helium production in nickel, J. Nucl. Mater. 115 (1983) 137–142.
[85] R.M. Boothby, The microstructure of fast neutron irradiated Nimonic PE16, J. Nucl. Mater. 230 (1996) 148–157.
[86] D. Leblanc, Molten salt reactors: a new beginning for an old idea, Nucl. Eng. Des. 240 (2010) 1644–1656.
[86a] HASTELLOY® N alloy—Haynes International, H-2052B, 2002. Available at: www.haynesintl.com/pdf/h2052.pdf.
[87] M.H. Yun, J.P. Ennis, H. Nickel, H. Schuster, The effect of high temperature reactor primary circuit helium on the formation and propagation of surface cracks in alloy 800H and Inconel 617, J. Nucl. Mater. 125 (3) (1984) 258–272.
[88] H.-Y. Lee, Y.-W. Kim, K.-N. Song, Preliminary application of the draft code case for alloy 617 for a high temperature component, J. Mech. Sci. Tech. 22 (5) (2008) 856–863.
[89] K. Natesan, A. Purohit, S.W. Tam, Materials behavior in HTGR environments, Report NUREG/CR-6824, Office of Nuclear Regulatory Research, Washington, 2003.
[90] W.L. Graham, Corrosion of metallic materials in HTR-helium environments, J. Nucl. Mater. 171 (1990) 76–83.
[91] C. Cabetand, B. Duprey, Long term oxidation resistance of alloys for gas-cooled reactors, Nucl. Eng. Des. 251 (2012) 139–145.
[92] J. Berka, M. Vilémová, P. Sajdl, Testing of degradation of alloy 800 H in impure Helium at 760°C, J. Nucl. Mater. 464 (2015) 221–229.
[93] L.R. Greenwood, B.M. Oliver, Comparison of Predicted and Measured Helium Production in US BWR Reactors, J. ASTM Int. 3 (3) (2005). Art. no. JAI13490.
[94] L.R. Greenwood, F.A. Garner, Hydrogen generation arising from the ^{59}Ni (n,p) reaction and its impact on fission-fusion correlations, J. Nucl. Mater. 233–237 (Part 2) (1996) 1530–1534.
[95] L.R. Greenwood, R.K. Smither, SPECTER: Neutron Damage Calculations for Materials Irradiations, ANL/FPP/TM-197, Argonne National Laboratory, 1985. https://www-nds.iaea.org/irdf2002/codes/index.htmlx.
[96] F.A. Garner, L.R. Greenwood, E.R. Gilbert, M. Griffiths, Impact of Ni-59 (n, alpha) and (n, p) reactions on DPA rate, heating rate, gas generation and stress relaxation in fast, light water and CANDU reactors, Proc. 14th Intern. Conf. on Environmental Degradation of Materials in Nuclear Power Systems, 2009.
[97] F.A. Garner, B.M. Oliver, L.R. Greenwood, M.R. James, P.D. Ferguson, S.A. Maloy, W.F. Sommer, Determination of helium and hydrogen yield from measurements on pure metals and alloys irradiated by mixed high energy proton and spallation neutron spectra in LANSCE, J. Nucl. Mater. 296 (2001) 66–82.
[98] L.R. Greenwood, F.A. Garner, B.M. Oliver, M.L. Grossbeck, W.G. Wolfer, Surprisingly large generation and retention of helium and hydrogen in pure nickel irradiated at high temperatures and high neutron exposures, in: M.L. Grossbeck, T.R. Allen, R.G. Lott, A.S. Kumar (Eds.), Effects of Radiation on Materials, ASTM International, West Conshohocken, PA, 2004, pp. 529–539, ASTM STP 1447.
[99] B.A. Oliver, F.A. Garner, S.A. Maloy, W.F. Sommer, P.D. Ferguson, M.R. James, S.T. Rosinski, M.L. Grossbeck, T.R. Allen, A.S. Kumar (Eds.), Retention of very high levels of helium and hydrogen generated in various structural alloys by 800 MeV protons and spallation neutrons, Proceedings of Symposium on Effects of Radiation on Materials, 20th International Symposium, ASTM STP 1405, American Society for Testing and Materials, West Conshohocken, PA, 2001, pp. 612–630.
[100] M. Griffiths, G.A. Bickel, S.A. Donohue, P. Feenstra, C.D. Judge, D. Poff, L. Walters, M.D. Wright, L.R. Greenwood, F.A. Garner, in: Degradation of Ni-alloy components in a CANDU reactor core, 16th Int. Symposium on Environmental Degradation in Materials, Asheville, NC, 2013.
[101] R. Bajaj, W.J. Mills, B.F. Kammenzind, M.G. Burke, Effects of neutron irradiation on deformation behaviour of Ni base fastener alloys, Ninth International Symposium on Environmental Degradation of Materials in Nuclear Power Systems-Water Reactors, TMMMS, 1999, p. 1069.
[102] ENDF/B nuclear library, Available from: http://t2.lanl.gov.
[103] F. Onimus, J.L. Bechade, Radiation effects in zirconium alloys, in: Comprehensive Nuclear Materials, 4, Elsevier, Amsterdam, 2012, pp. 1–31.
[104] C.D. Judge, M. Griffiths, L. Walters, M. Wright, G.A. Bickel, O.T. Woo, M. Stewart, S.R. Douglas, F. Garner, Embrittlement of nickel alloys in a CANDU reactor environment, in: T. Yamamoto (Ed.), Effects of Radiation on Nuclear Materials, Vol. 25, ASTM International, Anaheim, CA, 2012, pp. 161–175.

[105] C.D. Judge, N. Gauquelin, L. Walters, M. Wright, J.I. Cole, J. Madden, G.A. Botton, M. Griffiths, Inter-granular fracture in irradiated Inconel X-750 containing very high concentrations of helium and hydrogen, J. Nucl. Mater. 457 (2015) 165–172.

[106] N. Yonker and J. Williams, (Surry) In-core Flux Thimble Broke into Six Pieces during the Removal Process, Operating Experience Entry, (Nuclear Network), OE23922, 12/26/2006.

[107] H.K. Zhang, Z. Yao, G. Morin, M. Griffiths, TEM characterisation of neutron irradiated CANDU spacer material, Inconel X-750, J. Nucl. Mater. 451 (2014) 88–96.

[108] N.M. Ghoneim, G.L. Kulcinski, The use of the fully dynamic rate theory to predict void growth in metals, Rad. Effects 41 (1979) 81–89.

[109] R.S. Nelson, J.A. Hudson, D.J. Mazey, The stability of precipitates in an irradiation environment, J. Nucl. Mater. 44 (1972) 318–330.

[110] L.E. Thomas, B.H. Sencer, S.M. Bruemmer, Radiation-induced phase instabilities and their effects on hardening and solute segregation in precipitation strengthened alloy 718, Proc. Mater. Res. Soc. 650 (2001). R1.5.1.

[111] C. Howard, S. Parker, D. Poff, C. Judge, M. Griffiths, P. Hosemann, in: Characterization of neutron irradiated CANDU-6 Inconel X-750 Garter Springs via lift out three point bend tests, Proceedings of HOTLAB Conference: Leuven, Belgium, 2015.

[112] F. Garzarolli, D. Alter, P. Dewes, in: Deformability of austenitic stainless steels and Ni base alloys in the core of a boiling and pressurized water reactor, Second International Symposium on Environmental Degradation of Materials in Nuclear Power Systems-Water Reactors, TMMMS, 1985, p. 131.

[113] F. Garzarolli, D. Alter, P. Dewes, J.L. Nelson, in: Deformability of austenitic stainless steels and Ni base alloys in the core of a boiling and pressurized water reactor, Third International Symposium on Environmental Degradation of Materials in Nuclear Power Systems-Water Reactors, TMMMS, 1987, p. 657.

[114] F. Garzarolli, P. Dewes, R. Hahn, J.L. Nelson, in: Deformability of high purity stainless steels and Ni base alloys in the core of a PWR, Sixth International Symposium on Environmental Degradation of Materials in Nuclear Power Systems-Water Reactors, TMMMS, 1993, p. 607.

[115] L.E. Thomas, S.M. Bruemmer, in: Radiation-induced microstructural evolution and phase stability in Ni-base alloy 718, Eighth International Symposium on Environmental Degradation of Materials in Nuclear Power Systems, ANS, 1997, p. 772.

[116] T.S. Byun, K. Farrell, Tensile properties of Inconel 718 after low temperature neutron irradiation, J. Nucl. Mater. 318 (2003) 292–299.

[117] A.L. Ward, J.M. Steichen, R.L. Knect, Irradiation and thermal effects on the tensile properties of Inconel-718, ASTM STP 611, 1976. p. 156.

[118] W.J. Mills, Effect of irradiation on the fracture toughness of alloy 718 plate and weld, J. Nucl. Mater. 199 (1992) 68.

[119] F.A. Garner, Evolution of microstructure in face-centred cubic metals during irradiation, J. Nucl. Mater. 205 (1993) 98–117.

[120] S.A. Maloy, T.J. Romero, P. Hosemann, M.B. Toloczko, Y. Dai, Shear punch testing of candidate reactor materials after irradiation in fast reactors and spallation environments, J. Nucl. Mater. 417 (2011) 1005–1008.

[121] S.A. Maloy, M.R. James, W.F. Sommer, W.R. Johnson, M.R. Louthan, M.L. Hamilton, F.A. Garner, in: S.T. Rosinski, M.L. Grossbeck, T.R. Allen, A.S. Kumar (Eds.), The effect of high energy protons and neutrons on the tensile properties of materials selected for the target and blanket components in the accelerator production of tritium project, Proceedings of Symposium on Effects of Radiation on Materials, 20th International Symposium, ASTM STP 1405, American Society for Testing and Materials, West Conshohocken, PA, 2001, pp. 644–659.

[122] M.L. Hamilton, F.A. Garner, M.B. Toloczko, S.A. Maloy, W.F. Sommer, M.R. James, P.D. Ferguson, M.R. Louthan Jr., Shear punch and tensile measurements of mechanical property changes induced in various austenitic alloys by high energy mixed proton and neutron irradiation at low temperatures, J. Nucl. Mater. 283–287 (2000) 418.

[123] F. Carsughi, H. Derz, G. Pott, W. Sommer, H. Ullmaier, Investigations on Inconel 718 irradiated with 800 MeV protons, J. Nucl. Mater. 264 (1999) 78.

[124] M.R. James, S.A. Maloy, F.D. Gac, W.F. Sommer, J. Chen, H. Ullmaier, The mechanical properties of an alloy 718 window after irradiation in a spallation environment, J. Nucl. Mater. 296 (2001) 139.

[125] B.H. Sencer, G.M. Bond, F.A. Garner, M.L. Hamilton, B.M. Oliver, L.E. Thomas, S.A. Maloy, W.F. Sommer, M.R. James, P.D. Ferguson, Microstructural evolution of alloy 718 at high helium and hydrogen generation rates during irradiation with 600–800 MeV protons, J. Nucl. Mater. 283–287 ((2000) 324.

[126] B.H. Sencer, G.M. Bond, F.A. Garner, M.L. Hamilton, S.A. Maloy, W.F. Sommer, Correlation of radiation-induced changes in mechanical properties and microstructural development of alloy 718 irradiated with mixed spectra of high energy protons and spallation neutrons, J. Nucl. Mater. 296 (2001) 145.

[127] D.N. Braski, The Effect of Neutron Irradiation on Vanadium Alloys ORNL-CONF-860421 (Dec.), www.iaea.org/inis/collection/NCLCollectionStore/_.../17067690.pdf, 1986.

[128] J.E. Pawel, A.F. Rowcliffe, G.E. Lucas, S.J. Zinkle, Irradiation performance of stainless steels for ITER application, J. Nucl. Mater. 239 (1996) 126–131.

[129] J.D. Hunn, E.H. Lee, T.S. Byun, L.K. Mansur, Ion irradiation induced hardening in Inconel 718, J. Nucl. Mater. 296 (2001) 203.

[130] N. Hashimoto, J.D. Hunn, T.S. Byun, L.K. Mansur, Microstructural analysis of ion-irradiation-induced hardening in Inconel 718, J. Nucl. Mater. 318 (2003) 300–306.

[131] H.K. Zhang, Irradiation Induced Damage in Candu Spacer Materials Inconel X-750, (Ph. D. dissertation), Queen's University, 2013.

[132] H.K. Birnbaum, P. Sofronis, Hydrogen-enhanced localized plasticity—a mechanism for hydrogen-related fracture, Mat. Sci. Eng. A176 (1994) 191–202.

[133] P. Sofronis, I.M. Robertson, in: Viable mechanisms of hydrogen embrittlement—a review, 2nd Int. Symp. on Hydrogen in Matter, Amer. Inst. of Physics, 2006, pp. 64–70.

[134] F.A. Garner, E.P. Simonen, B.M. Oliver, L.R. Greenwood, M.L. Grossbeck, W.G. Wolfer, P.M. Scott, Retention of hydrogen in Fcc metals irradiated at temperatures leading to high densities of bubbles or voids, J. Nucl. Mater. 356 (2006) 122–135.

[135] W.J. Mills, M.R. Lebo, J.J. Kearns, Hydrogen embrittlement, grain boundary segregation and stress corrosion cracking of alloy X-750 in low and high-temperature water, Metall. Mater. Trans. A 30A (1999) 1579–1596.

[136] J. Prybylowski, R.G. Balinger, in: J.L. Nelson (Ed.), An overview of advanced high strength nickel-base alloys for LWR applications, Proceedings: 1986 Workshop on Advanced High-Strength Materials, 1989, EPRI NP-6363, Paper 15.

[137] N.M. Ghoneim, S. Sharafat, J.M. Williams, L.K. Mansur, Theory of helium transport and clustering in materials under irradiation, J. Nucl. Mater. 117 (1983) 96–105.

[138] M.B. Toloczko, F.A. Garner, V.N. Voyevodin, V.V. Bryk, O.V. Borodin, V.V. Melnychenko, A.S. Kalchenko, Ion-induced swelling of ODS ferritic alloy MA957 tubing to 500 dpa, J. Nucl. Mater. 453 (2014) 323.

[139] J.I. Bennetch, W.A. Jesser, Microstructural aspects of He embrittlement in type 316 stainless steel, J. Nucl. Mater. 103-104 (1981) 809–814.

[140] R.M. Boothby, Modelling grain boundary cavity growth in irradiated Nimonic PE16, J. Nucl. Mater. 171 (1990) 215–222.

[141] J.L. Scott, M.L. Grossbeck, P.J. Maziasz, Radiation effects for materials in fusion reactors, J. Vac. Sci. Technol. A 20 (4) (1982) 1297.

[142] D.J. Edwards, F.A. Garner, S.M. Bruemmer, P.J. Efsing, Nano-cavities observed in 316SS PWR flux thimble tube irradiated to 33 and 70 dpa, J. Nucl. Mater. 384 (2009) 249–255.

[143] L.K. Mansur, M.L. Grossbeck, Mechanical property changes induced in structural alloys by neutron irradiations with different helium to displacement ratios, J. Nucl. Mater. 155-157 (1988) 130–147.

[144] W.J.S. Yang, D. Gelles, J.L. Straalsund, R.J. Bajaj, Post-irradiation ductility loss of Fe-Ni-Base precipitation Hardenable alloys, J. Nucl. Mater. 132 (1985) 249–265.

[145] W.J.S. Yang, Grain boundary segregation in solution-treated Nimonic PE16 during neutron irradiation, J. Nucl. Mater. 108–109 (1982) 339–346.

[146] W.J.S. Yang, B.J. Makenas, in: F.A. Garner, J.S. Perrin (Eds.), Microstructure of irradiated Inconel 706 fuel pin cladding, Effects of Radiation on Materials: 12th International Symposium, vol. 1, American Society for Testing and Materials, Philadelphia, PA, 1985, pp. 127–138. ASTM STP 870.

[147] S. Vaidyanathan, T. Lauritzen, W.L. Bell, in: H.R. Brager, J.S. Perrin (Eds.), Irradiation embrittlement in some austenitic superalloys, 11th Int. Symp. on Effects of Radiation on Materials, American Society for Testing and Materials, Philadelphia, 1982, pp. 619–635.

[148] K. Aoki, O. Izumi, Flow and fracture behavior of Ni3(Ti,Al) single crystals tested in tension, J. Mater. Sci. 14 (1979) 1800–1806.

[149] W.G. Wolfer, Fundamental properties of defects in metals, in: Comprehensive Nuclear Materials, vol. 1, Elsevier, Amsterdam, 2012, pp. 1–45.

[150] A.D. Marwick, Fluctuations in defect concentration due to inhomogeneous production of point defects by collision cascades, J. Nucl. Mater. 116 (1983) 40–43.
[151] D.P. Dunne, J.H. Zhu, The role of vacancies on inhibition of reverse transformation in rapidly solidified Ni66Al34 alloy, Mat. Sci. Eng. A273–275 (1999) 690–696.
[152] G.S. Was, Fundamentals of Radiation Materials Science: Metals and Alloys, Springer-Verlag, New York, 2007.
[153] A.R. Causey, G.J.C. Carpenter, S.R. MacEwen, In-reactor stress relaxation of selected metals and alloys at low temperatures, J. Nucl. Mater. 90 (1980) 216–223.
[154] M.L. Grossbeck, L.K. Mansur, Low-temperature irradiation creep of fusion reactor structural materials, J. Nucl. Mater. 179–181 (1991) 130–134.
[155] M.L. Grossbeck, L.T. Gibson, S. Jitsukawa, L.K. Mansur, L.J. Turner, in: Irradiation creep at temperatures of 400°C and below for application to near-term fusion devices, 18th International Symposium on Effects of Radiation on Materials, ASTM STP 1325, 1999, pp. 725–741.
[156] R.E. Stoller, M.L. Grossbeck, L.K. Mansur, in: A theoretical model of accelerated irradiation creep at low temperatures by transient interstitial absorption, 15th International Symposium on Effects of Radiation on Materials; ASTM STP 1125, 1992, pp. 517–529.
[157] M. Griffiths, N. Wang, A. Buyers, S.A. Donohue, in: Effect of irradiation damage on the deformation properties of Zr-2.5Nb pressure tubes, Fifteenth International Symposium on Zirconium in the Nuclear Industry, Sun River, 2007.
[158] M. Briceno, J. Fenske, M. Dadfarnia, P. Sofronis, I.M. Robertson, Effect of ion Irradiation produced defects on the mobility of dislocations in 304 stainless steel, J. Nucl. Mater. 409 (2011) 18–26.
[159] T.R. Allen, J.T. Busby, Radiation damage concerns for extended light water reactor service, JOM 61 (7) (2009) 29–34.
[160] F.A. Garner, Void swelling and irradiation creep in light water reactor environments, in: P.G. Tipping (Ed.), Understanding and Mitigating Ageing in Nuclear Power Plants, Woodhead Publishing Limited, Cambridge, 2010, pp. 308–356 (Chapter 10).
[161] O.K. Chopra, W.J. Shack, Effect of LWR Coolant Environments on the Fatigue Life of Reactor Materials, NUREG/CR-6909, February, 2007.
[162] M.L. Grossbeck, K. Ehrlich, C. Wassilew, An assessment of tensile, irradiation creep, creep rupture and fatigue behavior in austenitic stainless steels with emphasis on spectral effects, J. Nucl. Mater. 174 (1990) 264.
[163] C. Wassilew, in: Mechanical behaviour and nuclear applications of stainless steel at elevated temperatures, Proceedings of the International Conference organized jointly by the CEC JRC Ispra Establishment and The Metals Society and held at the Villa Ponti, Varese, Italy on 20-22 May 1981; Book 280 published in 1982 by The Metals Society, London, 1982.

CHAPTER 10

LOW-ALLOY STEELS

Tim Williams*, Randy Nanstad†

39bhr Consulting (39bhr Ltd.), Derby, United Kingdom †R&S Consultants, Knoxville, TN, United States

CHAPTER OUTLINE

- 10.1 Composition, Fabrication, and Properties of LAS 412
  - 10.1.1 Types and Composition of LAS 412
  - 10.1.2 Design and Fabrication of LWRs 415
  - 10.1.3 Microstructure and Properties 423
- 10.2 Principal Applications of LAS 427
  - 10.2.1 Reactor Pressure Vessels 427
  - 10.2.2 Other Pressure Vessels 428
  - 10.2.3 Piping 428
- 10.3 Performance 429
  - 10.3.1 Regulatory Codes and Structural Integrity Assessment (SIA) 431
  - 10.3.2 In-Service Degradation 438
  - 10.3.3 Other Performance Issues 469
- 10.4 Current Developments and Future Prospects 470
  - 10.4.1 Improved Prediction of Through-Life Toughness 470
  - 10.4.2 Improved Materials 472
  - 10.4.3 Other Issues 473
- References 473

Low-alloy steels (LASs) are used in about 415[1] operating nuclear power plants (NPPs). Section 10.1 of this chapter describes the range of LAS used, the processes by which components are made from them, and their mechanical properties at beginning of life (BOL). Section 10.2 describes the main applications. The most demanding of these is the reactor pressure vessel (RPV) for light water reactors (LWRs) and some gas cooled reactors (GCRs). Although LASs are strong and have generally good resistance to service conditions, their resistance to fracture can be substantially degraded by irradiation damage,[2] potentially resulting in catastrophic failure. Section 10.3 provides a

---

[1]Estimate based on numbers of reactors, including 367 PWRs and BWRs, given in International Atomic Energy Agency (IAEA) Reference Data Series No. 2, 2018 Edition, Nuclear Power Reactors in the World, IAEA Vienna, 2018. IAEA-RDS-2/38, ISBN 978-92-0-101418-4.

[2]The terminology used in this chapter is as follows: *damage* refers to any change to the structure (generally at the nanometer scale) of the steel by irradiation; a *feature* is a nanostructural object induced or enhanced by the effect of damage; most, but not all, damage causes *hardening* (changes to the flow properties of the material). Hardening causes *embrittlement* (reduction in toughness), but embrittlement can also occur by nonhardening mechanisms (*nonhardening embrittlement*) when grain boundaries (and other internal surfaces) are weakened by the segregation of elements such as phosphorus to them. Hardening, embrittlement and nonhardening embrittlement may also be caused by thermal aging, but the detailed mechanisms may be different.

brief overview of the regulatory codes and structural integrity assessment (SIA) methods needed to ensure that failure of LAS RPVs is incredibly improbable. It then describes in detail the degradation processes that must be considered in SIA, with an emphasis on the challenges presented by the need to develop understanding of the physical mechanisms of irradiation damage. Finally, Section 10.4 speculates on prospects for use of this versatile, but complex, class of materials in future NPP. The main focus of this chapter is on Gen I to Gen III pressurized and boiling water reactor (PWR and BWR) RPVs, however, other applications for LAS, and issues specific to them, are discussed in Sections 10.2.2 and 10.2.3.

The objective of this chapter is to provide to the reader a sense of the complexity inherent in the application of LASs in nuclear reactors and how the understanding of how those materials react to the exposure conditions provides reliable long-time operation and enables development of improved materials for future applications.

10.1 COMPOSITION, FABRICATION, AND PROPERTIES OF LAS

Plain carbon steels (simple alloys of iron and carbon with small additions of deoxidants) provide adequate strength for many structural engineering purposes. However, when higher strength is required without loss of toughness (see Section 10.3.1) it is necessary to use LAS, which contain a total of up to about 5% of other alloying elements (Mn, Ni, Cr, etc.) (all compositions are given in mass%). Section 10.1.1 describes the types of LAS used in NPPs for the main pressure retaining components. Section 10.1.2 describes the methods used to fabricate LWR RPVs. Section 10.1.3 discusses the microstructure of RPV components, how this is influenced by composition and fabrication, and how it affects mechanical properties and their variability.

10.1.1 TYPES AND COMPOSITION OF LAS

Table 10.1 gives specified compositions of the major alloying elements and minimum mechanical properties for a selection of LAS currently in use in RPVs. These steels were based on those used for thick-walled vessels in boilers and in chemical plants. They were developed in the United States to provide the greater strength and hardenability[3] required for application in PWRs and BWRs, which had relatively thicker sections (≥ 200 mm) to accommodate the higher pressures. Initially ASME[4]-SA 302 Grade B (a Mn-Mo steel) was used because it had a higher yield strength (345 MPa minimum, cf $\approx$ 200 MPa) than the C-Mn steels (although two power reactor vessels, Indian Point-1 and Pathfinder, were fabricated with A212 Grade B plates with minimum yield strength of 262 MPa). A "Ni-modified" version with the same strength, ASTM[5] SA302 Grade BM, was subsequently introduced to provide a lower ductile-brittle transition temperature. This was a Mn-Mo-Ni steel, containing about 0.6% Ni and was later designated as ASTM SA533 Grade B Class 1. For forged components, ASTM SA508 Class 2 was used; later ASTM SA508 Class 3 was preferred to reduce the likelihood of underclad cracking (UCC), see Section 10.3.3. Similar steels (see Table 10.1) were adopted for PWRs and BWRs made in other countries that used designs similar to the US PWRs and BWRs. For the relatively much lower pressure Magnox reactors developed in the United Kingdom in the 1950s and 1960s, plain carbon-Mn (C-Mn) steels were used for the rolled plates that were welded together to form the relatively thin-walled (75–100 mm) RPVs. Much of the early work to develop an understanding of RPV steels embrittlement was carried out on these materials [1].

The PWRs developed independently in the former Soviet Union (FSU) were the WWER-440,[6] with two models V230 and (later) V213. The design of these required higher strength steels than for the US LWR steels (specification minimum 0.2% proof stress of 430 MPa, compared with 345 MPa, above), and this was achieved by using

---

[3]The depth through the thickness of a component over which the hardness (strength) of the component can be increased by heat treatment.
[4]American Society of Mechanical Engineering.
[5]American Society for Testing and Materials (now known as ASTM International).
[6]Also known as VVER-440. WWER (water-water energy reactor) is the preferred International Atomic Energy Agency (IAEA) description. The Russian name transliterates to Vodo-Vodyanoi Energetichesky Reactor (VVER) and this abbreviation is often preferred by the countries who now own these reactors.

10.1 COMPOSITION, FABRICATION, AND PROPERTIES OF LAS

Table 10.1 Compositions of Some Representative LAS Used for PWR and BWR RPVs

Country of Origin	Alloy (Specification Date)	Product Form (Reactor Type)	Carbon Min	Carbon Max	Silicon Min	Silicon Max	Manganese Min	Manganese Max	Molybdenum Min	Molybdenum Max	Nickel Min	Nickel Max	Chromium Min	Chromium Max	Vanadium Min	Vanadium Max
France	18MnD5 RCC-M2112 (1988)	F	0.00	0.20	0.10	0.30	1.15	1.55	0.45	0.55	0.50	0.80	0.00	0.25	0.00	0.01
FSU	15Kh2MFA	F (WWER-440)	0.13	0.18	0.17	0.37	0.30	0.60	0.60	0.80	0.00	0.40	2.50	3.00	0.25	0.35
FSU	Sv-10KhMFT +AN-42M	SAW	0.04	0.12	0.20	0.60	0.60	1.30	0.35	0.70	0.00	0.30	1.20	1.80	0.10	0.35
FSU	Sv13Kh2MFT +OF-6	ESW	0.11	0.16	0.17	0.35	0.40	0.70	0.17	0.37	NS	NS	1.40	2.50	0.17	0.37
FSU	15Kh2NMFA	F (WWER-1000)	0.13	0.18	0.17	0.37	0.30	0.60	0.50	0.70	1.00	1.50	1.80	2.30	0.00	0.10
FSU	Sv12Kh2N2MA +FC-16	SAW	0.05	0.12	0.15	0.45	0.50	1.00	0.45	0.75	1.20	1.90	1.40	2.10	NS	NS
Germany	20MnMoNi55 (1983, 1990)	F	0.17	0.23	0.15	0.30	1.20	1.50	0.40	0.55	0.50	0.80	0.00	0.20	0.00	0.02
Germany	22NiMoCr37 (1991)	F	0.17	0.23	0.15	0.35	0.50	1.00	0.00	0.60	0.50	1.20	0.25	0.50	0.00	0.02
UK	Mild, C-Mn, steel	P (Magnox)[a]	0.09	0.17	0.10	0.60	1.04	1.32	NS	NS	NS	NS	NS	NS	NS	NS
USA	ASTM A302B	P	0.00	0.25	0.15	0.30	1.15	1.50	0.45	0.60	NS	NS	NS	NS	NS	NS
USA	ASTM A508 Cl2 (1971)	F	0.00	0.27	0.15	0.35	0.50	0.90	0.55	0.70	0.50	0.90	0.25	0.45	0.00	0.05
USA	ASME A508 Cl3 (1989)	F	0.00	0.25	0.15	0.40	1.20	1.50	0.45	0.60	0.40	1.00	0.00	0.25	0.00	0.05
USA	ASME A533 GrB (1989)	P	0.00	0.25	0.15	0.40	1.15	1.50	0.45	0.60	0.40	0.70	NS	NS	NS	NS

F, forging; P, plate; SAW, submerged arc weld; ESW, electroslag weld; NS, not specified. Further details of these materials can be found from the specifications cited.
[a] Magnox composition ranges are typical, not specification ranges.

Cr-Mo steels; the later WWER-1000 series (models V-302 and V320) used Cr-Ni-Mo alloys, which were stronger still (490 MPa minimum). In many publications Mn-Mo-Ni and WWER steels are described as "western" and "eastern" steels, respectively. This once had some geopolitical basis but is geographically erroneous. In terms of brittle to ductile transition temperature (see Section 10.3.1), the FSU and US standards had similar requirements with a maximum RT_{NDT} or T_{K0}, respectively, around −10°C.

The welds used to join the plates and forgings in these steels have generally lower carbon content to reduce weldability problems, together with increased Si to offset the corresponding reduction in strength and to act as a deoxidant. Welds are generally "matching," that is they have a similar composition and strength compared to the adjoining base materials. However, there have been exceptions, where the manufacturers have used a significantly different composition weld.

The specifications and characteristics of individual steels have changed with time as understanding has developed and with changes to steel-making practice. A key specification change has been to progressively restrict the permitted level of Cu in the RPV beltline region[7] (in some cases to as low as 0.05%). Until about 1970, the adverse effect of Cu on irradiation sensitivity was unknown (see Section 10.3.2.1) and up to about 0.4% Cu could be present in steels due to the use of unsorted scrap (e.g., containing automotive wiring looms and electric motors) and, in particular, the practice of coating welding wire with Cu to increase electrical conductivity and for protection against corrosion. It is also notable that permitted levels of S and P have also been reduced from about 0.04% to 0.01%.

The effects of the alloying and impurity elements in LAS are complex and cannot be properly discussed in isolation due to interactions between them. The following provides a brief overview of the effect of the elements that are controlled in RPV steels, more specific information is provided in the later sections (10.1.2, 10.1.3, and 10.3.2):

Aluminum refines grain size, increasing strength
Carbon increases hardness but reduces toughness; free carbon can affect irradiation sensitivity
Chromium increases hardenability
Copper markedly increases irradiation sensitivity
Manganese increases hardenability and reduces the adverse effect of P (see further) and sulfur. It can increase the effect of Ni on irradiation sensitivity
Molybdenum forms stable carbides and inhibits grain growth, increasing toughness; it protects against temper embrittlement
Nickel increases hardenability and toughness, but increases irradiation sensitivity
Nitrogen (if in solution in the steel) can reduce toughness and increase irradiation sensitivity
Phosphorus can increase irradiation sensitivity and induce embrittlement during heat treatment, irradiation annealing and thermal aging.
Silicon increases strength
Sulfur reduces toughness and resistance to environmentally assisted cracking
Vanadium is a carbide former, increases strength and toughness.

There are significant differences between the compositions of the different LAS used in RPVs, and a significant variability even within a specific LAS. Fig. 10.1 [2] shows values from the US surveillance database; alloys are A302B, A533B and A508 types, and associated welds; the latter tends to have the greater variability. Each of the 660 data represents a single sample point from a unique heat or weld. Fig. 10.1A illustrates the range and variability of Mn and Ni contents in US PWRs and BWRs; Fig. 10.1B shows the variability in the P and Cu. There is a correlation between Cu and P contents in part due to the limitations placed on both these elements in later RPVs, but also may reflect their occurrence in nature. The variability within the US fleet of LWRs may not be representative of those of other nations that adopted nuclear energy later and were able to take benefit of the US experience. At that time, 920 sets of surveillance data had been obtained from the 660 different alloys; most heats or welds were irradiated in more than one capsule.

---

[7]The region of the RPV shell surrounding the reactor core and most heavily irradiated.

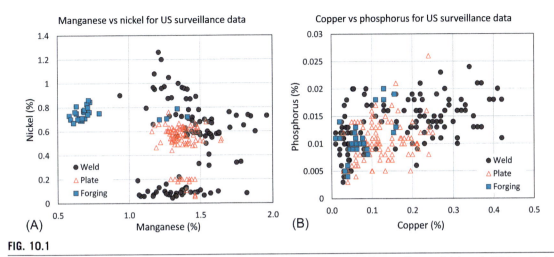

FIG. 10.1

Examples of compositional variability between RPV steels (values reported in US surveillance database): (A) the alloying elements Mn and Ni; and (B) the impurities Cu and P [2].

The variations in composition shown in Fig. 10.1 have several sources. The values of some elements may be deliberately restricted by different steel makers to narrower ranges from what the specification allows, for example, to make it easier to achieve the mechanical properties required; and other variations are due to inherent variabilities of the process, including source of the ore and scrap metal feedstock. In addition, the composition of a steel can vary significantly within a component and this can lead to sampling variations. Variations in composition (and microstructure) within a component are discussed in Section 10.1.2.

Chemical composition values can vary significantly within a component (see Section 10.1.2) but are generally determined from a single relatively small sample (a few grams) from a single location within the component. Normally, only values of the total amount of that element in the sample are given; as we will see later (in Section 10.3.2.1, under subsection "Effect of environmental and materials variables"), sensitivity to irradiation damage is dependent on the amount of an element in solution, which may be substantially lower. For example, Mn is partially sequestered in carbides. As discussed in Section 10.1.2, Mn and C can co-segregate in "ghost-lines" within forgings, thus reducing the Mn in solution relative to other regions. These factors can contribute to the uncertainties in irradiation damage models empirically fitted to data.

It should be noted that compositional variations, within specification limits and within individual components, can give rise to corresponding mechanical property variations. Composition can also vary significantly within a component. Fig. 10.2 gives an example from the destructive examination of a weld from the Midland RPV, which was not put into service due to site-specific civil engineering problems. The substantial variation in copper content through the vessel thickness for this submerged-arc weld resulted from varying thickness of copper coating on the multiple coils of welding wire [3, 4]. The effect of compositional variation on mechanical properties is also affected by the fabrication processes. These issues are discussed in Section 10.1.3.

10.1.2 DESIGN AND FABRICATION OF LWRs

Rules for the design, materials, fabrication, and inspection of RPVs are given in engineering codes. In the United States, the ASME BPVC (American Society of Mechanical Engineers Boiler and Pressure Vessel Code) [5] applies; Section III Division 1 provides rules for the construction of nuclear facility components. Other countries have adopted ASME or developed similar codes, for example, the German KTA (KernTechnischer Ausschuss) code, the Japanese JSME (Japanese Society of Mechanical Engineering) code, the Russian PNAE-G

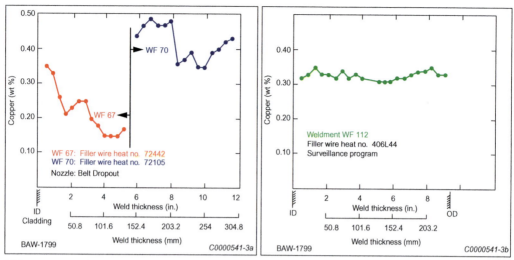

FIG. 10.2

Variability of Cu content through the thickness of two different welds. The weld on the left from the Midland RPV was made using different filler wire heats on each side of the double-V preparation weld; both show high variability but with differences in the trend vs depth. The weld on the right for a different reactor shows low variability. The differences within a wire heat are primarily associated with variations in the Cu coating on the welding wire.

(Federal Standards and Rules in the Field of Use of Atomic Energy) code, and the French RCC-M (Règles de Conception et de Construction des Matériels Mèchanique) code. Some code comparisons are given in Refs. [6, 7].

LWR RPVs are broadly similar to each other in design and construction. Fig. 10.3 [8] compares constructional details for a representative early US PWR with the two main VVER designs. The designs are broadly similar but the VVERs have two nozzle courses and a higher number of circumferential seam welds between the shell courses. The use of higher-strength steels, compared to US designs, allowed relatively small wall thicknesses. The VVER-440, in particular, is narrower to facilitate transportation to the reactor site, a factor that increases irradiation dose at the RPV wall. In comparison, BWRs are significantly taller than PWRs with the same power output and have a rather larger diameter (reduced irradiation dose). The operating pressure in BWRs is about half that of PWRs, which allows a reduction in wall thickness.

In the oldest US-designed RPVs, the main shell was constructed from rolled and formed plates, joined by longitudinal submerged arc seam welds, generally with a double V preparation, to form rings that are joined together by circumferential welds. The lower closure head is fabricated from a ring forging joined to a pressed plate. The upper closure head is a more complex construction fabricated from plates and forgings; it is subsequently bolted in place by high strength LAS studs and nuts. The head includes many complex welded joints associated with the control rods and other penetrations. The nozzles are forged and welded to the main shell. VVERs were constructed in a broadly similar manner, except that ring forgings were used for the main shell. This practice has been followed more generally since the 1960s and it eliminates the longitudinal seam welds. These were undesirable, particularly in the beltline because welds are more likely to contain flaws than base materials and potentially more limiting in terms of structural integrity (SI) (see Section 10.3.1).

RPV manufacturers use conventional steel-making and fabrication practices. The finished steel is poured into an ingot mold. These can have a capacity of 250 tons or more and take around 2 days to solidify. During this solidification there are complex interactions between thermodynamic, fluid dynamic, and gravitational processes [9]. This leads to a particular pattern of compositional and microstructural variations, on a scale of millimeter to meter. The macrostructural variations (at the upper end of this scale) are illustrated in Fig. 10.4 [10]. Although

10.1 COMPOSITION, FABRICATION, AND PROPERTIES OF LAS

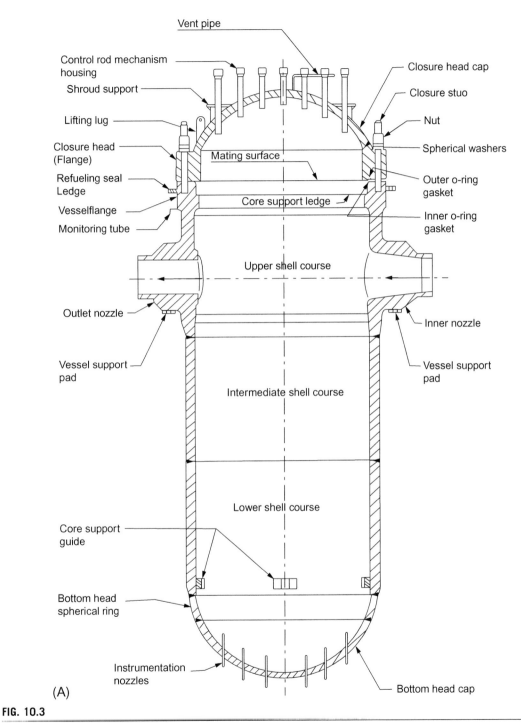

FIG. 10.3

Construction details of (A) a US-type PWR showing the major components in comparison with (B) WWER-440 and WWER-1000 designs (not to same scale) showing the higher aspect ratio [8].

(Continued)

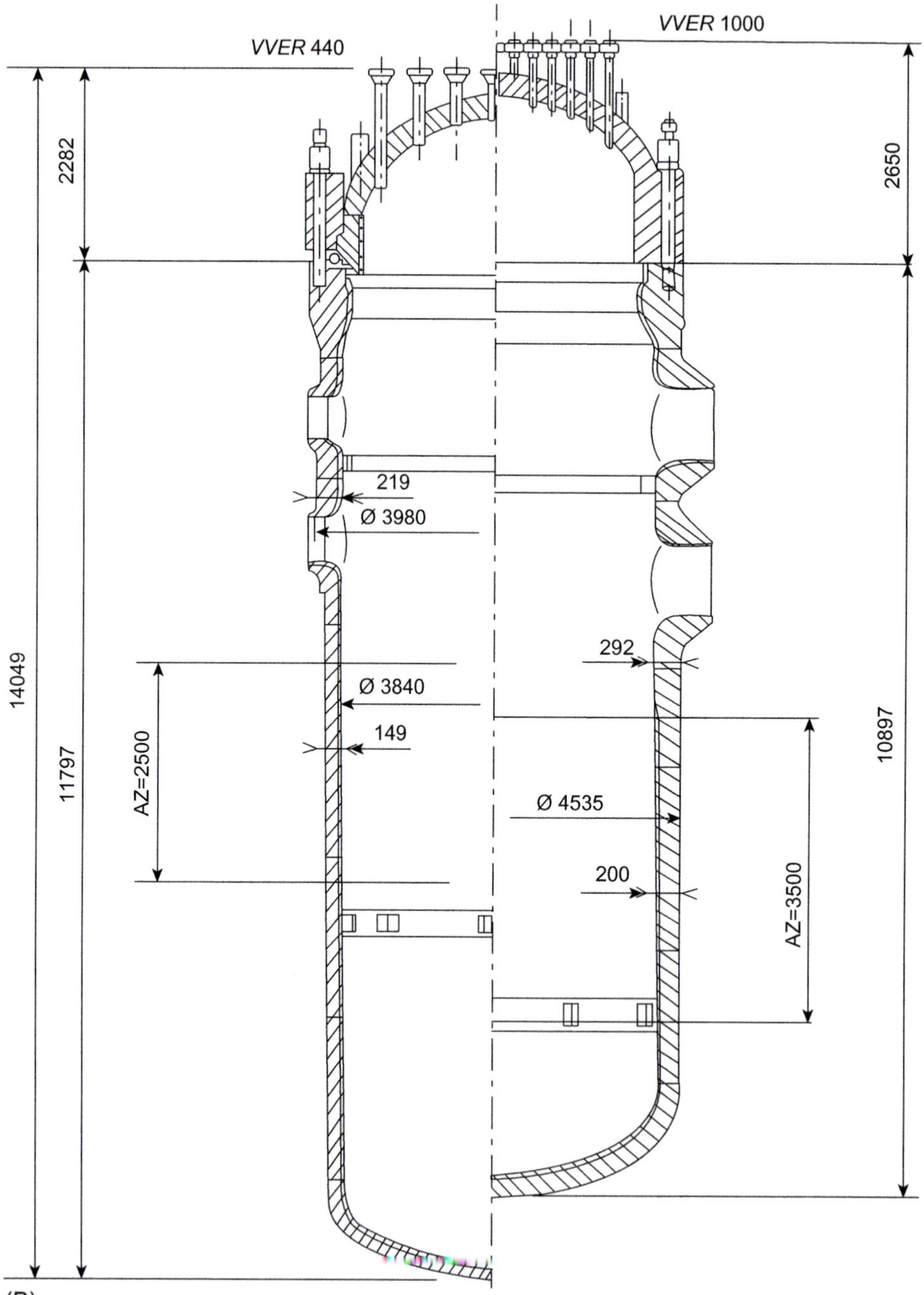

418 CHAPTER 10 LOW-ALLOY STEELS

(B)

FIG. 10.3 CONT'D

macrosegregation is defined by compositional variations, different types of nonmetallic inclusions may be associated with specific regions of macrosegregation.

The French Heavy Forge project found that the long-range C segregation that occurs from top to bottom and radially in large ingots can result in +20% to −10% deviations in C contents from the mean. These were accompanied by co-segregation of P and Cu. As discussed in Ref. [11], the variability of Cu, P, and C affects both irradiation sensitivity and the interpretation of surveillance test results.[8] Segregation of C, P, S, Mn, and other alloying elements can also be pronounced in local regions known as A segregates or "ghost lines" [12]. These form in regions toward the upper end and approximately mid-way between the center and the outside of the ingot. Segregation, and co-segregation, can potentially result in significant differences in irradiation sensitivity in different regions of a component. For example, as already noted (Section 10.1.1), Mn is sequestered by C, so the co-segregation of these elements in ghost lines can reduce the amount of Mn in solution (hence available for irradiation damage processes) relative to other regions in the same component. The amount and location of segregation depends on several factors, including alloy composition, ingot mold design, and casting techniques.

The formation of carbides and other nonmetallic inclusions in steels during the steel-making ingot solidification are an inherent and, for most types of inclusions, a desirable (or at least not undesirable) part of the process. In addition, less-desirable inclusions (often termed exogenous) arising from the casting process, for example, from slag on top of the molten metal, or erosion of the refractory materials lining the mold, can also be deposited. Both inherent (termed endogenous) and exogenous inclusions can be preferentially deposited in specific regions of the ingot. This can be due to local chemistry at the time of formation, or to complex processes involving the relative densities of the inclusions and the still-molten steel, convection currents and entrapment by the steel as it freezes [9, 10, 13].

The structures shown in Fig. 10.4 are modified in subsequent processing of the ingot. The gross defects and some of the macrosegregation at top and bottom of the ingot are removed by cropping. For ring forgings the V-segregates at the center of the ingot are removed when the central core[9] is punched out to allow it to be expanded to form the RPV shell. It is difficult to punch the core in exact alignment with the original axis of the ingot so that there can be variations in the effects of segregation around the circumference of the ring at a given height. The A-segregates are likely to remain, to eventually be located close to the inner surface of the finished component. In plates, the V-segregates also remain, but will be in the center of the finished component.

The forging and rolling operations used to produce ring forgings and plates ready for machining to final size, coupled with the austenitization, quenching and tempering heat treatments, break up clusters of inclusions, and refine the microstructure to make it stronger, tougher, and more homogeneous. However, working and heat treatment cannot remove the macroscopic compositional variations, and these will influence the effects of heat treatment, hence mechanical properties. The details of the processes used are particular to the individual forgemaster and may be proprietary. They also vary between different components.

The various forged or plate components of an RPV must be welded together and machined to final size individually or as subassemblies. Major welds in some vessels were produced by electroslag welding (e.g., Dresden Units 2 and 3, Quad Cities Units 1 and 2) [14] but submerged-arc welding (SAW) with double V-preparation gives better properties and is more common. Manual welding may also be required for smaller or less accessible joints, for example, those associated with penetrations. The inside surface of the RPV is protected from the coolant (and also to protect the coolant from corrosion products) by two or three layers of stainless steel cladding, which is deposited by multiwire or strip electrodes on individual components or subassemblies; manual welding may be required to fill the final gaps.

Fig. 10.5 shows a cross section of a welded joint. This is from a destructive examination of a decommissioned Magnox reactor to confirm irradiation damage predictions [15]. The heavy etching reveals the individual

---

[8]Surveillance testing is described in Section 10.3.1.
[9]Issues with segregation can now be significantly reduced by hollow ingot technology.

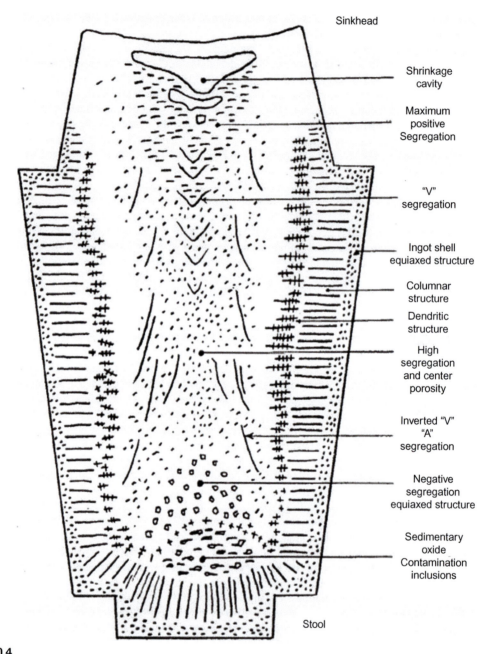

FIG. 10.4

Grain structure and segregation in a steel ingot [10].

weld beads and macrostructural inhomogeneity. This is because of the effect of the temperature from the molten weld bead on previously deposited beads (or the base material). The initial microstructures formed in this way are refined by the subsequent postweld heat treatment (PWHT) to improve properties. In some cases the welded joint may be re-austenitized, in that case it may be difficult to distinguish the weld from the base material.

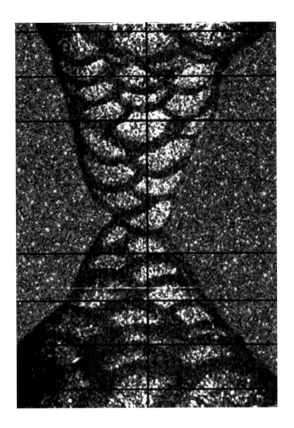

FIG. 10.5

Macrostructure of a double V-weld, section perpendicular to the welding direction, heavily etched to show the weld beads clearly (horizontal lines are from the original reference to indicate specimen sampling levels) [15].

Compositional variations within welds can come from variations in welding wire composition, between batches, or along the length of a single batch of wire, the thickness of the Cu coating (if there was one, see Section 10.1.1), variations in the welding flux and welding parameters, and from dilution of the molten weld pool from melting of the material on which it is deposited (base material or previous weld beads). The dilution of the weld bead from melting of the substrate can be 20%–30%. This is unimportant if the substrate has the same composition, but dilution effects can be of practical significance when a copper-coated welding wire was deposited onto plate or forging materials (which generally had a lower copper content) or when weld and substrate have different alloy compositions. As with forgings or plate production, welding practice differs between components and fabricators.

Fabrication processes are carefully controlled to avoid defects[10] and to develop the specified properties. For plates and forgings, this includes control of the amount and direction of the reduction of the initial ingot to the final shape, of the temperatures and cooling rates at which these and the final heat treatment are undertaken. It is not a production line operation, and the knowledge and expertise of the forgemaster can affect the quality of the

---

[10] A defect is a flaw (an imperfection or unintentional discontinuity that can be detected by nondestructive examination) with characteristics such that it might be rejected (dependent on size) by the relevant regulatory code.

final product. The joining operations are in the hands of the RPV fabricator. The weld preheat temperatures and the PWHT must be controlled to avoid cracking and provide the required properties. The welding process can introduce various defects, such as slag entrapment, these must be detected and removed as the welding proceeds. Overall RPV production is a long multistage process; the course of which is somewhat result—and circumstance—dependent at each stage. It is perhaps more akin to a large civil engineering project than a manufacturing operation. Because it is difficult to control, monitor, or specify precisely, the ultimate arbiters of fitness for purpose are the final nondestructive examination (NDE) for defects, and the results of mechanical property "acceptance tests." To enable the latter, components are made to be larger than required in the finished RPV so that acceptance test material can be cut off prior to assembly. Such material is given the same further heat treatment cycles, for example, those for stress relief after welding, as its parent so that it is as representative of the material in the finished RPV as possible. However, as discussed earlier, RPV components may have variations in chemical composition and mechanical properties on a scale of centimeter and greater. The effect of this may need to be taken into consideration in safety assessments. For circumferential welds, acceptance tests are carried out on a test weld made with the same batches of weld wire and flux using the same welding parameters. In addition to retaining surplus acceptance test material, RPV operators may require retention of tested samples and manufacturing offcuts from the RPV as a contingency against any future issues. Minor flaws or departures from specification requirements for fabrication may be allowed through concession procedures. These allow for the fact that codes and specifications cannot account for every eventuality, thus are developed on the basis that any infringement could be in a critical location and occur in combination with other adverse factors. Concessions may be granted when the effect of the infringement can be demonstrated to be inconsequential.

Heat treatment affects BOL properties and may also affect irradiation damage sensitivity. Details can vary between RPV vendor, designer, fabricator, production date, alloy specification, and component within an RPV, and are not usually published in detail. In broad terms, however, Mn-Mo-Ni plates and forgings are austenitized at 800–950°C and water quenched, then tempered in the range 600–675°C for a few hours, followed by controlled cooling. WWER-440 steels are austenitized at 960–1000°C and oil quenched, then tempered at 660–680°C; the corresponding values for WWER-1000 steels are 910–930°C and water quenched, and 640–660°C. PWHT is generally carried out in the range 590–625°C for PWR and BWR steels, but VVER-440 and VVER-1000 forging PWHTs are 660–680°C and 620–650°C, respectively, followed by slow cooling [16, 17]. As PWHT is generally carried out after each welding or cladding operation, the total PWHT time varies with component in the RPV depending on whether it was early or late in the construction sequence. Further, later RPVs have fewer, but larger, components. Although PWHT is generally carried out at significantly lower temperatures than tempering, the accumulated time for the earliest-in-sequence components can be potentially as significant to microstructural development as tempering itself. When material is procured for experimental characterization it may be difficult to exactly reproduce actual RPV production practices; for example, production heating and cooling rates may be much slower than the specified maximum due to the mass of the material.

The design and fabrication of RPVs have evolved considerably over the past 40 years. The designs themselves benefit from the use of modern computerized design techniques and past operational experience. Modern RPVs tend to be much larger and built with much larger and more massive forgings, as opposed to plate materials, thus they contain a smaller proportion of welds. In ring forgings, the welds are positioned outside the most heavily irradiated regions of the reactor shell. In addition, steels and fabrication processes are now understood, and controlled, much better generally resulting in more consistent and better-characterized end products. However, issues still emerge, particularly in the fabrication of large forgings [18], which can lead to a need to remake components, and thus costly delays. There is clearly a need to continue research and development of manufacturing and fabrication technologies.

Rarely, fabrication issues may not become apparent until after RPVs have entered service. These can call into question the safety of continuing operation and, since it is impractical to replace an RPV, may cause a major problem. Examples are given in Section 10.3.3.

10.1.3 MICROSTRUCTURE AND PROPERTIES

The microstructures of LAS are complex and vary between alloys. The relatively low-alloy content C-Mn steels used in Magnox RPVs base materials are, after quenching, generally pearlitic or ferritic-pearlitic; in contrast Mn-Mo-Ni and WWER base materials can be described as mixed bainite and ferrite or bainitic. Martensite may also be present, especially near the quenched surfaces. The significantly higher austenitization temperatures for WWER-440 steels (see Section 10.1.2) result in substantially larger prior-austenite grain sizes.

The microstructural complexity and inhomogeneity of commercial steels presents a problem for scientific studies. There are two main approaches. Carbon-free model alloys with a simple ferritic structure are used in some fundamental experiments, especially when the very fine-scale damage processes are to be observed. For studies intended to investigate the effect of damage processes on commercial steels, model steels may be used. These can use tailored compositions, and be processed and heat treated to represent commercial fabrication or to investigate variations. However, due to the small volumes of material involved or (intentional) deviations from chemical composition specifications, these may not be fully representative of commercial materials, notably the macroscopic variability discussed further.

The microstructures of specific LAS may be complex and inhomogeneous on a scale of microns. The ferritic regions and bainitic laths can have widely different dislocation densities and differences in carbide sizes and distributions (see Fig. 10.6 [19]). This scale is much larger than the scale of individual displacement cascades (in Section 10.3.2.1, under subsection "Microstructural evolution during irradiation"), which extend over about 0.005–0.01 µm for very high-energy primary recoil atoms. For very high-energy events, subcascades may form and increase the total volume over which a neutron damage event takes place.

Microstructures of LAS also vary over macroscopic length scales. As discussed earlier (Section 10.1.2), base materials can contain areas of segregation and nonmetallic inclusions such as Mn-Ni-Si (MNS) that affect microstructure on a scale of millimeter to meter. In addition, base materials (plates and forgings) are generally quenched and tempered and this results in a significant gradient in microstructure from the surfaces to the interior. The gradient decreases with the alloying element content, Ni, Cr, and Mn, as well as with decreasing quench severity (e.g., from water to oil).

In welds, which are not normally quenched and tempered, there can be compositional and microstructural variations on a scale of tenths of millimeter. As already noted, the melting of the substrate at each weld bead pass can cause compositional gradients. In addition, the rapid freezing of the molten metal adjacent to a previously deposited

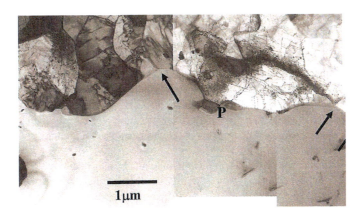

FIG. 10.6

Fine-scale microstructure showing a high degree of inhomogeneity. At the top are fine grains with high dislocation density, below is a single large grain with very low dislocation density, pinned at the precipitate P and at the points marked by the arrows. In comparison, a 10 keV displacement cascade (Microstructural evolution during irradiation section) has a size of about 0.01 µm [19].

bead or to the base material results in columnar growth, tending to be perpendicular to the surface on which is being deposited. Subsequently, the exposed surfaces of each weld bead that is not actually melted are heated sufficiently to cause recrystallization, producing a narrow layer of equiaxed grains in those regions. A heat-affected zone (HAZ) (about 3–5 mm wide) is formed in the adjacent base material. This has a complex structure of coarse and fine-grained regions, depending on the maximum temperature, the subsequent cooling rate and whether there is overlap between the HAZ immediately adjacent to the weld bead and the one deposited above it (see Fig. 10.5). In some welds there may also be variations in weld composition and microstructure at the root due to sealing either by manual welding or with a different wire.

These microstructural variations have a significant effect on mechanical properties and sensitivity to irradiation damage. The microscale variations are much smaller than the scale of most mechanical tests but can be detected by such techniques as micro and nano-indentation and their effects are evident in the scatter in fracture toughness tests. The macroscale variations are significant at the component level and of considerable engineering importance (see Section 10.3.1). They have been studied in several experiments, which have generally used plates, forgings, and welds made with the same practices as the RPV component, archive RPV material, and in a few instances [15, 20, 21] material from decommissioned RPVs. In all mechanical properties studies, test orientation (the axis of loading) and, for tests involving crack propagation, the direction of crack propagation (see e.g., Ref. [22]), may be important. This is due to anisotropic material properties arising from the microstructural variations caused by the fabrication processes (including welding and forging).

One of the earliest and most extensive of these programs was the Oak Ridge National Laboratory (ORNL) Heavy Section Steel Technology (HSST) program [23]. The HSST program investigated many 300-mm thick plates, with the first three being HSST-01, HSST-02, and HSST-03. Fig. 10.7 shows the variation in the drop weight, tensile and hardness values with distance from the surface through thickness. The hardness and drop weight values clearly shows the classical "bathtub" shape caused by the more rapid quenching of the surfaces compared to the interior (the slower cooling of the latter may also be important at subsequent stages of heat treatment). The strength properties have only been determined to $T/2$, but values from $T/2$ to T would have been the mirror image of those from 0 to $T/2$. The drop weight NDT profile is the inverse of the hardness and strength profiles due to the beneficial effect of quenching on refining the microstructure.

Fig. 10.8 [23] shows Charpy impact 44J transition temperature variation from the edge of the plate for samples taken at different distances from the surface.

Fig. 10.8 also shows the beneficial effect of quenching on toughness; in addition, there is clearly an effect of distance from the edge of the plate (where the quenching is on three surfaces, four at corners). It should be noted that the quench effect may vary along the length of the plate or forging depending on the design and efficiency of the quenching rig. The quench effect also varies between components due to differences in composition as well as quenching rate, thickness, and tempering treatments. Experimental programs usually use full-thickness material; as already noted, RPV components are machined after quenching, thus removing some of the effects of spatial variability. Investigations of regions of A-segregation have shown [12] that these could have substantially higher Charpy transition temperatures than adjacent material. However, these regions are small and narrow and there is evidence that their effect is satisfactorily accounted for in fracture toughness bounding curves provided that the test specimens are randomly located.

Toughness properties can also vary significantly within individual welds due to the material and process variations. There can be quite substantial differences when different wire batches are used for different parts of a weld. The toughness properties of the HAZ also vary but are difficult to investigate because of the problem of locating the crack tip in the HAZ (which is nonplanar) and the variability of HAZ structures. Even if the crack tips are located entirely in the HAZ, propagation normally deviates to the weld or base metal side, since HAZ toughnesses are generally better than those of the base material due to the lower temperature transformation microstructure.

The variability in mechanical properties within a component is, however, generally small compared to those between different alloys and components. Compositional differences amplify the effects of differences between fabricators and variability in the processes. For example, WWER-440 steels contain V, which produces smaller and more finely distributed carbides, hence greater strength.

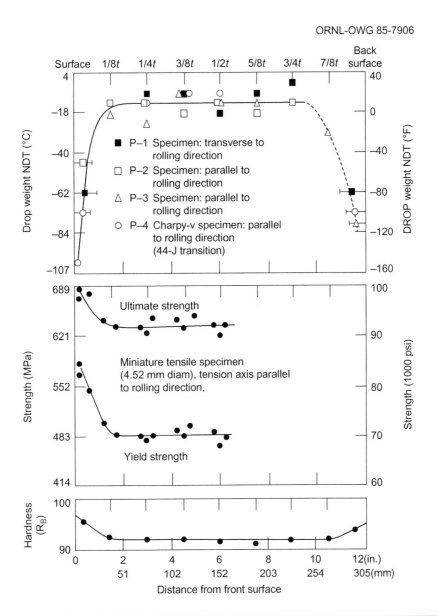

FIG. 10.7

Variations in strength through the thickness of the HSST01 plate, illustrating the classic "bathtub" shape in hardness caused by the higher degree of quenching at the surfaces. The same shape is found for strength properties but only half the profile was measured here. The drop weight NDT profile is an inverted bathtub due to the beneficial effect of quenching on refining the microstructure [23].

Fig. 10.9 shows the BOL Charpy 41J transition temperatures for materials in the US surveillance database. The scatter is much higher than the uncertainty in the mean values (the standard error on the mean can range between about 5°C and 10°C) and the total range of transition temperatures approaches 120°C, which is greater than the predicted end-of-life (EOL) irradiation shift in many cases. Thus, it is as important to determine and control BOL properties as it is for irradiation shift.

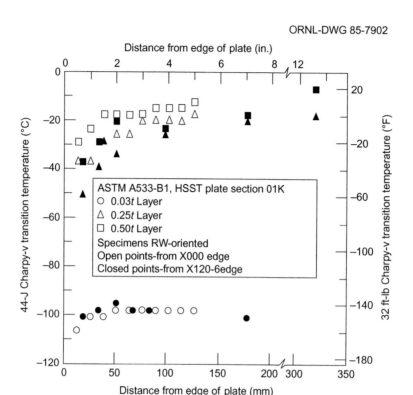

FIG. 10.8

Charpy transition temperature variation with distance from edge of plate and depth from the surface. The results reflect those in Fig. 10.7 showing the beneficial effect of quenching in improving toughness transition temperature; conversely the poorer properties at locations further from quenched surfaces [23].

Fig. 10.10 shows variability in toughness transition temperatures in a set of RPV steels. The data are from 27 unirradiated (8 plates, 7 forgings, 11 welds, and 1 HAZ) and 10 irradiated (1 plate and 9 welds) RPV materials, as measured by various specimen types with a wide range of sizes. The left-hand figure shows the raw data revealing a wide spread in the toughness transition temperature commensurate with the spread in the Charpy transition temperature shown in Fig. 10.9. In the center and right figures the data are normalized using the RT_{NDT} and RT_{T_0} parameters. These are described in detail in Section 10.3.1 but are, respectively, early [prefracture mechanics (FM)] and recent methods for normalizing toughness data to take into account the effects of heat-to-heat variations and specimen size on transition temperature. It can be seen that RT_{NDT} produces little or no improvement in the data spread. RT_{T_0} is much better; the scatter that remains is largely due to material inhomogeneity.

The prediction of the mechanical properties of a component, and property variation within it, on the basis of the alloy composition and the fabrication parameters, remains a largely empirical and inexact process [18]. The values used for SIA (see Section 10.3.1) are, therefore, based on results from samples taken from the individual RPV components in conjunction with previous data for the same heat and for similar applications. The SIA process includes large margins for uncertainty, for example due to variations in properties within the component, relative to the test sample location. The size of these margins could be potentially reduced by research to establish the ranges of microstructures and properties within prototypical components and to establish the sources of this variability.

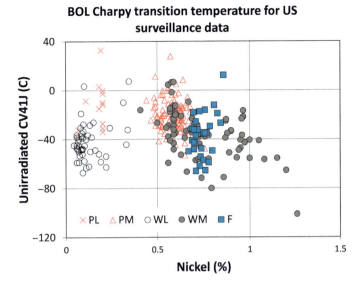

FIG. 10.9

Variability in unirradiated Charpy transition temperature, plotted vs Ni content. PL points are low Ni plates (A302B and A302BM with arbitrary upper limit of 0.35% Ni); PMs are medium Ni plates (A533B); WLs are low Ni welds; WMs are medium to high Ni welds, and Fs are forgings. As the case for Fig. 10.1, each point represents a single sample from a unique heat of weld [2].

10.2 PRINCIPAL APPLICATIONS OF LAS
10.2.1 REACTOR PRESSURE VESSELS

LASs have been the dominant material for Gen I to Gen III GCR and LWR RPVs. The C-Mn steels used in the early Gen I (≈ 100 MWe) Magnox GCRs were also used in the UK's AGR internals. Although these steels are now technologically less important, a significant amount of the pioneering experimental work done on them remains relevant. All existing in-service and in-build PWR and BWR RPVs are made from either one or more of the Mn-Mo-Ni variants or one of the WWER (Cr-Mo and Cr-Ni-Mo) steels. These materials have performed remarkably well. The few in-service problems have been mostly caused by the high irradiation sensitivity and the relatively poorer BOL properties of the earlier LAS. The higher irradiation sensitivity was primarily associated with Cu, though this, and the importance of other elements such as Ni and P, was unknown at the time these RPVs were made. For later RPVs, changes to compositional specifications and manufacturing processes have substantially improved the quality of LAS and minimized the chance of the recurrence of the problems that affected the production of some earlier RPVs. Overall, the extensive feedback available on the performance of LAS in production and service, and the need on occasion to resolve some challenging SIA issues (e.g., see Section 10.3.2.4) has resulted in a mature set of alloys for future use in these types of reactors. However, there are significant differences between LWR RPVs in terms of irradiation flux, fluence and temperature, and operational transients; thus, the performance of LAS in one reactor design may not necessarily be transferrable to another. Current experience is limited to operating temperatures less than about 300°C and fluences less than about 5×10^{19} n/cm^2 ($E > 1$ MeV). While more than 95% of the LWRs in the United States have been granted license extensions to 60 years, it is currently planned to extend the lifetime of many RPVs to 80 years (six plants have submitted applications), giving an EOL fluence of double this value.

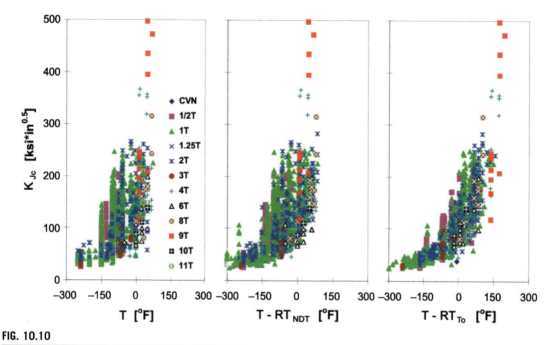

FIG. 10.10

Variability in toughness transition temperature. The left-hand figure shows the scatter in the raw data arising from heat to heat variations and size effect (CVN = precracked Charpy V-notch and, e.g., $1T = 1$-in-thick compact tension specimen). The middle figure shows the ineffectiveness of RT_{NDT} as a normalizing parameter relative to RT_{T_0} (right-hand figure), which significantly reduces the spread of the data relative to temperature [24].

10.2.2 OTHER PRESSURE VESSELS

LAS are also used for PWR pressurizer and steam generator (SG) shells. These can be made using essentially the same materials and fabrication techniques as RPVs. Pressurizers are not irradiated but operate at significantly higher temperatures (e.g., 340–350°C) than RPVs so they are, therefore, more likely to degrade by thermal aging (Section 10.3.2.2), and although this has not been a practical issue for a typical reactor operating time of 40 years, it has been identified as a potential issue for extended operation. SG shells are subject to secondary system pressures allowing a lower wall thickness. Since there are normally two or more SGs per RPV, which can be individually isolated from the primary circuit, and since SG shells are not part of the primary circuit and are not subject to irradiation damage, they have somewhat less stringent SI requirements than RPVs. There are no known SG shell degradation issues that have not been addressed for RPVs. Although there have been many problems with SGs, all have been due to fretting, corrosion, and cracking issues with SG tubes and internals; moreover, the SG is a replaceable component and many have been replaced in operating plants.

10.2.3 PIPING

Although cast austenitic stainless steel has been used for main coolant piping of PWRs and the recirculation piping in BWRs, carbon and LAS are used for many piping components in both. Carbon and LAS are often preferred because of their combination of relatively low cost, good mechanical properties in thick sections, good weldability, and generally high resistance to stress corrosion cracking (SCC). In many cases, the carbon and LAS are clad with a weld overlay on the inside wetted surface with corrosion-resistant materials, such as austenitic stainless steels or nickel-base alloys.

Typical reactor coolant piping system applications and specifications of carbon and low-alloy steels include, for example, SA-516 grade 70 carbon steel with internal stainless steel cladding for PWR piping, SA-106 grade B, and SA-333 grade 6 for BWR main steam and feedwater piping in the unclad condition. The chemical compositions of such steels are shown in Table 10.2 [25].

The experience with these materials in LWRs has been good for the most part, although there have been some occurrences resulting in technical issues identified for further research. Some of these issues associated with carbon and LASs for LWR piping include [25–28]:

1. *Low resistance to flow-accelerated erosion/corrosion (FAC)*, especially if the steel is low in chromium. Low chromium is common for carbon steels as chromium is simply a tramp element in most carbon steels, causing a high susceptibility to FAC/erosion-corrosion. Resistance to FAC/erosion corrosion can be increased with small additions of chromium, copper, and molybdenum in the steel. Additionally, control of the pH, temperature, chemistry, and flow rate of the coolant are also helpful in minimizing FAC effects.
2. *Susceptibility to general, crevice, and pitting corrosion in uncontrolled environments:* The effects of general corrosion are taken into account in the design basis for PWR and BWR primary pressure boundary carbon steel and LAS components by virtue of a corrosion allowance to account for the loss of material during service life, and this is not generally a significant issue except for consideration of extended operation of LWRs to 60 or 80 years.
3. *Boric acid corrosion* in some materials in PWRs, but mostly this occurs outside the primary pressure boundary. A highly concerning example was the corrosion of the Besse Davis RPV head from an external leak (see Section 10.3.3).
4. *Transgranular stress corrosion cracking (TGSCC)*; in general, carbon and low-alloy steels have a good resistance to TGSCC in the high-temperature conditions of LWRs, but there have been observations of TGSCC that compels examination of compositional limits for specific elements, such as aluminum and nitrogen.
5. *Corrosion fatigue* has been identified as a potential concern in piping applications but has not been observed to be a serious problem in operating PWRs and BWRs except for a few cases at connections and nozzles.
6. *Strain-aging embrittlement* can occur in the operating range for LWR piping but can be mitigated by proper heat treatment.
7. *Environmental fatigue* has not been observed in service under normal operating conditions, but there are research results published indicating that fatigue resistance in LWR coolants is lower than previously thought; thus, improved understanding of the issue is needed in consideration of extended operation.

In summary, the experience with carbon and LASs for piping applications in LWRs has been generally favorable for the original design life of about 40 years. However, there are some technical issues requiring vigilance and, in some cases, additional research especially associated with extended operation to 60 or 80 years [25].

10.3 PERFORMANCE

The failure of an RPV is potentially catastrophic in human, economic, and political terms. Accordingly, they are designed, constructed, and operated to regulatory codes that require safety margins sufficient to ensure that such failures are incredibly unlikely.[11] Central to these codes is SIA. Section 10.3.1 provides a background to the regulatory codes, standards, and SIA methods relevant to PWRs and BWRs. It focuses on US practice, but the approaches used elsewhere are broadly similar. The materials inputs required for the SIA of RPVs include the materials property values of the RPV components at BOL (see Section 10.1.3), and estimates of how these change

---

[11] Taken as a failure rate of 1 in 10^7 per vessel year. This, for example, is about 10 times less likely than the probability of a super-eruption at Yellowstone (estimated by Bill McGuire, Hazard Research Centre at University College London, reported in the Guardian newspaper, January 6, 2009).

Table 10.2 Compositions and Properties of Some Representative LAS Used for PWR and BWR Piping

ASME/ASTM Spec. Type Grade, UNS No.	C	Mn	P	S	Si	Cu	Ni	Cr	Mo	V	Nb
SA/A 105 CS Forgings K03504	0.035 max	0.60–1.05	0.035 max	.040 max	0.10–0.35	0.40 max (1)	0.40 max (1)	0.30 max (1)	0.12 max (1)	0.05 max	0.02 max
SA/A 106 Seamless CS Pipe Grade B, K03006	0.30 max	0.29–1.06	0.035 max	0.035 max	0.10 min	0.40 max (2)	0.40 max (2)	0.40 max (2)	0.15 max (2)	0.08 max (2)	—
SA/A 333 CS	0.3 max	0.29–1.06	0.035 max	0.035 max	0.1 max	—	—	—	—	—	—
SA/A 5_6 CS Plates Gr. 70, K02700	(3)	0.85–1.20	0.035 max	0.035 max	0.15–0.40	—	—	—	—	—	—

Notes:
1. Sum of Cu, Ni, Cr, and Mo shall be <1.00%; and sum of Cr and Mo shall not exceed 0.32%.
2. Limits for V and Nb may be increased to 0.1% and 0.05%, respectively.
3. C max. varies with thickness of plate; 0.5–2 in., 0.28% max; 2–4 in., 0.30% max; 4–8 in., 0.31% max.

during service due to irradiation and aging (see Section 10.3.2). Section 10.3.3 describes other performance issues for RPVs including structural degradation through mechanisms such as fatigue and corrosion. It also describes the SIA issues that emerge in cases where in-service inspection (ISI) has unexpectedly revealed manufacturing flaws.

10.3.1 REGULATORY CODES AND STRUCTURAL INTEGRITY ASSESSMENT (SIA)

10.3.1.1 Codes and regulations

Governments ensure that critical structures, such as RPVs, are safe through regulations policed by a regulatory or licensing authority. In the United States, the legal requirements for RPVs are defined in the US Code of Federal Regulations (CFRs), backed up by the US Nuclear Regulatory Commission (USNRC) Regulatory Guides (RG). Both of these incorporate by reference design and construction rules such as the ASME BPVC (see also Section 10.1.2) and materials standards, such as the ASTM International Standards and Guides. Both ASME and ASTM depend on work by expert volunteers to ensure that the requirements of the codes and standards appropriately reflect current knowledge and best practice; for example, there may be a requirement to obtain mechanical property values at distances greater than a quarter thickness from quenched surfaces to conservatively take account of the bathtub effect discussed in Section 10.1.3. Other countries have adopted US codes or developed their own domestic versions. Codes and standards used in the United States, France, Japan, Korea, Canada, and Russia are compared in Ref. [29].

There can be differences in regulatory frameworks. For example, while most regulations are prescriptive, the UK approach defines a set of safety principles [30]; licensees must convince the regulator that the methods and standards they have chosen to use meet these principles. However, international collaboration has brought increasing convergence on the technical details within regulatory codes. International regulatory cooperation is promoted through bodies such as the IAEA, the Organization for Economic Cooperation and Development Nuclear Energy Agency (OECD-NEA), and the Western European Nuclear Regulators Association (WENRA).

The safety and performance of an RPV, as any engineering structure, depends on:

- Good design and proven materials and manufacturing processes.
- Inspection and testing to ensure that the design and materials requirements have been met.
- Analysis to demonstrate that it will withstand its service loading during its lifetime.
- Appropriate maintenance and monitoring in service.

Underlying these requirements is the technical discipline of SIA.

10.3.1.2 Structural integrity assessment (SIA)

A general introduction to SIA is given in Ref. [31]. SIA is used to establish the ability of a component or structure to survive, by an adequate margin, its design lifetime without failure or unacceptable deterioration in performance. It must consider the properties of the materials, the possibility of defects, changes to the materials or to the structure due to in-service degradation, and the loadings on the structure due to normal operation and (within-design-basis) accidental service conditions. Design codes can incorporate SIA using a simplified conservative approach, for example, by specifying conservative flaw sizes and analysis methods. For complex or unusual design features, and when unexpected threats to SI arise, detailed finite element analysis (FEA) calculations may be needed.

SIA must consider all potential failure modes. In the case of LWR RPVs, brittle failure is the dominant concern. Temperatures are too low for creep to be an issue; corrosion is unlikely because the internal surface of the RPV is normally protected from the coolant by the stainless steel cladding, and the external surface is normally dry (but see Section 10.3.3); fatigue and fatigue crack growth are not normally significant due to the low number of significant stress cycles.

RPVs are designed such that at operating pressures, the stress on the material is well below the yield stress and they cannot fail by "plastic collapse." However, in the presence of a sharp ("crack-like") defect, ferritic steels can fail by a brittle cleavage mechanism or by ductile tearing at stresses lower than the yield stress. This behavior can be

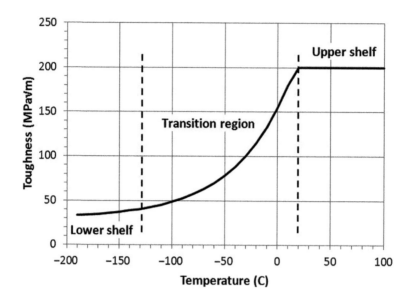

FIG. 10.11

Mean fracture toughness vs temperature curve showing the factor of about 6 increase in toughness over a temperature range of about 150°C (the "transition region"). At the lowest temperatures (the "lower shelf") LAS are brittle, and specimens fail by unstable cleavage fracture. At the highest temperatures (on the "upper shelf") crack initiation is followed by stable ductile tearing, which continues only if the stress is increased, a relatively benign condition [2].

predicted by FM [32]. Elastic FM defines a stress intensity factor, K_I, due to applied loads, to represent the "driving force" for the initiation of cleavage from a sharp crack. The K_I controls the stress and strain fields near the tip of a crack under so called small-scale yielding (SSY) conditions, provided that the local plastic zone that forms at a blunting crack is very small compared to the cracked body dimensions and overall plastic deformation is negligible. K_I is proportional to a characteristic loading stress multiplied by the square root of the crack length, so has units of MPa$\sqrt{m}$. A nondimensional coefficient accounts for the cracked-body geometry.

K_I is valid only when plastic yielding is localized to the crack tip. If the cracked body undergoes overall yielding before the elastic K_I is reached, plastic work is expended that must be accounted for. A major development in FM was the J-integral method of characterizing the crack tip state [32], which can in some circumstances account for the plastic work. The J-integral, J_I, values have units of MPa-m, but these are usually converted to the K_I equivalent values, K_{JI}. The J_I accounts for the plastic work that flows through the blunted crack tip to establish the local stress and strain fields for conditions of so called SSY, which is self-similar and dimensionally scales with K_I^2. Under SSY, the peak crack tip stress reaches values of 3–5 times the yield stress and the spatial extent of the stress concentration varies as K_I^4. The plastic J is related to the area under the elastic-plastic load-displacement curve. Values of K_I and J_I depend on the crack size, the cracked body geometry and loading details.[12] Modern FEM codes can provide the assessment of J, and the corresponding crack tip stress and strain fields, in three-dimensional geometries for arbitrary sizes and geometries.

While K_I (or K_{JI}) is the driving force for fracture, K_{Ic} (or K_{Jc}) is a material property, namely the critical stress intensity factor at which crack propagation initiates and is referred to as "fracture toughness" [33]. Values of K_I (or K_{JI}) are obtained by testing specimens with sharp fatigue cracks grown from machined notches. Fig. 10.11 schematically illustrates a mean fracture toughness vs temperature curve. Over a temperature range of about

---

[12] The suffix "I" in the K and J nomenclature refers to the "Mode I" crack loading in which the applied stress is perpendicular to the plane of the crack. If the flaw is not so aligned Mode II and Mode III loadings may also be important.

150°C (the "transition region") K_I or K_{Ij} increases by about a factor of 6, corresponding to a similar increase in the load on the specimen at initiation. At the lowest temperatures (the "lower shelf") LAS are brittle, and specimens fail by unstable cleavage fracture. At the highest temperatures (on the "upper shelf") ductile crack initiation is followed by stable crack tearing. The point of initiation of stable tearing is generally defined at 0.2 mm of stable growth, termed $K_{J0.2}$. Further growth can be characterized by J_R-da (J-resistance as a function of ductile crack extension) curves. The SIA of RPVs is normally dominated by the need to avoid brittle failure at temperatures below the onset of the upper shelf; thus, the low-to-mid transition region is generally the range of most practical concern.

Fig. 10.12, after Wallin [34], shows fracture toughness data from the very large Euro dataset [35]. Although all results are from a single heat of material there is considerable scatter; this is typical of all LAS and results from intrinsic material inhomogeneity at the microstructural scale (see Section 10.1.3). It is clear that larger specimens give lower mean values. Before the development of J-integral methods, very large specimens were used to maintain elastic fracture (minimal plastic deformation) conditions that are "valid" with respect to the requirements of the ASTM E399 standard [22]) K_{Ic}, even at the upper end of the transition region. Elastic-plastic fracture mechanics (EPFM) led to a major relaxation of these size requirements, since K_J could still uniquely characterize the crack tip SSY fields, at least for deep cracks below specified plastic deformation limits.

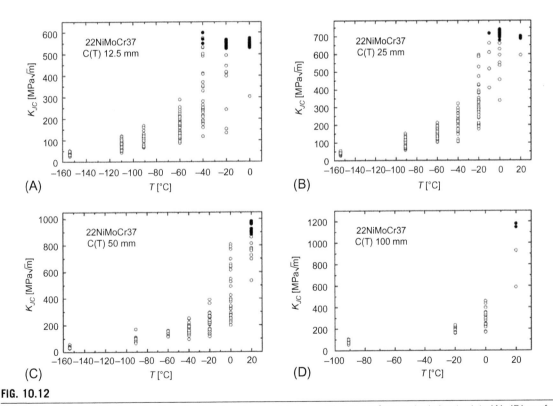

FIG. 10.12

Fracture toughness data from the Euro fracture toughness database. All data are from a single heat; plots (A)–(D) are for different specimen sizes and show the increase in transition temperature as size increases. Filled symbols represent specimens that failed by ductile tearing [34].

From K. R. W. Wallin, Distribution free statistical assessment of scatter and size effects in the Euro fracture toughness data set, Engineering Fracture mechanics 103 (2013) 69–78. Original data from K. Wallin, Fracture toughness of engineering materials – estimation and application, EMAS Publishing, Warrington (2011).

However, SSY fields occur only under limited conditions that requires deep cracks and limited plastic deformation. Deviations from these conditions lead to so-called loss of constraint. Low-constraint conditions associated with excessive deformation often occur during fracture testing, and commonly apply when near surface, or large, cracks in heavy section components, like RPVs, are assessed by SIA. Loss of constraint results in decreases in the amplitude of the crack tip stress fields relative to SSY conditions. The net result is that a larger J is needed to cause cleavage. In the limit, the crack tip stresses are not sufficient to initiate cleavage and fracture occurs by ductile tearing, which is generally not the limiting safety concern. Finally, as discussed further, cleavage fracture toughness also depends on the crack front length (B) due to weakest link statistical effects, related to the random distribution of trigger sites. The stresses needed to trigger cleavage sites can be described by Weibull statistics that rationalize the large intrinsic scatter in K_{Ic} or K_{Jc} measurements. Fortunately, loss of constraint can be accounted for by FEM simulations of the blunting crack tip stress fields and local fracture models of cleavage initiation [36]. Statistical size effects approximately follow a simple $K_{Jc} \propto (K_{Jc} - K_0)B^{-1/4}$ type scaling where K_0 is minimum toughness.

While EPFM allows fracture testing in the transition with relatively small specimens, development of $K_{Jc}(T)$ curves for both unirradiated and various irradiated conditions would require an impracticably large number of statistically valid fracture tests. This problem has now been largely resolved by the development of the master curve (MC) by Wallin [37]. The MC, Fig. 10.13 [38], has a nominally invariant $K_{Jc}(T)$ for a large range of alloys and alloy conditions, including following irradiation. However, the alloy-condition specific MC has different positions on an absolute temperature scale. Since their shape is constant, these differences can be accounted for by indexing the MC to a reference temperature (T_0) for a specific alloy condition. Thus, when plotted on a $K_{Jc}(T-T_0)$ scale the data form a single scatter band, with a median K_{Jc} of the MC shape. Here, T_0 is the temperature for a median $K_{Jc} = 100\,\mathrm{MPa}\sqrt{\mathrm{m}}$. Wallin also showed that the scatter in toughness about the mean curve is roughly the same for a wide range of ferritic steels and can be described by weakest link Weibull statistics. Specimens with longer crack front lengths are more likely to contain weak initiators than those with short ones, and thus tend to give K_{Jc} values at the bottom of the scatter band. Further, using the weakest-link statistics concept, the MC for various B can be normalized to a reference crack front length (or specimen thickness where the reference is 25 mm). The MC

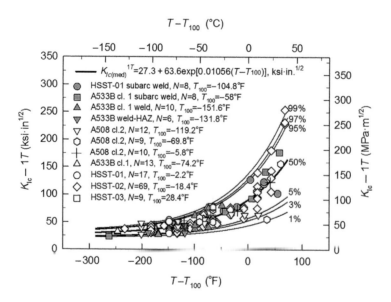

FIG 10.13

Demonstration of the effectiveness of the master curve in normalizing data with a wide range of transition temperatures [38].

concept has also been demonstrated for irradiated steels [39] and for a wide range of specimen geometries [36, 40]. The big advantage of the MC is that, with limitations associated with constraint and material homogeneity effects (see Ref. [41]), it allows the toughness distribution of a heat of material over a range of temperatures to be determined using a relatively small number of small specimens. Note, an alternative model, the unified curve [42], of which the MC is a special case (according to the Unified Curve author), proposes that the curve flattens with increasing T_0, hence with increasing irradiation damage. Such flattening has been long predicted and is under investigation. However, a more rigorous model that accounts for the influence of the alloy strength on the temperature dependence of the critical cleavage fracture stress rationalizes a constant MC shape for a wide range of T_0 [43].

Nuclear regulatory codes were developed before the development of modern FM. The relatively simple empirical methods that were then adopted to protect against RPV fracture remain influential. The pre-FM concept was that toughness could be characterized by a reference temperature RT_{NDT} determined from the most limiting value from two types of tests, both carried out under impact loading: the Pellini drop weight test and the Charpy impact test. The drop weight test determines the temperature, T_{NDT}, at which a cleavage fracture, initiated in a deliberately brittle crack-starter weld bead deposited on the test material, arrests in the test material. The Charpy test measures the total energy required to initiate and propagate a crack in a small blunt-notched specimen.[13] As shown in Fig. 10.14 [44], a number of Charpy tests are carried out at different temperatures to establish the temperature at which the mean energy absorbed is equal to a reference energy; the ASTM standard specifies the energy as 40.7 J (the metric equivalent of the originally specified 30 ft-lb). In practice, particularly in countries that use

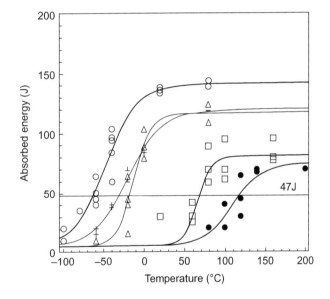

FIG. 10.14

Charpy transition curves for a WWER-1000 weld, illustrating the large irradiation shift and upper shelf drop at high fluences and the substantial data scatter that can occur at high fluences or with irradiation-sensitive materials (from left to right the curves are unirradiated and irradiated to 3.8, 5.9, 18.4, and 26.2 × 10^{23} n/m^2, $E > 0.5$ MeV, respectively) [44].

---

[13] Both the drop weight and Charpy tests were developed to provide an empirical measurement of the toughness of the material and were often described as toughness tests. However, they should not be confused with fracture toughness tests, which are founded on well-established theoretical models.

the metric system, the value is often rounded upwards and we term the reference temperature as CV41.[14] The Russian codes use a somewhat similar approach to define a "critical temperature of brittleness," T_{k0}; this is based on Charpy tests, but the reference energy depends on the material's (unirradiated) yield stress.

It was assumed that RT_{NDT} would normalize FT data from different heats. All data available at the time were plotted vs $(T-RT_{NDT})$ and the ASME K_{IC} curve was drawn as the lower bound curve to be used in SIA. Although all subsequent data have been reasonably well bounded by the ASME K_{Ic} curve, it is now known that RT_{NDT} can give overly conservative results (e.g., [45]). ASME now allows a hybrid approach [46], which uses a new normalizing parameter RT_{T0}, based on the MC. Guidelines for application of the MC approach to RPVs are described in Ref. [47].

SIA must demonstrate that there is a margin between K_I and the lower bound K_{Ic} adjusted for the irradiation shift. This must be the case for all potentially limiting locations (including nonirradiated regions) and all operational and potential accidental loadings. Fig. 10.15 shows the K_I transient from a pressurized thermal shock (PTS) event in which, following a loss of coolant accident (LOCA), relatively cold water is injected into the RPV with the internal pressure of the RPV remaining high. To ensure that the margin is maintained, RPVs are operated to pressure/temperature limits defined in a "P–T" diagram [8]; a simplified illustration of which is given in (Fig. 10.16). The P–T limit, the shape of the limiting curve on the left-hand-side, mirrors the fracture toughness lower bound. For normal operation, start-ups, cool-downs, and operation transients, all the factors involved in determining the margin (such as values of BOL toughness, and K_I transient) are calculated in a highly conservative (i.e., on the safe side) manner. For accident conditions, such as PTS, which are extremely rare, some relaxation of the conservatisms is allowed, and some regulatory codes permit use of probabilistic methods. USNRC 10CFR50.61a [49], for example, may permit the use of probabilistic FM analysis for their assessment. The USNRC report, NUREG-1874, [50] references and discusses the Fracture Analysis of Vessels-Oak Ridge (FAVOR) Code for carrying out the assessment. Nevertheless, as noted earlier in this section, failure probability targets of less than 1×10^{-7} per RPV year are sought.

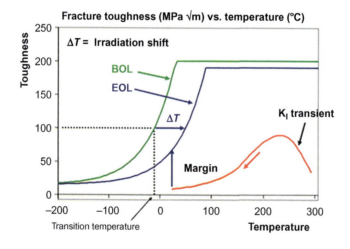

FIG. 10.15

To ensure safe operation a substantial margin must be maintained between the conservatively calculated stress intensity factor (K_I) at the tip of a conservatively sized defect in the RPV wall and the lower bound toughness curve for the material surrounding the crack tip. The bounding curve is defined by the beginning of life (BOL) toughness bound, shifted by a conservative estimate of the irradiation shift at the crack tip location [2].

---

[14] Various other nomenclatures are used in the literature.

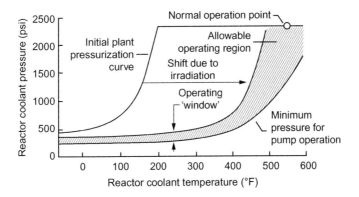

FIG. 10.16

Schematic diagram illustrating the ASME Section XI Appendix G operating requirements. The initial plant pressurization curve, and that curve shifted due to irradiation, reflect the shape of the toughness curves shown in Fig. 10.15. If the shift becomes too large, the operating window (shaded region) becomes too small for continued operation to be allowable [8].

From International Atomic Energy Agency, Integrity of Reactor Pressure Vessels in Nuclear Power Plants: Assessment of Irradiation Embrittlement Effects in Reactor Pressure Vessel Steels, IAEA Nuclear Energy Series NP-T-3.11, IAEA, Vienna (2009). Used with permission from IAEA.

Regulatory codes require ISI in critical regions to demonstrate that any flaws in the vessel have not grown. They also require surveillance testing to monitor irradiation embrittlement as determined by the irradiation-induced transition temperature shift, the difference between the transition temperature in the unirradiated and irradiated states. Note that the transition temperature corresponds to a specific mean toughness and the irradiation shift is related to the temperature increase associated with the effect of irradiation damage on yield stress (see discussion of Fig. 10.26 in Section 10.3.2.1, under subsection "Relationship between hardening features and hardening and embrittlement"). Capsules containing mechanical test specimens are positioned inside the RPV at BOL in a location where they are irradiated at a faster rate than the RPV itself by a lead factor, usually between about 2 and 5, but sometimes outside this range, particularly in early RPVs. The capsules are withdrawn at specified intervals and the specimens tested, to provide estimates of the irradiation shift of the vessel wall at a future point in service life, given by the effective service life at the time of specimen withdrawal multiplied by the lead factor. These values are used to confirm the predictions based on trend curves (see Section 10.3.2.1, under subsection "Applications modeling") that were used to define the pressure-temperature (P–T) limits. Significant deviations from predictions require investigation and adjustment of P–T limits as appropriate; accumulated surveillance data for a fleet of RPVs are used to produce improved trend curves. The ASTM requirements for surveillance testing are given in Ref. [51].

Although it would in principle be possible to do surveillance testing using fracture toughness specimens, most regulatory codes retain the use of pre-FM methods. However, it was impractical to irradiate drop-weight specimens because of their size, thus the effect of irradiation on ASME K_{Ic} curve, that is, the change in RT_{NDT} (ΔRT_{NDT}), was assumed to be the same as the shift in the Charpy curve at 41J (ΔT_c), the increase in CV41 from the unirradiated (BOL) to the irradiated condition. The validity of this assumption is intrinsically questionable, since the tests and what they measure are very different. For example, Charpy shifts are affected (relative to fracture toughness) by the decrease in the upper shelf energy (USE), and ultimately if the USE falls below the reference energy, the nominal shift would be infinite. Fortunately, the shelf drop is, in practice, not this large, and can be corrected for application with methods described in Ref. [52]. Further, as shown in Fig. 10.14, shift values can be imprecise due to material scatter. Nevertheless, although SIA requires values of fracture toughness, Charpy testing remains the standard method of determining irradiation shift. This is because most surveillance capsules (see further) were populated with Charpy specimens at build, the simplicity of Charpy testing and the resulting large historical database. As we see (in Section 10.3.2.1, under subsection "Relationship between hardening features and hardening and embrittlement"), fracture toughness shifts and Charpy shifts are generally, but not always, reasonably well correlated. It is noted that some surveillance programs have included small fracture toughness specimens. In the United States, most have not been tested awaiting regulatory guidelines for acceptability.

The trend curves used to predict fracture toughness shift during the lifetime of an RPV must correctly identify the limiting materials that should be included in surveillance capsules and they must be conservative. It is unavoidable that trend curve predictions must be extrapolated. For the oldest reactors, the lead factor ensures surveillance data are available for fluence values appropriate to future operation, but the data may be relatively sparse, and the predictions must be extrapolated in terms of time. This is particularly an issue when consideration is given to extending life beyond that originally planned, and there are no surveillance capsules available to cover the extension. Although surveillance date for older reactors can be used in the development of shift predictions for more recent ones the materials and manufacture of reactors, and even details of their operation, have changed over time. Thus it is important to have sufficient understanding of irradiation damage mechanisms to be confident that these extrapolations beyond current experience have a rigorous physical basis. These issues are discussed in Section 10.3.2.

There have been cases where it has not been possible to demonstrate with sufficient confidence that the margin between K_I and the lower bound K_{Ic} will be sufficient in future operation, and some RPVs have been taken out of service before planned end of life. This has generally occurred because the continued surveillance, ISI and other oversight processes embedded in regulatory procedures produced new information that potentially undermined the existing safety case. An example was the Yankee Rowe reactor in the United States, which was shut down in 1992 after only about ¾ of its planned life due to embrittlement concerns. A more recent case was the discovery of a population of small cracks in the Belgian reactors Doel 3 and Tihange 2 RPVs (see Section 10.3.3). It took 3 years of very intensive research and safety analysis work, and regulatory assessment before a restart was authorized. However, the decision to decommission an RPV following such discoveries is often made for political or economic reasons. Particularly in the case of older plants, when required data are not available, and which are nearing the end of their planned life, the risk that an adequate safety case might not be achieved may be judged to be too high. In several cases where surveillance data have revealed unacceptably high levels of embrittlement, RPV annealing has been used to recover most of the damage and enable continued operation (see Section 10.3.2.4).

10.3.2 IN-SERVICE DEGRADATION
10.3.2.1 Irradiation damage
Historical background

It was predicted by Wigner in 1946 [53] that any substance subjected to collision of neutrons would undergo atomic displacements, and further recognized from the early days of commercial nuclear power in the mid-1950s that irradiation could embrittle RPV steels [54], but the potential severity and complexity of the effects were underestimated. By the early 1960s some understanding of the mechanisms of embrittlement was developed by Brinkman [55], Seeger [56], and others, based on theoretical descriptions for the effect of neutron interactions creating displacement cascades and displaced atoms defects in the form of isolated and small clusters of vacancies and self-interstitial atoms (SIA) (see Chapter 5). It was empirically established that embrittlement (generally measured as irradiation shift, see Section 10.3.1) depends strongly on the "environmental" variables of fluence (Φ) and irradiation temperature (T_{irr}). Fluence is the total number of neutrons with energy above a specified threshold, normally assumed to be 1 MeV (0.5 MeV in WWER practice) passing through a unit area (generally 1 cm^2) of the material during the irradiation. Changes in mechanical properties, including transition temperature shifts (Λ) were modeled using expressions such as Eq. (10.1), which implicitly assume that there is a single damage mechanism with irradiation shift dependent on Φ raised to a fixed exponent. Some early models adopted a ⅓ exponent, but later a value of ½ became more widely used. Still later models have their exponents fitted to Δ data trends, leading to exponents greater than ½ in some cases. The fitting parameter, A, was fitted to take account of the influence of irradiation temperature and LAS variables, like plate and weld product forms as well as alloy type and composition.

Simple early irradiation shift model.

$$\Delta = A\Phi^n \tag{10.1}$$

where Δ is the mechanical property change due to irradiation, Φ is the fluence (n/cm^2, $E > 1$ MeV), and A and n are fitting parameters.

While fluence remains a standard dose unit, neutron spectrum effects can be approximately accounted for by displacement per atom (dpa) exposure units; dpa scales with the kinetic energy deposited onto LAS atoms by the high-energy neutrons (see "Microstructural evolution during irradiation" subsection below). As noted above, A depends on material, as indicated by considerable variability in sensitivity to irradiation damage between different alloys and between different heats of the same alloy. As described in "Applications modeling" section, in later models, A was replaced by more complex expressions incorporating chemistry terms and parameters. Its dependence on T_{irr} was also recognized in models applicable to reactors with significant temperature gradients. Similarly, in later models, the exponent, n, became a fitting parameter and in some cases defined by a more complex expression.

A key breakthrough in the development of models came in 1969 from a series of elegant studies by Potapovs and Hawthorne [57, 58] at the US Naval Research Laboratories, which found that Cu and P were dominant causes of irradiation embrittlement variability. As a result, specification limits were introduced for Cu and P for LAS used in highly irradiated regions in an RPV. Fig. 10.17 illustrates the effect of Cu on irradiation sensitivity.

The specification change mitigated the problem for most new RPVs, but it was apparent that irradiation shift might restrict the operational lifetime (at that time generally planned as 40 years) of some of those already in service (some with 30 or more years to go). Irradiation damage became a "materials problem" and attracted significant research funding. The high costs of irradiating LAS and postirradiation examination of activated specimens and the importance of the results promoted technical exchange and collaboration. For example, the ASTM Effects of Radiation on Materials Symposia, which had started in 1957 [59], became a major technical forum for RPV irradiation damage issues and, in 1971, the IAEA launched a series of coordinated research projects [60].

A key early technical objective was to identify the mechanism of the Cu effect. The problem was that the microstructural examination techniques [e.g., transmission electron microscopy (TEM)] available at the time could not clearly identify the difference between unirradiated and irradiated LAS microstructures. However, in 1981, proceedings papers by Lott et al [61] and Miller and Brenner [62] noted the formation of irradiation-induced copper-rich precipitates (CRPs), and their significance to radiation damage in an RPV steel, with the use of a field ion microscope-atom probe (FIM-AP). Subsequently, a seminal paper in 1983 [63] by Odette of the University of California Santa Barbara (UCSB) proposed the now-accepted mechanism, the formation of CRPs under radiation-enhanced diffusion (RED) (see "Microstructural evolution during irradiation" section). A broad outline of the same mechanism was proposed independently at about the same time by Fisher et al., but not published openly until later [64]. The Fisher model was highly influential in the development of radiation damage understanding, especially in the United Kingdom, where it was used for the development of the trend curves used in the SIA of C-Mn steel Magnox reactors, a major step in the use of mechanistic understanding to support shift predictions.

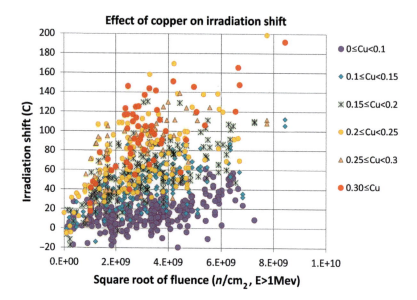

FIG. 10.17

Effect of Cu on irradiation shift; scatter is substantial even within a given group indicating that other variables are also important. Data from the US surveillance database plotted vs square root of fluence for clarity [2].

These early insights led to the development of new irradiation shift models based on a fit to the US surveillance database by both Odette [65] and Guthrie [66]. These early models were the basis of a revision of the US NRC Regulatory Guide on embrittlement, RG1.99-Rev 2 [67]. An important development in the new model was the inclusion of significant Ni effects that had previously not been accounted for. While this was somewhat controversial at the time, the evidence that Ni enhances embrittlement was shown to be overwhelming, leading to sustained studies of the responsible mechanisms [68]. Further, the recognition of possible flux, or dose rate, effects resulted in a decision to restrict the shift model fits to surveillance data (see Section 10.3.1 for a discussion of RPV surveillance programs).

Over the next several years Odette, Lucas and co-workers greatly expanded the theory [69–72] and developed the standard "two-feature" (2F) model of irradiation damage, Eq. (10.2) and Fig. 10.18. This model distinguishes between the hardening contributions from precipitation (the CRPs), ΔP, and those from so-called stable matrix features (SMFs),[15] ΔM, that form in low Cu steels. The 2F model remains the standard framework for irradiation shift correlation modeling for most Mn-Mo-Ni steels and irradiation environments (see "Applications modeling" section).

Standard two feature model of irradiation embrittlement.

$$\Delta T = \Delta M + \Delta P$$
$$\Delta M = A\Phi^n$$
$$\Delta P = B \tanh(\Phi/C)$$

(10.2)

where ΔT is total irradiation embrittlement, ΔM is the embrittlement contribution from matrix hardening, ΔP is the embrittlement contribution from precipitation hardening, and Φ is fluence.

A, B, C, n are fitting parameters or expressions that depend on other variables.

---

[15] This is now often described as stable matrix feature hardening (or stable matrix feature damage) to distinguish it from unstable matrix features (see Section 0).

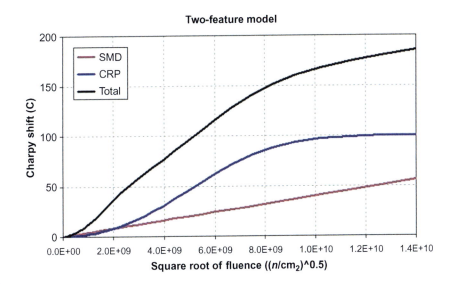

FIG. 10.18

Standard two feature irradiation shift model. The total shift is the sum of the stable matrix damage which has a square root dependency on fluence and the saturating copper-rich precipitate term [2].

The UCSB group, led by Odette, has been responsible for many major technical advances in embrittlement research [73]. Odette strongly advocated the need for mechanistic understanding, supported by good data, in shift model development. He cofounded, with Hawthorne, the International Group on Radiation Damage Mechanisms (IGRDMs) in RPV Steels and was the key figure in its development. IGRDM meetings, held about every 18 months, continue to provide a major forum for collaboration between the many groups involved in LAS irradiation damage studies worldwide.

Since the mid-1980s work on C-Mn, Mn-Mo-Ni, and WWER steels has produced several technical surprises, some of which have had potentially unwelcome implications for RPV lifetimes, even for RPVs with lower Cu steels. This work has shown that irradiation damage mechanisms in LAS are complex and scientifically interesting. The remainder of this section describes how LAS microstructures evolve during irradiation; the effect of environmental and materials variables on this evolution; and the relationship between the irradiation damage features produced by irradiation and hardening and embrittlement of the steels. Finally it discusses modeling of irradiation, including modern physical method of modeling and the earlier more empirical method that are still important in engineering applications. Currently overall knowledge of the mechanisms of irradiation damage derives from consideration of information gained in all these areas. However, each technique used for investigation has different strengths and limitations (see Chapter 4).

Microstructural techniques can identify nanostructural features (above a certain size), but sample only very small volumes; they may have artifacts and do not directly measure hardening or embrittlement. Standard mechanical property tests measure the total hardening or shift, not the contributions of the individual nanostructural features, but a combination of annealing and hardness tests has been used to resolve the various features [74]. Empirical relationships between nanostructure and hardening and embrittlement are obviously affected by the uncertainties in both techniques. Detailed physical models and computer simulations can provide information and insights that are not empirically accessible, but they must be informed by experiment (e.g., what output is to be modeled) and validated by comparing their predictions directly to relevant data. More importantly, simpler, reduced-order, physically motivated analytical expressions, derived from the more detailed physics models, can be statistically fit to large databases. The physical underpinning of the reduced order models can partly offset issues related to the limited quality of the available data.

It should be emphasized that understanding of the physical mechanisms of irradiation damage remains incomplete. This has potential implications for the extension of RPV lifetimes, the introduction of new materials, and the accuracy of predictions of irradiation shifts (see "Applications modeling" section).

Finally, we note that the nomenclature used in the literature on irradiation damage and embrittlement of LAS is inconsistent partly because of changes in the common understanding with time, and because of individual authors' varied perceptions and the methods that they used. In the rest of this chapter we have tried to use a consistent nomenclature but noted alternatives where appropriate.

Microstructural evolution during irradiation

Fig. 10.19 provides a simplified schematic view of nanostructural evolution during irradiation, showing the features that may be formed and the routes and processes through which they can evolve. These are described in the following paragraphs.

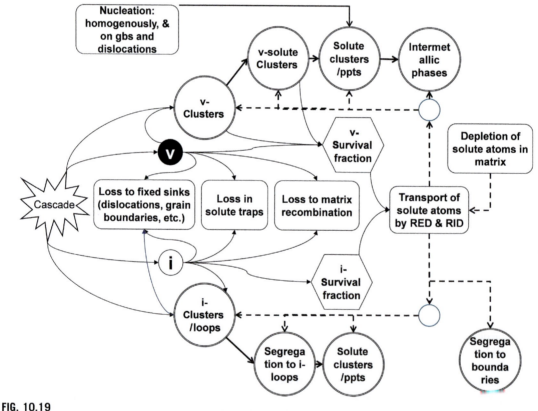

FIG. 10.19

Simplified illustrations of the possible processes and paths involved in nanostructural evolution. The circles are nanostructural features that originate from the cascade. The thin solid lines show the possible fates of these, either in processes that cause temporary or permanent removal from the matrix (rounded rectangles) or through evolution into more complex defects (via the thicker solid lines). The rate of microstructural evolution depends on the survival fractions (hexagons) of the species involved in radiation enhanced (or induced) diffusion of solute atoms (thick dashed lines) and potentially by nucleation rates [2].

Vacancies and self-interstitial atoms. Vacancies and SIAs, or *point defects (PDs)*, are initially produced in equal numbers by neutron interaction generated by high-energy primary recoil atoms (average energy in the tens of KeV) in displacement cascades as *Frenkel Pairs* (i.e., every SIA produced is necessarily accompanied by a vacancy). A fraction of the PD either recombines, or clusters, as the cascade cools to thermal equilibrium over the order of 10s of ps. The number of initially displaced atoms that survive immediate recombination can be calculated by simple Kinchen and Pease (K&P) type models [75], that account for the primary recoil energy partitioning between electronic and nuclear loss processes. The K&P model proposes that net displacements do not occur when the cascade recoil energies fall below a threshold energy of ≈ 40 eV in Fe. Primary defect production models, like Norgett, Robinson and Torrens (NRTs) [76], provide the basis to compute a *displacements per atom* (dpa) exposure, or dose, unit [77], which is the average number of times each atom of a solid is displaced from its crystal lattice site during an irradiation. Molecular dynamics (MD) simulations show that only about 1/3 of the primary displacements escape intracascade recombination [78]. Nevertheless, dpa is regarded as a good comparative measure of damage dose produced by neutrons with a wide and varying range of spectral energy distributions. Note that dpa is really a proper unit of dose, which scales directly with the kinetic energy deposited in solids by high-energy neutrons. Note also that thermal neutrons (and electrons and gamma rays) also produce dpa. However, the more isolated PD from these corresponding low energy recoils experience less recombination since there is no cascade containing vacancies and SIA. Thus, for heavy water and graphite moderated reactors, with a high flux of thermal neutrons (see "Effect of environmental and materials variables" subsection below–*Neutron Spectrum*), the dose is best expressed using weighted dpa units that account for the relative effects of recombination on the surviving PD. Thermal neutrons also generate local dpa from recoil products produced by n, α reactions with the impurity element B.

Single PDs do not themselves cause significant hardening but are fundamental to a wide range of irradiation effects. PDs cluster to form vacancy nanovoid and SIA dislocation loop obstacles. Furthermore, they play a vital role in the long-term evolution of nanostructural features by transporting solute atoms. Under irradiation, vacancies and interstitials reach a steady-state concentration when losses to sinks and by recombination ("Effect of environmental and materials variables" subsection below–*Neutron flux*) are equal to the net generation rate in cascades. RED is the result of the excess concentration of these defects, primarily by an enhanced vacancy exchange mechanism. However, the effect of excess interstitials on diffusion is generally very small, and does not add to the uncertainties for modeling RED, which involves complex effects, such as vacancy drag. As we see in "Effect of environmental and materials variables" subsection below–*Neutron flux*, PD concentrations depend on neutron flux. However, the dependence is not linear and can result in a flux, or "dose rate," effect on irradiation damage. Knowledge of the details of the formation and fate of PDs is essential to understanding and quantifying this effect (Computational modeling studies and Applications modeling sections).

SIA clusters. Modeling studies (e.g., Ref. [79]) suggest that very small *SIA clusters* are directly formed in cascades and are quite mobile in pure alloys [80]. Interactions with solutes can reduce their mobility. These SIA clusters are effectively proto prismatic $a/2<111>$ dislocation loops. Unresolvable, so-called black spot damage, has long been observed in irradiated steels. However, current techniques are generally unable to detect SIA cascade features at these sizes. This may be in part because solute atoms segregate to them, reducing both their mobility and strain contrast, hence visibility in TEM. The SIA-solute cluster complexes may be a stable matrix component in the 2F concept of irradiation embrittlement. Most importantly, SIA loops may serve as heterogeneous nucleation sites for both Cu- and MNS-rich precipitates under conditions where homogeneous nucleation is difficult [81–84]. Recently larger SIA clusters or dislocation loops have been observed in LAS steels at higher fluence. Presumably these loops have grown from the much smaller cascade clusters. Although the latest work remains unpublished at this time, advanced TEM methods suggest that loops may be far more numerous than previously thought. MD modeling studies [85] and many experiments [86] have demonstrated that loops are significant dislocation obstacles; therefore, they are a potential cause of matrix hardening. However, the dose, dose rate, and temperature dependence of both the smaller and larger loops has not been established, nor has their dependence on alloy chemistry and balance of the microstructure.

Vacancy clusters. Small vacancy clusters, often described as nanovoids, [87, 88] are extremely difficult to observe in TEM, but they have been detected by positron annihilation spectroscopy (PAS) in unalloyed Fe and simple binary Fe-Cu alloys [89]. They have also been observed in SANS studies of Fe and simple model alloys. Very early models suggested that they form in displacement cascades as Cu-coated nanovoids [90]. Such features have been detected by PAS [91]. Kinetic Lattice Monte Carlo studies [92] support this hypothesis and show that the cluster-solute complexes are mobile and can undergo coalescence reactions. However, nanovoids per se do not appear to form in LAS with higher solute contents.

Solute point-defect cluster complexes—SMF. Even though the total solute content of an LAS is less than about 5 atomic percent, PDs are never far from Ni, Mn, Si, and Cu atoms (at 2% solute the distance is ≈ 0.7 nm). The 1980s concept is that there are two distinct contributions to irradiation hardening; a matrix hardening feature, which is either some form of SIA (Fisher) or vacancy (Odette) cluster complex, and precipitation hardening has gradually been replaced by the concept of a continuum of features evolving toward local thermodynamic equilibrium. As noted above, SIA and vacancy clusters form cascade complexes with solute atoms that are more stable and less mobile than nanovoids. APT and other studies identify these complexes as solute clusters that are typically called SMFs. SMFs are a potential intermediate step between cascade PD cluster complexes and Ni, Mn, Si, and Cu precipitates. SMFs are distinguished from what are called unstable matrix defects (UMDs) (see further). The latter are less thermally stable and dissolve in typical low-flux, long-time irradiations.

Precipitates. The equilibrium solubility of Cu in iron at LWR operating temperatures of $\approx 290°C$ is around 0.01%. However, for final RPV heat treatment at typical temperatures of $\approx 620°C$, up to 0.25%–0.3% is retained in supersaturated solid solution during cooling to room temperature and is generally referred to as "matrix Cu." At higher Cu levels preprecipitation reduces the matrix Cu to the limits cited above. The Cu in solution can precipitate during thermal aging, or much more rapidly by RED. However, the precipitation of well-formed CRP is very slow for Cu levels below about 0.08%, although it can still play a role in defect-solute cluster complexes at lower concentrations. The matrix Cu minus the threshold Cu is sometimes described as the "effective Cu." The precipitates are pure Cu in simple Fe-Cu binary alloys. At small sizes (<2 nm) the precipitates are coherent, and the Cu has a bcc structure. These transform to the 9R structure and ultimately to fcc Cu at larger sizes. In more complex multiconstituent alloys and steels, CRPs incorporate Ni, Mn, Si, and P in amounts that depend on the corresponding solute composition of the steel, Fig. 10.20. These elements, particularly Ni, affect the number density, size, and volume fraction of the precipitates, hence the corresponding hardening. Precipitates nucleate homogenously within the grains as well as on dislocations, including loops, and grain boundaries. Under purely thermal conditions, diffusion is slow; at 320°C it takes about 10 years to reach peak hardening in a high Cu material (which is followed by overaging). Extrapolation of thermal aging data suggests that it would take in the order of hundreds of years to develop maximum hardness in a 0.3% steel at 290°C (see Section 10.3.2.2).

Under RED these processes are greatly accelerated. There is evidence from MD modeling studies that during cascade aging very small Cu clusters are mobile and can "getter" isolated Cu atoms and smaller clusters locally, accelerating nucleation. Alternatively, interstitial clusters may provide a nucleation route. Subsequently, the high concentration of irradiation-induced vacancies accelerates growth of the nucleated precipitate. The time to 95% full precipitation in an RPV wall depends on the material and flux but can be around 10 years according to the Eason Odette Nanstad and Yamamoto (EONY) model discussed in "Applications modeling" subsection below. This is at least an order of magnitude faster than precipitation under thermal aging conditions. At the high fluxes that can be achieved in a materials test reactor (MTR), precipitation times can be reduced to days. Following nearly complete depletion of matrix Cu the precipitates are often observed to coarsen, but there is little over aging effect on hardening. Indeed, it has been recently shown that the CRPs with predominantly Cu cores surrounded by an Mn, Ni, and Si shell continue to grow by the slow accumulation of the later elements, ultimately reaching volume fractions that are a large fraction of the 2Ni + Cu bulk content.

As discussed in "Historical background" subsection above, knowledge that copper and solute clusters were a major contributor to irradiation hardening in certain steels took many years to develop. Due to their size and other characteristics, no single experimental technique could provide all the information required. Field

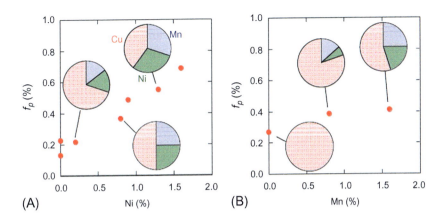

FIG. 10.20

Effect of Mn and Ni on volume fraction (f_p) and composition of clusters; (A) in a 0.4%Cu, 1.4%–1.6%Mn model MnMoNi steel, increasing Ni increases volume fraction and reduces the proportion of Cu; in a 0.4%Cu-0.8%Ni model MnMoNi steel, increasing Mn increases f_p slightly but has a marked effect on Cu content. Steels were irradiated to 3.4×10^{23} nm^{-2} at 290°C and 8×10^{11} ncm^{-2}s^{-1}; $E > 1$ MeV [93].

From E. D. Eason, G. R. Odette, R. K. Nanstad and T. Yamamoto, A physically-based correlation of irradiation-induced transition temperature shift for RPV steels, ORNL/TM-2006/530, Oak Ridge National Laboratory, 2007. Courtesy of Oak Ridge National Laboratory, U.S. Dept. of Energy.

Emission Gun Scanning Transmission Electron Microscopy (FEGSTEM), Small Angle Neutron Scattering (SANS), Field Ion Microscopy (FIM) and Atom Probe Tomography (APT) (see Chapter 4) have all been used extensively, most powerfully in combination with each other. Much of the information available on solute clusters has derived from APT. This has the immediate appeal of the ability to image individual atoms, but has several issues that make the interpretation of the images much less certain than might at first sight appear. In early studies, the volume of the material that could be examined was extremely small (approximately 1000–5000 nm^3). To increase the chance of finding features, many early studies were done on medium Ni steels, with high Cu and high fluence. The features observed were identified as CRPs, with Cu as the nucleating element and the other solutes being coprecipitants. Many workers reported that the precipitates contained a high proportion (up to about 50%) Fe, but this is inconsistent with other evidence and may be an artifact of the AP technique. Later AP studies revealed the importance of alloy composition on solute precipitate composition and alternative terminologies, such as copper-enriched clusters (CECs), Mn-Ni precipitate (MNP), or MNS precipitates, were adopted by various workers. These studies increasingly took advantage of the ability of the more recently developed local electrode atom probe (LEAP) to analyze a much larger volume of material. However, in this instrument a much larger percentage (about 60%) of atoms are "lost" and not imaged. This, together with the difficulty of defining a boundary between the solute cluster and the steel matrix, adds to the difficulty of identifying the cluster species and the mechanisms of their formation. However, it is now becoming clear that there is a continuum of solute complexes depending on several factors including local bulk chemical composition. It follows that even in high Cu, medium Ni steels, while early precipitates may be CRPs, later-formed ones are more likely to be MNPs.

Late blooming phases. In 1995 Odette [94] predicted on thermodynamic grounds that even with low or zero Cu contents Mn-Ni-Si phases (MNSPs) could be formed. These would have much lower nucleation and growth rates than the CRPs, would be favored by lower T_{irr}, lower flux and higher Ni and Mn. Since LAS contain much larger amounts of alloying elements Mn, Ni, and Si compared to the impurity Cu, volume fractions of MNSPs could be much larger than those of CRPs. They would also be potent hardening features and could lead to unexpectedly high rates of embrittlement late in life (even in high Cu steels once the available Cu has been exhausted) [95]. Such "late blooming phases" (LBPs) would have potentially serious implications to RPV lifetime extension and are not easy to characterize since the formation of LBPs for a combination of low flux and high fluence requires very long irradiation times. While there had been previous hints and glimpses, MNSPs were first found by Odette and coworkers in 2003 in a 270°C low flux test reactor irradiation of a Cu free, 1.6% Ni steel [96], as well as several simple model alloys. Since then, a large body of high to very high fluence test reactor data have shown the formation of large volume fractions of MNSPs in the form of near-stoichiometric G ($Ni_{16}Si_7Mn_6$) and Γ_2 ($Ni_3(Mn+Si)_3$) intermetallic phases, consistent with thermodynamic predictions [81] and X-ray diffraction studies [97]. Recently MNSPs have also been found in high 1.6% Ni weld surveillance steels of the Ringhals reactors [98]. Further, the initial results of a large intermediate flux-high fluence irradiation experiment, carried out in the ATR reactor at 290°C by Odette and co-workers, show that MNSPs also form over a wide range of intermediate to high Ni at extended life conditions of $\approx 10^{20}$ n/cm² in both Cu-free and Cu-bearing LAS. At full decomposition, the volume fractions of MNSPs can reach several %, depending on the alloy composition. Ni is the dominant factor in MNSP formation. The volume fractions are a significant fraction (e.g., $\approx$40% on average) of the alloy Ni content at $\approx 10^{20}$ n/cm² [83].

Unstable matrix defects. Work by Mader, Odette, and others [82, 99–101] has shown that an additional damage feature, UMDs, is formed at very high neutron fluxes. These are conceived as being thermally unstable small *vacancy solute cluster complexes* that continuously form and dissolve under irradiation in displacement cascades. The steady-state population is only high enough to cause significant hardening if the flux (hence the cascade generation rate) is high enough and the irradiation temperature (hence the dissolution rate) is low enough. This contrasts with SMFs, which continue to accumulate. UMDs are also sinks for vacancies, hence, decrease RED and delay precipitation hardening to higher fluence. Thus, depending on the overall combination of variables, UMDs may increase, decrease, or leave unaffected, hardening. UMDs are most easily detected by short time (5 hours) low temperature (350°C) annealing recovery measurements Fig. 10.21. UMDs are important and give rise to dose rate effects only at very high flux.

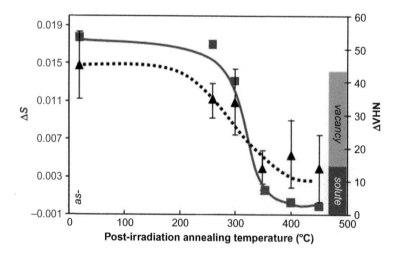

FIG. 10.21

Annealing of high Ni steel 124S285 between about 250°C and 350°C for 5 hours reduces hardness (triangles) and the PALA ΔS signal, which is associated with vacancies. However, while the "vacancy-related" damage (which may be attributed to UMDs) is removed, significant residual hardness (which can be attributed to solute atom hardening) remains [123].

Phosphides and Irradiation Hardening. It has long been recognized that P contributes to irradiation hardening and embrittlement of LAS, particularly at low Cu levels. P is highly insoluble in iron and even small amounts (<0.05%) can cluster and precipitate under RPV irradiation conditions due to RED and perhaps segregation enhancement [58, 103–105]. Precipitation of alloy (Mn, Mo, Cu) M_xP is most likely thermodynamic [106]. Strong P and Mn interactions have been observed in SANS studies with unusually low magnetic to nuclear scattering ratios, consistent with the presence of $Mn_{2-3}P$ precipitates for medium flux irradiations to $\sim 1.6 \times 10^{23}$ n/m^2 at 270°C and 290°C [93]. Phosphide contributions to embrittlement have generally been lumped into the SMF term in 2F models, although this is not strictly correct since this does not account for P depletion. Phosphorus associates with CRPs [104], so that in higher Cu steels P effects are diminished compare to those in low Cu steels. However, it has also been observed that P results in higher number densities of CRPs in a way that could enhance hardening, at least as a second-order effect. P also segregates to both dislocations and interfaces. Clustering and precipitation at dislocations may contribute to hardening. However, P atmospheres also likely result in some source hardening.

Other microstructural changes. *Segregation*: P also thermally segregates to grain and lath boundaries at temperatures around 400°C during heat treatment causing temper embrittlement. RED accelerates thermal P segregation at lower temperatures. Vacancy and SIA solute drag mechanisms [107] can enhance P segregation. P segregation (and C depletion) weaken prior austenite grain boundaries, resulting in nonhardening embrittlement (NHE) and an increase in intergranular fracture (IGF). This has been an issue for Magnox (C-Mn steels) RPVs operated above 300°C, but in Mn-Mo-Ni steels P does not appear to sufficiently segregate to grain boundaries to cause a significant embrittlement. The mechanism (and corresponding limits) for this relative immunity are not well-established, but it may be a result of competition from other elements segregating to boundaries, in combination with reductions in the effective P composition due to segregation to higher densities of dislocations and lath boundaries and enrichment at irradiation enhanced and induced features. Synergisms between hardening and NHE can result in very severe embrittlement.

Mn, Ni, and Si also segregate to dislocations and interfaces due to a combination of thermally and radiation driven processes (RED and RIS). There is emerging evidence that interstitial dislocation loops play a role in irradiation hardening and embrittlement of LAS at high fluence. Although publications have not yet emerged, modern TEM techniques have recently reported loops with sufficient density ($>10^{22}$/m^3) and size (≈ 4 nm) to produce significant hardening, perhaps up to 100 MPa or more, at 10^{20} n/cm^2. However, the general significance of such loops is an open question. Probably their most important role is that very small loops formed in displacement cascades are heterogeneous nucleation sites for precipitates. Larger loops, in the range of 5–20 nm, may also assist precipitate nucleation, as do network dislocations.

Other changes to the microstructure during irradiation include a reduction in solid solution strengthening due to incorporation of solute atoms in precipitates and other features, possible changes to the network dislocation structure, and refinement of the preirradiation carbides. As discussed in Section 10.3.2.3, there is also some evidence of strain aging in LAS. Although carbon and nitrogen migrate freely at RPV operating temperatures, radiation damage features may act as deep interstitial traps, thus reducing the amount of free nitrogen or carbon available for strain aging processes. For BWR and PWR conditions, however, the later effects are likely to be small.

Effect of environmental and materials variables.

Neutron spectrum. As described in the previous section, neutron spectrum affects PD production per unit fluence, usually determined above a threshold value of 1 MeV, but can be approximately taken into account by using dpa. Nevertheless, fluence (>1 MeV) is still the most common unit used in regulatory codes. In many irradiation locations of interest, the error induced by using fluence rather than dpa is modest compared with other uncertainties. A useful simple measure of spectral differences is the ratio of dpa/fluence. This ratio is typically about 1300 barns (1 barn = 10^{-28} m^2) in a test reactor core. But the spectrum "softens" as the neutron flux is attenuated and the corresponding fraction of damaging neutrons between 0.1 and 1 MeV increases. Ratios of dpa/fluence for typical test reactor irradiations, surveillance capsules and the RPV wall (at the inside surface) of a PWR range from about 1450 to

2150 barns[16] (i.e., by ±20% from the mean of these those values), see also [108]. However, as neutron flux attenuates through an RPV wall, the spectral softening becomes significant. For this reason, the USNRC RG 1.99 [67] provides a formula that calculates fluence attenuation with depth in the RPV wall adjusted by the dpa/fluence ratio.

Use of dpa to account for spectrum effects may also be required for making comparison between results or predictions from different types of reactors (or between different locations in the same reactor), to compare results with fluence quoted with different thresholds (e.g., $E > 0.1$ MeV), and to compare results using different bombarding species. Use of dpa can be recommended more generally to improve precision in reporting results and modeling based on them; imprecision fogs analysis.

For LWRs, dpa is a reasonably good exposure unit. However, dpa does not account for all potential displacement processes. In some reactor designs, notably heavy water and graphite moderated reactors, there can be a high flux of thermal neutrons, those with energy levels insufficient to displace atoms from their lattice positions (discussed above). However, thermal neutrons undergo nuclear reactions with iron and other nuclei to form isotopes, the subsequent decay of which produces a recoil reaction sufficient to form isolated surviving PDs at a greater efficiency (order 2×) than high-energy neutrons. This was found to be a significant effect for the UK Magnox reactors with high thermal neutron fluxes. Further information on thermal neutron effects is given in Refs. [109, 110].

Neutron flux. Nanostructural evolution requires the transport of solute atoms by vacancies and SIAs. The kinetics of microstructural evolution depends on the concentrations of these species and is usually modeled in terms of the excess vacancies. The excess vacancy concentration depends on the rates at which they are generated in cascades (or thermally) and balanced by the rates at which they are removed from the matrix. Fig. 10.19 illustrates the several ways in which these PDs can be lost. The primary cause of the flux (fluence rate or dpa rate) effect is recombination of the excess vacancies and SIA. This can be modeled in terms of the fraction of vacancies that survive recombination (with an SIA) as a function of flux. Model predictions of flux effects are shown in Fig. 10.22. Note this curve is for a specific set of parameters at 290°C but is generally consistent with observed data trends. At very low dose rates, vacancy concentration is dominated by thermal vacancies, and is independent of dose rate. Above an ill-defined dose rate, irradiation-induced vacancies become increasingly dominant and the curve becomes nonlinear. At these lower dose rates, all the vacancies remaining after the cascade event survive recombination (i.e., are lost to fixed sinks, primarily features such as grain boundaries and dislocations where they are annihilated). At some point the irradiation-induced vacancies become dominant (the thermal vacancies making a negligible contribution to the total). The vacancy concentration becomes proportional to dose rate, the rate of generation being matched by the rate of loss to fixed sinks, producing the central linear part of the curve. At higher-dose rates the curve becomes nonlinear again due to vacancy losses in solute traps and recombination with interstitials due to chance encounters in the matrix. At the highest dose rates there may also be losses to vacancy clusters induced by irradiation. The rate of diffusion of solute atoms to precipitates and other irradiation defects is proportional to vacancy concentration (note that under variable flux conditions it may take time to reach steady state). Hence the precipitation rate is increased with increasing dose rate by an amount that depends on dose rate (also temperature and sink density). In the linear part of the vacancy concentration vs dose-rate curve, if the dose rate doubled, the time to reach a given state of precipitation would halve, but the dose (dose rate x time) to reach the latter would be unchanged, and we would not observe a dose rate (flux) effect.[17] At the higher-dose rates, the curve is nonlinear and, to a given dose, there will have been fewer total vacancies available for diffusion. For these conditions a flux effect will potentially be observed (if a diffusion-dominated damage mechanism, such as copper precipitation, applies); the amount of precipitation damage at a given dose will be reduced for the higher-dose rate irradiations relative to lower-dose irradiations. This will, however, only apply if precipitation is incomplete, at least for the higher-dose rate irradiation.

---

[16] R E Stoller, private communication.

[17] Semantically, there is a flux effect, doubling the flux halves the time to reach a given level of damage. However, by convention we normally plot irradiation damage vs fluence.

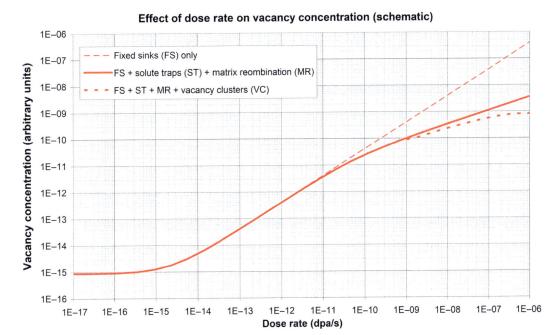

FIG. 10.22

Effect of flux on vacancy concentration. At higher-dose rates, irradiation becomes less effective in increasing the concentration of vacancies in the matrix, hence less effective in increasing diffusivity. This creates a potential dose-rate (flux) effect on irradiation damage in the upper nonlinear region. In the lower nonlinear region there is also a potential for such an effect due to the influence of thermal neutrons [2].

However, as flux increases, an increasing fraction of the vacancies produced is lost to recombination with SIA via the routes indicated in Fig. 10.19. Vacancies that become trapped by solute atoms (i.e., bound to them by potential attraction, but not so strongly they cannot escape thermally) are more likely to react with SIAs, as flux increases, resulting in mutual annihilation. With further flux increases, the increased concentration of PDs increases the chance of random encounters in the matrix. At very high fluxes significant numbers of UMDs are formed (depending on irradiation temperature) and this further reduces the survival fraction.

The survival fraction can be regarded as an efficiency factor. At low fluxes, irradiation is fully efficient; all vacancies reach fixed sinks and can achieve their maximum potential in terms of enabling solute atoms to diffuse. As flux increases, damage rates increase overall (vacancies are produced at a faster rate), but an increasing fraction of them are lost before they reach fixed sinks. As a result, high flux irradiations produce lower damage per unit fluence than low flux ones. Thus, accelerated irradiations can increase the fluence required to reach a given state of precipitation, resulting in nonconservative hardening and embrittlement relative to RPV wall fluxes.

A detailed model of flux effects by Eason, Odette, Nanstad, and Yamamoto (known as EONY) is given in [93, 111]; the first of these provides full details of the work, the second is a much shortened journal publication. EONY is calibrated to the IVAR database [112] and illustrated for two alloys in Fig. 10.23, based on the work described in [114–116]. The left-hand figures show that the effect of one order of magnitude difference in flux is equivalent to about a factor of two in effective fluence (or dpa). In the right hand figures, the "effective fluence" parameter, which models the survival fraction, and which is further discussed in "Applications modeling" subsection below, brings all the data together.

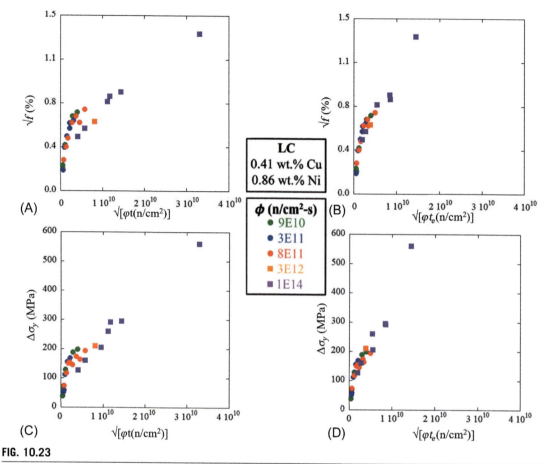

FIG. 10.23

Results from a high-Cu medium-Ni steel irradiated to various fluxes showing (left) the square root of the volume fraction of precipitates and the consequent yield stress increase as a function of the square root of fluence, and (right) these quantities as a function of effective fluence. Effective fluence corrects for the effect of high flux in reducing precipitation at a given fluence [113] based on the work described in Refs. [114, 115].

Fig. 10.24 shows the flux effect in a low Cu steel. Since MD hardening is small, the effect is numerically small compared to flux effects on CRPs. Nevertheless, the mean trend is significant, is observed in many alloys, and is consistent with the same effective fluence as for precipitates. This issue is not normally relevant to power reactors, which predominantly operate under base load conditions with constant flux over long periods of time, but may apply to some very high flux test reactor irradiations carried out over a small number of days, which is the time for PDs and UMDs to reach steady concentrations. Work by Stoller [117] has indicated that it can take some hours for vacancy concentrations to reach equilibrium and hence for RED to reach its maximum rate for that flux.

Flux effects are not always apparent in data. They will not be observed when the flux is high enough that the irradiation-induced vacancies generation rate is much higher than the thermal vacancy generation rate, but is not so high that the survival fraction falls significantly below unity. Flux effects may not be readily distinguishable from scatter (except in high-quality databases) in low-irradiation sensitivity materials or when the flux range is small. They should be most apparent when there is significant precipitation damage, but this only applies up to fluences

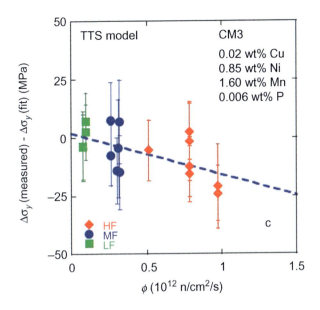

FIG. 10.24

Residuals in the EONY model predictions for a low copper commercial steel (CM3) (which was not included in the fitted database) showing a significant trend with flux that indicates that hardening at a given dose decreases with flux even when copper is very low [93].

From E. D. Eason, G. R. Odette, R. K. Nanstad and T. Yamamoto, A physically-based correlation of irradiation-induced transition temperature shift for RPV steels, ORNL/TM-2006/530, Oak Ridge National Laboratory, Courtesy of Oak Ridge National Laboratory, U.S. Dept. of Energy.

with nearly complete precipitation. Finally, when UMDs are important, the flux effect may be at least partly offset by the hardening caused by the UMDs.

Irradiation time. Due to thermal aging effects, time is also potentially important for very low flux irradiations where equilibrium and irradiation-induced vacancies have similar concentrations. If the flux is high enough for irradiation-induced vacancies to dominate, the effect of time is subsumed into the flux effect. The flux range over which an independent time effect might be important is not well established but may be relevant for some BWR vessels and regions outside the beltline. However, since thermal aging is sluggish for normal LWR operating temperatures even at the highest levels of Cu found in commercial RPVs (see Section 10.3.2.2), this is not formally a consideration for such conditions.

Irradiation fluence. While neutron fluence (or dpa) is the prime independent variable, damage and embrittlement do not depend on it in a simple way. In models using a simple chemistry factor (CF) times fluence raised to a power (e.g., Eq. (10.1)), reported exponents vary from around 0.3 to around 0.8 [118]. Different features have different fluence dependencies. For example, in typical 2F models developed to date (see "Applications modeling" subsection below), CRPs form rapidly (especially at low flux) and saturate due to Cu depletion at a fluence less than 10^{19} n/cm². The corresponding SMF term increases in rough proportion to the square root of fluence. The physically motivated EONY model has fluence dependence terms that depend on flux and composition. However, existing models, including EONY, are based on surveillance data that cover only a range of flux, temperature (see further), and fluence. They are not reliable at fluences greater than $\approx 5 \times 10^{19}$ n/cm², and do not fully account for flux effects. A large amount of data has been generated by accelerated irradiations in test reactors at higher fluence pertinent to extended life. Unfortunately, current low flux fluence models underpredict test reactor embrittlement trends at high fluence [82]. This is primarily due to the significant emergence of the LBP MNSPs described previously that are not included in regulatory models circa 2017. Therefore, there are large experimental and modeling efforts to characterize the fluence dependence up to about 10^{20} n/cm² or more by adding a MNSP term while modifying that for the SMF. The MNSP term also saturates at full intermetallic phase precipitation and can also be described by a saturating model and a flux-dependent indexing fluence. Both CRP and MNSP precipitation are well described by Avrami models as noted in "Applications modeling" section.

Irradiation temperature. Increasing irradiation temperature (T_{irr}) reduces irradiation shift in low Cu steels in which MF dominate irradiation hardening. However, the types of reactors that use LAS operate within a relatively narrow temperature range (~20°C), and the effect of temperature within this range is small or neglected. There is, therefore, little need to characterize the temperature effect very precisely, especially since surveillance specimens are normally irradiated at approximately the same temperature as the RPV wall. However, an understanding of temperature effects is important where there are significant temperature gradients in reactors and for developing models of irradiation shift that are to be applied to different reactors. A complication in this respect is that reactor temperatures can vary with location and, in a given location with power, the relationship between surveillance capsule and RPV wall temperatures may be different from RPV designs.

Temperature gradients were very significant in the UK Magnox reactors [1] resulting in a significant body of work on C-Mn steels (e.g., Barton et al. [119]). These data, together with data from Grounes [120], were analyzed by Jones and Williams [121] to demonstrate that the effect of temperature on matrix damage is reasonably linear over a wide range of temperatures, and can be represented by a sensitivity factor F_T as shown in Eq. (10.3), which is normalized to $F_T = 1$ at 190°C (a baseline for Magnox reactors). The effect on matrix damage of changing temperature from X to Y (all other variables constant) is estimated from $\Delta_Y/\Delta_X = F_{TY}/F_{TX}$. Thus, increasing irradiation temperature from 285°C to 295°C would reduce MF hardening by about 2.4%.

Jones and Williams temperature function.

$$F_T = 1.869 - 4.57 \times 10^{-3} T_{irr} \tag{10.3}$$

F_T is the temperature sensitivity factor ($F_T = 1$ at 190°C) and T_{irr} is the irradiation temperature (°C)

Eason et al. [111] has shown that, contrary to the orthodox view, embrittlement rates from PD can be significantly affected by irradiation temperature, identifying a number of factors that would reduce CRP with increasing irradiation temperature. The formation of UMDs has a classical dependence on time and temperature; they are formed in significant numbers only if the flux, hence generation rate, is high enough relative to their dissolution rate, which follows an Arrhenius dependence on temperature.

Although temperature effects have limited engineering importance, they may be significant in data comparisons and important in the context of development of mechanistic understanding. For this reason, they have been extensively studied in UCSB experiments [105]. Little work has been done at the ambient temperatures at which some RPV support structures operate. These experience low flux and fluence but may use less well-controlled materials and operate at temperatures closer to the DBTT. The evidence suggests that the adverse effect of reducing temperature saturates below about 100°C or higher, and that high-temperature models are not extrapolatable down to ambient due to mechanism changes.

Material composition. Irradiation hardening and embrittlement are strongly affected by alloy composition, Cu, Ni, P, Mn, and Si. C and N can also play a role especially in plain carbon steels. Composition effects are complex in part due to synergistic interactions between the various solutes and with other material and irradiation variables. The material variables include the thermomechanical processing history that mediate microstructures and dissolved solute contents and distributions. In some databases interpreting the effects of multiple solute compositions is confounded by strong correlations between them and their limited ranges (clumped) of values. In some cases, like P, solute contents were not recorded. The following paragraphs summarize the major compositional effects.

As already discussed, Cu is highly insoluble in steels and readily precipitates under both thermal and RED conditions. Its effect is magnified by its forming precipitates (CRPs) that include other elements, in particular Ni, Mn, and Si. Increasing Cu increases the precipitate volume fraction and results in about a roughly square-root dependency on Cu above a threshold of about 0.07%–0.08%. These values are well above the solubility limit at LWR operating temperatures but can be explained by the increased difficulty in nucleation of well-formed precipitates. At lower levels Cu is an important constituent of SMDs. An upper limit to the Cu effect is defined by its solubility at the final heat treatment temperature, typically at 600–650°C, of the vessel (some Cu preprecipitation may also take place during a slow cool down). There is also evidence [122] that matrix Cu can be reduced by S due to the formation of digenite ($Cu_{1.8}S$).

Nickel acts in synergism with Cu, increasing both precipitate number density and maximum volume fraction. This is expressed in several engineering models by Cu-Ni cross-terms. Nickel is observed to have an effect independent of Cu in low-Cu high-Ni steels, which generally have higher fluence exponents than their medium and low Ni equivalents. Note that Ni also affects the unirradiated, start-of-life, microstructure and properties of LAS which indirectly influences hardening and embrittlement.

Manganese is a constituent of CRPs, MNPs, and LBPs and has also been found to segregate to dislocations and SIA loops. Its influence is widespread, but it is seldom directly featured in engineering shift models because Mn-Mo-Ni steels tend to contain about the same Mn levels. The EONY model [111] includes an MnP term in its MF hardening component, representing the effect of phosphide formation. Burke et al. [123] have shown that low-Cu high-Ni steels without Mn have very low irradiation sensitivity. There is also substantial evidence with WWER-1000 steels that Mn plays a significant role in the embrittlement of Cr-Mo steels, especially some welds with relatively high Ni contents. This may be because Mn (an oversized atom) segregates to SIA loops enabling nucleation of Mn-Ni features. Since Mn forms carbides and other compounds in steels, matrix levels, available to contribute to features developed under irradiation, may be significantly lower than nominal bulk levels. Finally, it has been recently observed that for typical western RPV LAS there is a trade-off between the composition of the Ni-rich precipitates that reflect the alloy Mn and Si content. The precipitate compositions, and even their precursor MS, tend to have an Si+Mn for every Ni atom. This is related to the similar compositions in the intermetallic phases that form at high fluence, like G, $Ni_{16}Mn_6Si_7$, and G2, $Ni_3(Mn+Si)_2$, phases [81].

Phosphorus affects irradiation sensitivity by forming phosphides and by segregating to dislocations and interfaces, including CRPs and grain boundaries [102, 104, 124]. Segregation to grain boundaries, usually described as temper embrittlement, can potentially lead to NHE. As noted previously, this does not appear to be a significant issue for Mn-Mo-Ni steels [125] but may influence WWER and C-Mn steels behavior [126, 127]. However, Mn-Mo-Ni steels may be affected by P segregation following postirradiation thermal annealing [128].

Silicon is incorporated in CRPs, NMPs, and LBPs and therefore might be expected to increase shift. However, in some early engineering models [129–131], Si may have a negative coefficient that reduces shift. Again, Si is associated with MNS precipitates but trades off with Mn as noted earlier (e.g., Ref. [83]).

Nitrogen effects have been observed in Magnox steels. Si-killed steels had higher sensitivity than those killed with aluminum, which is more effective in gettering nitrogen during steel production. It is possible that the effect of nitrogen is to stabilize MFs, but little work has been done and free (uncombined) nitrogen values are very infrequently measured.

Carbon may play a similar role to N in governing the development of SIA features in some alloys. Carbon has been observed to have a small effect in LAS. Of course, C, and processing treatments, has a very large effect on the microstructure of LAS that varies significantly from simple ferritic model alloys. This rationalizes differences in the irradiation effects in model alloys vs steels, by mechanisms such as superposition effects, sink densities, and removal of elements in carbides (e.g., Mn) from solution.

Microstructure, product form, and categorical variables. Composition, fabrication, and processing histories affect the microstructure of unirradiated steels and in combination affect irradiation sensitivity. It would be expected that microstructure would affect the kinetics of RED (through sink density), the influence of

P segregation (through grain size), the effectiveness of hardening features (see the next section) due to strength superposition effects, and the relationship between hardening and embrittlement (through differences in the fracture process). However, the relevant microstructural data are generally not usually available and could not readily be deduced even if the necessary details of the compositional, fabrication, and processing variables were accurately known. For these reasons, it has been necessary to analyze and fit data using *categorical variables*, most notably *product form* (usually plate, weld and forging, but sometimes subdivisions of these). Either each category is fitted separately, or they may be co-fitted with most or all parameters and functional forms the same but with at least one individual fitting parameter. In some cases, there are physical variables that distinguish the categories, for example the differences in Si and C contents in base materials compared with welds; however, it cannot be assumed that the difference in sensitivity for categories can be attributed to these factors. The use of categorical variables also is implicit in the use of regional (e.g., French, Japanese) models.

Relationship between hardening features and hardening and embrittlement

The hardening features developed during irradiation obstruct dislocation movement and thereby increase the critical resolved shear stress (CRSS) for plastic flow. In studies of microstructural evolution with irradiation dose it is often found that there is a reasonably linear correlation between increase in yield stress (or shift) with the observed increase in the square root of the volume fraction of the features responsible. The strengthening produced by a number density, N, of obstacles in an otherwise empty matrix, can be approximated by

Simple irradiation hardening model.

$$\Delta\sigma_y = 0.55 T\alpha\mu b \sqrt{f}/r \tag{10.4}$$

where T is the Taylor factor, which relates CRSS to yield stress, α is the obstacle strength parameter ($0 \le \alpha \le 1$), μ is the shear modulus of iron, b is the Burgers vector, f is the volume fraction of precipitates, and r is the precipitate radius.

Here, $T \approx 3.06$, and values of α differ for different features; MFs are relatively weak ($\alpha \approx 0.05$–0.1) whereas CRPs are stronger with $\alpha \approx 0.1$–0.3. Fine scale carbides, like Mo_2C, which contribute to the initial yield strength of a steel, have $\alpha = 0.8$–1, since they cannot be cut by dislocations, thus are bypassed, at large bowing angles, by the Orowan looping mechanism. Values of α can be characterized experimentally by measuring the size and volume fraction (or number density) of the hardening features along with their individual contribution to the yield stress. Fig. 10.25 (left-hand figure) compares MD simulations by Bacon and Osetsky [132] to establish the critical bowing angle for a range of precipitate sizes and the resulting increase in yield stress per square root of the volume fraction of precipitates. The right-hand figure [from Ref. 111] compares the predictions of the MD simulations using with predictions made using the Russell and Brown analytical hardening model [133] for a range of materials.

A key issue is how the effects of the different strengthening features should be superposed. This has been modeled using computer simulations [103, 134, 135], which show that features with similar α superpose by root sum of squares (RSS) addition; those with very different (very weak and very strong) α superpose linearly whereas, when the total strengthening contribution is from a mixture (intermediate and large), α falls between these limits. The RSS superposition is explained by the distance between pinning points on a slip plane that scales with square root dependency on volume fraction. Linear superposition is explained on the basis that the large bowing angle from strong obstacles increases the number of weak obstacles that the dislocation segment encounters, where the latter act as a friction stress.

Most, but not all [136], engineering models of irradiation embrittlement (see "Applications modeling" subsection below) assume linear superposition between the preexisting dispersed barrier hardening precipitates and MF irradiation-induced hardening. Such simplifying superposition assumptions are unavoidable unless there is detailed

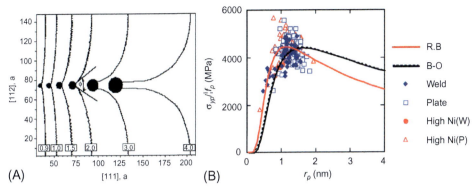

FIG. 10.25

(A) Molecular dynamics simulations by Bacon and Osetsky (BO) [132] of the critical bowing angle at which a coherent copper precipitate is cut for precipitate radii, r_p, from 0.9 to 4 nm; (B) the effect of r_p on hardening efficiency (increase in strength per square root of volume fraction of precipitates) for both BO and the fitted Russell and Brown modulus hardening model, compared with measurements based on SANS data and tensile tests [93].

(A) From Bacon, D. J. and Osetsky, Y. N., Hardening due to copper precipitates in α-iron studied by atomic-scale modelling, J. Nuc. Mar. 329–333, 1233, 2004. (B) From E. D. Eason, G. R. Odette, R. K. Nanstad and T. Yamamoto, A physically-based correlation of irradiation-induced transition temperature shift for RPV steels, ORNL/TM-2006/530, Oak Ridge National Laboratory, 2007.
Courtesy of Oak Ridge National Laboratory, U.S. Dept. of Energy.

knowledge of the hardening feature populations. However, the resulting errors are at least partially subsumed into the fitting parameters. While use of linear superposition may be of limited consequence to engineering models, it can distort estimates of the relative embrittlement contributions of different features. As an example, if an irradiation contribution to hardening of 200 MPa is added to an unirradiated dispersed barrier hardening of 200 MPa, the net yield stress increase would be the full 200 MPa contributions for a linear sum rule compared to only 83 MPa for an RSS rule.

The yield stress (more specifically the average flow stress between strain of 0 and 0.1) changes due to irradiation and correlates well with hardness changes [137]. The latter can, therefore, provide a relatively convenient, low-cost and reasonably accurate alternative to tensile testing, with the advantage that it can be used on small specimens, or to investigate local variations in irradiation damage. Care is needed to ensure that preirradiation and irradiation results are obtained in the same region. Neither yield stress nor hardness change can detect NHE.

Except for NHE (see further), irradiation shifts the DBTT through its effect on flow properties (any changes to the fracture initiators are generally considered insignificant for RPV irradiation conditions). The classical explanation is illustrated by the Davidenkov diagram in Fig. 10.26. This assumes that fracture occurs when the yield stress, multiplied by a factor to account for the increased stress at the crack tip due to triaxial constraint, reaches the critical (roughly temperature independent) internal stress for cleavage fracture caused by activating brittle trigger particles. More physically-detailed models require a critical stress over a microstructurally relevant volume containing cleavage trigger particles. Irradiation roughly increases yield stress by a constant amount that only weakly depends on temperature in association with the elastic modulus. The fracture stress is generally assumed to be independent of irradiation if the cleavage trigger particle populations are stable. However, for a constant MC toughness-temperature curve shape the critical stress must be weakly dependent on temperature [43]. In either case the result is that the shift in the fracture toughness curve is directly related to the increase in yield stress—or more precisely to the strain hardened flow stress in the fracture process zone. This has been modeled in detail [138] taking account of the effect of irradiation in reducing strain hardening. At very high irradiation damage levels that reduction decreases the ratio of the toughness shift to the yield stress increase. However, the effect was predicted to be small for typical LWR irradiation conditions [139]. The consistency of the correlations between hardening and yield stress [138] and between the latter and fracture toughness shift [139] suggests that hardness change could

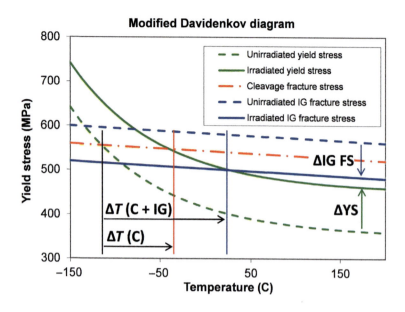

FIG. 10.26

Modified Davidenkov diagram. The DBTT corresponds to the intersection between the yield stress and the cleavage fracture stress. Irradiation increases the yield stress, but not the fracture stress, hence the DBTT is increased ($\Delta T(C)$). The intergranular (IG) fracture stress is normally higher than the cleavage fracture stress but can be reduced by irradiation. If it falls below the cleavage fracture stress, the DBTT shift can be increased ($\Delta T(C+IG)$). Note that this effect may have an irradiation dose threshold [2].

be a useful means of estimating embrittlement. Hardness change, rather than Charpy shift, has been used as the independent variable in empirically fitted irradiation damage models [130].

Similar arguments apply to the relationship between irradiation hardening and Charpy shifts. However, in this case changes in the Charpy USE (Fig. 10.27) also must be accounted for [52]. Further, traditional Charpy hardening shift models assumed that the cleavage fracture stress is independent of both irradiation and temperature. As a result, the hardening shift relation is nonlinear. Odette [52] has modeled this in some detail. Unexpectedly, given the significant differences between the two types of test, there is a reasonable correlation between CV41J and T_0 values (Fig. 10.28 [139]). Limited data available suggests that the correlation between shifts in these values is also reasonably linear as shown in Fig. 10.29 [139]. The correlation was found to depend on product form: for plate and forging (combined) the ratio of fracture toughness shift to Charpy shift was approximately 1.16:1, compared with 1:1 for welds.

The Davidenkov concept (Fig. 10.26) and other models can account for the effect of NHE [140]. As P segregation increases the IGF stress decreases; when it falls below the cleavage fracture stress, the fracture mode changes to intergranular. Normally NHE will be accompanied by yield stress/hardening increases from the dispersed irradiation damage features; hence NHE and other irradiation damage mechanisms are synergistic and result in larger shifts than due to either mechanism acting alone.

The effect of irradiation on upper shelf toughness (both Charpy and fracture toughness) has been of relatively low technological concern and much less studied since the levels are high and the ductile tearing failure mechanism relatively benign. A past concern was that the Charpy USE in some early welds fell below the 41J indexing level; however, these welds were demonstrated to have adequate initiation toughness and reasonable ductile tearing resistance.

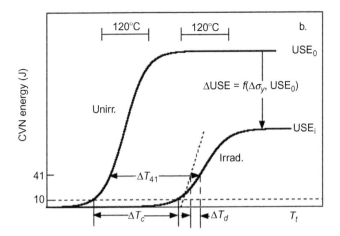

FIG. 10.27

Effect of upper shelf energy drop on Charpy shift. The cleavage fracture temperature corresponds approximately to the temperature of the mean Charpy curve at 10J. The width of the Charpy transition curve is about 120°C for all materials, independent of irradiation damage. For materials with low initial transition temperature and high upper shelf drop, the Charpy 41J shift overestimates the 10J shift, hence overestimates the shift in the fracture toughness curve [52].

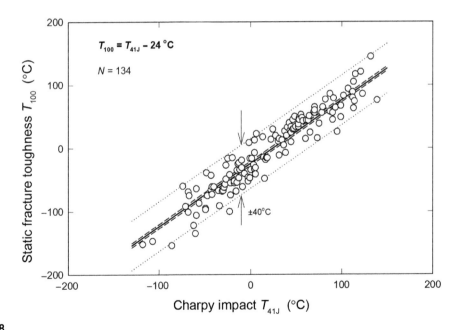

FIG. 10.28

The fracture toughness transition temperature (at $100\,\text{MPa}\sqrt{m}$) correlates linearly with the Charpy 41J transition temperature with an offset that corresponds to the relationship between the 41J temperature and the cleavage fracture temperature. There is however considerable scatter [139]

From M. A. Sokolov and R. K. Nanstad, Comparison of Irradiation-Induced Shifts of KJc and Charpy Impact Toughness for Reactor Pressure Vessel Steels, pp. 167–190 in Effects of Radiation on Materials: 18th International Symposium, ASTM STP 1325, R. K. Nanstad, M. L. Hamilton, F. A. Garner, and A. S. Kumar, Eds., American Society for Testing and Materials, West Conshohocken, PA, 1999.

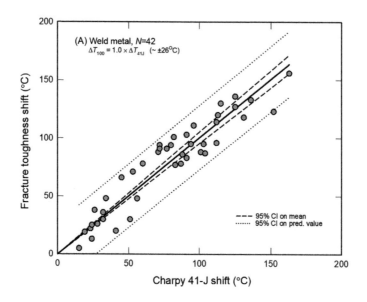

FIG. 10.29

Fracture toughness shift correlates reasonably well with Charpy shift. In the case of weld (shown here) there is a 1:1 relationship, however, for plates and forgings the toughness shift is 16% higher than the Charpy shift [139].

From M. A. Sokolov and R. K. Nanstad, Comparison of Irradiation-Induced Shifts of KJc and Charpy Impact Toughness for Reactor Pressure Vessel Steels, pp. 167–190 in Effects of Radiation on Materials: 18th International Symposium, ASTM STP 1325, R. K. Nanstad, M. L. Hamilton, F. A. Garner, and A. S. Kumar, Eds., American Society for Testing and Materials, West Conshohocken, PA, 1999.

Computational modeling studies

The ultimate challenge for irradiation damage modeling is to produce a predictive model—one that properly represents the physics of the phenomena and relies on universal physical parameters rather than calibration to irradiation data. Such a model could be reliably extrapolated outside current experience to make accurate predictions for new alloys or irradiation conditions.

The physical processes involved in irradiation embrittlement range in length scale from nanometers to centimeters, and from picoseconds to years; for structural integrity assessments the length scale extends to meters. This requires a multiscale modeling (MSM) approach [141, 142]. MSM[18] is employed in a wide range of technical fields and uses a sequence of linked physical models to span the scales required using appropriate computational methods for each. The basic framework for MSM of embrittlement was first introduced by Odette and collaborations in the early 1980s [143] and further developed and refined as summarized in a paper in the early 2000s [144]. The basic strategy was to link near atomic scale microstructural changes driven by irradiation to dislocation level hardening effects on the constitutive properties. These are then linked by local micromechanics models of cleavage by continuum finite element simulations of crack tip stress fields. Over the succeeding years much more detailed submodels of the individual stages of the overall process have been proposed. For example, Fig. 10.30 illustrates the model developed in the European Commission Euratom projects PERFECT [146] and PERFORM 60 [147]. These multimillion Euro programs were led by EDF (Electricité de France) and involved many other, mainly European, partners. They were successors to the pioneering EDF REVE project led by Jumel and Van Duysen [135, 145] with international collaboration.

---

[18] Also known as multiscale multiphysics modeling (MSMP) or multiscale materials modeling (MMM).

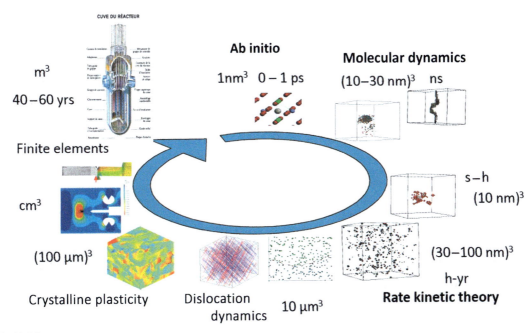

FIG. 10.30

Multiscale modeling—illustration of the approach used in the European Commission PERFORM 60 project and the time and length scales for each type of model. Compare with the simplicity of the analytical expressions, Eqs. (10.1)–(10.11) [145].

MSM of embrittlement remains work in progress although major advances are emerging, especially based on integrating modeling with experiment. However, this is a broad and diverse topic that cannot be treated with due justice in this chapter and the reader is directed to a large and growing literature for further details [148–174].

Applications modeling

Applications (or engineering) models are often developed by fitting rather simple equations to embrittlement data. They are used to predict irradiation embrittlement in operating RPVs or used to analyze data to identify and characterize the effects of embrittlement variables. Regulatory models are used in many cases by licensing authorities, as being sufficiently well validated or conservative, to extrapolate predictions to be confirmed by the ongoing surveillance programs. Many nonregulatory application models have also been developed over the past decades, each representing the status of knowledge and data available at that time and place. The earliest models were necessarily largely empirical; as understanding has developed, there has been an increasing movement toward mechanistically based correlation (MBC) models—models fitted to data but formulated on the basis of mechanistic understanding.

However, the development of a model able to accurately predict irradiation shift for all PWR or BWR designs has been elusive and attempts to do so have been controversial. The current ASTM model [175] provides a reasonably good fit to a wide range of data from different national databases. However, it is essentially an empirical model and, as the guide warns, the user has the responsibility to show that the conditions to which the model is to be applied are addressed adequately by the technical information on which it is based. Attention is drawn to the uneven distribution of variables and the need for concern for applications near the extreme of the data ranges on which the model is based, or where data are sparse. Similar caveats would apply to other such models.

Databases. The quality of applications models is highly dependent on the quality of the available databases. Models intended for regulatory application are generally fitted to surveillance databases, most often for a specific RPV fleet. These provide information as close as is practicable to predict the behavior of the RPV wall materials for that RPV. However, surveillance databases have limitations with respect to investigating irradiation phenomena: values of the variables may be clustered within the overall dataspace and may not be known to the accuracy that would be expected in laboratory experiments. It is also possible that values of some variables now known to be important were not originally determined. Further, the data represent only existing plants (with the older ones having heavier weighting due to higher fluence); values of the variables, or their combination, may be somewhat systematically different in future plants.

For these reasons, extensive use has been made of data generated from material irradiated in test reactor experiments for both scientific investigations and in a small number of regulatory models. Most countries with NPP have done such experiments; outstanding examples of which are the UCSB experiments IVAR [112] and ATR2 [176]. Such laboratory experiments often use commercial materials, but model alloys and model steels with tailored compositions are also used. These enable investigation of unpopulated regions of data space or can provide an optimum distribution of variables for analysis, avoiding such problems as correlations between the compositions of specific elements that can occur in commercial materials. Although test reactor databases can provide excellent information, an unresolved issue is the extent to which the findings can be extrapolated to the relatively low flux conditions within an RPV wall (see "Effect of environmental and materials variables" subsection above–*Flux effects*).

Unfortunately, databases are often proprietary or commercially sensitive. Not all countries have followed the US lead in openly publishing surveillance data; and when test reactor data are published it is often done so incompletely.

Early empirical Models. The earliest shift models were simple upper bound or mean curves fits using the limited amount of data available, often taking the form of Eq. (10.1). The discovery that irradiation embrittlement of RPV steels might be a significant issue (see "Historical background" subsection above) led to codification of shift prediction methods. A key development was the USNRC RG 1.99 (RG1.99), first issued in 1975, with Revision 1 ("Rev 1") issued in 1977 [177]. This provided upper bound shift predictions (Eq. 10.5), developed by finding an analytical expression that bounded the relatively small number of Charpy shift surveillance data then available. The assumption that there are no effects of Cu and P below threshold levels was based on observation. The square-root dependence of shift on fluence was consistent with the findings of theoretical and experimental work by Makin et al. [178].

USNRC RG1.99, Rev 1 Model (1977).

$$\Delta = [40 + 1000(\text{Cu} - 0.08) + 5000(\text{P} - 0.008)]\Phi^{1/2} \tag{10.5}$$

where Δ is the irradiation shift (°F) and Φ is the fluence $/1 \times 10^{19}$ n/cm^2, $E > 1$ MeV.
The copper and phosphorus functions have a minimum value of zero

---

The identification of Cu as the major source of heat to heat variability, in conjunction with increases in the size of the database and better access to (then mainframe) computers, enabled the development of more detailed statistical models and complex analytical forms. A good example is the 1979 model of Varsik and Byrne [179] (Eq. 10.6). This was fitted to about 100 data points from surveillance and MTR irradiations, normalized to a fluence of 3×10^{19} n/cm^2 ($E > 1$ MeV). A "Chemistry Ratio" (CR) was developed to express the shift at that fluence; the elements in the CR (except Mo) feature in many current models.

Varsik and Byrne model for weld metal (1979).

$$\Delta = [377.9 \log(CR) + 331.9]\Phi^{0.43}$$
$$CR \text{ is chemistry ratio} \tag{10.6}$$
$$CR = \{[1.5\text{Ni} + \text{Si} + 0.5\text{C} + 0.5(\text{Mn} - 0.05)]\text{Cu}\}/(0.5 + 0.5\text{Mo})$$

By 1984 the US surveillance database had grown to 177 data points (today there are around 1000), and it was evident that RG1.99 Rev 1 was nonconservative with respect to the effect of Ni. The US NRC commissioned Guthrie and the Electric Power Research Institute (EPRI) commissioned Odette to (independently) assess the existing database and to develop fitted models. Randall [180] combined the two fluence functions (FFs) into a worse case, and then calculated shifts for a wide range of Cu and Ni values (these being the only elements found to significantly affect irradiation sensitivity) at a fluence of 1×10^{19} ($E > 1$ MeV). These values were described as the CF and were presented as a table rather than as an analytical expression. RG1.99, Rev. 2 (Eq. 10.7) was issued in 1988; unlike RG1.99 Rev 1, it predicted mean shift, but required a margin to be added for uncertainties in shift and the start of life transition temperature [67].

USNRC RG1.99 Rev 2 model (1988).

$$\Delta = CF \Phi^{[0.28 - 0.010 \log(\Phi)]} \tag{10.7}$$

CF is chemistry function (values are tabulated and depend on Cu and Ni)

Other regulatory models adopted similar CF × FF equations, but with analytical expressions for the CF. Examples include the 1989 Russian PNAE model (Eq. 10.8 [181]), the 1987 French FIM (mean shift) (Eq. 10.9) and FIS (upper bound) formulae, described by Brillaud et al in 1987 [182] and the 1991 Japanese JEAC (Japan Electric Association Code) model (Eq. 10.10 [183]). In contrast, the German KTA 3203 code [184] retained a graphical presentation of bounding shift values.

Russian regulatory (PNAE) model (1989).

$$\Delta = [230(\text{Cu} + 10\text{P}) + 20]\Phi^{1/3} \tag{10.8}$$

French regulatory (FIM—mean shift) model (1987).

$$\Delta = \left[17.3 + 1537(\text{P} - 0.008) + 238(\text{Cu} - 0.08) + 191\,\text{Ni}^2\text{Cu}\right]\Phi^{0.35} \tag{10.9}$$

JEAC model (1991).

$$\Delta = \left(26 - 240 \cdot \text{Si} - 61 \cdot \text{Ni} + 301\sqrt{\text{Cu} \cdot \text{Ni}}\right) \times \Phi^{(0.25 - 0.1 \log \Phi)} \tag{10.10}$$

Eqs. (10.7)–(10.10) can give substantially different predictions even for the same materials and irradiation conditions. In the case of the PNAE model this is not surprising due to the substantial difference in the steels. In the case of the other models, which apply to very similar steels, the reason for difference in prediction is that they are empirically fitted by different people at different times to different surveillance databases; those for the specific reactor fleet within the jurisdiction of the regulator. Thus, differences may arise due to differences in the mean values and distributions of both materials variables (e.g., alloy composition and fabrication parameters) and irradiation variables (e.g., temperature, flux, and spectrum) all of which affect the empirical interpretation of the data and the statistical fit. For example, the US surveillance database contains a substantial amount of data from older RPVs which contain higher levels of copper, while French surveillance data are dominated by more recent low copper steels. There may also be biases between the databases, for example, between the surveillance capsule irradiation temperature for the different fleets. The differences in formulation of the models can reflect not only the differences in the databases but also differences in the scientific understanding of irradiation damage at the time they were formulated. There may also be differences in technical and regulatory opinion and in fitting procedures. At best empirical models can only describe the data to which they are fitted and may become less accurate outside the dataspace to which they are fitted (and to which they are applied).

MBC models. MBC models have the advantage over empirical models that the physical understanding used in their development makes them potentially more reliable in extrapolation. Odette, who is a strong advocate of such models, has developed several, including Refs. [111, 185, 186]. The processes involved in producing MBC models is described in Ref. [69] and have been used by many others (e.g., Refs. [6, 93, 130, 187–190]). Although the French and Japanese regulatory codes have incorporated MBC models [6, 191], RG1.99 still uses the 1988 Rev 2 model. The 2007 ASTM standard for predicting irradiation shift [192] used a MBC model based on work by Eason et al [93, 111] that later resulted in the model used in the USNRC PTS rule [49]. The first of these references presents the model fully, but the second gives a very comprehensive description of the understanding underlying the model and how it was developed. EONY is given in Eq. 10.11 and the remainder of this section explains some of its main features. As previously noted, the current (2015) ASTM standard [176] uses an empirical approach for shift prediction.

EONY is a 2F model with the matrix and precipitation terms linearly added and the implicit assumption that the BOL hardening features do not affect (Charpy 41J) shift (or are accommodated in the parameterization). The model was fitted to surveillance data, guided by mechanistic understanding and insights. The parameters A and B account for differences in categorical variables (see " Effect of environmental and materials variables" subsection above); the increased sensitivity of Combustion Engineering manufactured vessels is not understood, which serves to remind that RPVs may not all be made in the same way.

The matrix term was initially derived by fitting the low Cu subset of the database. This avoids the problem of "trade-off" between terms in a 2F model when both features are fitted simultaneously, and the same variable may affect both. The fitting of the two terms was done iteratively as described in Ref. [111]. The MF term includes a linear temperature effect; although other forms might be preferred on physical grounds, the linear function was more stable for fitting and sufficiently accurate within the narrow temperature band of the data. The MF term also includes the synergistic Mn-P effect, observed in IVAR irradiations of simple model alloys, and possibly a consequence of P and Mn segregation to, and stabilization of, matrix features. Although it is likely that other elements influence matrix embrittlement, it is usually very difficult to identify this from surveillance databases; the effects within the range of values of the element within the database may be small compared with data scatter, or values might not be available.

The CRP term also includes P; this can be rationalized because of the presence of P in CRPs. The dual role of P illustrates the difficulty of using mechanical property data alone to isolate irradiation effects. Similarly, Ni has more than one function; it increases the precipitation plateau, by producing a larger volume fraction of precipitates with a finer dispersion and shifts the plateau to higher fluence. Precipitation of Cu during heat treatment is considered through the Cu effective formulation. It was found to be unnecessary to include a temperature term in the CRP component. The dependence of CRP on effective fluence is modeled as a tanh function, which gives a good approximation of the classical Avrami model of precipitation kinetics and may be more tractable for statistical fitting.

A key aspect of both the matrix and the precipitation terms is the use of the effective fluence term to model flux effects. The formalism is based on a simplified rate theory model that considers the fraction of vacancies surviving recombination as a function of flux (discussed in "Effect of environmental and materials variables" section and illustrated in Fig. 10.19). However, this is only applied to BWR irradiations, even though flux effects would be expected for PWRs also. The rationale is that the BWR and PWR datasets form isolated clumps in the database but in the PWR clump most data fall within a relative narrow range of fluxes, and that surveillance fluxes for BWRs are one to two orders of magnitude lower than those for PWRs. The effective fluence term was co-fitted to the MF and CRP terms. This is justified on the basis that both require transport of solute atoms.

EONY model.

$$\Delta T = \Delta M + \Delta P$$

$$\Delta M = A(1 - 0.001718 T_i)(1 + 6.13 \text{PMn}^{2.47})\sqrt{\varphi t_e}$$

$$A = \begin{cases} 1.140 \times 10^{-7} \text{ for forgings} \\ 1.561 \times 10^{-7} \text{ for plates} \\ 1.417 \times 10^{-7} \text{ for welds} \end{cases}$$

$$\Delta P = B(1 + 3.77\text{Ni}^{1.191}) f(\text{Cu}_e, P) g(\text{Cu}_e, \text{Ni}, \varphi t_e)$$

$$B = \begin{cases} 102.3 \text{ for forgings} \\ 102.5 \text{ for plates in non-CE manufactured vessels} \\ 135.2 \text{ for plates in CE manufactured vessels} \\ 155.0 \text{ for welds} \\ 128.2 \text{ for standard reference material plates} \end{cases}$$

$$\text{Cu}_e = \begin{cases} 0 \text{ for Cu} \leq 0.072\text{wt}\% \\ \min[\text{Cu}, \text{Max}(\text{Cu}_e)] \text{ for Cu} > 0.072\text{wt}\% \end{cases}$$

$$\text{Max}(\text{Cu}_e) = \begin{cases} 0.242 \text{ for typical (Ni} > 0.5 \text{ Linde 80 welds)} \\ 0.301 \text{ for all other materials} \end{cases}$$

$$f(\text{Cu}_e, P) = \begin{cases} 0 \text{ for Cu} \leq 0.072\text{wt}\% \\ [\text{Cu}_e - 0.072]^{0.668} \text{ for Cu} > 0.072\text{wt}\% \text{ and } P \leq 0.008\text{wt}\% \\ [\text{Cu}_e - 0.072 + 1.359(P - 0.008)]^{0.668} \text{ for Cu} > 0.072 \text{ and } P > 0.008 \end{cases}$$

$$g(\text{Cu}_e, \text{Ni}, \varphi t_e) = \frac{1}{2} + \frac{1}{2}\tanh\left[\frac{\log_{10}(\varphi t_e) + 1.139\text{Cu}_e - 0.448\text{Ni} - 18.120}{0.629}\right]$$

$$\varphi t_e = \begin{cases} \varphi t \text{ for } \varphi \geq 4.39 \times 10^{10} \\ \varphi t \left(\frac{4.39 \times 10^{10}}{\varphi}\right)^{0.259} \text{ for } \varphi < 4.39 \times 10^{10} \end{cases} \quad (10.11)$$

where ΔT, ΔM, and ΔP are the total, matrix feature, and precipitation shift (°C), T_i is irradiation temperature (°C), elemental compositions are in weight percent, φ is neutron flux (n/cm^2, $E > 1$ MeV), t is irradiation time (s), φt_e is effective fluence, and Cu_e is effective copper content.

The precision of the EONY model was partly validated by extensive analysis of residuals, partly by using reserved data that had not been used in the fitting process, and partly by comparing its predictions with data from the IVAR database. The scatter (1SD) of the residuals to the fit ranged from about 8°C (for low Cu plate) to 15°C (for high Cu weld). These values are probably about the minimum achievable for surveillance data, which are generally relatively imprecise, compared to laboratory data.

It should be noted that even MBC models may not provide accurate predictions outside the database to which they are fitted. Soneda et al. [190] compared predictions from the CRIEPI model, which explicitly uses rate theory to account for flux effects, with predictions from the EONY model and data from the US surveillance database. There were large differences between the predictions of the two models for forgings and at low flux. Differences between CRIEPI model weld predictions and US weld results could be accounted for to a large extent by upper shelf drop (see "Relationship between hardening features and hardening and embrittlement" section). Differences between the CRIEPI model predictions and the US forging results could be largely compensated by the offset adjustment method. This feature of the CRIEPI models allows account to be taken of systematic variations that can exist in BOL toughness within a component, particularly forgings (see Section 10.1.3), which result in a systematic bias of all shift estimates based on a common BOL datum. Such adjustments are controversial, and there is no universal agreement on how they should be done. A French approach discussed in Ref. [11] recognizes the potential bias, but resolves it by reduction of the margin applied to the mean curve rather than by adjustment of the raw data. In the United Kingdom, Bayesian methods have been used to model the entire set of Charpy curves from a surveillance program, rather than the curves individually [193]. As in the CRIEPI approach, this explicitly recognizes, and considers, the uncertainty in the BOL curve.

Other comparisons between MBC models confirm the difficulty of producing a single model appropriate to all LWR designs. There are several issues associated with the empirical nature of these models. These include differences in the variables used in the models (some variables, such as microstructural factors are not quantified, others, such as start of life strength, may be confounded with other variables and others have too small a range within the database to which the model is fitted). Neutron spectrum effects may also be an issue if fluence rather than dpa is used (e.g., Refs. [194, 195]), and for BWRs vs PWRs.

So, despite 30 years work on MBCs, efforts to develop more accurate engineering models, suitable for life extension to 60 or more years continue. Issues include: the need to develop means of properly incorporating high flux data; identification of the regimes in which LBPs and other possible high fluence features become important; and the development of methods to combine advanced models based on self-consistent analysis of high-quality data with the relatively scattered and ill-characterized surveillance data [82]. Arrangements have been made to increase the residence time for some of the current surveillance capsules in selected reactors to increase the database at high fluences [196].

10.3.2.2 Thermal aging

Irradiation damage is the most significant mechanism of in-service degradation in LAS. However, microstructural changes can also take place in the absence of irradiation through thermal aging. Although the damage caused by thermal aging is generally lower, it may be significant to SIA if it occurs in regions which are more highly stressed than the beltline region, have inferior start of life toughness properties, or are found to contain significant flaws.

Three types of thermal aging may be considered:

- Formation of fine-scale precipitates that harden the material thereby causing shift.
- The weakening of grain and other microstructural boundaries by the segregation to them of solutes that promote IGF.
- Changes to the preexisting fine-scale microstructure, thereby affecting hardness or shift.
- Formation or coarsening of cleavage trigger particles

The first of these occurs by a mechanism similar to precipitation damage under irradiation. Since the thermal vacancy concentration is much lower than the irradiation-enhanced vacancy concentration, precipitation rates are extremely slow. Nevertheless, substantial damage (a hardness change of about 70 HV (kgf/mm^2), equivalent to a Charpy shift of about 150°C) was observed in Mn-Mo-Ni steels aged for 20,000 hours at a temperature of only 330°C [197]. This observation was, however, for a weld containing about 1.7% Ni and 0.4% Cu in solution (see "Microstructural evolution during irradiation" subsection above). Ni has a major effect on both irradiation damage and thermal aging and a weld with similar Cu, but a Ni level of about 0.3%, hardened by 15 HV (kgf/mm^2) (30°C) under the same conditions. The kinetics of Cu precipitation depends very strongly on temperature and Cu content and very little precipitation would be expected at an RPV wall temperature of 290°C over a 40-year lifetime. Confirmatory evidence is available from a number of thermal surveillance capsules. Nevertheless, there may be concern for components with significant levels of Cu operating at higher temperatures, for example in GCRs or in

LWR pressurizers, which operate at 345°C, as well as for long-term operation to 80 years and beyond for some LWRs

Temper embrittlement is a well-known hazard of heat treatment at temperatures in the range 400–500°C. It occurs through a similar mechanism to NHE, discussed earlier (Microstructural evolution during irradiation section). It is avoided by correct heat treatment but Mn-Mo-Ni RPV steels are, in any case, thought to be relatively immune. Concerns have been expressed that temper embrittlement might occur during irradiation damage annealing, but the evidence is mixed (see Section 10.3.2.4).

Other microstructural changes could occur though prolonged exposure to reactor operating temperatures. In general, however, diffusion rates are so small (unless enhanced by irradiation) that only the smallest features are likely to change significantly in size and such changes are unlikely to any measurable impact on mechanical properties.

WWER-440 steels are reported to be practically immune to thermal aging, but WWER-1000 forgings may exhibit thermal aging leading to shifts of up to 30°C [198]. The latter is due to a carbide precipitation process, but this is followed by coalescence that reduces the effect. WWER-1000 welds have a low carbon content and this effect is not apparent, but in welds used for off-beltline applications higher levels of phosphorus are permitted and there is evidence of aging by a thermal segregation temper embrittlement mechanism.

The thermal aging processes discussed earlier have similar, but faster, irradiation damage counterparts. It is, therefore, generally assumed that irradiation damage and thermal aging are not additive, the shift to be used in an SIA is, therefore, the maximum of the irradiation or thermal aging shift; as flux tends to zero, the irradiation shift tends to the thermal aging shift. For WWER RPVs, however, the PNAE Code adds 30°C to the ΔTT based on irradiation.

10.3.2.3 Strain aging

Strain aging is known to occur in LAS; the issue is whether it needs to be considered, and one of some interest in the United Kingdom [199]. Strain aging is caused by solute elements (predominantly C and N) segregating to dislocations. This resists the onset of plastic flow causing the well-known yield stress drop when the dislocations break free. Static strain aging (SSA) can occur when a material is strained beyond yield, creating new dislocations, is unloaded and then loaded again. If there is time for the solutes to segregate to the new dislocations before the material is reloaded, SSA will occur. Since the new dislocations are now pinned, the stress required to yield the material will be higher than that at which yielding previously took place. Dynamic strain aging (DSA) occurs at combinations of strain rate and temperature at which the (interstitial) solutes can just keep pace with the dislocations. This produces serrated yielding, elevates the flow stress and reduces ductility.

SSA has been observed to significantly increase Charpy transition temperature in C-Mn and Mn-Mo-Ni steels (e.g., [200, 201]). However, the relevance of SSA, normally investigated by straining blocks of material to (usually) around 5% before machining test specimens from them, to RPV material may be questioned. RPV material enters service in the stress relieved condition without significant prestrain and, in any case, initial toughness values are obtained in the condition in which the material enters service. Loads during service are well below yield, except at the tips of any flaws within the material of a large enough size for local yielding to take place. In principle, these very localized regions could experience SSA. However, FT specimens contain small yield regions due to fatigue precracking. Since interstitial elements diffuse rapidly even at ambient temperatures, the specimens are likely to be strain aged before testing. The great majority of in-service transients are for lower K_I than the fatigue precracking transient. The most significant in-service transients occur under PTS conditions. If the K_I were to exceed the bounding K_{IC} (which could only happen if a major error had been made in the SIA) it would most likely do so with K_I reducing (see Fig. 10.15). In this case protection against failure would be provided by the phenomenon of warm prestressing (WPS), whereby if a crack survives loading to a particular K_I at a higher temperature, it will not fail below that K_I at a lower temperature. The potential interactions between strain aging and WPS are complex and difficult to unravel experimentally. At the peak K_I, the interstitials are likely to be able to keep pace with the dislocation movements because temperatures are relatively high and strain rates low. At temperatures lower than that at the peak, the strains will reduce with a reversal in dislocation movement. In addition, WPS may be accompanied by crack tip blunting and decohesion of potential fracture initiators. Experimentally, the positive effects of WPS are seen to outweigh any negative effects of SSA/DSA, but it is difficult to isolate the components.

DSA reduces J_I and flattens the J–R curve. However, because of the speed of interstitial diffusion at RPV operating temperature, DSA is only observed at strain rates beyond those possible in service (except perhaps during a failure event).

There is evidence that nitrogen segregates to MF defects, particularly in some carbon steels, which reduces the amount available for strain aging. In the case of DSA there is evidence that this is the case. There is also evidence [202] that SSA and irradiation shifts are additive when irradiation follows SSA. As far as is known, strain aging is not explicitly accounted for in SIAs.

10.3.2.4 Irradiation damage annealing

LWR licenses in the United States are now being extended from 40 to 60 years by USNRC with the possibility of further extensions to 80 years and beyond. Extending operation from 40 to 80 years implies a doubling of the neutron exposure for the RPV, unless flux reduction methods are implemented. Thus, for the RPVs of PWRs expected to experience neutron fluences from 1 to 5×10^{19} n/cm^2 (>1 MeV) after 40 years, the exposures will be 2 to 10×10^{19} n/cm^2 after 80 years. Even for normal start-up and cool-down transients, the coolant-pressure-temperature (P–T) curve must be below the corresponding stress (from pressure) that could cause fracture for an assumed very large crack size. The mechanisms that cause irradiation-induced embrittlement of RPV steels are discussed in Section 10.3.2.1 and are discussed further here in the context of the mechanisms that take place during the thermal annealing process, which are inextricably linked to those which cause the embrittlement.

Various options are possible to mitigate the effects of irradiation embrittlement on the RPV. Many of these options are discussed by Planman, Pelli, and Torronen [203]. Recovery of the material toughness through thermal annealing is one method of increasing safety margins of the RPV. Thermal annealing involves heating the RPV beltline materials to temperatures ~50–200°C above the normal operating temperature for about 1 week, with the amount of recovery increasing with increasing annealing temperature. Two different procedures can be used to perform the thermal anneal, a wet anneal or a dry anneal. A wet anneal is performed with cooling water remaining in the RPV and is limited to the RPV design temperature (343°C for US LWRs). A dry anneal requires removal of the cooling water and internal components and would normally be performed at temperatures in the range of 430–500°C.

Annealing recovery is a function of irradiation temperature and flux, chemical composition of the material, annealing temperature and time, with annealing temperature being the dominant factor in recovery of fracture toughness. A key aspect of thermal annealing, in addition to demonstration that the annealing had been effective, is the rate of reembrittlement that occurs following the annealing treatment [204]. Fig. 10.31 shows schematic depictions of two different procedures to predict the postannealing reembrittlement of the material. The lateral shift method assumes the reembrittlement will occur at the same rate that occurs at the beginning of the original irradiation embrittlement, while the vertical shift method assumes the reembrittlement will occur at the rate of the original embrittlement from the point of annealing [205]. The lateral method is more appropriate to RPV steels at high enough fluence for Cu to be precipitated and re-irradiation approximately follows the MF hardening. However, the behavior at very high fluence is still being studied.

There are many examples of thermal annealing results on LASs [204], with one example for a high-copper weld from the Midland Unit 1 reactor shown below; this reactor did not operate but material was removed from the RPV for various material variability, mechanical property, and irradiation effects studies by the NRC-sponsored Heavy-Section Steel Irradiation (HSSI) Program [206]. Fig. 10.32 shows the beneficial effect of the 1-week high-temperature annealing at 454°C (850°F) compared with that for a 1-week anneal at 343°C (650°F). The higher temperature anneal resulted in a Charpy 41-J transition temperature recovery of about 80%, compared with about 50% for the lower temperature anneal.

As with irradiation effects mechanisms that cause RPV material embrittlement, the mechanisms of thermal annealing that result in recovery of the material toughness are quite complicated. For a copper-bearing material

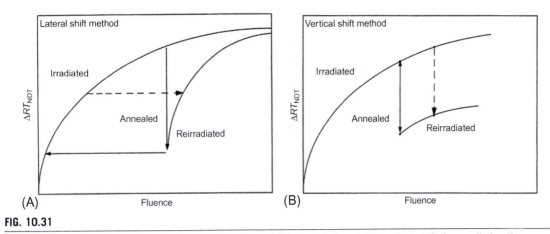

FIG. 10.31

Schematic depictions of two thermal annealing methods, (A) lateral shift and (B) vertical shift, for predicting the postannealing reembrittlement of RPV materials. The more-conservative lateral shift method assumes that reembrittlement follows the original embrittlement curve; the vertical shift method that it continues at the same rate as that just before annealing [205].

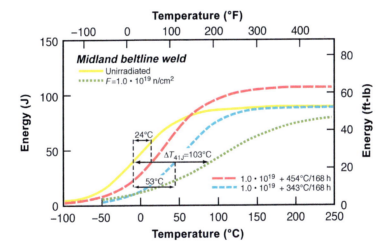

FIG. 10.32

Effects of thermal annealing at 343°C and 454°C on the high Cu Midland Unit 1 RPV beltline weld. At the higher annealing temperature, the transition temperature is almost fully recovered; the overrecovery of the upper shelf is a common observation.

Redrawn from Iskander, S. K., Sokolov, M. A., and Nanstad, R. K., Comparison of Different Experimental and Analytical Measures of the Thermal Annealing Response of Neutron-Irradiated RPV Steels, pp. 403–420 in Effects of Radiation on Materials: 18th International Symposium, ASTM STP 1325, R. K. Nanstad, M. L. Hamilton, F. A. Garner, and A. S. Kumar, Eds., American Society for Testing and Materials, 1999.

such as the example shown for the Midland weld, irradiation of the material caused the formation of CRPs, which are about 1–2 nm in diameter, and various forms of matrix damage defect-solute complex features (MF). The high-temperature thermal annealing treatment dissolves most of the small irradiation-induced defect clusters and complexes. Because of the differences in surface energy, the larger CRPs grow at the expense of the smaller ones and the mean size of the CRPs increases to to 3–5 nm. This is accompanied by a slight reduction in their volume fraction and a significant reduction in their number density. The overall results of these dissolution mechanisms are a significant decrease in hardening (i.e., yield strength), and recovery of toughness (i.e., decreases in Charpy impact and fracture toughness transition temperatures) depending on annealing temperature. There is generally small recovery at 343°C, at low flux, modest recovery at 373–400°C and almost complete recovery at 450°C. The amount of recovery at a given temperature depends on flux in combination with copper content [48, 100]. At high flux (i.e., ~1E13 n/cm^2/s or higher), UMDs may be formed that contribute to the embrittlement [101]. Due to this, thermal annealing recovery on test reactor irradiated material may be different from that experienced by the RPV under the same annealing conditions.

The available data on thermal annealing of LASs are relatively robust and well documented. However, there are only limited results available regarding the reembrittlement rate. An issue of investigation regarding effects of thermal annealing and reirradiation on the microstructural changes in RPV steels also involves the potential for temper embrittlement resulting in significant IGF. A study by McElroy et al. [207] clearly demonstrated the embrittlement susceptibility of grain-coarsened RPV steels. It is important to note that the study was performed on 11 different steels intentionally heat treated with temper embrittlement procedures to assess susceptibility to the phenomenon. Very coarse grains (e.g., ASTM 0–1) were obtained and, as stated by Nanstad et al. [124] in a similar study, such coarse grains may not truly represent the coarse-grain microstructure in RPV weld heat-affected zones. Moreover, IGF of low-alloy steels for PWRs has not been reported in commercial surveillance programs, although there are reported observations from VVER reactors [208]. Although significant IGF has not yet been observed in irradiated PWR steels, IGF has been observed in the HAZ of some steels in the postirradiation annealed condition after irradiation to a fluence about 1×10^{19} n/cm^2, raising concern regarding behavior after irradiation to even higher fluences [124].

In the United States, Title 10CFR 50.66 [209] permits the thermal annealing of light-water reactors. It refers to RG 1.162 [205], which describes the format and content of an acceptable Thermal Annealing Report and addresses the metallurgical and engineering issues that need to be addressed in an application to perform a thermal annealing. The reembrittlement trend in RG 1.162 is based on the "lateral" shift procedure shown in Fig. 10.32.

Thermal annealing of operating nuclear reactors has been performed at least 16 times, once in the United States, once in Belgium, and 14 times at Russian-designed VVER-440 plants. The first annealing operation was performed on the US Army SM-1A nuclear reactor in Fort Greely, Alaska in August 1967 [210]. The Belgian Reactor 3 (BR3) was the first commercial power reactor to be annealed, with a wet anneal being performed in 1984 (22 years after startup) using primary pump heat [211]. BR-3 operated at 260°C (500°F) and the thermal annealing operation was performed at 343°C (650°F), the RPV design temperature, with the Charpy impact transition temperature recovery estimated to be 50%.

Following the annealing of the BR3 reactor, many thermal annealing treatments were performed on Russian-designed VVER-440 reactors. Table 10.3 shows the name of each reactor, the year of the annealing, the annealing temperature and time, and whether the RPV was clad with austenitic steel on the inside surface [212]. Thus, there is experience with annealing and reoperation of nuclear plants to give credence to application of the technology [212]. In fact, an Annealing Demonstration Project (ADP) funded jointly by the US Department of Energy and the nuclear industry was performed at the uncompleted Marble Hill nuclear plant in Indiana in 1996/1997 and an independent evaluation concluded that "Successful completion of the ADP has demonstrated that functional requirements for in-place annealing of a US RPV can be met using existing equipment and procedures" [213, 214, 215].

Despite the technical success of the Marble Hill demonstration and the existence of RG 1.162, thermal annealing is not likely to be deployed for plants operating to 40 or 60 years based on the current understanding of RPV degradation (and resulting reduced uncertainty in safety margins) as well as the potential liability of permanently damaging a reactor vessel. However, in some cases, thermal annealing may be required to extend plant life to 80 years. Additional research, to address the issue of irradiation damage effects at high fluence under low flux conditions is needed to reduce the uncertainties, hence the potential liabilities and make this technique more acceptable for industry [213, 215].

Table 10.3 Thermal Annealing Performance of VVER-440 Type RPVs

Reactor	Year	Temp/Time	SS Clad	Country
Novovoronezh-3	1987	430 ± 20°C/168 h	No	Russia
Armeniya-1	1988	450 ± 50°C/168 h	No	Russia
Greifswald-1	1989	475 ± 15°C/150 h	No	Germany
Kola-1	1989	475 ± 15°C/150 h	No	Russia
Kola-2	1989	475 ± 15°C/150 h	No	Russia
Kozloduy-1	1989	475 ± 15°C/150 h	No	Bulgaria
Kozloduy-3	1989	475 ± 15°C/150 h	Yes	Bulgaria
Greifswald-2	1990	475 ± 15°C/150 h	No	Germany
Greifswald-3	1990	475 ± 15°C/150 h	Yes	Germany
Novo Voronezh-3	1991	475 ± 15°C/150 h	No	Russia
Novo Voronezh-4	1991	475 ± 15°C/150 h	No	Russia
Kozloduy-2	1992	475 ± 15°C/150 h	No	Bulgaria
J. Bohunice-2	1993	475 – 503°C/168 h	Yes	Slovakia
J. Bohunice-1	1993	475 – 496°C/168 h	Yes	Slovakia
Loviisa-1	1996	475 ± 15°C/100 h	Yes	Finland
Rovno-1	2010	475 ± 15°C/150 h	Yes	Ukraine

10.3.3 OTHER PERFORMANCE ISSUES

Although irradiation damage has generally been the dominant integrity issue for RPVs, there have been a few other problems in specific cases. In about 1980 UCC was found in several French RPVs during ISI. These were small (in the order of mm) defects emanating from the interface between the cladding and the LAS of the RPV wall. The cracks were caused by hydrogen cracking during the cladding process. They were not detected at the time of manufacture because the NDE techniques available at that time, were not able to do so. The ISI benefitted from significant development in NDE in the intervening years. Justification for continued operation of the cracked RPVs was achieved through extensive SIA. The solution for subsequent vessels was to increase the preheat temperature during the cladding process to allow the hydrogen to diffuse harmlessly out of the material.

A second problem with hydrogen cracking emerged in 2012 when a routine ISI of the Belgian Doel 3 was extended, as a precaution, to include detection of UCC in forging material.[19] No significant UCC was found, but instead there were some thousands of small laminar defects lying parallel to the vessel wall. Thorough investigation revealed these to be "hydrogen flakes" produced during manufacture, but not detected at the time. Doel 3 and its sister reactor Tihange 2, which was found to be similarly though less extensively affected, were shut down for several months, pending the development of a safety justification (SJ). Evaluations of this SJ are described in Refs. [216, 217]. The work to produce and evaluate the SJ was supported by many international experts.

Reactor operators and regulators around the world were kept informed of the problem and this led to reviews of other RPVs in service to determine whether they might be affected by hydrogen flaking. In the case of the Swiss Beznau Unit 1 reactor, the precautionary ultrasonic test (UT) inspection done in 2015 to determine whether hydrogen flakes were present in the RPV recorded several thousand UT indications.[20] After a two-and-a-half-year

---

[19] ISI inspections normally focus on welds, which are more likely to contain flaws than bases materials and are optimized for detection and sizing of specific types and orientations of defects (those most likely/harmful) and may have limited performance for others.

[20] A UT indication is "Evidence of a discontinuity that requires interpretation to determine its significance" (ASTM Standard Terminology for Non-Destructive Examinations ASTM E 1316-91b).

program of testing and analysis the licensee was able to justify that the indications were from agglomerates of alumina inclusions produced during manufacture of the RPV and that these did not adversely affect the safety of the RPV. As in the case of the Belgian reactors, many international experts were involved [217a].

Unexpected in-service issues such as the examples above can be highly expensive in terms of loss of revenue. The costs of the SIA work and materials investigations are very modest in comparison, but can take many months (one or more years if it is necessary to irradiate material). Limited understanding of materials behavior and lack of accurate BOL data or shift predictions for the specific RPV and the locations of interest can be major problems, which can be compounded by failure to retain sufficient well-qualified archive material. Assumptions about materials behavior, which were reasonable in the context of a defect-free RPV and a healthy SIA, and which were, therefore, never fully validated, can suddenly become important when the SIA margins are smaller. The work required to satisfy regulatory concerns is seldom quick and easy.

A further major service issue was the hole discovered in the Davis-Besse RPV top head in 2002 [218]. This was caused by leakage of boronated water onto the head, so is not directly comparable to the other degradation mechanisms discussed in this chapter. However, it does illustrate the vulnerability of RPV steels to corrosion and the need to ensure that protection is in place for all possible failure paths.

In principle, but never observed in practice, fatigue crack growth provides a possible RPV failure path. Cycling loads are generally highest in areas subject to high thermal stresses, particularly where there are discontinuities or changes in section, for example, the nozzles. However, even here they are generally relatively low, and these regions receive high levels of inspection such that any undetected defect would have low cycling stress intensity factors. RPV steels have generally good resistance to fatigue crack growth, however, growth rates are markedly increased by exposure to high-temperature pressurized water environments. This is not normally an issue since the cladding on the RPV protects the ferritic RPV material itself and the number of cycles is relatively low. This protection would be lost should the cladding be penetrated by a defect. EASCC effects are discussed in Chapter 6 of this book.

10.4 CURRENT DEVELOPMENTS AND FUTURE PROSPECTS
10.4.1 IMPROVED PREDICTION OF THROUGH-LIFE TOUGHNESS

Most current research on LAS is being driven by the economic and environmental needs to extend NPP lifetimes. The work is driven by the need to extrapolate beyond current experience and is, therefore, focused on the development of understanding, both by careful experiments and MSM. One trend has been convergence of MBC and MSM models. Insights from numerical physical models of individual parts of the irradiation damage have long been used (by some workers) to guide the development of statistically fitted, analytical, MBC models. However, the integration of physical submodels into full MSM models, as exemplified in the PERFORM 60 project, coupled with developments in computer technology and the growth of physical understanding and data at the scales of the processes involved, provide a potential for increasingly accurate MBC models. If this process of convergence of correlation and physical models continues, it will ultimately provide explanations for the differences between the shift predictions of correlation models fitted to surveillance data for different (national) RPV fleets. This would remove the unsatisfactory categorical variable of "geographical location"; similarly, there would be no need for variables such as "product form" or "vendor," or at least the reasons for the effect of these variables would be understood.

By developing more comprehensive, detailed, and physically demonstrable models, to remove or explain unquantified factors and anomalies, confidence in shift predictions for extended life will increase. It is a virtuous circle, the more irradiation damage is understood, the greater the confidence that there are no "hidden factors" that have not yet made themselves known. Confidence is essential, not from a safety viewpoint since embrittlement is monitored, but for economic and public confidence reasons. Life extension potentially provides great economic benefits, but requires major investment, which is at risk if shift predictions are underestimated.

10.4 CURRENT DEVELOPMENTS AND FUTURE PROSPECTS

Currently, the largest program of work to address the issue of life extension is the ongoing ATR-2 experiment [115, 177, 219, 220], funded by the US DOE, and led by UCSB and ORNL in collaboration with a number of US and international organizations. This is centered on an irradiation of a very large number of materials, to fluences up to those relevant for an 80-year RPV irradiation at a flux that bridges the gap between existing databases. Most of the irradiations are at typical PWR or BWR temperature, but there are some samples over a wider range to provide information on mechanisms. Samples are being tested, using several different techniques, to develop a better understanding of microstructural evolution and its effect on materials properties. The aim is to confirm existing physical models, some developed many years ago, and enable them to be developed as necessary. In this context, it is worth noting that some of the materials involved have compositions that extend beyond materials in current commercial use. It is hoped that these will provide valuable tests of the models and potentially new materials for future use (see Section 10.4.2).

It should be noted that it is not practicable to avoid extrapolation of test data to RPV conditions. To provide advanced warning of unexpected high irradiation sensitivity, surveillance data are irradiated at flux levels usually between about 2 and 5 higher than the RPV that they monitor, but sometimes outside this range. With originally planned RPV lifetimes generally of 40 years, results for surveillance program exposures of more than 20 years are relatively uncommon. It is generally impracticable to do test reactor irradiations for more than about 2 years, so the choice ranges between irradiating to the EOL fluence, but at high flux, or the RPV flux to low fluence. ATR-2 bridges the gap between the previous UCSB IVAR (low flux) and ATR-1 (high fluence) experiments. However, it will still be necessary to extrapolate to RPV fluxes and, hence, reliance on physical models to do so with confidence.

The commercial nuclear industry in the United States has developed programs to obtain additional data at high fluences to support extended operations to 80+ years, as some RPVs are projected to have peak vessel fluences approaching 1×10^{20} n/cm^2 [116, 196]. The Coordinated PWR Reactor Vessel Surveillance Program (CRVSP) and the PWR Supplemental Surveillance Program (PSSP) have been developed with a survey of the existing high fluence data and considering projected data to come from existing surveillance programs. The EPRI CRVSP will provide additional high fluence data by deferring some of the existing surveillance capsule withdrawals to later times, thus, providing higher fluences than originally planned. About 16 capsules will provide 38 new data from 6 to 9×10^{19} n/cm^2, with data becoming available from the current time until about 2025. The additional data from the CRVSP will be complemented by the EPRI PSSP which will generate an additional 27 data by reconstituting previously irradiated and tested Charpy surveillance specimens, and further irradiating them to higher fluence in a host PWR. The PSSP data will likely be available between 2025 and 2030. An important aspect of the industry programs is the link to the ATR2 program in that eight (five welds and three base metals) of the commercial surveillance materials included in the ATR2 experiment are included in the PSSP, with the intent to provide direct material correspondence for evaluating flux effects.

Increasing attention is also being given to the treatment of fracture toughness as of equal importance to irradiation shift. Originally, the two were often considered as two separate issues and disciplines. In reality they do not exist in isolation; what matters from a safety standpoint is the fracture toughness of irradiated material, not the individual components, BOL toughness and toughness shift, in isolation. However, it is still generally the case that safety cases are based on BOL toughness, determined from the empirical parameter RT_{NDT}, and irradiation shift determined by Charpy tests. It has been clearly shown that the MC is generally a more accurate basis for estimating toughness, and the physical basis for it is now much better understood (see Section 10.3.1). Similarly, the relationship between the microstructural changes produced by irradiation and the effects of these on materials flow properties and hence, ultimately, the MC have become reasonably well-established. In the PERFORM-60 project a preliminary attempt was made to provide a model that could be run through a relatively simple to use-interface to provide a prediction from the atomic to RPV length and time scales, from microstructural development, through to fracture toughness.

The above developments will help to provide confidence for investment in lifetime extension, by providing more accurate (reliable) predictions, and more accurate estimates of the uncertainties in prediction. The biggest challenge is to establish sufficient confidence in models that they gain regulatory acceptance. This is likely to be achieved only through continuing work to understand irradiation damage and fracture and produce validated physical models. However, a need for empirical validation of predictions through surveillance data will remain for the foreseeable future [221]. The benefits of this could be enhanced through shared surveillance programs, international pooling of data, and investigation of archive samples using modern techniques [222, 223]. There has been a little progress in these areas, but costs and commercial issues are barriers.

Given the continuation of the work and continued international collaboration, RPV toughness issues should not be an obstacle to license extension to 80 years for many RPVs. The same work will, as described in the next two sections, also contribute to developments in materials for new, long lifetime RPVs.

10.4.2 IMPROVED MATERIALS

For current LAS and RPVs designs, attention to BOL properties is at least as important as attention to irradiation shift, since the latter is relatively small in modern materials thanks mainly to control of Cu. The BOL properties of current materials are also better than those of a decade or two ago due to developments in steelmaking and fabrication practices. Continuing developments in these will potentially produce further improvements in both BOL fracture toughness and, importantly, better homogeneity within components. These improvements will be facilitated by a combination of improved understanding of the micromechanisms of fracture, coupled with better process models and improved process monitoring for plates and forgings. Historically, welds have long been regarded as potential areas of weakness and this has led to the development of very large forgings. However, developments in welding technology, in particular to eliminate the use of welding fluxes, may reverse this differential and enable use of smaller forgings, for which it is easier to achieve better properties. For smaller component, hot isostatic pressing (HIPping) might be used to make items that that are more homogeneous and defect-free, and nearer net shape, than possible by forging.

The compositions of the LAS used in modern reactors are probably reasonably well optimized. Further developments in understanding may enable improvements to BOL properties without detriment to irradiation sensitivity. Increased consistency of composition may not be easy to achieve but would be beneficial in the context of eliminating variability from databases, enabling better read-across between subcomponents and between RPVs, and enabling the effect of other factors, such as fabrication variability to be more clearly identified. However, it should be noted that while the LAS used in reactors must meet specified requirements for composition and materials properties, there is considerable leeway within the specifications. Some alloying elements are much more expensive than others and steelmakers and forgemasters can reduce costs by tailoring compositions (within the specification) and yet achieve the materials properties required. Likewise forging and heat treatment are expensive. Thanks to the development of knowledge, and improved process control, which can also result in much cleaner steels, these developments do not prejudice the achievement of minimum materials properties requirements. However, a potential downside is that historical toughness and irradiation shift data may not be fully representative for modern versions of the same specific grade of steel.

The industry has been slow to adopt new grades of steel, but there would be potential benefits in using the stronger and tougher grades of steel with higher Ni or Cr content. These would enable designs with higher stresses, higher pressures, temperatures, or thinner RPV walls with operational benefits in terms of reduced warm-up and cool-down times. The latter is because the steel would potentially be fully ductile at ambient temperatures.

One obstacle to using these higher strength grades of steel has been concerns about irradiation embrittlement. For example, as already discussed, increased Ni in MnMoNi steels can substantially increase irradiation sensitivity even with very low copper levels. However, the limited experimental data previously available suggested that this is not the case when Mn levels are low. Work currently underway in the ATR-2 experiment [115, 116, 220] has confirmed the hypothesis that this result is due to a Mn-starvation mechanism. With Mn <0.25%, the volume fraction of detrimental MNS features in a 3.5% Ni steel is greatly reduced.

The nuclear industry is necessarily extremely cautious about introducing radical new materials. However, performance improvements, albeit more modest, can potentially be achieved by better materials characterization from both BOL tests and expanded surveillance programs, supported by more accurate and precise models of fracture and irradiation damage firmly based on physical understanding.

10.4.3 OTHER ISSUES

Although LAS are not suitable for use at the high-temperatures of Gen-IV reactors, they are good candidates for future Gen-III reactor designs, and have been proposed for some Gen IV reactors with provisions for cooling of the RPV. Their advantage is that they are very well-known; a very large amount of knowledge and expertise exists on the design and manufacture of LAS RPVs and there is around 14,300 reactor year's operating experience.[21] With currently available technical knowledge of LAS, it is possible to envisage making further improvements to enable RPVs with lifetimes of well over 100 years, particularly if flux levels are reduced in future designs by increasing water gaps or by other now well-established methods. However, the aging of current experts in the through-life fracture behavior of LAS is a significant issue. The generation that was involved in the pioneering developments of the 1980s is now at or beyond retirement date. Further, a consequence of the downturn in the nuclear industry after the Three Mile Island and Chernobyl accidents was that the following generation of scientists and engineers with a deep understanding of RPV steels issues is much smaller. The recent revival of interest in nuclear power as a large-scale C-free energy source may reverse this trend.

An RPV is a passive component and its performance can essentially be defined in terms of through-life costs. For the most part, these are defined by manufacture, ISI, warm-up and cool-down times, surveillance testing and other materials support and decommissioning. There is a strong prospect that all these costs can be reduced through further development of the understanding of steels. In the case of decommissioning, the development of low activation steels will be the key factor for the future. In the shorter term, the biggest potential saving enabled by development of understanding is the avoidance of the technical surprises that have resulted in the premature decommissioning of reactors, extended outages, or the requirement for engineering measures such as the installation of neutron shields. It is important to develop such understanding broadly, to consider not just the known hazards, but also those that may become evident given the time and circumstances.

REFERENCES

[1] R.B. Jones, M.R. Wooton, Surveillance of reactor pressure vessel (RPV) embrittlement in Magnox reactors (Chapter 7), in: N. Soneda (Ed.), Irradiation Embrittlement of Reactor Pressure Vessels (RPVs) in Nuclear Power Plants, Woodhead Publishing, Cambridge, UK, 2015.
[2] Figures supplied by T. J. Williams (unpublished).
[3] R.K. Nanstad, D.E. McCabe, R.L. Swain, Evaluation of variability in material properties and chemical composition for Midland Reactor Weld WF-70, in: R.K. Nanstad, M.L. Hamilton, F.A. Garner, A.S. Kumar (Eds.), Effects of Radiation on Materials: 18th International Symposium, ASTM STP 1325, American Society for Testing and Materials, West Conshohocken, PA, 1999.
[4] K.E. Moore, A.S. Heller, B&W 177-FA, Reactor Vessel Beltline Weld Chemistry Study, BAW-1799, Babcock and Wilcox Company, Utility Power Generation Division, Lynchburg, Virginia, 1983.
[5] ASME Boiler and Pressure Vessel Codes, ASME, New York, NY.
[6] G. Baylac, J.M. Grandemange, The French Code RCC-M: design and construction rules for the mechanical components of PWR nuclear islands, Nucl. Eng. Des. 129 (3) (1991) 239–254.
[7] STP-NU-051, Code Comparison Report for Class 1 Nuclear Power Plant Components, Multinational Design Evaluation Program Codes and Standards Working Group, ASME Standards Technology, LLC, New York, 2012.

---

[21] Figure based on IAEA figures for reactor years' experience to end 2012, factored by percentage of PWRs and BWRs in total number of reactors operating in 2010–12 (Tables 4 and 18 of 2018 Edition of Nuclear Power Reactors in the World, IAEA, Vienna, 2018).

[8] IAEA, Integrity of reactor pressure vessels in nuclear power plants: assessment of irradiation embrittlement effects in reactor pressure vessel steels, in: IAEA Nuclear Energy Series No. NP-T-3.11, IAEA Vienna, 2009.
[9] E.J. Pickering, Macrosegregation in steel ingots: the applicability of modelling and characterization techniques, ISIJ Int. 53 (6) (2013) 935–949.
[10] K. Kajikawa, A. Mitchell, Solidification modelling for the manufacture of forging ingots, in: IFM Proceedings, Pittsburgh, 2008.
[11] P. Todeschini, Y. Lefebvre, H. Churier-Bossennec, N. Rupa, G. Chas, C. Benhamou, Revision of the irradiation embrittlement correlation used for the EDF RPV fleet, in: Proceedings of Fontevraud 7, 2010.
[12] C. Naudin, A. Pineau, J.-M. Frund, Toughness modeling of PWR vessel steel containing segregated zones, in: Tenth International Conference on Environmental Degradation of Materials in Nuclear Power Systems—Water Reactors, NACE International, 2002.
[13] J. Campbell, Complete Castings Handbook, second ed, Butterworth-Heinemann, Oxford, UK, 2016.
[14] L.E. Steele, Review and analysis of reactor vessel surveillance programs, in: L.E. Steel (Ed.), Status of USA Nuclear Reactor Pressure Vessel Surveillance For Radiation Effects, ASTM STP 784, ASTM, Philadelphia, 1983 (Chapter 8).
[15] C.J. Bolton, P.J.E. Bischler, M.R. Wootton, R. Moskovic, J.R. Morri, H.C. Pegg, A.B. Haines, R.F. Smith, R. Woodman, Fracture toughness of weld metal samples removed from a decommissioned Magnox reactor pressure vessel, Int. J. Pres. Ves. Pip. 79 (8–10) (2002) 685–692.
[16] B.T. Timofeev, M. Brumovsky, U. von Estorff, The Certification of 15Kh2MFA/15Cr2MoVA Steel and its Welds for WWER Reactor Pressure Vessels, Capture, JRC European Commission, EUR 24581 EN, 210.
[17] B.T. Timofeev, G.P. Karzov, A.A. Blumin, V.I. Smirnov, Determination of crack arrest toughness for Russian light water reactor pressure vessel materials, Int. J. Pres. Ves. Pip. 77 (2000) 519–529.
[18] Y. Tanaka, Reactor pressure vessel (RPV) components processing and properties (Chapter 2), in: N. Soneda (Ed.), Irradiation Embrittlement or Reactor Pressure Vessels (RPVs) in Nuclear Power Plants, Woodhead Publishing, Cambridge, UK, 2015.
[19] J.M. Titchmarsh, Private Communication.
[20] H.-W. Viehrig, M. Houske, K.S. Arora, U. Rindelhart, What one learns from reactor pressure vessels of decommissioned nuclear power plants, in: Proceedings of Fontevraud 7, 2010.
[21] E. Krasikov, A. Amaev, S. Kotlov, Decommissioned LWR pressure vessel materials properties studies, in: Proceedings of Fontevraud 7, 2010.
[22] ASTM, Standard E399 2012 Standard Test Method for Linear-Elastic Plane-Strain Fracture Toughness KIc of Metallic Materials, ASTM International, West Conshohocken, PA, 2013.
[23] W.J. Stelzman, R.G. Berggren, T.N. Jones Jr., ORNL Characterization of Heavy-Section Steel Technology Program Plates 01, 02 and 03, US NRC Report NUREG CR-4092, 1985.
[24] M. Kirk, R. Lott, W.L. Server, R. Hardies, S. Rosinski, Bias and precision of T_0 values determined using ASTM Standard E 1921-97 for nuclear reactor pressure vessel steels, in: M.L. Hamilton, A.S. Kumar, S.T. Kumar, S.T. Rosinski, M.L. Grossbeck (Eds.), Effects of Radiation on Materials: 19th International Symposium, ASTM STP 1366, ASTM, 2000.
[25] Electric Power Research Institute, Primary System Corrosion Research Program: EPRI Materials Degradation Matrix, Revision 2, EPRI Report No. 1020987, Electric Power Research Institute, Palo Alto, CA, 2010.
[26] NUREG-1801, Generic Aging Lessons Learned Report, Revision 1, U.S. Nuclear Regulatory Commission: September, 2005.
[27] J.X. Muscara, Expert Panel Report on Proactive Materials Degradation Assessment, NUREG/CR-6923, U.S. Regulatory Commission, 2007.
[28] P. Andresen, Expanded Materials Degradation Assessment (EMDA): Aging of Core Internals and Piping Systems (NUREG/CR-7153, Volume 2), NUREG/CR-7153, Vol. 3 (ORNL/TM-2013/532), Office of Nuclear Regulatory Research, U.S. Nuclear Regulatory Commission, Washington, DC, 2013.
[29] STP-NU-051-1, Code Comparison Report for Class 1 Nuclear Power Plant Components, ASME Standards Technology LLC, 2012.
[30] Safety assessment principles for nuclear facilities, UK Health and Safety Executive (HSE).
[31] Comprehensive Structural Integrity, Vol. 1 to 10, Elsevier Science Ltd, 2007.
[32] T.L. Anderson, Fracture Mechanics—Fundamentals and Applications, Taylor and Francis, ISBN 978 0 8493 1656 2.
[33] K.R.W. Wallin, Fracture Toughness of Engineering Materials, EMAS Publishing, 2012. ISBN 978 0 9552994 6 9.
[34] K.R.W. Wallin, Distribution free statistical assessment of scatter and size effects in the Euro fracture toughness data set, Eng. Fract. Mech. 103 (2013) 69–78.

[35] J. Heerens, D. Hellmann, Development of the Euro fracture toughness dataset, Eng. Fract. Mech. 69 (2002) 421–449.
[36] H.J. Rathbun, G.R. Odette, M.Y. He, T. Yamamoto, Influence of statistical and constraint loss size effects on cleavage fracture toughness in the transition—a model based analysis, Eng. Fract. Mech. 73 (2008) 2723–2747.
[37] K. Wallin, The master curve: a new method for brittle fracture, Int. J. Mater. Product Technol. 14 (2/3/4) (1999) 342–354.
[38] M.A. Sokolov, Statistical analysis of the ASME KIc database, J. Pres. Ves. Technol. Trans. ASME 120 (1998) 24–28.
[39] IAEA TECDOC 1631, Master Curve Approach to Monitor Fracture Toughness of Reactor Pressure Vessels in Nuclear Power Plants, International Atomic Energy Agency, 2009.
[40] H.J. Rathbun, G.R. Odette, T. Yamamoto, G.E. Lucas, Influence of statistical and constraint loss size effects on cleavage fracture toughness in the transition—a single variable experiment and database, Eng. Fract. Mech. 73 (2006) 134–158.
[41] ASTM E 1921, Test Method for the Determination of Reference Temperature, T_0, for Ferritic Steels in the Transition Range, Annual Book of ASTM Standards, ASTM International, West Conshohocken, PA, 2015.
[42] B. Margolin, B. Gurovich, V. Fomenko, A. Shvetsova, D. Gulenko, M. Zhurko, E. Korshunov, E. Kuleshova, Fracture toughness prediction for highly irradiated RPV materials: From test results to RPV integrity assessment, J. Nucl. Mater. 432 (1–3) (2013) 313–322.
[43] G.R. Odette, H.J. Rathbun, M. Hribernik, T. Yamamoto, M. He, P. Spätig, A multiscale approach to measuring and modeling cleavage fracture toughness in structural steels, in: V. Ghetta, D. Gorse, D. Mazière, V. Pontikis (Eds.), Materials Issues for Generation IV Systems. NATO Science for Peace and Security Series B: Physics and Biophysics, Springer, Dordrecht, 2008.
[44] Y. Nikolaev, A. Nikolaeva, A. Kryukov, V. Levit, Y. Korolyov, Radiation embrittlement and thermal annealing of Cr-Ni-Mo reactor pressure vessel materials, J. Nucl. Mater. 226 (1995) 144–155.
[45] E. Lucon, M. Scibetta, R. Gérard, Analysis of the Belgian surveillance fracture toughness database using conventional and advanced master curve approaches, J. Astron. Instrum. 6 (3) (2009), https://doi.org/10.1520/JAI101897.
[46] ASME Boiler and Pressure Vessel Code, Section XI, Rules for In-service Inspection of Nuclear Power Plant Components, Code Cases N-629 (1998 edition) and N-631 (1999 edition).
[47] IAEA, Guidelines For Application Of The Master Curve Approach To Reactor Pressure Vessel Integrity In Nuclear Power Plants, IAEA TRS 429, International Atomic Energy Agency, Vienna, 2005.
[48] E.D. Eason, J.E. Wright, E.E. Nelson, G.R. Odette, E.V. Mader, Embrittlement recovery due to annealing of reactor pressure vessel steels, Nucl. Eng. Des. 179 (1998) 257–265.
[49] U.S. Nuclear Regulatory Commission Code of Federal Regulations, 10 CFR Part 50.61a, Alternate Fracture Toughness Requirements for Protection Against Pressurized Thermal Shock Events, Volume 75, No. 1, dated January 4, 2010, effective February 3, 2010.
[50] M.T. EricksonKirk, T.L. Dickson, Recommended Screening Limits for Pressurized Thermal Shock (PTS), NUREG-1874, Office of Nuclear Regulatory Research, 2010.
[51] ASTM E2215-16 Standard Practice for Evaluation of Surveillance Capsules from Light-Water Moderated Nuclear Power Reactor Vessels, ASTM International.
[52] G.R. Odette, P.M. Lombrozo, R.A. Wullaert, The relation between irradiation hardening and embrittlement, in: 12th International Symposium on the Effects of Irradiation on Materials-12, ASTM-STP-870, 1985, p. 841.
[53] E.P. Wigner, Theoretical physics in the metallurgical laboratory of Chicago, J. Appl. Phys. 17 (1946) 857–863.
[54] G.D. Calkins, C.C. Woolsey, Introduction to Symposium on Radiation Effects of Materials, Vol. 2, 1958. ASTM STP 220, ASTM.
[55] J.A. Brinkman, On the nature of radiation damage in Metals, J. Appl. Phys. 25 (1954) 961.
[56] A. Seeger, The nature of radiation damage in metals, In: Proceedings of the Symposium on Radiation Damage in Solids and Reactor Materials held by the IAEA at Fondazione Cini, S. Giorgio Maggiore, Venice, 7-11 May, Part 11962, IAEA, Vienna, 101–125.
[57] U. Potapovs, J.R. Hawthorne, Nuclear applications, Nucl. Apl. 6 (1) (Jan 1969) 27–46.
[58] J.R. Hawthorne, Demonstration of improved radiation embrittlement resistance of A533B steel through control of selected residual elements, in: Irradiation effects of structural alloys for nuclear reactor applications, ASTM STP 484, ASTM, 1970, pp. 96–127.
[59] Symposium on Radiation Effects on Materials, Volume 1, ASTM STP 208, 1957.
[60] W.L. Server, R.K. Nanstad, International Atomic Energy Agency coordinated research projects on structural integrity of reactor pressure vessels, Journal of ASTM International 6 (7) (2009) 1–17, https://doi.org/10.1520/JAI102096.
[61] R.G. Lott, S.S. Brenner, M.K. Miller, A. Wolfenden, Development of radiation damage in pressure vessel steels, Trans. Am. Nucl. Soc. 38 (1981) 303–304, ISSN: 0003-018X.

[62] M.K. Miller, S.S. Brenner, An atom probe study of irradiated pressure vessel steels, in: Proceedings of International Field Emission Symposium, July 27-31, Portland, Oregon, 1981.
[63] G.R. Odette, On the dominant mechanism of irradiation embrittlement of reactor pressure vessel steels, Scr. Mater. 17 (1983) 1183.
[64] S.B. Fisher, J.E. Harbottle, N. Aldridge, Radiation hardening in Magnox pressure vessel steels, Phil. Trans. Roy. Soc. Lond. A 315 (1985) 301–332.
[65] G.R. Odette, et al., Physically Based Regression Correlations of Embrittlement Data from Reactor Pressure Vessel Surveillance Programs, Electric Power Research Institute, NP-3319, January(1984).
[66] G.L. Guthrie, Charpy trend curves based on 177 PWR data points, in: LWR Pressure Vessel Surveillance Dosimetry Improvement Program, NUREG/CR-3391, Vol. 2, prepared by Hanford Engineering Development Laboratory, HEDL-TME 83-22, April(1984).
[67] U.S. Nuclear Regulatory Commission, Radiation Embrittlement of Reactor Vessel Materials, Regulatory Guide 1.99 Rev. 2, 1988.
[68] G.E. Lucas, G.R. Odette, P. Lombrozo, J.W. Sheckherd, The effects of composition, temperature, and microstructure on the radiation hardening of pressure vessel steels, in: 12th International Symposium on the Effects of Irradiation on Materials Effects of Irradiation on Materials, ASTM STP-870, 1985 pp. 901–241.
[69] G.R. Odette, G.E. Lucas, Irradiation embrittlement of reactor pressure vessel steels: mechanism, models and data correlations, In: Radiation Embrittlement of Reactor Pressure Vessel Steels, ASTM STP 909, ASTM, 1986 206.
[70] G.R. Odette, G.E. Lucas, Recent advances in predicting embrittlement of reactor pressure vessel steels, in: Proceedings of the 2nd International Symposium on Environmental Degradation of Materials for Nuclear Power Systems-Water Reactors, ANS, 1986, p. 295.
[71] G.E. Lucas, G.R. Odette, H.R. Chen, Recent advances in understanding radiation hardening and embrittlement mechanisms in pressure vessel steels, in: Proceedings of the 2nd International Symposium on Environmental Degradation of Materials for Nuclear Power Systems-Water Reactors, ANS, 1986, p. 345.
[72] G.R. Odette, G.E. Lucas, The effect of heat treatment on irradiation hardening of pressure vessel steels: data trends and mechanisms, in: Proceedings of the 3rd International Symposium on Environmental Degradation of Materials for Nuclear Power Systems-Water Reactors, TMS, 1987, p. 95.
[73] G.E. Lucas, An evolution of understanding of reactor pressure vessel steel embrittlement, J. Nucl. Mat. 407 (2010) 59–69.
[74] E.V. Mader, G.E. Lucas, G.R. Odette, The effects of metallurgical and irradiation variables on the postirradiation annealing kinetics of pressure vessel steels, in: R.E. Stoller, A.S. Kumar, D.S. Gelles (Eds.), Effects of Radiation on Materials: 15th International Symposium ASTM STP 1125, American Society for Testing and Materials, Philadelphia, 1992.
[75] G.H. Kinchin, R.S. Pease, The displacement of atoms in solids by radiation, Rep. Prog. Phys. 18 (1955) 1–51.
[76] M.J. Norgett, M.T. Robinson, I.M. Torrens, A Method of Calculating the Number of Atom Displacements in Irradiated Metals, AERE Report TP/494; CEA Report No. 4389; ORNL Solid State Division Report No. 72-70(1972).
[77] ASTM Standard Practice E693. Standard Practice for Characterizing Neutron Exposures in Iron and Low Alloy Steels in Terms of Displacements Per Atom (DPA), ASTM.
[78] R.E. Stoller, Evaluation of neutron energy spectrum effects and RPV through-wall attenuation based on molecular dynamics cascade simulations, Nucl. Eng. Des. 195 (2000) 129–136.
[79] R.E. Stoller, Pressure vessel embrittlement predictions based on a composite model of copper precipitation and point defect clustering, in: D.S. Gelles, R.K. Nanstad, A.S. Kumar, E.A. Little (Eds.), Effects of Radiation on Materials: 17th. International Symposium, ASTM STP 1270, American Society for Testing and Materials, Philadelphia, 1996, pp. 25–58.
[80] R.E. Stoller, G.R. Odette, B.D. Wirth, Primary defect production in bcc iron, J. Nucl. Mater. 251 (1997) 49–60.
[81] G.R. Odette, B.D. Wirth, A computational microscopy study of nanostructural evolution in irradiated pressure vessel steels, J. Nucl. Mater. 251 (1997) 157–171.
[82] G.R. Odette, R.K. Nanstad, Predictive reactor pressure vessel steel irradiation embrittlement models: issues and opportunities, JOM 61 (7) (2009) 17–23.
[83] H. Ke, P. Wells, P. Edmondson, N. Almirall, L. Barnard, G.R. Odette, D. Morganet, Thermodynamic and kinetic modeling of Mn-Ni-Si precipitates in low-Cu reactor pressure vessel steels, Acta Mater. 138 (2017) 10–26.
[84] L. Messina, M. Chiapetto, P. Olsson, C.S. Becquart, L. Malerba, An object kinetic Monte Carlo model for the microstructure evolution of neutron-irradiated reactor pressure vessel steels, Phys. Status Solidi A 213 (2016) 2974–2980.

[85] D.J. Bacon, Y.N. Osetsky, D. Rodney, Dislocation-obstacle interactions at the atomic level, in: J.P. Hirsh, L. Kubin (Eds.), Dislocations in Solids, Elsevier, 2009.

[86] K. Fukuya, Current understanding of radiation-induced degradation in light water reactor structural materials, J. Nucl. Sci. Technol. 50 (3) (2003) 213–254, https://doi.org/10.1080/00223131.2013.

[87] M. Hernández-Mayoral, D. Gómez-Briceño, Transmission electron microscopy study on neutron irradiated pure iron and RPV model alloys, J. Nucl. Mater. 399 (2010) 146–153.

[88] E. Meslin, M. Lambrecht, M. Hernández-Mayoral, F. Bergner, L. Malerba, P. Pareige, B. Radiguet, A. Barbu, D. Gómez-Briceño, A. Ulbricht, A. Almazouzi, Characterization of neutron-irradiated ferritic model alloys and a RPV steel from combined APT, SANS, TEM and PAS analyses, J. Nucl. Mater. 406 (2010) 73–83.

[89] S.C. Glade, B.D. Wirth, G.R. Odette, P. Asoka-Kumar, P.A. Sterne, R.H. Howell, Positron annihilation spectroscopy and small-angle neutron scattering characterization of the effect of Mn on the nanostructural features formed in irradiated Fe-Cu-Mn alloys, Philosophical Magazine 85 (4-7) (2005).

[90] G.R. Odette, C.K. Sheeks, A model for cascade induced microvoid formation in dilute iron-copper alloys, TMS (1981) 415.

[91] B.D. Wirth, G.R. Odette, P. Asoka-Kumar, et al., G.S. Was, J.L. Nelson (Eds.), Proceedings of the 10th International Symposium on Environmental Degradation of Materials in Light Water Reactors, NACE, 2001 (CD).

[92] F.G. Djurabekova, L. Malerba, C. Domain, C.S. Becquart, Stability and mobility of small vacancy and copper-vacancy clusters in bcc-Fe: An atomistic kinetic Monte Carlo study, Nucl. Instrum. Methods Phys. Res. Sect. B Beam Interact. Mater. Atoms 255 (1) (2007) 47–51.

[93] E.D. Eason, G.R. Odette, R.K. Nanstad, T. Yamamoto, A physically-based correlation of irradiation-induced transition temperature shift for RPV steels, ORNL/TM-2006/530, Oak Ridge National Laboratory, 2007.

[94] G.R. Odette, Microstructural evolution during irradiation, in: MRS Symp. Proc. 373, Materials Research Society, Warrendale, PA, 1995, p. 137.

[95] C.L. Liu, G.R. Odette, B.D. Wirth, G.E. Lucas, A lattice Monte Carlo simulation of nanophase compositions and structures in irradiated pressure vessel Fe-Cu-Ni-Mn-Si steels, Mater. Sci. Eng. A 238 (1997) 202–209.

[96] G.R. Odette, T. Yamamoto, B.D. Wirth, Late blooming phases and dose rate effects in RPV steels: integrated experiments and models, in: N.M. Ghoniem (Ed.), Second International Conference on Multiscale Materials Modeling. Los Angeles, 2004, p. 355.

[97] D.J. Sprouster, J. Sinsheimer, E. Dooryhee, S.K. Ghose, P. Wells, T. Stan, N. Almirall, G.R. Odette, L.E. Ecker, Structural characterization of nanoscale intermetallic precipitates in highly neutron irradiated reactor pressure vessel steels, Scr. Mater. 113 (2016) 18–22. http://refhub.elsevier.com/S1359-6454(17)30571-2/sref42.

[98] M.K. Miller, K.A. Powers, R.K. Nanstad, P. Efsing, Atom probe tomography characterization of high nickel, low copper surveillance RPV weld irradiated to high fluences, J. Nucl. Mater. 437 (2013) 107–115.

[99] G.R. Odette, E.V. Mader, G.E. Lucas, W.J. Phythian, C.A. English, The effect of flux on the irradiation hardening and embrittlement of pressure vessel steels, in: A.S. Kumar et al., (Ed.), Effects of Radiation on Materials: 16th International Symposium, ASTM STP 1175, American Society for Testing and Materials, West Conshohocken, PA, 1994.

[100] E.V. Mader, Kinetics of Irradiation Embrittlement and the Post-Irradiation Annealing of Nuclear Reactor Pressure Vessel Steels, PhD Thesis, The University of California, Santa Barbara, 1995.

[101] R.K. Nanstad, G.R. Odette, R.E. Stoller, T. Yamamoto, Review of Draft NUREG Report on Technical Basis for Revision of Regulatory Guide 1.99, Oak Ridge National Laboratory ORNL/NRC/LTR-08/03; 2008, (2008).

[102] R.J. McElroy, C.A. English, A.J.E. Foreman, G. Gage, J.M. Hyde, P.H.N. Ray, I.A. Vatter, Temper embrittlement, irradiation induced phosphorus segregation and implications for post-irradiation annealing of reactor pressure vessels, in: R.K. Nanstad, M.L. Hamilton, F.A. Garner, A.S. Kumar (Eds.), Effects of Radiation on Materials: 18th International Symposium, ASTM STP 1325, American Society for Testing and Materials, West Conshohocken, Pennsylvania, 1999, pp. 296–316.

[103] G.R. Odette, G.E. Lucas, Recent progress in understanding reactor pressure vessel embrittlement, Radiat. Eff. Defects Solids 144 (1998) 189.

[104] M.K. Miller, M.G. Burke, An APFIM survey of grain boundary segregation and precipitation in irradiated pressure vessel steels, in: A.S. Kumar et al., (Ed.), Effects of Radiation on Materials: 16th International Symposium, ASTM STP 1175, American Society for Testing and Materials, West Conshocken, PA, 1994, p. 492.

[105] G.R. Odette, G.E. Lucas, R.D. Klingensmith, B.D. Wirth, D. Gragg, The Effects of Composition and Heat Treatment on Hardening and Embrittlement of Reactor Pressure Vessel Steels, NUREG/CR-6778, U.S. Nuclear Regulatory Commission, Washington, DC, 2003.

[106] R.B. Jones, J.T. Buswell, The interactive roles of phosphorus, tin and copper in the irradiation embrittlement of PWR pressure vessel steels, in: G.T. Theus, et al. (Eds.), Environmental Degradation of Materials in Nuclear Power Systems — Water Reactors, Proceedings of the 3rd International Symposium, Metallurgical Society, Warrendale, PA, 1988, p. 111.

[107] Z. Lu, R.G. Faulkner, P.E.J. Flewitt, The role of irradiation-induced phosphorus segregation in the ductile-to-brittle transition temperature in ferritic steels, Mater. Sci. Eng. A 437 (2006) 306–312.

[108] L.R. Greenwood, An evaluation of through-thickness changes in primary damage production in commercial reactor pressure vessels, in: S.T. Rosinski, M.L. Grossbeck, T.R. Allen, A.S. Kumar (Eds.), Effects of Radiation on Materials: 20th International Symposium, ASTM STP 1405, American Society for Testing and Materials, West Conshohocken, PA, 2001.

[109] G.R. Odette, G.E. Lucas, D. Klingensmith, Irradiation hardening of pressure vessel steels at 60°C: the role of thermal neutrons and boron, in: R.K. Nanstad, M.L. Hamilton, F.A. Garner, A.S. Kumar (Eds.), Effects of Radiation on Materials: 18th International Symposium, ASTM STP 1325, American Society for Testing and Materials, West Conshohocken, PA, 1999, pp. 3–13.

[110] R.B. Jones, D.J. Edens, M.R. Wootton, Influence of thermal neutrons on the hardening and embrittlement of plate steels, in: M.L. Hamilton, A.S. Kumar, S.T. Rosinski, M.L. Grossbeck (Eds.), Effects of Radiation on Materials: 19th International Symposium, ASTM STP 1366, American Society for Testing and Materials, West Conshohocken, PA, 2000.

[111] E.D. Eason, G.R. Odette, R.K. Nanstad, T. Yamamoto, A physically-based correlation of irradiation-induced transition temperature shifts for RPV steels, J. Nucl. Mater. 433 (2013) 240–254.

[112] G.R. Odette, T. Yamamoto, D. Klingensmith, The effect of dose rate on irradiation hardening of RPV steels: a comprehensive single variable database and model based analysis, University of California Santa Barbara Letter Report UCSB-NRC-03/1; 2003, (2003).

[113] G. R. Odette, Private communication.

[114] P.B. Wells, T. Yamamoto, B. Miller, T. Milot, J. Cole, Y. Wua, G.R. Odette, Evolution of manganese–nickel–silicon-dominated phases in highly irradiated reactor pressure vessel steels, Acta Mater. 80 (2014) 205–219.

[115] G.R. Odette, T. Yamamoto, P.B. Wells, N. Almirall, K. Fields, D. Gragg, R.K. Nanstad, J.P. Robertson, K. Wilford, N. Riddle, T. Williams, Summary of Progress on the ATR-2 Experiment Post-Irradiation Examination of Reactor Pressure Vessel Alloys, UCSB ATR-2 2017-1, University of California Santa Barbara, Santa Barbara, CA, 2017.

[116] W. Server, B. Burgos, T. Hardin, J.B. Hall, The EPRI PWR Supplemental Surveillance Program (PSSP) final design and implementation, in: Proc. of ASME 2017 PVP Conference, PVP2017, July, Waikoloa, Hawaii, USA, 2017. PVP2017-65307.

[117] R.E. Stoller, Modeling the influence of irradiation temperature and displacement rate on radiation-induced hardening in ferritic steels, in: A.S. Kumar, D.S. Gelles, R.K. Nanstad, E.A. Little (Eds.), Effects of Radiation on Materials: 16th International Symposium, ASTM STP 1175, American Society for Testing and Materials, Philadelphia, 1993, pp. 394–423.

[118] Y.A. Nikolaev, Radiation embrittlement of Cr-Ni-Mo and Cr-Mo steels, J. ASTM Int. 4 (8) (2007). Paper ID JAI100695.

[119] P.J. Barton, D.R. Harries, I.L. Mogford, Effects of neutron dose rate and irradiation temperature on hardening, J. Iron Steel Inst. (London) 203 (1965) 507–510.

[120] M. Grounes, Review of Swedish work on irradiation effects in pressure vessel steels and on significance of data obtained, In: Effects of Radiation on Structural Metals, ASTM STP 426, American Society for Testing and Materials, Philadelphia, 1967, pp. 224–259.

[121] R.B. Jones, T.J. Williams, The dependence of radiation hardening and embrittlement on irradiation temperature, in: D.S. Gelles, R.K. Nanstad, A.S. Kumar, E.A. Little (Eds.), Effects of Radiation on Materials: 17th International Symposium, ASTM 1270, ASTM, 1996, p. 569.

[122] S.B. Fisher, J.E. Harbottle, N.B. Aldridge, Microstructure related to irradiation hardening in pressure vessel steel, in: Proc. Conf. on Dimensional Stability and Mechanical Behaviour of Irradiated Metals and Alloys held in Brighton, UK, 1983, pp. 87–91. pub. BNES, London (1984), vol. 2, Paper 34.

[123] M.G. Burke, R.J. Stofanak, J.M. Hyde, C.A. English, W.L. Server, Microstructural aspects of irradiation damage in A508 Gr 4N forging steel: composition and flux effects, in: M.L. Grossbeck, R.G. Lott (Eds.), Effects of Radiation on Materials, ASTM STP 1447, American Society for Testing and Materials, West Conshohocken, PA, 2002.

[124] R.K. Nanstad, D.E. McCabe, M.A. Sokolov, C.A. English, S.R. Ortner, Investigation of temper embrittlement in reactor pressure vessel steels following thermal aging, irradiation, and thermal annealing, in: S.T. Rosinski, M.L. Grossbeck, T.R. Allen, A.S. Kumar (Eds.), Effects of Radiation on Materials: 20th International Symposium, ASTM STP 1405, American Society for Testing and Materials, West Conshohocken, Pennsylvania, 2001, pp. 356–382.

[125] C. English, Final report of the PISA project: phosphorus influence on steels ageing (PISA), Final Technical Report, European Commission, Contract No: FIKS-CT-2000-00080(2005).

[126] B.A. Gurovich, E.A. Kuleshova, Y.I. Shtrombakh, D.Y. Erak, A.A. Chernobaeva, O.O. Zabusov, Fine structure behaviour of VVER-1000 RPV materials under irradiation, J. Nucl. Mater. 389 (2009) 490–496.

[127] P.J.E. Bischler, V.M. Callen, R.C. Corcoran, P. Spellward, A microstructural examination of irradiated reactor pressure vessel weld samples, Int. J. Pres. Ves. Pip. 77 (2000) 629–640.

[128] R.K. Nanstad, G.R. Odette, Reactor pressure vessel issues for the light-water reactor sustainability program, Proc. Env. Deg. Conf. (2009).

[129] T.J. Williams, D. Ellis, C.A. English, J. Hyde, A model of irradiation damage in high nickel submerged arc welds, Int. J. Pres. Ves. Pip. 79 (8–10) (2002) 649–660.

[130] R. Ahlstrand, B. Margolin, I. Akbashev, L. Chyrko, V. Kostylev, E. Yurchenko, V. Piminov, Y. Nikolaev, V. Koshkin, V. Kharshenko, V. Bukhanov, TAREG 2.01/00 project, Validation of neutron embrittlement for VVER 1000 and 440/213 RPVs, with emphasis on integrity assessment, Progr. Nucl. Energy 58 (2012) 52–57.

[131] Japan Electric Association, JEAC4201-2004, Methods for Surveillance Tests for Structural Materials of Nuclear Reactors, December 21; 2004, (2004).

[132] D.J. Bacon, Y.N. Osetsky, Hardening due to copper precipitates in α-iron studied by atomic-scale modelling, J. Nucl. Mater. 329-333 (2004) 1233.

[133] K.C. Russell, L.M. Brown, A dispersion strengthening model based on differing elastic moduli applied to the iron-copper system, Acta Metall. 20 (1972) 969.

[134] S. Jumel, C. Domain, J. Ruste, J.-C. Van Duysen, C. Becquart, A. Legris, P. Pareige, A. Barbu, E. Van Walle, R. Chaouadi, M. Hou, G.R. Odette, R.E. Stoller, B.D. Wirth, Simulation of irradiation effects in reactor pressure vessel steels: the reactor for virtual experiments, REVE project, J. Test. Eval. 30 (1) (2002) 37.

[135] S. Jumel, J.C. Van Duysen, J. Ruste, C. Domain, Interactions between dislocations and Irradiation-induced defects in light water reactor pressure vessel steels, J. Nucl. Mater. 346 (2005) 79.

[136] N. Soneda, A. Nomoto, Characteristics of the new embrittlement correlation method for the Japanese reactor pressure vessel steels, J. Eng. Gas Turb. Power 132 (2010). 102918-1 to 101918-9.

[137] M.Y. He, G.R. Odette, A universal relationship between indentation hardness and flow, J. Nucl. Mater. 367 (2007) 556.

[138] G.R. Odette, M.Y. He, T. Yamamoto, On the relation between irradiation induced changes in the Master Curve reference temperature shift and changes in strain hardened flow stress, J. Nucl. Mater. 367 (2007) 561.

[139] M.A. Sokolov, R.K. Nanstad, Comparison of irradiation-induced shifts of KJc and Charpy impact toughness for reactor pressure vessel steels, in: R.K. Nanstad, M.L. Hamilton, F.A. Garner, A.S. Kumar (Eds.), Effects of Radiation on Materials: 18th International Symposium, ASTM STP 1325, American Society for Testing and Materials, West Conshohocken, PA, 1999, pp. 167–190.

[140] C.A. English, S.R. Ortner, G. Gage, W.L. Server, S.T. Rosinski, Review of phosphorus segregation and intergranular embrittlement in reactor pressure vessel steels, in: S.T. Rosinski, M.L. Grossbeck, T.R. Allen, A.S. Kumar (Eds.), Effects of Radiation on Materials: 20th International Symposium, ASTM STP 1405, American Society for Testing and Materials, West Conshohocken, Pennsylvania, 2001.

[141] C.S. Becquart, C. Domain, Modeling microstructure and irradiation effects, Metall. Mater. Trans. A 42A (April 2011) 852–870.

[142] J.A. Elliott, Novel approaches to multiscale modelling in materials science, Int. Mater. Rev. 58 (4) (2011) 207–225.

[143] G.R. Odette, G.E. Lucas, Irradiation embrittlement of reactor pressure vessel steels: mechanisms, models and data correlations, in: L.E. Steele (Ed.), Radiation Embrittlement of Reactor Pressure Vessel Steels—An International Review, ASTM STP 909, ASTM, Philadelphia, PA, 1986, pp. 206–241.

[144] G.R. Odette, B.D. Wirth, D.J. Bacon, N.M. Ghoniem, Multiphysics modeling of radiation-damaged materials: embrittlement of pressure-vessel steels, MRS Bull. (2001) 176–181.

[145] S. Jumel, RPV-1: a virtual reactor to simulate irradiation effects in pressurized water reactor pressure vessel steels, PhD thesis, Université de Lille, 2004.

[146] G. Adjanor, S. Bugat, C. Domain, A. Barbu, Overview of the RPV-2 and INTERN-1 packages: from primary damage to microplasticity, J. Nucl. Mater. 406 (2010) 175–186.

[147] A. Al Mazouzi, A. Alamo, D. Lidbury, D. Moinereau, S. Van Dyck, PERFORM 60: Prediction of the effects of radiation for reactor pressure vessel and in-core materials using multi-scale modelling—60 years foreseen plant lifetime, Nucl. Eng. Des. 24 (9) (2011) 3403–3415.

[148] G.J. Ackland, M.I. Mendelev, D.J. Srolovitz, S. Han, A.V. Barashev, Development of an interatomic potential for phosphorus impurities in α-iron, J. Phys. Condens. Matter. 16 (2004) S2629–S2642.
[149] P. Olsson, C. Domain, J. Wallenius, Ab initio study of Cr interactions with point defects in bcc Fe, Phys. Rev. B 75 (2007) 014110.
[150] L. Malerba, G.J. Ackland, C.S. Becquart, G. Bonny, C. Domain, S.L. Dudarev, C.-C. Fu, D. Hepburn, M.C. Marinica, P. Olsson, R.C. Pasianot, J.M. Raulot, F. Soisson, D. Terentyev, E. Vincent, F. Willaime, Ab initio calculations and interatomic potential for iron and iron alloys: achievements within the PERFECT Project, J. Nucl. Mater. 406 (2010) 7–18.
[151] R.E. Stoller, A.F. Calder, Statistical analysis of a library of molecular dynamics cascade simulations in iron at 100K, J. Nucl. Mater. 283–287 (2000) 746–752.
[152] C.S. Becquart, C. Domain, A. Legris, J.C. Van Duysen, Influence of the interatomic potentials on molecular dynamics simulations of displacement cascades, J. Nucl. Mater. 280 (2000) 73–85.
[153] D. Terentyev, D.J. Bacon, Y.N. Osetsky, Interaction of an edge dislocation with voids in α-iron modeled with different interatomic potentials, J. Phys. Cond. Matter 20 (2008) 445007.
[154] G. Monnet, Mechanical and energetical analysis of molecular dynamics simulations of dislocation-defect interactions, Acta Mater 55 (2007) 5081–5088.
[155] R.E. Stoller, Primary radiation damage formation, in: R.J.M. Konings, T.R. Allen, R.E. Stoller, S. Yamanaka (Eds.), Comprehensive Nuclear Materials, Elsevier Ltd., Amsterdam, 2012, pp. 293–332.
[156] P. Jung, Atomic displacement functions of cubic metals, Phys. Rev. B 23 (1981) 664–670.
[157] G. Wallner, M.S. Anand, L.R. Greenwood, M.A. Kirk, W. Mansel, W. Waschkowski, Defect production rates in metals by reactor neutron irradiation at 4.6 K, J. Nucl. Mater. 152 (1988) 146–153.
[158] S. Jumel, J.-C. Van Duysen, INCAS: an analytical model to describe displacement cascades, J. Nucl. Mater. 328 (2004) 151–164.
[159] B.D. Wirth, G.R. Odette, Kinetic Lattice Monte Carlo simulations of diffusion and decomposition kinetics in Fe-Cu alloys: embedded atom and nearest neighbor potentials, Mater. Res. Soc. Symp. Proc. 481 (1998) 151–156.
[160] R. Ngayam-Happy, C.S. Becquart, C. Domain, L. Malerba, Formation and evolution of MnNi clusters in neutron irradiated dilute Fe alloys modelled by a first-principle based AKMC method, J. Nucl. Mater. 426 (s1–3) (2012) 198–207.
[161] E. Vincent, C.S. Becquart, C. Domain, Ab initio calculations of self-interstitial interaction and migration with solute atoms in bcc Fe, J. Nucl. Mater. 359 (2006) 227–237.
[162] E. Vincent, C.S. Becquart, C. Domain, Microstructural evolution under high flux irradiation of dilute Fe-CuNiMnSi alloys studied by an atomic kinetic Monte Carlo model accounting for both vacancies and self-interstitials, J. Nucl. Mater. 382 (2008) 154–159.
[163] N. Soneda, S. Ishino, A. Takahashi, K. Dohi, Modeling the microstructural evolution in bcc-Fe during irradiation using kinetic Monte Carlo computer simulation, J. Nucl. Mater. 323 (2003) 169–180.
[164] A. Chatterjee, D.G. Vlachos, An overview of spatial microscopic and accelerated kinetic Monte Carlo methods, J. Comp. Aid. Mater. Des. 14 (2007) 253–308.
[165] G. Was, Fundamentals of Radiation Materials Science, Springer, New York, NY, 2007.
[166] R.E. Stoller, S.I. Golubov, C. Domain, C.S. Becquart, Mean field rate theory and object kinetic Monte Carlo: a comparison of kinetic models, J. Nucl. Mater. 382 (2008) 77–90.
[167] M. Hou, A. Souidi, C.S. Becquart, C. Domain, L. Malerba, Relevancy of displacement cascades features to the long term point defect cluster growth, J. Nucl. Mater. 382 (2008) 103–111.
[168] L.P. Kubin, G.R. Canova, The modelling of dislocation patterns, Scr. Metall. 27 (1992) 957–962.
[169] B. Devincre, V. Pontikis, Y. Brechet, G. Canova, M. Condat, L.P. Kubin, Three-dimensional simulations of plastic flow in crystals, in: M. Marechal, B.L. Holian (Eds.), Microscopic Simulations of Complex Hydrodynamic Phenomena, Plenum Press, New York, 1992, p. 413.
[170] G. Monnet, C. Domain, S. Queyreau, S. Naamane, B. Devincre, Atomic and dislocation dynamics simulations of plastic deformation in reactor pressure vessel steels, J. Nucl. Mater. 394 (2009) 174–181.
[171] A. Andrieu, A. Pineau, J. Besson, D. Ryckelynck, O. Bouaziz, Beremin model: Methodology and application to the prediction of the Euro toughness data set, Eng. Fract. Mech. 95 (2012) 102–117.
[172] S.R. Bordet, A.D. Karstensen, D.M. Knowles, C.S. Wiesner, A new statistical local criterion for cleavage fracture in steel. Part I: model presentation, Eng. Fract. Mech. 72 (3) (2005) 435–452.
[173] B. Margolin, V. Shvetsova, A. Gulenko, Radiation embrittlement modelling in multi-scale approach to brittle fracture of RPV steels, Int. J. Fract. 179 (2013) 87–108.

[174] K.B. Cochran, M. Erickson, P.T. Williams, H.B. Klasky, B.R. Bass, A dislocation-based cleavage initiation model for pressure vessel steels, in: Proc. ASME 2012 Pressure Vessels & Piping Division Conference, PVP2012-78564, July, 2012.
[175] ASTM E900-15 Standard Guide for Predicting Radiation-Induced Transition Temperature Shift in Reactor Vessel Materials, ASTM International, West Conshohocken, PA, USA.
[176] R.K. Nanstad, G.R. Odette, M.A. Sokolov, Ensuring the performance of nuclear reactor pressure vessels for long-time service, in: Proc. of ASME Pressure Vessels and Piping Conference, 2010, pp. PVP2010–25832.
[177] Regulatory Guide 1.99 Revision 1, Effects of residual elements on predicted radiation damage to reactor vessel materials, U.S. Nuclear Regulatory Commission, Washington DC, April, 1977.
[178] M.J. Makin, A.D. Whapham, F.J. Minter, The formation of dislocation loops in copper during neutron irradiation, Philos. Mag. 7 (1962) 285.
[179] J.D. Varsik, S.T. Byrne, An empirical evaluation of the irradiation sensitivity of reactor pressure vessel materials, in: J.A. Sprague, D. Kramer (Eds.), Effects of Radiation on Structural materials, 1979, pp. 252–266 ASTM STP 683, ASTM.
[180] P.N. Randall, Basis for Revision 2 of the US Regulatory Commission's Regulatory Guide 1.99, in: L.E. Steele (Ed.), Radiation Embrittlement of Nuclear Reactor Pressure Vessels, An International Review (Second Volume), ASTM STP 909, ASTM, 1986, pp. 149–162.
[181] Standards of Calculation for Strength of Nuclear Power Plant Equipment and Pipelines, PNAE G-7-002-86, M ENERGOATOMIZDAT, 1989.
[182] C. Brillaud, F. Hedin, B. Houssin, Comparison between French surveillance programme results and predictions of irradiation embrittlement, in: R.E. Stoller, F.A. Garner, C.H. Henager, N. Igata (Eds.), Effects of Radiation on Materials, 13th International Symposium, ASTM STP 956, ASTM, 1987, pp. 420–447.
[183] JEAC 4201-1991, Surveillance Tests of Structural Materials for Nuclear Reactors, Japan Electric Association, 1991 (in Japanese).
[184] Surveillance of the irradiation behaviour of reactor pressure vessel materials for LWR facilities, Nuclear Safety Standards Commission, KTA 3203, GRS, Issue 3/84.
[185] G.R. Odette, P. Lombrozo, Physically Based Regression Correlations of Embrittlement Data from Reactor Pressure Vessel Surveillance Programs, Electric Power Research Institute, NP-3319, 1984.
[186] E.E. Eason, J.E. Wright, G.R. Odette, Improved Embrittlement Correlations for Reactor Pressure Vessel Steels, US Nuclear Regulatory Commission, UREG/CR-6551, NRC, 1998.
[187] R. Chaouadi, An Engineering Radiation Hardening Model of Reactor Pressure Vessel Materials, SCK-CEN Report R-4235, 2005.
[188] N. Soneda, Multiscale computer simulations and predictive modelling of RPV embrittlement, MATGEN-IV, Cargese, Corsica, 2007.
[189] N. Soneda, K. Dohi, A. Nomoto, K. Nishida, S. Ishino, Embrittlement correlation method for the Japanese reactor pressure vessel materials, J. ASTM. Int. 7 (3) (2010), https://doi.org/10.1520/JAI102127.
[190] L. Debarberis, A. Kryukov, F. Gillemot, B. Acosta Iborra, F. Sevini, Semi-mechanistic analytical model for radiation embrittlement and re-embrittlement data analysis, Int. J. Pres. Ves. Pip. 82 (2005) 195–200.
[191] JEAC 4201-2007, Method of Surveillance Test for Structural Materials of Nuclear Reactors, Japan Electric Association, Chiyoda-ku, Tokyo, Japan.
[192] ASTM E900-02(2007) Standard Guide for Predicting Radiation-Induced Transition Temperature Shift in Reactor Vessel Materials.
[193] R. Moskovic, P.L. Windle, A.F. Smith, Modeling Charpy impact energy property changes using a Bayesian method, Metall. Mater. Trans. A 28 (5) (1997) 1181–1193.
[194] R.E. Stoller, L.R. Greenwood, An evaluation of neutron energy spectrum effects in iron based on molecular dynamics displacement cascade simulations, in: M.L. Hamilton, A.S. Kumar, S.T. Rosinski, M.L. Grossbeck (Eds.), Effects of Radiation on Materials: 19th International Symposium, STP1366, ASTM International, West Conshohocken, PA, 2000.
[195] S.R. Ortner, P. Styman, G. Hopkin, The Range of Applicability of Embrittlement Trend Curves Transactions, SMiRT-23 August 10-14, 2015, Division VIII, Paper ID, 866.
[196] W.L. Server, T.C. Hardin, J.B. Hall, R.K. Nanstad, U.S. High Fluence Power Reactor Surveillance Data-Past and Future, J. Pres. Ves. Technol. 136 (2014) 021603-1.
[197] G. Sha, A. Morley, S. Hirosawa, A. Cerezo, G.D.W. Smith, D. Ellis, T.J. Williams, Thermal Ageing Effect of Pressure Vessel Steels, Vacuum Nanoelectronics Conference, 2006 and the 2006 50th International Field Emission Symposium. IVNC/IFES, 2006. Technical Digest. 19th International.

[198] B.Z. Margolin, E.V. Yurchenko, A.M. Morozov, D.A. Chistyakov, Prediction of the effects of thermal ageing on the embrittlement of reactor pressure vessel steels, J. Nucl. Mater. 447 (2014) 107–114.

[199] An Assessment of the Integrity of PWR Pressure Vessels: 2nd Report By a Study Group Under the Chairmanship of Dr W Marshall, United Kingdom Atomic Energy Authority, 1982.

[200] M. Koçak, B. Petrovski, D.R.G. Achar, G.M. Evans, Fracture mechanics and wide plate tests for analysis of nitrogen and strain ageing effects of weld metal fracture properties, in: M.M. Salama, S.E. Webster, J.V. Haswel, E.A. Patterson, J. Haagensen (Eds.), Proceedings of the 3rd International Conference on OMAE, vol. 3, part B1993, pp. 741–751.

[201] B. Houssin, G. Slama, P. Moulin, Strain ageing sensitivity of pressure vessel steels and welds of nuclear reactor components in assuring structural integrity of steel reactor pressure vessels, in: Proceedings of the 1st International Seminar, held in Berlin, West Germany, August 1979 (in conjunction with the 5th International Conference on Structural Mechanics in Reactor Technology).

[202] IAEA, Analysis of The Behaviour Of Advanced Reactor Pressure Vessel Steels Under Neutron Irradiation Final Report Of IAEA Coordinated Research Programme 1977-1983, Technical Reports Series No. 265, IAEA, Vienna, 1986. ISBN: 92-0-155186-X.

[203] T. Planman, R. Pelli, K. Torronen, State of the Art Review on Thermal Annealing, AMES Report No. 2, European Commission, EUR 16278 EN, 1994.

[204] M.A. Sokolov, R.K. Nanstad, S.K. Iskander, Effects of thermal annealing on fracture toughness of low upper-shelf welds, in: D.S. Gelles, R.K. Nanstad, A.S. Kumar, E.A. Little (Eds.), Effects of Radiation on Materials: 17th International Symposium, ASTM STP 1270, American Society for Testing and Materials, 1996, pp. 690–705.

[205] U.S. Nuclear Regulatory Commission, Format and Content of Report for Thermal Annealing of Reactor Pressure Vessels, Regulatory Guide 1.162, U.S. Nuclear Regulatory Commission, Washington, DC, 1996.

[206] R.K. Nanstad, B.R. Bass, J.G. Merkle, C.E. Pugh, T.M. Rosseel, M.A. Sokolov, Heavy-section steel technology and irradiation programs—retrospective and prospective views, J. Pres. Ves. Technol. 132 (2010). 064001-1.

[207] R.J. McElroy, A.J.E. Foreman, G. Gage, W.J. Phythian, P.H.N. Ray, I.A. Vatter, Optimization of Reactor Pressure Vessel surveillance programmes and their analysis, contribution to IAEA CRP 3 Research Program, AEA-RS-2426, 1993.

[208] E.A. Kuleshova, B.A. Gurovich, Ya.I. Shtrombakh, D.Yu. Erak, O.V. Lavrenchuk, Comparison of microstructural features of radiation embrittlement of VVER-440 and VVER-1000 reactor pressure vessel steels, J. Nucl. Mater. 300 (2002) 127–140.

[209] Title 10, Code of Federal Regulations, Part 50.66, Requirements for thermal annealing of the reactor pressure vessel, United States Printing Office, 2014.

[210] U. Potapovs, G.W. Knighton, A.S. Denton, Critique of In-Place annealing of SM-1A Nuclear Reactor Vessel, Nucl. Eng. Des. 8 (1968) 39–57.

[211] M.A. Sokolov, R.K. Nanstad, W. Server, Thermal Annealing of Reactor Pressure Vessels: International Experience and U.S. Perspectives, Transactions of the American Nuclear Society, Vol. 110, Reno, Nevada, June 15-19, 2014.

[212] N.M. Cole, T. Friderichs, Report on Annealing of the Novovoronezh Unit 3 Reactor Vessel in the USSR, NUREG/CR-5760, MPR-1230, MPR Associates, Inc, 1991.

[213] C.B. Oland, B.R. Bass, J.W. Bryson, L.J. Ott, J.A. Crabtree, Marble Hill annealing Demonstration Evaluation, NUREG/CR-6552, (ORNL/TM-13446), Oak Ridge National Laboratory, Oak Ridge, TN, 1998.

[214] Marble Hill Demonstration Report—EPRI TR-104934, Electric Power Research Institute, Palo Alto, CA, 1998.

[215] R.K. Nanstad, W.L. Server, Reactor Pressure Vessel Task of Light Water Reactor Sustainability Program: Initial Assessment of Thermal Annealing Needs and Challenges, ORNL/LTR-2011/351, Oak Ridge National Laboratory, Oak Ridge, TN, 2011.

[216] Federal Agency for Nuclear Control, Flaw Indications in the reactor pressure vessels of Doel 3 and Tihange 2 Final Evaluation Report 2015, FANC-AFCN, 12-11-2015.

[217] B.R. Bass, T.L. Dickson, S.B. Gorti, H.B. Klasky, R.K. Nanstad, M.A. Sokolov, P.T. Williams, W.L. Server, ORNL Evaluation OF Electrabel Safety Cases for Doel 3/Tihange 2: Final Report (R1), ORNL/TM-2015/59349, Oak Ridge National Laboratory, Oak Ridge, TN, 2015.

[217a] ENSI Review of the Axpo Power AG Safety Case for the Reactor Pressure Vessel of the Beznau NPP Unit 1, ENSI 14/2573, 28 February 2018.

[218] NUREG BR-0353, Rev.1, Davis-Besse Reactor Pressure Vessel Head Degradation: Overview, Lessons Learned, and NRC Actions Based on Lessons Learned, August 2008.

[219] R.K. Nanstad, G.R. Odette, T. Yamamoto, M.A. Sokolov, Post-Irradiation Examination Plan for ORNL and University of California Santa Barbara Assessment of UCSB ATR-2 Irradiation Experiment, ORNL/TM-2013/598, Oak Ridge National Laboratory, Oak Ridge, TN, 2013.

[220] G.R. Odette, T. Yamamoto, P. Wells, N. Almirall, D. Gragg, Microstructural Characterization of Reactor Pressure Vessel Alloys: ATR-2 Experiment Update, UCSB ATR-2 2017-2, University of California Santa Barbara, Santa Barbara, CA, September 2017.

[221] W.L. Server, M. Brumovsky (Eds.), International Review of Nuclear Reactor Pressure Vessel Surveillance Programs, STP1603, ASTM International, West Conshohocken, PA, 2018, https://doi.org/10.1520/STP1603-EB.

[222] R.K. Nanstad, M.A. Sokolov, S.R. Ortner, P.D. Styman, Neutron and thermal embrittlement of rpv steels: an overview, in: W.L. Server, M. Brumovsky (Eds.), International Review of Nuclear Reactor Pressure Vessel Surveillance Programs, ASTM STP1603, ASTM International, West Conshohocken, PA, 2018, pp. 68–106, https://doi.org/10.1520/STP160320170063.

[223] T.M. Rosseel, M.A. Sokolov, X. Chen, R.K. Nanstad, Current status of the characterization of RPV materials harvested from the decommissioned zion unit 1 nuclear power plant, in: Proceedings of ASME 2017 Pressure Vessels and Piping Conference, PVP 2017, July 16–20, Waikoloa, Hawaii, 2017.

CHAPTER 11

FERRITIC AND TEMPERED MARTENSITIC STEELS

Philippe Spätig*, Jia-Chao Chen*, G. Robert Odette[†]

Laboratory for Nuclear Materials, Nuclear Energy and Safety, Paul Scherrer Institute, Villigen, Switzerland
[†]*Materials Department, University of California, Santa Barbara, CA, United States*

CHAPTER OUTLINE

11.1 Short Historical Development of the Ferritic/Martensitic Steel: Composition and Constitution 485
11.2 Applications of the Ferritic/Martensitic Steels in Generation IV Nuclear Systems and Fusion Reactors 488
11.3 Environmentally Assisted Cracking 489
11.4 Compatibility With Liquid Metal Coolants 490
11.5 Radiation Hardening and Softening, Embrittlement, Fatigue, and Thermal Creep 493
    11.5.1 Radiation Hardening and Softening 496
    11.5.2 Irradiation Embrittlement—Fast Fracture 501
    11.5.3 Fatigue 507
    11.5.4 Thermal Creep 509
11.6 Helium Effects 510
11.7 Void Swelling and Irradiation Creep 514
    11.7.1 Void Swelling 514
    11.7.2 Irradiation Creep 516
11.8 Future Prospects for Improved Performance 518
References 519

11.1 SHORT HISTORICAL DEVELOPMENT OF THE FERRITIC/MARTENSITIC STEEL: COMPOSITION AND CONSTITUTION

High-chromium ferritic steels (*ferritic* has to be understood here in the broad sense, such as ferritic, tempered bainitic, and tempered martensitic) have been used in power plants, as well as chemical industry, for many decades. The development of ferritic steels for nuclear applications was initiated in the mid-1970s after it was discovered that austenitic stainless steels for the in-core components of the liquid metal-cooled reactors (LMRs) were susceptible to irradiation-induced void swelling [1]. AISI 316 austenitic steel was initially chosen for the LMR applications owing to its good high-temperature properties, corrosion resistance, and ease of fabrication and welding. However, swelling leads to core component distortion such as axial and/or radial expansion of fuel cladding and wrappers constituting the fuel subassemblies. The main aim of the subsequent LMR structural materials research programs was to develop steels with reduced tendency for swelling while possessing adequate creep and tensile strength. Owing to their inherent high void swelling resistance up to large damage doses, typically <0.5% up to about 140 displacement per atom (dpa), ferritic steels subsequently emerged as one of the leading candidates for the LMR core applications. A wide range of ferritic steels were considered throughout the world, ranging from carbon free >12 Cr nontransformable ferritic alloys, to C-bearing transformable 9–12Cr tempered martensitic steels (TMSs) (e.g., 9Cr-1Mo, 9Cr-1MoVNb, etc.) [2]. All compositions here are in weight percent (wt%).

In parallel to the LMR structural material efforts, various national programs on alloy development for thermonuclear fusion applications included ferritic steels, due to the broad similarities between the fission and fusion neutron environments, with the exception of helium generation rates discussed below and in Chapters 3, 5, and 12. The US program initially focused on a conventional Cr-1Mo steel HT9 (12Cr-1Mo-0.25V-0.5W-0.5Ni-0.2C), a modified 9Cr-1Mo steel (9Cr-1Mo-0.2V-0.06Nb-0.1C), and a 2.25Cr-1Mo (2.5Cr-1Mo-0.1C), all in normalized and tempered conditions. The 9Cr and 12Cr steels have a tempered martensitic microstructure while the 2.25Cr steel has a tempered bainitic microstructure. In Europe, the 10Cr MANET alloy, containing 0.5Mo along with a small addition of Nb as a grain stabilizer [3], was extensively studied. A series of investigations were carried out to optimize the overall balance of mechanical properties, with the objectives of achieving a low ductile-to-brittle transition (DBT) temperature, along with high fracture toughness, tensile, and creep strength [4–7].

In order to obtain fully martensitic structures in the 12Cr steels, austenite-stabilizing C and Mn additions were added. However, neutron irradiation was found to induce severe degradation of the Mn-stabilized martensitic alloys [8]. Despite the austenite stabilization by Mn, the 12Cr steels contained up to 15% δ-ferrite [9] and suffered from grain boundary embrittlement, as reflected in a drastic degradation of the alloys mechanical properties. Embrittlement was due to the formation of Mn-rich intermetallic χ phase and elemental Mn segregation to grain boundaries [9].

Tempered bainitic 2.25Cr-1Mo steels were also irradiated over the temperature range 390–510°C from 74 to 116 dpa [10]. Neutron irradiation was found to induce carbide precipitation below 450°C and carbide coarsening above 500°C, raising serious concerns about their applicability for nuclear application [11]. Reduced activation 2Cr-V tempered bainitic steels were found to behave in a similar fashion [8].

The most promising *ferritic* steel composition, both for fast reactor and fusion applications, was found to be in the 7–9Cr range, owing to their superior resistance to degradation from neutron irradiation and significant corrosion and oxidation resistance [6]. While commercial 9Cr-1Mo steels were first investigated, alloy development efforts, involving compositional modifications were soon initiated. Indeed, in the early-1980s, the concept of reduced activation steels became the main pathway chosen by the fusion reactor materials programs to enhance public acceptance and minimize radioactive waste disposal. Fusion first wall and blanket structure become highly radioactive during operation, due to activation reactions with neutrons. Low activation requires removing the alloying elements that produce long-lived radioactive isotopes, such as Cu, Mo, Ni, Nb, and N. Thus, in reduced activation TMS, Mo is replaced by W on a 2/1 wt% basis (since the molar of W is about twice that of Mo). Strengthening provided by carbide forming Nb was replaced by V, Ti, and Ta carbides.

The concept of reduced-activation materials is important for several reasons. The first relates to possible release of radioactivity to the environment in case of an accident. Thus, materials with low decay heat, activity, and radiotoxicity are desirable. Second, plant maintenance and repair operations are facilitated if dose rates are sufficiently low to permit the use of robotic devices. Recycling and waste disposal of activated materials is the third consideration. Typically, remote recycling is possible if the surface dose rate is lower than 10 mSv/h, while hands-on recycling requires dose rate that must be <10 μSv/h. To classify a material as low-level waste, the dose rate and decay heat typically must be lower than 2 mSv/h and 1 W/m^3, respectively. The volume of radioactive waste, which can neither be cleared nor recycled, must meet low-level waste disposal criteria within several decades in order to avoid imposing complex burdens on future generations. Thus, the concept of reduced activation is critical to fusion materials waste management strategies.

Many different heats of so-called reduced activation ferritic/martensitic steels (RAF/M) 7–9Cr TMS have been produced in Japan, Europe, and more recently in China. Note that, while use for the term RAF/M is the convention, this is a bit of a misnomer, since these alloys are all in the TMS condition. Their compositions differ slightly, but the various TMS steels are otherwise generally similar. It is not our purpose to describe the details on all these steels, but the interested reader can find a good survey in Ref. [2]. In all cases, 7–9Cr reduced activation

11.1 SHORT HISTORICAL DEVELOPMENT OF THE FERRITIC/MARTENSITIC STEEL

FIG. 11.1

TEM observation of Eurofer97 microstructure.

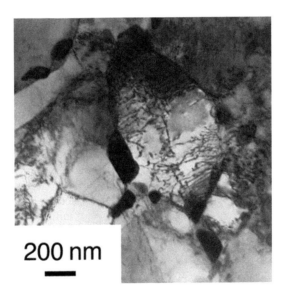

FIG. 11.2

$M_{23}C_6$ carbides along prior austenite grain boundaries.

TMSs exhibit a typical microstructure resulting from a two-step final heat treatment: (1) normalization at about 1000°C for about 0.5h, and (2) tempering around 750°C for about 1h. Of course, the specific temperatures must be optimized for each particular composition. The RAF/M steels are generally characterized by equiaxed prior austenitic grains (20–60 µm), containing multiple packets of martensitic laths (see Fig. 11.1). They have a moderately high dislocation density included in subgrain structures. The C and metallic solute composition are adjusted to produce various alloy carbides. The largest $M_{23}C_6$ carbides, ranging from 50 to 500nm, have a high Cr content, and are primarily located along the prior austenite grain boundaries, in some cases forming colonies (see Fig. 11.2), energy-dispersive X-ray spectroscopy (EDS) shows that the $M_{23}C_6$ carbides also contain some Fe, W, V and, in some cases, a small amount of Mn. A much higher number of fine-scale secondary hardening precipitates, mainly distributed in the matrix, such as VN, (CrV)N, and TaC. These fine-scale precipitates play a key role in increasing creep strength and are also crucial to refining the austenite grains by providing nucleation sites and impeding the grain growth.

For fusion reactor applications, the chemical composition and heat treatments applied to the unirradiated steels were designed to optimize the tensile and creep strength up to about 550°C, which as a rule can be regarded as their upper temperature limit. In principle, this practice differs from that necessary for fast breeder wrappers and duct applications, since these components are generally not highly stressed. Rather, the LMR focus was on good high-temperature ductility and a low initial DBT temperature, so that structural integrity is maintained during both irradiation and fuel handling operations. In contrast, for cladding applications strength is important. More generally, the specific requirements for a given final application can be achieved by optimizing the normalizing and tempering conditions. Note that, depending on the equivalent Cr and other alloying elements, 9–12Cr steels can have either a fully austenitic structure or a duplex austenitic-ferritic structure in the temperature range 850/950°C to 1200/1150°C. Ultimately, the final heat-treatment controls the microstructure, which in turn mediates the balance of tensile creep strength, fatigue resistance, and fracture toughness properties. RAF/M for fusion applications are discussed in detail in Chapters 2 and 3, while Chapter 12 addresses advanced oxide dispersion strengthened (ODS) alloys.

11.2 APPLICATIONS OF THE FERRITIC/MARTENSITIC STEELS IN GENERATION IV NUCLEAR SYSTEMS AND FUSION REACTORS

Generation IV (Gen-IV) reactors are discussed in Chapter 2. The Generation-IV International Forum (GIF) was created in 2000. It is a collective of 13 countries committed to the development of the next generation of nuclear reactors. Six reactor concepts have been selected by GIF, namely: gas fast reactor (GFR), sodium fast reactor (SFR), lead-cooled fast reactor (LFR), which are fast reactors; very high-temperature reactor (VHTR), molten salt reactor (MSR), which are thermal reactors; and supercritical water reactor (SCWR), which has a mixed thermal/fast neutron spectrum [11]. The core outlet temperature is significantly higher (up to several hundred °C) in all six concepts, than in the light water reactors in service today. These high temperatures, along with much higher dpa doses and more corrosive environments compared with current reactors, raise new challenges for the development of superior materials able to withstand such harsh conditions.

Therefore, within the framework of materials development programs for Gen-IV nuclear systems, as discussed in more detail in Chapter 2, ongoing investigations are focused on materials that exhibit good high-temperature mechanical properties (e.g., strength, ductility, creep, toughness, etc.), sufficient resistance to fast neutron irradiation (e.g., dimensional and microstructural stability, etc.), and compatibility between the fuel on the one hand, and the coolant on the other hand. These materials programs address issues related to fuel cladding, wrappers, and any other in-core components, such as ducts, as well as for pressure vessels. Note that, Zr alloys are the reference cladding in current light water reactors (see Chapter 7), while advanced austenitic stainless steels serve that function in SFRs (see Chapter 8). However, Zr alloys cannot be used beyond 400°C and they are not accident tolerant. Austenitic steels are limited to a moderate maximum dpa level, due to their intrinsic tendency for swelling. Historically, there has been a continuous increase in fuel burnup, which is the amount of energy extracted from the U, from about 20–25 GWd/tU for the Gen I reactors to about 50–60 GWd/tU in currently operating light water reactors (Gen II). Anticipated burnups for Gen III and IV reactors, burnups are up to 100–200 GWd/tU, requiring the development of even more irradiation-resistant materials. As noted previously, TMS steels are leading candidate core and cladding materials owing to their excellent dimensional stability under irradiation, as well as their good corrosion resistance in various environments. However, the relatively low creep strength at high temperature of TMS with 8%–12% Cr and a few weight percent of other alloying elements limits their use in high-temperature applications; thus their creep properties must be improved. ODS alloys, containing a fine dispersion of stable matrix oxides being developed to provide high creep strength, are very promising materials for cladding applications (see Chapter 12).

TMS, such as T91 (9Cr-1MoVNb), T92 (9Cr-0.5Mo2WVNb), and HT9 are being considered for both the primary and secondary circuits of the SFR as well as for the steam generator pipes as replacements for the austenitic steels [12]. This selection is motivated by their lower fabrication costs compared with austenitic stainless steels, and the fact that the TMS have better thermal conductivity and lower thermal expansion coefficients, which allows use smaller ducts, hence a reduction in the steel volume. T92 was proposed as material for the hexagonal tube component of the SFR fuel assembly. Use of the 9Cr TMS class is envisaged for the vessel of gas-cooled reactors. The pressure vessel of the VHTR is about twice as large as a pressurized water reactor (PWR) vessel. Forging and welding of such enormous components fabricated using 9Cr TMS will require a comprehensive research program; for example, to avoid hot-cracking during welding.

As mentioned in the previous section, the 9Cr TMS are also leading material candidate structural materials for thermonuclear fusion applications, in particular for the first wall and the blanket structures (see Chapter 3). The deuterium-tritium (D-T) fusion reaction produces a high-energy 14 MeV neutron, which is well above the threshold for (n,α) and (n,p) reactions in Fe, which are about 5 and 1 MeV, respectively. Thus, typical appm He/dpa and appm H/dpa ratios (atom parts per million/dpa) in a fusion reactor first wall are about 10 and 40, respectively. Other fusion irradiation conditions that differ from fission include: (i) the primary knock-on atom (PKA) spectrum in fusion that includes a high-energy component (>100 keV), which is absent in fission irradiations; (ii) neutron reactions transmutations that result in more chemical composition changes in fusion versus fission environments; and (iii) the first wall will be exposed to enormous heat fluxes that result in unprecedented thermomechanical challenges [13]. However, the first wall dpa rates will be similar, or less, than in LMRs (see Chapters 2 and 3). The stress, temperature, and environmental conditions in a fusion reactor depend on the first wall and blanket and coolant concept, typically helium or liquid lead-lithium eutectic alloy. Moreover, plasma-facing components will be subjected to significant time varying mechanical and electromechanical loading as well as severe heat fluxes and thermal stresses as noted above. Coupled with neutron irradiation, fusion clearly represents the most extreme service environment ever encountered in any source of sustained energy that requires long-lived components.

11.3 ENVIRONMENTALLY ASSISTED CRACKING

Environmentally assisted cracking (EAC) of metals and alloys is a very complex corrosion phenomenon that is influenced both by the service conditions and the material characteristics. Depending on the applied loading conditions, cracking phenomena include: stress corrosion cracking (SCC) with static loads, corrosion fatigue (CF) with cyclic loads, and strain-induced corrosion cracking (SICC) with slow monotonically rising, or very low-cyclic, loads. In an irradiation environment, SCC is typically exacerbated by irradiation-assisted stress corrosion cracking (IASCC). In aging light water reactors (see Chapter 6), IASCC has been observed to be a major mode of component failure of the austenitic steels. Owing largely to the water chemistry, differences between boiling water reactors (BWRs) and pressurized water reactors (PWRs), IASCC is a more serious issue in a BWRs. IASCC affects various components such as small bolts, springs, control blade handles, instrumentation tubes and core shrouds in a BWR, as well as baffle bolts in a PWR. IASCC susceptibility is greatly affected by the microchemistry at segregated, especially Cr depleted, grain boundaries. The susceptibility to IASCC is typically investigated by postirradiation slow strain-rate tensile, crack growth rate, and crack initiation tests in autoclaves that simulate nuclear water chemical environments.

While SCC in austenitic steels has been extensively investigated, there are far fewer corresponding studies for TMS. Thus, far less is known about corrosion and cracking phenomenon in a high-temperature pressurized water environment. Nonetheless, a number of experimental results have shown that 12Cr and 12Cr-1MoVW TMS in various normalized and tempered conditions are even more resistant to SCC than the austenitic steels in simulated BWR environments. For example, prolonged exposures of TMS alloys to a BWR environment with 6–9.5 ppm dissolved oxygen or a hydrogen water chemistry with 6–14 ppb oxygen and 170–190 ppb dissolved hydrogen, did not lead to environmental cracking [14]. More recently, crack propagation rate investigations on wrought ferritic steels containing 5%, 9%, 13%, and 17% Cr were carried out by Rebak et al. using compact tension specimens [15]. The tests were performed at 288°C in both pure water and water containing sulfuric acid (30 ppb of sulfate ions), as well as in normal water conditions (2 ppm of dissolved oxygen) and under charged hydrogen water conditions (63 ppb of dissolved hydrogen). The dissolved oxygen, hydrogen and H-ions control the corrosion potential of the loaded specimen that, in turn, mediates the crack growth rate. Notably, excellent crack growth resistance was observed in the TMS under typical high-temperature pressurized water conditions, with sustained cracking observed only above stress intensity factors of ≈ 43 MPam$^{1/2}$. Further, the postthreshold crack growth rates in these wrought martensitic/ferritic steels were at least two orders of magnitude lower than those in austenitic steels. Slow strain-rate SCC tests on several TMS were carried out in water loop, representing environment of supercritical water reactors (SCWR) [16]. The oxygen and steel composition were found to influence the SCC behavior. For typical SCWR highly corrosive operating conditions of 250 bar (2.5×10^4 kPa) at 500°C, higher Cr ODS steels are more resistant to SCC [17]. SCC behavior is discussed in more detail in Chapter 6.

Note that, radiolysis of the water by ionizing radiation leads to the formation of hydrogen peroxide which, in turn, increases the corrosion potential. For unirradiated reduced activation Eurofer97, SCC is enhanced by increases in strength and/or the electrochemical potential, as well as by trace chloride additions [18]. Vankeerberghen et al. showed the effect of irradiation with in-core electrochemical measurements in the BR2 reactor at SCK-CEN for both austenitic and Eurofer97 TMS steels. Notably, the corrosion potential was also found to increase with neutron flux [19].

11.4 COMPATIBILITY WITH LIQUID METAL COOLANTS

Corrosion phenomena in the ferritic/martensitic (F/M) steels also occur in other environments such as liquid metals like lithium and lead-lithium eutectic (Pb-17%Li), which are potential coolants and tritium breeders in fusion reactor blankets [20], or lead-bismuth eutectic, which is chosen as primary coolant and spallation source for the subcritical core of the accelerator-driven system [21]. Corrosion effects are often described by a simple mass loss rate in g/m^2/year, or in surface recession rate µm/year. Liquid metal corrosion causes degradation by three major processes: surface regression or component thinning, surface degradation, and bulk-phase intergranular attack. The latter mechanism is related to liquid metal embrittlement (LME) that is manifested in significant to severe reduction in ductility, and especially fracture resistance in the presence of cracks [22]. Thus, mass loss is only a partial metric of corrosion-induced damage, and it is necessary to consider surface and, especially, grain boundary degradation leading to LME [23]. As a consequence, liquid metal corrosion also leads to the loss of the mechanical strength. Corrosion tests can be done either under static exposures or flowing liquid conditions. Corrosion is generally enhanced under liquid flow conditions by accelerated chemical reactions, due to the removal of reaction products or erosion-corrosion. Dissolution-deposition mass transport in a flowing system under corrosion conditions is also a major issue.

Here, we can only present some simple illustrative examples of liquid metal corrosion phenomena. Kondo et al. carried out corrosion tests on the Japanese JLF-1 RAF/M (Fe-9Cr-2W-0.1C) steel at 600°C in liquid Li and Pb-17Li [24] under both static and flow conditions. This work showed that the concentration of nonmetallic impurities in the liquid metal is an important factor affecting corrosion. For example, corrosion rates increased with higher concentration of O or N in Li. Nitrogen promotes the formation of unstable nitrides such as Li$_3$FeN$_2$, Li$_9$CrN$_5$, and LiCrN. In this case, corrosion takes place preferentially at boundaries, including along laths, lath packets, and prior austenite grain boundaries, often causing grain exfoliation. The role of carbon depends on its concentration in the liquid, the composition of the steel, and the alloy carbide stability. The effect of carbon can be manifested either as carburization or decarburization. An example of nonmetallic impurity effects on the weight loss rate of JLF-1 in liquid Li is shown in Fig. 11.3.

Kondo et al. also reported the time dependence of Fe, Cr, and W in Pb-17Li (see Fig. 11.4). The dissolution of W and Cr are similar, and appear to be saturated, at 3000 h. However, dissolution of Fe continues beyond 500 h. Presumably, these trends can be attributed to the respective elements solubility limits in Pb-17Li, which is higher for Fe.

SEM observations of the surface morphology after 750 and 3000 h in static Pb-17Li, as well as after 250 h in flowing Pb-17Li, are shown in Fig. 11.5. Under static conditions, the surface is pitted at 750 h, prior to extensive

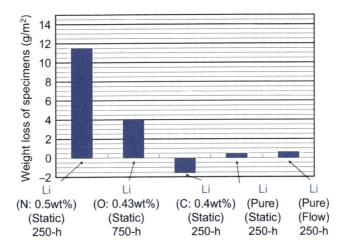

FIG. 11.3

The weight loss of JLF-1 steel showing the effect of nonmetallic impurities on Li corrosion [24].

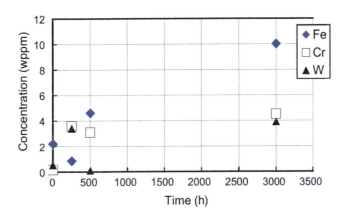

FIG. 11.4

The concentration of metallic elements in Pb-17Li as a function of time in a static corrosion test of JLF-1 steel [24].

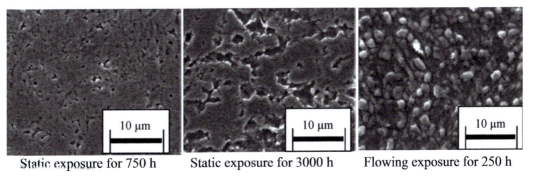

FIG. 11.5

SEM observations of JLF-1 steel after exposure to Pb-17Li [24].

grain boundaries attack at 3000 h. In contrast, after exposure to flowing Pb-17Li for only 250 h, a granular surface microstructure develops, probably as the result of a grain exfoliation mechanism; is that entire grains are attacked and peeled off by the liquid metal flow.

As a simple example, the time dependence of the concentration of a metallic element in a static liquid metal based on dissolution and mass transfer between the specimen and test container at a specified temperature is given by [25]

$$C = C_s \left(1 - \exp\left(-\frac{h(S_s + S_c)t}{V} \right) \right) \tag{11.1}$$

Here, C and C_s are the concentration of the metallic element in the liquid metal and the corresponding solubility limit, respectively, S_c and S_s are the surface areas of the test container and the specimen, respectively, and h is a fitted dissolution rate mass transfer coefficient in m/s. The other known variables are V the liquid volume and t the time. Clearly, this dissolution-deposition model does not account for liquid flow or chemical reactions involving nitrogen and carbon. Thus, more complex models taking these mechanisms into account, as well as the critical effect of temperature, are required.

Liquid-solid interaction induced erosion-corrosion must also be considered. However, detailed discussion of this topic is beyond the scope of this chapter. Hence, we simply present one example of an empirical engineering model. Clearly, metal loss rates must be related to the liquid metal temperature and fluid velocity. Over the operation range of interest from 400°C to 550°C, the following, fully empirical, expression has been proposed for flowing Pb-17Li [26]:

$$\Delta M_l = 8 \times 10^9 \exp - \left(\frac{25690}{1.98T} \right) v^{0.875} d^{-0.125} \tag{11.2}$$

Here ΔM_l is the metal recession rate (μm/y), v is the liquid flow velocity, T is the absolute temperature, and d is the flow channel hydraulic diameter in meters. Thus, for example, at 500°C, $v = 0.02$ m/s in a duct with $d = 0.05$ m, $\Delta M_l \cong 19$ μm/y.

However, even if such empirical treatments are complete and valid, which is generally impossible, there are additional complications in the case of fusion blankets. In particular, the high magnetic fields, in the range of 5–10 T, enhance corrosion rates in fusion blanket ducts, due to the interaction of the flow-induced electric currents with the applied plasma-confinement magnetic fields, resulting in a flow opposing Lorentz forces. These forces lead to high magneto-hydro-dynamic (MHD) pressure drops and suppression of turbulence, along with corresponding effects on heat and mass transfer coefficients, and other important MHD phenomena [27]. Pertinent to erosion-corrosion is the fact that the liquid metal velocity distributions are strongly modified, with a large, stable gradient in the thin boundary layer at the liquid-duct wall interface, which promotes mass transfer into the liquid. The occurrence of MHD-driven electric currents across the liquid/solid interface also influences corrosion processes. Efforts are being made to develop more comprehensive models of Pb-17Li corrosion mechanisms in TMS such as those proposed by Moreau et al. [28]. Clearly, the combination of such effects adds enormous complexity to accurately predict in-service corrosion rates. One solution is development of durable protective coatings for steels. Recent research has been mainly focused on sandwich coatings made of Al_2O_3 or Er_2O_3 together with W that act as both corrosion and tritium anti-permeation barriers [29].

LME refers to a degradation of the mechanical properties of normally ductile materials resulting from synergistic interactions between a liquid metal and subsurface microstructures, especially grain boundaries. LME is a very complex phenomenon that depends on a large number of variables, such as the grain boundary type and structure, grain size, the liquid metal properties and chemistry, applied stress, type of loading (monotonic vs. cyclic), loading rate, component geometry (smooth or with stress concentration), and temperature. LME is manifested by a loss of ductility provided that: (i) the applied stress is above the yield stress; and (ii) a close contact between the liquid metal and the steel (wetting) exists [30]. If these conditions are met, a DBT induced by the liquid metal can be in the temperature range 300–450°C. Note that LME differs from SCC in water environment, due to the fact that the former does not necessarily involve corrosion processes, which actually tend to prevent wetting and therefore embrittlement [31]. High oxygen content in the liquid metals promotes the formation of an oxide scale on the steel surface that acts as a protective coating. Indeed, from small punch tests performed on T91 steel in lead-bismuth eutectic, Ye et al. showed that at high displacement rate LME is not significant at low oxygen contents, while at low displacement rate a LME DBT was observed. It was proposed that the LME DBT was associated with a decrease in the local cohesion of the steel due to a reduction in the surface energy by the adsorbed liquid metal atoms [31]. Predicting LME is a highly challenging, especially if potential irradiation effects are considered. In particular, radiation-induced segregation may play an important role in LME. Defining the conditions where LME occurs is only one part of the problem, and the impact of LME on the component integrity largely remains to be investigated.

Finally, fuel cladding tubes are also subjected to the chemical attack by fission products. However, this complex and specialized topic is beyond the scope of this chapter.

In closing, it is important to emphasize that the causes and consequences of corrosion are very system specific. For example, corrosive dissolution rates in one location are affected by deposition rates in another part of the system, since both depend on the concentration (chemical potential) of the transported species in the liquid metal as well as the varying temperatures.

11.5 RADIATION HARDENING AND SOFTENING, EMBRITTLEMENT, FATIGUE, AND THERMAL CREEP

Before discussing irradiation-induced changes on the tensile, fracture, impact, and fatigue behavior of TMS steels, we briefly summarize their unirradiated properties. The tensile properties of RAF/M steels are close to those of the conventional TMS 9Cr-1Mo alloys. While these alloys differ slightly in chemical composition, production routes, and final heat treatment, all show relatively similar tensile properties. This is illustrated in Figs. 11.6 and 11.7

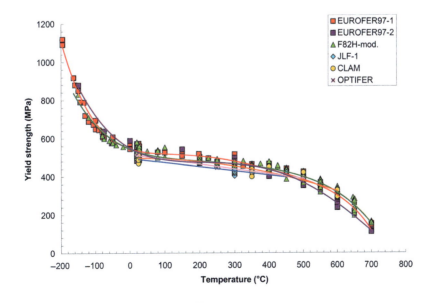

FIG. 11.6

Yield stress versus temperature of various unirradiated RAF/M steels [32].

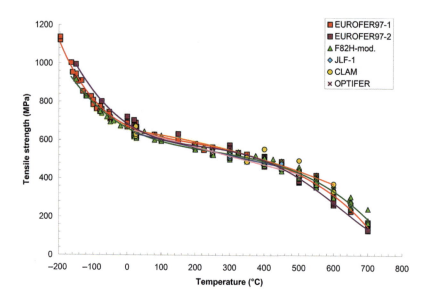

FIG. 11.7

Tensile strength versus temperature of various unirradiated RAF/M steels [32].

where the yield stress (σ_y) and ultimate tensile strength (σ_u as an engineering stress) versus temperature of six RAF/M steels are shown.

TMS lose their strength at higher temperatures/lower strain rates deformation. At typical tensile test strain rates, the transition begins at ≈400°C and is complete by 600°C [33,34]. The limits on creep strength of TMS led to development of other grades of these steel, like NF616 [35], as well as other classes of alloys, such as ODS steels (ODS—see Chapter 12). Below ambient temperature (≈20°C) and/or high strain rates σ_y increases with decreasing temperature. This is primarily due to the lattice Peierls stress contribution to strength contribution, which is thermally activated [36]. The σ_y and flow stress dynamics can be well described by a strain-rate ($\dot{\varepsilon}$) compensated temperature (T_{SRC}). For Arrhenius-type activated processes, the σ_y is the same at equivalent temperature adjusted for different ($\dot{\varepsilon}$) as

$$T_{SRC} = T\left(1 + C\ \ln\left(\frac{\dot{\varepsilon}_r}{\dot{\varepsilon}}\right)\right) \tag{11.3}$$

Here, $\dot{\varepsilon}$ is the actual strain rate and $\dot{\varepsilon}_r$ is a reference strain rate at the actual T, and C is a constant that depends on the activation energy of dislocation glide rate controlling mechanism [37]. A typical value of C is ≈0.03.

The strain-hardening capacity of the RAF/M steels can be inferred from the ratio, $R = \sigma_u/\sigma_y$. Over the temperature range from 20°C to 450°C, R lies between 1.1 and 1.4 for all the RAF/M steels considered in Figs. 11.6 and 11.7. In the same temperature range, the uniform elongation decreases and is limited to few percent (see Figs. 11.8 and 11.9). The total elongation is of the order of 10%–20% in this temperature range.

The impact and fracture toughness properties of most unirradiated TMS and RAF/M steels are sufficiently high so as to not to be of concern for design. Typically, the ductile-to-brittle transition temperature (DBTT) measured in Charpy impact tests are around −50°C, or less, and the Master-Curve reference temperature T_0 is around −100°C (see below for more details). The DBTT in some TMS weld conditions are higher than in base metals. Both

11.5 RADIATION HARDENING AND SOFTENING

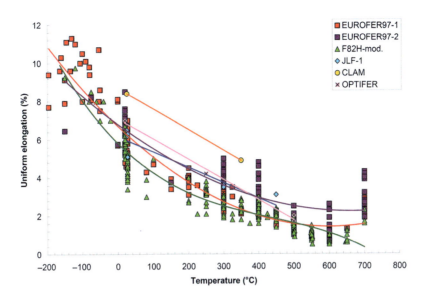

FIG. 11.8

Uniform elongation versus temperature of various unirradiated RAF/M steels [32].

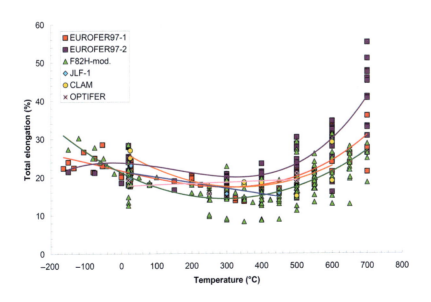

FIG. 11.9

Total elongation versus temperature of various unirradiated RAF/M steels [32].

conventional TMS and RAF/M steels have lower creep strength than austenitic stainless steels, particularly above 550°C.

11.5.1 RADIATION HARDENING AND SOFTENING

At irradiation below about $0.4T_m$ (T_m is the absolute melting temperature), irradiation results in hardening due to the accumulation of extended lattice defects and precipitation. The primary defects are vacancies, self-interstitial atoms (SIA), and small clusters of these defects created in the displacement cascade that are generated by a PKA following the interaction between an impinging neutron and a nucleus (see Chapter 5). Impurity atoms are also produced by neutron transmutation reactions. The most significant transmutation product is He, especially in fusion environments (see Chapter 3). Many of the vacancies and interstitials annihilate one another (recombine) during the cascade-cooling phase or long-range migration, especially at trapping sites [38,39]. However, point defects that survive recombination diffuse to sinks, and a fraction form extended like defect clusters that grow as interstitial dislocation loops and, in some cases, voids. Insoluble helium clusters to form bubbles. The excess mobile defects also accelerate diffusion rates and can induce solute segregation. This also leads to solute clustering, often at small loop or nano-void defect-solute complexes, as well as nonequilibrium phase precipitation (see Chapter 3). The evolutions of dislocation loops, precipitates, nanovoids, helium bubbles, and larger growing voids, that are induced or enhanced by irradiation, are strong functions of dpa and temperature. The feature sizes generally increase with irradiation temperature while their number density decreases. The features sizes and number densities increase with dpa up to a saturation levels, except in the case of voids, which can continue to grow leading to significant amounts of swelling.

Under displacement damage dominated conditions, these evolutions are largely governed by the excess defect concentrations. These concentrations are controlled by the defect generation rate balanced by their corresponding annihilation rates at sinks as well as by recombination. The high sink density in TMSs, composed of dislocations, various boundaries, a variety of precipitates, and helium bubbles, reduce excess defect concentrations, hence, the corresponding rate of evolution of the features cited above. Thus, one of the reasons for the void swelling resistance of TMS is the high sink density and lower excess defect concentrations compared with austenitic stainless steels. So, formation and accumulation of defects are responsible for the degradation of mechanical properties.

Increase in the yield stress, $\Delta\sigma_y$, or irradiation hardening, is caused by the irradiation-induced nano-features that act as obstacles to dislocations. Hardening is normally mainly due to the small $a\langle 100\rangle$ and $a/2\langle 111\rangle$ type interstitial loops and small precipitates induced by irradiation. The contribution of small <1 nm cavities to irradiation hardening appears to be limited in some cases, but larger voids can cause considerable hardening if present [40–43]. Large concentration of He, in the form of bubbles also contributes to irradiation hardening (see Section 11.6). The tensile curves reported by Dai et al. [44] shown in Fig. 11.10 were obtained from the ferritic-martensitic steel Optimax-A (9CrW0.2V) after irradiation in the Swiss spallation neutron source SINQ located at Paul Scherrer Institute. The tests were conducted at the same temperature as that of irradiation. The dose, helium content, and irradiation temperature are indicated for each curve. In the irradiation temperature range from 100°C to 350°C, a clear $\Delta\sigma_y$ increase is observed along with a strong reduction in ductility characterized by either the uniform elongation or the total elongation at failure. For these irradiation conditions, except for $T_{irr} = 350°C$, the uniform elongation is extremely small as the onset of necking practically corresponds to the yield point. Irradiation hardening in this case is primarily due to loops and with possible smaller contributions from precipitates and bubbles.

Yamamoto et al. have compiled and analyzed the existing irradiation hardening data on high-chromium TMS in terms of dose and temperature dependence [45]. They showed that the irradiation hardening can be effectively modeled by the Makin and Minter equation, initially derived in [46,47]

$$\Delta\sigma_y = \Lambda\sigma_{ys}\left(1 - \exp\left(-\frac{dpa}{dpa_0}\right)\right)^{1/2} \tag{11.4}$$

11.5 RADIATION HARDENING AND SOFTENING

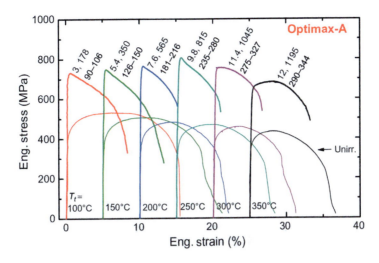

FIG. 11.10

Engineering tensile stress-strain curves of the Optimax-A steel irradiated in SINQ. The numbers for each curve refer to dpa, He (appm), and irradiation temperature range [44].

The irradiation hardening saturates with neutron fluence at $\Delta\sigma_{ys}$, and dpa_0 characterizes the dose transient before saturation. The behavior reflects that initially the number of isolated irradiation-induced features, acting as dislocation obstacles, increases linearly with neutron dose. However, as the number of features increases, there is an increasing probability that newly created defects are destroyed by or destroy the existing clusters. So, at saturation, the net rate of creation of new features decreases to zero. In Fig. 11.11, we show an example of the dose dependence of the irradiation hardening for TMS and ODS steels [48], which supports the saturation law.

Both $\Delta\sigma_y$ and dpa_0 are functions of irradiation temperature. At temperatures higher than about $0.4T_m$, the densities of dislocation loops and precipitates decrease, along with the corresponding hardening. Above about $0.4T_m$, irradiation-induced softening increases with increasing temperature due to various instabilities in the overall steel microstructure [49,50]. These instabilities are expected to slow or saturate in some softened state.

As noted above, $\Delta\sigma_y$ results from the accumulation of the small irradiation-induced features that act as the obstacles to dislocations glide. The moving dislocations bow out between the obstacles up to a critical angle, the breakaway escape angle. That critical angle characterizes the strength of the obstacles, typically denoted as the α parameter. Along with the number density N_V and the diameter d of the obstacles α mediates $\Delta\sigma_y$. The simplest relation for σ_{yj} associated with a specific obstacle type j, is

$$\sigma_{yj} = \alpha_j(d_j)\mu M b \sqrt{N_{Vj} d_j} \qquad (11.5)$$

Here, μ is the shear modulus, M is the Taylor factor, and b is the dislocation Burgers vector length. The strength parameter α, depends on the type j and size d of the obstacle, which increases up to a value of about 0.8–1 for strong obstacles that are bypassed by Orowan bowing rather than escaping or cutting. The most important obstacles are loops (l), α' precipitates (α'), cavities (c), and other, typically G-phase type, precipitates (p). Microstructural data can be used to calibrate the $\alpha(d)$ factors for various features. Typical values of various features are $\alpha_l \approx 0.2$–0.4, $\alpha_{\alpha'} \approx 0.03$–$0.06$, $\alpha_c \approx 0.5$–1, and $\alpha_p \approx 0.4$–0.6 [43,51].

The overall strength of an alloy is governed by contributions from a number of different obstacles, such as those that preexist irradiation and contribute to σ_{yu} (=yield stress of the unirradiated material). For the simple case where there is only one obstacle resulting in σ_{yi} (=yield stress of irradiated material) and one for σ_{yu}, their net contribution

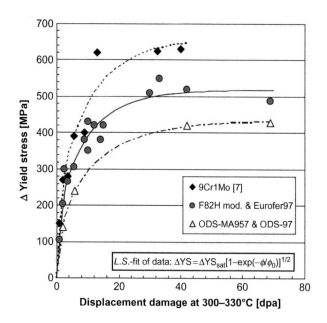

FIG. 11.11

Irradiation hardening versus dpa for high-chromium conventional and reduced activation steels, as well as for two ODS-steels after irradiation at 300–330°C to 80 dpa in the BOR-60 fast reactor [48].

is given by limiting superposition laws that gives the net $\Delta\sigma_y$ due to irradiation as either a linear sum superposition (LS) law,

$$\Delta\sigma_y = \sigma_{yi} \tag{11.6}$$

or a root sum square (RSS) superposition law,

$$\Delta\sigma_y = \sqrt{\sigma_{yi}^2 + \sigma_{yu}^2} - \sigma_{yu} \tag{11.7}$$

The LS model in Eq. (11.6) applies when the irradiation-induced features are weak and α_i is small. The RSS model in Eq. (11.7) is appropriate for larger α_i combining with a typically large α_u. Single dislocation computer simulations for various combinations of α's have been used to derive analytical expression to treat cases that fall between the LS and RSS limits [52]. However, the RSS model is a reasonable approximation for most radiation-induced hardening features.

The preceding expressions for $\Delta\sigma_y$ are clearly oversimplified and approximate. Complications include the large variety of obstacles, hardening contributions that are not from dispersed barriers: polycrystalline Hall-Petch and subgrain structure effects, long-range internal stress fields and back stresses, and so on. Another important complication is source hardening due to dislocation atmospheres that may exist both prior to and especially after irradiation. Specifically, a population of mobile interstitial dislocation loops has been observed to segregate to dislocations [53]. Despite, the simple dispersed models have proven to be very successful in relating microstructural evolution under irradiation to the corresponding hardening.

Dislocation loops are features contributing to irradiation hardening and play a number of important roles in irradiation effects on TMS. A complete discussion of this topic is beyond the scope of this chapter. Rather than attempting to describe many detailed observations in the literature, which are both scattered and incomplete, or discussing extensive modeling results, the following paragraphs attempt to paint a broad picture of the causes, character, and consequences of loop evolution in TMS. No attempt is made to fully reference the literature, but we note that a useful summary of much of the loop story is given in the introduction of Ref. [54] and references therein as well the discussion in an earlier paper [55].

Loops typically make a significant contribution to irradiation hardening below $\approx 430°C$, and hardening peaks at $\approx 300°C$ [45]. At this temperature, loop hardening begins at <1 dpa, and increases rapidly prior to saturating between ≈ 10 and 70 dpa [45,56]. Displacement damage hardening is minimal above $\approx 430°C$. The hardening dose and temperature dependence are qualitatively consistent with the observed evolution of the loop microstructure [55]. Even higher dose behavior remains to be explored.

Loops are born in the form of platelet-shaped clusters of SIA in displacement cascades [57,58]. At small sizes, these loops can be viewed as clusters of crowdions, which are defects involving two atoms sharing one lattice site along a close-packed direction. This configuration of SIA results in TMS loops with an $a/2\langle 111\rangle$ Burgers vector and an $\{110\}$-habit plane [58]. According to the molecular dynamics models, the small loops are highly mobile in a perfectly pure Fe bcc lattice, with a very low activation energy (<0.1 eV) for one-dimensional (1D) diffusion on their glide prism [58,59]. Note that some experimental estimates of the migration energy of loops are higher (up to 0.4 eV), but this may be due to the effects of impurity trapping (see below). Small loops can reorient from one glide prism ($\langle 111\rangle$ direction) to another, but 1D migration is typically dominant. This strongly affects the rate and mechanism by which the SIA clusters are annihilated at sinks [60]. Owing to strain field interactions, loops may raft and decorate dislocations, at least at lower dose and in pure iron [61]. Annihilation of the SIA clusters at dislocations can take place primarily by conservative climb on their habit plane, due to atomic diffusion at the loop periphery, thereby forming a super jog.

Loops interact with each other and can transform from having a glissile $a/2\langle 111\rangle$ character, to a sessile $a[100]$ configuration that occurs by complex $a/2[111]+a/2[1-1-1]$ reaction paths [62,63]. The sessile $a\langle 100\rangle$ type loops cause more hardening than $\langle 111\rangle$ loops [54,55,64]. Again, both types of loops grow due to their strong SIA bias. The population of $a\langle 100\rangle$ loops increases with dose and temperature, in part to their greater stability [54,55] and at intermediate dpa they typically outnumber $a\langle 111\rangle$ loops by ratios of about 2/1 or more. At higher dpa, some fraction of growing loops may transform to line dislocation segments with both types of Burgers vectors [54]. Fig. 11.12 illustrates the different loop configuration as observed with TEM by Yao et al.

Loop mobility decreases with increasing size (n), roughly scaling as $n^{-2/3}$ [43,65]. Loops are also trapped by both impurities, and especially solute additions, hence, they are much less mobile in alloys [66,67]. Since detrapping is a thermally activated process, the mobility of loops in alloys is expected to increase with increasing temperature. Loops are also sites for significant segregation of solutes such as C, P, Si, Mn, and Ni due to both thermal segregation, presumably due to the strain energy shielding (in the case of Cr), and radiation-induced vacancy drag segregation mechanisms for several other solutes [68,69]. Indeed, the clusters of these solutes seen in atom probe tomography studies are believed to be at small loops [68] that act as a heterogeneous nucleation sites for the formation of the corresponding well-formed intermetallic phases that develop at higher dpa [70]. Segregation of Cr to loops increases their strength as dislocation obstacles [71].

The most important practical characteristics of dislocation loops are their type, number densities (N), and diameters (d), since these factors directly mediate their corresponding sink strength and hardening contributions, that approximate scale with Nd and $\sqrt{Nd}$, respectively. The density of the loops decreases and their size increases with increasing temperature. The underlying mechanisms of the loop dependence on temperature relate to $\langle 111\rangle$ loop mobility, $\langle 100\rangle$ loop growth by absorption of $\langle 111\rangle$ SIA clusters, coarsening and growth of large loops at the expense of smaller ones [54]. The loop dose and temperature behavior are reflected in the corresponding hardening trends [45].

Unfortunately, data on the dose and temperature dependence of N and d are scattered and limited. This is partly due to the challenge of properly imaging loops by TEM, especially at very small sizes [54]. At $\approx 300°C$ and below the loops are so small that they are most often described as "black spot" damage. It is possible to characterize slightly larger loops using two beam, g·b visibility methods [54]. Very roughly characteristic values of N are $\leq \approx 1$, 5, and $30 \times 10^{22}/m^3$ at 500, 400, and 300°C, respectively. However, reported values vary by factors of at

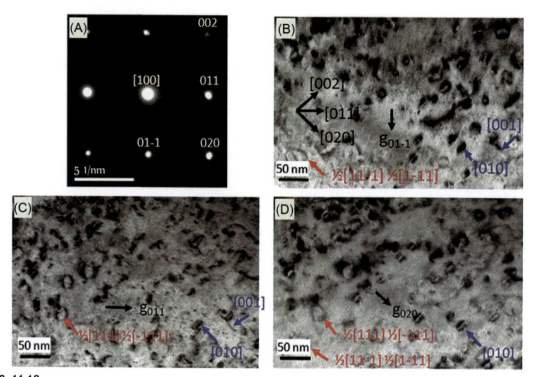

FIG. 11.12

TEM observation of different dislocation loops in RAF/M steel observed under [100] zone axis, according to Ref. [54].

least 3 from these prototypical estimates. The corresponding loop d depends more on dose and less on temperature, but typical averages are from 10 to 20 nm [54] although at higher dpa some may be much larger and smaller loops appear as solute clusters.

A crude, but perhaps practical, estimate of Nd can be extracted from the hardening trends [45]. Assuming that, in the absence of voids or bubbles, dislocations contribute $\approx 75\%$ of the hardening, independent of temperature, Fig. 11.13 shows the corresponding estimate of Nd as a function of $1/T$ between 300°C and 400°C. Here, we use a loop strength parameter of $\alpha = 0.3$ and an unirradiated dispersed barrier strength contribution of 200 MPa. Below 300°C Nd levels off or slightly decreases. There is also a TEM-based average Nd data point at 500°C [72]. Note that loops only partially offset any softening that is due to coarser scale recovery and microstructural instability effects. The error bars of ±75% reflect minimum expected uncertainties in this assessment of Nd.

It has been recently reported that the simultaneous injection of significant quantities of He, in combination with displacement damage, increases the loop N and d for irradiations at 500°C. This results in Nd that is a factor of 5–10 times higher than in the absence of He [72]. Once formed loops are rather stable, and do not readily anneal at temperatures below about 550°C. However, loops can undergo Ostwald-type coarsening by absorbing vacancies at smaller loops and emitting vacancies from larger loops, respectively [73,74]. Detrapped loops can coalesce or annihilate at sinks if and mobilized by glide and conservative climber mechanisms [75], again resulting in decreases in N. Discrete loops may also convert to network dislocation segments during annealing [54].

To close this section, we mention that, concomitantly with irradiation hardening, ferritic steels undergo both a reduction of tensile ductility and strain hardening, which are interrelated. The strain hardening reduction is generally attributed to mechanical erosion of the irradiation hardening features, that, in many metals and alloys,

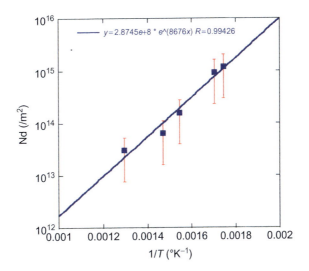

FIG. 11.13

Nd estimates versus $1/T$.

leads to flow localization and channeling [76,77]. However, the effect of even a modest loss of strain hardening on the reduction in uniform strain ductility due to a premature plastic instability, is greatly amplified by a corresponding increase of the flow stress [78], consistent with the Considère's criterion, $d\sigma/d\varepsilon = \sigma$, where σ and ε are the true stress and strain, respectively [79]. In the most severe, albeit frequent, case the uniform elongation nearly coincides with the 0.2% plastic strain at yield. However, even if the engineering stress/strain curve exhibits a very low uniform elongation and load reductions following almost immediate necking, the actual true stress/strain curve experiences softening only over a limited strain range (typically few percent) and is often followed by a regime of strain hardening; note modest softening or perfectly plastic behavior have also been observed [80]. The net effect is that the total true-stress irradiation hardening persists over the large increments of true-strain [37]. It is also notable that the increment of total engineering strain beyond the region of uniform elongation is not very sensitive to irradiation damage.

11.5.2 IRRADIATION EMBRITTLEMENT—FAST FRACTURE

One of the main challenges to engineering application of TMSs is related to their DBT as a function of temperature that is intrinsic to bcc metals and alloys. Below the DBT temperature (DBTT), the fracture mode typically changes from high toughness microvoid nucleation, growth, and coalescence to low toughness quasi-cleavage [81]. The DBT arises from the rapid increase in the bcc lattice strength below a strain-rate sensitive temperature, combined with the activation of low microscale toughness cleavage systems, like (100)⟨011⟩. Cleavage is activated by a sufficiently high internal stress, usually by breaking a brittle grain boundary trigger particle in the steel. This condition requires an internal critical microcleavage fracture stress acting over a sufficient volume of the steel microstructure to break a trigger particle. The critical microcleavage fracture stress is typically on the order of 2000 MPa [82]. Hence, in addition to low temperatures, triaxial stress concentration near notches and cracks are usually needed to induce cleavage. At the start of their life, the DBTTs of TMS components are low, typically <−50°C. However, neutron irradiation shifts the DBTT to higher temperatures, restricting the use of TMS to limited temperature and dpa windows in nuclear applications [83].

Historically, Charpy impact test has been used to characterize the fracture behavior of bcc alloys and their DBTT. Charpy tests measure the energy absorbed in breaking small shallow-notched bend bars under impact loading. A typical Charpy curve of absorbed energy versus temperature consists of a lower shelf, at lower temperature, where ferritic steels are brittle and the absorbed energy is very low, and an upper shelf at higher temperature, in the ductile regime, where even more energy is needed to break the specimen that is dissipated in plastic work. A gradual transition between the two shelves typically takes place over a temperature interval as large as 120°C or more, but this can also be very sharp, especially in the case of very small specimens. The Charpy DBTT, referred as to here as T_c, can be indexed by the absorbed energy-temperature curve in different ways. Standard $10 \times 10 \times 55 \, \text{mm}^3$ Charpy specimens are typically indexed at 41 J. For smaller subsized Charpy bars, the DBTT is often indexed at the absorbed energy is half-way between the upper and lower shelf energies. Indexing can also be based on the fraction of brittle versus ductile fracture on the broken surface. It is emphasized that T_c, whatever its definition, represents a specific test parameter, rather than a fundamental material property. Further, the DBTT depends on a number of extrinsic variables such as specimen size and geometry, as well as the loading rate. Once a definition and test procedure is adopted, the effect of irradiation on the indexed T_c can be quantified. In almost all cases, T_c increases ($\Delta T_c > 0$). In addition, irradiation *embrittlement* also results in a reduction of the Charpy upper shelf energy. These neutron irradiation effects on the Charpy curves are illustrated in Fig. 11.14 for four high-chromium TMSs after irradiation at 300–330°C to 80 dpa in the fast neutron BOR-60 test reactor [48]. Note that the large DBTT shifts, from 150°C to 240°C, depend on the steel. In this case, the upper shelf-energy decreases from about 50% in the reduced activation steels, to about 80% in the conventional 9Cr1Mo steel.

Irradiation-hardening results in embrittlement and ΔT_c because the internal stresses needed to reach the microcleavage stress occur at higher temperature than in the unirradiated condition. Detailed models relating ΔT_c to $\Delta \sigma_y$ have been developed and predict nonlinear relations between temperature shifts and hardening as well as effects of both the unirradiated T_c and upper shelf energy [45]. However, simple linear correlations between ΔT_c and the

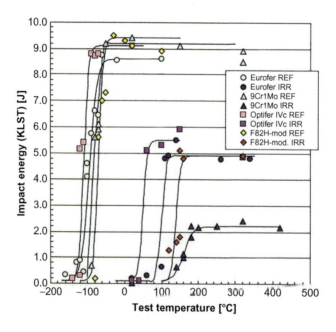

FIG. 11.14

Charpy impact energy-temperature curves for four high-chromium tempered martensitic steels before and after irradiation at 300–330°C to 80 dpa in the BOR-60 fast reactor [48].

ambient temperature $\Delta\sigma_y$ are often reported in literature, as $\Delta T_c = C_c \, \Delta\sigma_y$ [29,50,84,85]. A compilation and analysis of $\Delta\sigma_y$ and ΔT_c data for 8–10Cr TMS found that C_c ranges from about 0.25 to 0.55°C/MPa with an average value of about 0.4°C/MPa [45]. Notably, C_c increases with irradiation temperature above ≈ 400°C. Note that, there are cases of embrittlement where $\Delta\sigma_y < 0$ with $\Delta T_c > 0$. This behavior is a signature of nonhardening embrittlement, which is also caused by high concentrations of He (see below).

While widely used to characterize embrittlement, Charpy data cannot be used for quantitative integrity assessment of structures containing loaded cracks. In such cases, fracture mechanics methods and fracture toughness tests are needed to assess the stability of cracks under load. For very large specimens, or structures, and low toughness, where the crack tip plastic zone is very small with respect to all the other characteristic dimensions, the macroscopic loading is linear elastic, and the fracture toughness, K_{Ic} is related to the critical load for crack initiation P_c, and the crack length a by a nondimensional geometrical function Y that depends only on the specimen geometry [86] as

$$K_{Ic} = P_c Y \sqrt{a} \tag{11.8}$$

Under elastic fracture conditions, the near crack tip stress field is fully characterized by K_I. For the cases where the loading is not linear elastic, but the plastic zone at the crack tip remains much smaller than the specimen dimensions, for deep cracks, the J-integral loading parameter characterizes the crack tip field, and is related to a dimensionally equivalent nonlinear elastic-plastic stress intensity factor as [86]

$$K_{Jc} = \sqrt{\frac{JE}{(1-v^2)}} \tag{11.9}$$

Brittle cleavage fracture is triggered by particles in the process zone of a blunting crack tip. There is only a finite probability of a cleavage trigger-crack initiator within the crack tip stress field at an applied K_I. As a result, cleavage toughness is intrinsically scattered (see Fig. 11.15). Further, the longer the crack front, B, the higher the probability of activating a trigger particle. Since cleavage fracture toughness depends on B, it is not a geometry and dimensionally material independent property per se. The effects on measured toughness associated with B and other crack-specimen dimensions are often modeled using the concept of weakest link cleavage crack initiation, which

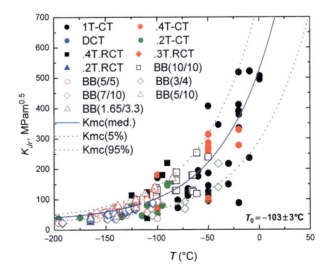

FIG. 11.15

F82H-mod steel CL and B-adjusted K_{Jc} data based on Odette's fracture local critical stressed volume model [94].

can be described by Weibull statistics [87]. Based on the simple assumption that equivalent stressed volumes yield the same fracture probability, the measured toughness can be adjusted for specimens having B_1 and B_2 respectively as [82]

$$K_{B_2} = K_{min} + (K_{B_1} - K_{min})\left(\frac{B_1}{B_2}\right)^{1/4} \tag{11.10}$$

The K_{min}, which is typically taken as ≈ 20 MPa $\sqrt{m}$, is the minimum required for cleavage crack macroscopic propagation in structural steels. Similar to Charpy tests, the fracture $K_{Jc}(T)$ curve is characterized by a lower brittle fracture shelf and an upper ductile tearing shelf, connected by a transition region where cleavage fracture is semibrittle. Wallin found that the shape of the $K_{Jc}(T)$ curve in the transition region is practically the same for all unirradiated and irradiated ferritic low-alloyed reactor pressure vessel (RPV) steels [88]. A large body of research suggests that, for engineering purposes, the median toughness, for a reference $B = 25.4$ mm thick specimen, can be described by the ASTM E1921 standard Master-Curve expressions [88]:

$$K_{Jc\,med}(T) = 30 + 70\exp(0.019(T - T_0)) \tag{11.11}$$

In principle, the only difference between different steel and alloy conditions is the so-called reference temperature T_0, at a median toughness of 100 MPa $\sqrt{m}$. That is, T_0 indexes the master shape of the $K_{Jc\,med}(T)$ curve on an absolute temperatures scale. Thus, the effect of irradiation on fracture toughness in the cleavage transition is reflected in changes in T_0 (ΔT_0). This observation and basic hypothesis led to the development of the so-called Master-Curve method for measuring K_{Jc} in the cleavage transition. The Master-Curve method was initially developed for RPV steels, as documented in the ASTM E1921, that establishes the requirements to determine the index T_0, as well as $K_{Jc}(T)$ curves at specified confidence (probability) intervals (e.g., 5% and 95%) that reflect the inherent statistical variability in measured cleavage toughness, for the same steel and test conditions.

The Master-Curve method is applicable to elastic-plastic fracture toughness when cleavage initiates beyond general yield, as indicated by a nonlinearity in the load-displacement curve. However, there are constraint loss (CL) limits for measuring K_{Jc} for a particular specimen size. For example, the CL limit for a 10-mm wide bend specimen, with a 5 mm precrack and $\sigma_y = 500$ MPa, is ≈ 130 MPa $\sqrt{m}$; and the limit is ≈ 80 MPa $\sqrt{m}$ for a corresponding 1/3 sized pre-cracked Charpy specimen. CL occurs when the plastic zone is no longer small compared with the specimen dimensions, so the crack tip fields are no longer self-similar, such that they can be described fully by the applied K_J, which only applies to deep crack and small-scale yielding. CL reduces the magnitude of the crack tip stress fields, hence, increases the measured toughness. Note that there is considerable evidence that CL effects occur at lower K_{Jc} values than the limit recommended in ASTM E1921 [89], but further discussion is beyond the scope of this chapter. ASTM E1921 also imposes a number of other standard restrictions, such as the number of valid specimens needed to establish T_0 and the crack length to specimen width ratio, a/W ($a/W \approx 0.5$) and testing in bending. However, as pointed out by Odette et al. [82], the Master-Curve method allows evaluating the median $K_{Jc}(T)$ curve based on a relatively small number of small specimens. This is enormously important, since otherwise, characterizing the effects of irradiation on fracture toughness would not be practically possible.

The Master-Curve approach, recognized in the ASTM-E1921 standard, is generally applicable to ferritic, bainitic, and TMSs. However, based on the fracture toughness dataset of several unirradiated and irradiated ferritic-martensitic steels, Lucon raised some doubt about the applicability of the Master-Curve method in this case, observing that intrinsic scatter of some toughness data in the transition region is greater than predicted [90]. Most of the available fracture toughness data for the TMSs were obtained with sub-sized specimens, that are expected to enhance the K_{Jc} scatter. The driver for using sub-sized specimen is mainly due to the limited volume of the current irradiation facilities.

In order to account for specimen geometry and size effects on toughness, local approach models for brittle fracture have been developed over the past 30 years [91]. These local approaches relate the micromechanisms of cleavage fracture, which is triggered when a critical crack tip stress field has a high probability of causing cleavage. Odette proposed a powerful method to quantify specimen size effects and to transfer the data from one specimen size to another [82]. This method takes into account both the statistical effect associated with B (statistically stressed volume SSV) and CL effects. Odette verified and adopted the procedure for the effect of B in ASTM-E1921 (see Eq. 11.10) through a basic scaling $(B_1/B_2)^{1/4}$, and to adjust T_0 to account for CL effects, that are not treated directly in ASTM-E1921, Odette's model computes the theoretical toughness K ratio of the large-scale yielding to small-scale yielding (K_{LSY}/K_{SSY}) levels that produces the same local crack tip cleavage conditions. It is based on a local fracture criterion that assumes the cleavage occurs when the normal stress σ_n is greater than a critical stress σ^* within a critical average in-plane area, A^*, characterizing the fracture process zone in front of the crack tip. The actual criteria is a critically stress volume (V^*) but this is accounted for in the SSV $B^{1/4}$ scaling. Calibrated values of σ^* and A^* are used to adjust the measured toughness from one specimen size to another by calculating the respective applied stress intensity factors, K_J, necessary to reach the local cleavage criterion.

As a first approximation, σ^* and A^* are calibrated by the unirradiated $K_{Jc}(T)$ curve assuming that they are temperature independent, in such a way that the Master-Curve can be reconstructed by σ^*/A^*. The σ^* represents the local fracture stress that is interpreted in terms of the Griffith-Orowan criterion [92], which in the case of penny-shaped cracks of radius r_0 is given by [93]

$$\sigma^* = \left[\frac{\pi E(\gamma_s + w_p)}{2(1-\nu^2)r_0}\right]^{\frac{1}{2}} \tag{11.12}$$

Here, E is the Young's modulus, ν is the Poisson's ratio, γ_s is the surface energy of the ferrite matrix, w_p ($>>\gamma_s$) is the plastic work necessary for trigger-particle microcrack propagation, and r_0 is the effective radius of the fractured brittle particle microcrack initiator. In principle, w_p and σ^* should be temperature dependent since they are mediated by dislocation mobility. The SSV and CL local fracture model was employed and adjusted to a 1T-specimen to assemble all the fracture toughness data gathered for the RAF/M F82H steel, with a large variety of geometries of mostly sub-sized specimens [94] (see Fig. 11.15).

Indeed, it has been shown that a moderate positive temperature dependence of σ^* has to be considered to model an invariant Master-Curve shapes associated with a large irradiation-induced ΔT_0 [95]. It is also emphasized that the calibrated σ^*/A^* values on the unirradiated material may not be representative of the material after irradiation, especially in case of nonhardening embrittlement (see below), if a change of fracture micromechanism is induced by irradiation.

Despite a somewhat larger scatter than predicted especially in the lower transition region, the data in Fig. 11.15 were found to be consistent with the ASTM-E1921 standard Master-Curve. Subsequently, the applicability of the Master-Curve method to the unirradiated TMS Eurofer97 was also assessed by Mueller et al. [96]. This study was based on the analysis of a fracture toughness database of Eurofer97 steel obtained with sub-sized 0.35T and 0.87T $C(T)$ specimens tested in the lower to middle transition region. The large number of data points of the analyzed database allowed recalculation of the Master-Curve coefficients. The data suggest an adjustment of the athermal part of Master-Curve equation (30 MPa $\sqrt{m}$ in the standard Master-Curve) is sufficient to represent the data with a slightly modified equation as [96]

$$K_{Jc\,med}(T) = A + (100 - A)\exp(0.019(T - T_0)) \tag{11.13}$$

An $A = 12$ MPa $\sqrt{m}$ provided the best fit to the $K_{Jc}(T)$ data for Eurofer97 down to the lower shelf as can be seen in Fig. 11.16. Without this adjustment, it would be difficult to define T_0 from data near the lower shelf.

It is important to recognize that the Master-Curve method was primarily developed for statically loaded deep cracks in thick-walled structures, which does not represent all the possible loading conditions and crack configurations. An important example is that of thin-walled fusion first wall and blanket structures, which would be expected to contain shallow cracks and to be susceptible to very fast loading rates in some cases as a consequence plasma disruptions. Fracture safety assessment in these cases can certainly be based on an approach similar to the Master-Curve but describing the necessary modifications is beyond the scope of this chapter.

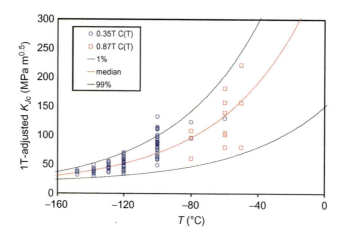

FIG. 11.16

Modified Master-Curve of Eurofer97 steel, $T_0 = 78°C$ [96].

Neutron exposures in the irradiation-hardening regime below ≈400°C induce upward shifts in both Charpy and fracture toughness-temperature curves (ΔT_c and ΔT_0) as well as a decrease of the upper shelf fracture tearing resistance. Fig. 11.17 illustrates how the Master-Curve method can be used to determine ΔT_0 with a small number of small specimens (in principle only six valid K_{Jc} points are sufficient). In this case, the K_{Jc} were measured using pre-cracked sub-sized $0.18T$ compact tension specimens of unirradiated and irradiated Eurofer97. The neutron-irradiated conditions, obtained in the experimental reactor at AEKI-KFKI in Budapest, were at a nominal dose of about 0.35 dpa at two different temperatures of 150°C and 350°C [97]. In spite of the low dose, the ΔT_0 were 98°C and 50°C at the lower and higher temperatures, respectively.

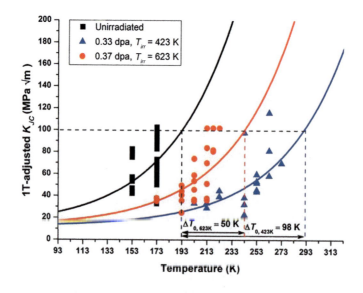

FIG. 11.17

Fracture toughness results and corresponding Master-Curves in the unirradiated and irradiated conditions [97].

11.5 RADIATION HARDENING AND SOFTENING

Similar to Charpy curves, it has been shown that $\Delta T_0 \approx C_y \Delta \sigma_y$. However, the average flow stress (σ_{fl}) between 0% and 10% plastic strain controls fracture in the process zone of a blunting crack [98]. Since highly irradiated TMS experience a significant loss of strain hardening, hence, changes in the flow stress, the $\Delta \sigma_{fl}$, are less than $\Delta \sigma_y$. Thus, a more rigorous hardening shift relation is given by.

$$\Delta T_0 \approx C_f \Delta \sigma_{fl} \qquad (11.14)$$

Both experiment and a fundamental model of the Master-Curve suggest that this relation is slightly nonlinear as illustrated by the dashed line in Fig. 11.18, with an average value of $\approx 0.8°C/MPa$ shown by the simple linear fit. The corresponding C_y is typically in the range of 0.5–0.6°C/MPa. Charpy C_c are generally even smaller with typical $C_c \approx 0.4°C/MPa$. Hence, Charpy ΔT_0 are a nonconservative measure of embrittlement, and should not be used as a surrogate for fracture toughness ΔT_0 [84]. As noted previously, the C's increase if there are nonhardening mechanisms. Nonhardening embrittlement occurs at higher temperatures and, as discussed below, especially in the presence of large quantities of He.

11.5.3 FATIGUE

Fatigue limits are almost always a challenge to structural materials for fossil and nuclear energy applications. TMS components, especially as foreseen for fusion devices, will experience cyclic stresses-strains such as mechanical and electromagnetic loads, as well as temperature changes induced by the plasma burn-on and plasma-off periods. The actual loading and unloading sequence will also include periods of static stress so that the interaction between creep and fatigue will be important [99]. Many strain-controlled fatigue studies have been conducted in the low cycle fatigue (LCF) regime, where the number of cycles to failure, N_f, is typically $<10^4$. For fusion applications, emphasis on LCF is motivated by the need to prevent failure generated by repeated thermal transients and corresponding thermal expansion stresses and strains that result in accumulating cyclic strain, and in the limit component ratcheting [100].

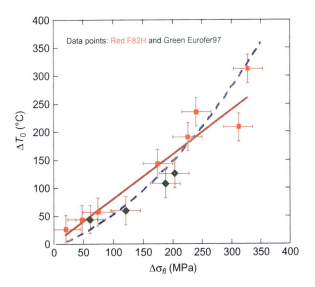

FIG. 11.18

Shift of the reference temperature T_0 versus average flow stress increase $\Delta \sigma_{fl}$ (Odette et al., to be published).

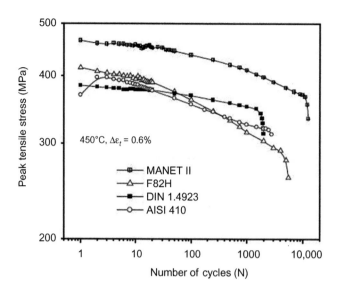

FIG. 11.19

Peak stress versus cycle number of tempered martensitic steels [101].

The attractive high-temperature static strength properties of the TMS that persist up to about 400–500°C, are to some extent attained at the expense of their corresponding fatigue (and fracture) properties. This is due to the stability of TMS microstructures again composed of: (1) the high dislocation densities; (2) the micrometer size of elongated grains laths originating from the normalized heat treatment; and (3) the dispersed distribution of small carbides in the matrix. These microstructures are prone to instabilities and softening during cyclic loading. An example of this phenomenon is shown in Fig. 11.19 for isothermal uniaxial fatigue tests on a number of TMS alloys, where the peak stress (maximum stress during a cycle, σ_p) is plotted against the cycle number at 450°C and a total strain amplitude of the strain range, $\Delta\varepsilon_t = 0.6\%$. Notably F82H, which is the only RAF/M steel, exhibits more pronounced softening compared with the three other variants. Based on the data shown in Fig. 11.19, Armas et al. suggested that the softening depends on the martensitic start temperature M_s [101]. The effects of temperature and $\Delta\varepsilon_t$ on the peak cyclic stress are shown in Fig. 11.20 for Eurofer97.

All the data show an approximate linear dependence on a log-log N scale after an initial transient period prior to softening. Thus, cyclic softening can be described by an expression of the type: $\sigma_p = BN^{-s}$, where s is the softening exponent that increases with temperature but is relatively independent of $\Delta\varepsilon_t$ at higher N. At 450°C, $s \approx 0.065$.

The current design of the test blanket modules for ITER has motivated research on the transferability of fatigue predictions from isothermal uniaxial to multiaxial conditions. For example, Aktaa et al. carried out two strain-controlled multiaxial experiments, one with a fixed principal strain direction and the other with cyclic axial and torsional components under nonproportional loading. The results showed that the multiaxial fatigue life was significantly shorter than for uniaxial loading [102]. Similar conclusions were drawn from corresponding thermomechanical fatigue experiments. This is the only limited information on TMS with respect to the many factors that influence fatigue life, such as: mean stress effects along with many other details of complex load cycles; non-isothermal thermomechanical fatigue; da/dN type fatigue crack growth; and creep-fatigue with hold time effects. Most of these studies have been conducted on 9Cr-1Mo type steels, like T91 in support of an ASME code case, and for fossil energy applications, rather than for RAF/M alloys with W entirely replacing Mo [103,104]. Likewise, there is a body of data for the higher strength P92 steel (NF616), that contains Mo and W, as well as solutes that form fine-scale MX type precipitates [105,106]. Notably, both T91 and P92 have established a considerable high-temperature experience base in worldwide fossil energy applications.

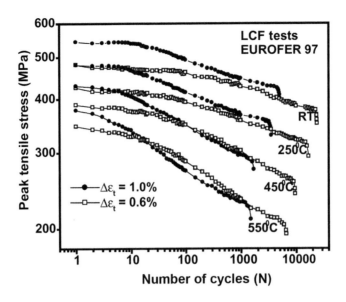

FIG. 11.20

LCF tests on Eurofer97 for various T and $\Delta\varepsilon_t$ [101].

The effects of neutron irradiation on LCF were investigated by several authors and for different conditions [107–109]. Nonetheless, the fatigue data on irradiated materials remain relatively scarce. As an example, we highlight the recent result of Materna-Morris et al. on Eurofer97 after irradiation in the High Flux Reactor in Petten, Netherlands to 16.3 dpa at 250°C, 350°C, and 450°C [110]. The fatigue tests were at the irradiation temperature. Cyclic softening was observed for all the steels and test conditions, for both unirradiated, irradiated and for varying strain amplitude and temperature. Notably, the fatigue life was not significantly affected, by irradiation at 350°C and 450°C. In contrast, as seen in Fig. 11.21, a significant increase in fatigue life was observed at 250°C, which was almost an order of magnitude larger at high N at a strain amplitude of 0.5%.

Helium also affects LCF properties as shown in Ref. [111]. Here, tests on helium implanted specimens of TMS F82H at ambient temperature in air were carried out for total strain ranges from 0.6% to 1.5%. A reduction of N_f was found at higher strain ranges from 1.0% to 1.5%, while no measurable effect was detected in the corresponding $\Delta\varepsilon_t$ range from 0.6% to 0.8%. Stress-controlled fatigue data are extremely limited. Most notably, Ullmaier and Schmitz performed a series of experiments on DIN 1.4914 Manet steel over a range of stress amplitudes, temperatures, mean stress, and hold time [112].

11.5.4 THERMAL CREEP

Creep and creep rupture behavior at high temperature is a key issue for TMS. Notably, RAF/M steels have a somewhat lower creep strength than a commercial mod. 9Cr-1Mo steels, with small additions of V and Nb that form MX precipitates, as can be seen in Fig. 11.22 comparing rupture times as a function of stress at 550°C [113]. Nonetheless, the creep strength of the RAF/M steels is comparable to that of other 9Cr-1Mo steels that are extensively used worldwide in fossil power plants. Some advanced TMS will allow increases in the operating temperature up to about 650°C [114], and ODS alloys extend this even further. For example, a number of ODS TMS variants were produced and creep tested in the range 600–700°C up to rupture times of about 10^4 h. The Larson-Miller plot presented in Fig. 11.23 shows the creep rupture stress for the ODS-Eurofer variants have creep strengths that are over twice that for the RAF/M F82H and Eurofer97 [115]. Boutard et al. suggested that the ODS ferritic steel 12YWT and commercial PM2000 can withstand operating temperature upto 750°C [116].

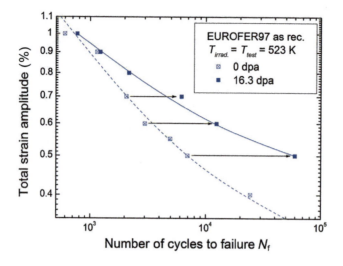

FIG. 11.21

Eurofer97 fatigue life evolution after 16.3 dpa at $T_{irr}=250°C$ [110].

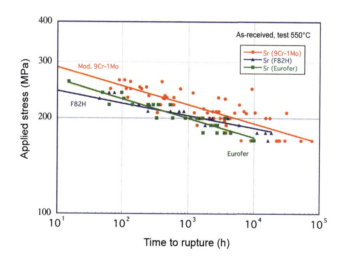

FIG. 11.22

Creep rupture properties at 550°C of F82H and Eurofer 97 RAF/M steels, and mod. 9Cr-1Mo steels [113].

11.6 HELIUM EFFECTS

Helium accumulation in ferritic steels leads to significant degradation of some mechanical properties. Since the helium is practically insoluble in metals, it precipitates as bubbles, even at very low concentrations. Bubbles preferentially form at various trapping sites such as grain boundaries, precipitate-matrix interfaces, dislocations, etc. [117,118]. Bulk helium is produced in service primarily by neutron (n,α) transmutation reactions that are

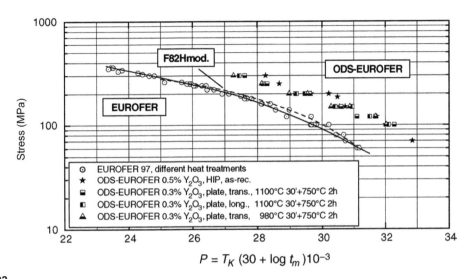

FIG. 11.23

A Larson-Miller plot of the creep rupture time (t_r) for F82H and Eurofer97 RAF/M steels, with ODS-Eurofer [115].

accompanied by displacement (dpa) damage. The He production rate in ferritic alloys ranges from ≈0.2 appm He/dpa in fission reactors to ≈10 appm He/dpa up to 100 appm He/dpa in fusion first walls and the spallation targets, respectively [119]. In addition to the He/dpa ratio, the transport, fate, and consequences of helium depend on the combination of a number of variables such as: dose rate (dpa/s), total dose (dpa), irradiation temperature, and the alloy start of life microstructure and microchemistry. These factors mediate the number density and size distribution of the bubbles, as well their location in the alloy microstructure. The fate of helium, in turn, determines its consequences to both microstructure and mechanical properties. For example, beyond a critical size, bubbles convert to unstably growing voids, hence, He affect incubation dpa for swelling [70]. Likewise, grain boundary bubbles are formation sites for creep cavities that control high-temperature creep rupture time and ductility. We first discuss the effects of helium on hardening ($\Delta\sigma_y$) and fast fracture (ΔT_0). To begin, however, we briefly describe the different experimental techniques that have been used to introduce helium and investigate its effects in steels. More details are provided in Ref. [119].

- Dual He plus heavy (self) and He dual ion irradiations (DII) are a useful and convenient technique to simultaneously generate dpa and accumulate He in a rather well-controlled manner for a range of He/dpa ratios and temperature up to high dpa. However, DII irradiations generally cannot directly simulate neutron damage due to: very high dpa rates, proximity of the specimen surface, effects of the injected heavy ion interstitial atom, and differences in the atomic recoil spectra.
- In mixed spectrum, fast-thermal fission reactor irradiations, Ni and B produce helium through the two-step reaction $^{58}Ni(n,\gamma)^{59}Ni(n,\alpha)^{56}Fe$ sequential reaction or the direct $^{10}B(n,\alpha)^{7}Li$ reaction, respectively [120]. However, issues associated with the utilization of Ni and B, include: a thermal neutron fluence is needed to form ^{59}Ni to produce the desired He/dpa; B burns out rapidly at high thermal neutron flux and is not uniformly distributed; and, typically neither Ni nor B is an alloying element in TMS, and both can themselves affect the alloy microstructure and properties both prior to and under irradiation.
- He can be implanted simultaneously with dpa production from fast neutrons in mixed spectrum fission reactors by exposing a specimen to a thin layer containing Ni, resulting in $^{59}Ni(n,\alpha)^{56}Fe$ reactions, using what is known as the in situ He injection (ISHI) technique [120]. He is uniformly deposited to a depth of several

microns. This limited range does not generally allow mechanical properties characterization, but can be used to extract useful information on the evolution of the microstructure and nanohardness.
- Isotopic tailoring of ferritic alloys with ^{54}Fe is an appealing technique to generate He effects in Fe-based materials by (n,α) transmutations [121]. However, ^{54}Fe is extremely costly and does not yield the levels characteristic of a fusion environment. Indeed, the He/dpa production ratio is about 2.3 for a fission neutron spectrum, compared with a value of around 10 for a fusion neutron spectrum.
- Irradiation with a mix of high-energy spallation protons and neutrons is a very attractive option to investigate the synergistic effects of dpa and He on bulk microstructures and properties [122]. Spallation neutron facilities can produce about 10–20 dpa/y with a range of 30–160 appm He/dpa. Note that the He/dpa ratio is quite large compared with those the fusion first wall (10 appm He/dpa). Owing to the high-energy spallation neutron spectrum tail spallation proton irradiations also produce some very high-energy primary recoil atoms, along with a significant amount of H, which might contribute to embrittlement, as well as a variety of solid transmutation products. However, retained H concentrations are generally low owing to its high mobility. Irradiations with spallation neutrons below about 25 dpa keep the solute transmutation concentration below 0.025%–0.25%.
- Helium implantation only is also a useful to study basic He mechanisms and, in some cases, the corresponding irradiation effects under He dominated conditions.

For example, the effects of He on the tensile properties of two 9%Cr TMSs (EM10 and T91) at concentrations up to 5000 appm implanted at temperatures from 150°C to 550°C were studied by Jung et al. [123]. Tensile tests at ambient and implantation temperatures showed very large amounts of hardening at high He levels. The displacement damage from the He implantations was relatively low, at ≈0.8 dpa/5000 appm He. Thus, most of the hardening was attributed to He bubbles, rather than displacement damage defects. Notably, helium bubble hardening was found after implantations at 550°C, in contrast to the absence of hardening (or even softening) observed in corresponding low He fission neutron irradiations. The ductility trends in the He-implanted steels were also in contrast to those for neutron irradiations. Neutron irradiations result in a nearly complete loss of uniform stain capacity at high hardening levels. However, considerable ductility (fracture strain and reduction in area) remained below 2500 appm He, even for the most brittle condition, for the 250°C implantations at both test temperatures. Fractography indicated that intergranular fracture occurred in the latter case [123]. In another study by Ando et al. [124], irradiations of F82H at 360°C by a single-ion beam (10.5 MeV Fe^{3+}) and dual ions (10.5 MeV Fe^{3+} and 1.05 MeV He^{2+}) beam, showed that He contributes to the hardening (measured by microhardness) at concentrations above ≈600 appm.

A large body of data has been generated for a wide range of alloys and irradiation conditions at the PSI (Switzerland) SINQ spallation proton-neutron source, for example, Ref. [125]. The most salient observations regarding the evolution of the ambient temperature tensile properties after spallation irradiations are illustrated in Fig. 11.24, for T91 and EM10 steels irradiated up to about 20 dpa and ≈1800 appm He at temperatures between 80°C and 350°C. The $\Delta\sigma_y$ versus $\sqrt{dpa}$ are compared with fitted curves derived from neutron irradiation data at low He concentrations [45]. The SINQ-irradiation hardening shows no indication of saturation, in contrast to neutron irradiation trends, where in the latter case $\Delta\sigma_y$ saturation occurs by about 10 dpa. Again, the extra increment of $\Delta\sigma_y$ is attributed to contributions. The fracture surfaces of the specimens irradiated at doses >15 dpa were found to be predominantly intergranular [125]. It was also observed that the tensile curves of the SINQ-irradiated specimens at about 525–560°C still exhibits a large amount of hardening, while corresponding neutron irradiations produce moderate softening. This constitutes more evidence of bubble hardening at high He content. Bubble hardening has also been demonstrated by annealing experiments that remove loop and precipitate contributions [119]. Overall, the data indicate that the contribution of He to hardening becomes increasingly significant for concentrations larger than ~500 600 appm.

The excess and extended temperature range of hardening due to He contributes directly to irradiation embrittlement. However, at high concentrations, He accumulation also weakens the prior austenite grain boundaries, leading to extremely brittle, intergranular fracture. The latter effect is reflected in a higher to much higher $C - \Delta T_0/\Delta\sigma_y$, that increases with He beyond a threshold, again currently estimated to be ≈500–600 appm [45,119]. The actual

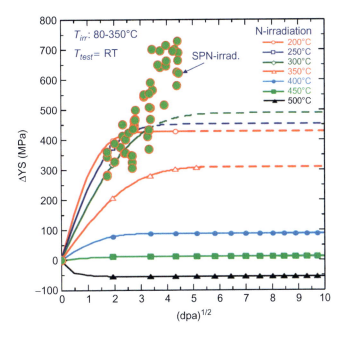

FIG. 11.24

Comparison of irradiation hardening induced by neutron and spallation proton-neutron (SPN) irradiation [125].

From Dai, Y, Henry, J, Tong, Z, Averty, X, Malaplate, J & Long, B 2011, 'Neutron/proton irradiation and He effects on the microstructure and mechanical properties of ferritic/martensitic steels T91 and EM10', Journal of Nuclear Materials, vol. 415, no. 3, pp. 306–310. Originally adapted from T. Yamamoto, G.R. Odette, H. Kishimoto, J.-W. Rensman, P. Miao, J. Nucl. Mater., 356 (2006), p. 27.

value of C depends on the test type. Estimated Master-Curve toughness shifts, ΔT_0, reach enormous levels of about 700°C as shown in Fig. 11.25.

It is well established that He can also adversely affect high-temperature creep rupture properties of many structural steels. Most previous research focused on austenitic stainless and other nickel bearing alloys [126–129]. High-temperature helium embrittlement (HTHE) causes a dramatic decrease of creep rupture time in these alloys, which generally have rather simple microstructures [128]. HTHE is due to the formation of He bubbles on the grain boundaries. When He bubbles reach a critical radius, determined by the normal stress on a grain boundary, they become unstable and grow continuously by the accumulation of a flux of vacancies [119]. The cavities experience stress-driven rapid grow until impingement, or rupture of the boundary ligaments occurs, leading to intergranular facture. While both inter and transgranular cavitation fracture can take place in the absence of He, the rupture times are greatly reduced when creep cavities nucleate on the grain boundary bubbles that have grown to a critical size. See further discussion of void swelling below.

In contrast to austenitic steels, however, TMS appear to have a very high resistance to HTHE. For example, implantation experiments on F82H by Yamamoto et al. showed that creep rupture remains transgranular and ductile at 550°C up to 1000 appm He [130,131]. Note that, larger cavities (bubbles or possibly some voids) have been observed in He-implanted steels in association with both Y_2O_3 oxides [132] and high-temperature heat treatments [117]. Explanations of the different HTHE behavior of austenitic versus TMS include that the high sink density and numerous trapping sites, in the latter case, distribute He in fine bubbles so as to protect the grain boundaries [119]. Further, the lower overall TMS creep strength, and corresponding reduced stresses on grain boundaries, may contribute to their higher HTHE resistance [119]. However, it must be emphasized that there are no lower-flux and longer-time neutron data on TMS that address the question of their resistance to HTHE under service-relevant irradiation conditions.

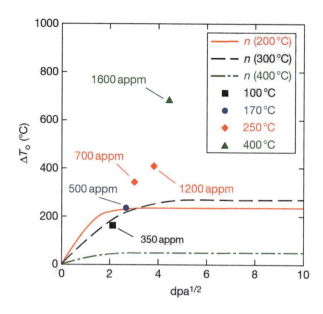

FIG. 11.25

Master-Curve reference temperature shift illustrating He embrittlement at high concentrations [119].

11.7 VOID SWELLING AND IRRADIATION CREEP
11.7.1 VOID SWELLING

As noted above, the remarkable resistance to void swelling of the ferritic steels makes them very attractive candidates for nuclear applications. The database on void swelling of this class of materials is even more limited than that for austenitic alloys. Despite, swelling in ferritic alloys has been evaluated after both neutron [133–144] and charged particle (heavy ions and HVEM-electrons) irradiations [145–150] for the last 40 years. Here, a brief summary of this literature is given, with special reference to review papers [151–153] that contain many more details. A typical displacement (dpa) dose dependence of swelling of ferritic alloys for fast neutron or heavy ion irradiations is shown in Fig. 11.26.

Three dose regimes can be distinguished: an extended incubation dpa_i, a transient regime of increasing swelling rate per dpa, and swelling at a constant rate. This behavior is qualitatively similar to that for austenitic steels, but ferritic alloys have much longer incubation and transient dpa regimes, as well as much lower steady-state swelling rates (about one-fifth of that in austenitic steels). As in the case of other classes of materials, void swelling of ferritic alloys depends on many variables. For instance, loops are strongly biased sinks for SIA [120], thus they are important to void swelling. Void formation and swelling occur in the temperature range from $\approx 400°C$ to $600°C$ in a fast neutron irradiations, mainly in the range from $\approx 400°C$ to $475°C$, often peaking at $\approx 425°C$ [133]. Peak swelling generally occurs at a temperature substantially higher under heavy ion irradiations, probably mainly due to the higher displacement damage rate. Sencer and Garner have assembled data suggesting that dose rate mainly influences incubation and transient periods, while the steady-state regime remains nearly the same at $\approx 0.2\%/dpa$ [154].

Fast reactor irradiations of a series of simple binary Fe-Cr alloys show that Cr plays a significant role in swelling behavior. Swelling increases with increasing Cr from 3 to 12 wt%, then decreases again at 15 wt%. Studies on Fe-Cr alloys demonstrated that Cr in solid solution is not probably responsible for the high swelling resistance of 9%–12% Cr TMS. Other elements, such as C, Mo, and others may play a role as well. For example, irradiations with C+ ions

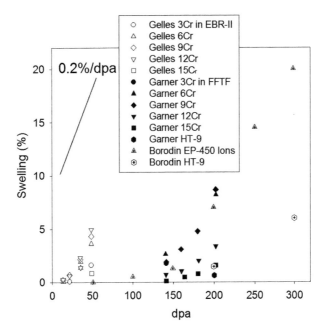

FIG. 11.26

Comparison of swelling of Fe-Cr alloys and ferritic-martensitic steels in EBR-II, data assembled from Refs. [150,152].

[155] and neutrons [154] both demonstrated that the addition of carbon to model Fe-12Cr alloys dramatically enhances swelling at higher temperature. Composition also affects the microstructure (ferritic vs. bainitic vs. tempered martensitic) and both play a role in complex multiconstituent, multiphase commercial TMS that have high swelling resistance, primarily associated with large incubation and transient regime doses of 100 dpa or more [136,151]. Thus, in large part, the effects of Cr, C, and other elements are associated with the formation of tempered martensite, with high sink densities, in contrast to simple ferritic microstructures.

The much shorter incubation dose for fast reactor irradiations in EBR-II versus FFTF, shown in Fig. 11.26, may suggest an important role of He mediated by the respective neutron spectra, that are ≈ 0.17 versus 0.02–0.08 He/dpa, respectively [156]. He effects on swelling in TMS are discussed in more detail in Chapters 3, 5, and 12 and more extensively in the literature [157,158]. Insight on He effects on swelling can be found in a series of papers by Yamamoto and co-workers, based on the ISHI and DII methods [120,159]. Both methods show that high He reduces the incubation dpa dose for swelling in a way that inversely scales with the He/dpa ratio.

Fig. 11.27 shows the effect of the He/dpa ratio in a DII on void swelling in RAF/M F82H steel. Odette and others, long ago established a framework for He effects in terms of the so-called critical bubble concept as discussed in recent reviews [33,119]. A large body of data is consistent with a heterogeneous void nucleation path involving the stable growth of bubbles by He additions up to near or at a critical size at which point they covert to unstably growing voids.

Irradiation creep (IC) is also enhanced by swelling rates that increase under stress, and growth of grain boundary cavities nucleated on bubble involve the same general type of mechanisms as void swelling [72].

The swelling resistance of ferritic alloys is not completely understood. A model of Sniegowski and Wolfer [160] suggests that the interstitial dislocation bias is intrinsically lower in bcc versus fcc iron, giving a theoretical steady-state swelling rates of 0.23%/dpa versus 1.4%/dpa, respectively, reasonably consistent with observation. However, and inherent bcc versus fcc bias difference is not consistent with data on various vanadium-binary

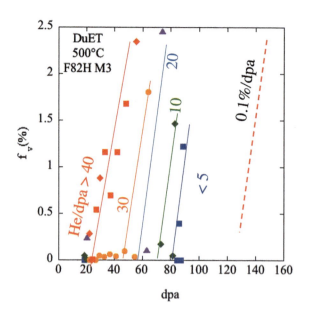

FIG. 11.27

Swelling versus dose for different He/dpa ratios in dual ion irradiations.

bcc alloys that were found to swell at rates that in some cases were even higher than fcc steels [161,162]. Rate theory rationalizes why He plays a prominent role in void swelling incubation dpa$_i$, based on rate the critical bubble model. Odette [153] also pointed out that in addition to low bias, intrinsically higher self-diffusion rates in bcc ferritic compared with fcc austenitic alloys increases dpa$_i$. Odette further noted that the He/dpa ratio is lower in TMS since, unlike austenitic alloys, they do not contain Ni that has a high (n,α) cross section compared with Fe.

11.7.2 IRRADIATION CREEP

Observation of IC, or time-dependent plastic deformation, in irradiation environments at stresses below σ_y, preceded the discovery of void swelling. Early IC studies were mainly focused on Zr alloys and austenitic steels in fission reactors [163–165]. Later, reactor even more limited IC data was obtained for TMS, mainly based on the pressurized thin-walled tube method, involving measuring diametric strains following (mostly) fast reactor irradiation increments [166]. Light ions irradiations, using accelerators with energies in the tens to >100 MeV range, have also been used to study IC, mostly in the lower dpa regime [167–171]. IC, which is significant and approximate athermal, over a wide range of temperatures, typically can be described by the simple relationship between stress (σ) and inelastic strain (ε_{ic}) as

$$\varepsilon_{ic} = B_0 \sigma \, \mathrm{dpa} \tag{11.15}$$

Here, B_0 is the so-called IC compliance. Typical B_0 ranges from ≈4 to $10 \times 10^{-6}\,\mathrm{dpa}^{-1}\,\mathrm{MPa}^{-1}$ for light ion irradiated ODS steels with different Cr-content as reported in Ref. [172] and shown in Fig. 11.28 for PM2000 (with a $B_0 \approx 4.5 \times 10^{-6}\,\mathrm{dpa}^{-1}\,\mathrm{MPa}^{-1}$). Corresponding values of B_0 typically range from ≈2 to $8 \times 10^{-7}\,\mathrm{dpa}^{-1}\,\mathrm{MPa}^{-1}$ in fast reactors [138,173] (the average B_0 is ≈$5.5 \times 10^{-7}\,\mathrm{dpa}^{-1}\,\mathrm{MPa}^{-1}$) [138,140,152]. Possible reasons for the differences in IC compliance under light ion and neutron irradiation are discussed in Ref. [174]. Notably, microstructure and compositional variations, as well as crystal structure have only modest influence on IC compliance. For example, Garner et al. concluded that B_0 for TMS is about half that for austenitic alloys [152].

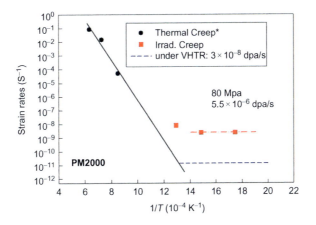

FIG. 11.28

Temperature dependence of creep rates of PM2000 without irradiation (solid line), under He-implantation (dash-dotted), and expected for VHTR conditions (dashed line).

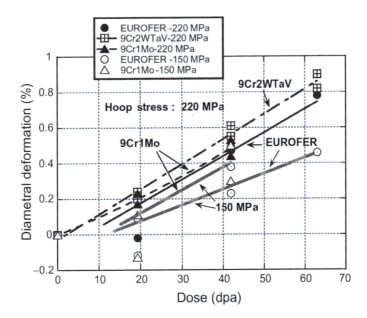

FIG. 11.29

Diametral strain of 9Cr1Mo(EM10), 9Cr2WTaV, and 9Cr1WTaV (EUROFER) steels as a function of dose for hoop stresses of 150 and 220 MPa, irradiated in BOR60 at 325°C [176].

Some low dose studies of IC suggest that an initial transient regime occurs prior to approximate linear increase in strain with dpa [172]. These transients may be due to microstructural changes that cause hardening [175]. However, as shown in Fig. 11.29, in the case of a BOR60 neutron irradiation at 325°C, either no transient is observed (9Cr-2W at 325°C [176]), or there is a threshold dpa prior to approximate constant ε_{ic} increase with dpa (EUROFER at 325°C [176], HT-9 and 9Cr-1Mo at 400°C [138]).

Linear stress dependence of the creep rate up to ≈250 MPa is generally well established for various structural alloys for most reactor [166,176] and light ion irradiations [172]. Note however that some neutron irradiations give stress exponents above one [138], in some cases in the absence of swelling and following a transient strain. Below the thermal creep regime, with stress exponents >>1, IC the weak temperature dependence presumably governed by defect recombination. In general, IC rates decrease with increasing dpa rate due to the corresponding increases temperature-dependent recombination, although systematic data are not available on this effect in TMS.

The coupling of IC compliance and stress-dependent swelling rate (S') per dpa has been widely observed in austenitic stainless steels as discussed comprehensively in a review by Garner [177]. Primarily due to the low swelling, only few systematic investigations on creep-swelling interactions have been reported for TMS. Toloscko observed a nonlinear stress exponent in HT9, which He attributed to swelling. This adds a DS' term to B_0, as.

$$\varepsilon_{ic} \approx (B_0 + DS')\sigma \, dpa \tag{11.16}$$

The D is estimated to be $\approx 7\text{-}10 \times 10^{-3}$ MPa^{-1} [138]. However, nonlinear stress dependence of swelling was not observed in the TMS T91 alloy.

For perspective, ignoring swelling, 200 MPa and 200 dpa result in $\varepsilon_{ic} \approx 0.02$, for $B_0 = 5 \times 10^{-7}$ MPa^{-1} dpa^{-1}, which may not be negligible for some application. If there is 2% swelling at 200 dpa, the IC strain could increase to $\varepsilon_{ic} \approx 0.05$. Finally, two important points are that IC does generally not contribute to damage that results in creep rupture [178] and, the stress relaxations accompanying IC may be beneficial in many cases.

11.8 FUTURE PROSPECTS FOR IMPROVED PERFORMANCE

In general, TMS are technologically very mature, primarily owing to their widespread use in fossil energy applications. For nonnuclear applications, the main TMS limitations are allowable stresses and temperatures, as well as corrosion-oxidation and compatibility issues. Significant increases in TMS upper use temperatures have been achieved by combinations of microalloying compositional modifications and microstructural control by thermomechanical treatment (TMT) processing paths that yield fine and stable dispersions of MC and MN-type phases. The RAF/M steels for fusion applications are not nearly as technologically mature, and their qualification for service is limited. As noted previously, the TMS temperature strength and corrosion limits are $\approx 550°C$, and this may be lower for service in neutron environments, where data are very limited for higher temperatures, longer times ($>>10^4$ h that are typical maximum times in creep tests) and for properties like for various manifestation of creep fatigue. Dimensional instabilities such as IC and void swelling (at high helium levels) are issues near and above 400°C; and synergistic irradiation-thermal instabilities are possible, perhaps above 450–500°C. Note that in general, microstructural instabilities in structural alloys are enhanced under stress, but this has not been explored for RAF/M variants of TMS. At <350–400°C degradation of toughness and loss of ductility are limiting, and embrittlement is a showstopper at high He levels over a much larger range of service temperatures.

Two paths have been identified to expand the irradiation service window of ferritic alloys.

- Nano-ODS ferritic (>12 Cr w/o C) and martensitic alloys (≤12 Cr w C), called ODS steels (nanostructured ferritic alloys) that are discussed in Chapter 12.
- Melt practice TMS dispersion strengthened with fine-scale MX precipitates where the M are V, Ta, and Nb, while the X are C and N. These developmental alloys have been widely discussed in the literature, such as a recent review by Zinkle et al. [179]. The earlier TMS path involved TMT deformation processing to form a high concentration dislocation nucleation sites for fine-scale MX precipitation [180] in wrought products. More recently, so-called cast nanostructured alloys, are being explored [181].

The ODS alloys have outstanding mechanical properties, are thermally and radiation stable and are highly irradiation tolerant, including easily managing very high levels of He [33]. The major ODS challenges are cost and production capacity. The melt-practice MX-TMS are still in a developmental stage, and their irradiation performance limits remain to be explored. The MX steels combine the positive characteristics of the TMS and ODS steels. MX alloys are designed and processed to contain a high number density of fine nitride and carbide precipitates, on size scales of order of 10 nm. While this is several times coarser than the nanooxides in advanced ODS steels, conventional processing greatly reduces manufacturing and fabrication costs. To meet reduced activation requirements, only limited variations in the TMS compositions are possible. For example, the nominal chemical composition of one of the new generation of MX RAF/M steels is targeted in the range of 8.3–9.0 wt% Cr, 1.0–1.5W, <1.0 Mn, 0.1–0.3V, 0.05–0.15Ta, <0.2Ti, <0.2Si, 0.08–0.15C, <0.06N, balance Fe [182]. Computational thermodynamics models have been used to tailor the compositions and TMT and to predict the fractions and stability of the various fine-scale carbides, nitrides, and carbonitride phases [183,184]. Again, the underlying idea is to produce a large number density of nanoscale precipitates, for example, by creating a large number of nucleation sites by hot-working to generate a high network dislocation density [114]. Probably, the biggest challenge the advanced MX steels face is the stability of their performance sustaining nano/microstructures over long periods of irradiation service at high temperature and under stress.

In closing, we mention the particular grand challenge posed by fusion applications (see Chapter 3). In addition to potentially severe irradiation effects arising from both dpa and He, the effects of potentially large concentrations of H, due to both transmutations and high fugacity plasma sources, must be considered. Potential synergistic effects of He and H on the dimensional instabilities, swelling, irradiation creep, thermal creep, fatigue, creep-fatigue, and so on are not known. It will be possible to gain experimental information about these phenomena as soon as a dedicated fusion neutron irradiation facility [185] will be available. In the meantime, sustained modeling efforts have to be done to better understand the role of helium in materials and, as a consequence to pave the road of future material development.

Fabricating complex and intricate structures, which require high and sustained tolerances, such as the breeding blanket of future fusion reactors is an enormous challenge. Joints and welds may be the weakest links and could limit, for example, the combination of product forms that are consistent with a joining practice. For example, wrought product forms and corresponding melt welding requiring postweld heat treatments may not be compatible. Conventional weld technology is basically applicable to TMS steels. However, effects of irradiation on the wide range fusion, fine grain, heat-affected and over-tempered zones microstructures present a great challenge, and emphasize the importance of developing solid-state joining method like friction stirred welding and diffusion bonding. The possibility of using highly integrated cast structures and/or near net shape processing may offer some advantages.

Finally, in the case of fusion event, if the enormous problems of irradiation and other sources of in service degradation are resolved along with the multitude of fabrication issues, the thermomechanical high-prolonged heat flux loading of plasma facing and blanket structures present an totally unique and unprecedented engineering challenge (see Chapter 3).

REFERENCES

[1] C. Cawthorne, E.J. Fulton, Voids in irradiated stainless steels, Nature 216 (1967) 575–576.
[2] R.L. Klueh, D.R. Harries, High-Chromium Ferritic and Martensitic Steels for Nuclear Applications, ASTM, West Conshohocken PA, 2001.
[3] D.L. Smith, R.F. Mattas, M.C. Billone, in: R.W. Cahn, P. Haasen, E.J. Kramer (Eds.), Materials science and technology: a comprehensive treatment, vol. 10B, Wiley-VCH, Weinheim, 1994, p. 263.
[4] T. Noda, F. Abe, H. Araki, M. Okada, Development of low activation ferritic steels, J. Nucl. Mater. 141–143 (1986) 1102–1106.
[5] R.L. Klueh, K. Ehrlich, F. Abe, Ferritic/martensitic steels: promises and problems, J. Nucl. Mater. 191–194 (1992) 116–124.
[6] D.S. Gelles, Development of martensitic steels for high neutron damage applications, J. Nucl. Mater. 239 (1996) 99–106.

[7] R.L. Klueh, Reduced-activation bainitic and martensitic steels for nuclear fusion applications, Curr. Opinion Solid State Mater. Sci. 8 (2004) 239–250.
[8] D.S. Gelles, Effects of irradiation on low activation ferritic alloys: a review, in: R.L. Klueh, D. Gelles, M. Okada, N.H. Packan (Eds.), Reduced Activation Materials for Fusion Reactors, ASTM-STP1047, Philadelphia, 1990, pp. 113–128.
[9] Y. Kohno, A. Kohyama, D.S. Gelles, K. Asakura, Radiation induced microstructural evaluation in ferritic/martensitic steels, Mater. Trans. Jpn. Inst. Metals 34 (11) (1993) 1018–1026.
[10] D.S. Gelles, in: F.A. Garner, N.H. Packan, A.S. Kumar (Eds.), Evaluation of ferritic alloy Fe-2 1/4Cr-1Mo after neutron irradiation: microstructural development, Radiation-Induced Changes in Microstructures:13th International Symposium, ASTM STP 955, ASTM, Philadelphia, 1987, pp. 560–587.
[11] K.L. Murty, I. Charit, Structural materials for Gen-IV nuclear reactors: challenges and opportunities, J. Nucl. Mater. 383 (2008) 189–195.
[12] Y. de Carlan, J. Henry, H. Burlet, A. Grosman, Les matériaux métalliques, une des clés de la 4e generation, Clefs CEA 55 (2007) 71–75.
[13] D. Stork, P. Agostini, J.L. Boutard, D. Buckthorpe, E. Diegele, S.L. Dudarev, C. English, G. Federici, M.R. Gilbert, S. Gonzalez, A. Ibarra, C. Linsmeier, A. Li Puma, G. Marbach, P.F. Morris, L.W. Packer, B. Raj, M. Rieth, M.Q. Tran, D.J. Ward, S.J. Zinkle, Developing structural, high-heat flux and plasma facing materials or a near-term DEMO fusion power plant: The EU assessment, J. Nucl. Mater. 455 (2014) 277–291.
[14] B.M. Gordon, C.W. Jewett, A.E. Pickett, M.E. Indig, in: J.W. Davis, D.J. Michel (Eds.), 'Corrosion resistance improvement of ferritic steels through hydrogen additions to the BWR coolant, Proceedings of Topical Conference on Ferritic Alloys for Use in Nuclear Energy Technologies, Metallurgical Society of AIME, New York, 1984, pp. 65–75.
[15] R.B. Rebak, P.L. Andresen, R.J. Blair, E.J. Dolley, in: Environmentally assisted cracking resistant ferritic steels for light water reactor applications, Corrosion 2013, Conference Proceedings, Document ID: NACE-2013-2599, NACE International, Houston, 2013. Available from: https://www.onepetro.org/conferences/nace.
[16] S.S. Hwang, B.H. Lee, J.G. Kim, J. Juang, SCC and corrosion evaluations of the F/M steels for a supercritical water reactor, J. Nucl. Mater. 372 (2008) 177–181.
[17] H. Je, A. Kimura, Stress corrosion cracking susceptibility of oxide dispersion strengthened ferritic steel in supercritical pressurized water dissolved with different hydrogen and oxygen contents, Corros. Sci. 78 (2014) 193–199.
[18] S. Van Dyck, R.W. Bosch, Environmentally assisted cracking of Eurofer 97 in water, Fusion Eng. Design 75–79 (2005) 973–977.
[19] M. Vankeerberghen, R.W. Bosch, R. Van Nieuwenhoven, In-pile electrochemical measurements on AISI 316 L(N) IG and EUROFER 97 – I: experimental results, J. Nucl. Mater. 312 (2003) 191–198.
[20] T. Muroga, T. Tanaka, M. Kondo, T. Nagasaka, Q. Xu, Characterization of liquid lithium blanket with RAFM and V-alloy for TBM and DEMO, Fusion Nucl. Technol. 56 (2009) 897–901.
[21] T. Sugawara, K. Nishihara, K. Tsujimoto, Transient analysis for lead-bismuth cooled accelerator-dirven system, Ann. Nucl. Energy 55 (2008) 238–247.
[22] B. Van den Bosch, R.W. Bosch, D. Sapundjiev, A. Almazouzi, Liquid metal embrittlement susceptibility of ferritic–martensitic steel in liquid lead alloys, J. Nucl. Mater. 376 (2008) 322–329.
[23] Corrosion Tests and Standards, in: R. Baboian (Ed.), Application and Interpretation, ASTM International, 2005, p. 473.
[24] M. Kondo, M. Takahashi, T. Tanaka, V. Tsisar, T. Muroga, Compatibility of reduced activation ferritic martensitic steel JLF-1 with liquid metals Li and Pb-17Li, Fusion Eng. Design 87 (2012) 1777–1787.
[25] M. Kondo, T. Muroga, T. Nagasaka, Q. Xu, V. Tsisar, T. Oshima, Mass transfer of RAFM steel in Li by simple immersion, impeller induced flow and thermal convection, J. Nucl. Mater. 417 (2011) 1200–1204.
[26] T. Flament, P. Tortorelli, V. Coen, H.U. Borgstedt, Compatibility of materials in fusion first wall and blanket structures cooled by liquid metals, J. Nucl. Mater. 191–194 (1992) 132–138.
[27] F. Barbier, A. Alemany, S. Martemianov, On the influence of a high magnetic field on the corrosion and deposition processes in the liquid Pb-17Li alloy, Fusion Eng. Design 43 (1998) 199–208.
[28] R. Moreau, Y. Bréchet, L. Maniguet, Eurofer corrosion by the flow of the eutecticalloy Pb-Li in the presence of a strong magnetic field, Fusion Eng. Design 86 (2011) 106–120.
[29] M. Abdou, N.B. Morley, S. Smolentsev, A. Ying, S. Malang, A. Rowcliffe, M. Ulrickson, Blanket/first wall challenges and required R&D on the pathway to DEMO, Fusion Eng. Design 100 (2015) 2–43.
[30] F. Balbaud, L. Martinelli, Corrosion issues in lead-cooled fast reactor (LFR) and accelerator driven systems (ADS), in: D. Féron (Ed.), Nuclear Corrosion Science and Engineering, Woodhead Publishing Limited, Oxford, 2012, pp. 807–841.

[31] C. Ye, J.B. Vogt, I. Proriol-Serre, Brittle fracture of T91 steel in liquid lead–bismuth eutectic alloy, J. Nucl. Mater. 280 (2014) 680–685.
[32] E. Lucon, W. Vandermeulen, Overview and critical assessment of the tensile properties of unirradiated and irradiated Eurofer97, in: Open Report SCK CEN-BLG-1042, 2007.
[33] G.R. Odette, M.J. Alinger, B.D. Wirth, Recent developments in irradiation-resistant steels, Annu. Rev. Mater. Res. 38 (2008) 471–503.
[34] G.R. Odette, D.T. Hoelzer, Irradiation-tolerant nanostructured ferritic alloys: transforming helium from a liability to an asset, J. Miner. Metals Mater. Soc. 62 (2010) 84–92.
[35] P.J. Ennis, O. Wachter, H. Nickel, in: Mechanical properties and microstructure of the 9% chromium steel NF616, Proceedings of the International Conference on Pressure Vessel Technology, ICPVT, vol. 1, 1996, pp. 461–466.
[36] P. Spätig, G.R. Odette, G.E. Lucas, Low temperature yield properties of two 7–9Cr ferritic/martensitic steels, J. Nucl. Mater. 275 (1999) 324–331.
[37] P. Spätig, G.R. Odette, E. Donahue, G.E. Lucas, Constitutive behavior and fracture toughness properties of the F82H ferritic/martensitic steel, J. Nucl. Mater. 283–287 (2000) 721–726.
[38] C.H. Woo, B.N. Singh, Production bias due to clustering of point defects in irradiation-induced cascades, Philos. Mag. 65 (4) (1992) 889–912.
[39] G.R. Odette, T. Yamamoto, D. Klingensmith, On the effect of dose rate on irradiation hardening of RPV steels, Philos. Mag. 85 (2005) 779–797.
[40] R. Kasada, A. Kimura, Modeling of microstructure evolution and mechanical property change of reduced-activation martensitic steel during varying-temperature irradiation, J. Nucl. Mater. 283–287 (2000) 188–192.
[41] C. Dethloff, E. Gaganidze, J. Aktaa, Quantitative TEM analysis of precipitation and grain boundary segregation in neutron irradiated EUROFER97, J. Nucl. Mater. 454 (2014) 323–331.
[42] O.J. Weiß, E. Gaganidze, J. Aktaa, Quantitative characterization of microstructural defects in up to 32 dpa neutron irradiated Eurofer 97, J. Nucl. Mater. 426 (2012) 52–58.
[43] Y.N. Osetsky, D.J. Bacon, Atomic-level dislocation dynamics in irradiated metals, in: Reference Module in Materials Science and Materials Engineering, Elsevier, 2016, pp. 333–356.
[44] Y. Dai, B. Long, Z.F. Tong, Tensile properties of ferritic/martensitic steels irradiated in STIP-I, J. Nucl. Mater. 377 (2008) 115–121.
[45] T. Yamamoto, G.R. Odette, H. Kishimoto, J.W. Rensman, P. Miao, On the effects of irradiation and helium on the yield stress changes and hardening and non-hardening embrittlement of 8Cr tempered martensitic steels: compilation and analysis of existing data, J. Nucl. Mater. 356 (2006) 27–49.
[46] M.J. Makin, F.J. Minter, Irradiation hardening in copper and nickel, Acta Metall. 8 (1960) 691–699.
[47] A.D. Whapman, M.J. Makin, The hardening of lithium fluoride by electron irradiation, Philos. Mag. 5 (1960) 237–250.
[48] B. van der Schaaf, C. Petersen, Y. De Carlan, J.W. Rensman, E. Gaganidze, X. Averty, High dose, up to 80 dpa, mechanical properties of Eurofer 97, J. Nucl. Mater. 386–388 (2009) 236–240.
[49] R.L. Klueh, J.M. Vitek, Tensile properties of 9Cr-1MoVNb and 12Cr-1MoVW steels irradiated to 23 dpa at 390 to 550 °C, J. Nucl. Mater. 182 (1991) 230–239.
[50] A. Kimura, Current status of reduced-activation ferritic/martensitic steels R&D for fusion energy, Mater. Trans. 46 (3) (2005) 394–404.
[51] F. Bergner, C. Pareige, M. Hernández-Mayoral, L. Malerba, C. Heintze, Application of a three-feature dispersed-barrier hardening model to neutron-irradiated Fe–Cr model alloys, J. Nucl. Mater. 448 (2014) 96–102.
[52] G.R. Odette, G.E. Lucas, Recent progress in understanding reactor pressure vessel steel embrittlement, Radiat. Effects Defects Solids 44 (1998) 189–231.
[53] B.N. Singh, N.M. Ghoniem, H. Trinkaus, Experiment-based modelling of hardening and localized plasticity in metals irradiated under cascade damage conditions, J. Nucl. Mater. 307–311 (2002) 159–170.
[54] D. Yao, D.J. Edwards, R.J. Kurtz, TEM characterization of dislocation loops in irradiated bcc Fe-based steels, J. Nucl. Mater. 434 (2013) 402–410.
[55] R. Schaeublin, D. Gelles, M. Victoria, Microstructure of irradiated ferritic/martensitic steels in relation to mechanical properties, J. Nucl. Mater. 307-311 (2002) 197–202.
[56] E. Gaganidze, C. Petersen, E. Materna-Morris, C. Dethloff, O.J. Weiß, J. Aktaa, A. Povstyanko, A. Fedoseev, O. Makarov, V. Prokhorov, Mechanical properties and TEM examination of RAFM steels irradiated up to 70 dpa in BOR-60, J. Nucl. Mater. 417 (2011) 93–98.
[57] R.E. Stoller, G.R. Odette, B.D. Wirth, Primary damage formation in bcc iron, J. Nucl. Mater. 251 (1997) 49–60.

[58] B.D. Wirth, G.R. Odette, D. Maroudas, G.E. Lucas, Dislocation loop structure, energy and mobility of self-interstitial atom clusters in bcc iron, J. Nucl. Mater. 276 (2000) 33–40.
[59] J. Marian, B.D. Wirth, A. Caro, B. Sadigh, G.R. Odette, J.M. Perlado, T.D. de la Rubia, Dynamics of self-interstitial cluster migration in pure α-Fe and Fe-Cu alloys, Phys. Rev. B 65 (2002). 144102/1–11.
[60] H.L. Heinisch, B.N. Singh, S.I. Golubov, The effects of one-dimensional glide on the reaction kinetics of interstitial clusters, J. Nucl. Mater. 283-287 (2000) 737–740.
[61] M. Wen, N.M. Ghoniem, B.N. Singh, Dislocation decoration and raft formation in irradiated materials, Philos. Mag. 85 (2002) 2561–2580.
[62] R. Bullough, M.H. Wood, E.A. Little, Microstructural explanation for the low swelling of ferritic steels, ASTM Special Technical Publication, 1981, pp. 593–609.
[63] J. Marian, B.D. Wirth, J.M. Perlado, Mechanism of formation and growth of <100> interstitial loops in ferritic materials, Phys. Rev. Lett. 88 (2002) 2555071–2555074.
[64] G. Bonny, D. Terentyev, J. Elena, A. Zinovev, B. Minov, E.E. Zhurkin, Assessment of hardening due to dislocation loops in bcc iron: overview and analysis of atomistic simulations for edge dislocations, J. Nucl. Mater. 473 (2016) 283–289.
[65] Y.N. Osetsky, D.J. Bacon, A. Serra, B.N. Singh, S.I. Golubov, Stability and mobility of defect clusters and dislocation loops in metals, J. Nucl. Mater. 276 (2000) 65–77.
[66] D. Terentyev, D. Klimenkov, L. Malerba, Confinement of motion of interstitial clusters and dislocation loops in BCC Fe–Cr alloys, J. Nucl. Mater. 393 (2009) 30–35.
[67] K. Yabuuchi, R. Kasada, A. Kimura, Effect of Mn addition on one-dimensional migration of dislocation loops in body-centered cubic Fe, Acta Mater. 61 (2013) 6517–6523.
[68] M. Bachhav, L. Yao, G.R. Odette, E.A. Marquis, Microstructural changes in a neutron-irradiated Fe-6at.%Cr alloy, J. Nucl. Mater. 453 (2014) 334–339.
[69] L. Messina, M. Nastar, T. Garnier, C. Domain, P. Olsson, Exact ab initio transport coefficients in bcc Fe-X (X=Cr, Cu, Mn, Ni, P, Si) dilute alloys, Phys. Rev. B 90 (2014) 104203/1–104203/15.
[70] H. Ke, P. Wells, P.D. Edmondson, N. Almirall, L. Barnard, G.R. Odette, D. Morgan, Thermodynamics and kinetic modeling of Mn-Ni-Si precipitates in low-Cu reactor pressure vessel steels, Acta Mater. 138 (2017) 10–26.
[71] D. Terentyev, F. Bergner, Y. Osetsky, Cr segregation on dislocation loops enhances hardening in ferritic Fe-Cr alloys, Acta Mater. 61 (2013) 1444–1453.
[72] H.J. Jung, D.J. Edwards, R.J. Kurtz, T. Yamamoto, Y. Wu, G.R. Odette, Structural and chemical evolution in neutron irradiated and helium-injected ferritic ODS PM2000 alloy, J. Nucl. Mater. 484 (2017) 68–80.
[73] S. Moll, T. Jourdan, H. Lefaix-Jeuland, Direct observation of interstitial dislocation loop coarsening in α-Iron, Phys. Rev. Lett. 111 (2013) 015503/1–015503/5.
[74] D. Mordehai, E. Clouet, M. Fivel, M. Verdier, in: Annealing of dislocation loops in dislocation dynamics simulations, IOP Conference Series: Materials Science and Engineering, vol. 3, 2009, p. 012001.
[75] T.D. Swinburne, K. Arakawa, H. Mori, H. Yasuda, M. Isshiki, K. Mimura, M. Uchikoshi, S.L. Dudarev, Fast, vacancy-free climb of prismatic dislocation loops in bcc metals, Sci. Rep. 6 (2016) 30596.
[76] M. Victoria, N. Baluc, C. Bailat, Y. Dai, M.I. Luppo, R. Schäublin, B.N. Singh, The microstructure and associated tensile properties of irradiated FCC and BCC metals, J. Nucl. Mater. 276 (2000) 114–122.
[77] N. Hashimoto, T.S. Byun, K. Farrell, S.J. Zinkle, Deformation microstructure of neutron-irradiated pure polycrystalline metals, J. Nucl. Mater. 329-333 (2004) 947–952.
[78] G.R. Odette, M.Y. He, E.G. Donahue, P. Spätig, T. Yamamoto, Modeling the multiscale mechanics of flow localization-ductility loss in irradiation damaged bcc alloys, J. Nucl. Mater. 307-311 (2002) 171–178.
[79] A. Considère, 1er semestre, 'L'emploi du fer et de l'acier dans les constructions', Annales des Ponts et Chaussées, Tome IX, 1885, pp. 574–775.
[80] S.A. Maloy, T.A. Saleh, O. Anderoglu, T.J. Romero, G.R. Odette, T. Yamamoto, S. Li, J.I. Cole, R. Fielding, Characterization and comparative analysis of the tensile properties of five tempered martensitic steels and an oxide dispersion strengthened ferritic alloy irradiated at 295 °C to 6.5 dpa, J. Nucl. Mater. 468 (2016) 232–239.
[81] G.R. Odette, On the ductile to brittle transition in martensitic stainless steels – mechanisms models and structural applications, J. Nucl. Mater. 212–215 (1994) 45–51.
[82] G.R. Odette, T. Yamamoto, H.J. Rathbun, M.Y. He, M.L. Hribernick, J.W. Rensman, Cleavage fracture and irradiation embrittlement of fusion reactor alloys: Mechanisms, multiscale models, toughness measurements and implications to structural integrity assessment, J. Nucl. Mater. 323 (2–3) (2003) 313–340.

[83] J.W. Rensman, J. van Hoepen, J.B.M. Bakker, R. den Boef, F.P. van den Broek, E.D.L. van Essen, Tensile properties and transition behaviour of RAFM steel plate and welds irradiated up to 10 dpa at 300 °C, J. Nucl. Mater. 307-311 (2002) 245–249.

[84] T. Yamamoto, G.R. Odette, D. Gragg, H. Kurishita, H. Matsui, W.J. Yang, M. Narui, M. Yamazaki, Evaluation of fracture toughness master curve shifts for JMTR irradiated F82H using small specimens, J. Nucl. Mater. 367–370 (2007) 593–598.

[85] P. Spätig, R. Stoenescu, P. Mueller, G.R. Odette, D. Gragg, Assessment of irradiation embrittlement of the Eurofer97 steel after 590 MeV proton irradiation, J. Nucl. Mater. 386–388 (2009) 245–248.

[86] T.L. Anderson, Fracture Mechanics Fundamentals and Applications, third ed., Taylor & Francis, Boca Raton, 2005.

[87] K. Wallin, The scatter in K_{Ic}-results, Eng. Fract. Mech. 19 (1984) 1085–1093.

[88] K. Wallin, Irradiation damage effects on the fracture toughness transition curve shape for reactor pressure vessel steels, Int. J. Press. Vessel. Pip. 55 (1993) 61–79.

[89] H.J. Rathbun, G.R. Odette, M.Y. He, T. Yamamoto, Influence of statistical and constraint loss size effects on cleavage fracture toughness in the transition – a model based analysis, Eng. Fract. Mech. 73 (2006) 2723–2747.

[90] E. Lucon, A closer look at the fracture toughness of ferritic/martensitic steels, J. Nucl. Mater. 367–370 (2007) 575–580.

[91] A. Pineau, Development of the local approach to fracture over the past 25 years: theory and applications, Int. J. Fract. 138 (1–4) (2006) 139–166.

[92] S.G. Roberts, S.J. Noronha, A.J. Wilkinson, P.B. Hirsch, Modelling the initiation of cleavage fracture of ferritic steels, Acta Mater. 50 (5) (2002) 1229–1244.

[93] D.A. Curry, J.F. Knott, Effect of microstructure on cleavage fracture toughness of quenched and tempered steels, Metal Sci. 13 (6) (1979) 341–345.

[94] G.R. Odette, T. Yamamoto, H. Kishimoto, M. Sokolov, P. Spätig, W.J. Yang, J.-W. Rensman, G.E. Lucas, A master curve analysis of F82H using statistical and constraint loss size adjustments of small specimen data, J. Nucl. Mater., Part B 329–333 (1–3) (2004) 1243–1247.

[95] G.R. Odette, M.Y. He, Cleavage toughness master curve model, J. Nucl. Mater. 283–287 (2000) 120–127.

[96] P. Mueller, P. Spätig, R. Bonadé, G.R. Odette, D. Gragg, Fracture toughness master-curve analysis of the tempered martensitic steel Eurofer97, J. Nucl. Mater. 386–388 (2009) 323–327.

[97] N. Ilchuk, P. Spätig, G.R. Odette, Fracture toughness characterization in the lower transition of neutron irradiated Eurofer97 steel, J. Nucl. Mater. 442 (2013) S58–S61.

[98] G.R. Odette, M.Y. He, T. Yamamoto, On the relation between irradiation induced changes in the master curve reference temperature shift and changes in strain hardened flow stress, J. Nucl. Mater. 367–370 (2007) 561–567.

[99] R. Vasina, P. Lukas, L. Kunz, V. Sklenicka, Interaction of high cycle fatigue and creep in 9%Cr-1%Mo steel at elevated temperature, Fatigue Fract. Eng. Mater. Struct. 18 (1995) 27–35.

[100] K. Mariappan, V. Shankar, R. Sandhya, K. Laha, Low cycle fatigue design data for India-specific reduced activation ferritic-martensitic (IN-RAFM) steel, Fusion Eng. Design 104 (2016) 76–83.

[101] A.F. Armas, C. Petersen, R. Schmitt, M. Avalos, I. Alvarez, Cyclic instability of martensite laths in reduced activation ferritic/martensitic steels, J. Nucl. Mater. 329–333 (2004) 252–256.

[102] J. Aktaa, M. Weick, C. Petersen, Reduced softening of Eurofer97 under thermomechanical and multiaxial fatigue loading and its impact on the design rules, J. Nucl. Mater. 386–388 (2009) 911–914.

[103] B. Fournier, M. Salvi, F. Dalle, Y. De Carlan, C. Caës, M. Sauzay, A. Pineau, Lifetime prediction of 9–12%Cr martensitic steels subjected to creep–fatigue at high temperature, Int. J. Fatigue 32 (2010) 971–978.

[104] F. Bassi, S. Foletti, A. Lo Conte, Creep fatigue crack growth and fracture mechanisms of T/P91 power plant steel, Mater. High Temp. 32 (2015) 250–255.

[105] Z. Zhang, Z. Hu, L. Fan, B. Wang, Low cycle fatigue behavior and cyclic softening of P92 ferritic-martensitic steel, J. Iron Steel Res. Int. 22 (2015) 534–542.

[106] R. Kannan, V.S. Srinivasan, M. Valsan, K.B.S. Rao, High temperature low cycle fatigue behaviour of P92 tungsten added 9Cr steel, Trans. Indian Inst. Metals 63 (2010) 571–574.

[107] Y. Miwa, S. Jitsukawa, M. Yonekawa, Fatigue properties of F82H irradiated at 523 K to 3.8 dpa, J. Nucl. Mater. 329–333 (2004) 1098–1102.

[108] S.W. Kim, H. Tanigawa, T. Hirose, A. Kohayama, Cyclically induced softening in reduced activation ferritic/martensitic steel before and after neutron irradiation, J. Nucl. Mater. 386–388 (2009) 529–532.

[109] N.V. Luzginova, J.W. Rensman, P. ten Pierick, J.B.J. Hegeman, Low cycle fatigue of irradiated and unirradiated Eurofer97 steel at 300 °C, J. Nucl. Mater. 409 (2011) 153–155.

[110] E. Materna-Morris, A. Möslang, H.C. Schneider, Tensile and low cycle fatigue properties of EUROFER97-steel after 16.3 dpa neutron irradiation at 523, 623 and 723 K, J. Nucl. Mater. 442 (2013) S62–S66.

[111] S. Nogami, M. Takahashi, A. Hasegawa, M. Yamazaki, Effect of helium on fatigue crack growth and life of reduced activation ferritic/martensitic steel, J. Nucl. Mater. 442 (2013) S43–S47.

[112] H. Ullmaier, W. Schmitz, Lifetime of Manet steel in load-cycling tests in vacuum at 20 and 550 °C, J. Nucl. Mater. 169 (1989) 233–240.

[113] A.A.F. Tavassoli, E. Diegele, R. Lindau, N. Luzginova, H. Tanigawa, Current status and recent research achievements in ferritic/martensitic steels, J. Nucl. Mater. 455 (2014) 269–276.

[114] R.L. Klueh, Ferritic/martensitic steels for advanced nuclear reactors, Trans. Indian Inst. Metals 62 (2009) 81–87.

[115] R. Lindau, A. Möslang, M. Rieth, M. Klimiankou, E. Materna-Morris, A. Alamo, A.A.F. Tavassoli, C. Cayron, A.M. Lancha, P. Fernandez, N. Baluc, R. Schäublin, E. Diegele, G. Filacchioni, J.W. Rensman, B. van der Schaaf, E. Lucon, W. Dietz, Present development status of EUROFER and ODS-EUROFER for application in blanket concepts, Fusion Eng. Design 75–79 (2005) 989–996.

[116] J.L. Boutard, in: V. Ghetta, D. Gorse, D. Mazière, V. Pontikis (Eds.), Materials Issues for Generation IV Systems – Status, Open Questions and Challneges, Springer, Dordrecht, The Netherlands, 2008, pp. 484–500.

[117] R. Coppola, M. Klimiankou, M. Magnani, A. Möslang, M. Valli, Helium bubble evolution in F82H-mod – correlation between SANS and TEM, J. Nucl. Mater. 329–333 (2004) 1057–1061.

[118] H. Ullmaier, The influence of helium on the bulk properties of fusion reactor structural materials, Nucl. Fusion 24 (8) (1984) 1039.

[119] Y. Dai, G.R. Odette, T. Yamamoto, The effects of helium in irradiated structural alloys, in: R.J.M. Konings (Ed.), Comprehensive Nuclear Materials, vol. 1.06, Elsevier, Amsterdam, 2012, pp. 141–193.

[120] Y. Yamamoto, G.R. Odette, P. Miao, D.T. Hoelzer, J. Bentley, H. Hashimoto, H. Tanigawa, R.J. Kurtz, The transport and fate of helium in nanostructured ferritic alloys at fusion relevant He/dpa ratios and dpa rates, J. Nucl. Mater. 367–370 (2007) 399–410.

[121] R.L. Klueh, D.S. Gelles, S. Jitsukawa, A. Kimura, G.R. Odette, B. van der Schaaf, M. Victoria, Ferritic/martensitic steels – overview of recent results, J. Nucl. Mater. 307-311 (2002) 455–465.

[122] Y. Dai, G.S. Bauer, Status of the first SINQ irradiation experiment, STIP-I, J. Nucl. Mater. 296 (2001) 43–53.

[123] P. Jung, J. Henry, J. Chen, J.-C. Brachet, Effect of implanted helium on tensile properties and hardness of 9% Cr martensitic stainless steels, J. Nucl. Mater. 318 (2003) 241–248.

[124] M. Ando, E. Wakai, T. Sawai, H. Tanigawa, K. Furuya, S. Jitsukawa, H. Takeuchi, K. Oka, S. Ohnuki, A. Kohyama, Synergistic effect of displacement damage and helium atoms on radiation hardening in F82H at TIARA facility, J. Nucl. Mater., Part B. 329–333 (1–3) (2004) 1137–1141.

[125] Y. Dai, J. Henry, Z. Tong, X. Averty, J. Malaplate, B. Long, Neutron/proton irradiation and He effects on the microstructure and mechanical properties of ferritic/martensitic steels T91 and EM10, J. Nucl. Mater. 415 (3) (2011) 306–310.

[126] H. Trinkaus, B.N. Singh, Helium accumulation in metals during irradiation – where do we stand? J. Nucl. Mater. 323 (2003) 229–242.

[127] H. Ullmaier, J. Chen, Low temperature tensile properties of steels containing high concentrations of helium, J. Nucl. Mater. 318 (2003) 228–233.

[128] H. Schroeder, W. Kesternich, H. Ullmaier, Helium effects on the creep and fatigue resistance of austenitic stainless steels at high temperature, Nucl. Eng. Design/Fusion 2 (1985) 65–95.

[129] H. Schroeder, P. Batfalsky, In-beam simulation of high temperature helium embrittlement of DIN 1.4970 austenitic stainless steel, J. Nucl. Mater. 103–104 (1981) 839–844.

[130] N. Yamamoto, J. Nagakawa, K. Shiba, Effects of helium implantation on creep rupture properties of low activation ferritic steel F82H IEA heat, J. Nucl. Mater. 283–287 (2000) 400–403.

[131] N. Yamamoto, Y. Murase, J. Nagakawa, K. Shiba, Creep behavior of reduced activation martensitic steel F82H injected with a large amount of helium, J. Nucl. Mater. 307–311 (2002) 217–221.

[132] J. Chen, P. Jung, W. Hoffelner, H. Ullmaier, Dislocation loops and bubbles in oxide dispersion strengthened ferritic steel after helium implantation under stress, Acta Mater. 56 (2008) 250–258.

[133] E.A. Little, D.A. Stow, Void-swelling in irons and ferritic steels: II. An experimental survey of materials irradiated in a fast reactor, J. Nucl. Mater. 87 (1) (1979) 25–39.

[134] J.J. Huet, P.H. Van Asnroek, W. Wandermeulen, in: Swelling of ferritic steels irradiated in fast reactors, Proceeding of the International Conference on Irradiated Behaviour of Metallic Materials for Fast Reactor Core Components, CEA-DMECN France, 1979, pp. 1–5.

[135] D.S. Gelles, Microstructural examination of several commercial ferritic alloys irradiated to high fluence, J. Nucl. Mater. 104 (1981) 975–979.

[136] D.S. Gelles, Microstructural examination of several commercial alloys neutron irradiated to 100 dpa, J. Nucl. Mater. 148 (2) (1987) 136–144.

[137] Y. Katoh, D.S. Gelles, A. Kohyama, Swelling and dislocation evolution in simple ferritic alloys irradiated to high fluence in FFTF/MOTA, J. Nucl. Mater. 225 (1995) 154–162.

[138] M.B. Toloczko, F.A. Garner, C.R. Eiholzer, Irradiation creep and swelling of the US fusion heats of HT9 and 9Cr-1Mo to 208 dpa at $\sim 400°C$, J. Nucl. Mater., Part 1 212–215 (1994) 604–607.

[139] M.B. Toloczko, F.A. Garner, Irradiation creep and void swelling of two LMR heats of HT9 at $\sim 400°C$ and 165 dpa, J. Nucl. Mater., Part 1 233–237 (1996) 289–292.

[140] M.B. Toloczko, D.S. Gelles, F.A. Garner, R.J. Kurtz, K. Abe, Irradiation creep and swelling from 400 to 600 °C of the oxide dispersion strengthened ferritic alloy MA957, J. Nucl. Mater., Part A 329–333 (2004) 352–355.

[141] J.M. Vitek, R.L. Klueh, Microstructure of 9 Cr-1 MoVNb steel irradiated to 36 dpa at elevated temperatures in HFIR, J. Nucl. Mater. 122 (1–3) (1984) 254–259.

[142] P.J. Maziasz, R.L. Klueh, J.M. Vitek, Helium effects on void formation in 9Cr-1MoVNb and 12Cr-1MoVW irradiated in HFIR, J. Nucl. Mater. 141–143 (2) (1986) 929–937.

[143] T. Okita, N. Sekimura, F.A. Garner, Effects of dpa rate on swelling in neutron-irradiated Fe–Cr and Fe–Cr–Mo alloys, J. Nucl. Mater. 417 (1–3) (2011) 944–948.

[144] J.L. Straalsund, R.W. Powell, B.A. Chin, An overview of neutron irradiated effects in LMFBR materials, J. Nucl. Mater. 108–109 (1982) 299–305.

[145] E. Kuramoto, N. Yoshida, H. Tsukuda, K. Karijima, N. Packan, M. Lewis, L. Mansur, Simulation irradiation studies on iron, J. Nucl. Mater. 104 (1981) 1091–1095.

[146] W.G. Johnston, T. Lauritzen, J.W. Rosolowski, A.M. Turkalo, Void swelling of ferritic alloys bombarded with nickel ions, in: Brager, Perrin (Eds.), Effects of Radiation on Materials: Proceedings of the Eleventh International Symposium, Vol. 782, ASTM STP, 1982, pp. 809–823.

[147] L.L. Horton, J. Bentley, K. Farrell, A TEM study of neutron-irradiated iron, J. Nucl. Mater. 108–109 (1982) 222–233.

[148] G. Ayrault, Cavity formation during single- and dual-ion irradiation in a 9Cr-1Mo ferritic alloy, J. Nucl. Mater. 114 (1) (1983) 34–40.

[149] L.L. Horton, L.K. Mansur, Experimental determination of the critical cavity radius in Fe-10Cr for ion irradiation, in: Garner, Perrin (Eds.), Effects of Radiation on Materials, Proceedings of Twelfth International Symposium, Vol. 870, ASTM STP, 1985, pp. 344–362.

[150] O.V. Borodin, V.V. Bryk, V.N. Voyevodin, A.S. Kalchenko, Y.E. Kupriyannova, V.V. Melnichenko, I.M. Neklyudov, A.V. Permyakov, Radiationswelling of ferritic-martensitic steels EP-450 and HT-9 under irradiation by metallic ions to super-higher doses, Prob. Atom. Sci. Technol. 2 (2011) 10–15.

[151] E.A. Little, Microstructural evolution in irradiated ferritic-martensitic steels: transitions to high dose behavior, J. Nucl. Mater. 206 (2–3) (1993) 324–334.

[152] F.A. Garner, M.B. Toloczko, B.H. Sencer, Comparison of swelling and irradiation creep behavior of fcc-austenitic and bcc-ferritic/martensitic alloys at high neutron exposure, J. Nucl. Mater. 276 (1–3) (2000) 123–142.

[153] G.R. Odette, On mechanisms controlling swelling in ferritic and martensitic alloys, J. Nucl. Mater., Part 2 155–157 (1988) 921–927.

[154] B.H. Sencer, F.A. Garner, Compositional and temperature dependence of void swelling in model Fe–Cr base alloys irradiated in the EBR-II fast reactor, J. Nucl. Mater., Part 1 283–287 (2000) 164–168.

[155] S. Ohnuki, H. Takahashi, T. Takeyama, Void swelling and segregation of solute in ion-irradiated ferritic steels, J. Nucl. Mater. 104 (1981) 1121–1125.

[156] F.A. Garner, D.S. Gelles, L.R. Greenwood, T. Okita, N. Sekimura, W.G. Wolfer, Synergistic influence of displacement rate and helium/dpa ratio on swelling of Fe–(9,12)Cr binary alloys in FFTF at $\sim 400°C$, J. Nucl. Mater. 329–333 (2004) 1008–1012.

[157] D.F. Pedraza, P.J. Maziasz, R.L. Klueh, Helium effects on the microstructural evolution of reactor irradiated ferritic and austenitic steels, Radiat. Effects Defects Solids 113 (1990) 213–228.

[158] X. Wang, A.M. Monterrosa, F. Zhang, H. Huang, Q. Yan, Z. Jiao, G.S. Was, L. Wang, Void swelling in high dose ion-irradiated reduced activation ferritic-martensitic steels, J. Nucl. Mater. 462 (2015) 119–125.

[159] T. Yamamoto, Y. Wu, G.R. Odette, K. Yabushi, S. Kondo, A. Kimura, A dual ion irradiation study of helium-dpa interactions on cavity evolution in tempered martensitic steels and nanostructured ferritic alloys, J. Nucl. Mater. 449 (2–3) (2014) 190–199.

[160] J.J. Sniegowski, W.G. Wolfer, in: J.W. Davis, D.J. Michel (Eds.), On the physical basis for the swelling resistance of ferritic steels, Proceedings of Topic Conference on Ferritic Alloys for Use in Nuclear Energy Technologies, AIME, 1984, pp. 579–586.

[161] F.A. Garner, D.S. Gelles, H. Takahashi, S. Ohnuki, H. Kinoshita, B.A. Loomis, High swelling rates observed in neutron-irradiated V-Cr and V-Si binary alloys, J. Nucl. Mater., Part B 191–194 (1992) 948–951.

[162] H. Matsui, D.S. Gelles, Y. Kohno, in: R.E. Stoller, A.S. Kumar (Eds.), Large swelling observed in a V-5at%Fe alloy after irradiation in FFTF, Effects of Radiation on Materials: 15th International Symposium, vol. 1125, ASTM STP, 1992, pp. 928–944.

[163] F.A. Nichols, Theory of the creep of zircaloy during neutron irradiation, J. Nucl. Mater. 30 (1969) 249–270.

[164] J.E. Habbottle, Anisotropic irradiation creep of zircaloy-2, Philos. Mag. A 38 (1978) 49–60.

[165] R.A. Holt, Effect of microstructure on irradiation creep and growth of Zircaloy pressure tubes in power reactors, J. Nucl. Mater. 82 (1979) 419–429.

[166] M.B. Toloczko, F.A. Garner, Variability of irradiation creep and swelling of HT9 irradiated to high neutron fluence at 400–600°C, in: ASTM Special Technical Publication, No 1325, 1999, pp. 765–779.

[167] D.J. Michel, P.L. Hendrick, A.G. Pieper, Transient irradiation-induced creep of nickel during deuteron bombardment, J. Nucl. Mater. 75 (1979) 1–6.

[168] T.C. Reiley, R.L. Auble, R.H. Shannon, Irradiation creep under 60 MeV alpha irradiation, J. Nucl. Mater. 90 (1980) 271–281.

[169] C.H. Henager, E.P. Simonen, E.R. Bradley, R.G. Stang, The stress dependence of creep in ni bombarded with 17 MeV deuterons, J. Nucl. Mater. 104 (1981) 1269–1273.

[170] H. Schroeder, High temperature helium embrittlement in austenitic stainless steels – correlations between microstructure and mechanical properties, J. Nucl. Mater. 155–157 (1988) 1032–1037.

[171] P. Jung, H. Ullmaier, Effects of light-ion irradiation on mechanical properties of metals and alloys, J. Nucl. Mater. 174 (1990) 253–263.

[172] J. Chen, P. Jung, J. Henry, Y. de Carlan, T. Sauvage, F. Duval, M.F. Barthe, W. Hoffelner, Irradiation creep and microstructural changes of ODS steels of different Cr-contents during helium implantation under stress, J. Nucl. Mater. 437 (1–3) (2013) 432–437.

[173] J.L. Seran, V. Levy, P. Dubuisson, D. Gilbon, A. Maillard, A. Fissolo, H. Touron, R. Cauvin, A. Chalony, E. Le Boulbin, in: Behavior under neutron irradiation of the 15-15Ti and EMI0 steels used as standard materials of the phenix fuel subassembly, Effects of Irradiation on Materials: Proceedings of 15th International Symposium, STP1125, ASTM, 1992, pp. 1209–1233.

[174] J. Chen, P. Jung, W. Hoffelner, Irradiation creep of candidate materials for advanced nuclear plants, J. Nucl. Mater. 441 (1–3) (2013) 688–694.

[175] P. Jung, H. Klein, Segregation in DIN 1.4914 martensitic stainless steel under proton irradiation, J. Nucl. Mater. 182 (1991) 1–5.

[176] A. Alamo, J.L. Bertina, V.K. Shamardin, P. Widemt, Mechanical properties of 9Cr martensitic steels and ODS-FeCr alloys after neutron irradiation at 325 °C up to 42 dpa, J. Nucl. Mater., Part A 367–370 (2007) 54–59.

[177] F.A. Garner, Irradiation performance of cladding and structural steels, in: R.W. Cahn, P. Haasen (Eds.), Liquid Metal Reactors in Materials Science and Technology, vol. 10A, VCH Verlag, Weinheim/New York, 1993, pp. 419–543.

[178] G.R. Odette, A model for in-reactor stress rupture of austenitic stainless steels, J. Nucl. Mater. 122 (1984) 435–441.

[179] S.J. Zinkle, J.L. Boutard, D.T. Hoelzer, A. Kimura, R. Lindau, G.R. Odette, M. Rieth, L. Tan, H. Tanigawa, Development of next generation tempered and ODS reduced activation ferritic/martensitic steels for fusion energy applications, Nucl. Fusion 57 (9) (2017) 092005.

[180] R.L. Klueh, N. Hashimoto, P.J. Maziasz, New nano-particle-strengthened ferritic/martensitic steels by conventional thermo-mechanical treatment, J. Nucl. Mater. 367-370 (2007) 48–53.

[181] L. Tan, Y. Katoh, A.A.F. Tavassoli, J. Henry, M. Rieth, H. Sakasegawa, H. Tanigawa, Q. Huang, Recent status and improvement of reduced-activation ferritic-martensitic steels for high-temperature service, J. Nucl. Mater. 479 (2016) 515–523.

[182] L. Tan, L.L. Snead, K. Yatoh, Development of new generation reduced activation ferritic-martensitic steel for advanced fusion reactors, J. Nucl. Mater. 478 (2016) 42–49.

[183] L. Tan, Y. Yang, J.T. Busby, Effects of alloying elements and thermomechanical treatment on 9Cr reduced activation ferritic–martensitic (RAFM) steels, J. Nucl. Mater. 442 (2013) S13–S17.

[184] S. Hollner, B. Fournier, J. Le Pendu, T. Cozzika, I. Tournié, J.-C. Brachet, A. Pineau, High-temperature mechanical properties improvement on modified 9Cr–1Mo martensitic steel through thermomechanical treatments, J. Nucl. Mater. 405 (2010) 101–108.

[185] S.J. Zinkle, A. Möslang, Evaluation of irradiation facility options for fusion materials research and development, Fusion Eng. Design 88 (2013) 472–482.

CHAPTER 12

NANO-OXIDE DISPERSION-STRENGTHENED STEELS

G. Robert Odette*, Nicholas J. Cunningham*, Tiberiu Stan*, M. Ershadul Alam*, Yann De Carlan[†]

Materials Department, University of California, Santa Barbara, CA, United States DEN, Department of Applied Metallurgical Research (SRMA), CEA, University of Paris-Saclay, Gif-Sur-Yvette, France[†]

CHAPTER OUTLINE

- 12.1 Introduction 530
- 12.2 A Brief History of ODS Alloys 531
- 12.3 Some Key Attributes of Nano-Oxide Dispersion-Strengthened (NODS) Iron-Based Alloys for Nuclear Applications—An Overview 532
  - 12.3.1 Unirradiated Mechanical Properties 532
  - 12.3.2 A Summary of Alloy Stability and Irradiation Effects 533
  - 12.3.3 Void Swelling and Helium Effects 535
  - 12.3.4 Other NODS Issues 535
- 12.4 Overview of the Composition and Processing Paths for NFA and NMS 536
  - 12.4.1 Alloy Compositions, Phase Diagrams, and Transformation Paths 536
  - 12.4.2 Preconsolidation Processing 538
  - 12.4.3 Consolidation 539
  - 12.4.4 Deformation Processing and Tube Fabrication 541
  - 12.4.5 Texturing and Damage Mechanisms During Deformation Processing 542
  - 12.4.6 Joining 545
  - 12.4.7 Alternative Compositions and Processing Paths 545
  - 12.4.8 Summary of Processing and Fabrication 547
- 12.5 Characteristics and Function of the NO 547
  - 12.5.1 NO Statistics 547
  - 12.5.2 Nature of the NO 548
  - 12.5.3 NO Functions and He interactions 550
  - 12.5.4 Summary 552
- 12.6 Mechanical Properties 552
  - 12.6.1 Static Tensile Strength and Ductility 552
  - 12.6.2 Creep 554
  - 12.6.3 Fast Fracture and Fatigue 556
- 12.7 Thermal Aging and Irradiation Effects 558
  - 12.7.1 Thermal Aging 559
  - 12.7.2 Overview Irradiation Effects on the Microstructure 560
  - 12.7.3 Irradiation Stability of the NOs 562

12.7.4 Dislocation Loops 562
12.7.5 Solute Segregation, Clustering, and Precipitation 564
12.7.6 Cavities and Swelling 566
12.7.7 Effects of Irradiation on Strength and Toughness 570
12.7.8 Effect of Irradiation on Other Properties 572
12.7.9 Summary of Thermal Aging and Irradiation Effects 572
12.8 Modeling 573
12.9 Future Prospects 574
References 575

12.1 INTRODUCTION

There is a growing worldwide interest in developing advanced sources of C-free fission and fusion energy. The viability of these sources of nuclear energy will ultimately depend on developing new, high-performance structural materials that can support extended service under extremely hostile conditions [1]. As described in more detail in Chapters 2, 3, 5, 6, and 11, conditions typical of advanced fission and fusion reactor service are characterized by various combinations of intense neutron radiation fields, large time-varying stresses, high temperatures, and aggressive chemical environments. The structural material of choice must provide high levels of reliability and large safety margins for extended periods of time. Neutron irradiation may produce up to more than 400 displacements-per-atom (dpa) in some cases (see Chapter 2). The displacement of atoms from their crystal lattice sites initially creates equal numbers of vacancy and self-interstitial atom (SIA) defects that result in complex microstructural and microchemical evolutions [1–7]. These include accumulation of extended vacancy and SIA defect clusters, increases in dislocation densities, solute segregation, radiation enhanced and induced precipitation, helium bubbles, and growing voids. These microstructural evolutions lead to changes in many performance-sustaining mechanical properties.

Displacement damage also interacts with hydrogen and, especially, helium transmutation products, which are generated in fusion reactor environments up to levels of a few tens of thousands to a few thousand atom parts per million (appm), respectively [2–5]. Helium is effectively insoluble in metals and diffuses to precipitate as second-phase gas bubbles. Under some conditions the bubbles act as heterogeneous nucleation sites for both: (a) growing voids in the alloy matrix, driving macroscopic swelling [3, 4]; and (b) stress-driven growing grain boundary cavities that can degrade the creep rupture time and ductility [3]. Helium accumulation on grain boundaries also synergistically interacts with irradiation hardening, including a contribution from the bubbles themselves, to severely degrade fracture toughness [3]. Other typical consequences of the microstructural-microchemical evolutions under irradiation include: frequent nearly complete loss of uniform tensile strain ductility (the necking strain) at low to intermediate irradiation temperatures; enhanced rates of environmentally induced cracking; and low-temperature irradiation creep [6]. Even absent irradiation effects, the thermomechanical structural challenges experienced by nuclear reactor structures will be daunting. For example, surface heat fluxes on some semipermanent plasma-facing fusion reactor components are comparable to those experienced by short-lived rocket nozzles [7]. Finally, greater energy conversion efficiency requires advanced materials with improved high-temperature properties.

The objective of this chapter is to summarize the status of developing an iron-based metallic alloy class that shows great promise in meeting many of these nuclear energy application challenges [3–5, 8–13]. This alloy class is traditionally known as oxide dispersion-strengthened (ODS) steels. However, the ODS designation includes a wide range of microstructures and properties, including alloys that contain dispersion strengthening features on the size scale of only a few nm. We emphasize these nano-dispersion-strengthened variants of ODS alloys, which we call nano-oxide dispersion-strengthened (NODS) steels. NODS steels have superior properties due to the presence of an ultrahigh density of the ultrafine nano-oxides [3–5, 8–13]. There are two primary variants of NODS steels: (a) low C, higher Cr (>12%) fully ferritic stainless steels, that fall outside the Fe-Cr γ-loop, which we call nanostructured ferritic alloys (NFAs); and (b) lower Cr (≈9%), C-bearing transformable Fe-based steels, that we call nanostructured tempered martensitic steels (NMS). We use the term NODS steels when both NFA and NMS ODS variants are being discussed.

NODS steels contain an ultrahigh density (typically of order $5 \times 10^{23}/m^3$) of 2–3 nm Y-Ti-O oxide (NO) features, represent a potentially transformative alloy class for nuclear applications, since they manifest a combination of very high tensile, creep and fatigue strengths, unique thermal stability, and remarkable irradiation tolerance [3–5, 8–13]. The nano-oxide (NO) impede dislocation climb and glide, enhance SIA-vacancy recombination (self-healing) and, most notably, trap helium in small, high-pressure interface gas bubbles. The nm-scale bubbles are sinks, reducing excess defect concentrations and the amount of He reaching grain boundaries, while remaining far too small to convert to growing voids or grain boundary creep cavities under stress. Thus the bubbles mitigate severe toughness loss, swelling, and degradation of creep rupture properties at lower, intermediate and higher temperatures, respectively. Since they are excellent defect sinks, small bubbles also mitigate other degrading radiation damage effects, such as grain boundary segregation, dislocation loop formation, enhanced and induced precipitation, that cause hardening and embrittlement, and low-temperature irradiation creep. The NOs also stabilize high dislocation densities and small grain sizes, which contribute significantly to the NODS steel strength. Note, not all ODS steels alloys contain such functionally effective NOs. As discussed in the following, coarser-scale oxides may even promote some manifestations of detrimental irradiation effects, such as void swelling [12].

Development of NODS steels is *work in progress*. Important scientific challenges that have seen major progress include establishing: (a) NO number densities, size distributions, compositions and structures, as well as the characteristics of their matrix interfaces; (b) how these NO characteristics are affected by alloy composition and processing variables; (c) the relationship between the character of the NOs and the corresponding mechanical properties and irradiation tolerance; and (d) the thermal and irradiation stability of the NOs and the balance of the alloy microstructures. More technology-oriented work in progress includes: (a) optimized alloy composition-synthesis and thermal-mechanical processing paths that provide a balanced mix of outstanding and isotropic properties; (b) chemical compatibility and corrosion-oxidation resistance; (c) ways to improve alloy homogeneity and reproducibility; and (d) methods for shape fabrication and joining that maintain optimal microstructures and yield defect-free structural components. Equally important, practical ODS challenges include reducing costs, establishing industrial scale supply sources, and qualifying new alloys for nuclear service.

The organization of this chapter is as follows: (a) a few remarks on the history of NODS steels; (b) a brief and broad and high level overview of key NODS steel characteristics; (c) a description of NODS steel processing and fabrication paths; (d) a series of slightly more detailed sections describing NODS microstructures, properties, thermal stability and irradiation effects, including in He-rich environments; (e) summary of related modeling studies; and (f) a short section on future prospects.

Accomplishing these goals presents a large challenge. First, there is a very large and rapidly growing literature that cannot be captured in a single chapter that aims to be comprehensive at a reasonable, but still limited, degree of depth. Second, the results in the literature on various topics are often highly disparate in the data that they present, and the conclusions they reach. In part, such disparities naturally derive from the large number of NODS alloy and test conditions involved, and in part the diversity of what has been probed, and the tools used to accomplish the measurements. So, what follows is based on the authors' best judgment of the most reliable, and hopefully mainly represents consensus opinions. But it is noted that there are exceptions to what is concluded on a topic that remains unsettled.

12.2 A BRIEF HISTORY OF ODS ALLOYS

The classical approach to strengthening structural alloys is to introduce a fine-scale dispersion of obstacles to dislocation climb and glide [4]. The use of highly stable oxides provides high-temperature strength. Oxide dispersion-strengthening (ODS) has been widely applied in alloy systems based on aluminum, copper, nickel, refractory metals, precious metals, intermetallics, and Fe-based steels. Perhaps the first application was the use of ThO_2 in W filaments, invented by Irving Langmuir around 1910. Subsequent research on ODS Ni (thoria-dispersed Ni, aka TD Ni) in the early 1960s demonstrated the viability of ODS for high-performance structural alloys. The first ODS steel papers listed in the Web of Science appeared in the late 1970s; and in 2016–17 ≈ 350 papers were listed in the NFA and ODS categories.

The roots of ODS steels trace to proprietary research in Belgium from in the late 1960s [14, 15]. Ferritic ODS alloys were then semicommercialized by Fischer, who patented a mechanical alloying (MA)-hot-extrusion powder processing route in 1978 [16]. These NFA, marketed in the 1980s by International Nickel Corporation (INCO) as MA956 and MA957, were in large part aimed at applications to liquid metal fast breeder reactor cladding and ducts. These high-performance NFAs are typically 12%–15%Cr ferritic steels micro-alloyed with Y, Ti, O and W, or Mo. Extensive early processing, fabrication and fast reactor irradiation studies, including high dose fast reactor irradiations, were carried out in the US, Europe, Russia, and Japan. Hamilton subsequently documented some of the early proprietary US breeder reactor research on NFA in 2000 [17]. However, the termination of the US breeder program in the late 1980s, and the absence of a large industrial market, led to a general decline in interest of NODS steels. Until recently, only Plansee has continued to produce the higher Cr and Al added (FeCrAl) heat resistant ODS steel, PM2000, largely for chemical industry applications. However, there is currently no industrial supplier of NODS steels.

Fortunately, beginning in the late 1980s an outstanding NODS steel research program began in Japan, that continues to this day, led by S. Ukai and coworkers [8, 9]. The initial Japanese effort focused on NFA, ultimately leading to the production of a commercial vendor Kobe 12Cr heat, often referred to as J12YWT. Since about 2000, a large body of international work has focused on similar 12-14Cr NFA, generally dubbed as 12-14YWT, and 9Cr transformable NMS, again with Japan in the lead [4, 5, 8–13, 18].

12.3 SOME KEY ATTRIBUTES OF NANO-OXIDE DISPERSION-STRENGTHENED (NODS) IRON-BASED ALLOYS FOR NUCLEAR APPLICATIONS—AN OVERVIEW

We begin with a brief overview of the key attributes of NODS steels, which make them potentially transformative structural materials for many nuclear applications. We also list their major liabilities. Additional details on the various topics are provided in the subsequent sections. For context, we compare NODS steels to conventional 9Cr tempered martensitic steels (TMS), since the latter are the current alloy class of choice for many nuclear applications (see Refs. [1, 19] and Chapter 11). Note, advanced melt practice TMS, strengthened by fine-scale alloy carbonitride precipitates, are also under development [19]. While superior to conventional TMS, the melt practice alloys generally have lower strength and upper use temperature limits than NODS steels.

12.3.1 UNIRRADIATED MECHANICAL PROPERTIES

The unirradiated mechanical properties of NODS steels are discussed in Section 12.6. Briefly, the static yield strength (σ_y) of NODS steels is generally much higher than TMS [2–5, 8–14, 16, 17, 19–21]. However, the NODS steel σ_y varies by large factors, depending on their composition and processing history. For example, the highest and lowest room temperature (RT) σ_y of NFA range from ≈ 600 to $1600\,MPa$ [12, 20, 21]. Notably, the strength advantage of NODS steels increases relative to TMS in the high-temperature creep regime. In the former case, creep rates and rupture times are governed by much higher threshold stresses [4, 5]. For example, at 700°C and 100 MPa, the rupture times for 9Cr TMS Eurofer97 and a French heat of NFA MA957 loaded in the extrusion direction are $\approx 40\,h$ and $5.8 \times 10^7\,h$, respectively, based on the Larsen-Miller plot in Refs. [4, 5]. The corresponding NODS steel general use temperatures limits are ≈ 150 (NMS) to 250°C NFAs higher than that for conventional

TMS, where the latter is typically used for nuclear service below 550°C [1]. Note that advanced 9-11Cr TMS have extended the use temperatures to above 550°C, mainly for fossil energy applications [19].

The creep strength of NFA is usually much lower transverse to the primary deformation (extrusion) directions (like the hoop direction of tubing) [4, 12, 17]. Note, Al-added ODS NiCrAl alloys, like PM2000 and MA956, are generally weaker than NFA [4, 21], since they typically have microstructures and oxides that are both coarser [4]. High-temperature re-crystallization heat treatments also reduce grain and dislocation contributions to NFA strength [8, 9, 22]. However, a second generation of NiCrAl low-carbon Zr-added steels has been developed with both high strength and ductility [23].

Transformable NMS are generally not as strong as NFAs, although the differences have been narrowing, especially at lower temperature [8, 9], and when compared to transverse NFA orientations. NMS creep properties are much more isotropic than for NFA following the austenite-to-ferrite/martensite transformation [8, 9]. Note, the lower and more isotropic strength is a major advantage of NMS in fabricating useful product forms, such as thin-walled tubing.

Unfortunately, the fracture toughness of NODS steels is generally substantially inferior to TMS, and is most often extremely anisotropic in NFA in deformation-processed conditions. This is due to texturing and formation of a brittle cleavage system texture component. In this case, the brittle orientation is for cracks running in the primary deformation directions (e.g., the axial direction for extrusions like tubing) [4, 8, 9, 24, 25]. The properties are more isotropic in recrystallized NFA [8, 9]. The fracture properties of NMS also vary widely, but again they are intrinsically more isotropic than NFA, and can be improved by optimizing thermal-mechanical processing paths [26, 27]. The fatigue properties of NODS steels are far superior to TMS ([12] and see Section 12.6), primarily due to their higher strength and considerable ductility.

12.3.2 A SUMMARY OF ALLOY STABILITY AND IRRADIATION EFFECTS

Irradiation and thermal aging effects in NODS steels are discussed in Section 12.7. A major advantage of NODS alloys is that they are able to operate above the displacement damage (vacancy and SIA defect) regime ($>\approx 0.45T_m$), where extended neutron irradiation-induced features, like SIA dislocation loops and vacancy clusters, do not accumulate [2, 4, 6]. The displacement damage resistance advantage of NODS steels versus TMS decreases at $<0.45T_m$, where dislocation loops, often born as cascade interstitial clusters, form in both systems, along with fine-scale precipitates ([3, 6], and see Section 12.7.2 and Chapter 11). In summary, NODS are even more resistant to void swelling than TMS. As noted in Section 12.7, the data on dislocation loops in NODS are extremely limited, and even the database on TMS is sparse, highly scattered, and not self-consistent. Thus it is difficult to provide quantitative details on dislocation loop behavior. However, the overall trend is that the dislocation loop density and total line length generally increase with decreasing temperature (see Section 12.7.4). At high He levels bubbles also contribute to hardening [3] and may enhance loop formation [28], but this has not been quantified for a wide range of conditions. Further, the corresponding differences, if any, between NFA and TMS have not been generally established. However, a ubiquitous observation is that small loops formed in displacement cascades, as well as network dislocations, are the sites of significant solute segregation due to both radiation-enhanced diffusion (RED) thermal and radiation-induced segregation (RIS) mechanisms and precipitate nucleation (see Section 12.7.5).

In principle, the high dislocation densities and fine grain sizes in optimized NODS steels should reduce excess point defect super saturations, and thus the related displacement damage effects, relative to TMS. In the absence of He bubbles, the effectiveness of NO depends on their binding energies with mobile defects, which control their effectiveness as vacancy-SIA recombination centers. However, this displacement damage mitigating effect of NO has not been fully confirmed. Note the irradiation creep rates in NFA are roughly half of those in TMS (see Section 12.7.8), and at low dpa the irradiation hardening rates are also less, although the mechanism may not be related to the irradiation-induced microstructure contribution to hardening per se; rather this is probably mainly due to superposition effects in adding new hardening features to an already highly dispersion-strengthened alloy (see Section 12.7.7). However, irradiation hardening at very high dpa is similar in TMS and NODS, suggesting that displacement damage effects, including on precipitation, are eventually qualitatively similar, perhaps in part due to saturation effects (see Section 12.7.2).

It is well established that the TMS are highly resistant to void swelling compared to austenitic stainless steels (see Refs. [1, 3, 4, 6, 9], Section 12.7 and Chapter 11). Nevertheless, growing voids are observed in TMS at sufficiently high dpa and He levels as discussed in [3–5, 29–32] and Section 12.7.2. Note that a small population of heterogeneously distributed cavities are observed in some NFA; however, they may be due to the milling gas, and in any event, they do not manifest a significant swelling potential. Ion irradiations confirm the high swelling resistance of NFA even at high He and dpa [32, 33], and at very high dpa [34].

Irradiation enhanced and induced nanoscale precipitates also form in both NODS steels and TMS (see Section 12.7.2). Enhanced precipitation is generally associated with higher solute diffusion rates under irradiation, or RED [35], mainly due to excess vacancies. Radiation induced precipitation reflects SIA and vacancy-driven radiation induced (solute) segregation, RIS [6]. Both RED and RIS work in tandem. Precipitation under irradiation is highly sensitive to alloy solute and impurity element concentrations (Cr, Ni, Si, Mn, P, etc.). In typical 14Cr NFA (14YWT), the primary precipitate phase is irradiation-enhanced coherent Cr-rich α' that forms at lower temperatures and times due to RED. Irradiation induced fine-scale clusters, or precipitates, containing Cr and impurities such as Si, Ni, Mn, and P, are also widely observed and are likely due to segregation at dislocation loops [35–38]. Note, solute segregation to network dislocations, loops, and grain boundaries may occur thermally, and does not necessarily require RIS. The precipitates, solute clusters and loops all contribute to irradiation hardening, usually characterized in terms of yield stress increases ($\Delta\sigma_y$) ([37], and see Section 12.7.2 and Chapter 11).

In the absence of large amounts of helium, significant hardening (yield stress increases, $\Delta\sigma_y$) in TMS and NMS occurs only at irradiation temperatures below $\approx 425°C$ ([39] and see Section 12.7.2). Hardening extends to higher temperatures in fission neutron irradiations of 12-14Cr NFA up to $\approx 500°C$, presumably primarily due to α' precipitation [17, 36, 40]. However, in spite of the fact that there is α' precipitation, under lower dose neutron irradiations at $\approx 320°C$, the $\Delta\sigma_y$ are generally less in NFAs compared to TMS [22, 41]. There is also much less reduction in the tensile uniform strain in NFA compared to TMS. For example, TMS F82H experiences almost complete loss of uniform strain ductility (<1%), while the necking strain remains at $\approx 5\%$ in NFA MA957 at a similar $\Delta\sigma_y$ of ≈ 200 MPa, that occur at ≈ 2 and 6 dpa in TMS and NFA, respectively [22, 41]. Unfortunately, for fast reactor irradiations to much higher dose up to ≈ 78 dpa at $\approx 320°C$, both TMS and NFA MA957 experience very large $\Delta\sigma_y$ up to 734 and 650 MPa, respectively, and loss of uniform strain (ε_u) to <1% [42, 43]. Other fast reactor irradiations at 410°C resulted in $\Delta\sigma_y \approx 250$ MPa in NFA MA957, decreasing to little or no hardening at 500°C [17].

The data in [42] suggest that hardening induced irradiation embrittlement measures in terms of subsized Charpy temperature shifts (ΔT) are generally larger in NFA (in the tough orientation) and NMS than for TMS. For example, the tough orientation of MA957 appears to suffer a Charpy transition temperature shift of $\approx 340°C$ for an irradiation to 42 dpa at $\approx 325°C$ [42], while a Eurofer NMS experiences an even larger shift, and a massive drop in the upper shelf energy. These measures of NODS embrittlement compare to a high dpa $\Delta T \approx 225°C$ for TMS [42].

Irradiation creep is the slow time-dependent strain (ε_{ic}) accumulation that occurs down to low temperatures under the combined effects of stress (σ) and dpa [6]. The irradiation creep rates in NFA MA957 are approximately half that for a TMS ([44] and see Section 12.7.2 and Chapter 11), likely due to NO-enhanced recombination.

Optimized microstructures must remain stable during extended service. This requirement often sets an upper use temperature for an alloy. At temperatures above $\approx 550–600°C$, TMS dislocation and lath structures recover, precipitates coarsen, and intermetallic phases form [39, 45–48]. Notably, in combination, these effects can lead to both softening and fast fracture embrittlement. In contrast, long-term thermal aging of MA957 at 850–1000°C for times up to 32,000 h (at some temperatures) showed that the NO are extremely stable at and below 900°C, while experiencing only slight, coarsening at 950°C and slow, but more systematic, coarsening at 1000°C [49]. The NO coarsening is accompanied by small reductions in NFA strength. Dislocation and grain structures are largely stable at 1000°C and $\approx 19,000$ h. These experimental observations, supplemented by higher temperature aging data, are discussed in the context of the controlling mechanisms and quantitative coarsening models in Section 12.7.1.

While there are a few and generally minor exceptions and a trend toward modest inverse coarsening at higher ion irradiation dose rates or very low irradiation temperatures, the preponderance of the evidence in the literature suggests that NO are also extremely stable under both neutron and charged particle irradiations up to high dpa at most service relevant temperatures and dpa rates [17, 50–53]. However, Wharry concludes that the NO experience inverse coarsening in NFA, while NMS undergo slight NO coarsening [52].

Finally as discussed in the following section, at high He levels, voids readily form on larger bubbles in the matrix and, in some cases, on coarser-scale precipitates that promote He bubble-to-void conversion [5].

12.3.3 VOID SWELLING AND HELIUM EFFECTS

As noted above TMS are highly resistant to void swelling in fission reactor irradiations. However, the displacement damage effects described earlier can be greatly modified and enhanced in the presence of high concentrations of helium [2–5, 12, 13, 29–33, 39, 47]. Some potentially severely degrading effects of high helium levels were noted in the introduction. In fission reactors, with low He/dpa ratios of less than ≈ 0.2 appm/dpa, He effects are moderate, and perhaps mainly manifested in promoting void nucleation, but only at very high dpa. However, fusion first wall spectra produce He/dpa ratios of order 10 appm/dpa, or 2000 appm He at power reactor lifetime goals of 200 dpa (see Chapter 3). Thus, void swelling may be an issue in this case.

He plays a key role in the incubation dose prior to rapid void swelling. The incubation dose is associated with precipitation of bubbles that grow slowly with the addition of He, but convert to unstable, rapidly growing voids above a critical size or He content [2–5]. The incubation dose varies widely, but is modest ($<\approx 100$ dpa) in austenitic fcc stainless steels that subsequently swell at high rates of up to $\approx 1\%$/dpa ([1, 2, 6, 9, 54] and see Chapter 8). TMS are much more inherently resistant to void swelling in fission reactors, with a greatly extended incubation dose [3–5, 55]. These reasons include higher self-diffusion rates and lower dislocation sink interstitial bias in bcc TMS, as well as lower He/dpa ratios, compared to nickel-bearing austenitic steels [48].

However, as noted above, sufficiently high He also results in the formation of growing voids in TMS with the potential for significant swelling at high dpa, at estimated post-incubation rates of order 0.1%–0.2%/dpa [3–5, 30, 31, 54]. As shown in Section 12.7.3, combinations of high hardening, including a contribution from bubbles, and grain boundary weakening due to He, can result in shifts of the brittle intergranular fracture temperature of TMS by up to $\approx 700°C$ [3–5, 39, 42, 47]. However, in contrast to TMS, very high He levels can be successfully managed in NFA [3–5, 29, 32, 33, 43, 47].

12.3.4 OTHER NODS ISSUES

NFA have generally superior corrosion resistance compared to NMS due to their higher Cr content. But adding at least 4% Al and higher Cr, up to 16%, may be needed to provide sufficient oxidation and corrosion resistance in NFA for many applications such as accident tolerant cladding and supercritical water reactors [8, 9, 56–60]. Al additions result in coarser-Al-Y-O oxides and a corresponding significant reduction of strength, but this is an acceptable trade-off in many applications. Further, additions of Hf or Zr help to refine the oxides [56]. However, the use temperature limits imposed by environmental attack also depend on many details of the type and chemistry of the coolant, including system level considerations. Candidate coolants include supercritical pressurized water [56, 57], liquid lead-bismuth [59], He (with O_2 impurities), and high-temperature steam, generated in light water reactor core melt accident conditions [60].

As noted previously, NFA and NMS present many outstanding challenges, like developing optimized alloy composition-synthesis designs and thermomechanical processing paths that provide a balanced suite of outstanding and isotropic properties [13]. Further, there is a need to improve alloy homogeneity and heat-to-heat reproducibility. A bigger challenge is developing methods for intricate shape fabrication and joining that maintain optimal microstructures and yield defect-free, functional structural components. Finally, important practical NFA and

NMS challenges include reducing costs, establishing industrial scale supply sources and qualifying new alloys for extended nuclear service [13].

12.4 OVERVIEW OF THE COMPOSITION AND PROCESSING PATHS FOR NFA AND NMS

12.4.1 ALLOY COMPOSITIONS, PHASE DIAGRAMS, AND TRANSFORMATION PATHS

As discussed earlier, there are two primary variants of Fe-based NODS alloys: low C, fully *ferritic* stainless alloys (NFAs) with >12%Cr; and C-bearing *ferritic-martensitic* transformable steels (NMS) with ≈9%–12%Cr. Both NFA and NMS are dispersion strengthened by a very high density of ultrafine Y-Ti-O NO. Typical alloy compositions are given in Table 12.1 in wt%. Note, along with some Ti, the O content must be in excess of that introduced by the Y_2O_3 to form $Y_2Ti_2O_7$ [4, 8, 9, 68]. Alloying with Cr provides corrosion-oxidation resistance. NFA typically contain 12%–16%Cr, while NMS have lower Cr and typically contain about 0.1 wt%C. Low neutron activation W, used in place of Mo, provides solid solution strengthening, while the Y, Ti, and O cluster to form NO dispersed barriers to dislocation glide and climb. The NO population also stabilizes fine grain and high dislocation structures. Notably, Ti is a necessary ingredient in forming NO as shown first by Ukai [69] and subsequently by many others such as in Refs. [4, 8, 9, 70, 71].

Fig. 12.1A and B show Fe-Cr [72] and Fe-9Cr-0.13C-0.2Ti [8, 9, 73, 74] phase diagrams that illustrate the key differences between the NFA and NMS alloy classes. The so-called ferritic stainless steels (>12%Cr), with low C, remain in a single bcc α-phase up to their solidus temperature, since they fall outside the Fe-Cr γ-loop shown in Fig. 12.1A (note Fig. 12.1A does not provide an accurate description of the $\alpha - \alpha'$ two-phase region as discussed in Section 12.7.5). Fig. 12.1B shows that lowering Cr and adding C in NMS with 0.2% Ti results in a classical high-temperature fcc γ (+TiC)-phase field. For example, at 0.1%C, the γ-phase exists between ≈900°C and 1100°C [8, 9, 73, 74]. Thus typical NMS normalizing heat treatments are to first austenitize in γ-phase region at ≈1050°C, prior to cooling at controlled rates to produce various transformation product microstructures, usually bcc α-ferrite or bct martensite, for slow and rapid cooling rates, respectively. Note, austenitizing followed by a moderate rate of cooling is often called normalizing. The slowly cooled ferritic variants are weak, so NMS are typically transformed

Table 12.1 Examples of the Nominal Key Element Contents for Some NFA and NMS (Balance Fe With Other Elements <0.5 wt%)

Alloy	Cr	W/Mo	Ti/Al	Y_2O_3/Y/O	C	Ref.
MA957	14	–/0.3	1/–	0.25/–/–	0.01	[16, 22, 44]
J12YWT	12	2/–	0.3/–	0.24/–/0.09	0.06	[8]
J15YWT	15	2/–	0.2/4	0.35/–/–	0.03	[61, 62]
Fr14YWT	14	1/–	0.4/–	0.3-0.6/–/–	0.02	[63]
NFA-1	14	3/–	0.3/–	–/0.2/0.125	0.03	[12, 20, 64]
ODS Eur.	9	1.1/–	–/–	0.3-0.5/–/–	0.11	[65]
J9CrNMS	9	2/–	0.2/–	0.35/–/–	0.13	[8, 66]
Fr9Cr NMS	9	1/–	0.2/–	0.3/–/–	0.1	[67]

J, Japanese; *Fr*, French; *YWT*, *NFA*; Note O content often not directly provided but can be found if atom probe tomography (APT) data are available.

12.4 OVERVIEW OF THE COMPOSITION AND PROCESSING PATHS

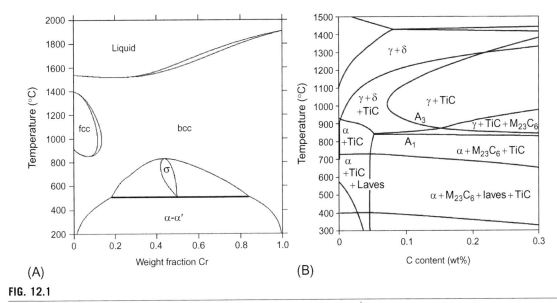

FIG. 12.1

(A) A standard-phase diagram for binary Fe-Cr showing the γ-loop, the $\alpha - \alpha'$ field below ≈500°C and the σ-phase region that forms at ≈50% Cr concentrations above ≈500°C [72]; and (B) the phase diagram for a 9Cr-0.2Ti-2W alloy as a function of the carbon (C) content [73].

to at least partially form martensite, that is then tempered at around 750–800°C [8, 9]. Tempering leads to reduced hardness and improved toughness. Again, standard 9Cr-C alloys (without NO) are called TMS.

Ukai reports that the presence of sufficient Y_2O_3 additions stabilizes a higher temperature $\gamma+\alpha$ two-phase field [8, 9, 74]. This is advantageous, since the residual ferrite retained upon cooling contains a much higher number density of smaller NO than in the transformed martensite phase. Thus the residual ferrite is much stronger than the martensite, especially at higher temperature. The ferrite fraction, in the form of elongated grains of ≈30–60 μm in length and 3–10 μm in width, comprises about 20% of the microstructure. Ukai proposed that the presence of residual ferrite at 1050°C is due to a combination of NO restricted grain γ-boundary mobility and dissolved O effects on the thermodynamic free energy differences driving the α-transformation [8, 9, 74]. As discussed in Section 12.6, analysis of data in [8] indicates that the creep rupture time at 700°C and 100 MPa is ≈1000 times longer for the dual-phase microstructure, compared to fully transformed tempered martensite. The creep strength in the dual-phase alloys is comparable to that in the weak direction of NFA. Even stronger dual-phase steels can be produced by deformation in the two-phase intercritical $\gamma+\alpha$ region illustrated in Fig. 12.1B [66].

These examples of the science of steel metallurgy lead to enormous practical differences between the NFA and NMS. In particular, the diversity of options for processing NMS is much greater than for NFA, since the fcc austenite ↔ bcc ferrite ↔ bct martensite transformation microstructures sensitively depend on, and can be manipulated by, many variables, including most simply the cooling rate and hot working conditions. Thus the transformation characteristics NMS are somewhat akin to those for ordinary 9Cr TMS, but modified by the detailed compositions and presence of NOs, as well as the overall powder metallurgy-based thermomechanical processing path.

The relative advantages and disadvantages of NFA versus NMS are further discussed in Sections 12.6 and 12.7. Briefly, NFA are generally stronger than NMS, at least in orientations in the primary deformation direction, and they are likely to be more He and damage tolerant. However, unrecrystallized extruded, or otherwise deformed, NFA are textured, with anisotropic properties, thus they are more difficult to fabricate into components, especially delicate product forms like thin-walled tubing (cladding). The superior NMS formability is due to the $\alpha \rightarrow \gamma \rightarrow \alpha$-phase transformation, which greatly mitigates the texturing and property anisotropy [8, 9, 67, 74]. Further, the phase transformation results in a ductile ferritic structure that can be shaped. In the duplex Japanese steels (ferrite + martensite) the initial ferrite does not completely transform, or recrystallize during austenization, hence, remains very strong. The austenite that subsequently transforms to a martensite phase is "ductile" and can be shaped. A good balance is needed between the untransformed strong ferrite and the transformed ductile martensite phase.

12.4.2 PRECONSOLIDATION PROCESSING

The processing paths for NFA and NMS prior to heat treatment and fabricating shapes are roughly similar [4, 8, 9, 74]. Yttrium is a very large atom and, along with O, has a vanishingly low equilibrium solubility in Fe [75]. Thus NODS cannot be processed by normal melt practice. As illustrated in Fig. 12.2, the standard NODS processing route begins with gas atomization of powders [77], that are either prealloyed with Cr, W, and Ti, or simply by blending elemental metallic powders. In both cases, the metallic powders are usually then mechanically alloyed (MA) with Y_2O_3 powders by high-energy ball milling [64, 78–80]. The MA step is followed by hot consolidation via hot isostatic pressing (HIPing) [64] or extrusion [78].

In gas atomization [64, 77], the powders are vacuum induction melted in a furnace lined with an alumina crucible that tilts to transfer the liquid alloy into a refractory pouring tundish (pot) with a ceramic nozzle at the

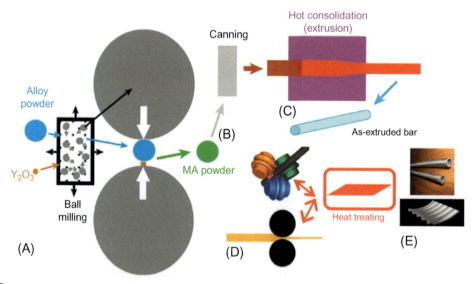

FIG. 12.2

NODS processing and fabrication paths: (A) ball milling; (B) MA and canning powders; (C) consolidation by hot extrusion (or HIPing or SPS not shown) [4]; (D) working by hot rolling and cold pilgering [76] with intermediate heat treatments to fabricate product forms like, (E) shaped plates and tubing [9].

Panels A–C republished with permission of Annual Reviews, From G.R. Odette, M.J. Alinger, B.D. Wirth, Recent Developments in Irradiation-Resistant Steels, Annu. Rev. Mater. Res. 38 (2008) 471–503 ; Panels D (pilgering) and E (tubing and shaped plate) courtesy of Y. De Carlan, (CEA) and the Cambridge University Advanced Structures Group, respectively.

bottom. The combined effects of gravity and applied pressure deliver the melt through the nozzle to a gas atomization die, typically consisting of coaxial, codirectional supersonic gas flow to disrupt the stream of molten metal. The gas flow breaks the melt stream into "atomized" spheroidal droplets that solidify quickly while moving through a cooling tower in a controlled atmosphere. Atomization is usually carried out in Ar, but He atmospheres, with greater thermal conductivity, and higher cooling rates, have also been used. The Y can be included in the melt [12, 64, 78]. In this case, a controlled Ar/O atmosphere can be used in order to increase the powder O content to an optimal value, since it is no longer being supplied by Y_2O_3 by MA. The atomized powders are then sieved into different large, medium, and small sizes. The powder compositions are relatively uniform, but the O content increases with decreasing powder size due to the larger fraction of oxide scale. Powders larger than $\approx 100\,\mu m$ are generally not included in the subsequent ball-milling step.

High-energy milling for MA (Fig. 12.2A) uses hardened steel balls that are accelerated to high velocities in a sealed, controlled atmosphere vessel containing a smaller charge of the powders [64, 78–80]. High velocities are imparted to the balls in a variety of ways, like "shaking" the can in a controlled pattern in a SPEX mill, or by stirring the balls with impeller arms in an attritor mill. The balls collide with each other and the walls of the can, producing a complex pattern of trajectories that frequently result of impacts between the balls and trapped powders (Fig. 12.2B). The ball-powder-ball and ball-powder-wall collisions produce several effects, including deforming the metal and fragmenting the ceramic powders, respectively; initially welding and strain hardening the metal powders; subsequently fragmenting the hardened metal powders; mixing the oxide and metal into multiphase composite powders; continued fragmentation of the embedded oxides, primarily by imposing large plastic strains producing dislocations slip that slice the Y and O-rich particles down to sub-nanometer sizes, thus eventually dissolving, or nearly dissolving these elements in the ferrite matrix [4]. In the mid-1990s that Okuda et al. proposed that oxides were dissolved by MA [81], as subsequently confirmed by Alinger et al. [4, 71]. Recently, Hary et al. showed that nano-oxides can also be dissolved by severe plastic deformation [82]. Ball milling is a severe deformation process that supplies more than roughly 24 eV of energy needed to fully dissolve Y and O from a Y_2O_3 reference state [82].

MA of Y and O requires proper milling conditions. There are a very large number of milling variables and parameters that interact with one another in a combinatorial fashion that are far too numerous to describe; a partial list includes [64, 78–80, 83–85]: the milling energy and tool; ball sizes and the ball to powder mass ratio; the milling environment; milling temperature and time; the oxide and metal powder sizes; and so on. The most important factors are probably the combination of milling energy and time. High-energy ZOZ attritor milling requires on the order of 40 h for proper Y mixing [4, 64, 78]. Powders pre-alloyed with Y form a cellular-phase separated $Fe_{17}Y_2$ as-atomized powder structures thus also require such long-time milling.

A typical set of attritor milling parameters is [78]: $\approx 70\,\mu m$ (average) prealloyed rapidly solidified metallic powders with 20 nm Y_2O_3 powders, at a 10/1 ball to powder mass ratio using 5 mm balls for 48 h in a dry Ar environment. MA by ball milling results in an enormous supersaturation of dissolved Y and O, which cluster and precipitate, along with Ti, during hot consolidation [4, 8, 9, 64, 70, 71]. The as-milled powders have very fine crystallite grain sizes (20–30 nm or less) and high dislocation densities. The hardness of as-milled powder is on the order of several GPa [4, 70, 71].

12.4.3 CONSOLIDATION

The milled powders are then typically canned and hot vacuum degassed below $\approx 400°C$ before HIPing or hot extrusion that yields high consolidated densities ($\geq 98\%$ theoretical), with little, or no, macroscopic porosity as illustrated in Fig. 12.2C [4, 8, 9, 64, 70, 71, 78]. Consolidation temperatures typically range from 850°C to 1150°C. HIPing is a consolidation process that deforms the cans and powders at high hydrostatic pressures of order $\approx 15\,MPa$ for hold times of a few hours, following heat-up and preceding a cool-down periods of about

1 h each [4, 64, 70, 71]. HIPing removes the porosity by a combination of plastic deformation and surface and grain boundary diffusion [64, 70, 71, 78]. Hot extrusion involves forcing the canned powders though a die resulting in a large reduction in cross-section and a corresponding increase in length [78, 80]. The MA powder can is rapidly heated and pushed through square or round cross-sections dies at extrusion ratios ranging from 5/1 to 10/1. Kilogram quantities laboratory NFA and NMS heats are common [78]. HIPing has also been used to consolidate multiple small batches of research powders simultaneously [4, 64, 68, 70, 71, 85]. Both consolidation methods can be scaled up to much larger sizes. Alloys extruded at lower temperatures (e.g., 850°C) are typically annealed for a short time at higher temperatures (e.g., 1 h at 1000°C) to relieve residual stresses and stabilize the NO and the balance of the microstructure [78].

The hot extruded or HIPing condition is usually the starting point for subsequent deformation processing (Fig. 12.2D) [76] into product form shapes like cross-rolled plates and thin-walled tube (Fig. 12.2E) [9]. However, near net shape HIPing may be possible in some cases. More recently, spark plasma sintering (SPS) has been used to consolidate the mechanically alloyed powders [63, 86], again with the possibility of a near net shape product. SPS involves using high die pressure and temperature in combination with current flow through the powders that generate spark plasmas within the green body. SPS consolidation is very rapid, which, in principle, may help to maintain small grain sizes.

Extruded NFAs usually have bimodal grain size distributions, and grains with a $<110>$ α-fiber texture [8, 9, 20, 25, 87–90]. The small grains are dominant in number, but not as much in area fraction [20, 25]. This texture is characteristic of deformed bcc ferrite [88, 89]. The bimodal grain size distribution is variously ascribed to differences in stored dislocation energy and boundary mobility related to inhomogeneous distributions of NO [8, 9, 91–93]. However, others find high numbers of NO in the large grains [64, 94]. Extrusion and HIPing at lower temperatures sometimes results in finer and more uniform grain sizes, but even this condition also more typically manifests some degree of bimodality [78]. NFA and NMS dislocation densities are high, on the order of $0.5–2 \times 10^{15}/m^2$ [4, 5, 20, 64, 88]. The preferred crystallographic orientations and elongated shapes of the grains lead to anisotropic mechanical properties [4, 5, 8, 9, 12, 17, 24, 27, 88, 95]. HIP consolidated NFA and NMS are more isotropic.

The property anisotropy in the as-extruded condition is manifested as significantly lower creep strength in the transverse direction (the hoop direction in a tube) and much lower toughness in the axial direction [4, 5, 8, 9, 12, 17]. Transverse orientations typically have only about $\approx 2/3$ the creep strength in the axial extrusion direction, and the difference may be larger for extruded tubing [8, 17]. In contrast to the creep strong and weak directions, the corresponding transverse fracture toughness of extruded MA957 has a much lower ductile to brittle cleavage transition temperature and higher ductile crack-tearing toughness [4, 5, 8, 9, 24, 25, 27, 95]. Thus high creep strength and higher toughness are favored in different orientations. NFA and NMS mechanical properties are discussed in more detail in Section 12.6.

For most applications the anisotropic grain structures and texture in extruded NFA must be modified by subsequent thermomechanical processing treatments (TMTs). These typically involve cycles of cold working (Fig. 12.2D and E), followed by softening recovery and, in some cases, recrystallization heat treatments [8, 9, 62, 76, 96–98]. Note manufacturing product forms like thin-walled cladding tubes is even more challenging than simpler shapes like as-extruded bars or plates [8, 9]. This topic is discussed further.

As a result of anisotropic properties and processing challenges transformable NMS are attractive alternatives to NFA, since they are more deformable and the austenite to martensite/ferrite transformation produces more equiaxed-isotropic grain structures, less texturing, and reduced mechanical property anisotropy [8, 9]. Further, NMS steels provide the opportunity for a much wider range of matrix microstructures, which are controlled by both composition and the time-temperature-phase transformations and posttransformation heat treatments. Transformable NMS single-phase steels are much weaker than fully ferritic NFA. However, as discussed in Section 12.6, the strength of dual-phase NMS is similar to the weak orientation of NFA, and can be even stronger in a hot-rolled condition.

12.4.4 DEFORMATION PROCESSING AND TUBE FABRICATION

The challenges of deformation processing and fabricating shapes depend on the product form. Producing defect-free thin-walled cladding with good isotropic properties illustrates the need for fundamental and physical-mechanical metallurgy principles to meet these challenges. Here, we focus on the status of leading efforts practiced at CEA in France and JAEA/Hokkaido University (JAEA/HU) in Japan. This remains work in progress, but recent advances have been significant and are summarized with further details, especially regarding the critical issue of texture evolution, in Refs. [8, 9, 62, 67, 76, 96–99].

12.4.4.1 NFA

Deformation processing of the US FCRD NFA-1 is discussed in Section 12.4.5. The composition of the CEA NFA is Fe14Cr0.25Ti0.2Y_2O_3 (wt%). The current CEA tube-making practice involves the following steps [9, 67, 98]. A thick-walled, textured mother tube extruded at 1100°C is first softened by annealing for 30 min at 1250°C, and slowly cooled, reducing the hardness from ≈ 500 to $300\,kg_f/mm^2$, while producing large grains; the initial texture remains unaffected. The initial softening is followed by three cycles of cold pilgering with a 20% wall thickness reduction per pass, each followed by intermediate softening anneals at 1200°C for 1 h, and subsequently very slow cooling. As illustrated in Fig. 12.2D, cold pilgering involves a mother tube on a mandrel that moves forward while being rolled by rotating ring dies, that move back and forth, resulting in a large decrease in the wall thickness and increase in the tube length [76]. The deformation is shear dominated and under nearly, but not completely, full compression, with only limited regions of tension [98]. Cold pilgering does not remove the texture, but elongates the large grains in the extrusion direction. Intermediate 1200°C anneals keep the hardness below a critical damage threshold of $\approx 400\,kg_f/mm^2$. A final cold pilger thickness reduction of $\approx 50\%$ is followed by a 750°C for 30 min heat treatment, that results in finer grains, but which are still textured and elongated grains. The final hardness is about $320\,kg_f/mm^2$. The CEA tubing has high tensile strength of ≈ 1000 and 250 MPa at RT and 700°C, respectively, in both longitudinal and circumferential orientations. However, the ductility, creep, and fracture properties are still anisotropic [67, 98, 99].

The most obvious approach to reducing microstructural and property anisotropy is a recrystallization heat treatment. Briefly, this involves introducing sufficient dislocation energy from cold work to drive the nucleation and growth of new grains that, to a great extent, loose their memory of the previous microstructures. An optimal recrystallization heat treatment preserves the NO and prevents excessive grain growth. However, the texturing and low boundary mobility that persists after the CEA extrusion and intermediate heat treatment (HT), coupled with the energies available from accessible amounts of cold-work, limit recrystallization to form new, more isotropic grains.

The composition of one Japanese NFA is Fe12Cr1.5W0.26Ti0.22Y_2O_3, with a minimum amount of Cr to make it just fully ferritic [8, 9, 100]. A second Japanese NFA are 12-15Cr alloys: Fe13Cr0.02C3W0.7Ti0.46Y_2O_3 and Fe15Cr0.03C2W0.2Ti0.3Y_2O_3 [8, 62, 96, 97]. The higher Cr is for corrosion resistance. In the latter case, a two-step tube fabrication path involves softening the 1100°C as-extruded mother mandrel mounted tube, followed by three pilger cold rolls and annealing cycles at a lower temperature of around 900°C to produce recovery while avoiding recrystallization. These intermediate steps form a (001) <110>-texture component, which is a brittle cleavage fracture system. The low-temperature recovery step allows sufficient dislocation energy storage to permit recrystallization. A final heat treatment at 1250°C recrystallizes the cold-worked microstructure, resulting in much more isotropic properties. The hardness of the recrystallized Japanese 15Cr NFA is $\approx 380\,kg_f/mm^2$.

12.4.4.2 NMS

The 9Cr NMS compositions are Fe9Cr1W0.1C0.2Ti0.3Y_2O_3-balance Mn, Ni, Si and Fe9Cr2W0.13C0.2-Ti0.35Y_2O_3 for the CEA and JAEA/HU steels, respectively [8, 9, 67, 74, 99]. Both tube-forming processes again involve cold pilgering. In the case of the CEA tubing, three cycles of 40% reduction mandrel pilger cold rolling plus full austenization at 1050°C for 60 min is followed by slow cooling at 0.03°C/s [67]. The final heat treatment involves the same austenitization conditions, but is followed by rapid cooling at ≈1°C/s, prior to 750°C, for 1 h temper. A roughly similar sequence is used in Japan [8, 9, 74]. However, there is only partial transformation to austenite, while ferrite grains are retained. The retained ferrite is attributed to reduction of the α/γ interface mobility, partly due to pinning by the NO, and O effects on the thermodynamic free energy differences driving the phase transformation [8, 9, 74]. More rapid cooling at 1.3°C/s then leads to a dual-phase structure with untransformed residual ferrite and martensite. A final 800°C, 1 h temper is used to improve the balance of properties. The final hardness levels of the normalized and tempered (NT) CEA and JAEA-HU NMS tubes are about 360 and 400 kg_f/mm^2, respectively. The tensile properties in these NMS conditions are very similar with ultimate tensile stresses of about 1200 and 450 MPa at RT and 700°C, respectively. Note, these static strength levels are actually higher than for some 14Cr ferritic NFA. The creep strength of the Japanese 9Cr NMS dual-phase steel is superior to that of its CEA counterpart, and at 700°C is comparable to that in many NFA in the weak orientation.

While the tensile properties are similar, the CEA and JAEA/HU ODS steels have very different microstructures. The CEA NMS contains a relatively uniform distribution of NO in a NT martensite matrix [9, 99]. In contrast, the martensite in the Japanese NMS steel contains a much lower number of somewhat larger oxides. The martensite phase is softer than the coarse-grained ferrite, which contains a high density of true 2–3 nm NO [8, 9, 74].

Ukai and colleagues at HU have recently developed advanced processing path that results in much higher strength levels in NMS plate [66]. This involves hot rolling between 1050°C and 1200°C, which is in the so-called intercritical two α-γ-phase temperature range region for the 9Cr-0.13C-0.2Ti alloy. The hot rolling produces different interfaces and α/γ interphase orientation relationships (ORs) and coarser-grain sizes than those that form during transformations from the cold-worked condition. This results in a remarkable tensile strength of about 540 MPa at 700°C, which is the highest that has been achieved to date, along with a creep strength advantage over dual-phase NMS. While the responsible mechanisms are not fully understood, it is hypothesized that the hot-rolled, intercritical microstructure, with coarser grains, increases the alloy strength by suppressing localized deformation in the martensitic phase [66]. The balance of properties of the advanced hot rolled plate has not been reported.

Low fracture toughness is a general issue that confronts many variants of NFA and NMS, reflected in high brittle to ductile (BDT) temperature and low upper-shelf-tearing toughness. Thus it is particularly notable that intercritical rolling has been used to improve the tearing toughness of NMS as discussed further in Section 12.6 [26, 27].

This section has focused on fabrication of thin-walled tubing. However, it is important to emphasize that other product forms, such as thicker plates and ducts, may be important in variety of applications. Other fabrication challenges include joining that is briefly discussed further.

12.4.5 TEXTURING AND DAMAGE MECHANISMS DURING DEFORMATION PROCESSING

A Fe-14Cr-3W-0.35Ti-0.25Y NFA 850°C extrusion that is cross-rolled at 1000°C was used to better understand the causes and consequences of texturing [12, 20, 25, 87, 88]. Following ball milling, gas atomized powders prealloyed with W, Ti, and Y were consolidated by extrusion at 850°C, and then annealed and cross-rolled at 1000°C to form an ≈10 mm-thick plate. Fig. 12.3A and B shows that the plate contains numerous microcracks running in planes parallel to the broad faces (and perpendicular to the thickness) that are visible on side and front faces. Fig. 12.3C shows the yield stress and total elongation of tensile tests with the axis in the through short (S) thickness direction compared to that for the in-plane axis extrusions (L) direction. The former orientation loaded normal to the microcrack planes is increasingly brittle below ≈100°C, while the latter is completely ductile.

This so-called NFA-1 heat has a bimodal distribution of pancake-shaped grains with thickness to diameter aspect ratios of ~0.35 in the middle of the plate [20, 25, 87, 88]. The grains have a strong (001)<110> α-fiber texture component in the extrusion direction and a weaker ε-fiber texture in the cross-rolling direction [87, 88].

12.4 OVERVIEW OF THE COMPOSITION AND PROCESSING PATHS 543

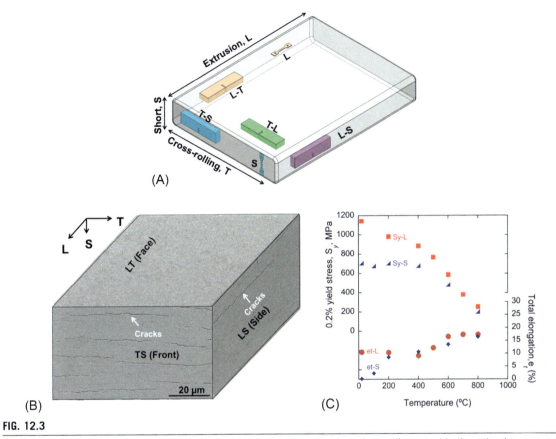

FIG. 12.3

(A) A schematic representation showing the NFA-1 plate specimen orientations with respect to the extrusion, cross-rolling, and short thickness directions (unpublished); (B) section views showing numerous microcracks running on planes normal to the thickness direction [20]; and (C) in-plane (L) and out-of-plane (S) 0.2% offset yield (or elastic fracture) stress and total elongation as a function of temperature, showing the reduced strength and low ductility associated with loading normal to the microcracks (unpublished). Note while such explicit and extensive microcracking is an extreme case, the brittle texture component generally manifests itself as very low toughness orientations in extruded NFA product forms.

The extrusion and cross-rolling deformation forms grains with (001) planes-oriented parallel to the plate faces, containing <110> directions, which is a low toughness ideal cleavage fracture system [47, 101]. The texture component leads to microcracking during cross-rolling. Microcracking is driven by residual stresses as the plate cools below the (001)<110> system BDT temperature.

Fig. 12.4 shows the sequence of texture development for extrusion and cross-rolling and illustrates the basic source of the microcracks [88]. Microscopically, dislocation slip under the plane strain deformation conditions leads to the brittle (001)<110> texturing due to the formation of sessile sub-boundaries via $a/2[111] + a/2[\bar{1}\bar{1}1] \rightarrow a[001]$-type reactions. The sessile dislocation arrays stack up on {100}-type planes to form a strong low-angle sub-boundary, that is, a barrier to further slip; subsequent deformation produces local dislocation pile-up stresses and nano-cracking at the sub-boundary. Again, the microcracking is subsequently driven by residual stresses as the

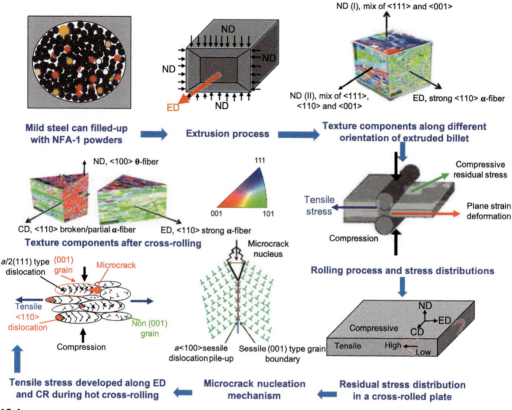

FIG. 12.4

A schematic illustration of deformation processing sequence-of-events in the 14YWT-NFA alloy, starting from the as-milled alloy powder and ending in an extruded billet bar cross-rolled plate [88]. ED, ND, and CD are extrusion, normal, and cross-rolling directions, respectively.

plate cools below the (001)<110>-system BDT temperature. The processing induced microcracks on loading lead to delaminations by cleavage propagation that improve toughness [25], but has only generally moderate effects on the tensile properties, except for loading in the brittle plate S thickness direction [20, 102], as shown in Fig. 12.3. However, these damage mechanisms make deformation processing very difficult. The sequence of deformation steps leading to the microcracking damage is illustrated in Fig. 12.4.

Shear dominated compressive stress states offer a superior approach to processing compared to the plane strain extrusion-rolling deformations described earlier. Hydrostatic extrusion, in which a fluid compressively loads the work piece in a pressurized die cavity, is one approach to producing shear-dominated conditions. To explore this option, a mother tube cut from the cross-rolled plate was hydrostatically extruded on a mandrel at 815°C [102]. This process produced tube sections with diameters of 5 mm and 0.5 mm wall thicknesses. The dominant plate texture component of brittle {001}<110>, gives way to easy shear texture components of {111}<110> and {111}<112> in the tube. The development of a shear dominated γ-fiber texture, lowers deformation stresses and facilitates relaxation of internal stress concentrations by easy slip. Notably, the hydrostatic extrusions are defect-free even in spite of the presence of microcracks in the cross-rolled mother tube.

12.4.6 JOINING

A major challenge to the use of ODS alloys is joining. Conventional fusion welding methods are not useful, since the NO dissolve during typical melting processes and Y phase separates during resolidification. Thus solid-state joining methods are required. For example, pressure resistive welding is used to join the caps to the tubes in the fast reactors [14]. There are many variants of solid-state joining and in the case of NODS, most attention has been paid to friction stir welding (FSW) and diffusion bonding (DB). FSW is a solid-state joining method that involves mechanical mixing of base metals with a rotating tool [103]. The tool advances to produce a nugget of mixed metal from the adjoining base pieces under severe deformation conditions. The FSW process is schematically illustrated in Fig. 12.5A. The key question is—can FSW provide high integrity, damage-free, joints in high strength ODS alloys? There are a huge number of issues and processing variables related to FSW. Thus a full discussion of this topic is far beyond the scope of this chapter.

However, the general conclusion derived from a growing literature, including dissimilar metal joints, is that it is possible to produce relatively high-quality NODS alloy FSW without severely compromising the enabling microstructure or high-performance mechanical properties [104–108], although in some other cases FSW efforts failed. Recognizing there are differences between the individual studies, in summary: (a) FSW modifies and redistributes the NO but they remain functionally similar to the baseline condition in their combined number density and size (Fig. 12.5B), the grain structure in the stirred weld nugget is refined and equiaxed, undergoing dynamic recrystallization (Fig. 12.5C) [109]; (c) the grain size in the heat affected zone of the base metal may coarsen, or become unstable, in NMS and dissimilar metal welds; and (d) the weld region is generally somewhat weaker compared the base metal, but the degradation may only be moderate. Indeed, higher strength in the joint has been observed as illustrated in Fig. 12.5D. FSW of dissimilar metal bonds is also possible [108, 110].

In principle, DB simply involves a high-temperature heat treatment for relatively short periods of time applied to the to-be-joined parts with clean and smooth mating surfaces under a significant pressure. For similar alloys, DB can create a seamless interdiffusion zone joint, which can be nearly indistinguishable from the adjoining base metals. Indeed ODS alloys are a nearly ideal candidate for DB due to their extraordinary thermal stability coupled with the fact that they do not undergo a phase transformation on cooling. The primary issues for dissimilar metal DB joints include: coefficient of thermal expansion mismatches, phase transformation volume changes, interdiffusion exchange of solvent and solute elements, especially C in the case of steels, formation of interface phases (often brittle), and Kirkendall porosity formation. Thus, dissimilar metal DB can alter the microstructure of TMS and NMS DB to NFA, primarily by C diffusion [111]. Liquid-phase and transient liquid-phase diffusion bonding (LPDB) is an alternative to the solid-state route [111]. However, a discussion of these issues is beyond the scope of this chapter.

12.4.7 ALTERNATIVE COMPOSITIONS AND PROCESSING PATHS

Space does not permit detailed discussion of alternative processing paths for NODS. A short list of other possibilities includes: atomization, or other solidification processes, that solute trap significant quantities of Y in solid solution and introduce reactive O from powder surface Cr scales during consolidation that reacts internally to produce NO [112]. The aim is to avoid, or minimize, ball milling. This approach has been described as gas atomization reaction synthesis (GARS). However, this approach has yet to produce true NO and very high strength NFA. Only the smallest powder sizes ($<5\,\mu m$) manifested significant Y-trapping that resulted in the formation of 3–12 nm NO at a number density of $\sim 3 \times 10^{22}\,m^{-3}$. Thus even in this case, the NO are fewer and coarser. Specialized high-velocity hot-spray methods have also been examined, but with a similar outcome. The underlying reason is that, while rapid compared to normal bulk melts, the solidification rates are simply too slow to produce sufficient solute trapping, and avoid phase separation in coarser-grain boundary matrix $Fe_{17}Y_2$ phases. It is likely that this will be the case for additive manufacturing approaches as well.

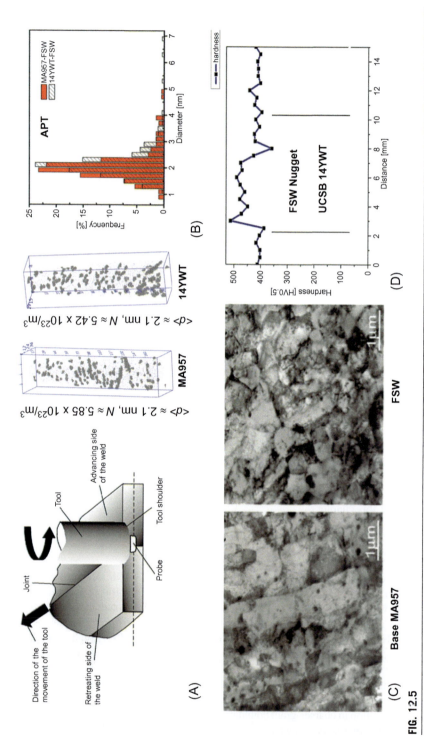

FIG. 12.5

(A) A schematic illustration of the FSW process; (B) atom probe tomography (APT) maps of NFA joints between both MA957 and a 14YWT heat, along with the corresponding size distributions, showing that the high number of nm-scale NOs are preserved; (C) the refined dynamically recrystallized grain structure in FSW MA957 [104]; and (D) a microhardness traverse showing the increased strength of the 14YWT FSW nugget in panel B.

Panel (A) is courtesy of Anand-FSW-Wikipedia and (B)–(D) is unpublished author research.

Limited studies of alternative NFA compositions have also been conducted. For example, substituting Hf or Zr for Ti forms finer and more stable NO in Al added NFA [56]. Further, substitution of Sc for Y has been examined [113]. Smaller Sc atoms have a much higher solid solubility than Y, raising the possibility of melt processing. However the cost of Sc would be prohibitive, and even if not, thermal instability of NO and NFA grain structures would make this option impractical.

Finally, here we focus on ferrite-martensite iron-based alloys. However, the principle of ODS applies to a variety of other structural materials, including austenitic stainless steels, refractory metals, and Ni-based alloys, including $\gamma - \gamma'$ super alloys, and intermetallics.

12.4.8 SUMMARY OF PROCESSING AND FABRICATION

NFA can be processed to high strength levels and recrystallized conditions manifest reasonably isotropic properties. However, they are generally less deformable and more difficult to fabricate into delicate product forms than lower Cr-C-bearing NMS. The NFA variant may have advantages of better: (a) corrosion-oxidation resistance; (b) higher long-time use temperature (associated with greater thermal stability); (c) high-temperature strength; (d) radiation damage tolerance; and (e) capability of managing high levels of He. However, NMS are much more flexible and take advantage of the innumerable options steel metallurgy provided by α/γ/martensite-ferrite-phase transformations. Further, the NMS 9Cr alloys are closing the strength gap and may have a path forward to higher toughness.

12.5 CHARACTERISTICS AND FUNCTION OF THE NO

Here, we focus on the fine-scale NO in Y-Ti-O-bearing NFA, without aluminum additions. After describing the general characteristic of the NO and how they depend on NFA composition and heat treatment variables, we will discuss their nature and how do they function to provide irradiation tolerance and effective He management.

12.5.1 NO STATISTICS

NO average diameters ($<d>$), number densities (N), and volume fractions (f) depend on the alloy composition and processing history. Typical values determined by atom probe tomography (APT), small-angle neutron scattering (SANS), and transmission electron microscope (TEM) are $<d> \approx 2.7 \pm 0.5$ nm, $N \approx 5(1-10) \times 10^{23}/m^3$, and $f \approx 0.5 \pm 0.3\%$ [4, 49, 64, 70, 71, 114]. NO in the martensitic phase of NMS are larger and fewer [8, 9]. The N and f increase with Y and the O as seen in Fig. 12.6A and B [64, 68]. Ti is necessary for fine-scale NO formation as first demonstrated by Ukai [69] and later by many others [4, 70, 71, 85]. Further, O levels of at least ≈ 0.1–0.15 wt % that is sufficient to balance the Ti+Y are important [69]. Fig. 12.6B shows the NO $<d>$, N, and f only weakly depend on the Ti/Y ratio from 1.57 to 3.14, except at the lowest Y [64]. Note, NO are typically somewhat heterogeneously distributed, with substantial local variations around the cited averages, especially in APT studies. While it is tempting to ascribe large grains to the absence of NO, this is not always the case [64, 94].

The $<d>$, N, and f characterized by various techniques, are reasonably consistent with one another, within the limits of the experimental uncertainties and methods as well as expected sample-to-sample variations [64, 115]. SANS generally yields the largest and TEM the smallest N and f, with APT falling in between. TEM tends to underestimate N for the smallest NO.

Adequate MA dissolves the Y and O into solution in the ferrite matrix and the NOs subsequently precipitate during high-temperature consolidation or annealing [4, 8, 9, 64, 70, 71, 85]. Fig. 12.6C shows that the N and f increase, while $<d>$ decreases, at lower heat treatment temperatures [4, 70, 71, 85], which is typical of classical precipitation behavior. Precipitation is also characterized by so-called C-curve kinetics, with the maximum transformation (nucleation and growth) rates peaking at a temperature at which the combination of free energy driving

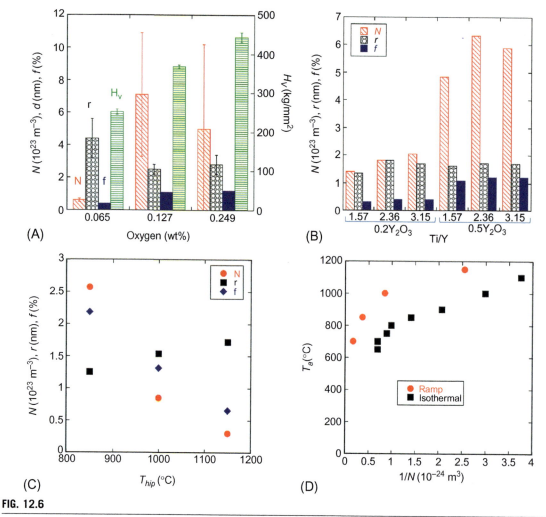

FIG. 12.6

(A) The effect of low, intermediate, and high O on the NO $<d>$, N, f, and NFA H_v [68]; (B) the effects of the Ti/Y ratio and the Y_2O_3 content of the NOs [64]; (C) the effect of consolidation temperature on $r(=<d>/2)$, N, and f [4]; and (D) quasi C-curve-type behavior shown by the variation of $1/N$ for both isothermal and ramped powder annealing temperatures [4].

forces and diffusion rates are optimized. Fig. 12.6D shows C-curve-type precipitation of NO, with increases in N, at a fixed time of 3 h, with decreasing temperature down to a diffusion limited nose at $\approx 700°C$ [4, 85]. Note at relevant consolidation temperatures >800°C, NO precipitation is extremely fast [4, 70, 71, 85].

12.5.2 NATURE OF THE NO

Bright-field-phase contrast and high-angle annular dark-field (HAADF) TEM studies have readily identified NO down to sizes of $d \approx 1$ nm under proper imaging conditions [104, 105, 114–117], as illustrated in Fig. 12.7A. While APT and TEM studies are in reasonably good agreement on N and $<d>$ [64, 115], the composition and phase

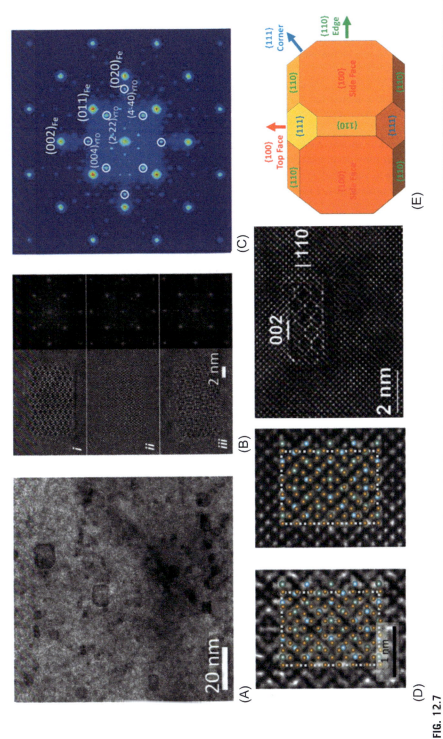

FIG. 12.7

(A) A lower magnification BF TEM image of the polyhedral NOs [116]; (B) high resolution through focal series images and FFT power spectra patterns for a larger NO, characterized in an exit-wave reconstruction analysis [116]; (C) the corresponding combined power spectra pattern showing the extra spots due to pyrochlore $Y_2Ti_2O_7$ [116]; (D) HRTEM and STEM reconstructions of the larger NO showing alternating Y, Y/Ti, and Ti atomic columns, and a 5 × 7 near coincidence lattice interface [116], and a higher magnification BF image of a smaller NO showing Moiré fringes due to coherency strains [118]; and (E) a plausible NO polyhedral shape construction bounded by low-index facets [119].

character of the NO has been a matter of long debate. In part, this was because APT compositional artifacts led to the conclusion that the NO contained far less Y and less O than would be associated with expected stoichiometric oxides, while being highly enriched in Fe [120, 121]. Thus as discussed and compared to other techniques in Refs. [4, 64], the NO was viewed by APT practitioners as being coherent solute clusters, or transition phases, with high Fe contents, rather than well-formed precipitates of second phase near stoichiometric oxides.

However, a large number of TEM studies have clearly shown that most NO, in NODS steels with proper amounts of Y, Ti, and O are near-stoichiometric fcc $Y_2Ti_2O_7$ pyrochlore complex oxides with specific interface OR variants; fewer and larger $Y_2Ti_2O_5$ have also been observed [114, 116, 122–127]. Note other Group IV elements, like Hf and Zr, can add to or replace Ti. Synchrotron X-ray diffraction (XRD), as well as combined SANS and small-angle X-ray scattering (SAXS) studies, have reached similar conclusions [128–130]. Some researchers propose the smallest NO have not yet fully developed into well-formed $Y_2Ti_2O_7$ pyrochlore [131, 132] although extraction studies find $Y_2Ti_2O_7$ down to very small sizes [114]. Fine details aside, however, most importantly understanding of the APT artifacts has evolved [64, 133–135], and their corrected compositions are now generally consistent with the TEM results.

The results of a recent TEM study are shown in Fig. 12.7 to illustrate these points [116]. Fig. 12.7A shows a bright-field image of embedded cuboidal NO over a very large range of sizes. Fig. 12.7B shows a high-resolution transmission electron microscope (HRTEM) focal series exit wave reconstruction of a relatively large $Y_2Ti_2O_7$ oxide along with the corresponding fast Fourier transform (FFT) power spectra. Fig. 12.7C shows the combined FFT diffraction pattern, with extra spots exactly corresponding to the $Y_2Ti_2O_7$ pyrochlore structure. Weaker, but similar diffraction patterns were observed in small NO (not shown) for a variety of MA957 conditions and an irradiated precursor alloy to the NFA-1 14YWT heat cited earlier. Fig. 12.7D compares the HRTEM and scanning TEM (STEM)-based reconstructions of the large NO structure, both showing a 5×7 near coincidence site lattice oxide-matrix interface and alternating pyrochlore columns of Ti, Y, and mixed T-Y. The coherency strains, as illustrated by the Moiré fringes in the small 1.8×3.3 nm NO shown in Fig. 12.7D, increase with decreasing NO size [116]. At larger sizes some NO interface facets become semicoherent [119]. In a parallel study, some of the interfacial ORs found in embedded NOs have also been observed for thin-film depositions of Fe on $Y_2Ti_2O_7$ substrates [136, 137]. The mesoscopic-scale interfaces provide insights into the structures, chemistries, and energies of the Fe-$Y_2Ti_2O_7$ system.

A plausible 3D polyhedral shape of a NO is shown in Fig. 12.7E [119] that is qualitatively consistent with a Wulff construction. However, since the relative energies of the low-index facets are not known, their relative sizes cannot be independently determined, thus the overall roughly cuboidal morphology of the NO is inferred from 2D TEM images. Further details of the NO morphology remain to be confirmed by 3D electron tomography.

Finally, while the NO cores are believed to be near stoichiometric $Y_2Ti_2O_7$ pyrochlore, unoxidized Cr and Ti segregate to their interfaces in what is best described as a core-shell structure [64, 125–127, 138, 139]. Notably, London showed that both Cr and Ti are needed to precipitate the small NO [127], likely in association with decreased interface energies. The Ti segregation increases with increasing alloy content of this element [64] and is very significant in MA957 with $\approx 1\%$Ti. Thus, in summary, the basic nature of the NOs is now well established.

12.5.3 NO FUNCTIONS AND HE INTERACTIONS

The key functions of NO are to provide high-temperature strength and stability of the NFA grains and dislocation structures, as well as irradiation tolerance, especially in managing He by trapping it in small nanometer-scale interface bubbles [3–5, 12, 13]. Thus, the NOs must be stable under prolonged high-temperature and intense irradiation conditions as discussed in Section 12.7.3. First principles calculations suggest that He is first trapped inside the NO [140–142]. Interface bubbles subsequently nucleate on favored facets [12, 119] when they reach a size such that the He energy is lower in the high-density fluid phase than in the oxide. The bubbles grow with the addition of He, but their size remains small, since the number density of high-pressure bubbles is approximately the same as the

12.5 CHARACTERISTICS AND FUNCTION OF THE NO

number of NOs (see Section 12.7). The NOs may also trap point defects, thus act as recombination sites with an efficiency mediated by the defect interface binding energies. A detailed analysis of interface structure and energetics is needed to understand possible NO vacancy-self interstitial recombination mechanisms, although this can be treated parametrically in rate theory models, in terms of sink strengths and defect binding energies. However, once a bubble has formed on a NO, it acts as a sink for the displacement damage defects [3, 4, 12]. This leads to lower excess defect concentration, hence, less clustering in loops and cavities, as well as, in principle, lower RED, RIS, and irradiation creep rates. Thus the ultrahigh bubble sink densities make NFA extremely irradiation tolerant. Indeed, it has been argued that NFA turn He from a liability to an asset [12]. Note, while T does segregate to He bubble interfaces in fusion structures, this is likely to be far less detrimental than T_2 formation in voids in TMS alloys [143].

Fig. 12.8 shows the results of an experiment to study details of the interface He bubble formation mechanisms and sites [119]. Here a 14YWT NFA was thermally aged at 1200°C for 228h to slightly coarsen the NO

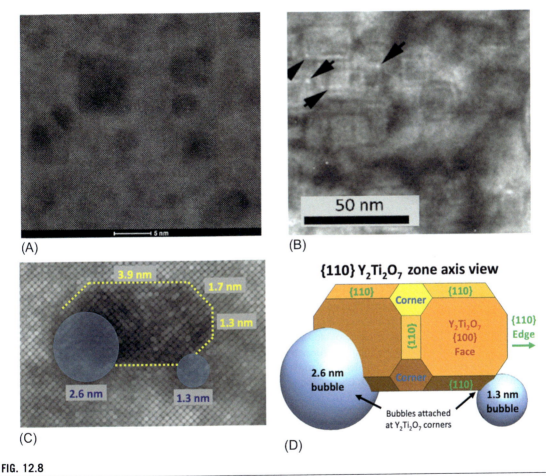

FIG. 12.8

(A) A high magnification STEM image of NOs and attached bubble seen as darker circular regions on the corners of the polyhedral NOs; (B) a lower magnification BF underfocus image of the NO with bubbles attached to small NO corner facets, seen as white circles; (C) a high magnification STEM image of a NO and two attached bubbles after filtering; and (D) a schematic reconstruction of the polyhedral NO and attached {111} corner facet bubbles [144].

so that they could be more easily characterized [119]. The aged alloy was then implanted with ≈4000 appm He at 700°C. Essentially all the He was collected in NO interface bubbles as shown by the He equation of state and TEM size distribution balances [145]. The bubble-NO associations were characterized in HRTEM studies on the TEAM 1 double aberration corrected microscope at the National Center for Electron Microscopy at the Lawrence Berkeley National Laboratory. Fig. 12.8A–D shows that the bubbles favorably form on the semicoherent (111) facets of the polyhedral NO [144]. The semicoherency is demonstrated by the observation of interface dislocations on the (111) facets. As the bubbles grow, they envelop other facets. The size of the He bubbles scales with the NO size. This may be due to the fact that larger oxides can store more He prior to forming interface bubbles; or this may simply be due to a higher He sink strength of larger NO. Further discussion of He bubble-NO associations can be found in Section 12.7. NOs also provide strength and irradiation tolerance in 9Cr NMS.

In general, there are fewer fine-scale NO in the tempered martensite phase of NMS [8, 9]. Thus, superior properties are associated with dual-phase 9Cr steels, which contain residual ferrite where the NOs preferentially form. The NOs in the ferrite phase of 9Cr dual-phase steels are qualitatively similar to those found in fully ferritic, higher Cr alloys [8, 9].

12.5.4 SUMMARY

There is a good understanding of the compositional (Y, O, Ti) and processing variables effects on forming a high density of nm-scale NO in C-free NFAs with 12 or more% Cr. The NOs are primarily $Y_2Ti_2O_7$ complex oxides with polyhedral low index facet-matrix interfaces with a reasonably established set of ORs. Details of how the NOs trap He and form nm-scale bubbles are also now reasonably well understood. The NOs in NMS primarily form in the residual ferrite phase in 9Cr transformable dual-phase steels that have approximately isotropic properties in contrast to fully ferritic NFA.

12.6 MECHANICAL PROPERTIES
12.6.1 STATIC TENSILE STRENGTH AND DUCTILITY

Fig. 12.9A shows that the RT yield stress (σ_y) of NFA varies over a very large range, even for the same nominal alloy, like the different variants of MA957 [12, 20, 21]. The strengthening microstructures are mediated by the alloy composition and its entire thermomechanical processing history. For example, in the case of MA957, σ_y is as low as ≈600 MPa in the fully recrystallized condition [22]. The σ_y is also lower in Al-bearing ODS alloys like PM2000 and MA956 (not shown), with coarser-scale oxides and larger grains [21]. Further, as noted previously, the tensile properties of extruded and otherwise highly deformed NFA are generally mildly anisotropic.

The high NFA static tensile strength derives from a combination of fine grains, high dislocation densities, and an ultrahigh number density (N) of NO. Except at the highest σ_y, the RT ultimate tensile stress (σ_u) of NFA is generally ≈100 MPa, or more, higher than σ_y, indicating considerable strain hardening and significant uniform strains (ε_u). Static RT σ_y up to ≈1200 MPa is observed in Ti-alloyed NMS with optimized dual-phase delta ferrite-martensite microstructures [8].

Fig. 12.9B shows the unirradiated tensile properties of NFA-1 as a function of temperature [20], which are reasonably representative of NFA in general. The σ_y decreases at a moderate rate from RT up to ≈400°C and then much faster between ≈400°C and 800°C. Of course, these nominal temperatures and magnitudes of σ_y depend on the strain rate, $\dot{\varepsilon}$. There is a transition from relatively athermal dislocation glide to the viscoplastic creep that starts at about 400°C [4]. The uniform tensile strains are typically $\varepsilon_u \approx 1\%$–8%, while the corresponding total strains (ε_t) range from $\varepsilon_t \approx 10\%$–25%. There are sometimes orientation effects on the tensile ductility of

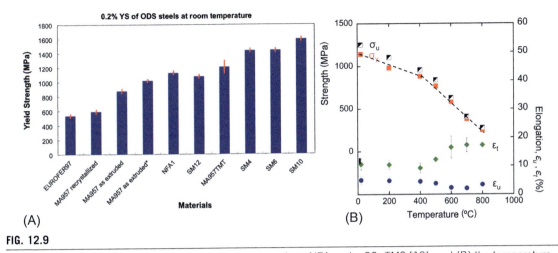

FIG. 12.9

(A) A comparison of the wide range of yield stresses in various NFA and a 9Cr TMS [12]; and (B) the temperature dependence of the NFA-1 yield and ultimate stresses as well as their uniform and total elongations [20].

NFA [24], but they are usually modest [20]. The NFAs ε_t are weakly temperature dependent, typically decreasing slightly up to about 400°C, then increasing up to a maximum between about 600°C and 800°C in the viscoplastic creep regime.

NFA strength generally increases with decreasing consolidation temperature as a result of both higher dislocation and NO number densities and smaller grain and subgrain sizes [70, 71, 85, 146, 147]. The highest strength of NFA as shown in Fig. 12.9A with $\sigma_y \approx 1600$ MPa is a very fine (average 0.16 μm) grained ORNL 14Cr ODS alloy, consolidated at 850°C [78].

The major contributions to the lower temperature ($< \approx 400$°C) strength on NFA alloys include [4, 68, 70, 71, 85, 146, 147]: (a) polycrystalline grain and subgrain hardening scaling with $\sigma_{hp} \approx k_{hp}/\sqrt{d_g}$, where d_g is the grain/subgrain dimension and k_{hp} is the Hall-Petch parameter ≈ 0.25–0.35 MPa $\sqrt{m}$; (b) dispersed barrier hardening roughly individually contributing as $\sigma_{db} \approx 1.69 \alpha_o G b \sqrt{f}/r$, where α_o, G, b, r, f obstacle strength factor, shear modulus, Burger's vector, NO radius, and volume fraction, respectively; (c) dislocation network hardening scaling with $\sigma_\rho \approx 3 \alpha_\rho G b \sqrt{\rho}$, where ρ is the dislocation density; and (d) the matrix strengthening, σ_m, including solid solution hardening contributions. Typical values of α_o and α_ρ are 0.1–0.2 and 0.2–0.4, respectively. The net $\sigma_y \approx \sqrt{(\alpha_o^2 + \alpha_\rho^2)} + \sigma_{hp} + \sigma_m$. For prototypic $f = 0.5\%$, $d_g = 1$ μm, $\rho = 5 \times 10^{14}/m^2$, and $\sigma_m = 200$ MPa, the direct and indirect NO contributions, due to dispersed barrier and grain refinement mechanisms, respectively, total roughly 500 MPa (individually 400 due to the NO and 300 MPa due to the fine grains); this is 50% of a total $\sigma_y = 1000$ MPa. However, smaller grains with $d_g \approx 0.16$ μm result in a much higher $\sigma_y \approx 1600$ MPa [146, 147].

An empirically calibrated σ_y model for these various processes is proposed in Refs. [147]. The σ_m and σ_{hp} significantly decrease with increasing temperature above 400°C, while σ_{db} and σ_ρ strength contributions are sustained up to more than 600°C [146, 147]. Further the deformation mechanism at higher temperatures changes from a mixture of athermal and thermally activated dislocation glide to a climb-controlled glide [148], threshold stress creep, although there is not a sharp boundary between these two processes. The σ_m contribution is also thermally activated, hence, increases rapidly below RT.

While the discussions of strength and ductility here have largely focused on NFA, similar concepts also apply to understanding the tensile deformation of NMS, although details must be modified to account for the wide range of microstructures and multiple phases with different strengths that can be created in this case.

12.6.2 CREEP

As noted above, NFA alloys transition to viscoplastic creep time-dependent deformation between ≈400°C and 600°C. NFA alloys that are stronger at low temperatures also tend to be stronger in the creep regime. The creep curves in Fig. 12.10A illustrate the large effect of orientation in an extruded MA957 thick-walled tube [12]. Here the transverse orientation of the tube is much weaker and less ductile than in the axial direction, as is typically observed for NFA [4, 8, 9, 12]. Typical creep parameters are the primary strain upon loading, typically less than ≈1%, a so-called minimum (or secondary) creep rate ($\dot{\varepsilon}_m$), where strain increases linearly with time, and an onset of tertiary creep strain and time, which in NFA coincides with the creep rupture time (t_r) and total useful strain. Thus the creep performance of NFA is largely controlled by $\dot{\varepsilon}_m$.

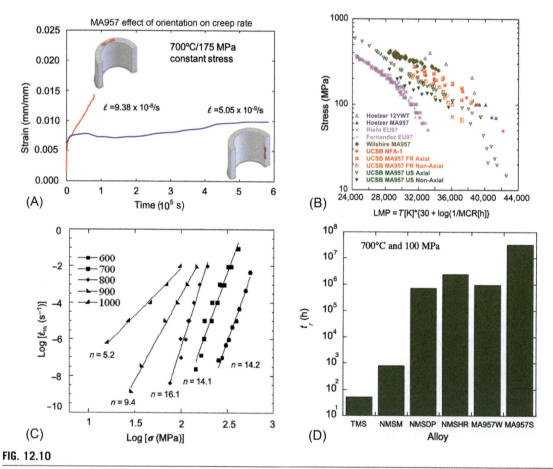

FIG. 12.10

(A) Creep time versus strain for thick-walled MA957 tubing at 125 MPa and 700°C showing highly anisotropic behavior [12]; (B) a Larson-Miller (LM) plot of $\dot{\varepsilon}_m$ of various NFA with a range of strengths that are generally much higher than for the two TMS that are also shown [the LM parameter is $T(K)(30+\log(1/\dot{\varepsilon}_m))$] [149–151,153]; (C) a log–log plot of σ versus $\dot{\varepsilon}_m$ for temperatures from 600°C to 1000°C, showing high stress exponents at 900°C and below due to large threshold stresses [151]; and (D) estimated creep rupture times based on LM plots for TMS, NMSM (martensitic phase), NMSD (dual phase), NMSHR (hot rolled), weak (W), and strong (S) directions of MA957 [21,149–151,153–155].

Panel C from M.C. Salston, G.R. Odette, A database and constitutive model for the static and creep strength of MA957 from room temperature to 1000°C, Trans. Am. Nucl. Soc. 98 (2008) 1146–1147. Copyright 2008 by the American Nuclear Society, LaGrange Park, Illinois.

12.6 MECHANICAL PROPERTIES

Larson Miller (LM) plots are a convenient way to present creep data ($\dot{\varepsilon}_m$ and t_r), where t_r or $1/\dot{\varepsilon}_m$ is related to the applied stress (σ) in terms of a combined temperature-time (T-t) LM parameter (LMP) as [152],

$$\text{LMP} = T(\text{°K})[C_{lm} + \log(1/\dot{\varepsilon}_m)],$$

Here C_{lm} is a fit constant that best collapses data at different σ and T onto a narrow stress versus LMP trend band. Fig. 12.10B shows a representative compilation of minimum creep rate data for $C_{lm} = 30$ [21,149–151,153–155]. Some conventional TMS and NMS data are shown for comparison [154, 155]. As in the case of tensile properties, there is again a very large range of NFA creep strengths, even for the single alloy like MA957. Such large differences are due to a combination of orientation, microstructure, and macro-scale defect effects. The TMS (× and + symbols) are much weaker than the NFA in all of the unrecrystallized conditions. The transverse orientations (open symbols) are systematically weaker than when the specimens are loaded in the axial extrusion direction (filled symbols). Similar LMP plots can be constructed for t_r. A very crude estimate of rupture time is $t_r \approx 0.05/\dot{\varepsilon}_m$, based on an average secondary creep strain. Note that in practice, t_r may also depend on and extrinsic variables, such as macro defects in tubing and the test environment. From a practical perspective, creep limitations are most likely be governed by maximum strain criteria (in structures, typically a maximum of a few %), rather than rupture times.

Another standard way to represent creep data is to plot $\log[\dot{\varepsilon}_m]$ versus $\log[\sigma]$ at different temperatures, to derive temperature-dependent stress exponents, $n(T)$, and effective stress-dependent activation energies, which are related to the creep mechanism. For example, for diffusion creep $n \approx 1$, for dislocation creep in simple single-phase metals $n \approx 3$–5, and for dislocation creep in dispersion strengthened alloys n is much larger, even reaching up to 30 or more in extreme cases [150]. Fig. 12.10C plots $\log[\dot{\varepsilon}_m]$ versus $\log[\sigma]$ at the specified temperatures for as-extruded MA957. In this case, n is ≈ 14–15 from 600°C to 800°C, decreasing to $n \approx 6$ at 1000°C. These high n is due to dispersion strengthening that results in an effective threshold stress (σ_t) for significant creep as discussed in Ref. [4], and, widely in the creep literature. Thus NFA creep behavior is best modeled as [4]

$$\dot{\varepsilon}_m = A \exp(-Q_c/RT)\{[\sigma - \sigma_t(T)]/G\}^n$$

Here the fitting parameters include a scaling coefficient A; the threshold stress σ_t at T which is an appreciable fraction of the static $\sigma_y(T)$; an activation energy for creep Q_c, as mediated by self-diffusion rates, and a temperature independent n that ideally related to dislocation creep mechanisms, theoretically less than around 5. For example, in the case of the as-extruded heat of MA957, assuming a typical self-diffusion value of $Q_c \approx 275$ kJ/mol and a fitted $n = 7$, the $\sigma_t(T)$ are about 0.5–0.6σ_y from 600°C to 700°C, decreasing to 0.4σ_y at 800°C, and approaching 0, along with σ_y, at 1000°C [151]. Similar σ_t/σ_y was found in a much stronger alloy at ≈ 1000°C [150]. This threshold stress creep behavior can be reasonably rationalized by a dislocation climb detachment model [147, 156, 157]. However, while the threshold stress-detachment model provides a basis to correlate NFA creep data, it is not clear that it rigorously applies to the coherent or semicoherent NO.

While these examples, and a number of other phenomenological models, have been reasonably successful in describing NFA creep behavior, a detailed understanding of the underlying mechanisms, and their relations to the microstructure are not yet fully understood. Further, very long-time creep data on NODS is very limited including in the low stress diffusion controlled regime with $n \approx 1$.

Creep data on NMS are more limited and also highly varied. This is not surprising given the wide range of MNS microstructures. Thus a detailed comparison of NFA and NMS creep strength is complex and beyond the scope of this chapter, but can be summarized as illustrated in Fig. 12.10D which plots the estimated t_r at 100 MPa and 700°C, based on LMP data in the literature for: TMS, martensitic NMS (NMSM), dual-phase NMS (NMSD), hot-rolled NMS (NMSHR), and both axial (strong) and transverse (weak) orientations of NFA MA957 (NFAA and NFAT) [109]. Note that NMS have approximately isotropic creep properties. In summary, (a) NMS are stronger than TMS steels; (b) stronger NFA orientations have much higher creep strengths than NMS; and (c) stronger NMS can have comparable, or slightly better, creep strengths than NFA in unfavorable transverse orientations. It is important to emphasize that these t_r are valued based on extreme extrapolations, and are not predictions, but rather a rough basis for comparison. Further the large differences in t_r are due to its extreme sensitivity to stress with a σ^n dependence, where n is very large (e.g., $n \approx 15$); the differences in the σ at the same $\dot{\varepsilon}_m$ are much smaller.

12.6.3 FAST FRACTURE AND FATIGUE

12.6.3.1 Fracture

Fracture resistance in NFA has been most often characterized by subsized Charpy impact tests. Here we focus on actual fracture toughness measurements on pre-cracked specimens [24, 25, 27]. A recent comprehensive review of the toughness databases for NFA and NMS can be found in Ref. [27]. As shown in Fig. 12.11A, NFAs manifest a very wide range of fracture toughness values. Here K_{Ic} and K_{Jc} fracture toughness data are plotted as a function of temperature, showing a classical increase in toughness from a cleavage lower shelf, up to a ductile-tearing upper shelf. Normally, the fracture properties of extruded NFA, with elongated and textured grains, are highly anisotropic. Fracture toughness is lowest for crack propagating in the extrusion (axial), or primary deformation, direction (most of the data in the dashed oval in Fig. 12.11A), while being higher, with a lower BDT, in the transverse or radial direction (most of the data in the dashed box in Fig. 12.11A). This is also illustrated in Fig. 12.11B as extruded MA957 [24]. Fine grains are believed to enhance toughness, and even a small fraction of larger grains, in typical bimodal size distributions, may be detrimental [158]. Thus NFA with uniformly distributed, very fine, nearly equiaxed, minimally textured grains are expected to have a lower BDT [159]. Fine grains are enhanced by extrusion at lower temperatures. HIP or SPS consolidated NFA also typically have more isotropic microstructures and properties than in extruded conditions, but this does not necessarily translate to higher toughness. Notably, transformable NMS are amenable to deformation-heat treatment processing paths that can substantially increase their toughness [26].

Fig. 12.11C shows the $K_{Jc}(T)$ curves for two orientations for a 14YWT plate (NFA-1) extruded at 850°C followed by annealing and cross-rolling at 1000°C [12, 25]. In both cases the NFA-1 crack front runs through the plate thickness dimension. Normally, the toughness is lower when the cracks propagate in the extrusion direction (TL). But in this case, the extrusion and transverse propagation direction (LT and TL) toughness values are similar. The BDT of $\approx -175°C$ is very low and similar in both orientations. The effective toughness is even higher for cracks initially propagating in the short thickness direction (TS and LS), due to crack deflection by $\approx 90°$ and subsequent Mode II propagation parallel to the plate surfaces.

The mechanism leading to high NFA-1 toughness in all of these cases is extensive delamination that occurs along planes perpendicular to the short direction (parallel to the plate faces), similar to those in the tough orientation of MA957 [24, 25]. In the case of LT and TL orientations, the delamination across the thickness direction suppresses cleavage by changing the internal crack-tip field from a near plane strain, to a near plane stress state, corresponding to much lower peak internal stresses that are insufficient for cleavage to occur [25]. Note that the extrusion and cross-rolling, and some other deformation processing paths, lead to texturing, with a large volume fraction of (100)<011>-cleavage system component formed as low-angle sessile dislocation boundaries [88]. Cleavage microcracks form and propagate along the low-angle grain boundaries.

Ductile crack-tearing toughness is another important measure of fracture resistance. Fig. 12.11D shows that the maximum load ductile-tearing toughness of NFA-1 is ≈ 100 MPa $\sqrt{m}$ at $-150°C$ [25]. The displacement at peak load is a measure of a cracked-body's ductility, analogous to the uniform strain in a tensile test, and the subsequent gradual fall-off of the load means the fracture process is very graceful, akin to a tensile total elongation ductility. Thus, ductile tearing is unlikely to be an important failure path in most structures with significant compliance. High compliance is associated with decreasing load with increased load point displacement. Thus in the absence of dead weight-type loading, high compliance leads to crack arrest.

Unfortunately, in NFA-tearing toughness and peak loads decrease significantly at higher temperatures [27], and ductile tearing ultimately transforms to slow creep crack growth that remains largely unexplored. The low

12.6 MECHANICAL PROPERTIES

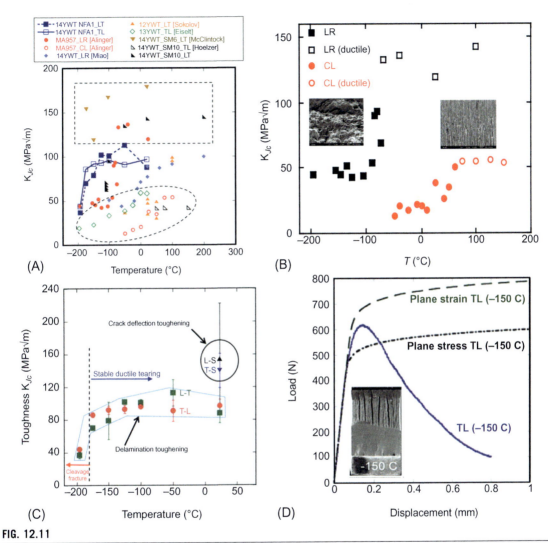

FIG. 12.11

(A) The huge spread of fracture toughness values for a variety of NFAs as a function of temperature, where the data enclosed by the dashed oval and square are generally for cracks propagating in the brittle deformation direction and normal to this direction, respectively, while the data connected by the solid lines for NFA-1 reflect the effect of delamination toughening [25]; (B) toughness-temperature data for NFA MA957 with cracks propagating in brittle extrusion direction (circles) and transverse to this direction (squares) showing cleavage and dimpled fracture surfaces, respectively [24]; (C) toughness-temperature data for NFA-1 showing a very low brittle to ductile transition temperature due to delamination toughening [25]; and (D) a load-displacement curve for NFA-1 showing extensive stable crack growth and high-effective ductility with a peak load falling between plane strain and stress states due to the delaminations seen in the insert [13, 25].

Panel C from S. Pal, M.E. Alam, G.R. Odette, S.A. Maloy, D.T. Hoelzer, J. Lewandowski, Microstructure, texture and mechanical properties of the 14YWT nanostructured ferritic alloy NFA-1, I. Charit, et al. (Eds.), Mechanical and Creep Behavior of Advanced Materials Ch 4, Springer (2017), pp. 43–54.

high-temperature toughness is due to the combination of decreasing σ_y and low blunting crack-tip opening ductility. Ductile tearing and creep crack growth are both likely promoted by a population of coarser (20–160 nm) Ti and Y ceramic inclusions [160]. Thus balancing Ti, W, and O, and avoiding impurities like C and N, may be especially important improving NFA properties at higher temperatures. In contrast to NFA, the tearing toughness in TMS and NMS do not decrease nearly as much at higher temperatures [27]. For example, higher tearing toughness, comparable to that in conventional 9Cr TMS, has been demonstrated in a low C NMS following hot rolling 20%–50% at 900°C in the two-phase α-γ inter-critical temperature region just outside the Fe-Cr gamma loop [26, 27]. The hypothesis is that this treatment improves the grain boundary strength.

While delamination toughening is beneficial in NFA-1, such damage may not be tolerable in many applications and for other properties. In the absence of delamination toughening, the fracture resistance of NFA can be improved by processing paths that lead to small and uniform grain sizes, minimal texturing, and the absence of microcracks and other features that promote cleavage. Hydrostatic extrusion has been identified as one such processing path [102]. Microcracks can also be partially healed by high-temperature ($\approx 1200°C$) heat treatments.

12.6.3.2 Fatigue

The higher strength and microstructural stability of NODS give them a significant fatigue resistance advantage over 9Cr TMS [161–166]. The main reason is that at a specified peak stress, the cyclic plastic strain component in the NODS case is a significantly smaller fraction of the total strain, leading to much greater fatigue strength as illustrated in Fig. 12.12A for RT stress-cycles to failure (S-N_f) tests [161]. Here the fatigue strength is about 2 times higher for the NODS steels versus TMS Eurofer97. Fig. 12.12B shows that the NODS steel fatigue strength advantage persists to high temperature and that, unlike TMS, these alloys do not cyclically soften [166]. This important NODS steel attributes to their microstructural stability, primarily provided by the NOs [163]. As expected, the NODS advantage also holds for creep-fatigue conditions [162, 164]. Note, there is no sign of a fatigue limit in the NODS steels. However, Fig. 12.12C [166] shows that in the elastic–plastic regime, a large number of fatigue cycles to failure (N_f) in NODS steels is limited to conditions with low plastic strain increments of order of $\approx 0.1\%$, again demonstrating the importance of the underlying alloy strength.

Limited data are available for very high cycle fatigue and fatigue crack, $da/dN(\Delta K)$, growth (fatigue crack growth, FCG) rates in NODS alloys. Notably, a comparison study of a set of very different alloys, including NODS in Fig. 12.12D, shows that the da/dN is very similar when plotted on an applied elastic-plastic $J_{a,pl}$ scale, as shown in Fig. 12.12D [167]. Unfortunately, like in the case of most other fatigue properties, there is not a significant database on the effects of many important fatigue variables, like mean stress, frequency, variable amplitude loading, creep hold times and so on. Finally, the critical effects of the service conditions, especially the chemical environment, remain largely unexplored. Irradiation effects on mechanical properties, including fatigue, are discussed in Section 12.7.

12.7 THERMAL AGING AND IRRADIATION EFFECTS

There are four key issues associated with the use of ODS alloys in fission and fusion service environments.

1. What is the thermal and irradiation stability of the start of life microstructure, especially the NO?
2. What new microstructures are introduced by irradiation service?
3. What are the consequences of these microstructural evolutions to mechanical properties and dimensional stability?
4. How does high He in fusion environments affect the irradiation response of NODS in comparison to the TMS alloys?

To begin, it is important to note that the effects of thermal aging and irradiation depend on many irradiation and material variables and their combinations. The data are also limited and the variables are often neither well

12.7 THERMAL AGING AND IRRADIATION EFFECTS

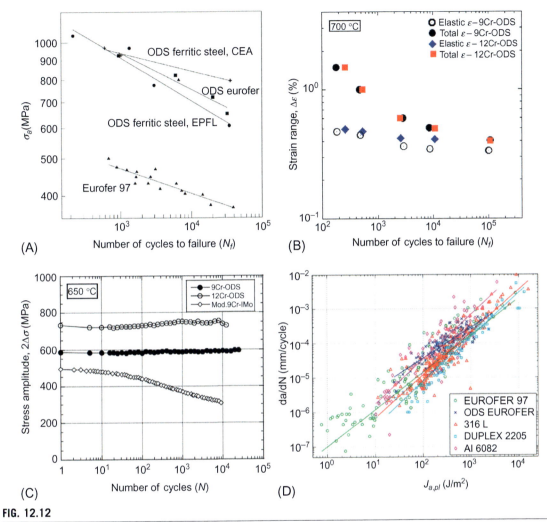

FIG. 12.12

(A) The cycles to failure (N_f) for 3 NODS compared to TMS Eurofer97 [161]; (B) N_f as a function of the tension-compression stress amplitude range ($2\Delta\sigma$) for NODS steels and TMS steels [166]; (C) the elastic and elastic plus plastic total strain range ($\Delta\varepsilon$) versus N_f for 2 NODS steels [166]; and (D) fatigue crack da/dN growth rates as a function of the peak elastic-plastic J-integral loading for a variety of alloys, including TMS and NMS [167].

controlled, nor systematically explored. Hence, data in the literature are scattered and not always self-consistent. The discussion below attempts to analyze the broad trends in thermal aging and irradiation effects on NODS that are judged to be most reliable and pertinent.

12.7.1 THERMAL AGING

Since coarsening reduces their beneficial contributions to alloy function, the thermal stability of the NOs is critical for temperatures more than 600°C where displacement damage effects are largely absent. A number of thermal aging studies on a variety of NODS alloys have been performed over a wide range of temperatures and times have yielded various results [4, 8, 9, 12, 49, 64, 70, 75, 85, 168, 169]. However, observations in the most extensive studies to date can be summarized as follows.

The NO are remarkably thermally stable below $\approx 950°C$ for NFA consolidated at higher temperatures around $\approx 1150°C$ [49, 64, 70, 75, 85, 169].

For lower consolidation temperatures of $\approx 850°C$, NO coarsen at annealing temperatures above $\approx 1100°C$ to sizes and number densities that are close to those that are observed at the same consolidation temperature [64].

At and above $\approx 1200°C$ NO coarsen rapidly (in a few to hundreds of hours), grains coarsen and dislocations recover, starting at lower temperatures [49, 70, 85, 168].

Recrystallization occurs in cold-worked NFA, with sufficiently high stored dislocation energy that depends on the alloy texture, at and above $\approx 1250°C$ [8, 9].

In order to develop improved NO thermal stability predictions closer to service conditions, Cunningham et al. extended thermal aging of MA957 consolidated at $\approx 1150°C$ to very long times at temperatures between 800°C and 1000°C up to 32.4 kh [49, 64, 75, 169]. Microhardness was used to track strength changes, while SANS, APT, and TEM were used to characterize changes in the NO $<d>$, N, and f, as well as the balance of the NFA microstructure and microchemistry. Fig. 12.13A shows that there are no significant changes in the microhardness during aging at temperatures at and below 900°C (not shown), and only modest softening at 950°C. Even at 1000°C the largest reduction in hardness is only $\approx 13\%$ at 19.5 kh. TEM showed the dislocation and grain structures were also stable at this aging condition [49, 64]. However, there is depletion of matrix Ti from 0.65 at.% to 0.05 at.% after aging at 1000°C for 19.5 kh, which was associated with the growth and coarsening of larger Ti-oxides that may have deleterious mechanical property effects. Note, the effects of TiO are mitigated in newer 14YWT alloys that contain less Ti (<0.4 wt% or less) than MA957 (≈ 1.0 wt%).

SANS APT and TEM showed that slight NO coarsening in long-term aged MA957 at 1000°C at 19.5 kh with an average increase in $<d>$ from 2.6 nm to 4.1 nm, a decrease in N from $4.5 \times 10^{23} \, m^{-3}$ to $1.2 \times 10^{23} \, m^{-3}$, and a small decrease in f from 0.60% to 0.51%.

The MA957 data from the long-term thermal aging and higher temperature (T_a) shorter (t_a) time data from 1150°C to 1400°C up to 480 h [112] were fitted with simple coarsening models as follows [71, 169]:

$$[<d(t)>^p - <d_o>^p] = k_o \exp(-Q_c/RT)t = K(T)t$$

Here $<d_o>$ is the average preaging NO diameter ≈ 2.8 nm, k_o and Q_c depend on the mechanism-dependent p. For example, $p=3$ for bulk diffusion controlled kinetics. The combined data were well fit by a $t^{1/5}$ pipe diffusion-coarsening model ($p=5$) for NO on dislocations. Fig. 12.13B shows the $\ln[K(T)]$ versus $1/T$ fit to the $<d>(T_a, t_a)$, yielding an effective activation energy of 673 kJ/mol and $k = 4.4 \times 10^{-25}$ nm^5/h. Fig. 12.13C shows (a) the predicted time needed to double the $<d>$, which is more than 5×10^6 h at 950°C; and (b) that the NO thermal coarsening will be negligible below $\approx 900°C$. TEM micrographs of the NO-dislocation association are shown in Fig. 12.13D. Notably, these results are consistent with conclusions derived from a cluster dynamics model, with first principles simulations showing that the solubility of Y at dislocations is the rate-controlling mechanism [75].

While most thermal aging studies reach generally similar conclusions, remaining NFA thermal stability issues include the effects of lower consolidation temperatures and possibly compositional factors, like the Ti/Y ratio and coherency state of the NO [169]. Finally, data are limited, but the properties of NMS have been shown to undergo only modest softening at 800°C up to 10^4 h [156].

12.7.2 OVERVIEW IRRADIATION EFFECTS ON THE MICROSTRUCTURE

NFA are highly irradiation damage tolerant [3–5, 12, 13]. In the case of displacement defect damage, this is due to: (1) the remarkable stability of NFA and NO even under extreme irradiation conditions; (2) the fact that NFA can operate at much higher temperatures than the displacement damage accumulation regime; (3) the possibility that

12.7 THERMAL AGING AND IRRADIATION EFFECTS

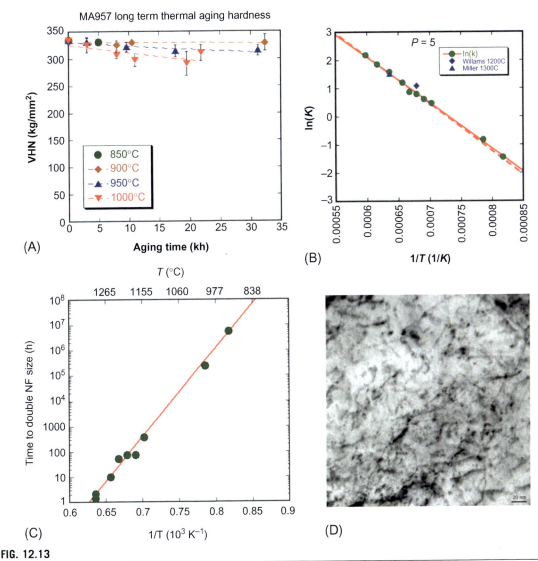

FIG. 12.13

(A) Vickers microhardness of MA957 as a function of annealing time at various temperatures; (B) the temperature dependence of $\ln[K(T)]$ fit by an effective activation energy of 652 kJ/mol for $p=5$; (C) the time (h) to double the size of the NOs at various temperatures; and (D) a BF TEM image of the association between dislocations and NOs under thermal annealing conditions [64].

NO defect trapping enhances SIA-vacancy recombination at lower temperature; and (4) NOs pin and stabilize high dislocation densities and fine grain sizes.

However, the most profound effect of NO is to sequester very large amounts of He in nanometer-scale interface bubbles that help to protect grain boundaries from embrittlement, suppresses void swelling, and enhances recombination. Managing He is absolutely critical in high He/dpa fusion environments, and may even play a role in lower He/dpa fission at very high dpa.

Here we consider the behavior of the NO, network dislocations, grain sizes, precipitates, solute segregation, and cavities including bubbles and voids. Given the greater complexity and diversity of NMS, we primarily focus on NFA in the following discussion.

12.7.3 IRRADIATION STABILITY OF THE NO

One key issue for NODS is the irradiation stability of the performance sustaining NO. There is a large literature on this topic with varied results as described in a recent review [52]. However, most studies find that NO are either relatively stable under both neutron and ion irradiations at elevated temperatures, or show a modest tendency toward inverse coarsening [50–52, 170]. The stability of NO (and other fine-scale precipitates) is mediated by two competing damage processes: (a) RED; and (b) athermal cascade ballistic mixing (BM) of precipitate and matrix atoms [115, 171].

The excess vacancy concentration (X_v) under irradiation results in a higher RED coefficient as $D^* \approx [X_v/X_{ve}] D_{th}$, where X_{ve} is the equilibrium vacancy concentration and D_{th} the thermal solute diffusion coefficient. The X_v can be determined from rate theory that accounts for vacancy production and annihilation both at sinks and by recombination, including at mobile defect trapping sites. In contrast, the BM effect is approximately athermal and is associated with dynamic injection of precipitate (NO in this case) solute atoms into the ferrite matrix, and vice versa, especially by displacement cascades. Thus BM tends to dissolve the NO while RED accelerates their self-repair.

For a typical damage rate of 5×10^{-7} dpa/s, and a sink strength of $5 \times 10^{14}/m^2$, the D^*/D_{th} increases rapidly below $\approx 475°C$ in ferritic crystal structures, while D^* itself decreases more slowly due to recombination and lower D_{th}. At higher temperatures $D^* \approx D_{th}$ increases exponentially with temperature, favoring self-repair. Alternatively, NO can be driven into a steady state, or patterned condition, at smaller sizes and higher number densities compared to the unirradiated condition. A decrease in precipitate size under irradiation is referred to as inverse coarsening, and involves both resolutioning of the solutes, in some cases coupled with NO renucleation on a finer scale.

Further, the rate of self-repair also scales with the chemical potential difference, which drives precipitation in the first place. These differences are huge for NO [75]. Thus complete NO dissolution is expected only at a combination of very low temperatures (low D^*) and high dose rates (high BM rates) consistent with observation [172]. While the review cited above identified a few examples of NO coarsening [52], most cases found them to be stable or to slightly inverse coarsen. One study of neutron-irradiated MA957 suggested a smaller and more numerous NO for an irradiation at 412°C [50]; however, the observed differences were within those associated with typical heterogeneous distributions of NO, and reflected in tip-to-tip variations found in APT studies [64, 115]. Notably, very high-dose neutron irradiation effects of the NO were not observed at higher neutron irradiation temperatures [50]. Other studies suggest that the NO are highly stable under neutron irradiation [17, 170].

Even at the extreme condition, shown in Fig. 12.14, for a 3.5 MeV ion irradiations up to a dose of ≈ 585 dpa at 1.7×10^{-3} dpa/s and 450°C, only slight inverse coarsening of the NO was observed by APT in a 14YWT NFA [51]. Note that TEM showed a slightly stronger, but still very modest, effect. Note the effects of BM are much more profound for Cr- and Cu-rich precipitates in Fe-Cr and Fe-Cu alloys, in which precipitation is prevented [173], or significantly modified [174], respectively. Perhaps what seems most surprising about NO stability is that irradiation does not enhance some coarsening that would be associated with higher D^* and interface roughening and resolutioning of Y. Finally irradiation can also induce solute segregation (RIS) that along with RED assists precipitation, and cascade mixing can amorphize, and thus, in principle, destabilize precipitates [6].

12.7.4 DISLOCATION LOOPS

There are two types of clearly identified dislocation loops in irradiated ferritic/martensitic steels, $a/2<111>$ and $a<100>$. They are the subjects of an extensive literature as discussed in Chapter 11. In addition, small interstitial loop cluster complexes (that may evolve to precipitates) and so-called black spot damage are observed at lower irradiation temperatures. Unfortunately, there are only a few systematic studies of the effects of irradiation

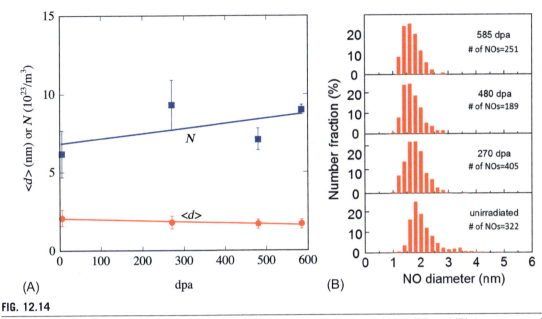

FIG. 12.14

(A) The NO N and $<d>$ as a function of dpa in ion irradiations at 400°C measured by APT; and (B) the corresponding size distribution at various dpa [51].

temperature, dpa, alloy type, and starting microstructure on dislocation loops in NODS steels; and even in TMS, the results are very limited. Hence we will not attempt to distinguish between the loop type or TMS versus NODS here, since for the latter at least one study found them to be similar at 500°C and 21 dpa [28]. Fast neutron irradiations loop densities (N_l) and sizes (d_l) depend on the Cr content, temperature, dpa, and He. However, reported results vary widely. The neutron irradiation of NODS and TMS at 500°C to 21 dpa cited above produced a significant populations of dislocation loops with identifiable <111> and, <100> character [28]. The overall loop diameter (d) and number densities (N) averaged ≈ 10 nm and $3 \times 10^{21}/m^3$, respectively, with a corresponding dislocation line length of $r_l \approx \pi dN \approx 10^{14}/m^2$. However, Gelles found no loops above 450°C in Fe-9 and 15Cr binary alloys [175]. Katoh found $r_l \approx 0.8 \times 10^{14}/cm^2$ at 423°C in the same Fe-15Cr alloy, decreasing to ≈$0.1 \times 10^{14}/cm^2$ at 9Cr [176]. Jiao reports $r_l \approx 1.5 \times 10^{14}/m^2$ at 376–415°C in a 9Cr TMS and none above 460°C [177]. Schäublin finds $r_l \approx 1.3 \times 10^{15}/m^2$ at ≈300°C [178]. At such low-irradiation temperatures, a high density of unresolved loops that initially form in cascades appear as black dots in unalloyed Fe, and also as small solute clusters formed by segregation at small immobilized loops. For example, Bergner found cluster πdN at ≈300°C that range from ≈$1-2.5 \times 10^{15}/m^2$, with a maximum at ≈9Cr [179], which is notably where there are far fewer identifiable larger loops.

Fig. 12.15 shows the general trend in ρ_l as a function of T, divided into low-to-intermediate (≈2–15) and higher dpa (>15). It appears that at higher temperature the loop populations continue to increase with dpa even up to 500°C. However, Weiss reports data for ≈335°C irradiations that peaks at 5 dpa with $\rho_l = 2 \times 10^{15}/m^2$, then decreasing to ≈$2 \times 10^{14}/m^2$ at >15 dpa [180]. Notably, these complex loop trends are generally able to predict only 50%, or less, of the observed irradiation hardening, $\Delta \sigma_y$, below 400°C, assuming an obstacle strength factor of $\alpha_l = 0.3$, combined with a preexisting obstacle strength contribution of 200 MPa [37, 42, 181, 182].

The key question of the effect of NO on the loop structure was addressed in an ion irradiation of a NMS with coarser-scale Y_2O_3 oxides [183]. The primary observation was that fewer loops formed in regions with higher densities of oxides, particularly at lower irradiation temperatures. In regions with low density of

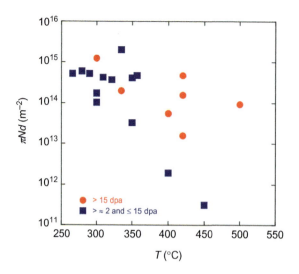

FIG. 12.15

Dislocation loop line length as a function of the irradiation temperature for low-to-intermediate (<15 dpa squares) and higher dose (>15 dpa circles). The data point at 500°C is for in situ He injection irradiations resulting in high 1200 appm He at 21 dpa [109].

oxides the loop populations resembled those in TMS. However, the issue of effect of NO on loops in neutron irradiations has not been resolved. Finally, limited data show that high ≈2000 appm He promotes the formation of dislocation loops at 500°C with average loop line length enhancement factor of by a factor of ≈5 [28].

12.7.5 SOLUTE SEGREGATION, CLUSTERING, AND PRECIPITATION

A variety of larger precipitates in the size range above 10–20 nm are observed in both unirradiated and irradiated NODS and other Fe-Cr alloys. Depending on the alloy composition, these include oxides, nitrides, and carbides of Ti, Y, Cr, W, and Mo. However, here we focus on nanoscale precipitates that form under irradiation, primarily α'- and G-phase-type precipitates ($Ni_{16}Mn_6Si_7$ at the stoichiometric composition), or their MnNiSi solute clusters precursors [35, 36, 38, 53, 179, 182, 184–187]. Note that P, Cu, and Cr are also found in association with these clusters. Further, Cr segregation and clustering observed even in the case of 3-6Cr subsaturated steels.

Cr-rich α' forms in Fe14Cr alloy down to low temperatures around 300°C, or less, due to RED accelerated kinetics. The NOs do not have a significant effect on α' precipitation [50], but less precipitation was observed in a NODS versus non-ODS FeCrAl steel after neutron irradiation at 300°C [188]. Fig. 12.16A shows an example of the high density of α' precipitates in an irradiated Fe-15Cr alloy [184]. Fig. 12.16B shows the Fe-Cr α' matrix composition data following irradiation compared [33] to the solvus line from a recent assessment, in part guided by first principles calculations [189]. The data are based on APT for neutron irradiations of Fe 3-18Cr binary alloys at 320°C and 455°C from ≈1.8 to 7 dpa as well as 9 and 12Cr TMS at 390°C and 410°C to 17 dpa [36]. The Fe18Cr alloy irradiated at 320°C to 1.8 dpa was also characterized after post irradiation annealing at 500°C and 600°C for 300 h. The predicted phase boundary and APT data are in nearly perfect agreement. Note, in order to avoid confusion, the corresponding α'-phase Cr-rich compositions are not shown in Fig. 12.16B, since they increase with the nm-scale precipitate radius, primarily due to APT artifacts; this information can be found, along with an explanation, in Ref. [36].

12.7 THERMAL AGING AND IRRADIATION EFFECTS

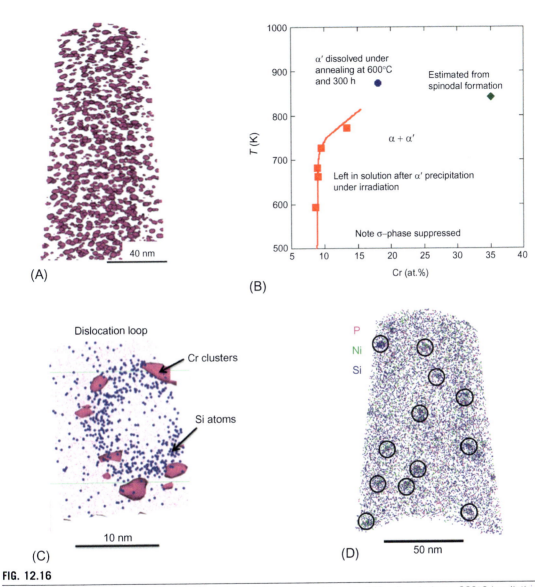

FIG. 12.16

(A) APT reconstruction of the high density of α' precipitates formed in a Fe-15Cr alloy during a 320°C irradiation to 1.8 dpa [184]; (B) the predicted Fe-Cr solvus boundary as a function of temperature (solid line) [189] along with APT data on the residual dissolved matrix Cr for various Fe-Cr alloys and irradiation conditions [173]; (C) a large dislocation loop with segregated Si and Cr clusters formed during a 1.8 dpa 320°C irradiation of 6Cr alloy (unpublished image from work describe in Ref. [185]); and (D) corresponding Ni, P, and Si clusters formed by RIS of solutes from highly under saturated solid solution.

However, the most reliable data for precipitates larger than ≈2 nm indicate that α' is ≈94%Cr at 455°C and ≈90%Cr at 500°C, while dissolving at 600°C, again very consistent with the predicted phase diagram. Thus, 14YWT NFA could contain up to several percent of α' up to ≈450°C decreasing to a negligible amount at 500°C. However, α' is weak dislocation obstacles limiting the associated irradiation hardening. For example,

α', dislocation loops and solute clusters in the Fe15%Cr alloy, in Fig. 12.16B, is estimated to contribute roughly equal amounts of hardening [37].

Irradiation also leads to a significant segregation of many solutes (Cr, Cu, Mn, Si, P, and Ni) to various microstructural features (dislocations, dislocation loops, grain boundaries, and precipitate interfaces) which in turn can lead to clustering and precipitation [35, 38, 53, 179, 182, 184–187]. Segregation mechanisms include both RED, which accelerates thermal processes, and RIS, which drives nonequilibrium solute enrichment at defect sinks. RED and RIS act in tandem to help nucleate and grow Cr-Cu-Mn-Ni-Si-P solute clusters (depending on the alloy composition) and G-phase like and other Ni silicides at dislocation loops (large and small) and network dislocations [35, 38, 53, 179, 182, 184–187, 190, 191]. Further, Cr clustering, along with the other solutes occurs even in the subsaturated 3Cr and 6Cr alloys [35, 184, 185]. Small loops created in cascades are immobilized by solute segregation, and, hence, are observed as clusters that can evolve into precipitates, as illustrated for the 15Cr alloy in Fig. 12.16C. These clusters also form on larger identifiable dislocation loops, as illustrated in Fig. 12.16D, as well as at network dislocations. The cluster number density (N_c) increases to very large values for an irradiation to ≈ 32 dpa at $\approx 335°C$ where they are expected to be the dominant source of a very large amount of irradiation hardening [182].

12.7.6 CAVITIES AND SWELLING

Swelling, or three-dimensional increases in the volume of a material ($\triangle V/V$), is driven by the growth of internal voids [1–6]. SIA and vacancies are created in equal numbers. The mobile SIA defects are preferentially annihilated at biased dislocation sinks, resulting in an excess flux of vacancies that can result in growing voids, if present. We first discuss TMS, with bcc crystal structures, which are much more inherently swelling resistant than fcc austenitic stainless steels [1–6, 9, 17, 19, 29–33, 54]. In large part this is due to the higher self-diffusion rates in bcc versus fcc crystal lattices, that make void nucleation more difficult [55]. Imagine that the excess vacancies are digging a hole in the sand (growing a void) in the lattice, while the self-diffusion rate controls how fast it fills in again with sand (atoms). Other factors that make TMS more swelling resistant than austenitic stainless steels include higher defect sink densities and lower dislocation biases, thus decreasing the net excess vacancy fluxes, as well as lower He/dpa [55]. However, once growing voids form, TMS can swell at what has been a proposed to be an eventual steady state rate of $\approx 0.2\%/dpa$, compared to $\approx 1\%/dpa$ in austenitic stainless steels [54].

Thus the swelling resistance of TMS is largely due to the long incubation times for void nucleation. The void nucleation occurs when helium bubbles reach a critical size [2–5, 55]. Cavities with less helium are stable bubbles with a positive gas pressure P_g balancing the negative surface capillary pressure as $P_g \approx 2\gamma/r_b$, where γ is the matrix surface energy and r_b is the bubble radius. Thus, bubbles grow only with the addition of He. However, due to the excess vacancy flux to unbiased (or less biased) void sinks, caused by the preferential annihilation of SIA at dislocations, cavities larger than the critical bubble size, grow unstably as voids ($P_g <$ to $<<2\gamma/r_v$), resulting in macroscopic swelling, $S = \triangle V/V$.

The dpa and He/dpa ratio (appm/dpa) that are required to evolve a significant fraction of a population of bubbles to the critical size for forming voids defines the swelling incubation dose, dpa_i. At low He/dpa levels, the dpa_i is very high, as is the case for TMS irradiated in fast fission reactors with low He/dpa ($<<1$) [55]. However, a high He/dpa (≈ 10) in fusion spectra results in much lower dpa_i [3–5, 12, 55]. Thus in fusion environments TMS have the potential for significant swelling at very high dpa. Ion irradiation studies and very limited neutron data (see below) suggest that for 10 appm He/dpa, the incubation $dpa_i \approx 50$ at 500°C [3–5, 12]. Thus at 200 dpa the swelling could be $\approx 30\%$.

Under similar irradiation conditions, NFA are much more swelling resistant than TMS due to the fact that the numerous NO trap helium in a very high density of nanometer-scale interface bubbles that are harmless, since they are far below the critical size [3, 4, 12]. In contrast, far fewer bubbles initially form in TMS, mostly on dislocations. Thus in NFA, He is partitioned to fewer and larger bubbles, which can reach the critical size for void conversion at a lower dpa_i. Note the high density of bubbles in NFA also act as sinks that lower vacancy excess vacancy fluxes, that further increase the dpa_i further relative to TMS. The sink strength of NO associated bubbles is estimated to be

12.7 THERMAL AGING AND IRRADIATION EFFECTS

about an order of magnitude higher than the sinks in TMS, potentially greatly reducing the displacement damage accumulation rates. Thus NFA may turn He from a liability to an asset [12].

Since fission reactor irradiations have low He/dpa ratios in TMS and NFA, special techniques are needed to experimentally study irradiation effects in a He-rich environments [3]. A number of studies have been carried out in dual charged particle ion irradiations (DII). In this case, one accelerator is used to implant He ions at MeV-range energies, while a second accelerator creates dpa, typically with high-energy self-ions (Fe^{+i}) in the energy range of 3–10 MeV. While many experimental details are significant, DII can generate high dpa at high He/dpa, in a relatively well-controlled way and in a short time, over the useful portion of the range of a heavy ion that avoids both near surface and end of ion range injected interstitial effects [3, 6]. The major drawback to DII is that the dpa rates are much higher than in neutron irradiations. In many cases, the accelerated dpa rate can have major effects on damage evolution.

Another technique to study He effects is to dope alloys with Ni. Ni generates He by a two-step thermal (th) neutron reaction: $^{58}Ni(n_{th},\gamma) \to {}^{59}Ni(n_{th},\alpha)$ [3]. The α-particle acquires two electrons and becomes a He atom when it reaches its end of range. However, Ni is an active alloying element that influences displacement damage in steels, thus doping confounds isolating He effects. Thus a preferable approach, first proposed by Odette in the mid-1980s, is to inject He into a sample from an external Ni source of (n,α) reactions [192]. An example of this so-called in-situ He injection (ISHI) method is shown in Fig. 12.17A [193]. Here, a 3-mm TEM disc, with a thin μm-scale NiAl coating, is paired with another TEM disc. The 4.8 MeV α particles implant a uniform concentration of He to a depth of approximately 8 μm, minus the coating thickness. Fast neutrons generate the dpa, while He is produced by thermal neutron (n,α) reactions, both at fusion reactor relevant rates. The He/dpa ratio depends on, and can be controlled by, the thermal to fast neutron flux ratio and the Ni coating thickness and composition. Extensive ISHI experiments have been carried out in the High Flux Isotope Reactor (HFIR) at ORNL to generate He/dpa from $\approx$5–60 appm/dpa at 300–500°C and up to 21 dpa. The results of ISHI and DII irradiations have been reported in a series of papers that are summarized below [5, 12, 28–33, 195].

A comparison of typical TMS versus NFA microstructures is shown in Fig. 12.17B for ISHI irradiation at 500°C in the ORNL HFIR to 21 dpa and 2100 appm He [5, 33, 193, 195]. The NFA MA957 contains only small He bubbles, while the TMS Eurofer97 contains a bimodal distribution of small bubbles and larger polyhedral voids. Fig. 12.17C shows that the reason for these differences is the fine distribution of bubbles revealed in a 3D correlative microscopy tomographic reconstruction of a He injected NFA [194, 196]. The bubbles, imaged in under focus TEM, are shown in blue, while NO, characterized by APT, are shown in red. Clearly, the He in the NFA is trapped and widely distributed, with at least one bubble on each NO. The NO bubbles also keep most of the He from reaching grain boundaries where it can promote creep rupture embrittlement at high temperature, as well as severe fast fracture embrittlement at low temperatures.

Fig. 12.18A and B show the systematic effect of higher dual ion irradiation He/dpa on the incubation dose, dpa_i, in a TMS F82H-Mod3 at 500°C [12, 32, 197]. Fig. 12.18C compares swelling in a DII TMS versus NFA. Here the displacement damage dose is plotted on a normalized scale as $dpa - dpa_i + 40$. The dpa_i collapses the data into a broad trend band, while the 40 is the incubation dpa_i at the average He/dpa ratio in the experiment at $\approx$30 appm/dpa. The DII data trend band indicates at a void swelling rate of $S \approx 0.1\%$/dpa. In contrast, the NFA shows a much lower swelling, which is almost entirely due to He bubbles and the NO precipitates (note the nm-scale oxides and attached bubbles are hard to distinguish by TEM), rather than growing voids. These results demonstrate the capacity for the NO in NFA to fully manage high levels of He. Fig. 12.18C also shows data points for 500°C HFIR ISHI irradiations with a $dpa_i \approx 5$. The ISHI data also fall near the DII trend band, but higher dpa (and presumably higher S) data are needed to confirm this trend. These results suggest that He begins to have major effects on void swelling in TMS at $\approx$500 appm He, while NFA will remain swelling resistant under all conceivable service conditions, since the bubbles are far below the critical size for conversion to voids.

Another important role of NOs He trapping in bubbles is to protect grain boundaries from He accumulation resulting in both low- and high-temperature embrittlement. Further while some He does reach grain boundaries, it is also locally trapped in tiny and harmless bubbles on slightly larger grain boundary NOs.

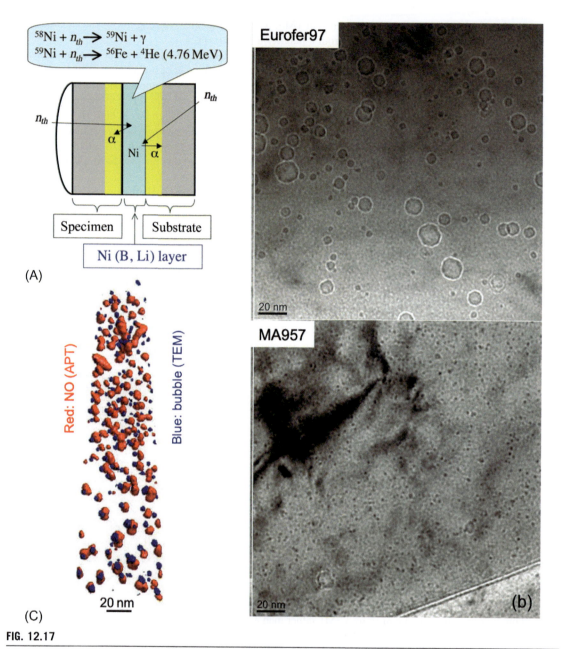

FIG. 12.17

(A) Schematic description of an ISHI approach to studying the effects of irradiation in He-rich environments [193]; (B) a comparison of cavity evolution in TMS Eurofer97 and NFA MA957 showing a bimodal distribution of voids and bubble versus just bubbles, respectively [193]; and (C) a 2D cross-section image of a 3D correlative APT-TEM tomogram showing the essentially one-to-one association of He bubbles (blue) and NOs (red) [194].

12.7 THERMAL AGING AND IRRADIATION EFFECTS

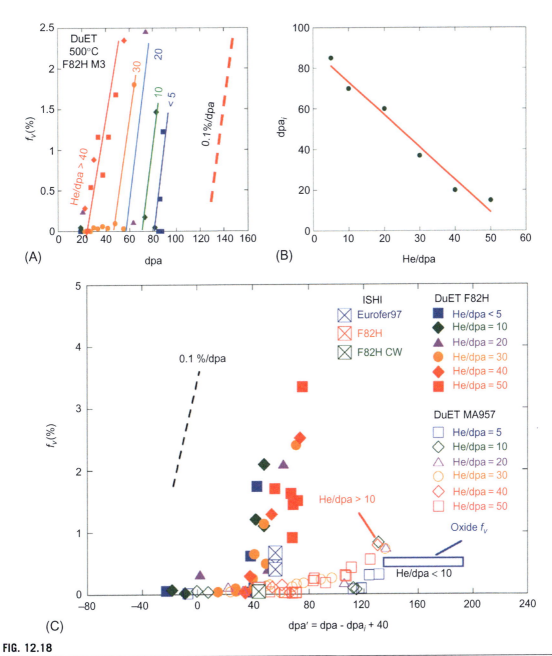

FIG. 12.18

(A) Void swelling for DII at 500°C versus dpa at various He/dpa ratios; (B) the corresponding incubation dpa_i for void swelling as a function of the He/dpa ratio; and (C) a comparison of void swelling in TMS F82H and the bubble volume fraction in NFA MA957 plotted on a $dpa-dpa_i$ dose scale normalized to a He/dpa ratio of 40 appm He/dpa [195].

12.7.7 EFFECTS OF IRRADIATION ON STRENGTH AND TOUGHNESS

Fig. 12.19A shows the effect of an advanced test reactor (ATR) irradiation to ≈ 7 dpa at $\approx 320°C$, on the engineering stress-strain, $\sigma(\varepsilon)$, curves of the TMS Eurofer97 and 2 NFA (MA957 and the high strength SM6) [12, 41]. As is generally observed, the increases in yield stress ($\triangle \sigma_y$) are smaller for the NFAs compared to TMS Eurofer 97 at this relatively low dpa. Similar results were observed after irradiation in the Osiris reactor [198]. This difference is probably in large part due to the effect combining the dispersed barrier hardening in the NFA before (NO and dislocations) and after irradiation (NO, dislocations, dislocations loops, solute clusters/precipitates and α'). As a simple example, if the dispersed barrier hardening before irradiation is 200 MPa in a TMS and 400 MPa in a NFA, an isolated irradiation hardening contribution of 400 MPa would produce a net yield stress increase of ≈ 250 MPa (TMS) versus 165 MPa (NFA), respectively, with the difference being entirely due to the root sum square strength superposition rule. In addition, however, high sink strengths would be expected to suppress displacement damage effects on hardening feature accumulation [23, 188].

Another significant difference between the NFA and TMS, shown in Fig. 12.19A, is that the uniform strain (ε_u) is very low for the irradiated TMS Eurofer97, while the post yield strain hardening and ε_u remain significant in the NFA MA957. Indeed, the irradiated uniform strain even appears to increase slightly in the SM6 following irradiation, although it is low in both cases in this ultrahigh strength NFA [41]. In summary, NFA are more resistant to low-dose irradiation effects in terms of their tensile properties. Unfortunately, at a much higher dpa, NFA loose much of their lower $\triangle \sigma_y$ and higher ε_u advantage with respect to TMS [42]. This is shown in Fig. 12.19B which plots $\triangle \sigma_y$ and ε_u versus dpa for irradiations at $\approx 300°C$ to $330°C$ for NFA MA957 and TMS Eurofer97 for data taken from Ref. [42] and the UCSB database.

Fig. 12.20A shows the $\triangle \sigma_y$ for TMS following spallation proton irradiations (SPI) with high He/dpa ratios of up to 50 appm/dpa (symbols) [3]. The SPI data are compared to the curves fitted to the low He fission neutron $\triangle \sigma_y$ database at various temperatures up to the dpa indicated by the end of the solid lines. The excess hardening at high He is primarily due to bubbles, as well as He enhanced dislocation loop populations. At very high He, even some TMS tensile tests fail elastically in association with a transition from ductile microvoid coalescence to brittle intergranular fracture [152]. Excess hardening in TMS at high He also persists up to much higher irradiation temperatures, where softening would normally occur at low levels [3]. For example, a $\triangle \sigma_y \approx 240$ MPa is observed in a

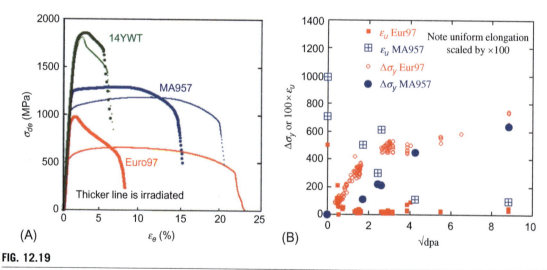

FIG. 12.19

(A) Engineering stress strain curves for 2 NFA and a TMS [12]; and (B) $\triangle \sigma_y$ and ε_u versus dpa for 300–330°C irradiations of NFA MA957 and TMS Eurofer97 [109].

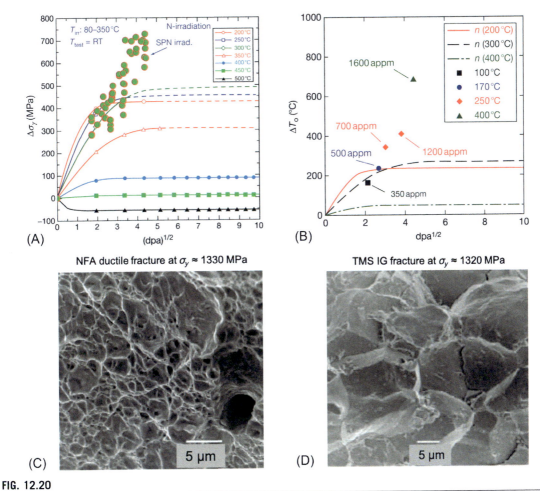

FIG. 12.20

(A) High He spallation proton irradiation (SPI) $\Delta\sigma_y$ data (filled circles) versus the $\sqrt{dpa}$ for TMS over a range of temperatures showing continued hardening compared to lines for fits to low He, neutron irradiation data (neutron data not shown); (B) ΔT_o shifts in the 100 MPa $\sqrt{m}$ master curve fracture temperature for SPI showing the large embrittling effect of high He compared to fits to low He, neutron irradiation data (neutron data not shown); and the effect of high He leading to a transition for ductile dimple (C) to intergranular fracture (D) along prior austenite grain boundaries [109].

TMS, at a test temperature of 450°C, following a spallation source irradiation program (STIP) irradiation at 560°C to 20 dpa and 1700 appm He [199].

Fig. 12.20B shows the enormous effect of high He SPI on the fast fracture toughness embrittlement of TMS compared to low He neutron irradiations conditions [3]. Here estimated shifts (ΔT_o) in the master toughness-temperature curve as a function of dpa at various He levels in SPI are compared to fitted ΔT_o curves for low He neutron irradiations. The enormous ΔT_o at high He are again accompanied by a transition from ductile to intergranular fracture as shown in Fig. 12.20C and D, respectively. This ultrasevere embrittlement in the spallation proton irradiations is due to a synergistic interaction between high $\Delta\sigma_y$ and grain boundary weakening by He accumulation.

Note, He contained in the NO interface would have a smaller effect on the irradiation hardening than when dispersed in the matrix of TMS, although the combined obstacle strength is yet to be determined. Further, in contrast to TMS, tensile tests of NFA MA957 continue to show fully ductile fracture up to very high He [43].

Finally, we note that the threshold for significant effects of He on fast fracture in TMS irradiated at low temperature is estimated to be ≈ 500 appm. Since less He collects on grain boundaries in NFA, they are expected to be highly resistant to such hardening-helium embrittlement effects.

12.7.8 EFFECT OF IRRADIATION ON OTHER PROPERTIES

Space does not permit a detailed discussion of the effect of irradiation on other NFA properties; however, we can briefly summarize a large body of research as follows. Irradiation creep (IC) in TMS and NODS occurs down to low temperatures, well below the thermal creep regime, and follows a stress (σ)-strain (ε)-dose (dpa) relation $\varepsilon \approx B\sigma$dpa. The NFA MA957 creep compliance, $B \approx 0.5 \times 10^{-6}$ MPa^{-1} dpa^{-1}, is approximately athermal from 400°C to 500°C, and is roughly half that of TMS [44]. However, above 500°C the linear stress dependence of IC breaks down in both alloys, and the creep rates increase nonlinearly with both temperature and stress. Thermal creep is dominant at 600°C in both alloy systems, but creep rates are much higher in the TMS. Note IC rates in NFA with high He may have lower IC rates, since the bubbles provide an ultrahigh sink strength for annihilating diffusing defects.

Fatigue behavior in TMS is not greatly affected by irradiation [200]. Indeed, it might be expected that irradiation would manifest some increased resistance to high cycle, stress controlled, fatigue and decreased resistance to low cycle, strain controlled, and fatigue. Since irradiated NFA are stronger and more ductile, they are likely to have a fatigue advantage over TMS following irradiation. There is no systematic data on da/dN versus $\triangle K_I$ crack growth behavior in irradiated NFA. However, these properties are not expected to be highly sensitive to irradiation.

High-temperature creep crack growth is an important property but has been largely unexplored even in the unirradiated condition. TMS may be subjected to He embrittlement at higher temperatures at high He in association with grain boundary bubbles conversion to growing creep cavities [3]. NFA are expected to be more He tolerant for reasons discussed previously.

Finally, environmental effects on fatigue and cracking TMS and NODS may be significant, but are highly complex and system specific beyond the scope of this chapter. Again, however, NFA would be expected to be less sensitive to environmental effects than lower Cr TMS and NMS.

12.7.9 SUMMARY OF THERMAL AGING AND IRRADIATION EFFECTS

The effects of thermal aging and irradiation on NODS can be summarized as follows.

- NFA microstructures and mechanical properties are remarkably thermally and irradiation stable for use temperatures up to at least 850°C at 300 dpa. The stability range for NMS is not as well established, but is likely to approach 800°C at high dpa.
- NFA-irradiated properties (creep, fast fracture embrittlement, hardening and tensile ductility, fatigue, swelling, and irradiation creep) are superior to TMS, to higher dpa. However, at higher dpa the corresponding irradiation hardening and tensile ductility loss degradation are more comparable. Further, the unirradiated fracture properties of TMS are often better than NFA. Optimally processed NMS have unirradiated prosperities that are similar to TMS and are expected to have roughly similar irradiation responses as NFA.
- In the presence of high He NFA (and likely NMS) the irradiated properties of NFA are far superior to those for TMS.

12.8 MODELING

Space does not permit a discussion of the extensive modeling activities relevant to NODS. However, in summary they fall into several categories including:

- Microstructure-property models as illustrated in a few examples previously cited, but that also include a far more extensive literature [75, 83, 140–142, 201–205].
- First principles and thermodynamic models of NO energetics [83] leading both to construction of phase diagrams, and to their use to evaluate free energies in precipitation kinetics models [75, 142, 201]. For example, Fig. 12.21A shows an oxide-phase selection diagram for $x_{Ti}/(x_{Ti}+x_Y)$ ratio and the O partial pressure (P_{O2}) [75]

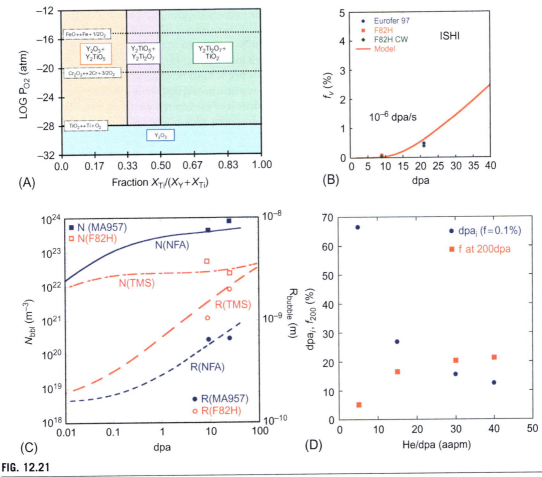

FIG. 12.21

(A) A $x_{Ti}/(x_{Ti}+x_Y)$—P_{O2} diagram oxide-phase selection at 1000°C [75]; (B) void swelling (f_v) predicted by the cluster dynamics He transport, fate, and consequences model for a 9Cr TMS as a function of dpa at 500°C and 40 appm He/dpa compared with ISHI data [206]; (C) cluster dynamics predicted and measured (ISHI irradiations) cavity N and average r as a function of dpa at 500°C for both MA957 and TMS F82H [206]; and (D) the corresponding cluster dynamics model predicted void formation incubation dose (dpa_i) and swelling at 200 dpa as a function of He/dpa [109].

at 1000°C. These thermodynamic insights provided by these models are very useful in guiding alloy design, for example in optimizing the alloy O content and Ti/(Y+Ti) ratio and temperatures to produce fine NO while avoiding coarser-scale detrimental Ti-oxide inclusions.
- A cluster dynamics model of the kinetics of NO precipitation and coarsening at high temperature, that is in good agreement with experimental observations [75]. The model shows very rapid nucleation and growth of NOs, which coarsen at very high temperatures by a dislocation pipe diffusion mechanism with annealing time (t_a) with $t_a^{1/5}$ kinetics.
- First principles and thermodynamic models of $Y_2Ti_2O_7$ and Y_2TiO_5 properties [201] and embedded NO interface energies, structures and termination chemistries, that also show great sensitivity to P_{O2} [142, 201, 202].
- First principles models show that He is deeply trapped in $Y_2Ti_2O_7$ and Y_2TiO_5, with very large binding energies [140–142]. The large trapping energy ensures that the NOs are able to manage large amounts of He up to high temperatures. These results also suggest that He is initially trapped in NOs before it forms interface bubbles.
- First principles models that show H has higher energies in $Y_2Ti_2O_7$ than in the Fe matrix [203]. Thus H (and thus T) is not sequestered inside NO but it segregates at their interfaces and at the surface of associated bubbles. Note trapping of T is a major issue in fusion technology, since it must be readily removed from the reactor structures in order to achieve tolerable inventories. Thus it is worth noting that voids would be a near permanent trap for T in the form of T_2 molecules in equilibrium with dissolved concentrations of this and other H isotopes.
- Perhaps somewhat counter intuitively, due to its large size, first principles model that shows Y is a relatively fast diffuser [204, 205].
- A cluster dynamics master model of He generation, transport and fate, and the corresponding effects on bubble evolution and void swelling have been generally successful in predicting the evolution of He bubbles in TMS and NFA, and their associations with various microstructural features, like precipitates, dislocations as well as grain and lath boundaries. Fig. 12.21B shows good agreement swelling predicted by a cluster He transport, fate, and consequences model for a 9Cr TMS as a function of dpa, at 500°C and 50 appm He/dpa. There is no corresponding void swelling NFA MA957 and the cavity volume is entirely made up of bubbles. Fig. 12.21C compares predicted and measured cavity N and the average r as a function of dpa at 500°C for both MA957 and TMS F82H. The MA957 bubbles are much smaller than the F82H voids (see Fig. 12.18) [206]. Fig. 12.21D shows the model predictions of the void formation incubation dose at $f_v=0.1$, and the magnitude of void swelling (f_v) at 200 dpa at 500°C predicted by the model as a function of the He/dpa ratio. The predicted 200 dpa swelling is $\approx 10\%$ at a fusion relevant He/dpa ≈ 10.

12.9 FUTURE PROSPECTS

In a recent viewpoint paper [13], Odette described the development of a sound science base for developing NFAs. The potential role for advanced manufacturing routes, like additive manufacturing using pre-alloyed atomized powders, is a natural question. However, based on what we know at this time, it seems clear that the highest performance NFA, with finely dispersed NOs, will require powder processing by MA and hot consolidation routes, since Y phase separates when melted and resolidified. Cold-spraying milled NODS powders are difficult because they are so hard. However, these processes may be applicable to ODS alloys with coarser oxides and lower strength and irradiation tolerance.

As noted above, deformation processing NFAs defect-free product forms, such as thin-walled tubing, is an unresolved challenge. But this challenge can be met with additional fundamental and applied research. Further, at some possible loss of performance, 9Cr transformable heat-treated NMS can be more readily fabricated into high-quality components, while maintaining a good balance of properties.

Producing NODS steels in raw billet form will be expensive, and the question remains about who will produce the alloys. Application of NODS cannot be limited to nuclear fusion and fission systems if lower costs due to large-scale commercialization are to be realized. Fortunately, energy, chemical, transportation, and aerospace technologies may be other applications for powder metallurgy of nano-dispersion strengthened steels. Perhaps the most difficult issue is how does a supplier appear without a wide base of applications; and how do widespread applications develop without a reliable supplier.

The lead author of this chapter believes that a NODS steel supplier will come forth, to serve a market for a wider range of NODS alloys including those based on austenitic stainless steels, Ti, Ni Al, Mg, and refractory metals, like W, including niche applications in common products. Note expensive, high strength NFAs likely will not be used in large amounts, such as heavy section structures. Rather, they would be used in small volume, high challenged components of complex multi-material hybrid systems. Further discussion of this important issue can be found in Ref. [13].

REFERENCES

[1] S.J. Zinkle, G.S. Was, Materials challenges in nuclear energy, Acta Mater. 61 (2013) 735–758.
[2] L.K. Mansur, Theory and experimental background on dimensional changes in irradiated alloys, J. Nucl. Mater. 216 (1994) 97–123.
[3] Y. Dai, G.R. Odette, T. Yamamoto, The effects of helium in irradiated structural alloys, in: Compr. Nucl. Mater., Elsevier, 2012, pp. 141–193.
[4] G.R. Odette, M.J. Alinger, B.D. Wirth, Recent developments in irradiation-resistant steels, Annu. Rev. Mater. Res. 38 (2008) 471–503.
[5] G.R. Odette, D.T. Hoelzer, Irradiation-tolerant nanostructured ferritic alloys: transforming helium from a liability to an asset, JOM 62 (2010) 84–92.
[6] G.S. Was, Fundamentals of Radiation Materials Science: Metals and Alloys, second ed., Springer, New York, 2016.
[7] V.P. Budaev, Results of high heat flux tests of tungsten divertor targets under plasma heat loads expected in ITER and tokamaks (review), Phys. At. Nucl. 79 (2016) 1137–1162.
[8] S. Ukai, Oxide dispersion strengthened steels, in: R. Konings (Ed.), Compr. Nucl. Mater., first ed., Elsevier, Atlanta, GA, 2012, pp. 241–271.
[9] S. Ukai, S. Ohtsuka, T. Kaito, Y. de Carlan, J. Ribis, J. Malaplate, Oxide dispersion-strengthened/ferrite-martensite steels as core materials for Generation IV nuclear reactors-Ch 10, in: P. Yvon (Ed.), Struct. Mater. Gener. IV Nucl. React., Woodhead Publishing, Elsevier, 2017, pp. 357–414.
[10] T.K. Kim, S. Noh, S.H. Kang, J.J. Park, H.J. Jin, M.K. Lee, J. Jang, C.K. Rhee, Current status and future prospective of advanced radiation resistant oxide dispersion strengthened steel (ARROS) development for nuclear reactor system applications, Nucl. Eng. Technol. 48 (2016) 572–594.
[11] A. Kimura, W. Han, H. Je, K. Yabuuchi, R. Kasada, Oxide dispersion strengthened steels for advanced blanket systems, Plasma Fusion Res. 11 (2016) 2505090.
[12] G.R. Odette, Recent progress in developing and qualifying nanostructured ferritic alloys for advanced fission and fusion applications, JOM 66 (2014) 2427–2441.
[13] G.R. Odette, On the status and prospects for nanostructured ferritic alloys for nuclear fission and fusion application with emphasis on the underlying science, Scr. Mater. 143 (2018) 142–148.
[14] A. De Bremaecker, Past research and fabrication conducted at SCK•CEN on ferritic ODS alloys used as cladding for FBR's fuel pins, J. Nucl. Mater. 428 (2012) 13–30.
[15] J.-J. Huet, Possible fast-reactor canning material strengthened and stabilized by dispersion, Powder Metall. 10 (1967) 208–215.
[16] J.J. Fischer, Dispersion strengthened ferritic alloy for use in liquid-metal fast breeder reactors (LMFBRS), US Patent 4.075.010, Washington, DC, 1978.
[17] M.L. Hamilton, D.S. Genes, G.D. Johns, W.F. Brown, Fabrication Technological Development of the Oxide Dispersion Strengthened Alloy MA957 for Fast Reactor Applications. Technical Report-PNNL 13168, (2000).

[18] R.L. Klueh, P.J. Maziasz, I.S. Kim, L. Heatherly, D.T. Hoelzer, N. Hashimoto, E.A. Kenik, K. Miyahara, Tensile and creep properties of an oxide dispersion-strengthened ferritic steel, J. Nucl. Mater. 307–311 (2002) 773–777.

[19] S.J. Zinkle, J.L. Boutard, D.T. Hoelzer, H.A. Kimura, R. Lindau, G.R. Odette, M. Rieth, L. Tan, H. Tanigawa, Development of next generation tempered and ODS reduced activation ferritic/martensitic steels for fusion energy applications, Nucl. Fusion 57 (2017) 92005.

[20] M.E. Alam, S. Pal, K. Fields, S.A. Maloy, D.T. Hoelzer, G.R. Odette, Tensile deformation and fracture properties of a 14YWT nanostructured ferritic alloy, Mater. Sci. Eng. A 675 (2016) 437–448.

[21] R.L. Klueh, J.P. Shingledecker, R.W. Swindeman, D.T. Hoelzer, Oxide dispersion-strengthened steels: a comparison of some commercial and experimental alloys, J. Nucl. Mater. 341 (2005) 103–114.

[22] A. Alamo, V. Lambard, X. Averty, M.H. Mathon, Assessment of ODS-14%Cr ferritic alloy for high temperature applications, J. Nucl. Mater. 329–333 (2004) 333–337.

[23] C.P. Massey, S.N. Dryepondt, P.D. Edmondson, K.A. Terrani, S.J. Zinkle, Influence of mechanical alloying and extrusion conditions on the microstructure and tensile properties of low-Cr ODS FeCrAl alloys, J. Nucl. Mater. 512 (2018) 227–238.

[24] M.J. Alinger, G.R. Odette, G.E. Lucas, Tensile and fracture toughness properties of MA957: implications to the development of nanocomposited ferritic alloys, J. Nucl. Mater. 307–311 (2002) 484–489.

[25] M.E. Alam, S. Pal, S.A. Maloy, G.R. Odette, On delamination toughening of a 14YWT nanostructured ferritic alloy, Acta Mater. 136 (2017) 61–73.

[26] T.S. Byun, J.H. Yoon, S.H. Wee, D.T. Hoelzer, S.A. Maloy, Fracture behavior of 9Cr nanostructured ferritic alloy with improved fracture toughness, J. Nucl. Mater. 449 (2014) 39–48.

[27] T.S. Byun, D.T. Hoelzer, J.H. Kim, S.A. Maloy, A comparative assessment of the fracture toughness behavior of ferritic-martensitic steels and nanostructured ferritic alloys, J. Nucl. Mater. 484 (2017) 157–167.

[28] H.J. Jung, D.J. Edwards, R.J. Kurtz, G.R. Odette, Y. Wu, T. Yamamoto, Microstructural Summary of ODS Ferritic Alloys (14YW, 14YWT, 12YWT, MA957FR, PM2000) and RAFM Steels (F82H Mod.3-CW, Eurofer97) From JP27 In-Situ He Injection (ISHI) Experiment at 500°C, Fus. Mater. Semiann. Prog. Rep. (2015) 55–65. DOE/ER-0313/58.

[29] T. Yamamoto, G.R. Odette, P. Miao, D.T. Hoelzer, J. Bentley, N. Hashimoto, H. Tanigawa, R.J. Kurtz, The transport and fate of helium in nanostructured ferritic alloys at fusion relevant He/dpa ratios and dpa rates, J. Nucl. Mater. 367 (2007) 399–410.

[30] T. Yamamoto, G.R. Odette, P. Miao, D.J. Edwards, R.J. Kurtz, Helium effects on microstructural evolution in tempered martensitic steels: in situ helium implanter studies in HFIR, J. Nucl. Mater. 386–388 (2009) 338–341.

[31] R.J. Kurtz, G.R. Odette, T. Yamamoto, D.S. Gelles, P. Miao, B.M. Oliver, The transport and fate of helium in martensitic steels at fusion relevant He/dpa ratios and dpa rates, J. Nucl. Mater. 367–370 (2007) 417–422.

[32] T. Yamamoto, Y. Wu, G.R. Odette, K. Yabuuchi, S. Kondo, A. Kimura, A dual ion irradiation study of helium-dpa interactions on cavity evolution in tempered martensitic steels and nanostructured ferritic alloys, J. Nucl. Mater. 449 (2014) 190–199.

[33] G.R. Odette, P. Miao, D.J. Edwards, T. Yamamoto, R.J. Kurtz, H. Tanigawa, Helium transport, fate and management in nanostructured ferritic alloys: in situ helium implanter studies, J. Nucl. Mater. 417 (2011) 1001–1004.

[34] E. Aydogan, S.A. Maloy, O. Anderoglu, C. Sun, J.G. Gigax, L. Shao, F.A. Garner, I.E. Anderson, J.J. Lewandowski, Effect of tube processing methods on microstructure, mechanical properties and irradiation response of 14YWT nanostructured ferritic alloys, Acta Mater. 134 (2017) 116–127.

[35] M. Bachhav, G.R. Odette, E.A. Marquis, α' precipitation in neutron-irradiated Fe–Cr alloys, Scr. Mater. 74 (2014) 48–51.

[36] E.R. Reese, M. Bachhav, P. Wells, T. Yamamoto, G. Robert Odette, E.A. Marquis, On α' precipitate composition in thermally annealed and neutron-irradiated Fe- 9-18Cr alloys, J. Nucl. Mater. 500 (2018) 192–198.

[37] D. Bhattacharyya, T. Yamamoto, P. Wells, E. Marquis, M. Bachhav, Y. Wu, J. Davis, N. Cunningham, A. Xu, G.R. Odette, Microstructural changes and their effect on hardening in neutron irradiated Fe-Cr alloys, J. Nucl. Mater. 519 (2019) 274–286.

[38] C. Pareige, V. Kuksenko, P. Pareige, Behaviour of P, Si, Ni impurities and Cr in self ion irradiated Fe–Cr alloys—Comparison to neutron irradiation, J. Nucl. Mater. 456 (2015) 471–476.

[39] T. Yamamoto, G.R. Odette, H. Kishimoto, J.-W. Rensman, P. Miao, On the effects of irradiation and helium on the yield stress changes and hardening and non-hardening embrittlement of ~8Cr tempered martensitic steels: compilation and analysis of existing data, J. Nucl. Mater. 356 (2006) 27–49.

[40] H.S. Cho, R. Kasada, A. Kimura, Effects of neutron irradiation on the tensile properties of high-Cr oxide dispersion strengthened ferritic steels, J. Nucl. Mater. 367–370 (2007) 239–243.

[41] S.A. Maloy, T.A. Saleh, O. Anderoglu, T.J. Romero, G.R. Odette, T. Yamamoto, S. Li, J.I. Cole, R. Fielding, Characterization and comparative analysis of the tensile properties of five tempered martensitic steels and an oxide dispersion strengthened ferritic alloy irradiated at ≈295°C to ≈6.5 dpa, J. Nucl. Mater. 468 (2016) 232–239.

[42] J. Henry, X. Averty, A. Alamo, Tensile and impact properties of 9Cr tempered martensitic steels and ODS-FeCr alloys irradiated in a fast reactor at 325°C up to 78 dpa, J. Nucl. Mater. 417 (2011) 99–103.

[43] J. Henry, X. Averty, Y. Dai, J.P. Pizzanelli, J.J. Espinas, Tensile properties of ODS-14%Cr ferritic alloy irradiated in a spallation environment, J. Nucl. Mater. 386–388 (2009) 345–348.

[44] M.B. Toloczko, D.S. Gelles, F.A. Garner, R.J. Kurtz, K. Abe, Irradiation creep and swelling from 400 to 600°C of the oxide dispersion strengthened ferritic alloy MA957, J. Nucl. Mater. 329–333 (2004) 352–355.

[45] P. Miao, G.R. Odette, D. Klingensmith, T. Yamamoto, On the thermal stability of 9% Cr tempered martensitic steels and nanostructured ferritic alloys, Trans. Am. Nucl. Soc. 98 (2008) 1150–1151.

[46] H. Sakasegawa, T. Hirose, A. Kohyama, Y. Katoh, T. Harada, K. Asakura, T. Kumagai, Effects of precipitation morphology on toughness of reduced activation ferritic/martensitic steels, J. Nucl. Mater. 307–311 (2002) 490–494.

[47] G.R. Odette, T. Yamamoto, H.J. Rathbun, M.Y. He, M.L. Hribernik, J.W. Rensman, Cleavage fracture and irradiation embrittlement of fusion reactor alloys: mechanisms, multiscale models, toughness measurements and implications to structural integrity assessment, J. Nucl. Mater. 323 (2003) 313–340.

[48] R.L. Klueh, K. Shiba, M.A. Sokolov, Embrittlement of irradiated F82H in the absence of irradiation hardening, J. Nucl. Mater. 386–388 (2009) 191–194.

[49] N. Cunningham, Y. Wu, D. Klingensmith, G.R. Odette, On the remarkable thermal stability of nanostructured ferritic alloys, Mater. Sci. Eng. A 613 (2014) 296–305.

[50] N.A. Bailey, E. Stergar, M. Toloczko, P. Hosemann, Atom probe tomography analysis of high dose MA957 at selected irradiation temperatures, J. Nucl. Mater. 459 (2015) 225–234.

[51] E. Aydogan, N. Almirall, G.R. Odette, S.A. Maloy, O. Anderoglu, L. Shao, J.G. Gigax, L. Price, D. Chen, T. Chen, F.A. Garner, Y. Wu, P. Wells, J.J. Lewandowski, D.T. Hoelzer, Stability of nanosized oxides in ferrite under extremely high dose self ion irradiations, J. Nucl. Mater. 486 (2017) 86–95.

[52] J.P. Wharry, M.J. Swenson, K.H. Yano, A review of the irradiation evolution of dispersed oxide nanoparticles in the b.c.c. Fe-Cr system: current understanding and future directions, J. Nucl. Mater. 486 (2017) 11–20.

[53] M.J. Swenson, J.P. Wharry, Nanocluster irradiation evolution in Fe-9%Cr ODS and ferritic-martensitic alloys, J. Nucl. Mater. 496 (2017) 24–40.

[54] F.A. Garner, M.B. Toloczko, B.H. Sencer, Comparison of swelling and irradiation creep behavior of fcc-austenitic and bcc-ferritic/martensitic alloys at high neutron exposure, J. Nucl. Mater. 276 (2000) 123–142.

[55] G.R. Odette, On mechanisms controlling swelling in ferritic and martensitic alloys, J. Nucl. Mater. 155–157 (1988) 921–927.

[56] A. Kimura, R. Kasada, N. Iwata, H. Kishimoto, C.H. Zhang, J. Isselin, P. Dou, J.H. Lee, N. Muthukumar, T. Okuda, M. Inoue, S. Ukai, S. Ohnuki, T. Fujisawa, T.F. Abe, Development of Al added high-Cr ODS steels for fuel cladding of next generation nuclear systems, J. Nucl. Mater. 417 (2011) 176–179.

[57] H.S. Cho, A. Kimura, S. Ukai, M. Fujiwara, Corrosion properties of oxide dispersion strengthened steels in super-critical water environment, J. Nucl. Mater. 329–333 (2004) 387–391.

[58] M.A. Montealegre, J.L. González-Carrasco, M.A. Morris-Muñoz, J. Chao, D.G. Morris, The high temperature oxidation behaviour of an ODS FeAl alloy, Intermetallics 8 (2000) 439–446.

[59] P. Hosemann, H.T. Thau, A.L. Johnson, S.A. Maloy, N. Li, Corrosion of ODS steels in lead-bismuth eutectic, J. Nucl. Mater. 373 (2008) 246–253.

[60] K.A. Terrani, Accident tolerant fuel cladding development: promise, status, and challenges, J. Nucl. Mater. 501 (2018) 13–30.

[61] S. Ukai, W. Izawa, N. Oono, S. Hayashi, Y. Kohno, S. Ohtsuka, T. Kaito, Charpy impact property related to {100} cleavage fracture in 15Cr ODS steel, Mater. Sci. Technol. 30 (2014) 1709–1714.

[62] T. Narita, S. Ukai, B. Leng, S. Ohtsuka, T. Kaito, Characterization of recrystallization of 12Cr and 15Cr ODS ferritic steels, J. Nucl. Sci. Technol. 50 (2013) 314–320.

[63] X. Boulnat, D. Fabrègue, M. Perez, S. Urvoy, D. Hamon, Y. de Carlan, Assessment of consolidation of oxide dispersion strengthened ferritic steels by spark plasma sintering: from laboratory scale to industrial products, Powder Metall. 57 (2014) 204–211.

[64] N.J. Cunningham, Study of the Structure, Composition, and Stability of Y-Ti-O nm-Scale Features in Nano-Structured Ferritic Alloys, (PhD thesis), University of California, Santa Barbara, USA, 2012.
[65] R. Lindau, A. Möslang, M. Schirra, P. Schlossmacher, M. Klimenkov, Mechanical and microstructural properties of a hipped RAFM ODS-steel, J. Nucl. Mater. 307–311 (2002) 769–772.
[66] S. Ukai, R. Miyata, S. Kasai, N. Oono, S. Hayashi, T. Azuma, R. Kayano, E. Maeda, S. Ohtsuka, Super high-temperature strength in hot rolled steels dispersing nanosized oxide particles, Mater. Lett. 209 (2017) 581–584.
[67] L. Toualbi, C. Cayron, P. Olier, J. Malaplate, M. Praud, M.H. Mathon, D. Bossu, E. Rouesne, A. Montani, R. Logé, Y. De Carlan, Assessment of a new fabrication route for Fe-9Cr-1W ODS cladding tubes, J. Nucl. Mater. 428 (2012) 47–53.
[68] N.J. Cunningham, Y. Wu, A. Etienne, E.M. Haney, G.R. Odette, E. Stergar, D.T. Hoelzer, Y.D. Kim, B.D. Wirth, S.A. Maloy, Effect of bulk oxygen on 14YWT nanostructured ferritic alloys, J. Nucl. Mater. 444 (2014) 35–38.
[69] S. Ukai, M. Harada, H. Okada, M. Inoue, S. Nomura, S. Shikakura, K. Asabe, T. Nishida, M. Fujiwara, Alloying design of oxide dispersion strengthened ferritic steel for long life FBRs core materials, J. Nucl. Mater. 204 (1993) 65–73.
[70] M.J. Alinger, G.R. Odette, D.T. Hoelzer, The development and stability of Y–Ti–O nanoclusters in mechanically alloyed Fe–Cr based ferritic alloys, J. Nucl. Mater. 329–333 (2004) 382–386.
[71] M.J. Alinger, G.R. Odette, D.T. Hoelzer, On the role of alloy composition and processing parameters in nanocluster formation and dispersion strengthening in nanostuctured ferritic alloys, Acta Mater. 57 (2009) 392–406.
[72] http://www.calphad.com/iron-chromium.html (last accessed 20.11.18).
[73] M. Yamamoto, S. Ukai, S. Hayashi, T. Kaito, S. Ohtsuka, Reverse phase transformation from α to γ in 9Cr-ODS ferritic steels, J. Nucl. Mater. 417 (2011) 237–240.
[74] S. Ukai, Microstructure and high-temperature strength of 9Cr ODS ferritic steel, metal, ceramic and polymetric composites for various uses, IntertechOpen 14 (2011) 283–302.
[75] L. Barnard, N. Cunningham, G.R. Odette, I. Szlufarska, D. Morgan, Thermodynamic and kinetic modeling of oxide precipitation in nanostructured ferritic alloys, Acta Mater. 91 (2015) 340–354.
[76] E.V. Márqueza, K. Mocellin, L. Toualbi, Y. De Carlan, R.E. Logé, Finite element simulation of cold pilgering of ODS tubes, Journées Annu. la SF2M Hal-007594 (2012) 1–3.
[77] A. Lefebvre, V. McDonell, Atomization and Sprays, CRC Press, 2017.
[78] D.T. Hoelzer, K.A. Unocic, M.A. Sokolov, T.S. Byun, Influence of processing on the microstructure and mechanical properties of 14YWT, J. Nucl. Mater. 471 (2016) 251–265.
[79] C. Suryanarayana, Mechanical alloying and milling, Prog. Mater. Sci. 46 (2001) 1–184.
[80] A. Upadhyaya, G.S. Upadhyaya, K.I. Takagi, Powder Metallurgy: Science, Technology and Materials, Universities Press. Chicago, 2011.
[81] T. Okuda, M. Fujiwara, Dispersion behaviour of oxide particles in mechanically alloyed ODS steel, J. Mater. Sci. Lett. 14 (1995) 1600–1603.
[82] B. Hary, R. Logé, J. Ribis, M.-H. Mathon, M. Van Der Meer, T. Baudin, Y. de Carlan, Strain-induced dissolution of Y-Ti-O nano-oxides in a consolidated ferritic oxide dispersion strengthened (ODS) steel, Materialia 4 (2018) 444–448.
[83] L. Barnard, G.R. Odette, I. Szlufarska, D. Morgan, An ab initio study of Ti–Y–O nanocluster energetics in nanostructured ferritic alloys, Acta Mater. 60 (2012) 935–947.
[84] N.J. Cunningham, G.R. Odette, The effects of ball milling parameters on the homogeneity of Y-Ti-O nano-feature distribution in nano-structured ferritic alloys, Trans. Am. Nucl. Soc. 98 (2008) 1093.
[85] M.J. Alinger, On the Formation and Stability of Nanometer Scale Precipitates in Ferritic Alloys during Processing and High Temperature Service, (PhD thesis), University of California, Santa Barbara, USA, 2004.
[86] I. Hilger, X. Boulnat, J. Hoffmann, C. Testani, F. Bergner, Y. De Carlan, F. Ferraro, A. Ulbricht, Fabrication and characterization of oxide dispersion strengthened (ODS) 14Cr steels consolidated by means of hot isostatic pressing, hot extrusion and spark plasma sintering, J. Nucl. Mater. 472 (2016) 206–214.
[87] E. Aydogan, S. Pal, O. Anderoglu, S.A. Maloy, S.C. Vogel, G.R. Odette, J.J. Lewandowski, D.T. Hoelzer, I.E. Anderson, J.R. Rieken, Effect of tube processing methods on the texture and grain boundary characteristics of 14YWT nanostructured ferritic alloys, Mater. Sci. Eng. A 661 (2016) 222–232.
[88] S. Pal, M.E. Alam, S.A. Maloy, D.T. Hoelzer, G.R. Odette, Texture evolution and microcracking mechanisms in as-extruded and cross-rolled conditions of a 14YWT nanostructured ferritic alloy, Acta Mater. 152 (2018) 338–357.
[89] D. Raabe, Overview on basic types of hot rolling textures of steels, Steel Res. Int. 74 (2003) 327–337.

[90] S. Takajo, C.N. Tomé, S.C. Vogel, I.J. Beyerlein, Texture simulation of a severely cold rolled low carbon steel using polycrystal modeling, Int. J. Plast. 109 (2018) 137–152.

[91] X. Boulnat, N. Sallez, M. Dadé, A. Borbély, J.-L. Béchade, Y. de Carlan, J. Malaplate, Y. Bréchet, F. de Geuser, A. Deschamps, P. Donnadieu, D. Fabrègue, M. Perez, Influence of oxide volume fraction on abnormal growth of nanostructured ferritic steels during non-isothermal treatments: an in situ study, Acta Mater. 97 (2015) 124–130.

[92] N. Sallez, X. Boulnat, A. Borbély, J.L. Béchade, D. Fabrègue, M. Perez, Y. de Carlan, L. Hennet, C. Mocuta, D. Thiaudière, Y. Bréchet, In situ characterization of microstructural instabilities: recovery, recrystallization and abnormal growth in nanoreinforced steel powder, Acta Mater. 87 (2015) 377–389.

[93] X. Boulnat, FAST high-temperature consolidation of Oxide-Dispersion Strengthened (ODS) steels: process, microstructure, precipitation, properties, (PhD thesis), INSA de Lyon, France, 2014.

[94] C.C. Eiselt, M. Klimenkov, R. Lindau, A. Möslang, Characteristic results and prospects of the 13Cr–1W–0.3Ti–0.3Y_2O_3 ODS steel, J. Nucl. Mater. 386–388 (2009) 525–528.

[95] A. Das, H.W. Viehrig, F. Bergner, C. Heintze, E. Altstadt, J. Hoffmann, Effect of microstructural anisotropy on fracture toughness of hot rolled 13Cr ODS steel—the role of primary and secondary cracking, J. Nucl. Mater. 491 (2017) 83–93.

[96] B. Leng, S. Ukai, Y. Sugino, Q. Tang, T. Narita, S. Hayashi, F. Wan, S. Ohtsuka, T. Kaito, Recrystallization texture of cold-rolled oxide dispersion strengthened ferritic steel, ISIJ Int. 51 (2011) 951–957.

[97] B. Leng, S. Ukai, T. Narita, Y. Sugino, Q. Tang, N. Oono, S. Hayashi, F. Wan, S. Ohtsuka, T. Kaito, Effects of two-step cold rolling on recrystallization behaviors in ODS ferritic steel, Mater. Trans. 53 (2012) 652–657.

[98] T. Louise, O. Patrick, R. Elodie, D. Bossu, Y. de Carlan, On the influence of cold rolling parameters for 14CrW-ODS ferritic steel claddings, Key Eng. Mater. 554–557 (2013) 118–126.

[99] P. Dubuisson, Y. de Carlan, V. Garat, M. Blat, ODS ferritic/martensitic alloys for sodium fast reactor fuel pin cladding, J. Nucl. Mater. 428 (2012) 6–12.

[100] T. Narita, S. Ukai, T. Kaito, Development of two-step softening heat treatment for manufacturing 12Cr—ODS ferritic steel tubes, J. Nucl. Sci. Technol. 41 (2004) 1008–1012.

[101] G.R. Odette, H.J. Rathbun, M. Hribernik, T. Yamamoto, M. He, P. Spätig, A multiscale approach to measuring and modeling cleavage fracture toughness in structural steels, in: V. Ghetta, D. Gorse, D. Mazière, V. Pontikis (Eds.), Mater. Issues Gener. IV Syst., Springer Netherlands, Dordrecht, 2008, pp. 203–226.

[102] S. Pal, M.E. Alam, G.R. Odette, S.A. Maloy, D.T. Hoelzer, J. Lewandowski, Microstructure, texture and mechanical properties of the 14YWT nanostructured ferritic alloy NFA-1, in: I. Charit, et al. (Eds.), Mechanical and Creep Behavior of Advanced Materials, Springer, 2017, pp. 43–54 (Chapter 4).

[103] Y. Hovanski, J.E. Carsley, K.D. Clarke, P.E. Krajewski, Friction-stir welding and processing, JOM 67 (2015) 996–997.

[104] P. Miao, G.R. Odette, J. Gould, J. Bernath, R. Miller, M. Alinger, C. Zanis, The microstructure and strength properties of MA957 nanostructured ferritic alloy joints produced by friction stir and electro-spark deposition welding, J. Nucl. Mater. 367–370 (2007) 1197–1202.

[105] A. Etienne, N.J. Cunningham, Y. Wu, G.R. Odette, Effects of friction stir welding and post-weld annealing on nanostructured ferritic alloy, Mater. Sci. Technol. 27 (2011) 724–728.

[106] G.J. Tatlock, K. Dawson, T. Boegelein, et al., High resolution microstructural studies of the evolution of nanoscale yttrium-rich oxides in ODS steels subjected to ball milling, selective laser melting or friction stir welding, Mater. Today Proc. 3 (2016) 3086–3093.

[107] B. Mazumder, X. Yu, P.D. Edmondson, C.M. Parish, M.K. Miller, H.M. Meyer, Z. Feng, Effect of friction stir welding and post-weld heat treatment on a nanostructured ferritic alloy, J. Nucl. Mater. 469 (2016) 200–208.

[108] H. Serizawa, M. Murakami, Y. Morisada, H. Fujii, S. Nogami, T. Nagasaka, H. Tanigawa, Influence of friction stir welding conditions on joinability of oxide dispersion strengthened steel/F82H ferritic/martensitic steel joint, Nucl. Mater. Energy 9 (2016) 367–371.

[109] G.R. Odette et al., Unpublished work, (2002–2018).

[110] D.T. Hoelzer, K.A. Unocic, M.A. Sokolov, Z. Feng, Joining of 14YWT and F82H by friction stir welding, J. Nucl. Mater. 442 (2013) S529–S534.

[111] S. Noh, R. Kasada, A. Kimura, Solid-state diffusion bonding of high-Cr ODS ferritic steel, Acta Mater. 59 (2011) 3196–3204.

[112] J.R. Rieken, I.E. Anderson, M.J. Kramer, G.R. Odette, E. Stergar, E. Haney, Reactive gas atomization processing for Fe-based ODS alloys, J. Nucl. Mater. 428 (2012) 65–75.

[113] L. Li, W. Xu, M. Saber, Y. Zhu, C.C. Koch, R.O. Scattergood, Long-term stability of 14YT–4Sc alloy at high temperature, Mater. Sci. Eng. A 647 (2015) 222–228.

[114] Y. Wu, E.M. Haney, N.J. Cunningham, G.R. Odette, Transmission electron microscopy characterization of the nanofeatures in nanostructured ferritic alloy MA957, Acta Mater. 60 (2012) 3456–3468.

[115] G.R. Odette, N.J. Cunningham, E.A. Marquis, S. Lozano-Perez, V. De Castro, P. Hosemann, E. Stergar, S. Liu, C.U. Segre, Atom-probe tomography, small angle neutron scattering, transmission electron microscopy, positron annihilation spectroscopy and X-ray absorption spectroscopy characterization of nano-scale features in nanostructured ferritic alloys, Microsc. Microanal. 15 (2009) 244–245.

[116] Y. Wu, J. Ciston, S. Kramer, N. Bailey, G.R. Odette, P. Hosemann, The crystal structure, orientation relationships and interfaces of the nanoscale oxides in nanostructured ferritic alloys, Acta Mater. 111 (2016) 108–115.

[117] P. Miao, G.R. Odette, T. Yamamoto, M. Alinger, D. Klingensmith, Thermal stability of nano-structured ferritic alloy, J. Nucl. Mater. 377 (2008) 59–64.

[118] J. Ciston, Y. Wu, G.R. Odette, P. Hosemann, The structure of nanoscale precipitates and precipitate interfaces in an oxide dispersion strengthened steel, Microsc. Microanal. 18 (2012) 760–761.

[119] Y. Wu, T. Stan, T. Yamamoto, G.R. Odette, J. Ciston, On the association between bubbles and nano-oxides in annealed and helium implanted 14YWT, Fus. Mater. Semiann. Prog. Rep. DOE/ER-0313/61 (2017) 50–56.

[120] M.K. Miller, E.A. Kenik, K.F. Russell, L. Heatherly, D.T. Hoelzer, P.J. Maziasz, Atom probe tomography of nanoscale particles in ODS ferritic alloys, Mater. Sci. Eng. A 353 (2003) 140–145.

[121] M.K. Miller, K.F. Russell, D.T. Hoelzer, Characterization of precipitates in MA/ODS ferritic alloys, J. Nucl. Mater. 351 (2006) 261–268.

[122] S. Yamashita, S. Ohtsuka, N. Akasaka, S. Ukai, S. Ohnuki, Formation of nanoscale complex oxide particles in mechanically alloyed ferritic steel, Philos. Mag. Lett. 84 (2004) 525–529.

[123] M. Klimiankou, R. Lindau, A. Möslang, Energy-filtered TEM imaging and EELS study of ODS particles and Argon-filled cavities in ferritic–martensitic steels, Micron 36 (2005) 1–8.

[124] J. Ribis, Y. de Carlan, Interfacial strained structure and orientation relationships of the nanosized oxide particles deduced from elasticity-driven morphology in oxide dispersion strengthened materials, Acta Mater. 60 (2012) 238–252.

[125] K. Dawson, G.J. Tatlock, Characterisation of nanosized oxides in ODM401 oxide dispersion strengthened steel, J. Nucl. Mater. 444 (2014) 252–260.

[126] V. Badjeck, M.G. Walls, L. Chaffron, J. Malaplate, K. March, New insights into the chemical structure of $Y_2Ti_2O_7$-δ nanoparticles in oxide dispersion-strengthened steels designed for sodium fast reactors by electron energy-loss spectroscopy, J. Nucl. Mater. 456 (2015) 292–301.

[127] A.J. London, S. Santra, S. Amirthapandian, B.K. Panigrahi, R.M. Sarguna, S. Balaji, R. Vijay, C.S. Sundar, S. Lozano-Perez, C.R.M. Grovenor, Effect of Ti and Cr on dispersion, structure and composition of oxide nano-particles in model ODS alloys, Acta Mater. 97 (2015) 223–233.

[128] A.J. London, B.K. Panigrahi, C.C. Tang, C. Murray, C.R.M. Grovenor, Glancing angle XRD analysis of particle stability under self-ion irradiation in oxide dispersion strengthened alloys, Scr. Mater. 110 (2016) 24–27.

[129] T. Stan, D.J. Sprouster, A. Ofan, G.R. Odette, L.E. Ecker, I. Charit, X-ray absorption spectroscopy characterization of embedded and extracted nano-oxides, J. Alloys Compd. 699 (2017) 1030–1035.

[130] M. Ohnuma, J. Suzuki, S. Ohtsuka, S.-W. Kim, T. Kaito, M. Inoue, H. Kitazawa, A new method for the quantitative analysis of the scale and composition of nanosized oxide in 9Cr-ODS steel, Acta Mater. 57 (2009) 5571–5581.

[131] M.A. Thual, J. Ribis, T. Baudin, V. Klosek, Y. de Carlan, M.H. Mathon, Relaxation path of nanoparticles in an oxygen-enriched ferritic oxide-dispersion-strengthened alloy, Scr. Mater. 136 (2017) 37–40.

[132] J. Ribis, M.A. Thual, T. Guilbert, Y. de Carlan, A. Legris, Relaxation path of metastable nanoclusters in oxide dispersion strengthened materials, J. Nucl. Mater. 484 (2017) 183–192.

[133] C.A. Williams, G.D.W. Smith, E.A. Marquis, Quantifying the composition of yttrium and oxygen rich nanoparticles in oxide dispersion strengthened steels, Ultramicroscopy 125 (2013) 10–17.

[134] A.J. London, S. Lozano-Perez, S. Santra, S. Amirthapandian, B.K. Panigrahi, C.S. Sundar, C.R. MGrovenor, Comparison of atom probe tomography and transmission electron microscopy analysis of oxide dispersion strengthened steels, J. Phys. Conf. Ser. 522 (2014) 12028.

[135] C. Hatzoglou, B. Radiguet, P. Pareige, Experimental artefacts occurring during atom probe tomography analysis of oxide nanoparticles in metallic matrix: quantification and correction, J. Nucl. Mater. 492 (2017) 279–291.

[136] T. Stan, Y. Wu, G.R. Odette, K.E. Sickafus, H.A. Dabkowska, B.D. Gaulin, Fabrication and characterization of naturally selected epitaxial Fe {111} $Y_2Ti_2O_7$ mesoscopic interfaces: some potential implications to nano-oxide dispersion-strengthened steels, Metall. Mater. Trans. A 44 (2013) 4505–4512.

[137] T. Stan, Y. Wu, P.B. Wells, H.D. Zhou, G.R. Odette, Epitaxial Fe thin films on {100} $Y_2Ti_2O_7$: model interfaces for nano-oxide dispersion strengthened steels, Metall. Mater. Trans. A 48 (2017) 5658–5666.

[138] E.A. Marquis, Core/shell structures of oxygen-rich nanofeatures in oxide-dispersion strengthened Fe–Cr alloys, Appl. Phys. Lett. 93 (2008) 181904.

[139] S. Liu, G.R. Odette, C.U. Segre, Evidence for core-shell nanoclusters in oxygen dispersion strengthened steels measured using X-ray absorption spectroscopy, J. Nucl. Mater. 445 (2014) 50–56.

[140] L. Yang, Y. Jiang, G.R. Odette, T. Yamamoto, Z. Liu, Y. Liu, Trapping helium in $Y_2Ti_2O_7$ compared to in matrix iron: a first principles study, J. Appl. Phys. 115 (2014) 143508.

[141] Y. Jin, Y. Jiang, L. Yang, G. Lan, G.R. Odette, T. Yamamoto, J. Shang, Y. Dang, First principles assessment of helium trapping in Y_2TiO_5 in nano-featured ferritic alloys, J. Appl. Phys. 116 (2014) 143501.

[142] L. Yang, Y. Jiang, Y. Wu, G.R. Odette, Z. Zhou, Z. Lu, The ferrite/oxide interface and helium management in nano-structured ferritic alloys from the first principles, Acta Mater. 103 (2016) 474–482.

[143] S.J. Zinkle, L.L. Snead, Designing radiation resistance in materials for fusion energy, Annu. Rev. Mater. Res. 44 (2014) 241–267.

[144] T. Stan, The Role of Oxides in Nanostructured Ferritic Alloys and Bilayers: Interfaces, Helium Partitioning and Bubble Formation, (PhD thesis), University of California, Santa Barbara, USA, 2017.

[145] Y. Wu, G.R. Odette, T. Yamamoto, J. Ciston, P. Hosemann, An electron energy loss spectroscopy study of helium bubbles in nanostructured ferritic alloys, Fus. Mater. Semiann. Prog. Rep (2013) 173–179. DOE-ER-0313/54.

[146] J.H. Kim, T.S. Byun, D.T. Hoelzer, S.W. Kim, B.H. Lee, Temperature dependence of strengthening mechanisms in the nanostructured ferritic alloy 14YWT: Part I-Mechanical and microstructural observations, Mater. Sci. Eng. A 559 (2013) 101–110.

[147] J.H. Kim, T.S. Byun, D.T. Hoelzer, C.H. Park, J.T. Yeom, J.K. Hong, Temperature dependence of strengthening mechanisms in the nanostructured ferritic alloy 14YWT: Part II—Mechanistic models and predictions, Mater. Sci. Eng. A 559 (2013) 111–118.

[148] M. Praud, F. Mompiou, J. Malaplate, D. Caillard, J. Garnier, A. Steckmeyer, B. Fournier, Study of the deformation mechanisms in a Fe–14% Cr ODS alloy, J. Nucl. Mater. 428 (2012) 90–97.

[149] D.T. Hoelzer, J.P. Shingledecker, R.L. Klueh et al., Fus. Mater. Semiann. Prog. Rept. (2008) 53. DOE/ER-0313/44.

[150] B. Wilshire, T.D. Lieu, Deformation and damage processes during creep of Incoloy MA957, Mater. Sci. Eng. A 386 (2004) 81–90.

[151] M.C. Salston, G.R. Odette, A database and constitutive model for the static and creep strength of MA957 from room temperature to 1000°C, Trans. Am. Nucl. Soc. 98 (2008) 1146–1147.

[152] J. Pelleg, Mechanical Properties of Materials, Springer Science & Business Media, Chicago, 2012.

[153] G.R. Odette, The data labeled as UCSB is from a database that has not been fully published (2002–2018).

[154] M. Rieth, M. Schirra, A. Falkenstein, P. Graf, S. Heger, H. Kempe, R. Lindau, H. Zimmermann, EUROFER 97. Tensile, charpy, creep and structural tests, in: No. FZKA-6911, 2003. Forschungszentrum Karlsruhe GmbH Technik und Umwelt (Germany), Inst. fuer Materialforschung.

[155] P. Fernández, A.M. Lancha, J. Lapeña, R. Lindau, M. Rieth, M. Schirra, Creep strength of reduced activation ferritic/martensitic steel Eurofer'97, Fusion Eng. Des. 75–79 (2005) 1003–1008.

[156] E. Arzt, D.S. Wilkinson, Threshold stresses for dislocation climb over hard particles: the effect of an attractive interaction, Acta Metall. 34 (1986) 1893–1898.

[157] J. Roesler, E. Artz, A new model-based creep equation for dispersion strengthened materials, Acta Metall. Mater. 38 (1990) 671–683.

[158] P. Miao, G.R. Odette, T. Yamamoto, M. Alinger, D. Hoelzer, D. Gragg, Effects of consolidation temperature, strength and microstructure on fracture toughness of nanostructured ferritic alloys, J. Nucl. Mater. 367–370 (2007) 208–212.

[159] D.A. McClintock, D.T. Hoelzer, M.A. Sokolov, R.K. Nanstad, Mechanical properties of neutron irradiated nanostructured ferritic alloy 14YWT, J. Nucl. Mater. 386–388 (2009) 307–311.

[160] S. Pal, M.E. Alam, G.R. Odette, S. Maloy, D.T. Hoelzer, Characterization of processing induced impurity phase precipitates in the as-processed FCRD-NFA-1, Fus. Mater. Semiann. Prog. Rep. (2015) 26–38. DOE-ER-0313/59.

[161] I. Kubena, B. Fournier, T. Kruml, Effect of microstructure on low cycle fatigue properties of ODS steels, J. Nucl. Mater. 424 (2012) 101–108.

[162] M. Kimura, K. Kobayashi, K. Yamaguchi, Creep and fatigue properties of newly developed ferritic heat-resisting steels for ultra super critical (USC) power plants, J. Soc. Mater. Sci. Jpn. 52 (2003) 50–54.

[163] A. Chauhan, M. Walter, J. Aktaa, Towards improved ODS steels: a comparative high-temperature low-cycle fatigue study, Fatigue Fract. Eng. Mater. Struct. 40 (2017) 2128–2140.

[164] A. Chauhan, L. Straßberger, U. Führer, D. Litvinov, J. Aktaa, Creep-fatigue interaction in a bimodal 12Cr-ODS steel, Int. J. Fatigue 102 (2017) 92–111.
[165] P. Hutař, I. Kuběna, M. Ševčík, M. Šmíd, T. Kruml, L. Náhlík, Small fatigue crack propagation in Y_2O_3 strengthened steels, J. Nucl. Mater. 452 (2014) 370–377.
[166] S. Ukai, S. Ohtsuka, Low cycle fatigue properties of ODS ferritic-martensitic steels at high temperature, J. Nucl. Mater. 367–370A (2007) 234–238.
[167] P. Hutař, J. Poduška, M. Šmíd, I. Kuběna, A. Chlupová, L. Náhlík, J. Polák, T. Kruml, Short fatigue crack behaviour under low cycle fatigue regime, Int. J. Fatigue 103 (2017) 207–215.
[168] S.Y. Zhong, J. Ribis, N. Lochet, Y. de Carlan, V. Klosek, M.H. Mathon, Influence of nano-particle coherency degree on the coarsening resistivity of the nano-oxide particles of Fe–14Cr–1W ODS alloys, J. Nucl. Mater. 455 (2014) 618–623.
[169] N.J. Cunningham, M.J. Alinger, D. Klingensmith, Y. Wu, G.R. Odette, On nano-oxide coarsening kinetics in the nanostructured ferritic alloy MA957: a mechanism based predictive model, Mater. Sci. Eng. A 655 (2016) 355–362.
[170] J. Ribis, S. Lozano-Perez, Nano-cluster stability following neutron irradiation in MA957 oxide dispersion strengthened material, J. Nucl. Mater. 444 (2014) 314–322.
[171] M.-L. Lescoat, J. Ribis, Y. Chen, E.A. Marquis, E. Bordas, P. Trocellier, Y. Serruys, A. Gentils, O. Kaïtasov, Y. de Carlan, A. Legris, Radiation-induced Ostwald ripening in oxide dispersion strengthened ferritic steels irradiated at high ion dose, Acta Mater. 78 (2014) 328–340.
[172] C.M. Parish, R.M. White, J.M. LeBeau, M.K. Miller, Response of nanostructured ferritic alloys to high-dose heavy ion irradiation, J. Nucl. Mater. 445 (2014) 251–260.
[173] E.R. Reese, N. Almirall, T. Yamamoto, S. Tumey, G.R. Odette, E.A. Marquis, Dose rate dependence of Cr precipitation in an ion-irradiated Fe18Cr alloy, Scr. Mater. 146 (2018) 213–217.
[174] S. Shu, N. Almirall, P.B. Wells, T. Yamamoto, G.R. Odette, D.D. Morgan, Precipitation in Fe-Cu and Fe-Cu-Mn model alloys under irradiation: dose rate effects, Acta Mater. 157 (2018) 72–82.
[175] D.S. Gelles, Microstructural examination of neutron-irradiated simple ferritic alloys, J. Nucl. Mater. 108–109 (1982) 515–526.
[176] Y. Katoh, A. Kohyama, D.S. Gelles, Swelling and dislocation evolution in simple ferritic alloys irradiated to high fluence in FFTF/MOTA, J. Nucl. Mater. 225 (1995) 154–162.
[177] Z. Jiao, S. Taller, K. Field, G. Yeli, M.P. Moody, G.S. Was, Microstructure evolution of T91 irradiated in the BOR60 fast reactor, J. Nucl. Mater. 504 (2018) 122–134.
[178] R. Schäublin, D. Gelles, M. Victoria, Microstructure of irradiated ferritic/martensitic steels in relation to mechanical properties, J. Nucl. Mater. 307 (2002) 197–202.
[179] F. Bergner, C. Pareige, M. Hernández-Mayoral, L. Malerba, C. Heintze, Application of a three-feature dispersed-barrier hardening model to neutron-irradiated Fe–Cr model alloys, J. Nucl. Mater. 448 (2014) 96–102.
[180] O. Weiß, E. Gaganidze, J. Aktaa, Quantitative characterization of microstructural defects in up to 32 dpa neutron irradiated EUROFER97, J. Nucl. Mater. 426 (2012) 52–58.
[181] F. Bergner, J. Aktaa, E. Altstadt, Effect of thermal ageing treatments and ion irradiation on the irradiation microstructure and mechanical behavior of ODS steel bars and plates, MATISSE Rep. D (2017) 4–13.
[182] S. Rogozhkin, A. Nikitin, N. Orlov, A. Bogachev, O. Korchuganova, A. Aleev, A. Zaluzhnyi, T. Kulevoy, R. Lindau, A. Möslang, P. Vladimirov, Evolution of microstructure in advanced ferritic-martensitic steels under irradiation: the origin of low temperature radiation embrittlement, MRS Adv. 2 (2017) 1143–1155.
[183] M. Klimenkov, R. Lindau, U. Jäntsch, A. Möslang, Effect of irradiation temperature on microstructure of ferritic-martensitic ODS steel, J. Nucl. Mater. 493 (2017) 426–435.
[184] M. Bachhav, G.R. Odette, E.A. Marquis, Microstructural changes in a neutron-irradiated Fe–15at.%Cr alloy, J. Nucl. Mater. 454 (2014) 381–386.
[185] M. Bachhav, L. Yao, G.R. Odette, E.A. Marquis, Microstructural changes in a neutron-irradiated Fe-6 at.%Cr alloy, J. Nucl. Mater. 453 (2014) 334–339.
[186] V. Kuksenko, C. Pareige, P. Pareige, Cr precipitation in neutron irradiated industrial purity Fe–Cr model alloys, J. Nucl. Mater. 432 (2013) 160–165.
[187] V. Kuksenko, C. Pareige, C. Genevois, F. Cuvilly, M. Roussel, P. Pareige, Effect of neutron-irradiation on the microstructure of a Fe-12at.%Cr alloy, J. Nucl. Mater. 415 (2011) 61–66.

[188] C.P. Massey, P.D. Edmondson, K.G. Field, D.T. Hoelzer, S.N. Dryepondt, K.A. Terrani, S.J. Zinkle, Post irradiation examination of nanoprecipitate stability and α' precipitation in an oxide dispersion strengthened Fe-12Cr-5Al alloy, Scr. Mater. 162 (2019) 94–98.

[189] G. Bonny, D. Terentyev, L. Malerba, On the α–α' miscibility gap of Fe–Cr alloys, Scr. Mater. 59 (2008) 1193–1196.

[190] G. Bonny, A. Bakaev, D. Terentyev, E. Zhurkin, M. Posselt, Atomistic study of the hardening of ferritic iron by Ni-Cr decorated dislocation loops, J. Nucl. Mater. 498 (2018) 430–437.

[191] J.-H. Ke, H. Ke, G.R. Odette, D. Morgan, Cluster dynamics modeling of Mn-Ni-Si precipitates in ferritic-martensitic steel under irradiation, J. Nucl. Mater. 498 (2018) 83–88.

[192] G.R. Odette, New approaches to simulating fusion damage in fission reactors, J. Nucl. Mater. 141–143 (1986) 1011–1017.

[193] G.R. Odette, T. Yamamoto, Y. Wu, E. Stergar, N.J. Cunningham, R.J. Kurtz, D.J. Edwards, Helium Effects in Advanced Structural Alloys: Progress on Solving a Grand Challenge to Fusion Energy, Fus. Mater. Semiann. Prog. Rep. (2011) 90–100. DOE/ER-0313/51.

[194] P.B. Wells, S. Krämer, Y. Wu, S. Pal, G.R. Odette, T. Yamamoto, Nanoscale 3D correlative atom probe-electron tomography: characterization of microstructures in dual ion irradiation NFA MA957, Fus. Mater. Semiann. Prog. Rep. (2015) 57–63. DOE/ER-0313/59.

[195] G.R. Odette, T. Yamamoto, Y. Wu, On the effects of helium-dpa interactions on cavity evolution in tempered martensitic steels and nanostructured ferritic alloys under dual ion-beam irradiation, Fus. Mater. Semiann. Prog. Rep. (2014) 8–13. DOE/ER-0313/57.

[196] S. Kramer, P. Wells, C. Oberdorfer, G.R. Odette, Correlative TEM and atom probe tomography—a case study on structural materials for fusion reactors, Microsc. Microanal. 23 (2017) 654–655.

[197] T. Yamamoto, Y. Wu, G.R. Odette, K. Yabuuchi, S. Kondo, A. Kimura, On the effects of helium-dpa interactions on cavity evolution in tempered martensitic steels under dual ion-beam irradiation, Fus. Mater. Semiann. Prog. Rep. (2015) 12–17. DOE/ER-0313/58.

[198] Y. De Carlan, X. Averty, J.-C. Brachet, J.-L. Bertin, F. Rozenblum, O. Rabouille, A. Bougault, Post-irradiation tensile behavior and residual activity of several ferritic/martensitic and austenitic steels irradiated in Osiris reactor at 325°C up to 9 dpa, in: T.R. Allen (Ed.), Eff. Radiat. Mater. 22nd Int. Symp. ASTM STP, 2004.

[199] Z. Tong, Y. Dai, The microstructure and tensile properties of ferritic/martensitic steels T91, Eurofer-97 and F82H irradiated up to 20dpa in STIP-III, J. Nucl. Mater. 398 (2010) 43–48.

[200] C. Petersen, A. Povstyanko, V. Prokhorov, A. Fedoseev, O. Makarov, M. Walter, Tensile and low cycle fatigue properties of different ferritic/martensitic steels after the fast reactor irradiation 'ARBOR 1', J. Nucl. Mater. 386–388 (2009) 299–302.

[201] Y. Jiang, L. Yang, Y. Jin, G.R. Odette, Surface stabilites and helium trapping of nano-sized oxide phases in nanostructured ferritic alloys: a first principles study, in: TMS 2014, 143rd Annu. Meet. Exhib., Springer International Publishing, Cham, 2014, pp. 163–170.

[202] L. Yang, Y. Jiang, G.R. Odette, W. Zhou, Z. Liu, Y. Liu, Nonstoichiometry and relative stabilities of $Y_2Ti_2O_7$ polar surfaces: a density functional theory prediction, Acta Mater. 61 (2013) 7260–7270.

[203] B. Tsuchiya, T. Yamamoto, K. Ohsawa, G.R. Odette, First-principles calculation of formation energies and electronic structures of hydrogen defects at tetrahedral and octahedral interstitial sites in pyrochlore-type $Y_2Ti_2O_7$ oxide, J. Alloys Compd. 678 (2016) 153–159.

[204] D. Murali, B.K. Panigrahi, M.C. Valsakumar, C.S. Sundar, Diffusion of Y and Ti/Zr in bcc iron: a first principles study, J. Nucl. Mater. 419 (2011) 208–212.

[205] J.-L. Bocquet, C. Barouh, C.-C. Fu, Migration mechanism for oversized solutes in cubic lattices: the case of yttrium in iron, Phys. Rev. B 95 (2017) 214108.

[206] T. Yamamoto, G.R. Odette, A master model of helium transport, fate and consequences in tempered martensitic and nanostructured ferritic steels, Fus. Mater. Semiann. Prog. Rep. (2019) to be published. DOE/ER-0313/66.

CHAPTER 13

REFRACTORY ALLOYS: VANADIUM, NIOBIUM, MOLYBDENUM, TUNGSTEN

Lance L. Snead*, David T. Hoelzer[†], Michael Rieth[‡], Andre A.N. Nemith[‡]

Department of Materials Science and Chemical Engineering, State University of New York at Stony Brook, Stony Brook, NY, United States Materials Science and Technology Division, Oak Ridge National Laboratory, Oak Ridge, TN, United States[†] Institute for Advanced Materials, Karlsruhe Institute of Technology, Karlsruhe, Germany[‡]*

CHAPTER OUTLINE

13.1 Introduction	585
13.2 Practical Routes for Refractory Alloy Production	587
13.2.1 Vanadium	587
13.2.2 Niobium	590
13.2.3 Fabrication of Nuclear-Grade Molybdenum	591
13.2.4 Practical Routes of Tungsten and Tungsten Alloy Production	596
13.3 As-Fabricated Mechanical Properties	600
13.3.1 Vanadium	600
13.3.2 As-Fabricated Mechanical Properties of Niobium	606
13.3.3 As-Fabricated Mechanical Properties of Molybdenum	607
13.3.4 As-Fabricated Mechanical Properties of Tungsten	611
13.4 As-Irradiated Mechanical Properties	615
13.4.1 As-Irradiated Mechanical Properties of Vanadium	615
13.4.2 As-Irradiated Mechanical Properties of Niobium	622
13.4.3 As-Irradiated Mechanical Properties of Molybdenum	625
13.4.4 As-Irradiated Mechanical Properties of Tungsten	629
13.4.5 Summary and Conclusions	634
References	635

13.1 INTRODUCTION

Refractory metals as a class of materials are understood to share the common properties of very high melting temperature and mechanical properties and wear resistance. A narrowly defined class of refractory metals would include metals with melting points >2000°C: niobium, chromium molybdenum, tantalum, tungsten, and rhenium [1], while a wider class would also include those with melting points above 1850°C: vanadium, hafnium, titanium,

CHAPTER 13 REFRACTORY ALLOYS

zirconium, ruthenium, osmium, rhodium, and iridium. The current practical application of refractory metals is relatively widespread (though arguably for specialty application) with examples being casting molds, wire filaments, reactant vessels for corrosive materials, hard tooling, and a myriad of applications where high density is desired.

This chapter focuses on a subset of refractory materials that are of particular interest for current and next-generation nuclear applications, specifically vanadium, niobium, molybdenum, and tungsten. It is noted that tantalum has historically been studied for space reactor applications, though is not discussed here, in part due to its similarity in behavior to niobium. A snapshot of how their physical properties compare is given in Tables 13.1 and 13.2. This chapter is supportive of content to Chapter 2 (Generation IV nuclear reactors), Chapter 3 (Fusion Reactors), and Chapter 5 (Radiation Effects and Thermomechanical Degradation Processes). Historically, refractory alloys have been considered in a wide array of nuclear applications including terrestrial, maritime, and space fission reactors as well as current and next-generation fusion reactors.

Because refractory metals are a class of materials possessing extraordinary high-temperature properties, they are perennial contenders for high-temperature nuclear applications. However, their use to date has been limited, due in part to the difficulty in fabricating high-performance refractory parts, and their environmental degradation

Table 13.1 Summary of Physical and Chemical Properties of Selected Refractory Metals

	Vanadium	Niobium	Molybdenum	Tungsten
Atomic number	23	41	42	74
Atomic volume in cm^3/mol	8.78	10.87	9.4	9.53
Average atomic mass g/mol	50.9415	93.2	95.94	183.85
Naturally occurring isotopes in %	^{50}V(0.25%), ^{51}V(99.75%)	93 (~100%)	92(14.84%); 95(9.25%); 96(16.68%); 97(9.55%); 98 (24.13%); 100 (9.63%)	180(0.135%); 182(26.4%); 183(14.4%); 184(30.6%); 186(28.4%)
Pauling electronegativity	1.63	1.6	2.16	2.36
Electronic shell	3d^3, 4s^2	4d^4 5s^1	[Kr] 5s1 4d5	[Xe] 4f14 5d4 6s2
Oxidation states	5, 4, 3, 2, 0	5, 4, 3, 2, 1, −1, −3 (a mildly acidic oxide)	6, 5, 4, 3, 2, 1, -1, -2 (strongly acidic oxide)	−2, −1, 0, +2, +3, +4, +5, +6
Heat of vaporization in kJ/mol	0.452	689.9 kJ/mol	598	824
Heat of fusion in kJ/mol	21.5	30	37.48	35.4
Specific heat in J/gK	490	265	24.06	0.13
Electrical resistivity in μΩcm (@ 20°C)	19.7	0.00152	0.0534	5.5
Thermal conductivity in W/cmK (@ 20°C)	0.307	53.7	138	1.74
Melting point in °C	1890	2477	2623	3410
Boiling point in °C	3407	4744	4639	5660
Density in g/cm^3 (@ 20°C)	5.8	8.57	10.28	19.3

Table 13.2 Typical Values for Single-Crystal Elastic Constants at Room Temperature

	Vanadium	Niobium	Molybdenum	Tungsten
Young's modulus—GPa	128	105	329	390–410
Shear modulus—GPa	47	38	126	156–177
Bulk modulus—GPa	160	170	230	305–310
Poisson's ratio	0.37	0.40	0.31	0.280–0.30

including irradiation effects. The following sections will discuss the current processing routes being taken to produce nuclear-grade refractory alloys, a general discussion of their properties, and the effects of irradiation on the materials.

13.2 PRACTICAL ROUTES FOR REFRACTORY ALLOY PRODUCTION
13.2.1 VANADIUM

Vanadium-base alloys emerged as a candidate material for nuclear energy applications during the 1960s primarily due to the low neutron-induced activation characteristics (particularly for fast neutron spectra), reasonable high-temperature mechanical properties, and high thermal stress factors mainly due to high thermal conductivity [2–5]. The early development of vanadium-base alloys was for fuel cladding for space and liquid metal fast breeder reactors (LMFBR's) during the 1960s in the United States [6] and related efforts in Europe [7]. The investigations conducted in this time frame led to the early awareness of physical metallurgy issues and the development of well-defined processing steps for producing small experimental heats (<10 kg) of vanadium-base alloys. Significant interest for the development of vanadium-base alloys grew in the US fusion power program (FPP) during the 1980s for first wall and blanket structures in fusion energy reactors, such as the international thermonuclear experimental reactor (ITER). Numerous laboratory-scale heats (<30 kg) of vanadium-base alloys were produced during the early phase of this program, mainly from the V, V-Ti, V-Cr, V-Cr-Ti, and V-Ti-Si systems with Ti and/or Cr concentrations within the range of ~3–15 wt% and Si levels <1 wt%. The goal of the research on these heats was to determine the role that processing and alloy composition had on the physical and mechanical properties and the microstructure of the vanadium alloys [6, 8, 9]. A substantial amount of research was conducted on these alloys before and after irradiation and the results obtained from the investigations identified a V-4Cr-4Ti alloy as the US reference alloy for subsequent development in the United States [8, 9]. This alloy composition was subsequently selected in similar fusion energy R&D efforts in countries such as Japan, Russia, and China. A common goal shared with each of the international countries with active R&D programs on vanadium alloys was to obtain the processing experience and physical metallurgy knowledge for producing large industrial-scale heats of the V-4Cr-4Ti alloy with better chemical homogeneity and reduced impurity levels and to establish a comprehensive database of physical and mechanical properties before and after fast neutron irradiations and in lithium environments.

During the 1990s, significant progress in the development of vanadium-base alloys occurred in the US fusion program led by a joint campaign between Argonne National Laboratory and Teledyne Wah Chang (Albany, Oregon). The fabrication steps are outlined in Fig. 13.1. This effort resulted in the first productions of industrial-scale heats of the V-4Cr-4Ti alloy; first, a 500 kg heat (#832665) that was used mainly for gaining valuable experience and knowledge in production, testing, and microstructural characterization studies and, second, a large 1200 kg heat (#832864) that was selected by General Atomics for the manufacture of the Radiative Divertor for the DOE DIII-D tokamak. Much of the experience for producing the large V-4Cr-4Ti heats was obtained from previous studies of fabrication, mechanical properties testing, and microstructural examination of laboratory scale heats (<30 kg) of V-(4-5)Cr-(3-5)Ti alloys [5, 6, 8, 9]. The specification for the chemical composition of the large heats of V-4Cr-4Ti called for careful control of impurities since numerous studies had shown that the mechanical properties, such as

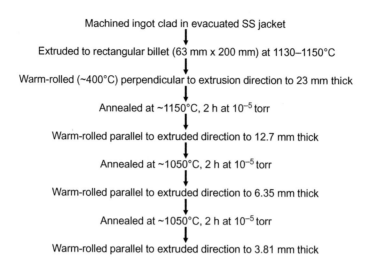

FIG. 13.1

Illustration of the fabrication schedule for production of plates from V-4Cr-4Ti cast ingots developed by Teledyne Wah Chang.

Developed by Allegheny Technologies Inc., Specialty Alloys and Components. Used with permission from Allegheny Technologies Inc.

tensile strength and ductility and ductile-to-brittle transition temperature (DBTT), were very sensitive to the impurity content and most notably to the interstitial O, C, and N concentrations [10]. Particular attention was given to optimization of the Si content of 400–1000 wppm to suppress swelling during neutron irradiation; minimization of Nb, Mo, and Ag to achieve low neutron activation; and limitation of other impurities (O, C, N, S, P, Ca, Na, etc.) to minimize grain boundary segregation and precipitation of second phase precipitates responsible for causing embrittlement of the V-based alloys. For production of the large heats of V-4Cr-4Ti, V ingots were first processed by electron beam melting that were then machined into large chips and consolidated with high-purity Cr and Ti chips by double vacuum arc melting. For the V stock, a purity level of 99.95 wt% was necessary and was typically obtained by an aluminothermic reduction method using high-purity V_2O_5 and additional electro-refining steps [2, 11]. Due to the high chemical reactivity of Ti, the recommendation was to use double- or triple-arc melting to achieve high-purity Ti stock. The fabrication practice developed for producing plates of V-4Cr-4Ti from the cast ingot is illustrated in Fig. 13.1. Once the cast ingot was produced, it was canned in stainless steel, heated for one or more hours at ~1150°C, and then extruded to form rectangular-shaped billets. The billets were then decanned and subjected to a series of heat treatments and rolling operations to form plates and sheets. It was a common practice to perform a final rolling to <50% thickness reduction followed by annealing at 1050°C for 2h in a high-vacuum furnace. The chemical analysis performed on both large heats of V-4Cr-4Ti after extrusion to the rectangular-shaped billet showed no additional pickup of impurities from the cast ingot indicating that the production and fabrication procedures were overall successful.

Ensuing characterization studies of the large V-4Cr-4Ti heats produced in the United States led to many advances in the understanding of the physical metallurgy of V-based alloys. In many sets of specimens prepared from the 500 kg V-4Cr-4Ti for mechanical properties evaluations, inhomogeneous microstructures appearing as alternating bands of coarse- and fine-grain regions were reported [12, 13] as shown in Fig. 13.2. Detailed microstructural analysis of the banded grain-size regions showed that higher concentrations of 0.1–0.3 μm size precipitates of the Ti(OCN) phase were associated with bands of finer grains aligned along the rolling direction of plates [13]. It was determined that the Ti(CON) precipitates would form any time the temperature decreased below the solvus during fabrication of plates from the cast ingot of V-4Cr-4Ti. This temperature was suggested to be ~1125°C [12, 14].

13.2 PRACTICAL ROUTES FOR REFRACTORY ALLOY PRODUCTION

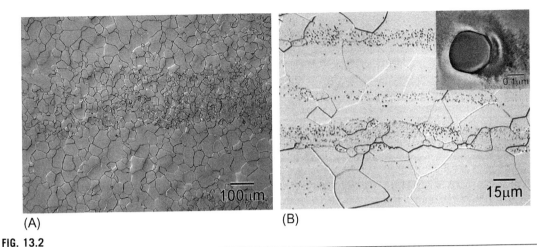

FIG. 13.2

Optical metallographic micrographs showing the (A) alternating bands of coarse- and fine-grain regions and (B) bands of high concentration of Ti(OCN) precipitates associated with the fine-grain regions. The inset is a TEM micrograph of the Ti(OCN) precipitate.

Understanding the precipitation behavior of the Ti(OCN) phase was important since studies showed that removal of interstitial O, C, and N atoms from the bcc matrix lowered the DBTT and increased the lower shelf energy in V-4Cr-4Ti [12, 14, 15]. Crystallographic investigations determined the Ti(OCN) precipitates (inset of the TEM micrograph shown in 13.2b) were consistent with the fcc Ti-x phase, where x = O, C, and N (~50% atomic percent total) and each possessing the $Fm\bar{3}m$ space group, which meant that each precipitate could accommodate different stoichiometries of the interstitial elements [16]. The globular-shaped Ti(OCN) precipitates nucleated preferentially in high dislocation density regions that formed during the fabrication steps. The inhomogeneous precipitation of Ti(OCN) complicated efforts to control formation of the banded microstructures of large V4Cr-4Ti heats assince it was not easy to redistribute them in the microstructure during thermal-mechanical treatments (TMT). In one study, it was shown that cross-rolling at room temperature produced a more uniform dispersion of Ti(OCN), but these results suggested that much more needed to be learned about the kinetics of recovery, recrystallization, and precipitation in V-4Cr-4Ti before optimal processing procedures could be developed.

Following the pioneering efforts of producing large V-4Cr-4Ti heats in the United States, similar efforts occurred in Japan, Russia, and China on the same reference V-4Cr-4Ti composition. In the early 2000s, several 30–160 kg ingots of NIFS-HEAT-1 and NIFS-HEAT-2 were produced in the Japanese program [17–19]. The goal of the development efforts was to reduce the total interstitial impurity (O, C, N) content of V-4Cr-4Ti to below 300 wppm, while maintaining low concentrations of Al (<200 wppm) and of Mo and Nb (<10 wppm). Detailed characterization studies were conducted on each production step to determine the most effective methods for reducing impurity pick-up for producing pure vanadium ingots. Several small heats of pure vanadium (25 kg) were successfully prepared with total interstitial concentrations (O, N, and C) of ~290 wppm. Alloying techniques were also reviewed and evaluated to identify promising methods for reducing impurity pick-up. In the Japanese program, the production method that was developed resulted in total interstitial concentrations of ~340 wppm for the NIFS heats of V-4Cr-4Ti [9]. The significant difference between the NIFS heats and the two US heats was the lower O concentration levels of ~150–180 wppm (NIFS) compared to ~330–360 wppm (US), which allowed for investigations of the oxygen effects of various physical and mechanical properties to be conducted. In Russia, a similar objective was employed for development of high-purity, chemically homogeneous V-Cr-Ti alloy ingots, mainly of the reference V-4Cr-4Ti and V-5Cr-5Ti. Ingots of 30–110 kg, that is, RF-VVC-2 and VM-DPCH-9 heats, were produced by vacuum-arc melting from 2000 to 2013 [20–22]. The investigations of the Russian heats focused on mainly lowering the oxygen impurity level and optimizing the TMT for improving the chemical homogeneity of solute-alloying elements, mainly Ti, and Ti(OCN) precipitate distribution to enhance the properties of the V-Cr-Ti alloys. The maximum allowed concentration of O, N, and C was specified at 200, 100, and 150 wppm, respectively, which was achieved in the production of the V-4Cr-4Ti and V-5Cr-5Ti heats. Analysis of the V-4Cr-4Ti (42 kg) and V-5Cr-5Ti (45 kg) ingots showed good macroscopic homogeneity in the Cr and Ti alloying elements, but detailed analysis by a microscopy X-ray spectral analyzer showed Ti segregation fluctuations within 2%–7% compared to the average concentration of 4.38% Ti. Subsequent TMT studies indicated that the composition variations in Ti could be minimized to lower fluctuation values. Development of V-Cr-Ti alloys in China remained at the laboratory scale until a 30 kg V-4Cr-4Ti heat (SWIP-30) was produced around 2013 [23–25]. The significant outcome of this development effort was achieving the desired limiting sum of O, C, and N impurity solute concentrations to about 400 wppm and identifying several TMT routines that resulted in high number densities of Ti(OCN] precipitates and dislocations that enhanced the high-temperature strength of V-4Cr-4Ti. The achievements obtained in the United States and international programs have demonstrated that it is feasible to fabricate V-Cr-Ti alloys in large-scale heats with the present industrial bases.

13.2.2 NIOBIUM

Interest arose during the 1950s to develop niobium (Nb)-based alloys for proposed aerospace and terrestrial nuclear reactor applications due to the low thermal neutron absorption cross-section for Nb, good compatibility with alkali liquid metals, acceptable high-temperature mechanical properties, and good fabricability [26–29]. Around this period, significant Nb ore deposits were discovered in the Western hemisphere that contributed to this interest [30]. Although early research focused on development of a wide range of Nb alloys, the only alloy that emerged on a commercial production status by the end of the 1950s was Nb-1Zr [31]. For most of the Nb alloys, the research focused on improving the high-temperature strength and oxidation resistance without degrading the inherently good properties of ductility and fabricability. The strategies that led to improvements in creep and oxidation resistance of Nb alloys consisted of alloying with W, Mo, Ta, and V for solid-solution strengthening and with Ti, Zr, and Hf for precipitation strengthening by reacting with interstitial O, C, and N atoms. By the early 1960s, the research identified Zr additions as essential for achieving acceptable oxidation resistance and compatibility with liquid metals such as Li. Alloying elements that promoted carbide formation, such as Zr, also emerged as the most effective method for precipitation strengthening. However, for many of the Nb alloys it became evident that improving the high-temperature strength properties reduced their fabricability, which resulted in further acceptance of Nb-1Zr for many applications due to the overarching importance of easily fabricable products. One notable derivative of Nb-1Zr was Nb-1Zr-0.1C, which was known as PWC-11. This alloy possessed higher strengths with minimum degradation of room-temperature ductility and fabricability. By the late 1960s, program shifts away from space reactor concepts resulted in significant reductions in the availability of Nb alloys and commercial producers, from over 20 alloys and 7 vendors in the mid-1960s to 5 alloys and 4 vendors by the late 1970s. Table 13.3 summarizes the commercially available Nb alloys at the end of the 1970s [32]. Resurgence in Nb alloy development occurred in the mid-1980s during the SP-100 space powder program, but effectively stopped when the program was terminated in 1994. Nb-1Zr and C-103 (Nb-10Hf-1Ti-0.7Zr) are the only alloys currently produced by commercial vendors.

The early production routes of Nb alloys involved casting and powder metallurgy. However, poor control of composition specifications and extensive thermo-mechanical processing that led to unacceptable interstitial O and C contamination levels in final product forms impeded the early alloy development programs. Better control of chemical composition with lower levels of impurities occurred with changes in casting techniques in the 1960s from arc-melting to electron-beam melting of the starting ingots. Further production improvements were achieved with consumable electrode arc melting methods during the 1970s and 1980s. This is illustrated in Fig. 13.3, which shows the effect of casting method on the impact toughness of unalloyed Nb [33, 34]. The higher purity levels achieved with electron beam casting showed the lowest DBTT. The typical process flow chart for the production

13.2 PRACTICAL ROUTES FOR REFRACTORY ALLOY PRODUCTION

Table 13.3 Summary of Nb Base Alloys Commercially Available by the Late 1970s [32]

Alloy	Vendor	Solute Concentration (wt. %)						
		W	Mo	Ta	Ti	Zr	Hf	Y
C-103	Fansteel, Kawecki Berylco Industries (KBI), Wah Chang	0.5		0.5	1	0.7	10	
C-129Y	Wah Chang	10					10	0.1
FS-85	Fansteel	10		28		0.8		
Nb-1Zr	Fansteel, KBI, Wah Chang, NRC					1		
SCb-291	Fansteel, KBI, Wah Chang	10		10				

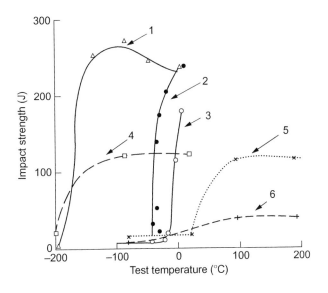

FIG. 13.3

Effect of temperature on impact properties of cast niobium [32, 33]. Curves 1–3 are from Charpy V-notch tests of (1) electron beam, and (2 and 3) of arc-melted samples. Curves 4–6 are from modified IZOD tests of (4) electron beam, (5) arc-melted, and (6) arc-melted and cold-worked materials.

of Nb alloys is shown in Fig. 13.4 [29]. Following the initial alloy melting by an electron beam, consumable electrode arc melting was used to introduce final alloying additions to the Nb alloy. Primary ingot breakdown to improve chemical homogeneity involved high-temperature extrusion or forging. Secondary final working was usually carried out at room temperature through rolling, drawing, or swaging. Thus, stringent process controls and materials specifications are necessary for the successful production of Nb alloys.

13.2.3 FABRICATION OF NUCLEAR-GRADE MOLYBDENUM

Because molybdenum possesses one of the highest melting temperatures (2623°C) and volatilizes or produces scale at a relatively low temperature [35], traditional alloy fabrication methods such as smelting are not possible. Instead, molybdenum can be produced through a series of grinding and separation steps leading to, as example, one of three

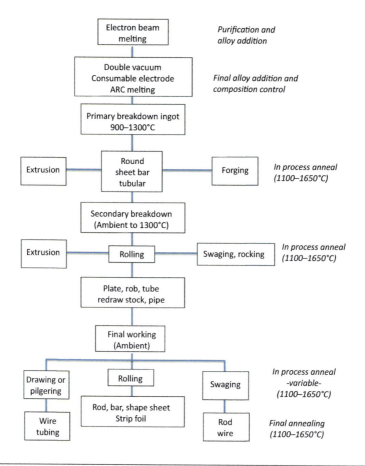

FIG. 13.4

Process flow diagram for the production of Nb base alloys [30].

types of product forms: sintered products, which are most of the commercial products available, arc cast products, which are also commercially available even though significantly more costly, and electron beam melted products, which are also quite costly and essentially laboratory-grade materials. Since polycrystalline molybdenum possesses inherently weak grain boundaries, grain boundary strengthening and refinement techniques have been pursued to mitigate embrittlement [36–40]. Significant improvement in ductility and fracture toughness has been achieved thus far mainly through impurity control, reduction of the grain size, and through addition of alloying elements such as Re, Ti, Zr, Al, B, and C, or oxide and carbide particles. In many cases, the control of impurities in both the matrix and grain boundary is crucial to obtain a product with good ductility. This is especially true for oxygen. Moreover, the Mo crystal shares the same room-temperature embrittlement issue as other Group VIA metals (tungsten, chromium), which has been long realized to be relieved through direct solid-solution alloying with Re. The underlying mechanisms of the so-called *Re*-effect [41], which is technologically important for both W and Mo, have been heavily debated in the literature with possible mechanisms including: (1) solid-solution softening attributed to improved $\frac{1}{2}\langle 110 \rangle$ screw dislocations [42–47] and (2) a change in the predominant slip plane from 110 to 112. While the Mo-Re alloys are technologically important, with alloys such as Mo-5Re, Mo-41Re, and Mo-47.5 Re being readily available, they are not currently considered for nuclear application in part due to the extreme cost of Re, and in part due to the poor irradiation performance of the alloys (see Section 13.4.3).

13.2 PRACTICAL ROUTES FOR REFRACTORY ALLOY PRODUCTION

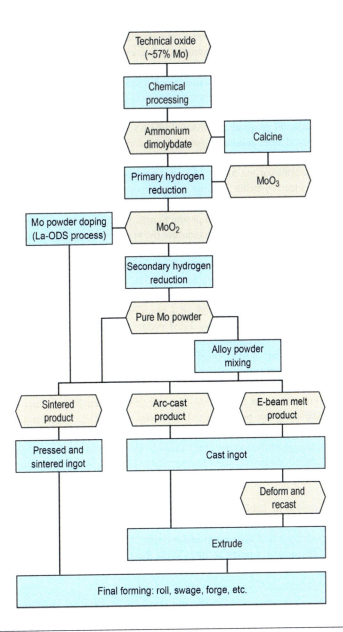

FIG. 13.5

Production route to common forms of molybdenum.

A schematic representation of the process for producing the three generic Mo products discussed above is given in Fig. 13.5. This process starts with a series of grinding and separation steps to isolate MoS_2, which is then roasted in air to product MoO_3 and SO_2. The SO_2 is then converted to sulfuric acid for commercial use, leaving a "technical oxide" containing approximately 57% Mo with ~0.1% S. This material is then chemically purified to produce Mo metal products. Pure Mo products can then be formed by sintering, arc casting, or through electron beam melting the purified powder, or alloys can be produced through standard powder mixing or through ball milling. The bulk of

Mo powder produced is sold in pressed and sintered mill products such as sheet, plate, rod, etc. These pure or alloyed powders are typically placed in an elastomeric mold, evacuated, and undergo a cold isostatic pressing (CIP). Typical pressures used are 200 MPa followed by a hydrogen sinter in the 1700–1800°C range. The hydrogen cover gas plays the crucial role of reducing the final oxygen content of the product to <20–30 wppm. A lesser amount of materials is produced following the press-sinter-melt (PSM) or arc-melt process originally developed by Climax Molybdenum. This semicontinuous process starts feeding the pure or alloyed product into a vacuum arc furnace heated by either consumable graphite electrodes (which suppress oxygen) or nonconsumable tungsten electrodes. Standard arc-cast molybdenum produces materials of very low oxygen content (<5 wppm) though at relatively high carbon content (~300 wppm.) A low-carbon arc-cast (LCAC) molybdenum is available with ASTM B386 specification of <100 wppm carbon and <15 wppm oxygen. Even higher purity levels or pure or alloyed material can be obtained by electron-beam melting in a copper crucible. For e-beam melted alloys uniformity can be an issue, typically addressed by repeated deformation and remelting.

In both the arc-cast product and electron-beam product casts, grain sizes are quite large leading to low ductility and poor fracture toughness. This is mitigated by extruding the products at relatively high temperature to refine the grain size, which in many cases causes highly anisotropic grain structures and associated mechanical properties.

Materials of specific interest to nuclear applications are unfortunately not the pressed and sintered materials most commonly available. As will be discussed in Section 13.4.3, this is attributed to the issue of the irradiation instability of grain-boundaries that are more difficult to control in the pressed and sintered materials. For this reason, the focus of attention over the past decade has been on the development and study of three types of Mo: high-purity LCAC Mo, the commercial alloy TZM produced through arc melting, and a lanthanum oxide particle containing oxide dispersion-strengthened "ODS-Mo."

Table 13.4 provides a comparison of the typical chemistry of a pressed and sintered bar product with the materials of most recent focus for nuclear structural applications [48]. The LCAC Mo and TZM products of Table 13.4 were produced through an arc-melt process, though it is noted that a sintered TZM product is also commercially available. The clear difference between the LCAC Mo and the sintered Mo product is the suppression of the oxygen, promoting strengthened grain boundaries. It is noted that TZM, which is the most widely available carbide-strengthened alloy, utilizes its alloying aids primarily to induce the formation of ZrC and TiC on the grain boundaries, which then inhibit grain-boundary grain-growth and grain-boundary failure. The presence of these carbides results in approximately twice the high-temperature (1095°C) strength as compared to pure Mo and a significant increase in material recrystallization temperature, potentially expanding the operating temperature window of this alloy.

The ODS Mo alloy discussed in Table 13.4 is an extremely pure alloy with the exception of the intentional addition of La. Specifically (see Fig. 13.5), wet doping of the MoO_2 powder with an La-nitrate solution takes place followed by pyrolysis, producing a fine dispersion of La-oxide "ODS" particles in the compact powder, which is then consolidated by hot extrusion and warm rolling. With the exception of the additional steps to produce the

Table 13.4 Chemical Composition of Nuclear-Grade Molybdenum Currently Under Study [48]										
	Chemical Composition (wppm)									
Material	C	O	N	Ti	Zr	Fe	Ni	Si	La
Sintered bar[a]	50	70	n/a	20	n/a	50	20	30	n/a
LCAC Mo	90	3	4	n/a	n/a	10	<10	n/a	n/a
ODS Mo	40	n/a	n/a	<10	n/a	27	<10	<10	1.08w/%
TZM	223	17	9	5000	1140	<10	<10	n/a	n/a

[a] Max Quantities, H.C. Starck. PD-7009, Issue 1–2009-08-10. Typical bar product typically contains 20–30 wppm.

13.2 PRACTICAL ROUTES FOR REFRACTORY ALLOY PRODUCTION

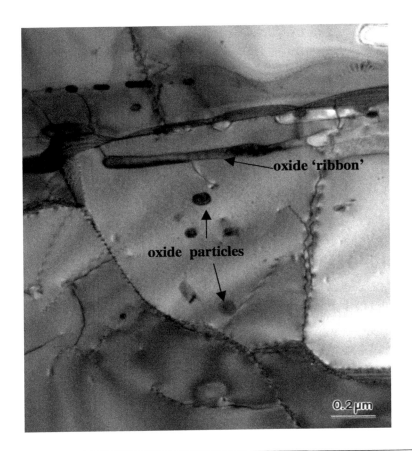

FIG. 13.6

Representative microstructure of ODS molybdenum microstructure [48].

From R. Baranwal and M. B. Burke. "Transmission Electron Microscopy of Oxide Dispersion Strengthened (ODS) Molybdenum: Effects of Irradiation on Material and Microstructure" Report B-T-3462. Bechtel Bettis 08/2007.

lanthanum oxide "nodules," the fabrication is more typical of the press-and-sinter route to production. For this material, the presence of the lanthanum in the final form serves to suppress oxygen segregation at the grain boundaries. The relative size and location of the ODS particles in this form is shown in the transmission electron micrograph in Fig. 13.6.

Molybdenum and its alloys can be welded through a number of methods including conventional gas tungsten arc welding (GTAW.) However, such welds are considered nonstructural as the heat-affected zone (HAZ) tends to have a much-reduced brittle-to-ductile transition (BDT) temperature and degraded strength that concentrates triaxial stresses in the weld region leading to (typically brittle) joint failure. Alternative methods to GTAW such as electron beam welding are often chosen in order to limit the absorbed heat and extent of the melt region. The underlying mechanism for weld embrittlement is oxygen contamination and segregation to grain boundaries and, for this reason, Mo materials are typically welded under controlled atmosphere or vacuum. Arc cast molybdenum, and in particular materials with higher carbon contents, have improved weldability as compared to typical powder metallurgical forms of molybdenum. Moreover, the reactive elements in TZM interact with oxygen during the welding process, mitigating its ability to embrittle grain boundaries, and making TZM a fairly robust alloy with regard to joining. In a similar vein, molybdenum alloys including Zr, B, C, and Al have been demonstrated to have the ability to preclude oxygen segregation to the grain boundaries (as determined though atom probe tomography studies [38, 39, 49]) during fabrication and welding, resulting in improved as-welded properties [40, 51].

13.2.4 PRACTICAL ROUTES OF TUNGSTEN AND TUNGSTEN ALLOY PRODUCTION

In this section, the practical fabrication route for tungsten and tungsten materials is described. The production route utilizes powder-metallurgy techniques due to the high melting point of tungsten (3422°C) and its susceptibility to produce volatile oxide products that makes traditional smelting processes impractical. Tungsten fabrication involves powder production, blending, compacting, sintering, and a hot- and cold-forming process. In addition, a near net-shape production route is introduced.

13.2.4.1 Powder production

Tungsten is usually prepared by hydrogen reduction of high-purity tungsten oxides. Such tungsten oxides can either be tungsten trioxide (WO_3) or tungsten blue oxide (WO_{3-x}) and, in rare cases, tungstic acid (H_2WO_4). Reduction is performed in rotary or pusher furnaces. The operating temperature lies between 600°C and 1000°C. The hydrogen gas has the additional purpose of carrying away the water vapor that is formed during the reduction. Tungsten particles form as a result of vaporization and later deposition processes. Therefore, due to a variation of temperature and water vapor partial pressure processes summarized in Fig. 13.7, tungsten particle sizes between 0.1 and 100 μm can be easily achieved. The powder metallurgical production of tungsten materials usually uses particle sizes between 2 and 6 μm.

13.2.4.2 Blending

For the production of tungsten alloys, tungsten powder is mixed with a powder of the alloying element. Common alloying elements include Re, Ta, or Mo. Most elements that form a solid solution form brittle phases at one point. Therefore, the alloying element has to be very evenly distributed. One approach is to mix rare oxides and co-reduce them to metal powder. In addition, often a binder is added.

13.2.4.3 Compacting

The fabricated powder is consolidated into a compact form. Tungsten has a relatively high hardness and strength and therefore the deformation of the particles is not easy. This compacting process however is performed without a lubricant to exclude impurities that would affect the mechanical properties of the end product. Two main routes

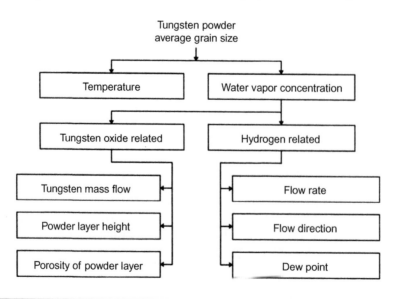

FIG. 13.7

Summary of tungsten oxide processing parameters that influence grain size.

have been established to compact tungsten powder: die pressing and isostatic pressing. After compacting, the component is called a green compact.

13.2.4.4 Die pressing
Die pressing of the power(s) is performed from the top or from the bottom and the top using rigid dies in mechanical or hydraulic presses. Mechanical presses can apply loads up to 1 MN, while hydraulic presses have a pressing force of up to 30 MN. Usually, pressures in the range of 200–400 MPa are needed in order to achieve a compact density of 55%–65% of the theoretical density. The pressure is also dependent on particle size and shape, as can be seen in Fig. 13.8.

13.2.4.5 Cold isostatic pressing
Cold isostatic pressing (CIP) is more common than die pressing. The raw powder(s) are filled into flexible molds (rubber or elastomers) and compressed. Most common is the wetbag technique where powder is filled in flexible molds that are immersed in water (in rare cases other liquids) and pressure is applied isostatically. With this process, tungsten ingots of up to 1 ton can be pressed and more complex components can be compacted. Another subtype of CIP is drybag pressing. This form is used for simple shapes. The powder-filled mold is sealed and the compression occurs between the mold and the pressure vessel.

13.2.4.6 Presintering
In some cases, the green compound has to be presintered to have a sufficient strength after the compacting. After compression, parts are then heat treated at 1100–1300°C under a hydrogen atmosphere.

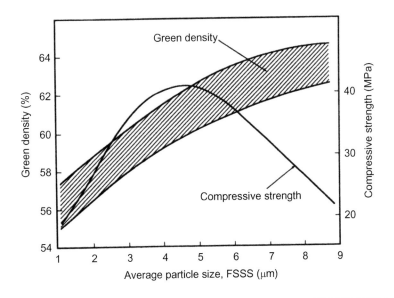

FIG. 13.8
Relationship between average tungsten grain size and green density as well as compressive strength for a constant compacting pressure (241 MPa) [52].

From Yih HW, Wang CT. Tungsten. Sources, Metallurgy, Properties, and Applications, 1979, reproduced with permission of SNCSC.

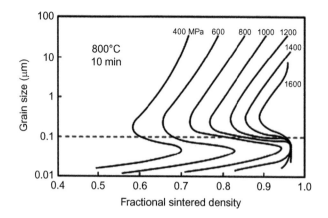

FIG. 13.9

Example grain size and sintered density contours for a variety of tungsten particle sizes processed at 800°C for 10 min over a range of possible compaction pressures, showing that extremely high compaction pressures are needed to deliver a grain size below 100 nm and nearly dense material. The best strength combinations result from tungsten particles below 100 nm [53].

13.2.4.7 Sintering

The sintering of tungsten is very challenging, as evidenced by the extraordinary pressures required to achieve a relatively modest as-sintered density at 800 C (see Fig. 13.9). For this reason, very high temperatures and pressures are generally used, with sintering temperatures between 2000°C and 3050°C commonplace. The high temperature leads to a volatilization of impurities in addition to particle growth. The temperature is either applied with self-resistance heating (several thousand amperes) or with the help of a resistance element heating system. The former option is called direct sintering, while the latter is called indirect sintering. Both methods are carried out under flowing hydrogen in order to remove the oxygen coating of the tungsten particle surface. The final density of the tungsten component is around 92%–98% of the theoretical density. The final grain size in as-sintered ingots is about 10–30 μm.

In order to achieve high sintering densities in relatively short times, the temperature has to be above 1900°C. At 2400°C, the average time to obtain a density of 92% lies at about 1 or 2 h. At 3000°C, the average time is already <30 min. Particle size has a great influence on the sintering time as well. The smaller the particle size, the shorter is the average time to obtain a high density at a given temperature.

Sintering can be accelerated if small amounts (0.5%–1%) of alloying elements like Ni or Pd that enhance the grain boundary diffusion in tungsten are present. This, however, has a negative influence on the mechanical properties.

13.2.4.8 Direct sintering

The direct sintering process uses electric current to control the temperature and, therefore, the sintering process. At the beginning stages, the current is kept relatively low. Due to a limited contact between the tungsten particles/grains, the electric resistance of the tungsten product is relatively high, particularly during the initial sintering period. The temperature is gradually raised up to 3000°C, with holding points in between. These hold times are essential to allow outgassing of undesired elements. The average holding time is approximately 30–60 min. Direct sintering has the advantage of having higher temperatures in the middle of the part. This favors diffusion processes and leads to an efficient evaporation of impurities.

13.2.4.9 Indirect sintering

Indirect sintering usually does not require presintering. The green compact is placed in a basket-shaped heating element of a furnace. Uniform shrinkage is achieved by placing the compacts on green tungsten shims [52]. The heating occurs at a slow rate leading to a hold at 2000–2700°C. Sintering times may range up to 24 h.

13.2.4.10 Forming process

In order to form a preshape, to improve the properties, and to achieve a completely dense material, a complex, multistage, hot- and cold-forming process is applied. Working is usually performed underneath the recrystallization temperature and therefore all forming is, strictly speaking, a cold-working process.

Some typical techniques are rolling (sheets) or swaging for rods, forging for larger parts, or drawing for wire and tubes. A detailed description of the most common forming processes can be found in the book "Tungsten," by Yih and Wang [52].

The first part of the forming process is carried out at around 1500–1700°C. If tungsten is worked at lower temperatures, cracks are reported to develop. With increasing degree of deformation, the grain size and shape changes significantly and anisotropy develops. This change in microstructure goes along with a change in mechanical properties. Due to this change, the forming process during the latter stages can be carried out at lower temperatures between 1100°C and 1300°C. Table 13.5 summarizes a typical progressive hot-rolling treatment and intermediate stress relief annealing schedule for the fabrication of tungsten sheet.

13.2.4.11 Near net-shape and mass fabrication

A time and cost-effective near-net-shape-forming process with the advantage of shape complexity, material utilization, and high final density is powder injection molding (PIM). This process was adapted and developed at KIT for tungsten and promising results have already been achieved. The key steps in PIM are kneading or extrusion

Table 13.5 Example of Rolling Schedules for Tungsten Sheet

	Thickness After Pass (mm)	Preheating Temperature (°C)
Starting thickness 24.4 mm		
1	20.3	1450
2	15.9	1400
3 (Cross roll)	12.7	1400
4	10.2	1400
Anneal 5min at 1300–1350°C		
5	8.26	1350
6	6.86	1300
7 (Cross roll)	5.46	1300
8	4.45	1300
Anneal 5min at 1250–1300°C; caustic pickling and condition		
9	3.51	1250
10	2.82	1200
11	2.26	1200
12	1.80	1150
13	1.63	1150

Source: Data from G.C. Bodine Jr., Tungsten Sheet Rolling Program, Final Report, Fansteel Metallurgical Corp., North Chicago, Illinois, 1963.

of a suitable feedstock (combination of powder and binder), where the heated powders (80°C) are mixed with a 50 vol% wax/thermoplastic binder system in a kneader at 120°C. The design and engineering of a tool (including filling simulation) is followed by injection molding of the so-called "green parts" at a feedstock temperature of 160°C and a mold temperature of 50–60°C. At the first step, the premold thimble is replicated; after that, the tool is moved around 180° and in the second step the tile is molded on top of the thimble. This process is very quick and effective. The de-binding (solvent and thermal) is carried out in n-hexane, followed by a thermal debinding step in dry hydrogen atmosphere. During the debinding step, not only are the binder and the impurities (oxygen and carbon) removed, but also the high residual stresses generated during injection molding are released. The last step of the process is a final heat-treatment process. The PIM process for tungsten is shown in Fig. 13.10. The process allows for an easy, reproducible, cost-effective mass production of parts of general shape. The process does not depend on the particular composition of the tungsten material. Further, the sintering process can be varied in different ways. For example, the HIP step can be replaced by a simple sintering step in a hydrogen furnace. Usually, one-component PIM (the produced parts consist of the same tungsten material) is sufficient for many applications.

13.3 AS-FABRICATED MECHANICAL PROPERTIES
13.3.1 VANADIUM

Several studies have shown that the mechanical properties and deformation behavior of V-Cr-Ti alloys are sensitive to composition and thermomechanical processing variables [4, 5, 8, 12, 15, 19, 25, 54]. The alloying additions of Cr and Ti form solid solutions in the bcc V matrix at elevated temperatures and provide strengthening mainly by solid solution (Cr) and precipitation (Ti). The main purpose of the Ti addition is to serve as a gettering agent for the interstitial O, C, and N atoms that are present as impurities in the bcc V matrix, and to form Ti(OCN) precipitates that provide some alloy strengthening [12, 15, 55]. The thermomechanical treatments performed during processing of V-Cr-Ti alloys also contribute to the sensitivity factor since they influence the amount of pickup of interstitial impurities as well as their removal from the bcc V matrix by precipitation processes and the average size, size distribution, and uniformity of grains.

The effect of solute Cr and Ti additions on the tensile properties of V-Cr-Ti alloys is shown in Fig. 13.11 [56]. Elevated levels of Cr and Ti were added to the V-4Cr-4Ti (#832665) heat and the processing conditions used resulted in similar O, C, and N concentrations of 368 ± 14, 118 ± 19, and 102 ± 19 wppm, respectively. Compared to V-4Cr-4Ti, the increase to 10%Cr and to 10% and 15%Ti resulted in continuously increasing yield and ultimate tensile stresses with essentially no effect on the uniform elongation and only a slight decrease in total elongation. Since the interstitial O, C, and N atoms are removed from the bcc V matrix by the formation of Ti9(OCN) precipitates, ductility is not affected significantly with higher Ti additions to the V-Cr-Ti alloy at room temperature.

The fracture toughness and Charpy impact properties of V-Cr-Ti alloys depend on composition and processing variables as well as numerous experimental factors, such as specimen geometry, notch characteristics, and test conditions, that is, strain rate and temperature [4, 15, 40, 41]. The results shown in Fig. 13.12 illustrate these factors. In Fig. 13.12A, data obtained from Charpy V-notched specimens show monotonic increases in the DBTT for a variety of Cr and Ti solute additions as the annealing temperature is increased above 1000°C, and monotonic increases in DBTT for each annealing temperature as the solute Cr and Ti concentrations are increased. The Charpy impact results shown in Fig. 13.12B for V-4Cr-4Ti (#832665) indicate that annealing at 1000°C resulted in no detectable DBTT down to −196°C (i.e., extremely good resistance to embrittlement down to very low test temperatures). Conversely, annealing at 1125°C resulted in a DBTT near −125°C with very low value of lower shelf energies. This poor Charpy impact behavior in the 1125°C annealed material was significantly improved by a subsequent anneal at 890°C. The increase in DBTT from annealing at 1125°C was attributed to a larger grain size and larger percentage of interstitial O, C, and N in solid solution, as opposed to finer grain size and Ti(OCN)

13.3 AS-FABRICATED MECHANICAL PROPERTIES

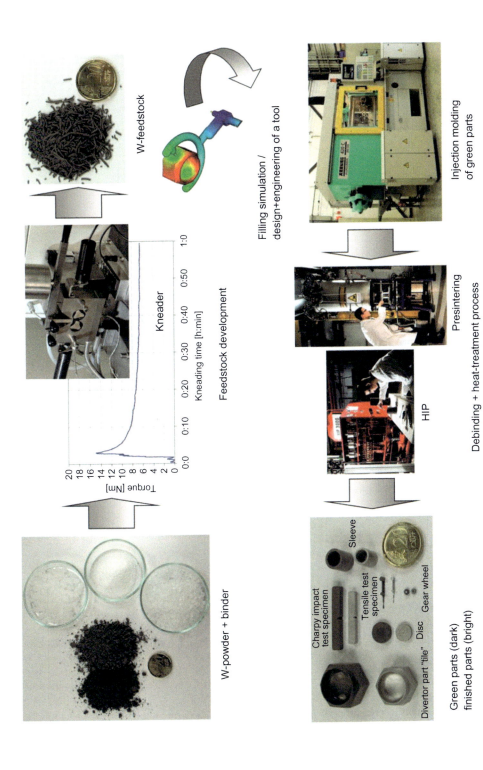

FIG. 13.10

The PIM process for tungsten as developed at KIT.

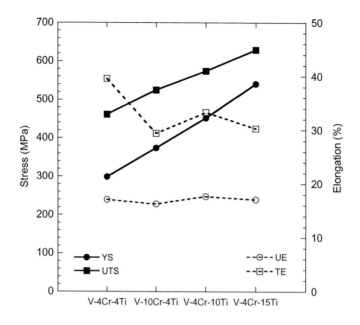

FIG. 13.11

Comparison of tensile properties between the reference V-Cr-4Ti alloy with increases in the Cr content (V-10Cr-4Ti) and in the Ti content (V-4Cr-10 and V-4Cr-15Ti). The tensile tests were conducted at room temperature using a strain rate of 10^{-3}/s.

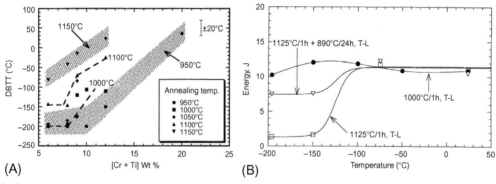

FIG. 13.12

Effect of annealing temperature on (A) the DBTT of V-Cr-Ti containing different solute additions [13, 57] and (B) the impact energy of the reference V-4Cr-4Ti (#832665) [14].

From R.J. Kurtz, M.L. Hamilton and H. Li, "Grain Boundary Chemistry and Heat Treatment Effects on the Ductile-To-Brittle Transition Behavior of Vanadium Alloys", Journal of Nuclear Materials, Vol. 258–263, (1998), p. 1375.

precipitates that formed during annealing at 1000°C. The annealing at 890°C increases the volume fraction of Ti(OCN) precipitates that lowers the O, C, N interstitial solutes in solid solution.

The temperature-dependent tensile properties of V-4Cr-4Ti (heat #832665) are shown in Fig. 13.13 [58]. The yield and ultimate tensile strengths (UTS) and total elongation decrease from room temperature to 300°C are followed by a nearly temperature-independent regime for strength and ductility up to 700°C. Above 700°C,

13.3 AS-FABRICATED MECHANICAL PROPERTIES

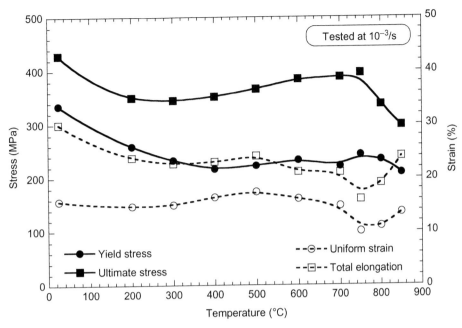

FIG. 13.13

Temperature dependence of tensile properties for the V-4Cr-4Ti (heat 832,665) alloy.

the strength increases to a local maxima and ductility decreases to a minima near 750°C, followed by relatively pronounced decreases in strength and increases in ductility at higher temperatures due to the onset of thermal creep processes.

The tensile curves obtained for V-4Cr-4Ti from 20°C to 850°C with a strain rate of 10^{-3}/s are shown in Fig. 13.14 [59, 60]. The results illustrate the influence that interstitial O, C and N have on the mechanical properties [4, 5, 8, 12, 15, 19, 25, 54, 57, 59]. The tensile curves are offset from the stress and strain axes for clarity. For most of the curves, the upper yield stress, σ_y, is preceded by a small plastic pre-strain in the elastic regime followed by a Lüders extension ranging from 0.5% to 1.0%. Prior to the Lüders extension, a prominent load drop occurs at 20°C, but is barely detectable at higher temperatures. Following the Lüders extension, deformation occurs homogeneously throughout the specimen with decreasing strain-hardening rate until reaching the ultimate tensile stress, σ_u, which defines the point of plastic instability. For temperatures between 300°C and 750°C, and at sufficiently low strain rates, serrations occur in various regions of the flow curves. These serrations are related to the diffusion of solute atoms and the locking of dislocations during the test, that is, the phenomenon of dynamic strain aging (DSA) [59, 60]. As evident in Fig. 13.14, the DSA is most pronounced near 600°C and for plastic elongations of ~10%–20% for the tensile tests performed at a 10^{-3}/s strain rate.

The DSA behavior of V-4Cr-4Ti was investigated using varying strain rates in tensile tests conducted over a temperature range from 20°C to 850°C [59, 60]. From the stress-strain curves, the strain rate sensitivity (SRS) parameter defined as [61]

$$m = \frac{1}{\sigma} \frac{\delta \sigma}{\delta \ln \dot{\epsilon}} \bigg|_{\epsilon, T} \quad (13.1)$$

was determined from the slope of the yield stress (σ_y), stress at 8% strain ($\sigma_{8\%}$), and ultimate tensile stress, σ_u plotted against the logarithm of strain rate. The temperature dependence of the ultimate stress, σ_u, as a function

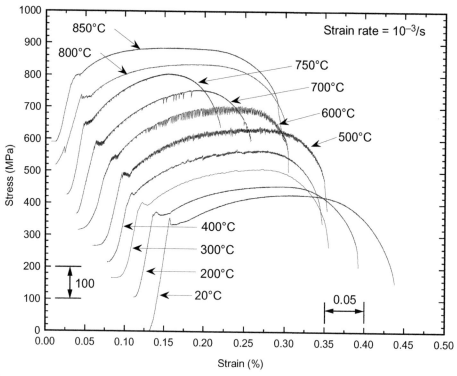

FIG. 13.14

Temperature-dependent tensile curves for V-4Cr-4Ti at a strain rate of 10^{-3}/s illustrating the dynamic strain aging regime at 400–700°C.

of strain rate and the calculated values of m as a function of temperature for V-4Cr-4Ti are shown in Fig. 13.15. Several different flow regimes are evident, with the SRS $m>0$ at low and high temperatures whereas $m<0$ at intermediate temperatures. During normal deformation processes, the flow strength increases with strain rate and m is positive, which generally occurred below 300°C and above 700°C for V-4Cr-4Ti, depending on the strain rate and the flow stress parameters. However, within the DSA regime, the increase in strain rate causes a decrease in the flow strength and m is negative. This regime generally occurred between 300°C and 700°C for V-4Cr-4Ti. Since deformation occurs by thermally activated slip processes at elevated temperatures, the ability of solute atoms to diffuse and form atmospheres around mobile dislocation cores will influence the type of deformation behavior. At low temperatures, solute mobility is too low to form sufficient solute concentrations while at high temperatures, solute concentrations are too mobile to exert sufficient drag on mobile dislocations. For the DSA regime, the increase in strain rate decreases the time available for solute diffusion to dislocations, which decreases the drag effect, resulting in a lowering of the flow strength, that is, negative SRS behavior. For the V-4Cr-4Ti, the interaction between Ti and interstitial O, C, and N impurities to form Ti-OCN precipitates removes the interstitial impurities from the bcc matrix and reduces the DSA effect. Thus, the DSA behavior observed for V-Cr-Ti alloys is indicative of the concentration of interstitial impurities that are in solution.

The high-temperature thermal creep properties of V-4Cr-4Ti are sensitive to changes in the interstitial concentrations due to the high affinity of V for interstitial solute atoms, especially O, and experimental factors such as long test times, high temperatures, and surrounding environment of specimens during the creep test [13, 19, 62–64]. The creep properties of V-4Cr-4Ti investigated in high vacuum using uniaxial tensile and biaxial pressurized tube

13.3 AS-FABRICATED MECHANICAL PROPERTIES

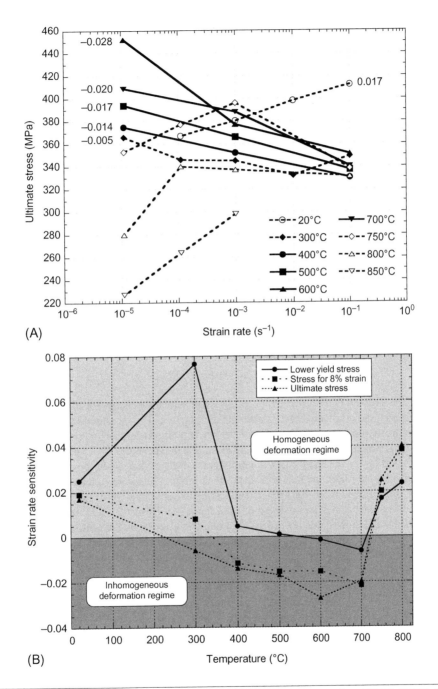

FIG. 13.15

(A) Ultimate stress as a function of strain rate of V-4Cr-4Ti over the temperature range from 20°C to 850°C. (B) Strain rate sensitivity as a function of temperature for the three measured flow stress parameters.

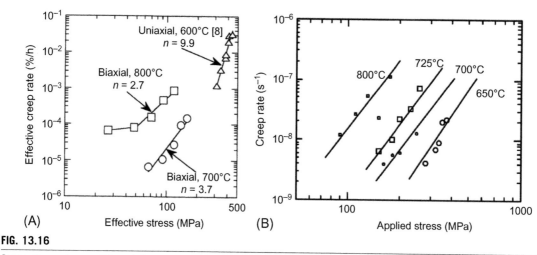

FIG. 13.16

Stress dependence of the minimum creep rate of V-4Cr-4Ti tested in vacuum at 600–800°C by (A) biaxial pressurized tube specimens [10] and (B) uniaxial tensile specimens [63].

specimens at temperatures >600°C are shown in Fig. 13.16 [13, 62, 64]. The results showed minimum creep rates that were consistent with the power-law creep expression:

$$\dot{\varepsilon} = A\sigma^n \exp\left(-\frac{Q}{RT}\right) \tag{13.2}$$

where σ is the applied stress, n is the stress exponent, Q is the activation energy for the controlling creep mechanism, R is the universal gas constant, and T is absolute temperature. The measured values of n were within the range of 2.7–4.0 for the two creep test methods. The measured activation energies for creep were ~260–330 kJ/mol for the two sets of data, which was close to the self-diffusion of pure vanadium, or ~270 kJ/mol. The predominant creep mechanism based on the data analysis was climb-assisted dislocation glide; however, thermally activated solute drag mechanisms could limit the rate of dislocation climb past obstacles in the microstructure at temperatures of 600–800°C and stresses of 50–400 MPa. Creep testing in liquid Li at 700°C and 800°C was conducted on V-4Cr-4Ti prepared from the United States (#832665) and Japan (NIFS-Heat-2) heats using biaxial pressured tube specimens [64]. The results showed that the creep strain rate was affected by the heat, tubing production method, and changes in the interstitial O and N levels that occurred during the creep test. The O levels decreased, while the N levels increased in the specimens during exposure in liquid Li, as expected from V-Li thermochemistry considerations (N and O being present in both the V alloy and molten Li as an impurity] [65]. The predominant creep behavior was an inverted primary transient creep period followed by steady state, or accelerating creep to rupture. Unlike the creep tests of V-4Cr-4Ti in vacuum, creep in liquid Li does not follow the classical three-stages of creep exemplified by primary, secondary, and tertiary creep.

13.3.2 AS-FABRICATED MECHANICAL PROPERTIES OF NIOBIUM

While niobium is BCC and therefore subject to low-temperature embrittlement and highly susceptible to impurity embrittlement (e.g., oxygen, nitrogen, carbon, hydrogen), the BDT temperature in the nonirradiated state is quite low (see Fig. 13.3). For this reason, niobium has very good formability and therefore is a desirable alloy for a number of practical applications. This enhanced fabricability comes at the expense of strength, which is typically lower than the other refractory metals being discussed in this chapter. As example, the UTS of a number of niobium alloys

13.3 AS-FABRICATED MECHANICAL PROPERTIES

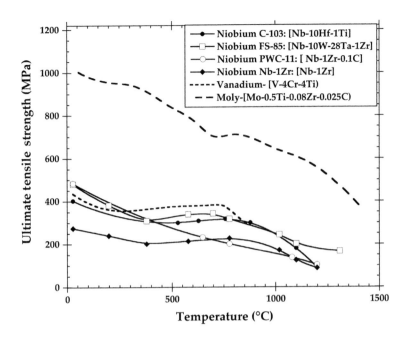

FIG. 13.17

Ultimate tensile strength of niobium alloys compared with vanadium and molybdenum.

studied for nuclear and aerospace applications (including the two alloys in current commercial production, Nb-1Zr and C-103) are provided in Fig. 13.17. Also provided in the figure for comparison purpose is the vanadium alloy V-4Cr-4Ti and the TZM molybdenum commercial alloy.

As drawn from Fig. 13.17, the beneficial combination of ductility and fabricability comes at the cost of alloy strength, with the suite of niobium alloys all similar to or lower than that of vanadium, and substantially lower than the strain-relieved TZM alloy depicted or tungsten alloys (not depicted). In comparison with tantalum alloys, which are routinely considered alongside niobium for space and nuclear applications, the niobium alloys have similar strength at low-to-intermediate level. For example, Nb-1Zr is very similar to the tantalum alloy T-111 (Ta-8 W-2Hf) to approximately 800°C, though due to a coarsening of carbides in the niobium alloy, its tensile and fracture properties degrade at higher temperatures [66, 67]. As pointed out by Leonard [68], such microstructural instability in alloys such as Cb-752 and Cb-753 at elevated temperature leads to a degraded creep strength, as evidenced by a negatively sloping Larson-Miller plot and embrittlement (in the case of FS-85) of grain boundaries.

13.3.3 AS-FABRICATED MECHANICAL PROPERTIES OF MOLYBDENUM

The transition metal molybdenum has the second highest melting temperature after tungsten, of the materials considered in this chapter. This high melting temperature, combined with the second commonality with tungsten-a BCC crystal structure, contributes to the limited low-temperature ductility of molybdenum. As discussed in the previous section, the primary Mo alloys of industrial interest are the pressed and sintered materials. The mechanical properties for each of these alloys, as with the higher purity alloys of more interest to nuclear applications, are strongly dependent on the amount of process deformation and annealing applied in order to gain some ductility. Fig. 13.17 provides a comparison of the temperature-dependent tensile yield stress for Plansee commercial pure sintered molybdenum and high-strength TZM alloy in the stress-relieved and recrystallized conditions. It can be seen that stress-relieved materials exhibit significantly higher strengths than recrystallized materials up to high temperatures, and that TZM alloy has much higher strength than pure Mo. In addition to increased yield strength, TZM possesses other beneficial engineering qualities such as approximately twice the stress rupture strength and a much lower steady-state creep rate. Inset within the figure are specific temperatures determined to provide 1 h recrystallization for 90% deformed molybdenum alloys (for lower deformation levels, longer times or higher temperature would be required). The 1-h recrystallization annealing times increase from ~1100°C for pure sintered Mo up to 1500°C for the specialty alloys such as the Mo-Re alloys, TZM, and the hafnium carbide grain boundary stabilized Mo-MHC. Essentially independent of processing route, the work deformation required in the production of molybdenum produces material with very limited room-temperature ductility, mandating some level of heat treatment to allow use in structural applications. As seen in Fig. 13.18, the primary result of recrystallization for pure Mo and Mo alloys is a product with significantly lower strength. A more modest reduction in strength is obtained by applying a stress-relieving anneal, which also yields modest improvements in ductility. The relative stress relief, or full recrystallization of the material can have a large impact on the ductility, strongly dependent on the initial interstitial content of the material and any interstitial pick-up during the process. For materials of interest to nuclear applications, the interstitial content will be well controlled and very low precluding the sweeping of oxygen to grain boundaries and associated reduced fracture toughness and embrittlement upon recrystallization. As example, in Cockeram's paper [51] on the development of weldable molybdenum alloys, both pure LCAC molybdenum and a number of "*weldable*" alloys improved from an as-worked total elongation of 2%–4% in the 100–1000°C range to 2%–18% in the stress-relieved condition to 14%–42% in the recrystallized condition.

The tensile behavior for Mo materials of current developmental interest for nuclear structural application is given in Fig. 13.19. These materials include the low carbon arc-cast Mo, arc-cast TZM (contrasting with the sintered TZM of Fig. 13.18), a lanthiated (ODS) Mo, and a molybdenum alloy developed as a high-purity weldable alloy containing B-11, C, and Zr as alloying elements to scavenge oxygen [51]. The yield strength for all these

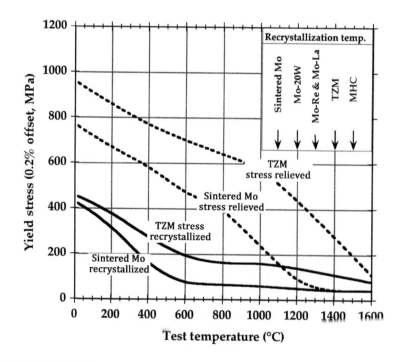

FIG. 13.18

Yield stress of commercial Mo alloys (Plansee technical data sheet).

13.3 AS-FABRICATED MECHANICAL PROPERTIES

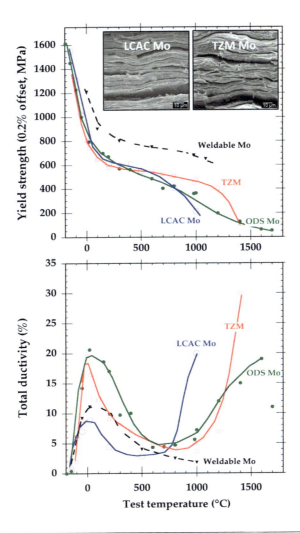

FIG. 13.19

Comparison on non-irradiated tensile properties of various forms of interest to nuclear application in stress-relieved conditions along the rolling direction.

Figure drawing upon data from [51,70,69] Inset, LCAC and TZM Mo fracture surface following ambient tensile test; Inset from Cockeram BV, Smith RW, Snead LL. The Influence of Fast Neutron Irradiation and Irradiation Temperature on the Tensile Properties of Wrought LCAC and TZM Molybdenum. Journal of Nuclear Materials. 2005;346:145–164.

tested Mo alloys decreases monotonically with increasing test temperature, with the weldable Mo alloy exhibiting the highest strength and the other three materials exhibiting strengths comparable to each other. In each case, the alloys possess moderate room-temperature ductility, which significantly decreases in the intermediate temperature range and then increases rapidly as test temperatures consistent with stress-relief annealing temperatures are achieved.

Due to the relatively large amount of reduction molybdenum undergoes during the forming process, it develops a highly anisotropic lamellar or "pancake-like" microstructure. Depending on the size of the grains and strength of

Table 13.6 Comparison of Tensile Properties of Stress Relieved Nuclear Grade Molybdenum in Longitudinal and Transverse Orientations [70]

	Yield Strength at RT (MPa)	Total Ductility at Room Temp.(%)	Yield Strength at 600°C (MPa)	Total Ductility @600°C (%)
LCAC—Longitudinal	751	17.6±1.5	475.6±33.4	3.8±0.5
LCAC—Transverse	809.3±9.7	13.1±2.8		
TZM—Longitudinal	803	19.6	546.5±26.9	4.7±0.2
TZM—Transverse	895	11	637.4±9.3	2.5±0.4
ODS—Longitudinal	735±32	14.2±2.6	445±20	4.3±0.4
ODS—Transverse	823±24	9.6±2		

the grain boundaries, this microstructural anisotropy can have a large effect on the mechanical properties. For the typical wrought alloys being discussed here, at temperatures near or above the DBTT, fracture initiates at grain boundaries (or perhaps at ODS particles) producing a separation of the adjoining lamella. An example of the typical microstructure in the as-deformed condition is provided as an inset to Fig. 13.19, showing the decohesion between lamellae in LCAC Mo and TZM following ambient temperature tensile testing. From studies on in-situ fractures, where fracture toughness was observed directly in a SEM, a crack-divider type of toughening mechanism for these materials has been inferred [69]. Given this failure microstructure, it is understandable that the longitudinal and transverse ductility of the materials would be significantly different. This is clearly seen in the data summarized in Table 13.6 for the longitudinal and transverse tensile ductility of LCAC Mo, TZM, and ODS Mo, which indicates a significant reduction in ductility in the transverse (across lamella) direction as compared to the longitudinal direction. The fracture toughness of nonirradiated molybdenum, while strongly affected by grain size, is also affected by orientation. Typically, the finest grain materials such as ODS Mo would possess the highest fracture toughness, while LCAC Mo with the largest grains exhibits the lowest fracture toughness and highest DBTT. Materials such as TZM, of intermediate grain size, fall in between. This can be seen by inspection of Fig. 13.20, along with the anisotropy in fracture toughness.

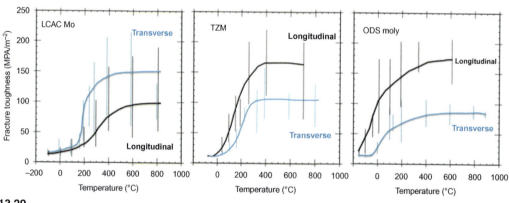

FIG. 13.20

Fracture toughness and anisotropy in fracture toughness of different types of molybdenum.

13.3.4 AS-FABRICATED MECHANICAL PROPERTIES OF TUNGSTEN

The outstanding mechanical properties of tungsten are the high tensile strength, including high thermal creep resistance at elevated temperatures. The biggest disadvantage, especially for the use as a structural material, is the intrinsic brittleness (measured by tensile, Charpy impact, or fracture mechanics tests) and the low fracture toughness, particularly near room temperature and below. In addition, the DBTT is especially high. Mechanical properties and therefore also fracture toughness of polycrystalline tungsten are strongly dependent on the underlying microstructure and the chemical composition. Table 13.7 compares the large difference in Vickers Hardness HV30 for polycrystalline tungsten at room temperature under recrystallized and work-hardened conditions [52].

Various tungsten materials and alloys have been developed in the past for structural applications, mainly in order to increase the ductility of tungsten.

The listed values for tungsten alloy mechanical properties in this chapter should be considered as general tendencies due to the strong dependence of mechanical properties on microstructural features and the fact that in most cases the precise microstructure, type and concentration of impurity elements, and their distribution are not detailed in published articles.

13.3.4.1 Structured tungsten materials

While tungsten is accepted as a very brittle material, deformation of high-purity single-crystal tungsten exhibits some ductility even at the low temperature of liquid helium. All tungsten materials show a distinct temperature dependence of the flow stress [71], typically explained by the thermally activated interaction between dislocations and interstitial impurities that are in solution. Above the critical temperature, a strong increase of the flow stress can be observed. For pure tungsten, this critical temperature is well above room temperature, at ~330–530°C [72].

The stress-strain curves for high-purity tungsten tested as a single crystal and a polycrystalline material swaged and subsequently annealed for 1 h at 2000°C are shown in Fig. 13.21 [73] for tests performed at room temperature and 400°C. The polycrystalline material shows good ductility at 400°C but is completely brittle at room temperature, indicating that the BDT temperature for this tungsten material lies between 20°C and 400°C. The precise value of the BDT temperature is dependent on the material condition and the strain rate, along with other experimental variables. For tungsten, the reported tensile DBTT usually lies between 100°C and 200°C, but in cases of extreme high strain rates this value can rise up to 1000°C [74].

After fabrication, tungsten products can exhibit different microstructures; typical examples are shown in Fig. 13.22A and B. While tungsten rods (Fig. 13.21B), due to the uniform radial stress applied during extrusion, have resulting elongated tendril grains along the extrusion (rod) axis, (Fig. 13.22A) the in-plain stress applied during the rolling of a plate results in a lathe-like (staked pancake) structure for the plates. While a number of factors can contribute to the strength, elongation, and fracture properties of toughness, the processing-derived structure is a major factor. For example, Fig 13.22C provides the absorbed fracture energy as a function of temperature as dependent on the orientation of the sample with respect to plate rolling (note: the plate lathe structure corresponds to Fig 13.22A.) As seen from the figure, for the two directions in which a crack is forced to propagate normal to the pancake structures (LS and LT) a reasonable measure of energy is absorbed in the sample upon failure. As the notch and therefore the crack is oriented in the plane of the pancakes, little energy is absorbed in crack propagation.

Table 13.7 Macro-Vickers Hardness for Polycrystalline Tungsten is Given as (HV30) [52]

Recrystallized	300
Work hardened	>650

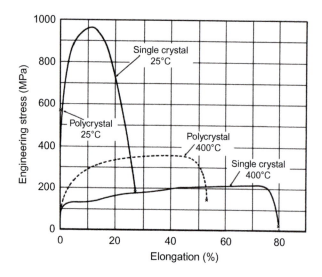

FIG. 13.21

Stress-strain curves of polycrystalline and single-crystal tungsten performed at a strain rate of 2%/min [72].

From Allen BC, Maykuth DJ, Jaffee RI. The Recrystallization and Ductile-Brittle Transition Behaviour of Tungsten-Effect of Impurities on Polycrystals Prepared from Single Crystals. Journal of the Institute of Metals. 1961;90(120–128).

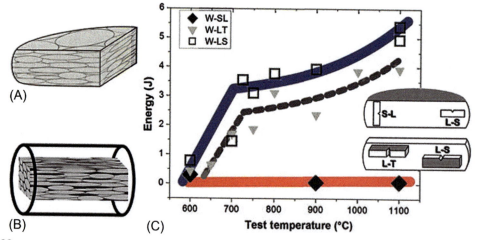

FIG. 13.22

(A) Pancake microstructure of plates, (B) fiber-like microstructure of rods. (C) Influence of the microstructure orientation on the Charpy properties of pure tungsten blanks.

The high sensitivity of fracture toughness regarding grain shape and orientation is due to the weak grain boundary cohesion in tungsten. Tungsten specimens are tougher when they are stressed parallel to the long side of the grains (fiber-like or "pancakes"). Stresses, acting perpendicular to the grain boundaries, cause easy crack initiation as well as easy crack propagation.

Charpy specimens oriented parallel to the axis of tungsten rods do not show a direct transition from brittle to ductile fracture with raising test temperatures. Moreover, a broad range of the so-called delamination fracture can be observed. Specimens fabricated from plates or round blanks behave in a similar manner.

13.3.4.2 Dispersion strengthened tungsten

Dispersion strengthening and precipitation hardening are the most effective ways to increase the high-temperature strength and creep resistance of tungsten. Such tungsten materials are often called ODS tungsten. Small amounts of insoluble elements like oxide or carbide powders are added to tungsten prior to the sintering process; the resulting material will consist of the typical tungsten microstructure but with an additional second phase at or around the grain boundaries. Either potassium solute additions in the ppm range (WVM) or oxides and carbides such as La_2O_3 (WL10), CeO_2 (WC20), and ThO_2 (WT20) are used. Other additions could be ZrO_2, CeO_2, ThO_2, TiC, and HfC. Due to high activation, Th and Hf have to be excluded for most nuclear applications. Adding lanthanum oxide (WL10) results in a shift of around 100K in the onset of brittle fracture to lower temperatures as shown in Fig. 13.23.

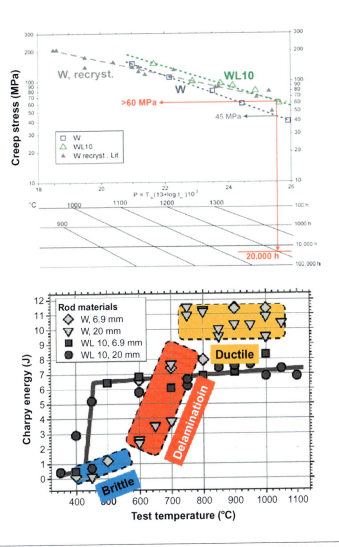

FIG. 13.23

Top: Creep strength prediction showing tungsten and tungsten lanthanum oxide (WL10). Bottom: Charpy test results of pure tungsten and tungsten lanthanum oxide (WL10).

At higher temperatures, oxides completely suppress ductile fracture, even at temperatures as high as 1100°C. It is very likely that the use of other oxides leads to similar results. In potassium-doped tungsten (WVM), the recrystallization temperature shifts by 200 K to higher temperatures. Since interstitially soluble elements like, for example, O, N, C, P, S, Al, or B, tend to segregate at grain boundaries, the addition of their small amounts would be another possibility for affecting the grain boundaries. While most of these elements will weaken the grain boundaries even more, only B is supposed to show a positive influence. Dispersion strengthening and precipitation hardening increase creep strength and recrystallization can be improved with only minor decrease in thermal conductivity by the use of dispersed oxides such as lanthanum that stabilize the grains. The intrinsic brittleness of tungsten (measured by Charpy tests), however, cannot be improved by oxide dispersion. On the contrary, intercrystalline fracture is enhanced even more.

13.3.4.3 Solution-strengthened tungsten

Only rhenium is rigorously known to improve the ductility of tungsten by solid solution and is known to shift the BDT to lower temperatures. Unfortunately, its use for nuclear energy applications has been ruled out for reasons of cost and irradiation embrittlement. Iridium could also have a positive effect, though it is even more expensive than Re. Other elements forming solid solution with tungsten are: Ta, V, Nb, Ti, Nb, and Cr. Of these elements, only Ta, V, Nb, and Mo form solid solutes in the complete practical alloying range. Since Nb and Mo transmute to very long-living radioactive isotopes, they are considered unattractive for fusion application. This leaves only Ta and V as possible candidate solutes for fusion applications. However, both mechanical properties measurement of W-(x)Ta and W-(y)V and supporting theoretical density functional theory calculation show that these alloys are more brittle than high-quality pure tungsten. The apparent cause for the inferior performance of the Ta and V bearing is the formation of highly brittle intermetallic phases. Charpy tests in the temperature range of 400–1100°C have shown that in comparison to pure tungsten the BDT temperature for solid solutions in all ranges studied is increased (i.e., worse.) This is seen from the right-most image of Fig. 13.24, in particular from inspection of the significant increase in the W-5Ta DBTT. However, strength was improved for very small additions of Ta as an alloying element and only at temperatures higher than 1000°C. As an additional note, these alloying elements can suppress the tungsten thermal conductivity significantly.

With the exception of W-Re, alloying tungsten in terms of establishing substitution solid solution does not improve the ductility. Grain boundary strengthening might improve the ductility. But so far it has not worked on the whole temperature range. B and TiC are candidates that still need to be tested. ODS tungsten materials like WL10 do not show fully ductile fracturing in Charpy tests at temperatures up to 1100°C.

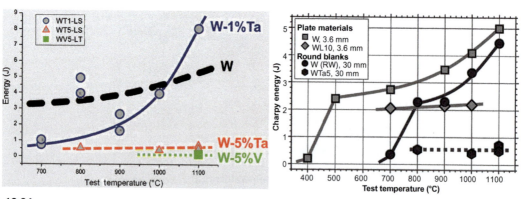

FIG. 13.24

Left: Charpy test results of pure tungsten, W-Ta, and W V alloys. Right: Charpy test results of different discussed tungsten materials.

In ODS tungsten as well as potassium-doped tungsten (WVM), recrystallization is shifted to higher temperatures by at least 200 K. The creep strength of ODS tungsten is slightly increased. The thermal conductivity of ODS tungsten is comparable to pure tungsten parallel to the rolling direction. In plates, it is slightly reduced perpendicular to the rolling surface. As shown in the left graph of Fig. 13.24, pure tungsten with an optimal microstructure seems still to be best suited for structural applications. As demonstrated in the left-most image of Fig. 13.24, the W-5Ta alloy performs considerably worse than the W-1Ta alloy, presumably due to the higher concentration of intermetallic present.

13.4 AS-IRRADIATED MECHANICAL PROPERTIES
13.4.1 AS-IRRADIATED MECHANICAL PROPERTIES OF VANADIUM

The effects of neutron irradiation on the mechanical properties of V-(4-5%)Cr-(4-5%)Ti alloys have been shown in several studies to be very detrimental at relatively low temperatures (<0.3T_M, where T_M is the melting temperature) even for low-dose (0.1–5 dpa) irradiation exposures [4, 6, 8, 12, 15, 19, 75–83]. In terms of tensile properties, the tensile curves shown in Fig. 13.25 following neutron irradiation in the High Flux Beam Reactor (HFBR) to 0.5 dpa at 110–420°C and to 0.1 dpa at 505°C reveal pronounced radiation hardening with concurrent severe reduction in strain-hardening capacity for irradiations at temperatures <400°C [78]. In this regime, significant increases in yield stress occur with yield peaks at uniform elongations <0.2%, followed by severe plastic instability (monotonically decreasing stress with increasing strain) for higher tensile strain levels. The slope of the tensile

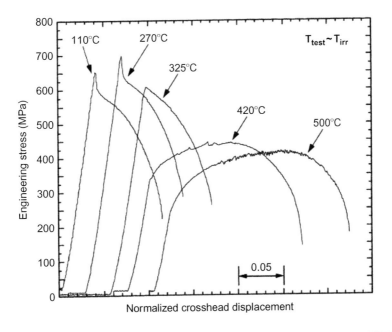

FIG. 13.25

Tensile stress vs strain curves for V-4Cr-4Ti following neutron irradiation in HFIR to 0.5 dpa at 110–420°C and to 0.1 dpa at 505°C [78]. The tensile curves are offset on the strain axis for clarity.

From Snead LL, Zinkle SJ, Alexander DJ, Rowcliffe AF, Robertson JP, Eatherly WP. Summary of the investigation of low temperature, low dose radiation effects on the V 4Cr 4Ti alloy. Fusion Materials Semiannual Progress Report for Period Ending Dec 31, 1997. DOE/ER-0313/231997. p. 81–98. Original from L.L. Snead et al., Presented at 8th Int. Conf. on Fusion Reactor Materials, Sendai, (1997) to be publ. in Fusion Materials semiann. Prog Rep. for period ending Dec. 31 1997.

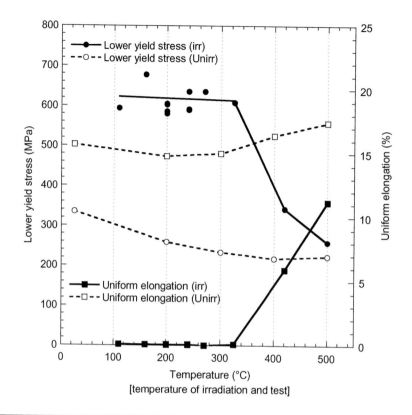

FIG. 13.26

Temperature dependence of yield strength and uniform elongation for unirradiated V-4Cr-4Ti and after neutron irradiation in HFBR to 0.5 dpa at 110–420°C and to 0.1 dpa at 505°C [59, 78]. Tensile tests were performed at the irradiation temperatures.

stress-strain curves changes abruptly at elongations of ~0.2%–0.4% for irradiations at temperatures <325°C. Conversely, for irradiations at temperatures >400°C, yielding occurs with no yield drop and the ensuing plastic deformation occurs with extensive strain hardening to the point of plastic instability at the ultimate tensile stress. Fig. 13.26 shows the temperature dependence of yield stress and uniform elongation for unirradiated V-4Cr-4Ti before and after irradiation in HFBR to 0.5 dpa [59, 78]. The data indicate pronounced radiation hardening and loss of uniform elongation for temperatures up to 325°C, whereas progressive recovery of the irradiation-hardening effects in V-4Cr-4Ti occurs for irradiations at temperatures >400°C. Above this irradiation temperature, the yield stress decreases and the uniform elongation increases with increasing irradiation temperatures until similar values occur for unirradiated and irradiated V-4Cr-4Ti at ~500°C.

Charpy impact test results of neutron-irradiated V-4Cr-4Ti also showed enhanced embrittlement behavior at relatively low temperatures and low-dose irradiation exposures. This is illustrated in Fig. 13.27 with absorbed energy data obtained from impact testing of machine-notched Charpy (MCVN) and pre-cracked Charpy (PCVN) specimens of V-4Cr-4Ti before and after neutron irradiation in HFBR to 0.5 dpa at 110–420°C [78]. In these figures, the tested specimens were miniaturized 1/3 CVN specimens with dimensions of ~3.3 × 3.3 × 25 mm. In the unirradiated condition, the DBTT was near −200°C. However, neutron-irradiation exposures at ≤420°C resulted in severe embrittlement of V-4Cr-4Ti by shifting the DBTT to higher temperatures and causing brittle cleavage and intergranular fracture of MCVN specimens. The magnitude of the DBTT shift was 200–350°C (Fig. 13.27A). The

13.4 AS-IRRADIATED MECHANICAL PROPERTIES

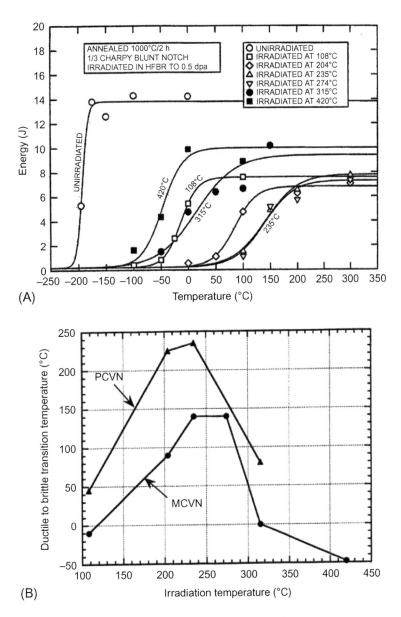

FIG. 13.27

Charpy impact absorbed energy curves of V-4Cr-4Ti before and after low-dose (0.5 dpa) neutron irradiation at 110–420°C [78].

notch acuity effect observed for unirradiated and irradiated V-4Cr-4Ti led to even higher DBTT shifts of ~50–120°C for PCVN specimens compared to MCVN specimens (Fig. 13.27B). One important issue that was resolved in the irradiation-embrittlement studies of V-4Cr-4Ti was that post-irradiation annealing at temperatures ≥650°C in vacuum restored the ductile failure behavior in specimens that had been embrittled by low-temperature neutron irradiation. Fig. 13.28 shows results obtained from post-irradiation annealing of V-4Cr-4Ti following neutron

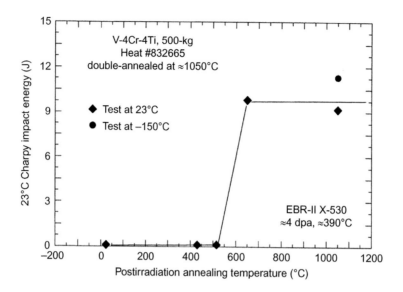

FIG. 13.28

Impact energy measured on specimens of 500 kg heat of V-4Cr-4Ti after irradiation at ~390°C to ~ 4 dpa and postirradiation annealing at 23–1000°C [82].

irradiation at ~390°C to 4 dpa in the Experimental Breeder Reactor-II (EBR-II) [63]. The neutron-irradiation exposure caused embrittlement of V-4Cr-4Ti by reducing the impact energy to <0.3 J in tests conducted at −150°C to +300°C (maximum tested temperature) in these miniaturized 1/3 CVN specimens. Following post-irradiation annealing at ≥650°C, the Charpy impact energy increased from <0.3 J to levels of ~9–11 J for tests at 23°C, which was near the impact energy of ~14 J for unirradiated V-4Cr-4Ti at 23°C (Fig. 13.27A).

In microstructural investigations of irradiated V-(4-5)Cr-(4-5)Ti alloys, significant changes were observed at doses as low as 0.1 dpa over temperatures ranging from 110°C to 500°C [84–86]. Fig. 13.29 shows two examples of the radiation-induced microstructural changes that occurred in HFBR neutron-irradiated V-4Cr-4Ti specimens at 268°C and 390°C. Transmission electron microscopy (TEM) showed the formation of a high number density of ~3-nm diameter faulted dislocation loops ($b = a/2 \langle 110 \rangle$ Burgers vectors) in specimens irradiated at <275°C (Fig. 13.29A). The faulted loops were found to have moderate dispersed barrier strengths ($\alpha < 0.5$). TEM analysis of tensile tested specimens following irradiation at <275°C showed cleared dislocation channels ~50 nm wide on [87] planes (Fig. 13.29A). Collectively, these microstructural results account for the pronounced radiation hardening and decrease in strain-hardening capacity to uniform elongations of <0.2% in tensile specimens at low irradiation temperatures (Fig. 13.25). For irradiation temperatures ranging from 275°C to 415°C, the dislocation loops changed from faulted loops to small perfect $a/2 \langle 111 \rangle$ loops with concomitant formation of large defect clusters lying on [001] habit planes at >315°C (Fig. 13.29B). TEM analysis of V-(4-5)Cr-(4-5)Ti alloys irradiated in EBR-II at 400°C also revealed the presence of a high number density of particles <6 nm diameter shown in Fig. 13.30 [88]. Diffraction from the particles showed radial streaks in the $\langle 200 \rangle$ direction near 3/4 $\langle 200 \rangle$ spots and tangential streaks near 2/3 $\langle 222 \rangle$ of the bcc matrix. Composition line profiles with electron energy loss spectroscopy showed Ti-enrichments of the particles. The diffraction characteristics of the particles were similar to those attributed to plate-shaped Ti-OCN ($Fm\bar{3}m$ space group) precipitates showing the Baker-Nutting Orientation Relationship, $\langle 001 \rangle_m // \langle 011 \rangle_p$ and $[200]_m // [200]_p$, between the matrix and precipitate [89]. For irradiation temperatures >415°C, the microstructural features were predominantly large [001] defect clusters that were enriched with Ti solute.

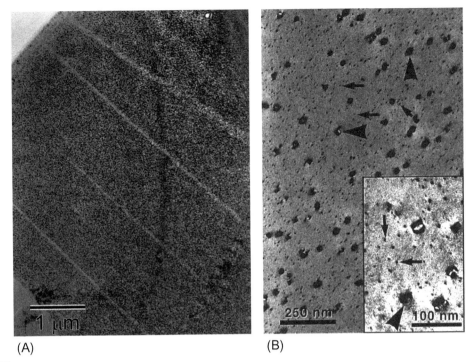

FIG. 13.29

TEM micrographs showing the microstructure of V-4Cr-4Ti irradiated to (A) 0.5 dpa at 268°C and tensile tested at 20°C and (B) 0.1 dpa at 390°C [85] In (B), the small arrows point to a/2<111> dislocation loops and the large arrows point to the [001] precipitates. The inset figure in (B) shows a magnified view of the 390°C microstructure.

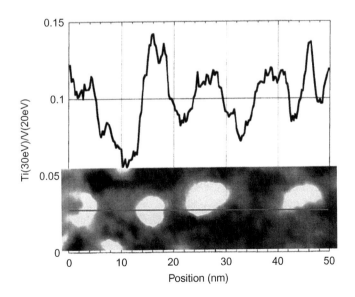

FIG. 13.30

Parallel electron energy loss spectroscopy line profile showing the correlation of Ti concentrations with the small particles in V-4Cr-4Ti irradiated at 400°C to 4 dpa [86].

From D.S. Gelles, P.M. Rice, S.J. Zinkle and H.M. Chung, "Microstructural Examination of Irradiated V (4 6%)Cr-(4–5%)Ti", Journal of Nuclear Materials, Vol. 258–263, (1998), p. 1380.

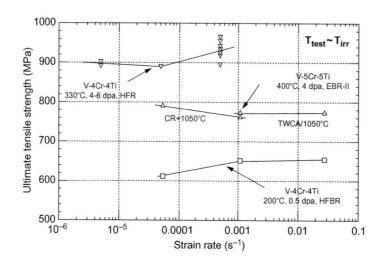

FIG. 13.31

Ultimate tensile strength as a function of strain rate for neutron irradiated V-4Cr-4Ti and V-5Cr-5Ti at 200–400°C to 0.5–6 dpa [79].

The strain rate dependence of the tensile properties of neutron-irradiated V-(4-5)Cr-(4-5)Ti alloys were investigated and correlated with the DSA behavior of unirradiated V-4Cr-4Ti [59, 77, 79]. Fig. 13.31 shows a compilation of UTS data vs strain rate for V-(4-5)Cr-(4-5)Ti alloys irradiated to 0.5–6 dpa at 200–400°C. For V-4Cr-4Ti specimens irradiated to 0.5 dpa at 200°C, the UTS increased slightly with increasing strain rate, consistent with a positive strain rate exponent that is commonly observed in irradiated materials. With irradiations to 4–6 dpa at 330°C, the UTS showed a weak, or slightly positive, strain rate exponent. Conversely, a slightly negative strain rate exponent for UTS was measured for V-5Cr-5Ti specimens irradiated to 4 dpa at the higher irradiation temperature of 400°C. The negative strain rate exponent (along with accompanying electrical resistivity measurements) suggests that a sizable concentration of interstitial O, C, and N solutes remained in the matrix of the irradiated V-Cr-Ti alloys and were not contained in Ti-OCN precipitates or defect clusters. The freely bound interstitial solutes were then able to interact with migrating dislocations during tensile testing causing the appearance of serrated flow in irradiated tensile curves, which is consistent with DSA [83]. Fig. 13.32 shows the SRS parameters obtained from tensile data for V-4Cr-4Ti before and after neutron irradiation in HFBR experiments to 0.1–4.0 dpa at 160°C to 500°C. The results showed that the irradiated strain rate exponent of the lower yield stress (LYS) was positive from 160°C to 500°C, but that the UTS changed from positive to negative values between ~300°C and 400°C. The magnitude of the negative strain rate exponent of UTS was reduced after irradiation at >300°C compared to that of unirradiated V-4Cr-4Ti. The decrease in magnitude of the strain rate exponent for UTS indicated that a greater concentration of interstitial O, C, and N solutes were bound to small defect clusters or Ti-OCN precipitates in specimens irradiated at temperatures ≥400°C compared to unirradiated V-4Cr-4Ti (although a significant concentration of O, C, and N solutes was dissolved in the matrix of the irradiated samples at >300°C, as evident from the negative SRS parameter).

A limited number of studies on the irradiation creep of V-4Cr-4Ti have been conducted and therefore understanding of effects of irradiation is still emerging [90–94]. Fig. 13.33 shows the creep strain rate (%/dpa units) plotted against applied stress for reactor irradiations performed in the ATR-A1 (to 5 dpa at 200°C and 300°C in lithium-filled capsules) [91–93], HFIR-12J (to 6 dpa at 500°C in He filled capsules) [92], BR-10 (to 3 dpa at 445°C) [90], and HFIR-17J experiments (to 3.7 dpa at 425°C and 600°C in lithium-filled capsules). Pressurized

13.4 AS-IRRADIATED MECHANICAL PROPERTIES

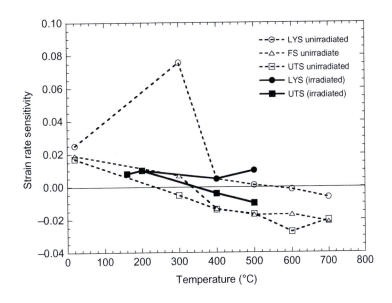

FIG. 13.32

Comparison of the strain rate sensitivity parameter as a function of temperature for unirradiated V-4Cr-4Ti before and after fission neutron irradiation in HFBR to 0.5 dpa at 160–500°C [16, 78, 79]. Values for LYS and UTS were similar at the 160°C and 200°C irradiation temperatures.

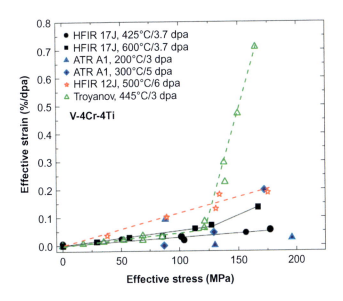

FIG. 13.33

Comparison of irradiation creep data for V-4Cr-4Ti [94].

creep tube specimens were used in the ATR-A1, HFIR 12J, and HFIR-17J experiments and thin-walled tube specimens loaded in torsion were used in the BR-10 experiment. The creep data from the ATR-A1 experiment showed no linear stress dependence of creep strain due to the large scatter in results, which may have been the result of nonuniform irradiation temperatures and neutron flux gradients experienced by the specimens. Linear stress dependence of creep strain was observed for HFIR-12J, HFIR-17J, and BR-10, although bilinear creep behavior was also observed in the BR-10 experiment and a slightly nonlinear relationship was observed in the HFIR-17J experiment at 600°C above ~110 MPa. The nonlinear relationship may suggest that thermal creep mechanisms begin to dominate at >110 MPa and 600°C. The discrepancies with data showing linear vs bilinear creep behavior could not be explained. The creep coefficients measured from the data varied from 1.4×10^{-6} to 11×10^{-6}/MPa dpa for irradiation doses of 3–6 dpa. However, the data showed no clear dependence on irradiation temperature or neutron dose for temperatures of ~400–600°C and doses of ~3–6 dpa. In general, the understanding of irradiation creep behavior of V-4Cr-4Ti remains unclear due to complexities in the production and fabrication of the alloys and neutron-irradiation experiments including the effects of dose, temperature, and environment on the specimens. Additional higher dose data over a wide range of temperatures are needed to improve understanding of irradiation creep in V alloys.

13.4.2 AS-IRRADIATED MECHANICAL PROPERTIES OF NIOBIUM

The studies of neutron-irradiation effects on the mechanical properties of Nb alloys have been primarily based on temperature and neutron fluence ranges projected for space reactor operating conditions, along with limited studies for sodium-cooled fast reactor applications. The space reactor operating goals were for 2–7 years at temperatures of 1100°C to 1400°C (1373–1673 K) to fast fluences of $0.4–4 \times 10^{22}$ n/cm² in liquid (Li, K) or gas (He) environments [95, 96]. The main concerns for Nb alloys consist of radiation hardening at all temperatures up to ~0.5 T_m, irradiation-assisted creep at ~0.2–0.5 T_m, void or cavity swelling at <0.5 T_m, and He embrittlement at >0.5 T_m, where the melting temperature $T_m = 2741$ K [95, 97, 98].

From the early development of Nb alloys in the mid-1950s to the present, there have been limited studies of irradiation effects and most have been on examinations of the microstructure and tensile properties of unalloyed Nb, Nb-1Zr, and Nb-1Zr-0.1C (PWC-11) [95]. Irradiation studies of Nb and Nb-1Zr showed rapid hardening occurs with complete loss of strain-hardening capacity due to rapid necking at temperatures <423 K [99–102]. Fig. 13.34 shows the reductions in uniform and total elongations that occurred for Nb-1Zr after irradiation to ~0.005–12.3 dpa at ~343 K [99]. Although the uniform elongation decreased to zero for irradiation doses >0.02 dpa, the total elongation remained at ~10% up to 12.3 dpa and the tensile fracture modes were ductile with high reduction in area. In the radiation-hardening regime, the localized plastic deformation after yielding occurs by dislocation channeling of accumulated radiation-induced defect clusters and dislocation loops, which is shown in Fig. 13.35 [32, 99]. Fig. 13.36 shows tensile data obtained from several studies of Nb-1Zr and Nb-1Zr-0.1C exposed to neutron irradiation to 0.1–5 dpa and tensile tested at the irradiation temperatures ranging from ~300 to 1573 K [32, 95, 103–105]. The irradiated UTS increases significantly below ~800 K and then decreases monotonically with increasing irradiation temperature to ~1000 K where it converges with the unirradiated UTS data. Since Nb-1Zr can become embrittled due to pickup of O, C, and N from the surrounding atmosphere at temperatures above ~800 K, special care must be taken in irradiation experiments to minimize such contamination effects. The irradiated uniform elongation is nearly zero up to ~723 K and begins to slowly increase at higher irradiation temperatures. The irradiated uniform elongation approaches acceptable "ductile" values of ~5% for irradiation and test temperatures of 1070 K or higher [105].

Irradiation temperatures of ~0.3–0.6 T_m can cause void swelling, He embrittlement, and irradiation creep in Nb alloys depending on the neutron fluences. For Nb and Nb-1Zr, swelling associated with void formation is <2% for fluences up to ~2–5 × 10^{26} n/m², E > 0.1 MeV [32, 99, 105]. Fig. 13.37 shows the temperature range for void

13.4 AS-IRRADIATED MECHANICAL PROPERTIES

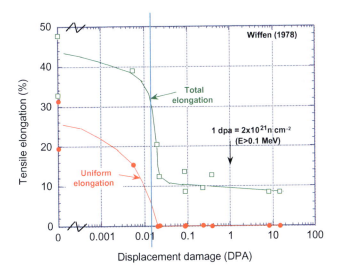

FIG. 13.34

Dose dependence of the total and uniform elongation of Nb-1Zr neutron irradiated at a temperature near 70°C and tested at room temperature [32, 98].

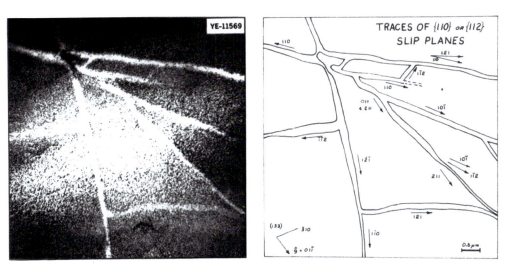

CHANNELING DEFORMATION IN Nb-1 Zr

Irradiated to 7.5×10^{20} n/cm^2 (0.1 MeV) at ~ 70°C   [0.36 dpa]

Tensile test at ~35°C, 0.1% uniform elongation, 12.7% total elongation

FIG. 13.35

Deformation microstructure of Nb-1Zr neutron irradiated at a temperature near 70°C and tensile tested at room temperature [32, 98].

624 **CHAPTER 13** REFRACTORY ALLOYS

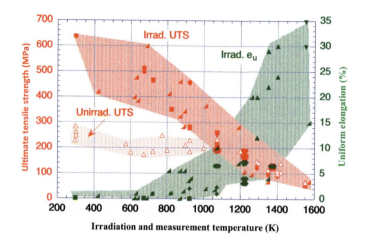

FIG. 13.36

Ultimate tensile strength (UTS) and uniform elongation (e_u) of Nb-1Zr and Nb-1Zr-0.1C neutron irradiated to 0.1–5 dpa and tensile tested at the irradiation temperature [98].

Adapted from Zinkle SJ, Wiffen FW. Radiation effects in refractory alloys. In: El-Genk MS, editor. Space Technology and Applications International Forum-STAIF 2004 (AIP Conf Proc vol 699). Melville, NY: American Institute of Physics; 2004. p. 733–40. Figure 2.

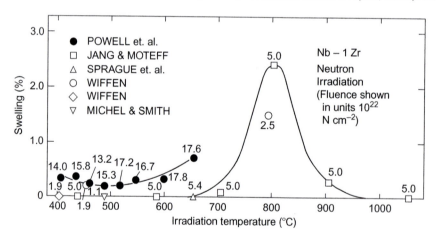

FIG. 13.37

Summary of void swelling behavior of neutron-irradiated Nb-1Zr [31, 99].

From Buckman RW, Jr. A review of tantalum and niobium alloy production. In: Cooper RH, Jr., Hoffman EE, editors. Proc Symp on Refractory Alloy Technology for Space Nuclear Power Applications, CONF-8308130: Oak Ridge National Lab; 1984. p. 86–97.101. Original from R. W. Powell, D. T. Peterson, M. K. Zimmerschiec, and J. F. Bates, Swelling of Several Commercial Alloys Following High Fluence Neutron Irradiation, J. NucL Mater., 103–104: 969–374 (1981).

swelling in Nb-1Zr is ~800–1200 K for doses up to ~35 dpa, with maximum swelling occurring near 1073 K [99]. Neutron irradiation [n, α] reactions can also cause embrittlement of Nb alloys by formation of He bubbles on grain boundaries. However, experimental studies have reported no severe embrittlement for He ions implanted to concentrations up to 100–500 appm, indicating Nb and Nb-1Zr have relatively good resistance to He embrittlement [87, 106, 107]. Experimental data on irradiation creep of Nb-1Zr are very limited, but the results suggest that irradiation creep rate would only produce minor deformations of <0.3% based on projected space reactor operating conditions [32, 104, 108].

The effect of irradiation on the DBTT and fracture properties of Nb-1Zr and PWC-11 is nonexistent. However, estimates for Nb-1Zr fracture behavior based on irradiated tensile strengths at temperatures below ~600–700 K suggests fracture toughness would be a concern for irradiation temperatures <800 K [32].

13.4.3 AS-IRRADIATED MECHANICAL PROPERTIES OF MOLYBDENUM

As with other BCC metals, molybdenum is subject to severe base metal embrittlement for typical temperature of interest in nuclear applications. Molybdenum, and for example tungsten, also share similar methods of fabrication and grain boundary susceptibility to neutron-irradiation embrittlement leading to material failure. Also, in materials such as some powder metallurgically derived and ODS molybdenum developed for radiation resistance, significant reduction during process rolling leads to anisotropic nonirradiated mechanical properties and response to irradiation. Unlike tungsten, in the nuclear application temperature range the defects responsible for embrittlement appreciably evolve through a homogeneous nucleation and growth process following the initial neutron-induced cascade [109] and therefore embrittlement mitigation strategies such as alloying or second phase additions can be considered. However, while molybdenum has historically been a popular conceptual material for nuclear energy applications leading to extensive R&D, limited success toward development of ductile nuclear-grade structural molybdenum has been achieved. It is noted that improved irradiation performance has been achieved by developing higher purity materials (particularly minimizing O, C, and N impurities) and materials with cleaner and therefore more irradiation-resistant grain boundaries. It is noted that this is in contrast to the common strategy of adding grain boundary precipitates to enhance thermal creep resistance and irradiation stability for nuclear alloys. However, this underscores the particular susceptibility of refractory materials to grain boundary impurities. For much the same reason, the most commonly produced materials, specifically hot-pressed and sintered powders, are not discussed here because of the inherent impurity content and as-irradiated embrittlement of their grain boundaries.

In fast and mixed spectrum reactors, 98-Mo (isotopically the most abundant) transmutes through a (n,γ) reaction first to 99-Mo and then to 99-Tc, a relatively long lived isotope that eventually produces 101-Ru. The majority of Mo transmutations simply create other Mo isotopes. Chain reactions with other isotopes produce similar, minor amounts of Zr and Nb, as discussed [110]. For doses <10 dpa, the transmutant levels are on the order of <1000 ppm and have negligible effect on hardening, although precipitates can be observed at higher irradiation temperatures [48].

Given the high melting temperature of molybdenum (2623°C), the homologous temperature range for its practical nuclear structural application temperatures (300–1000°C) is relatively low: 0.20–0.44T_m. For temperatures as low as 80°C, dislocation loops, rafts, and cavities (voids) are the major irradiation-induced defects observed in LCAC molybdenum microstructure [111]. At that temperature, loops were discernable by TEM at damage levels as low as 0.1 dpa with cavities detected by positron annihilation at levels <0.01 dpa, though higher temperature irradiation is required for cavities to become visible through TEM. Rafts of aligned dislocation loops emerge at 80°C irradiation at doses >0.01 dpa [111]. With increasing temperature (300°C), voids on the order of 1 nm become visible in TEM and are of similar size for LCAC and ODS molybdenum. Fig. 13.38 provides an overview schematic representation of the microstructural evolution in Mo as a function of dose and temperature as taken from the literature [48]. The void microstructure increases in number density with increasing dose and at first increases in size, then decreases within the limits of the current published work. At sufficiently high dose, the curious phenomenon of the ordered void lattice, a phenomenon first observed in molybdenum, can occur [112–115]. Increasing irradiation temperature results in larger voids of lower number density with a weak dependence on material microstructure: finer particle size materials such as ODS molybdenum result in somewhat finer scale void formation [48, 115]. Very high dose irradiation at high temperature leads to the formation of a low concentration of fine transmutation-induced precipitates [48].

As summarized in Fig. 13.38, the dominance of voids in the irradiated microstructure of molybdenum at intermediate to high temperatures raises the issue of irradiation-induced swelling. Void swelling under V+ ion irradiation has been studied to a damage level of 50 dpa (~900–1510°C), suggesting the potential for significant

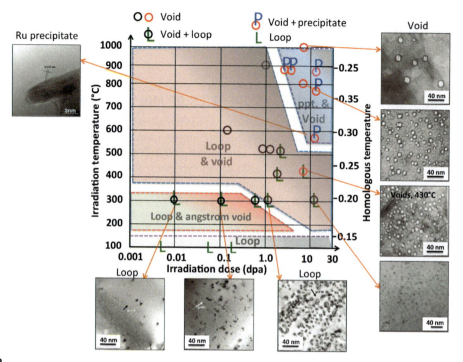

FIG. 13.38

Microstructural evolution map for irradiated molybdenum.

swelling [116]. Near the lower temperatures studied, a saturation level for void swelling (uncommon in most nuclear materials, and perhaps related to the propensity for void lattice formation in Mo alloys) approached 4%, though the saturation void swelling was closer to 10% at the upper end of the irradiation temperatures studied [116]. A compilation of the neutron-irradiation data has been assembled by Leonard [117] indicating <4% swelling for Mo irradiated to 8×10^{26} n/m^2 (E>0.1 MeV, ~35 dpa) in the 400–900°C range [113]. Therefore, void swelling does not appear to produce unacceptably large-dimensional changes in neutron-irradiated Mo, at least up to moderately high doses (≤100 dpa) that have been studied to date. The highest dose data, such as that replotted in Fig. 13.39, suggests that a peak in swelling occurs in a temperature range from 550°C to 650°C, perhaps with a slight shift depending on the material type.

For the lower end of the nuclear application temperature range, the irradiation-induced dislocation loops and tiny cavities result in significant matrix hardening. It is noted that at lower irradiation temperature and dose, and specifically as demonstrated in the systematic 80°C neutron-irradiation studies of Li [111], very low dose yield softening (doses <0.005 dpa) precedes the yield-hardening regime for Mo. However, for neutron irradiation in the range of 300–600°C on Mo materials selected and developed for irradiation resistance, substantial increased hardness occurs with increasing dose. Fig. 13.40 compares the change in room-temperature hardness for LCAC Mo, ODS Mo, and TZM irradiated in a mixed spectrum reactor at ~300°C, 600°C, and 900°C [36]. Mo materials irradiated at nominally 300°C and 600°C are seen to rapidly harden with a very weak temperature dependence and exhibit an approach to saturation at a hardness value approximately twice that of the initial value. For higher temperature irradiation (900°C in Fig. 13.40), fewer loops and small cavities are produced and the void, which is less effective hardening agent, becomes the dominant microstructural feature. Moreover, irradiation hardening at temperatures approaching 0.4 of the homologous temperature competes with anneal-softening of cold work,

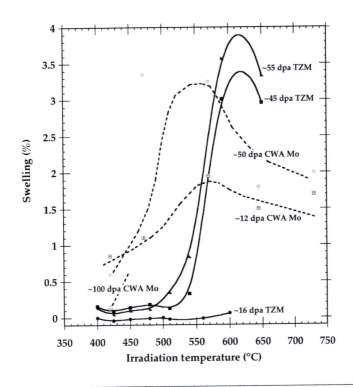

FIG. 13.39

Swelling of cold-worked and aged (CWA) and TZM molybdenum neutron irradiated to high dose as a function of temperature [116].

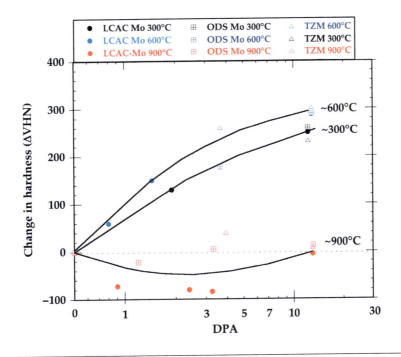

FIG. 13.40

Relative change in hardness of stress-relieved neutron-irradiated molybdenum and molybdenum alloys [36].

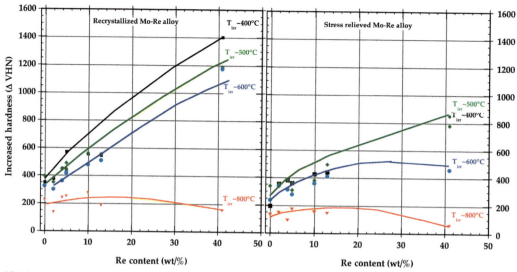

FIG. 13.41

Relative change in hardness as a function of Re content for recrystallized and stress relieved Mo. Rhenium transmutation produced through pre-irradiation alloying (FFTF-MOTA irradiation) [118].

yielding a much-reduced relative effect of irradiation. At irradiation temperatures in the 900–1100°C range, recrystallization of stress-relieved materials would be expected. A similar trend can be seen for Mo-Re alloys irradiated in the 400–600°C range (significant hardening) and at 800°C (anneal softening), as shown in Fig. 13.41 [118–120]. In this figure, rather than having a ductilizing effect such as has been observed in unirradiated Mo alloys, the presence of Re tends to increase the irradiation hardening of the matrix in part through the promotion of the embrittling σ and χ phases. While such second phases are not in equilibrium for the Mo-Re alloy Re contents of less than ~61%-eutectic temperature for the σ phase of ~1130°C, and ~77%-for the χ phase, under nonequilibrium irradiation conditions the undersized Re solute precipitates into the *Re*-rich phases. This phenomenon, known as Radiation Induced Segregation, is briefly summarized in Chapter 5 in this book.

As expected, the significant matrix hardening results in increased tensile strength and reduced ductility of Mo. Many approaches have been explored to mitigate this issue including alloying, ultra-pure alloys, and incorporation of nano-oxides, with moderate success. However, irradiation embrittlement remains a central issue to molybdenum in the practical temperature ranges for nuclear application. In contrast with the limited data on the effect of irradiation on the deformation and fracture of tungsten and tungsten alloys, a fair body of literature exists for modern molybdenum alloys including both fundamental irradiation effects studies and more practical engineering data on alloys developed for radiation damage resistance. Fig. 13.42 provides a simple example of the progressive hardening and loss of tensile ductility for relatively low-temperature irradiated Mo (nominally 80°C) over a range of very modest irradiation doses up to 0.3 dpa. As seen, this low-carbon arc-cast molybdenum, which is currently a leading irradiation-tolerant form of molybdenum, possesses over 50% total elongation prior to irradiation. The data clearly show progressive hardening (increased yield/ultimate stress) and rapid loss of ductility with near-complete loss of strain tolerance by 0.072 dpa (2×10^{24} n/m^2, $E > 0.1$ MeV.) As would be expected given the evolution of microstructure presented in Fig. 13.38 and the previous discussion on hardening, as the irradiation temperature is increased from ~80°C molybdenum embrittlement becomes less severe. However, for LCAC molybdenum and high-purity ODS molybdenum the transition temperature from ductile to brittle behavior for un-notched (smooth) tensile specimens shifts from well below room temperature in the nonirradiated condition to hundreds of degrees Celsius in the as-irradiated condition [121]. This shift in temperature is somewhat mitigated as the

13.4 AS-IRRADIATED MECHANICAL PROPERTIES

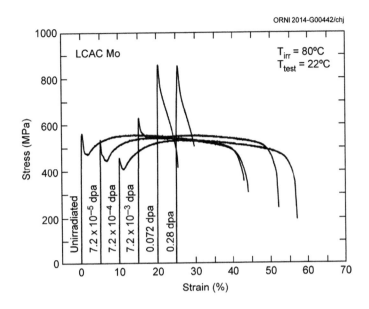

FIG. 13.42

The effect of mixed-spectrum reactor irradiation at 80°C on the tensile behavior of low-carbon arc-cast molybdenum [111]. Mixed spectrum High Flux Isotope Reactor neutron irradiation.

From Li M, Byun TS, Snead LL, Zinkle SJ. Defect Cluster Formation and Radiation Hardening in Molybdenum Neutron-Irradiated at 80°C. Journal of Nuclear Materials. 2007;367–370:817–22.

irradiation temperature is increased, though tensile embrittlement remains an issue for Mo for irradiation temperatures up to several 100°C.

Fig 13.43 provides data comprising a broad transition point from brittle (lower left) to ductile (upper right) behavior for smooth tensile specimens as a function of irradiation temperature for a range of nominally radiation-resistant forms of molybdenum: TZM, LCAC Mo, and ODS (La oxide) Mo. As seen from the figure, the nonirradiated ductile-brittle transition for nonirradiated specimens is approximately −100°C, increasing to a broad band between nominally 400–800°C in the as-irradiated condition. For practical applications, it is important to note that this data describes tensile ductility, which should not be confused with the material's ability to resist crack propagation, for example, fracture toughness. It has been demonstrated by Cockeram [122, 123], as typical of the fracture of most metallic systems, that tensile ductility of irradiated molybdenum is likely an optimistic measure of material performance (e.g., the brittle transition would be lower for a fracture toughness specimen as compared to a tensile specimen). An irradiation of the same materials depicted in Fig. 13.43 has been carried out on disc compact tension specimens and the fracture toughness was subsequently measured. In the nonirradiated condition, the fracture toughness of TZM, LCAC, and ODS forms of molybdenum in various orientations and recrystallized and strain-relieved conditions dipped below $30\,\text{MPa-m}^{1/2}$ fracture toughness in the temperature range between ∼−200°C and 0°C. For materials neutron irradiated at 386–574°C, the point associated with very low fracture toughness ($<30\,\text{MPa-m}^{1/2}$) was in the range of 300–600°C [122].

13.4.4 AS-IRRADIATED MECHANICAL PROPERTIES OF TUNGSTEN

As discussed in Section 13.2.4, the tungsten fabrication route has profound influence on the resulting ductility and mechanical performance of the finished product in the absence of irradiation. Similarly, the response of tungsten or tungsten alloys to neutron irradiation is also strongly influenced by the processing route for fabrication.

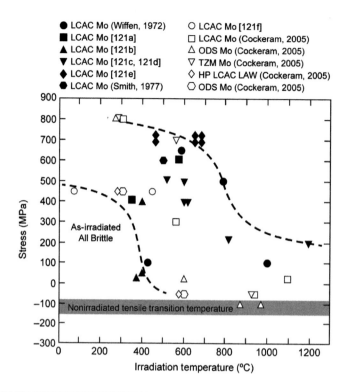

FIG. 13.43

The effect of irradiation on the brittle transition temperature for smooth tensile specimens of Mo and TZM [70]. Mixed spectrum High Flux Isotope Reactor neutron irradiation in the 0.23–44 × 10^{25} n/m^2 (E > 0.1 MeV) [121a–121f, 142].

Experimental studies reinforced with molecular dynamics simulations indicate that in metals with FCC crystal structure such as Cu, Ni, and Au, the formation of immobile "sessile" defect clusters primary occurs directly within the neutron-induced displacement cascade. This is in contrast with the behavior of BCC iron, for which defect clusters primarily form through a nucleation and growth process following the initial cascade event. In the latter case, such as in many steels, it is possible to tailor the alloy behavior under irradiation through mechanisms such as cold work, varying the type and level of interstitial solutes, and/or inclusion of secondary phases. Interestingly, as the atomic number of BCC metals increases, the relative impact of sessile defect formation and therefore the irradiation embrittlement increases. Molybdenum, which has mass between iron and tungsten, contains a mixture of sessile in-cascade and evolving defect clusters [69] while in-cascade sessile defect clusters are assumed to dominate the more massive tungsten. Hardening centers such as dislocation loops have been observed to form directly within tungsten displacement cascades [124].

As previously mentioned, rhenium has a ductilizing effect on unirradiated molybdenum, effectively reducing the BDT temperature to below room temperature. The current understanding of this effect is that, at low concentrations, rhenium softens the material through reduction of the Peierls stress while at higher concentrations it beneficially scavenges oxygen that would otherwise migrate to and embrittle grain boundaries. Upon neutron irradiation, and especially in mixed and thermal spectrum fission reactors, copious amounts of rhenium are built into tungsten through transmutation. This can be seen through inspection of Fig. 13.44, giving the atomic percent of transmuted rhenium as a function of dpa for a mixed spectrum fission reactor (in this case HFIR, a water-cooled thermal reactor used for materials irradiation studies), a fast reactor (a previous fast spectrum materials irradiation reactor), and a D-T fusion first wall spectrum (STARFIRE study). For the dose range of the plot, which can be

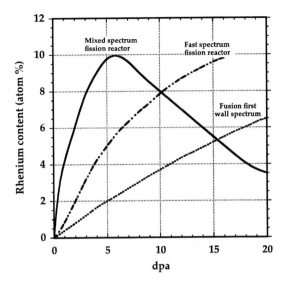

FIG. 13.44

Neutron-induced transmutation concentration of rhenium from tungsten in various neutron spectra.

Based on L.R. Greenwood, F.A. Garner, Transmutation of Mo, Re, W, Hf and V in various irradiation test facilities and STARFIRE, J. Nucl. Mater. 212–215 (1994) 635–639.

roughly interpreted as the yearly equivalent in a fission materials test reactor or fusion power system, several to ten atom percent of rhenium is produced and subsequently transmuted to osmium, which also accumulates within the tungsten. The effect of rhenium on irradiated microstructure has been historically studied [125, 126] and continues to be evaluated [127–129]. While the overall mechanism is somewhat in dispute, it is clear that the matrix embrittlement due to fast neutrons far outweighs any ductilizing effect of rhenium, especially at low rhenium levels (<1%) [125]. It is noted that Hasegawa [130] has reported that percent levels of rhenium in as-irradiated alloy do marginally reduce the Vickers hardening (ΔH) for low damage levels (<0.4 dpa). In a separate work, for higher levels of rhenium, the formation of the embrittling χ phase has been observed in W-26 Re irradiated in the temperature range of 373–800°C at doses of 2–9.5 dpa [131]. In any event, the use of rhenium as a route toward enhancing the ductility of as-irradiated tungsten seems ineffective. However, alloying or build-in of rhenium through transmutation does affect other thermophysical properties and the microstructural evolution. Rhenium is understood to suppress the growth of voids and therefore the amount of swelling [126]. In one of the few systematic studies on swelling, Matolich [132] compared the swelling in pure sintered tungsten to W-25Re irradiated in a fast spectrum (to 5.5×10^{25} n/m^2 ($E > 0.1$ MeV)) in the temperature range of 700–1200 K. Over this irradiation temperature range, pure tungsten yields a bell-curve behavior in swelling with a maximum swelling of about 1.6% at ~1100°C while the W-25% Re sample exhibits negligible swelling (~0.1%) at all temperatures. Finally, the presence of metallic rhenium will have a large negative impact on thermal conductivity, outweighing any impact of fast neutron damage on conductivity degradation. For the maximum concentrations schematically shown in Fig. 13.43 (~10%), tungsten would be expected degrade in thermal conductivity to approximately one-third of its original value [133].

Given the extremely high melting temperature of tungsten (3423°C), the homologous temperature range for its practical nuclear structural application temperatures (300–1000°C) is relatively low: 0.15–0.35 T_m. Given this, at the lower end of this range one would expect the irradiated tungsten microstructure to consist of finely dispersed defects of ill-defined character, which would coarsen as irradiation temperature and dose increased. Fig. 13.45 provides a map for the evolution in microstructure vs dose for pure tungsten spanning this temperature range, though at a relatively modest dose level (<2 dpa) compared to component lifetimes that would be expected in a nuclear power application. At the lowest temperature, the dense population of tiny defects is difficult to resolve,

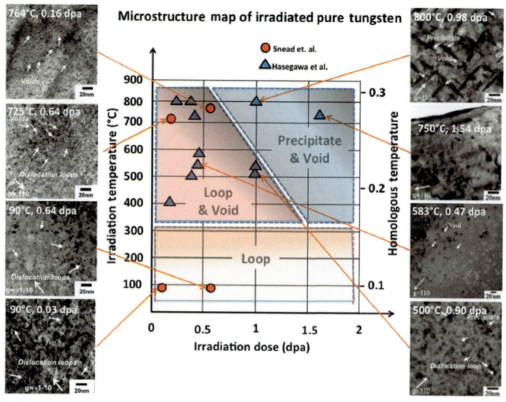

FIG. 13.45

Microstructural evolution of pure tungsten as a function of mixed-spectrum neutron irradiation.

Based on A. Hasegawa, T. Tanno, S. Nogami, T. Tanaka, Property change mechanism in tungsten under neutron irradiation in various reactors, J. Nucl. Mater. 417 (2011) 491–494.

though with increasing temperature and dose both loops and small voids become evident. As dose and temperature are further increased, voids coarsen and void-lattice can occur. Tanno [134] has observed the formation of a void lattice by 1.54 dpa at 750°C, significantly lower than previously reported [135]. For very high dose, consistent with high build-in of rhenium, or for alloys initially containing rhenium, needle-like precipitates of χ-phase rhenium become evident [126, 131, 134, 136].

The impact of irradiation in this relatively low homologous-temperature regime is to produce significant hardening in the tungsten crystal as can be seen through inspection of Fig. 13.46 [137]. In this data set, the Vicker's hardness of single-crystal W is provided for a mixed spectrum neutron irradiation spanning the temperature range of 90–750°C over a few orders of magnitude fluence. Within the significant statistical variation, there does not appear to be a strong function of irradiation temperature in this temperature/dose range. Moreover, the irradiation hardening of sintered, polycrystalline tungsten appears to be similar to that of the single-crystal W.

As mentioned in the introduction section, the properties of tungsten are a strong function of both the grain orientation and grain boundaries, with the mechanical properties of grain boundaries profoundly influenced by the fabrication route and post-fabrication processing such as deformation and annealing to produce strain relief or recrystallization. Moreover, it is assumed (given the dearth of data) that sintered forms of tungsten with boundaries weakened by the presence of impurities such as oxygen will undergo exaggerated irradiation-induced degradation. For the single-crystal materials shown in Fig. 13.46, a significant increase in yield strength occurs accompanied by loss of ductility for both [100, 110] orientations. This impact on ductility occurs by $\sim 1 \times 10^{24}$ n/m^2 (E > 0.1 MeV.) Fig. 13.47 provides a snapshot of the limited available tensile data on sintered pure polycrystalline tungsten

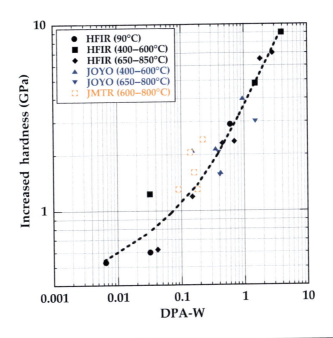

FIG. 13.46

Effect of mixed spectrum neutron irradiation on the hardness of single-crystal and sintered tungsten [137].

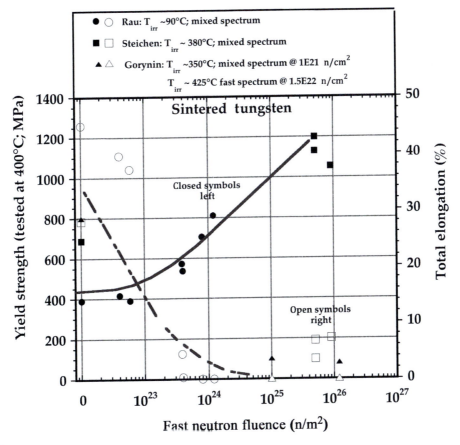

FIG. 13.47

Tensile behavior of polycrystalline tungsten under neutron irradiation [138–140].

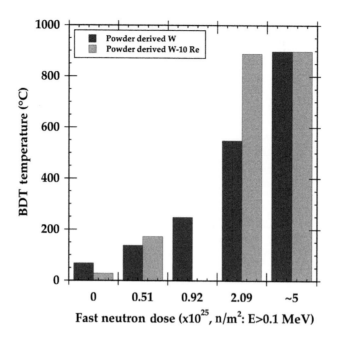

FIG. 13.48

Brittle-to-ductile transition temperature for relatively low-temperature mixed spectrum neutron-irradiated sintered tungsten and tungsten-10% rhenium alloys in the range of 250–300°C [141].

irradiated in mixed and fast spectrum reactors at 90–425°C. As expected given the typical hardening of Fig. 13.46, yield strength also exhibits a rapid increase with dose. This hardening, and likely a weakening of the grain boundaries, also results in a rapid loss of ductility with a complete loss of uniform and total elongation by $\sim 1 \times 10^{24}$ n/m² ($E > 0.1$ MeV), similar to that noted in single-crystal material. The work of Krautwasser [141] summarized in Fig. 13.48 for non-precracked bars provides the irradiation-induced shift in DBT temperature of both pure sintered tungsten and W-10Re indicating the DBT shifts to well above room temperature even for the lowest dose studied, 0.52×10^{25} n/m² ($E > 0.1$ MeV), and approaches 900°C for a dose an order of magnitude higher, with no apparent effect of rhenium.

13.4.5 SUMMARY AND CONCLUSIONS

Refractory metals have historically been, and continue to be, perennial candidates for high-temperature nuclear systems. However, their application has been predominantly notional or limited to nonstructural applications due largely to the issues highlighted in this chapter: challenging manufacturing issues associated with sensitivity to O, C, N pickup, a structure with inherently low as-fabricated ductility in most cases (particularly for Mo and W alloys), and pronounced embrittlement due to neutron irradiation. While the irradiation-induced embrittlement at the crystal level is extremely challenging to mitigate for this class of materials, strides are being made in our ability to both fabricate complicated (e.g., graded composition layers) refractory structures and in our understanding of how to design structures with low-ductility materials. Given this coupled processing and design understanding, it is becoming more likely that refractory materials will find more widespread and important application within nuclear systems in the near future.

REFERENCES

[1] Fang ZZ. (Editor in Chief) International Journal of Refractory Metals and Hard Materials.
[2] O.N. Carlson, F.A. Schmidt, E.N. Dalder, T. Grobstein, C.S. Olsen (Eds.), Vanadium Alloy Development in Evolution of Refractory Metals and Alloys, The Minerals, Metals, and Materials Society, 1994.
[3] Ross RB. Metallic Materials Specification Handbook (Springer US). 1980.
[4] H.M. Chung, B.A. Loomis, D.L. Smith, Development and testing of vanadium alloys for fusion applications, J. Nucl. Mater. 239 (1996) 139–156.
[5] D.L. Smith, M.C. Billone, K. Natesan, Vanadium-base alloys for fusion first-wall application, Int. J. Refract. Met. Hard Mater. 18 (2000) 231.
[6] H.R. Thresh, A.G. Hins, D.L. Smith, The fabrication of vanadium base alloys as canditate fusion reactor materials, J. Nucl. Mater. 155–157 (1988) 608.
[7] H. Böhm, W. Dienst, H. Hauck, H.J. Laue, Irradiation effects on the mechanical properties of vanadium-base alloys, in: W.C. Leslie (Ed.), Effects of Radiation on Structural Materials, American Society for Testing and Materials, Philadelphia, 1967, pp. 95–106. ASTM STP 426.
[8] D.L. Smith, H.M. Chung, B.A. Loomis, H.-C. Tsai, Reference vanadium alloy V-4Cr-4Ti for fusion application, J. Nucl. Mater. 233–237 (1996) 356–363.
[9] H.M. Chung, H.-C. Tsai, D.L. Smith, R. Peterson, C. Curtis, C. Wojcik, et al., Fabrication of 500 kg heat of V-4Cr-4Ti. Fusion Materials Semiannual Progress Report for Period ending Sept 30, 1994, DOE/ER-0313/17, Oak Ridge National Lab, 1994, pp. 178–182.
[10] B.A. Loomis, O.N. Carlson, Investigation of the Brittle-Ductile transition in vanadium, in: W. Clough (Ed.), Reactive Metals, Interscience, London, 1959, pp. 227–243.
[11] R.W. Buckman, Consolidation and fabrication of vanadium and vanadium-base alloys, Int. Metals Rev. 25 (1980) 158.
[12] R.J. Kurtz, K. Abe, V.M. Chernov, V.A. Kazakov, G.E. Lucas, H. Matsui, et al., Critical issues and current status of vanadium alloys for fusion energy applications, J. Nucl. Mater. 283–287 (2000) 70–78.
[13] A.F. Rowcliffe, D.T. Hoelzer, Microstructural inhomogeneities in V-4Cr-4Ti mechanical property specimens prepared for irradiation experiments in HFBR, BOR-60, and HFIR. Fusion Materials Semiannual Progress Report for Period ending December 31, 1998, DOE/ER-0313/25, Oak Ridge National Lab, 1998, pp. 42–58.
[14] R.J. Kurtz, M.L. Hamilton, H. Li, Grain boundary chemistry and heat treatment effects on the ductile-to-brittle transition behavior of vanadium alloys, J. Nucl. Mater. 258–263 (1998) 1375–1379.
[15] S.J. Zinkle, H. Matsui, D.L. Smith, A.F. Rowcliffe, E. van Osch, K. Abe, et al., Research and development on vanadium alloys for fusion applications, J. Nucl. Mater. 258–263 (1998) 205–214.
[16] D.T. Hoelzer, Structural Analysis of Ti-Oxycarbonitrides in V-Cr-Ti Based Alloys. In Fusion Materials Semiannual Progress Report for Period ending December 31, 1998, DOE/ER-0313/25, p. 59.
[17] T. Muroga, T. Nagasaka, A. Iiyoshi, A. Kawabata, S. Sakurai, NIFS program for large ingot production of a V-Cr-Ti alloy, J. Nucl. Mater. 283–287 (2000) 711–715.
[18] T. Muroga, T. Nagasaka, in: Development of manufacturing technology for high purity low activation vanadium alloys, 18th IAEA Fusion Energy Conference, Sorento, Italy, 2000.
[19] T. Muroga, T. Nagasaka, K. Abe, V.M. Chernov, H. Matsui, D.L. Smith, Vanadium alloys-overview and recent results, J. Nucl. Mater. 307–311 (2002) 547.
[20] A.K. Ikov, V.M. Chernov, M.M. Potapenko, I.N. Gubkin, C.A. Drobyshev, V.S. Zurabov, Development of production process and study of low-activity V-(4-5)%Ti-(4-5)%Cr structural alloys for thermonuclear reactors, Metal Sci. Heat Treat. 46 (2004) 497.
[21] A.N. Tyumentsev, I.A. Ditenberg, K.V. Grinyaev, V.M. Chernov, M.M. Potapenko, Multi-directional forge molding as a promising method of enhancement of mechanical properties of V-4Cr-4Ti alloys, J. Nucl. Mater. 329–333 (2011) 429.
[22] V.M. Chernov, M.M. Potapenko, V.A. Drobyshev, D. Blokhin, N. Budylkin, E. Mironova, et al., in: Low activation alloy V-4Cr-4Ti for fusion and fission power reactors-the RF results, Presented O2B2:11th ISFNT, Barcelona, Spain, 2013.
[23] H.Y. Fu, J.M. Chen, P.F. Zheng, T. Nagasaka, T. Muroga, Z. Li, et al., Fabrication using electron beam melting of a V-4Cr-4Ti alloy and its thermo-mechanical strengthening study, J. Nucl. Mater. 442 (Suppl. 1) (2013) S336.
[24] J. Chen, P. Wang, H. Fu, Z. Xu, in: Research of low activation structural materials for fusion power reactor in SWIP, 24th IAEA Fusion Energy Conference, 2012.
[25] J.M. Chen, C.M. Chernov, R.J. Kurtz, T. Muroga, Overview of the vanadium alloy researches for fusion reactors, J. Nucl. Mater. 417 (2011) 289.

[26] E.M. Sherwood, Columbium, in: C.A. Hampel (Ed.), Rare Metals Handbook, Robert Krieger Publishing Co., Huntington, NY, 1971, pp. 149–177.
[27] L.J. Pionke, J.W. Davis, Technical assessment of niobium alloys data base for fusion reactor applications. McDonnell Douglas, 1979 August, 1979. Report No.: C00-4247-2.
[28] J.R. Stewart, W. Lieberman, G.H. Rowe, Recovery and recrystallization of columbium-1.0% zirconium alloy, in: D. L. Douglass, F.W. Kunz (Eds.), Columbium Metallurgy, Interscience Publishers, New York, 1961, pp. 407–434.
[29] E.S. Bartlett, J.A. Houck, Physical and Mechanical Properties of Columbium and Columbium-Base Alloys, Defense Metals Information Center, Battelle Memorial Institute, Coloumbus, OH, 1960. Report No.: DMIC Report 125.
[30] R.G. Frank, Recent advances in columbium alloys, in: I. Machlin, R.T. Begley, E.D. Weisert (Eds.), Refractory Metal Alloys Metallurgy and Technology, Plenum Press, New York, 1968, pp. 325–372.
[31] R.W. Buckman Jr., in: R.H. Cooper Jr., E.E. Hoffman (Eds.), A review of tantalum and niobium alloy production, Proc Symp on Refractory Alloy Technology for Space Nuclear Power Applications, CONF-8308130, Oak Ridge National Lab, 1984, pp. 86–97.
[32] S.J. Zinkle, F.W. Wiffen, J.R. DiStefano, Historical Basis for Selection of Nb-1Zr Cladding for Space Reactor Applications, Oak Ridge National Lab, Oak Ridge, TN, 2004 September 30, 2004. Report No.: ORNL/LTR/NR/04-05.
[33] T.E. McGreevy, D.T. Hoelzer, S.J. Zinkle, C.E. Duty, K.J. Leonard, F.W. Wiffen, Outstanding Issues Concerning the Mechanical Behavior of Nb-1%Zr Alloys, Oak Ridge National Lab, Oak Ridge, TN, 2004. September 30, 2004. Report No.: ORNL/LTR/NR-JIMO/04-02.
[34] C.A. English, in: H. Stuart (Ed.), The physical, mechanical and irradiation behavior of niobium and niobium-base alloys, Proc Symp on Niobium, The Metallurgical Society of AIME, Warrendale, PA, 1984, pp. 239–324.
[35] E.A. Gulbransen, K.F. Andrew, F.A. Brassard, Oxidation of molybdenum 550°C to 1700°C, J. Electrochem. Soc. 110 (9) (1963) 952–959.
[36] B.V. Cockeram, R.W. Smith, T.S. Byun, L.L. Snead, The change in the hardness of LCAC, TZM, and ODS molybdenum in the post-irradiated and annealed conditions, J. Nucl. Mater. 393 (2009) 12–21.
[37] A.J. Bryhan, Welding Research Council Bulletin 312, United Engineering Center, 1986.
[38] M.K. Miller, A.J. Bryhan, Effect of Zr, B and C additions on the ductility of molybdenum, Mater. Sci. Eng. A 327 (1) (2002) 80–83.
[39] M.K. Miller, E.A. Kenik, M.S. Mousa, K.F. Russell, A.J. Bryhan, Improvement in the ductility of molybdenum alloys due to grain boundary segregation, Scr. Mater. 46 (2002) 299–303.
[40] N.N. Morgunova, L.N. Demina, N.I. Kazakova, Met. Sci. Heat Treat. 28 (12) (1986) 895–899.
[41] Y.N. Gornostyrev, M.I. Katsnelson, G.V. Peschanskikh, S.V. Trefilov, Phys. Staus Solid. 164 (1991) 185.
[42] L. Romaner, C. Ambrosch-Draxl, R. Pippan, Effect of rhenium on the dislocation core structure in tungsten, Phys. Rev. Lett. 104 (2010). 195503(1-4).
[43] M. Garfinkle, Room-Temperature Tensile Behavior of 100 Oriented Tungsten Single Crystals with Rhenium in Dilute Solid Solution, 1966, (1966)NASA TN D-3190 1966:17.
[44] A. Gilbert, M.J. Klein, J.W. Edington, Investigation of Mechanical Properties of Chromium, Chromium-Rhenium, and Derived Alloys, NASA CR 81225, (1966).
[45] P.L. Raffo, Yielding and Fracture in Tungsten and Tungsten-Rhenium Alloys, J Less Comm. Metals 17 (1969) 133.
[46] D.R. Trinkle, C. Woodward, The chemistry of deformation how solutes soften pure metals, Science 310 (2005) 1665.
[47] N.I. Medvedeva, Y.N. Gornostyrev, A.J. Freeman, Electronic origin of solid solution softening in BCC molybdenum alloys, Phys. Rev. Lett. 94 (2005) 136402.
[48] B.V. Cockeram, R.W. Smith, N. Hashimoto, L.L. Snead, The swelling, microstructure, and hardening of wrought LCAC, TZM, and ODS molybdenum following neutron irradiation, J. Nucl. Mater. 418 (1–3) (2011) 121–136.
[49] V.V. Bukhanovskii, N.G. Kartyshov, E.G. Polishchuk, V.K. Karchenko, M.I. Chikunov, Effect of heat treatment on the mechanical properties and nature of failure for sheet molybdenum alloys, Stren. Mater. 382 (1987) 229–241.
[50] Deleted in review.
[51] B.V. Cockeram, E.K. Ohriner, T.S. Byun, M.K. Miller, L.L. Snead, Weldable ductile molybdenum alloy development, J. Nucl. Mater. 382 (2008) 229–241.
[52] H.W. Yih, C.T. Wang, Tungsten, Sources, Metallurgy, Properties, and Applications, Plenum Press, New York, 1979.
[53] R.M. German, E. Olevesky, Int. J. Refract. Met. Hard Mater. 23 (2005) 294–300.
[54] H.M. Chung, B.A. Loomis, D.L. Smith, Creep properties of vanadium-base alloys, J. Nucl. Mater. 212–215 (1994) 772–777.

[55] T. Schober, D.N. Braski, The microstructure of selected annealed vanadium-base alloys, Metall. Trans. A. 20 (1989) 1927–1932.
[56] D.T. Hoelzer, A.F. Rowcliffe, L.T. Gibson, Processing and fabrication of experimental vanadium alloys for HFIR RB-17J experiment. Fusion Materials Semiannual Progress Report, DOE/ER-0313/34, Oak Ridge National Lab, 2003, pp. 22–26.
[57] D.T. Hoelzer, A.F. Rowcliffe, Investigating solute interactions in V-4Cr-4Ti based on tensile deformation behavior of vanadium, J. Nucl. Mater. 307–311 (2002) 596.
[58] A.F. Rowcliffe, D.T. Hoelzer, S.J. Zinkle, High Temperature Tensile Properties and Deformation Behavior of V-4Cr-4Ti. Fusion Materials Semiannual Progress Report DOE/ER-0313/26, p. 25.
[59] A.F. Rowcliffe, S.J. Zinkle, D.T. Hoelzer, Effect of strain rate on the tensile properties of unirradiated and irradiated V-4Cr-4Ti, J. Nucl. Mater. 283–287 (2000) 508–512.
[60] A.F. Rowcliffe, et al., Investigating solute interactions in V-4Cr-4Ti, J. Nucl. Mater. 508 (2000) 283–287.
[61] P. Haasen, Physical Metallurgy, Cambridge Univ. Press, New York, 1978, pp. 26–27.
[62] R.J. Kurtz, Biaxial thermal creep of V-4Cr-4Ti at 700°C and 800°C, J. Nucl. Mater. 283–287 (2000) 628.
[63] K. Natesan, W.K. Soppett, A. Purohit, Uniaxial creep behavior of V-4Cr-4T- alloy, J. Nucl. Mater. 307–311 (2002) 585–590.
[64] M. Li, T. Nagasaka, D.T. Hoelzer, M.L. Grossbeck, S.J. Zinkle, T. Muroga, et al., Biaxial thermal creep of two heats of V4Cr4Ti at 700 and 800°C in a liquid lithium environment, J. Nucl. Mater. 367–370 (2007) 788.
[65] D.L. Smith, K. Natesan, Influence of nonmetallic impurity elements on the compatibility of liquid lithium with potential CTR containment materials, Nucl. Technol. 22 (1974) 392–404.
[66] R.L. Stephenson, The creep-rupture properties of some refractory metal alloys I. Effect of heat treatment on the properties of the niobium-base alloy D-43, J Less Comm. Metals 15 (3) (1996) 395–402.
[67] D.M. Farkas, A.M. Mukherjee, Creep behavior and microstructural correlation of a particle-strengthened Nb-1Zr-0.1C alloy, J Mat Res. 11 (9) (1996) 2198–2205.
[68] K.J. Leonard, J.T. Busby, D.T. Hoelzer, S.J. Zinkle, Nb-base FS-85 alloy as a candidate structural materials for space reactor applications: effects of thermal aging, Met Mater Trans A 40A (2009) 838–855.
[69] B.V. Cockeram, R.W. Smith, K.J. Leonard, T.S. Byun, L.L. Snead, Irradiation hardening in unalloyed and ODS molybdenum and ODS molybdenum during low dose neutron irradiation at 300°C and 600°C, J. Nucl. Mater. 382 (2008) 1–23.
[70] B.V. Cockeram, R.W. Smith, L.L. Snead, The influence of fast neutron irradiation and irradiation temperature on the tensile properties of wrought LCAC and TZM molybdenum, J. Nucl. Mater. 346 (2005) 145–164.
[71] H. Schultz, Defect-sensitive properties of tungsten and other BCC transition-metals, Z. Metallkd. 78 (1987) 469–475.
[72] G. Bar, D. Gras, V. Hasse, System-Nr 54, Tungsten, Suppl In Gmelin Handbook of Inorganic Chemistry, Springer-Verlag, Heidelberg, 1989.
[73] B.C. Allen, D.J. Maykuth, R.I. Jaffee, The recrystallization and Ductile-Brittle transition behaviour of tungsten-effect of impurities on polycrystals prepared from single crystals, J. Inst. Metals 90 (1961) 120–128.
[74] D.B. Snow, Dopant observations in thin foils of annealed tungsten wire, Metall. Trans. A. 2 (1972) 2553.
[75] B.A. Loomis, H.M. Chung, L.J. Nowicki, D.L. Smith, Effects of neutron irradiation and hydrogen on ductile-brittle transition temperatures of V-Cr-Ti alloys, J. Nucl. Mater. 212–215 (1994) 799–803.
[76] B.A. Loomis, D.L. Smith, Vanadium alloys for structural applications in fusion systems: a review of vanadium alloy mechanical and physical properties, J. Nucl. Mater. 191–194 (1992) 84–91.
[77] S.J. Zinkle, D.J. Alexander, J.P. Robertson, L.L. Snead, A.F. Rowcliffe, L.T. Gibson, et al., Effect of Fast Neutron Irradiation to 4 dpa at 400°C on the properties of V-(4–5)Cr-(4–5)-Ti Alloys. Department of Energy Fusion Reactor Semiannual Period Ending December 31, 1996 DOE/ER-0313/21, Oak Rige National Lab, 1996, pp. 73–78.
[78] L.L. Snead, S.J. Zinkle, D.J. Alexander, A.F. Rowcliffe, J.P. Robertson, W.P. Eatherly, Summary of the investigation of low temperature, low dose radiation effects on the V-4Cr-4Ti alloy, in: Fusion Materials Semiannual Progress Report for Period Ending Dec 31, 1997, pp. 81–98 DOE/ER-0313/231997.
[79] S.J. Zinkle, L.L. Snead, J.P. Robertson, A.F. Rowcliffe, Strain rate dependence of the tensile properties of V-(4–5%)Cr-(4–5%)Ti irradiated in EBR-II and HFBR, in: Fusion Materials Semiann Prog Rep for period ending Dec 31 1997, DOE/ER-0313/23, Oak Ridge National Lab, 1997, pp. 77–80.
[80] S.J. Zinkle, Thermophysical and mechanical properties of V-(4–5%)Cr-(4–5%)Ti alloys, in: Fusion Materials Semiann Prog Report for period ending Dec 31 1997, DOE/ER-0313/23, Oak Ridge National Lab, 1997, pp. 99–108.
[81] E. Donahue, G.R. Odette, G.E. Lucas, J.W. Sheckherd, A.F. Rowcliffe, Effect of irradiation on toughness-temperature curves in V-4Cr-4Ti, in: Fusion Materials Semiann Prog Report for period ending Dec 31 1998, DOE/ER-0313/25, Oak Ridge National Lab, 1998, pp. 32–41.

[82] H.M. Chung, D.L. Smith, Tensile and impact properties of vanadium-base alloys irradiated at <430°C, J. Nucl. Mater. 258–263 (1998) 1442–1450.
[83] T. Muroga, Vanadium alloys for fusion blanket applications, Mater. Trans. 45 (2005) 405.
[84] J. Gazda, M. Meshii, H.M. Chung, Microstructure of V-4Cr-4Ti after low-temperature irradiation by ions and neutrons, J. Nucl. Mater. 258–263 (1998) 1437–1441.
[85] P.M. Rice, S.J. Zinkle, Temperature dependence of radiation damage microstructure in V-4Cr-4Ti neutron irradiated to low dose, J. Nucl. Mater. 258–263 (2) (1998) 1414–1419.
[86] D.S. Gelles, P.M. Rice, S.J. Zinkle, H.M. Chung, Microstructural examination of irradiated V-(4-5%)Cr-(4-5%)Ti, J. Nucl. Mater. 258–263 (1998) 1380–1385.
[87] F.W. Wiffen, Creep and tensile properties of helium injected Nb-1%Zr, in: J.S. Watson, F.W. Wiffen (Eds.), Radiation Effects and Tritium Technology for Fusion Reactors, CONF-750989. II, USERDA, Gatlinburg, TN, 1976, pp. 344–361.
[88] H. Kayano, S. Yajima, Mechanical properties of neutron-irradiated niobium, J. Nucl. Sci. Technol. 15 (2) (1978) 135–140.
[89] D.T. Hoelzer, S.J. Zinkle, Structural analysis of plate-shaped precipitates in neutron irradiated V-4Cr-4Ti, in: Fusion Materials Semiannual Progress Report for Period ending December 31 2000, DOE/ER-0313/29, Oak Ridge National Lab, 2000, pp. 19–25.
[90] V.M. Troyanov, M.G. Bulkanov, A.S. Kruglov, E.A. Krjuchkov, M.P. Nikulin, J.M. Pevchykh, et al., Irradiation creep of V-Ti-Cr alloy in BR-10 reactor core instrumented experiments, J. Nucl. Mater. 233–237 (1996) 381–384.
[91] H. Tsai, I.C. Gomes, D.L. Smith, A.J. Palmer, F.W. Ingram, F.W. Wiffen, Status of the irradiation test vehicle for testing fusion materials in the advanced test reactor, in: Fusion Materials Semiann Prog Report for period ending June 30, 1998, DOE/ER-0313/24, Oak Ridge National Lab, 1998, pp. 278–282.
[92] H. Tsai, T.S. Bray, H. Matsui, M.L. Grossbeck, K. Fukumoto, J. Gazda, et al., Effects of low-temperature neutron irradiation on mechanical properties of vanadium-base alloys, J. Nucl. Mater. 283–287 (2000) 362.
[93] D.S. Gelles, Microstructural Examination of Irradiated and Unirradiated V-4Cr-4Ti Pressurized Creep Tubes, J. Nucl. Mater. 386–388 (2002) 393.
[94] M. Li, D.T. Hoelzer, M.L. Grossbeck, A.F. Rowcliffe, S.J. Zinkle, R.J. Kurtz, Irradiation creep of the US Heat 832665 of V-4Cr-4Ti, J. Nucl. Mater. 386–388 (2009) 618.
[95] F.W. Wiffen, in: R.H. Cooper Jr., E.E. Hoffman (Eds.), Effects of irradiation on properties of refractory alloys with emphasis on space power reactor applications, Proc Symp on Refractory Alloy Technology for Space Nuclear Power Applications, CONF-8308130, Oak Ridge National Lab, 1984, pp. 252–277.
[96] J.H. DeVan, J.R. DiStefano, E.E. Hoffman, in: R.H. Cooper Jr., E.E. Hoffman (Eds.), Compatibility of refractory alloys with space reactor system coolants and working fluids, Proc Symp on Refractory Alloy Technology for Space Nuclear Power Applications, CONF-8308130, Oak Ridge National Lab, 1984, pp. 34–85.
[97] M. Kangilaski, Comparison of niobium (columbium) alloys for SP-100 space reactor applications (unpublished), General Electric Company, Advanced Nuclear Technology Operations, Sunnyvale, CA, 1984. November 1984. (Report No).
[98] S.J. Zinkle, F.W. Wiffen, in: M.S. El-Genk (Ed.), Radiation Effects in Refractory Alloys, Space Technology and Applications International Forum-STAIF 2004 (AIP Conf Proc, vol. 699), American Institute of Physics, Melville, NY, 2004, pp. 733–740.
[99] F.W. Wiffen, Plastic instability in neutron-irradiated niobium alloys, in: Alloy Development for Irradiation Performance Quarterly Progress Report, January–March, 1978, DOE/ET-0058/1, Oak Ridge National Lab, 1978, pp. 142–152.
[100] T.T. Claudson, H.J. Pessl, Irradiation effects on high-temperature reactor structural metals, in: C.E. Weber (Ed.), Flow and Fracture of Metals and Alloys in Nuclear Environments, American Society for Testing and Materials, Philadelphia, 1965, pp. 156–170. ASTM STP 380.
[101] R.P. Tucker, S.M. Ohr, M.S. Wechsler, Radiation hardening and transmission electron microscopy in niobium, in: Radiation Damage in Reactor Materials. I, IAEA, Vienna, 1969, pp. 215–233.
[102] F.W. Wiffen, The tensile properties of fast reactor neutron irradiated BCC metals and alloys, in: R.J. Arsenault (Ed.), Proc Int Conf on Defects and Defect Clusters in BCC Metals and their Alloys, Nuclear Metallurgy, vol. 18, National Bureau of Standards, Gaithersburg, MD, 1973, pp. 176–196.
[103] V. Kazakov, A.S. Pokrovsky, R. Melder, in: Mechanical properties of niobium alloys after neutron irradiation at 350–1300°C, Presented at 6th International Conference on Fusion Reactor Materials, Stresa, Italy, September 1993 (Unpublished Preprint), 1993.
[104] J.A. Horak, M.L. Grossbeck, M.M. Paxton, Effects of irradiation on the tensile properties of Nb-1Zr, in: M.S. El-Genk, M.D. Hoover (Eds.), 11th Symp on Space Nuclear Power and Propulsion, AIP Conf Proc 301, American Institute of Physics, Woodbury, NY, 1994, pp. 413–420.

[105] K.J. Leonard, J.T. Busby, S.J. Zinkle, Influence of thermal and radiation effects on microstructural and mechanical properties of Nb-1Zr, J. Nucl. Mater. 411 (2011) 286–302.
[106] D.G. Atteridge, L.A. Charlot, J.A.B. Johnson, J.F. Remark, Effects of helium implanted by tritium decay on the high temperature mechanical properties of niobium, in: F.W. Wiffen, J.S. Watson (Eds.), Radiation Effects and Tritium Technology for Fusion Reactors, 1976 (Gatlinburg, TN).
[107] A.A. Sagues, J. Auer, Mechanical behavior of Nb-1%Zr implanted with He at varioius temperatures, in: F.W. Wiffen, J.S. Watson (Eds.), IInd Radiation Effects and Tritium Technology for Fusion Reactors, 1976, pp. 331–343 (Gatlingburg, TN).
[108] M.M. Paxton, C.R. Eiholzer, Biaxial creep of rhenium and Nb-1Zr Re-bonded pressurized type specimens at SP-100 service conditions. Westinghouse Hanford Co., 1994 September, 1994, (1994) Report No.: WHC-SP-1078.
[109] M. Li, T.S. Byun, L.L. Snead, S.J. Zinkle, Defect cluster formation and radiation hardening in molybdenum neutron-irradiated at 80°C, J. Nucl. Mater. 367–370 (2007) 817–822.
[110] L.R. Greenwood, F.A. Garner, Transmutation of Mo, Re, W, Hf and V in various irradiation test facilities and STARFIRE, J. Nucl. Mater. 212–215 (1994) 635–639.
[111] M. Li, T.S. Byun, L.L. Snead, S.J. Zinkle, Low-temperature thermally-activated deformation and irradiation softening in neutron-irradiated molybdenum, J. Nucl. Mater. 377 (2008) 409–414.
[112] F.W. Wiffen, The effect of alloying and purity on the formation and ordering of voids in bcc metals, in: J.W. Corbett, L.C. Iannello (Eds.), Radiation Induced Voids in Metals, CONF-710601, National Technical Information Service, Springfield, VA, 1972, pp. 386–396.
[113] J.H. Evans, Observations of a regular void array in high purity molybdenum and TZM irradiated at high temperatures with 2 MeV nitrogen ions, Radiat. Eff. 10 (1971) 55–60.
[114] R.C. Rau, J. Moteff, R.L. Ladd, Comparison of radiation damage in molybdenum and some molybdenum-base alloys, J. Nucl. Mater. 40 (1971) 233–235.
[115] V.K. Sikka, J. Moteff, Damage in neutron-irradiated molybdenum, J. Nucl. Mater. 54 (1974) 325–345.
[116] F.A. Garner, J.F. Stubbins, Saturation of swelling in neutron-irradiated molybdenum and its dependence on irradiation temperature and starting microstructural state, J. Nucl. Mater. 212–215 (1994) 1298–1302.
[117] K.J. Leonard, Radiation effects in Refractory Metals and Alloys, in: R.J.M. Konings (Ed.), Comprehensive Nuclear Materials, Elsevier Publishing, Amsterdam, 2012, pp. 181–213.
[118] Y. Nemoto, A. Hasegawa, M. Satou, K. Abe, Y. Hiraoka, Microstructural development and radiation hardening of neutron irradiated Mo–Re alloys, J. Nucl. Mater. 324 (1) (2004) 62–70.
[119] J.T. Busby, K.J. Leonard, S.J. Zinkle, Radiation-damage in molybdenum-rhenium alloys for space reactor applications, J. Nucl. Mater. 366 (2007) 388–406.
[120] S.A. Fabritsiev, A. Pokrovsky, The effect of rhenium on the radiation damage resistivity of Mo-Re alloys, J. Nucl. Mater. 252 (1998) 216–227.
[121] B.V. Cockeram, R.W. Smith, L.L. Snead, The influence of fast neutron irradiation and irradiation temperature on the tensile properties of wrought LCAC and TZM molybdenum, J. Nucl. Mater. 346 (2005) 145–164.
[121a] V. Chakin, V. Kazakov, J. Nucl. Mater. 233–237 (1996) 570–572.
[121b] A. Hasegawa, K. Abe, M. Satou, C. Ueda, C. Namba, J. Nucl. Mater. 233–237 (1996) 565–569.
[121c] K. Abe, T. Takeuchi, M. Kikuchi, S. Morosumi, J. Nucl. Mater. 99 (1981) 23–37.
[121d] K. Abe, M. Kikuchi, K. Tate, S. Morozumi, J. Nucl. Mater. 122–123 (1984) 671–675.
[121e] T.H. Webster, B.L. Eyre, E.A. Terry, in: Proceedings of BNES conference on irradiation embrittlement and creep in fuel cladding and core components, British Nuclear Energy Society, London, 1972, pp. 61–80.
[121f] M. Schibetta, R. Chaouadi, J.L. Puzzolante, J. Nucl. Mater. 283–287 (2000) 455–460.
[122] B.V. Cockeram, T.S. Byun, K.J. Leonard, J.L. Hollenbeck, L.L. Snead, Post-Irrdiation fracture toughness of unalloyed molybdenum, ODS molybdenum, and TZM molybdenum following irradiation at 244°C to 507°C, J. Nucl. Mater. 440 (2013) 382–413.
[123] K. Babinsky, S. Primig, W. Knabi, A. Lorich, R. Stickler, H. Clemens, Fracture behavior and delamination toughening of molybdenum in Charpy impact tests, J. Mater. 68 (11) (2016) 2854–2863.
[124] D.R. Mason, X. Yi, M.A. Kirk, S.L. Dudarev, Elastic trapping of dislocation loops in cascades in ion-irradiated tungsten foils, J. Phys. Condens. Matter 26 (2014) 375701.
[125] L.K. Keys, J. Moteff, J. Nucl. Mater. 34 (1970) 260.
[126] R.K. Williams, F.W. Wiffen, J. Bentley, J.O. Stiegler, Irradiation induced precipitation in tungsten based, W-Re alloys, Mettal. Trans. A 14A (1983) 655–666.

[127] M. Fukuda, K. Yabuuchi, S. Nogami, A. Hasegawa, T. Tanaka, Microstructural development of tungsten and tungsten-rhenium alloys due to neutron irradiation in HFIR, J. Nucl. Mater. 455 (2014) 450–463.
[128] P.D. Edmondson, A. London, A. Xu, D.E.J. Armstrong, S.G. Roberts, Small-scale characterisation of irradiated nuclear materials: part 1-microstructure, J. Nucl. Mater. 462 (2015) 369–373.
[129] T. Koyanagi, N. Kumar, T. Hwang, L.M. Garrison, X. Hu, L.L. Snead, Y. Katoh, Microstructural evolution of pure tungsten neutron irradiated with a mixed energy spectrum, J. Nucl. Mater. 490 (2017) 66–74.
[130] A. Hasegawa, T. Tanno, S. Nogami, T. Tanaka, Property change mechanism in tungsten under neutron irradiation in various reactors, J. Nucl. Mater. 417 (2011) 491–494.
[131] Y. Nemoto, A. Hasegawa, M. Satou, K. Abe, Microstructural development of neutron irradiated W-Re alloys, J. Nucl. Mater. 283-287 (2000) 1144–1147.
[132] J. Matolich, H. Nahm, J. Moteff, Swelling in neutron irradiated tungsten and tungsten-25 percent rhenium, Scr. Metall. 8 (7) (1974) 837–842.
[133] M. Fujisuka, B. Tsuchiya, I. Mutoh, T. Tanabe, T. Shikama, Effect of neutron irradiation on thermal diffusivity of tungsten-rhenium alloys, J. Nucl. Mater. 283–287 (2000) 1148–1151.
[134] T. Tanno, A. Hasegawa, J.-C. He, M. Fujiwara, S. Nogami, M. Satou, et al., Effects of transmutation elements on neutron irradiation hardening of tungsten, Mater. Trans. 48 (9) (2007) 2399–2402.
[135] K. Krishan, Ordering of voids and gas bubbles in radiation environments, Radiat. Eff. 66 (1982) 121–155.
[136] R. Herschitz, D.N. Seidman, Radiation-induced precipitation in fast neutron irradiated W-Re alloys: an atom probe field microscopy study, Nucl Instrum Meth Phys Res B 7/8 (1985) 137.
[137] X. Hu, T. Koyanagi, M. Fukluda, N.A.P. Kiran Kumar, L.L. Snead, B.D. Wirth, Y. Katoh, Irradiation hardening of pure tungsten exposed to neutron irradiation, J. Nucl. Mater. 480 (2016) 235–243.
[138] R.C. Rau, J. Moteff, R.L. Ladd, Comparison of microstructure with mechanical properties of irradiated tungsten, J. Nucl. Mater. 24 (1967) 164–173.
[139] J.M. Steichen, Tensile properties of neutron irradiated TZM and tungsten, J. Nucl. Mater. 60 (1976) 13–19.
[140] I.V. Gorynin, A.A. Ignatov, V.V. Rybin, S.A. Fabritsiev, V.A. Kazakov, V.P. Chakin, V.A. Tsykanov, V.R. Barabash, Y.G. Prokofyev, Effects of neutron irradiation on properties of refractory metals, J. Nucl. Mater. 191–194 (1992) 421–425.
[141] P. Krautwasser, H. Derz, E. Kny, Influence of fast neutron fluence on the DBTT of tungsten, W-10Re and W-3.4Ni-1.6Fe, High Temp. High Pressures 22 (1) (1990) 25–32.
[142] H.H. Smith, D.J. Michel, Journal of Nuclear Materials, 66 (1) (1977) 125–142.

Index

Note: Page numbers followed by *f* indicate figures and *t* indicate tables.

A

ACAR. *See* Angular correlation of annihilation radiation (ACAR)
Accelerator production of tritium (APT) facility, 385
Acceptance tests, 421–422
Advanced fission energy systems, 52
AISI 316 austenitic steel, 294, 485–486
AISI 304 stainless steels, 294
ALFRED/ELFR system
   clad temperatures, 33
   flow conditions, 31, 32*t*
   materials, 34–35, 35*t*
Alloy D9, 337
American Society of Mechanical Engineers (ASME), 135
American Society of Mechanical Engineers Boiler and Pressure Vessel Code (ASME BPVC), 415–416
American Society of Testing and Materials (ASTM), 135
Amorphization. *See* Irradiation-induced amorphization
Angular correlation of annihilation radiation (ACAR), 130–131
Anisotropic nonirradiated mechanical properties, 625
Annealed austenitic stainless steel 1.4970, 321
Annealing temperature effect, 600–602, 602*f*
AOA. *See* Axial offset anomaly (AOA)
APT. *See* Atom probe tomography (APT)
ARIES tokamak power plant, components in, 63–64, 63*t*
ASME. *See* American Society of Mechanical Engineers (ASME)
ASME BPVC. *See* American Society of Mechanical Engineers Boiler and Pressure Vessel Code (ASME BPVC)
ASTM. *See* American Society of Testing and Materials (ASTM)
ASTM-E1921 standard, 504–505
Atomic displacements, 297
Atom probe tomography (APT), 444–445
   austenitic stainless steels, 123–125, 124*f*
   Cu-stabilized microvoids, 118–119
   field ion microscopes (FIMs), 118–119
   generation IV and fusion materials, 125
   interpretation, 125
   principles, 117–118, 118*f*
   reactor pressure vessel (RPV) embrittlement, 119–120, 119*f*, 121*f*
   ultrafine copper-enriched clusters and precipitates, 119
   zircaloys, 120–123
ATR-A1, HFIR 12J, and HFIR-17J experiments, 620–622
Austenitic stainless steels, 13–15, 14*t*, 27, 29, 36
   AISI 304 and AISI 316 stainless steels, 294
   atomic displacements, 297
   chemical compatibility with coolants, 321–323, 322*f*
   crystalline defects, 297
   dislocation microstructure, 302–303, 303*f*
   fatigue, 317, 317*f*
   fracture toughness and embrittlement, 309–311, 310*f*
   high-temperature He embrittlement, 311, 312*f*
   in-reactor creep rupture properties, 318, 319*f*
   intergranular stress corrosion cracking (IGSCC), 323–326, 324–325*f*
   irradiation-assisted stress corrosion cracking (IASCC), 123–124, 124*f*
   irradiation creep, 315–317, 315–316*f*
   irradiation hardening, 308–309, 309*f*
   in light water reactors, 294–295
   neutron irradiation, 297
   PCI/FCCI effects with fission fuels, 318–321
   perspectives and prospects, 337–340
   phase stability, 303–305, 304*f*, 305*t*
   radiation-definedquasiequilibrium state, 297
   radiation-induced segregation (RIS), 297–302
   in sodium-cooled fast reactors, 295–297
   stress-corrosion cracking (SCC), 123, 124*f*, 216, 217*f*, 323
   transmutation, 305–308, 307*f*
   void swelling, 311–314, 313–314*f*, 337, 339*f*
   water environment and radiation, combined effects of, 326–337
Axial offset anomaly (AOA), 259–260

B

BDT temperature. *See* Brittle-to-ductile transition (BDT) temperature
Beginning of life (BOL)
   mechanical properties, 411–412, 422
   unirradiated, 437–438
Boiling water reactors (BWRs), 2–5, 353
   axial splits in, 283
   components, 16–17, 18*t*
   coolant chemistry, 261–262, 261*f*
   fuel assemblies in, 377
   fuel assembly, 248, 250*f*
   liner cladding, 251
   schematic representation of, 5–6, 5*f*
   stress-corrosion cracking, austenitic stainless steels, 323
BOL. *See* Beginning of life (BOL)
Boric acid corrosion (BAC), 11–12
Brittle cleavage fracture, 503–504
Brittle-to-ductile transition (BDT) temperature
   molybdenum, 595
   tungsten, 632–634, 634*f*
Brittle transition temperature, smooth tensile specimens, 629, 630*f*
BWRs. *See* Boiling water reactors (BWRs)

C

Calculation of phase diagram (Calphad) method, 144–145
Calphad method, 144–145, 147, 149–150
CANDU reactors
    radiation damage and gas production, 369, 371–373, 374–376f, 375
    radiation hardening/softening and loss of ductility (see Inconel X-750 spacer)
    unirradiated vs. irradiated fracture surfaces, 377, 378f
Carbon fiber composites (CFCs)
    challenges, 90–91
    cracking, 91
    fragmentation, 91
    neutron dose and irradiation temperature, 91–92, 91f
    neutron irradiation, 91–92
    superior mechanical and physical properties, 90
    thermal fatigue, 91
    threshold ion energy, 90–91
    tritium retention, 92
Carbon-free model alloys, 423
Carburization, VHTR environment, 227
CASS. See Cast-austenitic stainless steels (CASS)
Cast-austenitic stainless steels (CASS), 15–16, 16t
Casting, niobium alloys, 590–591, 591f
Cavity swelling
    bubble, 188
    dose-dependent swelling behavior, cold-worked Type 316 stainless steel, 188–189, 189f
    isotropic volumetric expansion, 188
    neutron-irradiated-type 304-L austenitic (FCC) stainless steel vs. 9–12 Cr ferritic/martensitic (BCC) steel, 189, 190f
    spallation proton-neutron irradiation, 191
    steady-state swelling rate, 188–189
    void nucleation, 190
    void swelling, 188, 190–191
CFCs. See Carbon fiber composites (CFCs)
Charpy impact absorbed energy curves, 616–618, 617f
Charpy impact properties, vanadium alloys, 600–602
Charpy impact test, 435–436, 502
Charpy shift, 455–456, 457f
Charpy testing, 437–438
Charpy transition curves, WWER-1000 steel, 435–436, 435f
Charpy transition temperature, 424, 426, 426–427f, 465–466
Charpy V-Notch (CVN) impact tests, 138–139, 139f
Charpy V-notch (CVN)-indexed transition-temperature shifts, 19–21
Chromium, stress corrosion cracking, 216
CILC. See Crud-induced localized corrosion (CILC)
CIP. See Cold isostatic pressing (CIP)
CIPS. See Crud-induced power shift (CIPS)
C-Mn steels, 423
Coble creep, 168–169
Coincidence Doppler broadening (CDB), 130–131
Cold isostatic pressing (CIP)
    molybdenum, 593–594
    tungsten, 597
Cold-worked and aged (CWA) and TZM molybdenum neutron swelling, 625–626, 627f
Cold-worked (CW) austenitic steels, 295
Computational alloy design and optimization
    alloy optimization, 145–147, 145–146f
    alloy selection and design, 147, 148f
    Calphad method, 144–145
    computational thermodynamics, 144
    Gibbs energies, 144–145
    kinetics and mechanical property simulations, 147–149, 149f
    multiscale modeling methods, 149–150
Computational thermodynamics, 106–107, 144
Constraint loss (CL) limits, 504
Control rod drive mechanisms (CRDMs), 356
Copper-rich precipitates (CRPs), 439, 462
Corrosion, 11–12
    between alloy and coolant water, 264
    in ferritic/martensitic steels, 490
    in halide salts, 43
    in helium-cooled reactors, 224–229
    liquid lead alloys, 230, 239–242
    liquid metal corrosion, 230
    liquid sodium, 230, 236–239
    molten salts, 230–235
    in nuclear systems, 211–212
    types, 212, 213f
    in water cooled reactors, 212–223
    zirconium alloys, 258
Corrosion fatigue, austenitic stainless steels, 332–333, 333f
CRDMs. See Control rod drive mechanisms (CRDMs)
Creep fatigue
    damage, 172
    deformation, 400–401, 401f
    reactor structural alloys, 172, 173f
Creep strength
    nanostructured ferritic alloys (NFAs), 555
    nanostructured tempered martensitic steels (NMS), 555
    Ni-based alloys, 362–364, 363–364f
Creep testing, 604–606
Crevice corrosion, 11–12
9Cr ferritic/martensitic steel T91, 29
9-12Cr F/M steels, 35
CRIEPI model, 463–464
Critical microcleavage fracture stress, 501
Critical resolved shear stress (CRSS), 454
9Cr-1Mo steels, 486
CRPs. See Copper-rich precipitates (CRPs)
CRSS. See Critical resolved shear stress (CRSS)
Crud deposition
    fuel failures, 262, 263f
    on fuel rods, 258–260, 259t, 260f
Crud-induced localized corrosion (CILC), 259–260
Crud-induced power shift (CIPS), 259–260
Cu effect mechanism, 439
Cu-stabilized microvoids, 118–119

D

Davidenkov concept, 455–456, 456f
DBTT. *See* Ductile-to-brittle transition temperature (DBTT)
Debris fretting, 279, 280f
Decarburization, VHTR environment, 226–227, 226t, 228f
Decohesion, 609–610
Delayed hydride cracking (DHC), 268–269, 270f
Deuterium-tritium (D-T) fusion reaction, 489
DHC. *See* Delayed hydride cracking (DHC)
DII. *See* Dual ion irradiations (DII)
Direct age (DA), 355
Dislocation loops, 498–500, 500f
  nano-oxide dispersion-strengthened (NODS) steels, 562–564, 564f
Dispersion strengthened tungsten, 613–614, 613f
Displacement cascades, 56
Displacement damage, 530
Displacement dose, 56
Displacements-per-atom (dpa), 530
Dissolution-deposition mass transport, 490
Distinctive crud pattern, 262, 263f
Divertor/limiter application, materials for, 63
  carbonfiber composites, 90–92
  liquid walls, 92–93
  tungsten and tungsten alloys, 85–90
Double-step aging, 355
DSA. *See* Dynamic strain aging (DSA)
Dual ion irradiations (DII), 511
Ductile-to-brittle transition temperature (DBTT)
  ferritic/martensitic steels, 486, 501
  niobium alloys, 625
  vanadium-base alloys, 587–588
Duplex cladding, 251
Dynamic strain aging (DSA), 75, 465, 603–604

E

EAC. *See* Environmentally assisted cracking (EAC)
Eason Odette Nanstad and Yamamoto (EONY) model, 444, 449, 451–452, 462–463
ECP. *See* Electrochemical potential (ECP)
ECR. *See* Equivalent cladding reacted (ECR)
EDX spectroscopy. *See* Energy-dispersive X-ray (EDX) spectroscopy
EELS. *See* Electron energy loss spectroscopy (EELS)
Effective Cu, 444
Elastic-plastic fracture mechanics (EPFM), 433–435
Electrochemical potential (ECP), 323
Electron energy loss spectroscopy (EELS), 109–111
Electron microscopy
  aberration corrected microscopes, 113–114
  advanced specimen techniques, 117
  corrosion-resistant components in LWRs, 114
  fast reactors, 112–113
  field emission gun scanning transmission electron microscopy (FEGSTEM), 113–114
  high-angle annular dark field (HAADF) image, open crack, 116, 116f
  IG attack tip area, stereoscopic image, 115, 115f
  irradiation-assisted stress corrosion cracking (IASCC), 113
  oxide dispersion strengthened (ODS) steels, 112–113
  polyhedral voids, 113, 114f
  principles, 109–111
  quantitative information, 116
  reactor pressure vessel (RPV) steels, 113
  scanning transmission electron microscopy (STEM), 109–112
  solute redistribution and clustering, 116–117
  stress corrosion cracking (SCC), 115
  3D FIB slicing and electron tomography, 116
  transmission-electron backscatter diffraction (t-EBSD), 114–115
  transmission electron microscopy (TEM), 109, 111–112
Energetic neutrons, 62
Energy-dispersive X-ray (EDX) spectroscopy, 109–111
Environmentally assisted cracking (EAC), 11–12, 489–490
EONY model. *See* Eason Odette Nanstad and Yamamoto (EONY) model
EPFM. *See* Elastic-plastic fracture mechanics (EPFM)
Equivalent cladding reacted (ECR), 272
Erosion corrosion, 11–12
Eurofer97 steel
  modified Master-Curve, 505, 506f
  transmission electron microscopy (TEM) observation, 486–488, 487f
EUROPEAN LFR reactor design, 31, 32f
European Pb cooled ALFRED concept, 31, 32f
Ex-reactor corrosion, 258

F

FAC. *See* Flow-accelerated corrosion (FAC)
Fast-breeder reactors, 13
Fast flux test facility (FFTF), 25–26
Fast reactors, nickel alloys, 383–385, 385f
Fast-thermal fission reactor irradiations, 511
Fatigue
  austenitic stainless steels, 317, 317f
  corrosion, 332–333, 333f
  ferritic/martensitic steels, 507–509, 508–510f
  nano-oxide dispersion-strengthened (NODS) steels, 558, 559f, 572
  nickel alloys, 399–400, 400f
  reactor structural alloys, 171–172, 171f, 173f
Fatigue crack growth, LAS, 470
Fatigue damage-accumulation process, 172
FC. *See* Fuel cladding (FC)
Fe-Cr-Ni alloys, 45, 182–184, 183f, 301
FEGSTEM. *See* Field emission gun scanning transmission electron microscopy (FEGSTEM)
Ferritic/martensitic steels, 29–30
  compatibility with liquid metal coolants, 490–493
  composition and constitution, 485–488
  environmentally assisted cracking, 489–490
  fatigue, 507–509, 508–510f

Ferritic/martensitic steels *(Continued)*
  future prospects, 518–519
  in generation IV nuclear systems and fusion reactors, 488–489
  helium effects, 510–513
  irradiation creep, 516–518
  irradiationembrittlement, fast fracture, 501–507
  radiation hardening and softening, 496–501
  reduced activation ferritic/martensitic (RAF/M) steels, 493–496, 493–495f
  supercritical water (SCW), 223
  tempered martensitic steels (TMS) lose, 494
  thermal creep, 509, 510–511f
  void swelling, 514–516
Ferrules, 318
F82H-mod steel CL, 503–504, 503f
FHR. *See* Fluoride salt-cooled high-temperature reactor (FHR)
F82H, RAF/M steels
  Eurofer97 and, 70
  long-term aging of, 69–70
  thermal annealing, 69–70
FIB. *See* Focused ion beam (FIB)
Field emission gun scanning transmission electron microscopy (FEGSTEM), 444–445
Field ion microscopy (FIM), 118–119, 444–445
Filler metals (FMs), 356
FIM. *See* Field ion microscopy (FIM)
First-wall/blanket structure
  elements, 66
  functions, 61
  helium effects, 80–84
  nanostructuredferritic alloys, 72–75
  reduced activation ferritic/martensitic (RAF/M) steels, 66–72
  SiC_F/SiC composites, 77–80
  vanadium alloys, 75–77
Fission fuels, PCI/FCCI effects with, 318–321
Fission reactor irradiations, 567
Flow-accelerated corrosion (FAC), 11–12
Fluoride salt-cooled high-temperature reactor (FHR), 43, 229–230
Focused ion beam (FIB)
  milling, 108–109
  3D FIB slicing and electron tomography, 116
Forgings and plates, 419, 421–422
  Mn-Mo-Ni, 422
  V-segregates, 419
Fracture toughness, 138–141
  austenitic stainless steels, 309–311, 310f, 334–337, 336f
  ferritic/martensitic steels, 503, 506, 506f
  molybdenum alloys, 609–610, 610f
  shift, 456, 458f
  structural integrity assessment, 433, 433f
  transition temperature, 456, 457f
  vanadium alloys, 600–602
French CW15-15Ti, 296
Fuel assembly
  boiling water reactor (BWR), 248, 250f
  circular cross-sectional short assemblies, 248
  pressurized water reactor (PWR), 248, 249f
Fuel cladding (FC), 27, 248
  ferritic/martensitic steels, 493
  vanadium alloys, 587
  zirconium alloys (*see* Zirconium cladding alloys)
Fuel-cladding chemical interaction (FCCI) effects, 318–321
Fuel failures, zirconium alloys
  axial splits in BWR cladding, 283
  causes, 278–279, 279f
  corrosion and crud deposition, 262, 263f
  debris fretting, 279, 280f
  failure rates, 278, 278f
  grid-to-rod fretting (GTRF), 280, 281f
  oxide spallation, 283
  pellet-cladding interaction-stress corrosion cracking, 282, 282–283f
  pellet-cladding mechanical interaction (PCMI), 280–282, 281f
  primary failure, 278
  weld failures, 283
Fuel rod, 248, 249f
Fusion power reactor applications, 52. *See also* *See also* Nuclear fusion
Fusion reactors, ferritic/martensitic steels in, 488–489

G

Galvanic corrosion, 11–12
Gas-cooled fast reactor (GFR)
  candidate core materials, 41
  concept, 39–40, 40f
  flow conditions, 40
  predicted dose rates and total fluences, 41
Generation-IV (Gen-IV) reactors, 4, 23–24
  ferritic/martensitic steels in, 488–489
  helium cooled reactors, 36–41
  liquid metal cooled fast reactors, 24–36
  molten salt cooled reactor, 43
  molten salt fueled reactor (MSR), 41–43
  Ni-based alloys, 365–367, 367f, 368t
  super-critical water reactor (SCWR), 43–45
Gen II reactors, 2–4
Gen III reactors, 2–4
γ-phase, 351
Gen I reactors, 2–4
German DIN 1.4970 alloy, 297
GFR. *See* Gas-cooled fast reactor (GFR)
Gibbs energies, 144–145
Gibbs-Thomson effect, 83
Grid-to-rod fretting (GTRF)
  failure mechanisms, 280, 281f
  fuel leaking failures, 252
Griffith-Orowan criterion, 505
GTRF. *See* Grid-to-rod fretting (GTRF)
Guinier approximation, 126–127

Index

H

HAZ. *See* Heat-affected zone (HAZ)
Heat-affected zone (HAZ), 423–424, 595
Heavy Section Steel Technology (HSST) program, 424
He/dpa ratio, 566–567, 569f
Helium-cooled reactors
  alloy chemistry, optimization of, 229
  gas-cooled fast reactor (GFR), 39–41
  internal oxidation, 227–228, 229f
  oxidation potential gradients, 229
  temperature gradients, 229
  very high-temperature reactors (VHTR), 36–39, 224–227
Helium effects, ferritic/martensitic steels
  bubbles, 510–512
  dual He plus heavy, 511
  fast-thermal fission reactor irradiations, 511
  He/dpa ratio, 510–512
  He dual ion irradiations (DII), 511
  helium implantation, 512
  high-energy spallation protons and neutrons, 512
  high-temperature helium embrittlement (HTHE), 513
  in situ He injection technique (ISHI), 511
  isotopic tailoring of ferritic alloys, 512
  Master-Curve reference temperature shift, 512–513, 514f
  SINQ spallation proton-neutron source, 512, 513f
  tensile tests, 512
Helium effects, fusion structural materials
  annealing, 83
  bubble density, 81–82
  cavity nucleation, 84
  chemicallyvapor-deposited SiC bend bars, 83
  creep cavity, 84
  creep-rupture time, 84, 85f
  degradation, 80, 83
  fusion-relevant neutron source, 80–81
  Gibbs-Thomson effect, 83
  He and H generation rates, 80
  He-induced hardening and embrittlement, 82
  ion irradiation, 80–81
  neutron irradiation, 80–81, 83
  strengthening parameter, 81–82, 81f
  swelling, 83–84
  temperature dependence, V alloys preimplanted with He, 82–83, 82f
  time-dependent creep, 84
  vacancies and interstitials, 83
  vacancy clusters, 83
Helium embrittlement
  austenitic stainless steels, 311, 312f
  nickel alloys, 391–395
Helium transmutation products, 530
Hexagonal close packed (HCP) lattice structure, 252–253, 252f
HFBR. *See* High flux beam reactor (HFBR)
High-energy spallation protons and neutrons, 512
High flux beam reactor (HFBR), 615–616

High-temperature helium embrittlement (HTHE), 513
  austenitic stainless steels, 311, 312f
  cracking, 199
  creep rupture lifetime, Type 316 stainless steel, 197–198
  He-vacancy clusters, 197–198
  interstitial impurity, 195
  material parameters, 198–199
  matrix precipitates/dispersoids, 199
  migration, 199
  normalizedductilities, 195, 196f
  normalized tensile elongation ratio, vanadium alloys, 195–197, 197f
  predominant fracture mode, 195
  rupture lifetime *vs.* helium content reduction, 195, 196f
  strain rates, 197, 198f
Homogenization, 354
HT9 cladding, 29
HTHE. *See* High-temperature helium embrittlement (HTHE)
Hydride reorientation, 268, 270
Hydriding and mechanical properties
  hydrogen on post-accident transient mechanical properties, 272
  on irradiated mechanical properties, 266–271
  loss of coolant accident (LOCA), 262
  post-quench ductility, 262
  radiation damage, 262
  onunirradiated mechanical properties, 264–265
Hydrogen cracking, 469
Hydrogen embrittlement
  austenitic stainless steels, 334
  nickel alloys, 390–391
Hydrogen flakes, 469–470
Hydrogen isotopes, 59–60
Hydrogen pickup (HPU) fraction, 264
Hydrogen water chemistry (HWC), 261

I

IAC. *See* Irradiation-accelerated corrosion (IAC)
IASCC. *See* Irradiation-assisted stress-corrosion cracking (IASCC)
IC. *See* Irradiation creep (IC)
ICF. *See* Inertial confinement fusion (ICF)
IFMIF. *See* International Fusion Materials Irradiation Facility (IFMIF)
IGSCC. *See* Intergranular stress-corrosion cracking (IGSCC)
IHX. *See* Intermediate heat exchanger (IHX)
Immobile sessile defect clusters, 630
Impurity-driven corrosion, 234, 236f
Incipient cracking, 267–268
Inconel 718
  accelerator production of tritium (APT) facility, 385
  effective shear strength, 385–387, 386f
  fractured area of bent-beam samples, 385–387, 387f
  γ' and γ'' precipitates, 388–390
  from Los Alamos Neutron Science Centre (LANSCE), 385–387
  microhardness of, 385–387, 386f
  neutron irradiation, 385

Inconel 718 *(Continued)*
  postirradiation examination, 385
  room temperature stress-strain curves, 387–388, 388f
  time-temperature-transformation diagram, 355, 355f
  yield strength, 387–388, 389f
Inconel X-750 spacer
  cavity structure in, 377–378, 379f
  cavity structure near grain boundaries in, 378–380, 380f
  dislocation structure in, 377–380, 379f
  flexural stress *vs.* strain curves, 380, 381f
  HAADF and EELS spectra, 380, 382f
  and Inconel 600 flux detectors, 377–378
  load-displacement curves, 380, 381f
  time-temperature-transformation diagram, 355, 355f
Inertial confinement fusion (ICF)
  conceptual power plant designs, 64–65
  instantaneous displacement damage rate, 55
  and magnetic confinement fusion, similarities, 60
  peak neutron wall loads and gas production, 54, 54t
  plasma confinement, 52, 53f
Ingots
  steel, 419, 420f
  vanadium alloys, 587–590
  zirconium alloys, 251–252
In-pile tests, 320
In-reactor corrosion rate, 258
In-situ He injection (ISHI)
  ferritic/martensitic steels, 511
  nano-oxide dispersion-strengthened (NODS) steels, 567, 568f
Integral fast reactor (IFR), 29
Intergranular failure, embrittlement, 376–377
Intergranular stress-corrosion cracking (IGSCC), 11–12, 18, 184–185
  austenitic stainless steels, 323–326, 324–325f
  boiling water reactor (BWR), 261
  Cr depletion, 358
  subcritical water, 214
  water-cooled nuclear reactors, 184–187
Intermediate heat exchanger (IHX), 25
Internal oxidation, helium-cooled reactors, 227–228, 229f
International Fusion Materials Irradiation Facility (IFMIF), 135
International regulatory cooperation, 431
Inverse Kirkendall effect, 299
Ion irradiation facilities, nickel alloys, 390
Irradiated materials, characterization of, 106–107
Irradiated mechanical properties
  delayed hydride cracking (DHC), 268–269, 270f
  hoop strengths, recrystallized Zircaloy-4 G, 267, 268f
  hydride reorientation, 268, 270
  incipient cracking, 267–268
  Nb-containing alloys, 266
  recrystallized Zircaloy-4, ductility of, 267
  secondary hydriding, 267–268, 269, 270f
  stress-relief-annealed (SRA) and recrystallized annealed (RXA) conditions, 266–267
  uniform elongation (UE) and total elongation (TE), 267, 267f

Irradiated V-(4-5)Cr-(4-5)Ti alloys microstructure, 618, 619f
Irradiation-accelerated corrosion (IAC), 219, 332
Irradiation-assisted stress-corrosion cracking (IASCC), 184–185
  austenitic stainless steels, 326–332
  crack propagation, 108–109
  electron microscopy, 113
  ferritic/martensitic steels, 489–490
  subcritical water, 217–219, 220t
  supercritical water, 223, 224f
  susceptibility, 337, 339f
  water-cooled nuclear reactors, 184–185, 187
Irradiation creep (IC), 10
  austenitic stainless steels, 315–317, 315–316f
  ferritic/martensitic steels, 516–518
  nano-oxide dispersion-strengthened (NODS) steels, 572
  radiation-induced dimensional instability, 191–193
  vanadium alloys, 620–622, 621f
  zirconium alloys, 276–277
Irradiation damage, LAS
  annealing, 466–468, 467f
  ASTM model, 459
  computational modeling studies, 458–459, 459f
  databases, 459–460
  early empirical models, 460–462
  environmental and materials variables, 447–454
  hardening features and hardening and embrittlement, 454–457
  historical background, 438–442
  mechanistically based correlation (MBC) models, 462–464
  microstructural evolution, 442–447
  nanostructural evolution, processes and paths in, 442, 442f
  nonregulatory application models, 459
  regulatory models, 459
Irradiation effects, light water reactors, 10–11, 15–16, 19–21
Irradiation effects, nano-oxide dispersion-strengthened (NODS) steels
  environmental effects, 572
  fatigue behavior, TMS, 572
  high-temperature creep crack growth, 572
  irradiation creep, 572
  on microstructure, 560–562
  NFA MA957 creep compliance, 572
  on strength and toughness, 570–572, 570–571f
Irradiation effects, Zr-alloys
  on corrosion resistance, 274–275
  creep, 276–277
  dislocation-loops, 272–273, 273–274f
  growth, 275–276
  hardening and embrittlement, 275
  vacancies and interstitials, 272
Irradiation embrittlement
  ASME regulations, 139–140
  ASTM1921, 140
  Charpy V-Notch (CVN) impact tests, 138–139, 139f
  cleavage fracture, 140
  ferritic/martensitic steels, 501–507

fracture toughness, 139
invariant MC cleavage toughness, 140
large-scale yielding (LSY), 140–141
limiting conditions, 140
measured K_{Jm} values, unirradiated 9Cr tempered martensitic steel, 140–141, 142f
structural alloys, 138
unirradiated and irradiated K_{Jc}(T) master curves, 140, 141f
Irradiation fluence, 451–452
Irradiation growth
reactor structural alloys, 194–195
zirconium alloys, 275–276
Irradiation hardening
austenitic stainless steels, 308–309, 309f
embrittlement relations, 141–142
phosphides and, 447
Irradiation-induced amorphization
in ceramics and materials, 179–180
of complex ceramics, 181
dose *versus* temperature, 180, 180f
intermetallic compounds, 180
Irradiation-induced copper-rich precipitates, 439
Irradiation-induced microstructure, 106–109
Irradiation shifts, 455–456
Irradiation temperature, 452
Irradiation time, 451
ISHI. *See* In-situ He injection (ISHI)

J

Japan Atomic Energy Agency (JAEA), 38
JLF-1 steel
surface morphology, 490–492, 491f
weight loss of, 490, 491f

K

Kairos Power, 43
Kearns factors, 256
Kroll reduction process, 248

L

LANSCE. *See* Los Alamos Neutron Science Centre (LANSCE)
Lanthanide fission products, 25–26
Larson-Miller parameter (LMP), 73, 74f, 169, 337, 340f, 555
LASs. *See* Low-alloy steels (LASs)
Late blooming phases, 445–446
LBEs. *See* Licensing basis events (LBEs)
LCAC molybdenum microstructure, 625
Lead cooled fast neutron reactors. *See* Lead fast reactor (LFR)
Lead fast reactor (LFR)
accelerator-driven systems (ADS), 30, 33–34
candidate metal alloys and materials issues, 34–36, 35t
classical fast reactors, 30
design parameters, 30–31, 31t
eutectic Pb-Bi melt (LBE), 30

flow conditions, 31, 32–33t
fuel assembly geometry, 31, 32f
fuel pin designs, 33
in-core components, 33
objectives, 30
out-of-core components, 33
oxide and nitride fuels, 33–34
pool concept, 30, 30f
predicted dose rates and total fluence, 34
pure liquid lead (Pb), 30
LFR. *See* Lead fast reactor (LFR)
Licensing basis events (LBEs), 43
Light water reactors (LWRs)
austenitic stainless steels, 294–295
construction materials, 12–21
corrosion, 211–212
design and fabrication, 415–422
irradiation, 10–11
Ni alloys, 380–382
nuclear systems, corrosion in, 211–212
thermal aging and fatigue, 9–10
in United States, 1
water as coolant, 211–212
water environment, 11–12
Zr-alloys (*see* Zirconium alloys)
Linear sum superposition (LS) law, 497–498
Liner cladding, 251
Liquid breeders, 62
Liquid lead alloys, corrosion in
advantages, 239–240
corrosion control techniques, 241–242
general corrosion, 240–241, 240–241f
mechanical properties, 242
Pb/Pb-Bi alloy-cooled reactors, 239
Liquid metal-cooled fast reactors (LMRs), 485–486
ferritic/martensitic steels, 485–486
lead fast reactor (LFR), 30–36
sodium fast reactors (SFRs), 24–30
Liquid metal corrosion, 490
Liquid metal embrittlement (LME), 490, 492–493
Liquid sodium, corrosion in
corrosion and mass transfer, control factors, 238–239, 239f
corrosion mechanism, 236
oxygen concentration, 237, 238f
sodium-cooled reactors, 236
solubility, alloying elements, 236, 237f
temperature gradient-driven corrosion, 237–238
Liquid-solid interaction induced erosion-corrosion, 492
Liquid walls, 92–93
LME. *See* Liquid metal embrittlement (LME)
LMP. *See* Larson-Miller parameter (LMP)
LMRs. *See* Liquid metal-cooled fast reactors (LMRs)
Local electrode atom probe, 117, 118f
Localized corrosion modes, 11–12
Los Alamos National laboratory (LANL), 372

Los Alamos Neutron Science Centre (LANSCE), 385–387
Loss of coolant accident (LOCA), 262, 272
Low-alloy steels (LASs), 18–21, 20t
   codes and regulations, 431
   design and fabrication, 415–422
   Gen-III reactor designs, 473
   improved materials, 472–473
   in-service degradation, 438–468
   microstructure and properties, 423–426
   performance issues, 469–470
   piping, 428–429, 430t
   pressurized water reactor (PWR) pressurizer, 428
   reactor pressure vessels, 427, 429–431, 473
   steam generator (SG) shells, 428
   structural integrity assessment (SIA), 431–438
   through-life costs, 473
   through-life toughness prediction, 470–472
   types and composition, 412–415, 413t, 415–416f
Low-temperature crack propagation (LTCP), 11–12
Low-temperature radiation embrittlement, 176–179, 178f
LTCP. See Low-temperature crack propagation (LTCP)
LWRs. See Light water reactors (LWRs)

M

Macrosegregation, 416–419
Macro-Vickers hardness, polycrystalline tungsten, 611, 611t
Magnetic confinement fusion (MCF)
   conceptual power plant designs, 61–64
   and inertial confinement fusion, similarities, 60
   instantaneous displacement damage rate, 55
   peak neutron wall loads and gas production, 54, 54t
   plasma confinement, 52, 53f
Magneto-hydro-dynamic (MHD) phenomena, 492
Magnox reactor, 419–421
Makin and Minter equation, 496
Martensite, 423
Master-Curve method, 504, 506, 506f
Master-Curve reference temperature shift, 512–513, 514f
Material composition, low-alloy steels
   carbon, 453
   composition effects, 452
   manganese, 453
   nickel, 453
   nitrogen, 453
   phosphorus, 453
   silicon, 453
Materials degradation
   nuclear fusion environment, 55–60
   nuclear power reactors, 7–8
Materials test reactor (MTR), 444
Matrix Cu, 444
MBC. See Mechanistically based correlation (MBC)
$M_{23}C_6$ carbides, 486–488, 487f
MCF. See Magnetic confinement fusion (MCF)

Mechanical properties, as-fabricated
   molybdenum alloys, 607–610
   niobium alloys, 606–607, 607f
   tungsten and tungsten alloys, 611–615
   vanadium alloys, 600–615
Mechanical properties, as-irradiated
   molybdenum alloys, 625–629
   niobium alloys, 622–625
   tungsten and tungsten alloys, 629–634
   vanadium alloys, 615–622
Mechanical properties, nano-oxide dispersion-strengthened (NODS) steels
   creep, 554–556, 554f
   fatigue, 558, 559f
   fracture, 556–558, 557f
   static tensile strength and ductility, 552–553, 553f
Mechanical properties, vanadium alloys, 587–588
Mechanical property characterization techniques, 106
   aging phenomena, 134–135
   irradiationembrittlement tests, 138–141
   irradiation hardening embrittlement relations, 141–142
   mechanical tests at nanoscale, 142–143
   microhardness tests, 137, 138f
   nanoindentation, 144
   oxidation and corrosion, 134–135
   space limitation, 144
   space limits, 135
   subsized tensile tests, 135–137, 136f
Mechanistically based correlation (MBC), 459, 462–464
Metal injection molding (MIM), 86
Microbially induced corrosion (MIC), 11–12
Microhardness tests, 137, 138f
Microstructural characterization techniques, 106. See also Microstructural tools
Microstructural tools
   atom probe tomography (APT), 117–125
   direct techniques, 109
   electron microscopy, 109–117
   focused ion beam (FIB), 108–109, 134
   indirect techniques, 109
   irradiation damage, 109, 110t
   irradiation-induced microstructure, 107–109
   positron annihilation (PA) techniques, 130–134
   small-angle neutron scattering (SANS), 125–130
MIM. See Metal injection molding (MIM)
Missing pellet surface (MPS), 282
Mixed spectrum neutron irradiation, 632, 632–633f
Mixed-spectrum reactor irradiation effect, 628–629, 629f
Mn-Mo steel, 412
Mn-Ni-Si phases/precipitates (MNSPs), 120, 121f, 445–446
MNSPs. See Mn-Ni-Si phases/precipitates (MNSPs)
Molten salt cooled reactor, 43, 187
Molten salt fueled reactor (MSR)
   characteristic of, 41
   denatured, 42

Index

flow conditions, 42–43
fluoride and chloride-based salts, 42
from US Aircraft Reactor Experiment program, 41–42
Molten salts, corrosion in
  dissimilar-metal and temperature gradient-driven corrosion, 233, 234–235f
  fission product-driven SCC, 234–235
  fluid fuel, 229–230
  fluoride salt-cooled high-temperature reactor (FHR), 229–230
  general corrosion, 231–233, 231t, 232f
  impurity-driven corrosion, 234, 236f
  in nuclear systems, 230–231
Molybdenum (Mo) alloys
  as-fabricated mechanical properties, 607–610
  as-irradiated mechanical properties, 625–629
  nuclear-grade molybdenum fabrication, 591–595
  types, 594
MSM. *See* Multiscalemodeling (MSM)
MSR. *See* Molten salt fueled reactor (MSR)
MTR. *See* Materials test reactor (MTR)
Multiscalemodeling (MSM), 458–459, 470
Multivariate statistical methods (MVSA), 111
MVSA. *See* Multivariate statistical methods (MVSA)

N

Nabarro-Herring creep, 168–169
Nanofeatures (NFs), 72
Nanoindentation, 143
Nano-oxide (NO)
  functions and He interactions, 550–552, 551f
  nature of, 548–550, 549f
  statistics, 547–548, 548f
  $Y_2Ti_2O_7$ complex oxides, 552
Nano-oxide dispersion-strengthened (NODS) steels
  alloy stability and irradiation effects, 533–535
  cavities and swelling, 566–569
  vs. 9Cr tempered martensitic steels (TMS), 532
  dislocation loops, 562–564, 564f
  future prospects, 574–575
  irradiation effect, 560–562
  irradiation stability, 562, 563f
  mechanical properties, 552–558
  modeling, 573–574, 573f
  NFA and NMS, 536–547
  nm-scale bubbles, 531
  primary variants, 530–531
  scientific challenges, 531
  solute segregation, clustering, and precipitation, 564–566, 565f
  technology-oriented work, 531
  thermal aging, 559–560, 561f
  ultrafinenano-oxides, 530–531
  ultrahigh density, 530–531
  unirradiated mechanical properties, 532–533
  void swelling and helium effects, 535
Nanostructured ferritic alloys (NFAs), 125
  alloy compositions, 536, 536t
  alternative compositions and processing paths, 545–547
  challenges, 535–536
  consolidation, 539–540
  creep-rupture strength, 73, 74f
  creep strength, 555
  deformation processing and tube fabrication, 541
  displacement damage, 74–75
  fracture, 556–558, 557f
  irradiation effects, 74
  irradiationembrittlement effects, 74–75
  joining, 545, 546f
  phase diagrams, 536–537, 537f
  preconsolidation processing, 538–539, 538f
  process sequence, 73, 73f
  static tensile strength, 552
  vs. tempered martensitic steels (TMS), 567, 568–569f
  texturing and damage mechanisms, 542–544, 543–544f
  thermal stability, 73, 560
  transformation paths, 536–538
  viscoplastic creep time-dependent deformation, 554
  yield stresses, 552–553, 553f
Nanostructured tempered martensitic steels (NMS)
  alloy compositions, 536, 536t
  alternative compositions and processing paths, 545–547
  challenges, 535–536
  consolidation, 539–540
  creep strength, 555
  deformation processing and tube fabrication, 541–542
  fracture, 556–558, 557f
  joining, 545, 546f
  phase diagrams, 536–537, 537f
  preconsolidation processing, 538–539, 538f
  texturing and damage mechanisms, 542–544, 543–544f
  transformation paths, 536–538
Nb-1Zr, 590
  neutron, 622, 623f
Nd estimates *vs.* $1/T$, 500, 501f
Neutron flux, 443, 448–451, 449–451f
Neutron irradiation, 19–21, 297, 385, 530
Neutron spectrum, 371–373, 372f, 443
  low-alloy steels (LASs), 447–448
NF616, 29–30
NFAs. *See* Nanostructured ferritic alloys (NFAs)
NHE. *See* Nonhardeningembrittlement (NHE)
Ni-based superalloys, 353, 353f
Nickel alloys, 350–351
  chemical compatibility with coolants, 368–369
  fatigue, 399–400, 400f
  generation-IV reactors, 365–367, 367f, 368t
  heliumembrittlement, 391–395
  hydrogenembrittlement, 390–391
  irradiation creep and stress relaxation, 396–399
  point defects, 395–396

Nickel alloys *(Continued)*
  radiation damage and gas production, 369–375
  radiation hardening/softening and loss of ductility, 376–390
Nickel-based alloys, 16–18, 17t, 17f
  chemical compatibility with coolants, 368–369
  creep-fatigue deformation, 400–401, 401f
  deformation mechanisms, 362–364, 363–364f
  fracture modes, 360–362, 360–361f
  generation-IV reactors, 365–367, 367f, 368t
  joining processes, 356–358, 357t
  mechanical properties, 358–360, 359f
  minor elements/impurities, 350–351, 350t
  nominal compositions, 350–351, 350t
  in nuclear reactor cores and primary coolant circuits, 350–351, 350t
  physical metallurgy, 351–353
  stress corrosion cracking (SCC), 216, 364–365, 366f
  thermomechanical processing, 354–355
Ni+Cr alloys, 337
NIFS-HEAT-1, 589–590
NIFS-HEAT-2, 589–590
Niobium (Nb) alloys
  as-fabricated mechanical properties, 606–607, 607f
  as-irradiated mechanical properties, 622–625
  production, 590–591
NMS. *See* Nanostructured tempered martensitic steels (NMS)
Noble metal chemical application (NMCA), 261
NODS steels. *See* Nano-oxide dispersion-strengthened (NODS) steels
Nodular corrosion, 258, 259f
Non-carbon-emitting power generation, 1
Nonhardeningembrittlement (NHE), 447, 455–456
Nonirradiated tensile properties, molybdenum alloys, 608–609, 609f
Nonlinear elastic plastic stress intensity factor, 503
Nonmetallic elements, 27
Normal water chemistry (NWC), 261
Nuclear fusion
  basic reaction, 52
  divertor/limiter application materials, 85–93
  DT fuel, 52
  electrostatic repulsion, 52
  first-wall/blanket structural material options, 66–84
  vs. fission environment, 58–60, 58f
  hot DT plasma, 52
  inertial confinement fusion (*see* Inertial confinement fusion (ICF))
  magnetic confinement fusion (*see* Magnetic confinement fusion (MCF))
  magnet structural materials, 96
  materials degradation, environment, 55–60
  neutron and thermal loads, 54–55, 54t, 55f
  vacuum vessel (VV) materials, 94–96
Nuclear generation capacity, 1, 3t
Nuclear-grade molybdenum fabrication
  arc-cast product, 594–595

chemical composition, 594–595, 594t
coldisostatic pressing, 593–594
electron-beam product casts, 594
grinding and separation steps, 591–592
oxide dispersion-strengthened (ODS) molybdenum microstructure, 594–595, 595f
production route, 593–594, 593f
Re-effect, 591–592
semicontinuous process, 593–594
TZM products, 594–595
welding methods, 595
Nuclear power reactors
  component service and material performance, 7–8
  distribution of, 1–2, 2f, 4f
  materials degradation, 7–8
Nuclear systems, corrosion in, 212, 214t
NWC. *See* Normal water chemistry (NWC)

O

ODS F/M steel, 36
Optimax-A steel, engineering tensile stress-strain curves, 496, 497f
Orbital electron momentum spectra (OEMS) measurements, 134
Organization for Economic Cooperation and Development Nuclear Energy Agency (OECD-NEA), 431
Out-of-pile tests, 320
Oxidation, VHTR environment, 225, 226f
Oxide dispersion-strengthened (ODS) steels, 33–34, 112–113
  history, 531–532
  issues, 558
  nano-dispersion-strengthened variants, 530–531
  NODS steels (*see* Nano-oxide dispersion-strengthened (NODS) steels)
  radiation hardening and softening, 497

P

PALA. *See* Positron annihilation lineshape analysis (PALA)
PALS. *See* Positron annihilation lifetime spectra (PALS)
Pb-17Li, metallic element concentration, 490, 491f
Pb/Pb-Bi alloy-cooled reactors, 239
PCI/FCCI effects with fission fuels, 318–321
PCI-SCC. *See* Pellet-cladding interaction-stress corrosion cracking (PCI-SCC)
PCMI. *See* Pellet-cladding mechanical interaction (PCMI)
Pebble bed FHR design, 43
Peierls stress, 630–631
Pellet-cladding interaction-stress corrosion cracking (PCI-SCC), 282, 282–283f
Pellet-cladding mechanical interaction (PCMI), 280–282, 281f
Pellini drop weight test, 435–436
PFCs. *See* Plasma-facing components (PFCs)
PFZs. *See* Precipitate-free zones (PFZs)
Phosphides, 447
PHWRs. *See* Pressurized heavy water reactors (PHWRs)

Index   651

Physical metallurgy, Ni-based alloys
  body-centered tetragonal (BCT) structure, 353
  face-centered cubic (FCC) crystal structure, 351–352, 352f
  Ni-based superalloys, 353, 353f
  phases in, 353, 354t
  ternary diagram, Fe-Cr-Ni system, 351, 351f
  yield stress, temperature dependence of, 351–352, 352f
Physical metallurgy, vanadium alloys, 588–589
Pitting, 11–12
Plasma-facing components (PFCs), 85, 88
PNC316 and PNC1520 swelling, 295–296
Point defects (PDs), 443
Point defects, nickel alloys, 395–396
Poloidal field (PF) coils, 61
Polycrystalline tungsten, 611, 611t
  tensile behavior of, 632–634, 633f
Positron annihilation lifetime spectra (PALS), 130, 132
Positron annihilation lineshape analysis (PALA), 130–131
Positron annihilation spectroscopy-based techniques
  angular correlation of annihilation radiation (ACAR), 130–131
  application, 132–133, 133f
  characterization methods, 130
  coincidence Doppler broadening (CDB), 130–131
  γ-ray energy spectral peak, 130–131, 131f
  interpretation, 134
  orbital electron momentum spectra, 130–131
  orbital electron momentum spectra (OEMS) measurements, 134
  positron annihilation lifetime spectra (PALS), 130, 132
  positron annihilation lineshape analysis (PALA), 130–131
  resultant γ-rays, 130
  variable energy positron annihilation spectroscopy (VEPAS), 132
Post-irradiation annealing, vanadium alloys, 616–618, 618f
Post-weld heat treatment (PWHT), 77
Powder metallurgy (PM)
  niobium alloys, 590–591
  tungsten, 85–86
Power plant piping, 12
Precipitate-free zones (PFZs), 360
Precipitates, irradiation damage, 444–445, 445f
Pressurized heavy water reactors (PHWRs), 2–7, 7f
Pressurized water reactors (PWRs), 2–5, 19
  cladding alloys, 248–249
  coolant chemistry, 260–261
  cracking, 17–18
  duplex cladding, 251
  fuel assemblies in, 377
  fuel assembly, 248, 249f
  intergranular stress corrosion cracking (IGSCC), 323–326, 324–325f
  materials of service, 8, 9f
  schematic representation of, 6, 6f
Pressurized water stress corrosion cracking (PWSCC), 11–12, 185–187
Pressurizer shells, 19
Primary knock-on atom (PKA) spectrum, 59, 489

Primary water stress corrosion cracking (PWSCC), 261
Proton irradiation facilities, nickel alloys, 385–390
PWC-11, 590
PWRs. See Pressurized water reactors (PWRs)
PWSCC. See Pressurized water stress corrosion cracking (PWSCC); Primary water stress corrosion cracking (PWSCC)

Q
Quenching, low-alloy steels (LASs), 424

R
Radiation- and stress-modified corrosion and cracking phenomena
  Alloy 600 steam generator U-tubes, 185–187
  chronological summary, water-cooled nuclear reactors, 184–185, 185t
  intergranular stress-corrosion cracking (IGSCC), 184–187
  irradiation-assisted stress-corrosion cracking (IASCC), 184–185, 187
  IRSCC, 184–185
  in molten salt-cooled reactor, 187
  nickel- and iron-based alloys, simplified Pourbaix diagram, 185–187, 186f
  primary water stress corrosion cracking (PWSCC), 185–187
  radiation dose, influence of, 187, 188f
Radiation damage processes, 10
Radiation-enhanced and induced segregation
  in austenitic alloys, 182
  Fe-Cr-Ni alloys, 182–184, 183f
  high-energy particle irradiation on precipitates, 184
  temperature-dependent diffusion behavior, 181–182
  temperature effect and dose rate, sink-dominant conditions, 181–182, 181f
Radiation-enhanced diffusion (RED), 57
Radiation-enhanced precipitation (REP), 57
Radiation hardening and ductility reduction
  dose dependence, 173–175, 174–175f
  temperature dependence, 176, 176–177f
Radiation hardening and softening, ferritic/martensitic steels, 496–501
Radiation-induced dimensional instability
  cavity swelling, 188–191
  irradiation creep, 191–193
  irradiation growth, 194–195
Radiation-induced embrittlement, 11
Radiation-induced metallurgical changes
  atomic displacements, 297
  crystalline defects, 297
  dislocation microstructure, 302–303, 303f
  neutron irradiation, 297
  phase stability, 303–305, 304f, 305t
  radiation-defined quasiequilibrium state, 297
  radiation-induced segregation (RIS), 297–302
  transmutation, 305–308, 307f
Radiation-induced precipitation (RIP), 57

Radiation-induced segregation (RIS), 57
  binary, 50A–50B system, 297–299, 299f
  compositional profiles, 299, 300f
  enrichment and depletion, 299
  Fe-Cr-Ni alloys, 301
  grain boundary chromium concentration, dose dependence of, 301–302, 301f
  inverseKirkendall effect, 299
  irradiation temperature, dose, and dose rate, 299–301, 300f
  minor alloying elements and impurities, 301–302
  neutron dose, 301
  oversize solutes, 302
  and phase transformations, 10
  solute drag, 299
  vacancies and interstitials, 297–299
  vacancy exchange mechanism, 301
Radiation-induced swelling and creep effects, 10–11
RAF/M steels. *See* Reduced activation ferritic/martensitic (RAF/M) steels
Reactor coolant piping, 19
Reactor pressure vessels (RPVs), 5–6, 11, 19
  applications in LAS, 427
  embrittlement, atom probe tomography (APT), 119–120, 119f, 121f
  irradiation damage issues, 439
  safety and performance of, 431
Reactor pressure vessel (RPV) steels
  atom probe tomography (APT), 119
  compositional variability, 414, 415f
  electron microscopy, 113
Reactor structural alloys
  BOR-60 reflector assembly duct, embrittlement and fracture, 164, 165f
  high-temperature helium embrittlement, 195–199
  irradiated fuel assembly, fuel rod bowing in, 164, 164f
  radiation- and stress-modified corrosion and cracking phenomena, 184–187
  radiation hardening and embrittlement, 173–179
  radiation-induced degradation mechanisms, 164–166, 165f
  radiation-induced dimensional instability, 187–195
  radiation-induced phase and microchemical changes, 179–184
  thermomechanical degradation processes, 166–172
RED. *See* Radiation-enhanced diffusion (RED)
Reduced activation ferritic/martensitic (RAF/M) steels, 29–30
  advantages, 66–67
  chemical composition specifications, 66–67, 67t
  corrosion and compatibility, 72
  creep-fatigue interaction, 71
  7–9Cr TMS, 486–488
  dimensional stability, 69
  fatigue, 71
  F82H, 69–70
  first-wall structures and components, 71
  impact and fracture toughness properties, 494–496
  irradiation creep, 69
  irradiation hardening, 67
  irradiation temperature, 68–69, 68f, 71
  operating temperature window, 67
  PbLi flow velocity, 72
  quantitative corrosion models, 72
  strain-hardening capacity, 494
  tensile properties, 493–494
  tensile strength *vs.* temperature, 493–494, 494f
  tertiary creep behavior, 71
  total elongation *vs.* temperature, 494, 495f
  transition temperature shifts (TTS), 68–69
  uniform elongation *vs.* temperature, 494, 495f
  void swelling, 69
  yield strength with irradiation temperature, 67, 68f
  yield stress *vs.* temperature, 493–494, 493f
Reduced-activation materials, 486
Reference temperature T_0 *vs.* average flow stress shift, 507, 507f
Refractory alloys
  melting points, 585–586
  molybdenum (*see* Molybdenum alloys)
  niobium (*see* Niobium (Nb) alloys)
  physical and chemical properties, 586, 586t
  tungsten (*see* Tungsten and tungsten alloys)
  vanadium (*see* Vanadium alloys)
REP. *See* Radiation-enhanced precipitation (REP)
Research tools
  charged particle irradiation (CPI), 104–106
  computational alloy design and optimization, 144–150
  damage level *vs.* irradiation temperature, 104, 105f
  irradiated materials, characterization of, 106–107
  irradiation parameter space, 104, 105t
  material test reactors (MTRs), 104
  microstructural tools, 107–134
  radiation damage processes, 104, 106f
  subsized specimen tests, 134–144
Rhenium, 630–631, 631f
RIP. *See* Radiation-induced precipitation (RIP)
RIS. *See* Radiation-induced segregation (RIS)
Root sum of squares (RSS) superposition law, 454, 498
RPVs. *See* Reactor pressure vessels (RPVs)
Russian ChS-68, 297
Russian Pb-cooled fast reactor (BREST), 33

S

SANS. *See* Small-angle neutron scattering (SANS)
SAW. *See* Submerged-arc welding (SAW)
Scanning transmission electron microscopy (STEM), 109–112
SCC. *See* Stress corrosion cracking (SCC)
SCW. *See* Supercritical water (SCW)
SCWR. *See* Super-critical water-cooled reactor (SCWR)
Secondary hydriding, 267–268, 269–270f
Second-phase particle (SPP), zirconium alloys, 252
Self-interstitial atoms (SIA)
  clusters, 443

and vacancies, 566
vacancies and, 443
SFRs. *See* Sodium-cooled fast reactors (SFRs); Sodium fast reactors (SFRs)
SIA. *See* Structural integrity assessment (SIA)
SiC/SiC cladding, 41
Silicon carbide (SiC$_F$/SiC) composites
    attributes, 77
    brittlefiber fracture, 77
    hermeticity, 79–80
    point defect-induced swelling, 77–78, 78f
    radiation damage, 79
    thermal conductivity degradation, 77–78, 78f
    time-dependent deformation models, 79–80
    time-dependent deformation processes, 79–80
Single step aging, 355
SINQ spallation proton-neutron source, 512, 513f
Small-angle neutron scattering (SANS), 444–445
    application, 128–129, 128–129f
    direct and indirect information, 129
    principles, 125–127, 126f
    small solute clusters, 130
Small-scale yielding (SSY) conditions, 431–432, 434
SMFs. *See* Stable matrix features (SMFs)
Sodium-cooled fast reactors (SFRs), austenitic stainless steels
    cold-worked (CW) austenitic steels, 295
    compositions, 297, 298t
    creep rupture strength, 295, 296f
    French CW15-15Ti, 296
    French SFR fuel pins, diametral increase, 296, 296f
    German DIN 1.4970 alloy, 297
    PNC316 and PNC1520 swelling, 295–296
    radiation-induced void swelling, 295
    Russian ChS-68, 297
    UK FV548, 297
Sodium-cooled reactors
    corrosion, 236
    liquid sodium corrosion, 236
Sodium fast reactors (SFRs)
    candidate alloys and materials, 27–30, 27–28t
    experimental and demonstration reactors, 24, 25t
    fuel pin bundle assembly and duct in, 25, 26f
    in-core components, 25–27, 29–30
    out-of-core components, 25, 27–28, 27–28t
    pool and loop designs, 24, 24f, 25t
    predicted dose rates and total fluences, 27
Sodium flow conditions
    in-core components, 25–27
    out of core components, 25
Solid breeder blankets, 62
Solute point-defect cluster complexes, 444
Solute redistribution, 179
Solution anneal (SA), 355
Solution annealed austenitic stainless steel 1.4988, 321
Solution-strengthened tungsten, 614–615, 614f

Stable matrix features (SMFs), 440–441, 444
Static strain aging (SSA), 465–466
Steam generator (SG) shells, 19, 428
Steam generator tube sheets, 19
Steel ingot, 419, 420f
STEM. *See* Scanning transmission electron microscopy (STEM)
STEM X-ray mapping, 111
Strain aging, low-alloy steels, 465–466
Strain hardening, ferritic/martensitic steels, 500–501
Strain rate sensitivity (SRS) parameter, vanadium alloys, 603–604, 620, 621f
Stress corrosion cracking (SCC), 11–12, 45, 115
    austenitic stainless steels, 323
    electron microscopy, 115
    ferritic/martensitic steels, 489–490
    fission product-driven SCC, 234–235
    Ni-based alloys, 364–365, 366f
    subcritical water, 212–219
    supercritical water, 221–222, 223f
Stress intensity factor, 431–433, 436, 436f
Structural integrity assessment (SIA), 431–438
Structured tungsten materials, 611–612, 612f
Subcritical water, corrosion in
    coolant temperature, 212
    corrosion degradation, 212–214
    environment, 212
    irradiation-accelerated corrosion, 219
    irradiation-assisted stress corrosion cracking (IASCC), 217–219, 220t
    stress corrosion cracking (SCC), 214–217
Submerged-arc welding (SAW), 419
Subsized specimen tests, irradiated mechanical properties
    aging phenomena, 134–135
    irradiationembrittlement tests, 138–141
    irradiation hardening embrittlement relations, 141–142
    mechanical tests at nanoscale, 142–143
    microhardness tests, 137, 138f
    nanoindentation, 144
    oxidation and corrosion, 134–135
    space limitation, 135, 144
    subsized tensile tests, 135–137, 136f
Subsized tensile tests, 135–137, 136f
Superconductors, 63
Supercritical water (SCW)
    behavior variation, 219–220, 221f
    corrosion, 220–221, 222f
    ferritic-martensitic alloys, 223
    irradiation-assisted stress corrosion cracking (IASCC), 223, 224f
    stress corrosion cracking (SCC), 221–222, 223f
Super-critical water-cooled reactor (SCWR), 489–490
    candidate metal alloys, 45
    diagram of, 43–44, 44f
    flow conditions, 45
    predicted dose rates and total fluences, 45
Synergisms, irradiation damage, 447

T

t-EBSD. *See* Transmission-electron backscatter diffraction (t-EBSD)
TEM. *See* Transmission electron microscopy (TEM)
Tempered bainitic 2.25Cr-1Mo steels, 486
Tempered martensitic steels (TMS), 485–489
    fatiguebehavior, 572
    radiation hardening and softening, 497
    swelling resistance of, 566–567
Tensile ductility, ferritic/martensitic steels, 500–501
Tensile tests, helium effects, 512
Tensile yield stress, 137, 138f
TGSCC. *See* Transgranular stress corrosion cracking (TGSCC)
Thermal aging
    low-alloy steels, 464–465
    nano-oxide dispersion-strengthened (NODS) steels, 559–560, 561f
    reactor structural alloys, 166–167, 167f
Thermal annealing, 466–468
Thermal creep
    "Ashby" map, 169–171
    behavior, 168–169, 169f
    deformation mechanism map, 169–171, 170f
    deformation$vs.$ exposure time behavior, 167–168, 168f
    diffusional creep mechanisms, 168–169
    dislocation (power law) creep, 168–169
    ferritic/martensitic steels, 509, 510–511f
    Larson-Miller parameter, 169
    resistance, 625
    rupture data, 169, 170f
    time-dependent degradation phenomena, 168
Thermal nuclear reactor cores, 369
Thermomechanical processing, Ni-based alloys, 354–355
Thermomechanical treatment (TMT) processing paths, 518
TMS. *See* Tempered martensitic steels (TMS)
Tokamak
    ARIES-RS (Reversed Shear) power core, 61, 61f
    generic elements, radial build, 61, 62f
Toroidal field (TF) coils, 61
Total elongation (TE), 267, 267f
Toughness transition temperature variability, 426, 428f
Transgranular stress corrosion cracking (TGSCC), 11–12, 214
Transmission-electron backscatter diffraction (t-EBSD), 114–115
Transmission electron microscopy (TEM), 109–112
TRISO fuel particles, 43
Tungsten and tungsten alloys
    advantages and disadvantages, 85
    as-fabricated mechanical properties, 611–615
    as-irradiated mechanical properties, 629–634
    blending, 596
    coldisostatic pressing, 597
    compacting, 596–597
    direct sintering, 598
    forming process, 599, 599t
    indirect sintering, 599
    irradiation hardening, dose dependence of, 87, 88f
    jointdebonding, 86
    maximum sputtering rates, 88
    metal injection molding (MIM), 86
    nanostructured layer thickness, time dependence of, 89–90, 90f
    near net-shape and mass fabrication, 599–600, 601f
    neutron irradiation effects, 86
    nonequilibriumχ-phase, 87
    pertinent irradiation data, 87
    plasma-facing component (PFC) applications, 85, 88
    powder metallurgy (PM), 85–86
    powder production, 596, 596f
    presintering, 597
    sintering, 598, 598f
    thermal shock response, 89, 89f
    toughness, 85–86
    transmutation rates, 86, 87f
    tritium (T) retention, 89
    "W-fuzz" growth rate obeys diffusion-controlled kinetics, 89–90

U

UK FV548, 297
UMDs. *See* Unstable matrix defects (UMDs)
Uniform corrosion, 11–12, 258
Uniform elongation (UE), 267, 267f
Uniform oxide layer, 258, 258f
Unirradiated mechanical properties
    corrosion reaction, 264
    fuel cladding integrity with hydriding, 265, 266f
    hydride morphology and redistribution, 264–265, 265f
    hydrogen pickup (HPU) fraction, 264
Unstable matrix defects (UMDs), 446, 446f
Upper shelf energy (USE), 437–438
Uranium oxycarbide (UCO) material, 43
US Code of Federal Regulations (CFRs), 431
USE. *See* Upper shelf energy (USE)
US National Clad and Duct (NCD) program, 29
US Nuclear Regulatory Commission (USNRC) Regulatory Guides (RG), 431

V

Vacancy clusters, 444
Vacuum vessel (VV) materials, 62–63, 94–96
Vanadium alloys, 75–77
    as-fabricated mechanical properties, 600–615
    as-irradiated mechanical properties, 615–622
    production, 587–590
Variable energy positron annihilation spectroscopy (VEPAS), 132
VEPAS. *See* Variable energy positron annihilation spectroscopy (VEPAS)
Very high-temperature reactor (VHTR)
    carburization in, 227
    decarburization in, 226–227, 226t, 228f
    equilibrium Cr-O-C stability diagram, 224, 225f
    experimental and demonstration reactors, 36, 37t
    flow conditions, 38–39

grain boundary migration and sliding, 224–225
helium impurities, 224
hydrogen-generation processes, 38
metallic materials, 39, 39t
outlet gas temperatures, 36
oxidation in, 225, 226f
pebble bed reactors, 36–37, 37f
predicted dose rates and total fluences, 39, 39t
prismatic reactors, 36–37, 38f
thermochemical processes, 38
VHTR. See Very high-temperature reactor (VHTR)
Vicker's hardness, 137, 138f, 632
Vodo-VodyanoiEnergetichesky Reactor (VVER) designs, 416
Void swelling
  austenitic stainless steels, 311–314, 313–314f, 337, 339f
  DII TMS vs. NFA, 567, 569f
  ferritic/martensitic steels, 514–516
  and helium effects, 535
  radiation-induced, 295

W

Waste disposal, 60
Water cooled reactors
  subcritical water, corrosion in, 212–219
  supercritical water, corrosion in, 219–223
Water-cooled reactors
  boiling water reactors (BWRs), 2–6, 5f
  design features and operational characteristics, 7, 8t
  light water reactors (LWRs), 9–12
  material properties, 7
  pressurized heavy water reactors (PHWRs), 2–7, 7f
  pressurized water reactors (PWRs), 2–6, 6f
Waterside corrosion, cladding, 257
Welding process, low-alloy steels (LASs), 419–422
Western European Nuclear Regulators Association (WENRA), 431
WWER-440 steels, 412–414, 465

X

XT-ADS
  clad temperatures, 33
  flow conditions, 31, 33t

Y

Yield stress, 455
  molybdenum alloys, 607–608, 608f

nano-oxide dispersion-strengthened (NODS) steels, 552–553, 553f
Ni-based alloys, 362–364, 363–364f

Z

Zircaloys. See Zirconium alloys
Zirconium alloys
  atom probe tomography (APT), 120–123
  fuel cladding (see Zirconium cladding alloys)
  ingots, 251–252
  oxidation rate, 339–340
Zirconium-based alloys, 13
Zirconium cladding alloys
  anisotropy, 255
  boiling water reactor (BWR) coolant chemistry, 261–262, 261f
  cladding alloy compositions, 249, 250t, 251
  as core internals, 248
  corrosion, 257–258
  crud deposition, 258–260, 259t, 260f, 262, 263f
  duplex cladding, 251
  failure mechanisms, 278–283
  fuel cladding integrity with hydriding, 265
  fuel rods, 248, 249–250f
  grid-to-rod fretting (GTRF) fuel leaking failures, 252
  hydriding and mechanical integrity, 264–265
  ingots, 251–252
  irradiation effects, 272–277
  irradiation growth, 164, 164f
  Kroll reduction process, 248
  lattice structure and second-phase particles, 252–253, 252–253f
  liner cladding, 251
  mechanical properties, 251
  in nuclear industry, 284, 285f
  pilgering, 251–252
  processing and fabrication, 253–255
  pressurized water reactor (PWR) coolant chemistry, 260–261
  second-phase particle (SPP) size control, 252
  vs. stainless steel cladding, 318
  vs. stainless steels, 248
  strip fabrication, 252
  texture, 255–256, 255t, 256–257f
  Zircaloy-4, 248–249, 258, 258f
  Zircaloy-2 cladding, 251
  ZIRLO and M5, 249